Continued on back end papers

*Now available in a lower priced paperback edition in the Wiley Classics Library.

Statistical Factor Analysis
and Related Methods

Statistical Factor Analysis and Related Methods

Theory and Applications

ALEXANDER BASILEVSKY

Department of Mathematics & Statistics
The University of Winnipeg
Winnipeg, Manitoba
Canada

A Wiley-Interscience Publication

JOHN WILEY & SONS, INC.

New York • Chichester • Brisbane • Toronto • Singapore

Copyright © 1994 by John Wiley & Sons, Inc.

Library of Congress Cataloging in Publication Data:
Basilevsky, Alexander.
   Statistical factor analysis and related methods : theory and
applications / by Alexander Basilevsky.
     p. cm. -- Wiley series in probability and mathematical
statistics. Probability and mathematical statistics)
   "A Wiley-Interscience publication."
   Includes bibliographical references and index.
   ISBN 0-471-57082-6 (acid-free)
   1. Factor analysis. I. Title. II. Series.
QA278.5.B37 1993
519.5′354--dc20   93-2323

To my mother Olha
and
To the memory of my father Mykola

If one is satisfied, as he should be, with that which is to be probable, no difficulty arises in connection with those things that admit of more than one explanation in harmony with the evidence of the senses; but if one accepts one explanation and rejects another that is equally in agreement with the evidence it is clear that he is altogether rejecting science and taking refuge in myth.

– Epicurus (Letter to Pythocles, Fourth Century B.C.)

Physical concepts are free creations of the human mind, and are not, however it may seem, uniquely determined by the external world. In our endeavour to understand reality we are somewhat like a man trying to understand the mechanism of a closed watch. He sees the face and the moving hands, even hears its ticking, but he has no way of opening the case. If he is ingenious he may form some picture of a mechanism which could be responsible for all the things he observes, but he may never be quite sure his picture is the only one which could explain his observations. He will never be able to compare his picture with the real mechanism and he cannot even imagine the possibility of the meaning of such a comparison.

– A. Einstein, *The Evolution of Physics*, 1938

Preface

More so than other classes of statistical multivariate methods, factor analysis has suffered a somewhat curious fate in the statistical literature. In spite of its popularity among research workers in virtually every scientific endeavor (e.g., see Francis, 1974), it has received little corresponding attention among mathematical statisticians, and continues to engender debate concerning its validity and appropriateness. An equivalent fate also seems to be shared by the wider class of procedures known as latent variables models. Thus although high-speed electronic computers, together with efficient numerical methods, have solved most difficulties associated with fitting and estimation, doubt at times persists about what is perceived to be an apparent subjectiveness and arbitrariness of the methods (see Chatfield and Collins, 1980, p. 88). In the words of a recent reviewer, "They have not converted me to thinking factor analysis is worth the time necessary to understand it and carry it out." (Hills, 1977.)

Paradoxically, on the more applied end of the spectrum, faced with voluminous and complex data structures, empirical workers in the sciences have increasingly turned to data reduction procedures, exploratory methods, graphical techniques, pattern recognition and other related models which directly or indirectly make use of the concept of a latent variable (for examples see Brillinger and Preisler, 1983). In particular, both formal and informal exploratory statistical analyses have recently gained some prominence under such terms as "soft modeling" (Wold, 1980) and "projection pursuit" (Huber, 1985; Friedman and Tukey, 1974). These are tasks to which factor analytic techniques are well suited. Besides being able to reduce large sets of data to more manageable proportions, factor analysis has also evolved into a useful data-analytic tool and has become an invaluable aid to other statistical models such as cluster and discriminant analysis, least squares regression, time/frequency domain stochastic processes, discrete random variables, graphical data displays, and so forth although this is not always recognized in the literature (e.g. Cooper, 1983).

Greater attention to latent variables models on the part of statisticians is now perhaps overdue. This book is an attempt to fill the gap between the mathematical and statistical theory of factor analysis and its scientific practice, in the hope of providing workers with a wider scope of the models than what at times may be perceived in the more specialized literature (e.g. Steward, 1981; Zegura, 1978; Matalas and Reicher, 1967; Rohlf and Sokal, 1962).

The main objections to factor analysis as a bona fide statistical model have stemmed from two sources—historical and methodological. Historically, factor analysis has had a dual development beginning indirectly with the work of Pearson (1898, 1901, 1927), who used what later becomes known as principal components (Hotelling, 1933) to fit "regression" planes to multivariate data when both dependent and independent variables are subject to error. Also, Fisher used the so-called singular value decomposition in the context of ANOVA (Fisher and Mackenzie, 1923). This was the beginning of what may be termed the statistical tradition of factor analysis, although it is clearly implicit in Bravais' (1846) original development of the multivariate normal distribution, as well as the mathematical theory of characteristic (eigen) roots and characteristic (eigen) vectors of linear transformations. Soon after Hotelling's work Lawley (1940) introduced the maximum likelihood factor model. It was Spearman (1904, 1913), however, who first used the term "factor analysis" in the context of psychological testing for "general intelligence" and who is generally credited (mainly in psychology) for the origins of the model. Although Spearman's method of "tetrads" represented an adaptation of correlation analysis, it bore little resemblance to what became known as factor analysis in the scientific literature. Indeed, after his death Spearman was challenged as the originator of factor analysis by the psychologist Burt, who pointed out that Spearman had not used a proper factor model, as Pearson (1901) had done. Consequently, Burt was the originator of the psychological applications of the technique (Hearnshaw, 1979). It was not until later however that factor analysis found wide application in the engineering, medical, biological, and other natural sciences and was put on a more rigorous footing by Hotelling, Lawley, Anderson, Joreskog, and others. An early exposition was also given by Kendall (1950) and Kendall and Lawley (1956). Because of the computation involved, it was only with the advent of electronic computers that factor analysis became feasible in everyday applications.

Early uses of factor analysis in psychology and related areas relied heavily on linguistic labeling and subjective interpretation (perhaps Cattell, 1949 and Eysenck, 1951 are the best known examples) and this tended to create a distinct impression among statisticians that imposing a particular set of values and terminology was part and parcel of the models. Also, questionable psychological and eugenic attempts to use factor analysis to measure innate (i.e., genetically based) "intelligence," together with Burt's fraudulent publications concerning twins (e.g., see Gould, 1981) tended to further alienate scientists and statisticians from the model. Paradoxically, the rejection has engendered its own misunderstandings and confusion amongst statisticians (e.g., see Ehrenberg, 1962; Armstrong, 1967; Hills, 1977), which seems to have prompted some authors of popular texts on multivariate analysis to warn readers of the "...many drawbacks to factor analysis" (Chatfield and Collins, 1980, p. 88). Such misunderstandings have had a further second-order impact on practitioners (e.g., Mager, 1988, p. 312).

Methodological objections to factor analysis rest essentially on two criteria. First, since factors can be subjected to secondary transformations of the coordinate axes, it is difficult to decide which set of factors is appropriate. The number of such rotational transformations (orthogonal or oblique) is infinite, and any solution chosen is, mathematically speaking, arbitrary. Second, the variables that we identify with the factors are almost never observed directly. Indeed, in many situations they are, for all practical intents and purposes, unobservable. This raises a question concerning exactly what factors do estimate, and whether the accompanying identification process is inherently subjective and unscientific. Such objections are substantial and fundamental, and should be addressed by any text that deals with latent variables models. The first objection can be met in a relatively straightforward manner, owing to its somewhat narrow technical nature, by observing that no estimator is ever definitionally unique unless restricted in some suitable manner. This is because statistical modeling of the empirical world involves not only the selection of an appropriate mathematical procedure, with all its assumptions, but also consists of a careful evaluation of the physical-empirical conditions that have given rise to, or can be identified with, the particular operative mechanism under study. It is thus not only the responsibility of mathematical theory to provide us with a unique statistical estimator, but rather the arbitrary nature of mathematical assumptions enables the investigator to choose an appropriate model or estimation technique, the choice being determined largely by the actual conditions at hand. For example, the ordinary least squares regression estimator is one out of infinitely many regression estimators which is possible since it is derived from a set of specific assumptions, one being that the projection of the dependent variable/vector onto a sample subspace spanned by the independent (explanatory) variables is orthogonal. Of course, should orthogonality not be appropriate, statisticians have little compunction about altering the assumption and replacing ordinary least squares with a more general model. The choice is largely based on prevailing conditions and objectives, and far from denoting an ill-defined situation the existence of alternative estimation techniques contributes to the inherent flexibility and power of statistical/mathematical modeling.

An equivalent situation also exists in factor analysis, where coefficients may be estimated under several different assumptions, for example, by an oblique rather than an orthogonal model since an initial solution can always be rotated subsequently to an alternative basis should this be required. Although transformation of the axes is possible with any statistical model (the choice of a particular coordinate system is mathematically arbitrary), in factor analysis such transformations assume particular importance in some (but not all) empirical investigations. The transformations, however, are not an inherent feature of factor analysis or other latent variable(s) models, and need only be employed in fairly specific situations, for example, when attempting to identify clusters in the variable (sample) space. Here, the coordinate axes of an initial factor solution usually represent mathematically arbitrary frames of references which are chosen on grounds of convenience and east of computation, and which may have to be altered because of interpretational or substantive requirements. The task is much simplified, however, by the existence of well-defined statistical criteria which result in unique rotations, as well as by the availability of numerical algorithms for their implementation. Thus once a criterion function is selected and optimized, a unique set of estimated coefficients (coordinate axes) emerges. In this sense the rotation of factors conforms to general and accepted statistical practice. Therefore, contrary to claims such as those of Ehrenberg (1962) and Temple (1978), our position on the matter is that the rotation of factors is not intrinsically subjective in nature and, on the contrary, can result in a useful and meaningful analysis. This is not to say that the rotational problem represents the sole preoccupation of factor analysis. On the contrary, in some applications the factors do not have to be rotated or undergo direct empirical interpretation. Frequently they are only required to serve as instrumental variables, for example, to overcome estimation difficulties in least squares regression. Unlike the explanatory variables in a regression model, the factor scores are not observed directly and must also be estimated from the data. Again, well-defined estimators exist, the choice of which depends on the particular factor model used.

The second major objection encountered in the statistical literature concerns the interpretation of factors as actual variables, capable of being identified with real or concrete phenomenon. Since factors essentially represent linear functions of the observed variables (or their transformations), they are not generally observable directly, and are thus at times deemed to lack the same degree of concreteness or authenticity as variables measured in a direct fashion. Thus, although factors may be seen as serving a useful role in resolving this estimation difficulty or that measurement problem, they are at times viewed as nothing more than mathematical artifacts created by the model. The gist of the critique is not without foundation, since misapplication of the model is not uncommon. There is a difficulty, however, in accepting the argument that just because factors are not directly observable they are bereft of all "reality." Such a viewpoint seems to equate the concept of reality with that of direct observability (in principle or otherwise), a dubious and inoperative criterion at best, since many of our observations emanate from indirect sources. Likewise, whether factors correspond to real phenomena is essentially an empirical rather than a mathematical question, and depends in practice on the nature of the data, the skill of the practitioner, and the area of application. For example, it is important to bear in mind that correlation does not necessarily imply direct causation, or that when nonsensical variables are included in an analysis, particularly under inappropriate assumptions or conditions, very little is accomplished. On the other hand, in carefully directed applications involving the measurement of unobservable or difficult-to-observe variables—such as the true magnitude of an earthquake, extent and/or type of physical pain, political attitudes, empirical index numbers, general size and/or shape of a biological organism, the informational content of a signal or a two-dimensional image—the variables and the data are chosen to reflect specific aspects which are known or hypothesized to be of relevance. Here the retained factors will frequently have a ready and meaningful interpretation in terms of the original measurements, as estimators of some underlying latent trait(s).

Factor analysis can also be used in statistical areas, for example, in estimating time and growth functions, least squares regression models, Kalman filters, and Karhunen–Loève spectral models. Also, for optimal scoring of a contingency table, principal components can be employed to estimate the underlying continuity of a population. Such an analysis (which predates Hotelling's work on principal components—see Chapter 9) can reveal aspects of data which may not be immediately apparent. Of course, in a broader context the activity of measuring unobserved variables, estimating dimensionality of a model, or carrying out exploratory statistical analysis is fairly standard in statistical practice and is not restricted to factor models. Thus spectral analysis of stochastic processes employing the power (cross) spectrum can be regarded as nothing more than a fictitious but useful mathematical construct which reveals the underlying structure of correlated observations. Also, statisticians are frequently faced with the problem of estimating dimensionality of a model, such as the degree of a polynomial regression or the order of an ARMA process. Available data are generally used to provide estimates of missing observations whose original values cannot be observed. Interestingly, recent work using maximum likelihood estimation has confirmed the close relationship between the estimation of missing data and factor analysis, as indicated by the EM algorithm. Finally, the everyday activity of estimating infinite population parameters, such as means or variances, is surely nothing more than the attempt to measure that which is fundamentally hidden from us but which can be partially revealed by careful observation and appropriate theory. Tukey (1979) has provided a broad description of exploratory statistical research as

> ...an attitude, a state of flexibility, a willingness to look for those things that we believe are not there, as well as for those we believe might be there... its tools are secondary to its purposes.

This definition is well suited to factor and other latent variable models and is employed (implicitly or explicitly) in the text.

The time has thus perhaps come for a volume such as this, the purpose of which is to provide a unified treatment of both the theory and practice of factor analysis and latent variables models. The interest of the author in the subject stems from earlier work on latent variables models using historical and social time series, as well as attempts at improving certain least squares regression estimators. The book is also an outcome of postgraduate lectures delivered at the University of Kent (Canterbury) during the 1970s, together with more recent work. The volume is intended for senior undergraduate and postgraduate students with a good background in statistics and mathematics, as well as for research workers in the empirical sciences who may wish to acquaint themselves better with the theory of latent variables models. Although stress is placed on mathematical and statistical theory, this is generally reinforced by examples taken from the various areas of the natural and social sciences as well as engineering and medicine. A rigorous mathematical and statistical treatment seems to be particularly essential in an area such as factor analysis where misconception and misinterpretations still abound. Finally, a few words are in order concerning our usage of the term "factor analysis," which is to be understood in a broad content rather than the more restricted sense at times encountered in the literature. The reason for this usage is to accentuate the common structural features of certain models and to point out essential similarities between them. Although such similarities are not always obvious when dealing with empirical applications, they nevertheless become clear when considering mathematical-statistical properties of the models. Thus the ordinary principal components model, for example, emerges as a special case of the weighted (maximum likelihood) factor model although both models are at times considered to be totally distinct (e.g., see Zegura, 1978). The term "factor analysis" can thus be used to refer to a class of models that includes ordinary principal components, weighted principal components, maximum likelihood factor analysis, certain multidimensional scaling models, dual scaling, correspondence analysis, canonical correlation, and latent class/latent profile analysis. All these have a common feature in that latent root and latent vector decompositions of special matrices are used to locate informative subspaces and estimate underlying dimensions.

This book assumes on the part of the reader some background in calculus, linear algebra, and introductory statistics, although elements of the basics are provided in the first two chapters. These chapters also contain a review of some of the less accessible material on multivariate sampling, measurement and information theory, latent roots and latent vectors in both the real and complex domains, and the real and complex normal distribution. Chapters 3 and 4 describe the classical principal components model and sample-population inference; Chapter 5 treats several extensions and modifications of principal components such as Q and three-mode analysis, weighted principal components, principal components in the complex field, and so forth. Chapter 6 deals with maximum likelihood and weighted factor models together with factor identification, factor rotation, and the estimation of factor scores. Chapters 7–9 cover the use of factor models in conjunction with various types of data such as time series, spatial data, rank orders, nominal variables, directional data, and so forth. This is an area of multivariate theory which is frequently ignored in the statistical literature when dealing with latent variable estimation. Chapter 10 is devoted to applications of factor models to the estimation of functional forms and to least squares regression estimators when dealing with measurement error and/or multicollinearity.

I would like to thank by colleagues H. Howlader of the Department of Mathematics and Statistics, as well as S. Abizadeh, H. Hutton, W. Morgan, and A. Johnson of the Departments of Economics Chemistry, and Anthropology, respectively, for useful discussions and comments, as well as other colleagues at the University of Winnipeg who are too numerous to name. Last but not least I would like to thank Judi Hanson for the many years of patient typing of the various drafts of the manuscript, which was accomplished in the face of much adversity, as well as Glen Koroluk for help with the computations. Thanks are also owed to Rita Campbell and Weldon Hiebert for typing and graphical aid. Of course I alone am responsible for any errors or shortcomings, as well as for views expressed in the book.

Alexander Basilevsky

Winnigep, Manitoba
February 1994

Contents

Statistical Factor Analysis
and Related Methods

CHAPTER 1

Preliminaries

1.1 INTRODUCTION

Since our early exposure to mathematical thinking we have come to accept the notion of a variable or a quantity that is permitted to vary during a particular context or discussion. In mathematical analysis the notion of a variable is important since it allows general statements to be made about a particular member of a set. Thus the essential nature of a variable consists in its being identifiable with any particular value of its domain, no matter how large that domain may be. In a more applied context, when mathematical equations or formulas are used to model real life phenomena, we must further distinguish between a deterministic variable and a probabilistic or random variable. The former features prominently in any classical description of reality where the universe is seen to evolve according to "exact" or deterministic laws that specify its past, present, and future. This is true, for example, of classical Newtonian mechanics as well as other traditional views which have molded much of our comtemporary thinking and scientific methodology.

Yet we know that in practice ideal conditions never prevail. The world of measurement and observation is never free of error or extraneous, nonessential influences and other purely random variation. Thus laboratory conditions, for example, can never be fully duplicated nor can survey observations ever be fully verified by other researchers. Of course we can always console ourselves with the view that randomness is due to our ignorance of reality and results from our inability to fully control, or comprehend, the environment. The scientific law itself, so the argument goes, does not depend on these nuisance parameters and is therefore fixed, at least in principle. This is the traditional view of the role of randomness in scientific enquiry, and it is still held among some scientific workers today.

Physically real sources of randomness however do appear to exist in the real world. For example, atomic particle emission, statistical thermodynamics, sun spot cycles, as well as genetics and biological evolution all exhibit random behavior over and above measurement error. Thus randomness does not seem to stem only from our ignorance of nature, but also constitutes an important characteristic of reality itself whenever natural or physical processes exhibit instability (see Prigonine and Stengers, 1984). In all cases where behavior is purely or partially random, outcomes of events can only be predicted with a probability measure rather than with perfect certainty. At times this is counterintuitive to our understanding of the real world since we have come to expect laws, expressed as mathematical equations, to describe our world in a perfectly stable and predictable fashion. The existence of randomness in the real world, or in our measurements (or both), implies a need for a science of measurement of discrete and continuous phenomena which can take randomness into account in an explicit fashion. Such a science is the theory of probability and statistics, which proceeds from a theoretical axiomatic basis to the analysis of scientific measurements and observations.

Consider a set of events or a "sample space" S and a subset A of S. The sample space may consist of either discrete elements or may contain subsets of the real line. To each subset A in S we can assign a real number $P(A)$, known as "the probability of the event A." More precisely, the probability of an event can be defined as follows.

Definition 1.1. A probability is a real-valued set function defined on the closed class of all subsets of the sample space S. The value of this function, associated with a subset A of S, is denoted by $P(A)$. The probability $P(A)$ satisfies the following axioms.*

(1) $P(S) = 1$

(2) $P(A) \geq 0$, all A in S

(3) For any r subsets of S we have $P(A_1 \cup A_2 \cup \cdots \cup A_r) = P(A_1) + P(A_2) + \cdots + P(A_r)$ for $A_i \cap A_j = \emptyset$ the empty set, $i \neq j$

From these axioms we can easily deduce that $P(\emptyset) = 0$ and $P(S) = 1$, so that the probability of an event always lies in the closed interval $0 \leq P(A) \leq 1$. Heuristically, a zero probability corresponds to a logically impossible event, whereas a unit probability implies logical certainty.

Definition 1.2. A real variable X is a real valued function whose domain is the sample space S, such that:

(1) The set $\{X \leq x\}$ is an event for any real number x

(2) $P(X = \pm\infty) = 0$

This definition implies a measurement process whereby a real number is assigned to every outcome of an "experiment." A random variable can

---

* Known as the Kolmogorov axioms.

therefore be viewed as either a set of discrete or continuous measurements. At times a finer classification is also employed, depending on whether the random variable is ordinal, nominal, or consists of a so-called ratio scale (Section 1.5).

Once events or outcomes of experiments are expressed in terms of numerical values they become amenable to arithmetic computation, as well as algebraic rigor, and we can define functions for random variables much in the same way as for the "usual" mathematical variables. A random variable is said to have a probability function $y = f(x)$ where x is any value of the random variable X and y is the corresponding value of the function. It is convenient to further distinguish between continuous and discrete probability functions.

Definition 1.3. Let X be a continuous random variable. Then $f(x)$ is a continuous probability function if and only if

(1) $f(x) \geq 0$

(2) $\int f(x)\, dx = 1$

over values x for which X is defined. If A represents an interval of x, then A is an event with probability

(3) $P(A) = P(X \in A) = \int_A f(x)\, dx$

Definition 1.4. Let X be a discrete random variable. Then $f(x)$ is a discrete probability function if and only if

(1) $f(x) \geq 0$

(2) $\Sigma_x f(x) = 1$

over values (finite or countably infinite) for which X is defined. If A represents an interval of x, then A is an event with probability

(3) $P(A) = P(x \in A) = \Sigma_A f(x)$

Functions of the type $f(x)$ are known as univariate probability functions or distributions since they depend on a single random variable X. Johnson and Kotz (1969, 1970), for example, have tabulated many such functions. A probability distribution can be given either algebraically in closed form or numerically as a table of numbers. In both cases a distribution relates a single value of $f(x)$ with some value x of X. A distribution can also be characterized by a set of parameters, and we frequently require such parameters, which determine important characteristics of $f(x)$ such as location, spread, skewness, and other "shape" properties. Such parameters, if they exist, are known as moments. The kth moment of a probability function $f(x)$, about the origin, is defined as

$$\mu_k = \int_{-\infty}^{\infty} x^k f(x)\, dx \tag{1.1}$$

for a continuous probability function and

$$\mu_k = \sum_x x^k f(x) \tag{1.2}$$

for a discrete probability function. Usually only the first few moments are of interest, and in practice k rarely exceeds 4. The first moment about the origin, known as the expectation or the mean of the random variable, is given by

$$E(X) = \int_{-\infty}^{\infty} x f(x)\, dx \tag{1.3}$$

and

$$E(X) = \sum_x x f(x)\, dx \tag{1.4}$$

for continuous and discrete probability functions, respectively. The following properties of the mean value can be derived from Eqs. (1.3) and (1.4). Let X be any random variable and let c and k denote constants. Then

$$E(c) = c \tag{1.5a}$$

$$E(cX) = cE(X) \tag{1.5b}$$

$$E(k + cX) = k + cE(X) \tag{1.5c}$$

and E is a linear operator. Equation (1.5) generalizes to any number of random variables.

The second moment about the origin of a random variable, obtained by setting $k = 2$ in Eqs. (1.1) and (1.2) is also of major importance. It is usually adjusted for the mean value yielding the expression

$$E[x - E(X)]^2 = \int_{-\infty}^{\infty} [x - E(X)]^2 f(x)\, dx \tag{1.6}$$

known as the second moment about the mean, or the variance of a continuous random variable X, with summation replacing the integral for a discrete random variable. The variance is usually denoted as $\mathrm{var}(X)$ or simply σ^2. The square root of σ of Eq. (1.6) is then known as the standard deviation of X. From Eq. (1.6) we obtain the useful identity

$$\sigma^2 = E[x - E(X)]^2$$
$$= E(X^2) - E(X)^2 \tag{1.7}$$

where $E(X^2)$ is the second moment about the origin and $E(X)^2$ is the square of the expected value. Analogously to Eq. (1.5) we also have

$$\mathrm{var}(c) = 0 \tag{1.8a}$$

$$\mathrm{var}(cX) = c^2\,\mathrm{var}(X) \tag{1.8b}$$

$$\mathrm{var}(k + cX) = c^2\,\mathrm{var}(X) \tag{1.8c}$$

As implied by the term, the variance determines the variability or spread of a random variable in the sense that it measures the expected or average squared distance of a random variable from its mean $E(X)$.

Example 1.1. Consider the normal probability function

$$f(x) = \frac{1}{\sigma\sqrt{2\pi}} \exp\left[-\frac{1}{2}\left(\frac{x - \mu}{\sigma}\right)^2 \right] \tag{1.9}$$

which plays a major role in statistical sampling theory. It can be shown that Eq. (1.9) satisfies Definition 1.3 and is therefore a probability function. Also $E(X) = \mu$, $\mathrm{var}(X) = \sigma^2$, which in turn determine the location and spread of the normal distribution. The normal probability function appears in Figure 1.1 for two alternative values of μ and σ.

The normal probability function (Eq. 1.9) as well as its multivariate version (Section 2.8) play a central role in statistical inference, construction of confidence intervals, and estimation theory. In turn normality gives rise to the chi-squared, F, and t distributions. $\qquad\square$

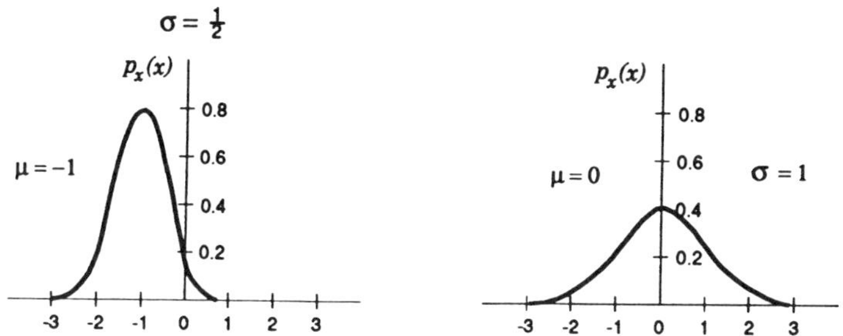

Figure 1.1 The normal probability function for two alternative values of μ, σ.

1.2 RULES FOR UNIVARIATE DISTRIBUTIONS

1.2.1 The Chi-Squared Distribution

Consider n normal random variables $X_1, X_2, \ldots, X_n$ which are independent of each other and which possess zero mean and unit variance. Then the sum of squares

$$\chi = X_1^2 + X_2^2 + \cdots + X_n^2$$

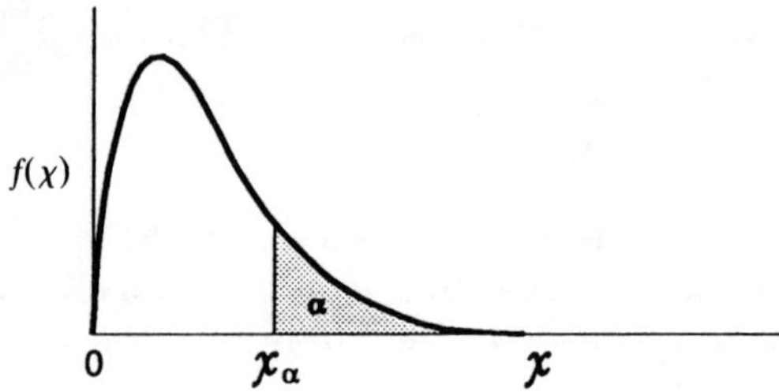

Figure 1.2 The chi-squared probability function for $n = 10$ degrees of freedom.

is distributed as the (central) chi-squared distribution

$$f(\chi) = \frac{1}{2^{n/2}\Gamma\left(\dfrac{n}{2}\right)} x^{(n-2)/2} \exp\left(-\frac{1}{2}x\right) \qquad 0 \le \chi < \infty \qquad (1.10)$$

with n degrees of freedom, where $E(\chi) = n$ and $\mathrm{var}(\chi) = 2n$. As is illustrated in Fig. 1.2 for $n = 10$ degrees of freedom the chi-squared distribution is skewed, but as n increases, χ tends to the standard normal distribution. When the random variables X_i $(i = 1, 2, \ldots, n)$ do not have unit variance and zero mean, they can be standardized by the transformation $Z_i = (X_i - \mu)/\sigma$ $(i = 1, 2, \ldots, n)$ and the sum of squares $Z_1^2 + Z_2^2 + \cdots + Z_n^2$ has chi-squared distribution with n degrees of freedom. When μ is not known (but σ is) and a normal sample $x_1, x_2, \ldots, x_n$ of size n is available, then

$$\sum_{i=1}^{n} \frac{(x_i - \bar{x})^2}{\sigma^2}$$

is distributed as a chi-squared random variable with $n - 1$ degrees of freedom.

1.2.2 The F Distribution

Let χ_1 and χ_2 be independent chi-squared random variables with m, n degrees of freedom respectively. Then the ratio

$$F = \frac{\chi_1/m}{\chi_2/n} \qquad (1.11)$$

is distributed as the F distribution:

$$f(F) = \frac{\Gamma[(m+n)/2]}{\Gamma(m/2)\Gamma(n/2)} \left(\frac{m}{n}\right)^{m/2} F^{(m-2)/2} \left(1 + \frac{m}{n}F\right)^{-(m+n)/2} \qquad (1.12)$$

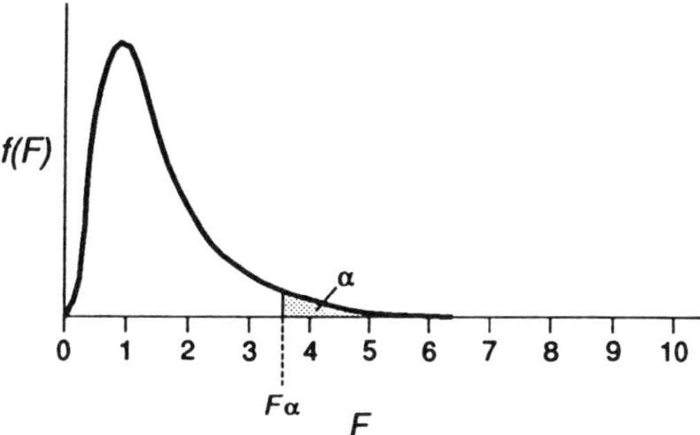

Figure 1.3 The F probability function for $m = 10$, $n = 10$ degrees of freedom.

with m and n degrees of freedom. We have

$$E(F) = \frac{n}{n-2}, \qquad \mathrm{var}(F) = \frac{2n^2(m+n-2)}{m(n-2)^2(n-4)}$$

The F distribution is also skewed (Fig. 1.3) and when m is fixed and n increases, Eq. (1.12) approaches the chi-squared distribution with m degrees of freedom. Since the chi-squared itself approaches the standardized normal, it follows the limit of Eq. (1.12) as m, $n \rightarrow \infty$ is also the normal distribution.

1.2.3 The t Distribution

Let $\bar{x}$, s^2 be the sample mean, variance of a normally distributed random variable. Then the standardized expression

$$t = \frac{(\bar{x} - \mu)}{s/\sqrt{n}}$$

is distributed as the t distribution (Fig. 1.4) with n degrees of freedom where $E(t) = 0$, $\mathrm{var}(t) = n/(n-2)$. The t distribution is a special case of the F distributions since for $m = 1$, $F = t^2$ in Eq. (1.12) and we obtain

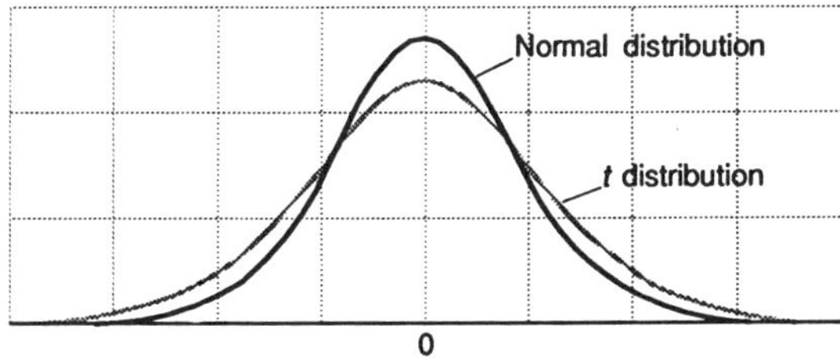

Figure 1.4 The Student t and the normal probability functions for $n = 6$ degrees of freedom and $\mu = 0$, $\sigma = 1$.

$$f(t) = \frac{\Gamma[(n+1)/2]}{\Gamma\left(\frac{1}{2}\right)\Gamma\left(\frac{n}{2}\right)\sqrt{n}}\left(1 + \frac{t^2}{n}\right)^{-(n+1)/2} \tag{1.13}$$

When z is a standard normal variate, the quotient

$$t = \frac{z}{\sqrt{\chi^2/n}}$$

is also distributed as Eq. (1.13), which can be seen by letting $m = 1$ and $F = t^2$ in Eq. (1.12).

1.3   ESTIMATION

Probability provides the basis for the notion of a random variable. One way in which randomness can be introduced is by random sampling. For example, in scientific experimentation where random variables are continuous and the sample space is thus infinite, a finite set of experiments can only represent an estimate of the underlying reality. Because of further random factors such as measurement error, variation in materials and/or subjects, and other diverse but minor external influences beyond the experimenter's control, any measured experimental quantity will necessarily possess some random variation. Even when the population is finite, pragmatic considerations may prohibit a total analysis of that population owing to practical factors such as cost or human effort. This is the case, for example, in sample surveys of human or inanimate populations where randomness is introduced, over and above imprecision of measurement and natural variation, by the artificial random sampling process. Thus, given the presence of random variation, a methodology of inference is required, which permits statements to be made concerning unobserved populations once a finite set of sample measurements is available. The situation is fundamentally unavoidable since in empirical work, unlike that of mathematical algebra or logic, the process of (logical) deduction is replaced by that of induction. Whereas logical deduction proceeds from the superset to a subset, induction works in the opposite direction by attempting to infer properties of the superset from those of a subset. All that can be done here is to ensure that the subset chosen, (i.e., the sample) "represents" any subset in that it does not differ significantly or systematically from any other subset which could have been chosen. The theoretical study of such equivalence between sample subsets, and thus between sample subsets and supersets (population), is the realm of sampling and estimation theory.

1.3.1 Point Estimation: Maximum Likelihood

Consider n observations made for a random variable X, such that its probability function $f(x)$ depends on r unknown parameters $(\theta_1, \theta_2, \ldots, \theta_r)$. This can be expressed by writing $f(x; \theta_1, \theta_2, \ldots, \theta_r)$. Our purpose is to estimate the parameters θ_i, given n sample values of X.

Definition 1.5. Any function of the sample observation is known as a statistic or an estimator. A particular value of an estimator is called an estimate.

Given a population parameter there exists, in principle, a large number of estimators from which we can usually choose. We wish to select those estimators that possess optimal precision or accuracy, defined in some sense. A well-known optimality principle which is frequently used is that of *maximum likelihood*.

Definition 1.6. The likelihood function of n random variables X_1, $X_2, \ldots, X_n$ is the joint distribution $L(X_1, X_2, \ldots, X_n; \theta)$ where θ is a set of unknown parameters. Clearly, when a set of observations $x_1, x_2, \ldots, x_n$ is given, the likelihood function depends only on the unknown parameters θ.

A likelihood function resembles a probability function except that it is not sufficiently restricted in the sense of Definitions 1.3 and 1.4.

Example 1.2. Consider the normal probability function Eq. (1.9) and a set of random independent observations $x_1, x_2, \ldots, x_n$. We have two parameters to estimate, $\theta_1 = \mu$ and $\theta_2 = \sigma^2$. For the ith observation we have

$$f(x_i) = \frac{1}{\sigma\sqrt{2\pi}} \exp\left\{-\frac{1}{2}[(x_i - \mu)/\sigma]^2\right\} \qquad (i = 1, 2, \ldots, n)$$

and the joint likelihood function of all the observations is

$$L(\mu,\sigma) = \frac{1}{\sigma\sqrt{2\pi}} \exp\left\{-\frac{1}{2}[(x_1 - \mu)/\sigma]^2\right\} \frac{1}{\sigma\sqrt{2\pi}} \exp\left\{-\frac{1}{2}[(x_2 - \mu)/\sigma]^2\right\}$$

$$\cdots \frac{1}{\sigma\sqrt{2\pi}} \exp\left\{-\frac{1}{2}[(x_n - \mu)/\sigma]^2\right\}$$

$$= \frac{1}{\sigma^n(2\pi)^{n/2}} \exp\left\{-\frac{1}{2}\sum_{i=1}^{n}\left(\frac{x_i - \mu}{\sigma}\right)^2\right\} \tag{1.14}$$

Equation (1.14) depends only on the unknown parameters μ and σ. The principle of maximum likelihood estimation is to find those values of μ and σ, that maximize Equation (1.14). Thus, in an a posteriori sense, the likelihood of the sample $x_1, x_2, \ldots, x_n$ is maximized. Given that the sample is taken randomly (and independently) by assumption, it should therefore represent the unknown population of values of μ and σ as best as is possible, in the sense that the sample estimates $\hat{\mu}$ and $\hat{\sigma}$ are the most probable or the most likely to appear. If the likelihood function possesses a derivative at μ and σ, it can be maximized in a straightforward fashion by setting these derivatives equal to zero. Actual numerical solutions however may not exist in closed form, and thus require successive iteration to obtain an adequate numerical approximation. As well, the function Eq. (1.14) may not possess a global maximum point and we then say that maximum likelihood estimators do not exist. $\qquad\square$

Example 1.3. Consider a normally distributed population where we wish to find the maximum likelihood estimators of μ and σ. Taking natural logarithms to facilitate differentiation, Eq. (1.14) becomes

$$L(\mu,\sigma) = -n \ln \sigma - \frac{n}{2} \ln(2\pi) - \frac{1}{2} \sum_{i=1}^{n} [(x_i - \mu)/\sigma]^2 \qquad (1.15)$$

which possesses the same maxima as Eq. (1.14) since a logarithmic transformation is monotonic, that is, order preserving. The partial derivatives of Eq. (1.15) with respect to μ,σ are

$$\frac{\partial L}{\partial \mu} = -\frac{2}{2} \sum_{i=1}^{n} [(x_i - \mu)/\sigma](-1/\sigma)$$

$$\frac{\partial L}{\partial \sigma} = -\frac{n}{\sigma} - \frac{2}{2} \sum_{i=1}^{n} [(x_i - \mu)/\sigma][-(x_i - \mu)/\sigma^2]$$

and setting to zero yields the estimators

$$\hat{\mu} = \frac{\sum_{i=1}^{n} x_i}{n} = \bar{x}, \qquad \hat{\sigma}^2 = \sum_{i=1}^{n} \frac{(x_i - \mu)^2}{n}$$

Since μ is not known it is replaced by the estimator $\hat{\mu} = \bar{x}$, and n by $n-1$ to ensure the unbiasedness of $\hat{\sigma}^2$. $\qquad\square$

Maximizing the likelihood function is but one procedure that can be used to estimate unknown parameters. Different procedures yield competing estimators, but we wish to pick out that estimator which possesses optimal accuracy. The concept of accuracy consists of two independent estimation criteria:

1. *Unbiasedness, consistency.* An estimator $\hat{\theta}$ of θ is said to be unbiased if and only if $E(\hat{\theta}) = \theta$, irrespective of sample size. For a sample size n, the

estimator $\hat{\theta}_n$ of θ is consistent when it converges, in probability, to the true value θ as $n \to \infty$. We write plim $\hat{\theta}_n = \theta$.

2. *Efficiency*. An estimator $\hat{\theta}$ of θ is said to be efficient if it possesses minimum variance. It is relatively efficient if for some other estimator $\hat{\Omega}$, we have $\text{var}(\hat{\theta}) \leq \text{var}(\hat{\Omega})$.

The criterion of consistency can also be described heuristically as unbiasedness in large samples, or assymptotic unbiasedness. Thus although an unbiased estimator will always be consistent, the converse is not necessarily true. When a choice of estimators is available, a common two-stage strategy is to consider only unbiased estimators, if they exist, and then to select those that possess minimum variance. The strategy however is not always optimal since a biased estimator can possess small variance and thus be more precise or accurate than an unbiased estimator with high variance (Fig. 1.5). Thus since variance is only defined for an unbiased estimator, a more general strategy is to consider the so-called mean squared error (MSE) criterion, where $\text{MSE} = \text{variance} + (\text{bias})^2$ (Fig. 1.6). The MSE provides a simulta-

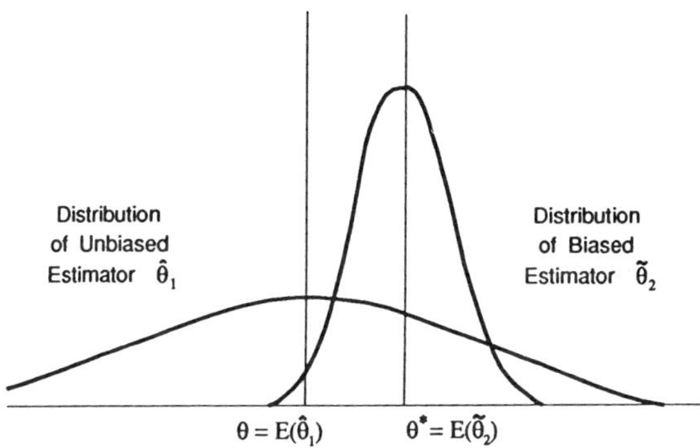

Figure 1.5 Sampling distributions of an unbiased, inefficient estimator $\hat{\theta}$ and a biased efficient estimator $\tilde{\theta}$.

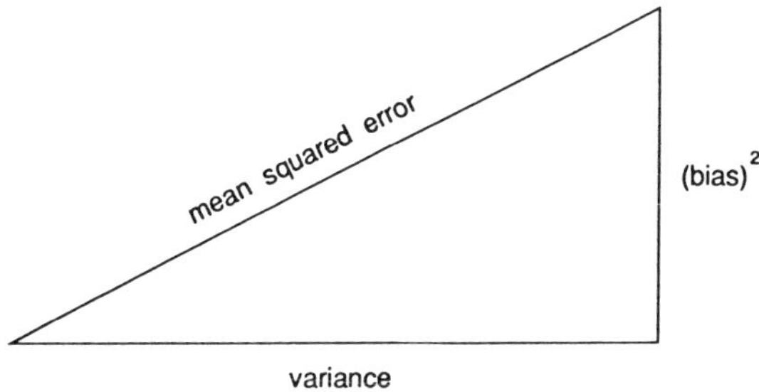

Figure 1.6 Mean squared error as a sum of squares of two orthogonal components—the squared standard deviation and the square of the bias.

neous measure of accuracy for an estimator both in terms of variance as well as bias.

The classical approach to estimation however is to avoid biased estimators. This is due in part to a reluctance to use estimators which, on the average, "miss the mark." Also, the bias may not be a decreasing function of the sample size but may increase as the sample size increases. In any case, it is often impossible to estimate the magnitude of the bias. However, if the bias is small, a biased estimator may be superior to one that is unbiased but is highly inefficient. Thus in Figure 1.5, for example, the biased estimator $\tilde{\theta}$ has a higher probability of being closer to the true value θ than the unbiased estimator $\hat{\theta}$. The reason lies in their variances, and this gives $\tilde{\theta}$ a smaller MSE, and thus better overall accuracy.

1.3.2 The Likelihood Ratio Criterion

A problem closely related to estimation is that of determining decision rules concerning parameters of probability functions. For example, given a random sample $x_1, x_2, \ldots, x_n$ we wish to test the null hypothesis $H_0:\mu = c$ against the composite alternative $H_a:\mu \neq c$.

More generally we may be interested in testing r parameters $\theta_1, \theta_2, \ldots, \theta_r$ belonging to a known probability function. Statistically powerful tests can be constructed by the generalized likelihood ratio criterion

$$\lambda = \frac{L(\hat{\Omega}_0)}{L(\hat{\Omega})} \tag{1.16}$$

where Ω_0 is the parameter space defined by a range of possible values under H_0, Ω_a the disjoint parameter space under H_a and $\Omega = \Omega_0 \cup \Omega_a$. Here $L(\hat{\Omega}_0)$ is the maximum of the likelihood function in Ω_0, that is, the likelihood function for which the unknown parameter(s) have been replaced by their maximum likelihood estimators. The term $L(\hat{\Omega}_a)$ is defined in a similar manner. We accept H_0 if $\lambda > k$, where k is a predetermined arbitrary constant and $0 < \lambda < 1$. Thus small values of λ indicate the likelihood of the sample is small under H_0 and we tend to accept H_a at some probability level α. In other words, we choose k such that $P(\lambda < c|H_0 \text{ is true}) \leq \alpha$, the probability of Type I error.

Example 1.4. Consider a sample of size n taken from a normal population with known variance σ^2 and unknown mean μ. The likelihood function Eq. (1.14) can be expressed as

$$L(\mu) = \left(\frac{1}{\sigma\sqrt{2\pi}}\right)^n \exp\left[-\frac{1}{2\sigma^2}\sum_{i=1}^{n}(x_i - \mu)^2\right]$$

$$= \left(\frac{1}{\sigma\sqrt{2\pi}}\right)^n \exp\left[-\frac{1}{2\sigma^2}\sum_{i=1}^{n}(x_i - \bar{x} + \bar{x} - \mu)^2\right]$$

$$= \left(\frac{1}{\sigma\sqrt{2\pi}}\right)^n \exp\left[-\frac{1}{2\sigma^2}\left\{\sum_{i=1}^{n}(x_i - \bar{x})^2 + \sum_{i=1}^{n}(\bar{x} - \mu)^2\right\}\right]$$

$$= \left(\frac{1}{\sigma\sqrt{2\pi}}\right)^n \exp\left[\frac{1}{\sigma^2}\left\{-\frac{1}{2}\sum_{i=1}^{n}(x_i - \bar{x})^2 - \frac{n}{2}(\bar{x} - \mu)^2\right\}\right]$$

Here the space Ω_0 consists of the point $\mu = c$, whereas Ω consists of the whole real μ axis. Thus

$$L(\hat{\Omega}_0) = \left(\frac{1}{\sigma\sqrt{2\pi}}\right)^n \exp\left[\frac{1}{\sigma^2}\left\{-\frac{1}{2}\sum_{i=1}^{n}(x_i - \bar{x})^2 - \frac{n}{2}(\bar{x} - c)^2\right\}\right]$$

and

$$L(\hat{\Omega}) = \left(\frac{1}{\sigma\sqrt{2\pi}}\right)^n \exp\left[\frac{1}{\sigma^2}\left\{-\frac{1}{2}\sum_{i=1}^{n}(x_i - \bar{x})^2\right\}\right]$$

where μ is replaced by c (under H_0) and by its maximum likelihood estimator $\bar{x}$ respectively. The likelihood ratio Eq. (1.16) is then

$$\lambda = \exp\left[-\frac{n}{2\sigma^2}(\bar{x} - c)^2\right] \tag{1.17}$$

The proper critical region for testing H_0 is the interval $0 < \lambda < k$, where $k < 1$ is chosen to yield the desired level of α. Let $\alpha = 0.05$. Then we chose k such that

$$\int_0^k h(\lambda | H_0 \text{ is true}) \, d\lambda = 0.05 \tag{1.18}$$

where $h(\lambda | H_0 \text{ is true})$ is the probability function of λ when $\mu = c$. From Eq. (1.17) we have

$$-2 \ln \lambda = \frac{(\bar{x} - c)^2}{\sigma^2 / n} \tag{1.19}$$

which, when H_0 is true ($\mu = c$), is distributed as the chi-squared distribution with one degree of freedom. The degrees of freedom are equal to the number of parameters determined by H_0. When the sample is taken from a normal population, Eq. (1.19) is an exact chi-squared random variable, and when the population is other than normal, the distribution of Eq. (1.19) approaches the chi-squared probability function as a limit, with increasing n. Let $\gamma = -2 \ln \lambda$, a chi-squared random variable. Then using tables the critical region is given by

$$\frac{(\bar{x} - c)^2}{\sigma^2 / n} > 3.84$$

and the 95% confidence interval for the normal sample mean is

$$\bar{x} > c + \frac{1.96}{\sigma/\sqrt{n}}, \qquad \bar{x} < c - \frac{1.96}{\sigma/\sqrt{n}} \qquad (1.20)$$

Alternatively the right-hand tail of the chi-squared distribution can be employed directly. We compute $-2 \ln \lambda$ and compare it with the critical value for $\alpha = 0.05$ and 1 degree of freedom. $\qquad \square$

Example 1.5. The likelihood ratio Eq. (1.17) is based on the assumption that σ^2 is known. We now consider the case when σ^2 is replaced by its maximum likelihood estimator (Example 1.3). We have

$$\hat{\mu} = \frac{1}{n} \sum_{i=1}^{n} x_i = \bar{x}$$

$$\hat{\sigma}^2 = \frac{1}{n} \sum_{i=1}^{n} (x_i - \bar{x})^2$$

and substituting these values in $L(\hat{\Omega}_0)$ and $L(\hat{\Omega})$ we obtain

$$L(\hat{\Omega}) = \left[\frac{n \sum_{i=1}^{n} (x_i - \bar{x})^2}{2\pi} \right]^{n/2} \exp(-n/2)$$

$$L(\hat{\Omega}_0) = \left[\frac{n \sum_{i=1}^{n} (x_i - c)^2}{2\pi} \right]^{n/2} \exp(-n/2)$$

where in Ω_0 we set $\mu = c$ in accordance with H_0. The likelihood ratio is then given by

$$\lambda = \frac{L(\hat{\Omega}_0)}{L(\hat{\Omega})} = \left[\frac{\sum_{i=1}^{n} (x_i - \bar{x})^2}{\sum_{i=1}^{n} (x_i - c)^2} \right]^{n/2} \qquad (1.21)$$

To determine a critical value k for which the critical region $0 < \lambda < k$ corresponds to a Type I probability, we proceed as follows. Using the identity

$$\sum_{i=1}^{n} (x_i - c)^2 = \sum_{i=1}^{n} (x_i - \bar{x})^2 + n(\bar{x} - c)^2$$

λ can be expressed as

$$\lambda = \left[\cfrac{1}{1 + \left[n(\bar{x} - c)^2 \Big/ \sum_{i=1}^{n} (x_i - \bar{x})^2 \right]} \right]^{n/2} \tag{1.22}$$

The critical region is then $\lambda < k$ so that

$$\cfrac{1}{1 + n(\bar{x} - c)^2 \Big/ \sum_{i=1}^{n} (x_i - \bar{x})^2} < k^{2/n}$$

or

$$t = \frac{(\bar{x} - c)}{s/\sqrt{n}} > \sqrt{(n-1)k'} \tag{1.23}$$

where

$$s^2 = \sum_{i=1}^{n} (x_i - \bar{x})^2 / n - 1 \quad \text{and} \quad k' = 1/k^{2/n} - 1 \qquad \square$$

Equation (1.23) is the t test with $(n-1)$ degrees of freedom. Most well-known tests can also be derived using the likelihood ratio and this provides a unifying approach to the theory of statistical testing. When the distribution of λ is not known, we know that for a large sample $-2 \ln \lambda$ tends to be distributed as chi-squared, with degrees of freedom equal to the number of fixed constants assigned by H_0.

The normal probability function figures prominently in a discussion when a random sample $x_1, x_2, \ldots, x_n$ is assumed to be drawn from a normal population. This need not always be the case. It is known however from the Central Limit Theorem that sums of independent, standardized random variables tend to be distributed approximately as the standard normal probability function. If a process is multiplicative, the Central Limit Theorem can still be invoked, but in terms of logarithms. Since much of statistical testing tends to involve sample means, the Central Limit Theorem makes the univariate normal probability function of great practical importance. The univariate normal in turn can be generalized to that of a multivariate normal probability function which plays a central role in statistical multivariate analysis (Section 2.8). The multivariate normal however is a special case of the so-called multivariate probability function.

1.4 NOTIONS OF MULTIVARIATE DISTRIBUTIONS

The preceding section dealt briefly with several key concepts relating to univariate probability functions of the form $y = f(x)$. More frequently, however, we observe phenomena that are outcomes of several random variables operating jointly. This suggests a generalization of the univariate distribution $f(x)$ to a multivariate distribution of the form $y = f(x_1, x_2, \ldots, x_p)$.

Definition 1.7. Let $X_1, X_2, \ldots, X_p$ be a set of p continuous random variables. Then $f(x_1, x_2, \ldots, x_p)$ is a continuous multivariate probability function if and only if

(1) $f(x_1, x_2, \ldots, x_p) \geq 0$

(2) $\int\int \cdots \int f(x_1, x_2, \ldots, x_p)\, dx_1\, dx_2 \cdots dx_p = 1$

over values $x_1, x_2, \ldots, x_p$ for which $X_1, X_2, \ldots, X_p$ are defined. If $A_1, A_2, \ldots, A_p$ represent intervals of $X_1, X_2, \ldots, X_p$, that is, $A_i = a_i \leq x_i \leq b_i$ $(i = 1, 2, \ldots, p)$, then $A = (a_1 \leq x_1 \leq b_1; a_2 \leq x_2 \leq b_2; \ldots, a_p \leq x_p \leq b_p)$ is termed an event and has probability

$$P(A) = P(a_1 \leq x_1 \leq b_1; a_2 \leq x_2 \leq b_2; \ldots, a_p \leq x_p \leq b_p)$$

$$= \int_{a_p}^{b_p} \cdots \int_{a_2}^{b_2} \int_{a_1}^{b_1} f(x_1, x_2, \ldots, x_p)\, dx_1\, dx_2 \cdots dx_p$$

$P(A)$ represents "volume" in pth dimensional space.

Definition 1.8. Let $X_1, X_2, \ldots, X_p$ be a set of p discrete random variables. Then $f(x_1, x_2, \ldots, x_p)$ is a discrete probability function if and only if

(1) $f(x_2, x_2, \ldots, x_p) \geq 0$

(2) $\sum_{x_p} \cdots \sum_{x_2} \sum_{x_1} f(x_1, x_2, \ldots, x_p) = 1$

A multivariate distribution captures the joint or simultaneous effect of a set of interrelated random variables. At times it is necessary to know the distribution of a subset of these variates, irrespective of the distribution of the remaining variables. Such distributions are known as marginal distributions.

Definition 1.9. Consider a set of p continuous random variables $X_1, X_2, \ldots, X_r, X_{r+1}, \ldots, X_p$. The marginal distribution of $X_1, X_2, \ldots, X_r$ is defined as

$$g(x_1, x_2, \ldots, x_r) = \int_{-\infty}^{\infty} \cdots \int_{-\infty}^{\infty} \int_{-\infty}^{\infty} f(x_1, x_2, \ldots, x_p)\, dx_{r+1}\, dx_{r+2} \cdots dx_p$$

Marginal distributions of discrete random variables are defined in a similar way, with summations replacing integrals. In particular, the marginal distribution of a single variate, say the first, can be obtained by integrating the multivariate distribution over the remaining $p - 1$ random variables.

A marginal distribution of a subset of random variables $X_1, X_2, \ldots, X_r$ is thus a probability function which does not involve the remaining set X_{r+1},

$X_{r+2}, \ldots, X_p$, that is, it describes the probabilistic behavior of the r variates irrespective of the remaining $p - r$ variables.

When dealing with a multivariate set of random variables questions of independence often arise, which are of crucial importance in theoretical and applied work.

Definition 1.10. Let $f(x_1, x_2, \ldots, x_p)$ be a multivariate distribution with marginal distributions $g_1(x_1)$, $g_2(x_2), \ldots, g_p(x_p)$. Then the random variables $X_1, X_2, \ldots, X_p$ are said to be independent if and only if

$$f(x_1, x_2, \ldots, x_p) = \prod_{i=1}^{p} g_i(x_i) \tag{1.24}$$

for all values within the range of the random variables for which the distribution $f(x_1, x_2, \ldots, x_p)$ is defined.

Both the concepts of joint and marginal distributions lead to yet a third type of probability distribution, that of a conditional probability function.

Definition 1.11. Consider a marginal distribution of a set of random variables $X_1, X_2, \ldots, X_r$. The conditional probability function of X_{r+1}, $X_{r+2}, \ldots, X_p$, given that the complementary set $X_1, X_2, \ldots, X_r$ takes on values $x_1, x_2, \ldots, x_r$, is given by

$$h(x_{r+1}, x_{r+2}, \ldots, x_p | x_1, x_2, \ldots, x_r) = \frac{f(x_1, x_2, \ldots, x_p)}{g(x_1, x_2, \ldots, x_r)} \tag{1.25}$$

A conditional distribution describes the probabilistic behavior of a set of random variables, when a complementary set is held fixed. When the two sets are independent, Eq. (1.25) can be written as

$$h(x_{r+1}, x_{r+2}, \ldots, x_p | x_1, x_2, \ldots, x_r) = \frac{f(x_1, x_2, \ldots, x_p)}{g(x_1, x_2, \ldots, x_r)}$$

$$= \frac{g(x_1, x_2, \ldots, x_r) h(x_{r+1}, x_{r+2}, \ldots, x_p)}{g(x_1, x, \ldots, x_r)}$$

$$= h(x_{r+1}, x_{r+2}, \ldots, x_p)$$

that is, the conditional and marginal distributions are identical only when the two sets are independent. Here $x_1, x_2, \ldots, x_r$ contribute no information toward the distribution of $x_{r+1}, x_{r+2}, \ldots, x_p$.

Notions of dependence (independence) play an important role in multivariate analysis. Using Definition 1.10 we see that a set of random variables is independent if and only if the joint distribution can be factored into a product of marginal distributions. Thus it is unnecessary to consider multivariate distributions if each random variable is independent. What defines "multivariateness" is therefore not the availability of more than a single random variable, but the fact that variables are interdependent.

In multivariate analysis the two most widely used measures of dependence are the covariance and its standardized version the correlation coefficient, defined as follows. Let X and Y be continuous random variables. The rth and the sth product moment about the origin of their joint distribution is defined as

$$\mu'_{r,s} = E(X^r Y^s) = \int_{-\infty}^{\infty} \int_{-\infty}^{\infty} x^r y^s f(x, y)\, dx\, dy \qquad (1.26)$$

When X and Y are discrete the double integral in Eq. (1.26) is replaced by double summation signs. Let μ_x and μ_y be the means of the marginal distributions of X and Y as defined by Eq. (1.3). Then

$$\mu_{r,s} = E[(X - \mu_x)^r (Y - \mu_y)^s]$$
$$= \int_{-\infty}^{\infty} \int_{-\infty}^{\infty} (x - \mu_x)^r (y - \mu_y)^s f(x, y)\, dx\, dy \qquad (1.27)$$

is known as the rth and sth product moment about the mean. The covariance between X and Y is obtained from Eq. (1.27) by setting $r = s = 1$. Symbolically the covariance is denoted by σ_{xy} or $\mathrm{cov}(X, Y)$ and can be evaluated in terms of moments about the origin as

$$\sigma_{xy} = E(XY) - E(X)E(Y) \qquad (1.28)$$

Since $-\sigma_x \sigma_y \le \sigma_{xy} \le \sigma_x \sigma_y$ we also have

$$\rho_{xy} = \frac{E[(X - \mu_x)(Y - \mu_y)]}{[E(X - \mu_x)^2 E(Y - \mu_y)^2]^{1/2}} = \frac{\sigma_{xy}}{\sigma_x \sigma_y} \qquad (1.29)$$

where $-1 \le \rho_{xy} \le 1$. The standardized quantity ρ_{xy} is known as the correlation coefficient. The covariance has the following two properties:

$$\mathrm{cov}[(X + c), (Y + k)] = \mathrm{cov}(X, Y)$$
$$\mathrm{cov}(cX, kY) = ck\, \mathrm{cov}(X, Y)$$

Let $Y = c_1 X_1 + c_2 X_2 + \cdots + c_p X_p$ be a linear combination of p random variables. Then the variance of Y is

$$\mathrm{var}(Y) = \sum_{i=1}^{p} \sum_{j=1}^{p} c_i c_j \sigma_{ij} \qquad (1.30)$$

where σ_{ij} is the covariance between X_i, X_j. The covariance between two linear combinations $Y_1 = c_1 X_1 + c_2 X_2 + \cdots + c_p X_p$, $Y_2 = d_1 X_1 + d_2 X_2 + \cdots + d_p X_p$ is the bilinear form

$$\mathrm{cov}(Y_1, Y_2) = \sum_{i=1}^{p} \sum_{j=1}^{p} c_i d_j \sigma_{ij} \tag{1.31}$$

From Eq. (1.28) when X and Y are independent $\sigma_{xy} = 0$, but the converse is not true.

1.5 STATISTICS AND THE THEORY OF MEASUREMENT

As we have discussed, random variables can be grouped into two broad categories, depending on whether they are continuous or discrete. At times it is useful to partition the two categories into still smaller subdivisions to reflect certain specific properties that are being measured. For consistency and clarity it is desirable to base such a classification on the theory of mathematical transformations. A useful byproduct of the exercise is that we are able to define the process of "measurement" in more general terms than is at times thought possible, which permits a uniform treatment of diverse scientific processes, both in the social and the natural sciences. We are also in a better position to define clearly what constitutes a measurement and whether certain data can legitimately be used in conjunction with some particular statistical model.

It is usual to consider two basic types of measurement processes; the so-called fundamental measurement and derived measurement. In this section we restrict ourselves to the former since derived measurement constitutes a major topic of factor analysis and is considered in other chapters. One of the earliest attempts to define fundamental measurement in mathematical terms is that of Russell (1937) who defined measurement as follows:

> Measurement of magnitudes is, in its most general sense, any method by which a unique and reciprocal correspondence is established between all or some of the magnitudes of a kind and all or some of the numbers, integral, rational, or real, as they may be.

Russell's definition is essentially that which is used today, albeit in modified form. Given a physical or a conceptual "object," we can always define properties of that object, a particular choice of properties being determined by the needs of the problem at hand. For instance, in an opinion poll we can discern properties of human beings such as "age," "sex," "occupation," "voting preference," or "income." Or given a chemical material we may be interested in properties such as atomic composition, molecular weight, whether the product is a fluid or a solid, viscosity if the former or melting point (if it exists) if the latter, and so forth. Similarly a biological experiment may be carried out to study the effect(s) of a parasitic worm on mammal tissue under various conditions. Measurement is then a process whereby such properties are related to sets of numbers by certain predetermined rules. Before considering these rules in more detail we develop the notions of a relation and that of a mapping or function.

1.5.1 The Algebraic Theory of Measurement

A measurement process concerns itself with a numerical description* or "quantification" of properties of objects, and as such implies a process of comparison whereby every element of a set is compared with other elements of the set. This in turn introduces the important notion of a relationship. Consider a nonempty set A consisting of a collection of objects possessing a common property. The inclusion of an element within a set is denoted by $a \in A$, whereas $a \notin A$ denotes "a is not in the set A." We can define three operations for sets in terms of their elements.

Definition 1.12. Let A and B be any two sets. Then for some element x

(1) The union of two sets A and B, written as

$$A \cup B = \{x : x \in A \quad \text{or} \quad x \in B\}$$

consists of all elements that are either in A or in B or both.

(2) The intersection of two sets A and B, written as

$$A \cap B = \{x : x \in A \quad \text{and} \quad x \in B\}$$

consists of all elements that are in both A and B.

(3) The difference between two sets A and set B, written as

$$A - B = \{x : x \in A \quad \text{and} \quad x \notin B\}$$

consists of all elements x that are in A but not in B.

The empty set $\emptyset$ is defined as that set which contains no elements and is therefore contained in every set. A set contained in another set is called a subset. Two sets A and B are equal if and only if they contain identical elements, that is, if A is contained in B and vice versa. When A is contained in B but is not equal to B we write $A \subset B$ and say that A is a proper subset of B.

The elements of a set need not stand in any particular order, and these sets are known as ordered sets.

---

* It can be argued that measurement can also proceed without the use of numbers (see Roberts, 1979).

Definition 1.13. Let a,b denote any two elements of a set A. Then the pair a,b is ordered if and only if

$$(a,b) = (c,d)$$

implies $a = c$ and $b = d$.

Definition 1.13 can be extended to any number of elements. Note that it is insufficient for ordered sets to simply possess equal elements in order to be considered equal, since the equal elements must also appear in an identical order.

Definition 1.14. Let A and B be any two sets. Then the Cartesian product $A \times B$ is the ordered set

$$A \times B = \{(a, b); a \in A \quad \text{and} \quad b \in B\}$$

The notion of a Cartesian product can be extended to any number of sets.

Definition 1.15. A binary relation R in a set A is a subset of $A \times A$, that is, $R \subset A \times A$. We write aRb for any two elements $(a, b) \in R \leq A$ where $a \in A$ and $b \in A$.

The elements of the relation $R \subset A \times A$ are ordered in pairs. Conversely, if some (or all) elements of A are ordered in pairs we obtain a relation R. For this reason we speak of a relation within a set. A special type of relationship which is important in measurement theory is that of an equivalence relation.

Definition 1.16. A relation R in A is an equivalence relation if and only if the following three properties hold:
(1) R is reflexive, that is, aRa for all $a \in A$.
(2) R is symmetric, that is, aRb implies bRa for all $a \in A$ and $b \in A$.
(3) R is transitive, that is, aRb and bRc implies aRc for all $a, b, c \in A$.

The set A, together with one (or more) relation R, is also known as a relational system. Thus the relation "equal to" is an equivalence relation but "less than" is not.

An equivalence relation enables elements of A to be partitioned into nonoverlapping and exhaustive sets known as equivalence classes. Let R be an equivalence relation of a set A. Then the equivalence class of $a \in A$, with respect to R, is the set

$$[a] = \{b \in A : aRb\} \tag{1.32}$$

The equivalence class of any element $a \in A$ is thus the set of all elements of

A for which the equivalence relation *R* holds. The following theorem can be proved.

THEOREM 1.1.   Let *R* be an equivalence relation on the set *A*. Then the set of equivalence classes forms a partition of *A*, that is, *R* induces a partition of *A* into mutually exclusive and exhaustive subsets.

So far we have considered relations defined on sets that do not necessarily contain numbers, since a set *A* can consist of any objects so long as they possess some common property. To define a scale of measurement, however, we must be able to relate the elements of a relation *R* with a set of numerical values such that they capture the essential property of these elements. Evidently a numerical representation of this type cannot, in general, be unique. For example, sample points can consist of discrete outcomes such as "heads" or "tails" of a coin-tossing experiment, denoted by 1 and 0, or at the other extreme we may be interested in continuous outcomes such as agricultural yield, temperature, or money incomes earned by individuals. In each of these examples we are concerned with a number associated with an outcome. Since such outcomes are generally stochastic, or contain stochastic components, we are in fact dealing with values of random variables.

Definition 1.17.   A relational system (or a relational structure) consisting of a set *A* and relations $R_1, R_2, \ldots, R_m$ is known as an algebra if and only if all relations are arithmetic operations. It is a *k*-dimensional numerical relational system if and only if $A = R^k$ where *R* is the set of real numbers and R^k is the *k*-dimensional plane. It is an empirical relational system if the elements of *A* are physical objects (events).

A relational system thus consists of a set *A* with one or more relations defined on that set. A numerical relational system consists of mathematical operations defined over the multidimensional set of real numbers $A = R^k$. There exist certain relations known as functions (mappings, transformations) which are of special interest in statistical work.

Definition 1.18.   A function *f* from set *A* into another (or the same) set *B* is a subset of the Cartesian product $A \times B$ such that
   (1) For any element $x \in A$ there exists an element $y \in B$ such that $(x, y) \in f$
   (2) $(x, y) \in f$ and $(x, z) \in f$ imply $y = z$

A function *f* thus represents a collection of ordered pairs (x, y) which satisfy Definition 1.18; that is, for every element *x* of *A* there is at least one corresponding element *y* in *B*. Given an element *x*, the element *y* associated with it must therefore be unique. A function can be represented by an

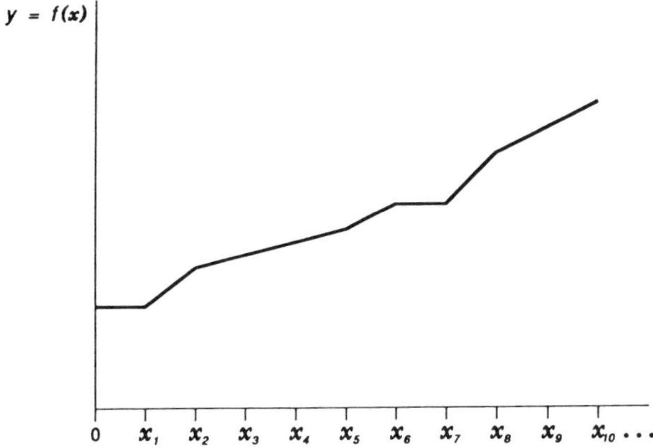

Figure 1.7   An arbitrary function $y = f(x)$ observed at discrete points $x_1, x_2, \ldots$.

algebraic equation $y = f(x)$, by a table of numbers, or by a graph (Fig. 1.7). Here x is known as the independent variable and y as the dependent variable. The element y is also known as the image of x under f, and x is known as a preimage of y under f. Definition 1.18 can be extended to a function of k variables, that is, to the Cartesian product $A_1 \times A_2 \times \cdots A_k$. In terms of the sets A, B we can write $f : A \to B$, indicating that f is a function (mapping) of A into B as shown in Figure 1.8. The range of $f : A \to B$ need not equal the codomain B since in general $y \in B$ need not have a preimage in A under f.

Definition 1.19.   Let A, B be the domain and codomain, respectively, of some function $f : A \to B$. Then the transformation $f : A \to B$, which preserves certain operative rules, is known as a homomorphism.

A homomorphism can be either "into" or "onto." If it is both, that is, if f is one-to-one (and onto), it is known as an isomorphism. More specifically, consider a binary operation $\circ$ defined on set A. Then the transformation

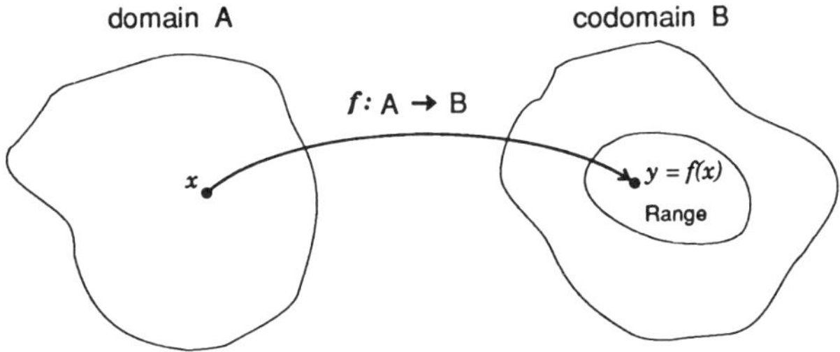

Figure 1.8   A transformation of elements of set A (domain) into set B (codomain).

$f: A \rightarrow A$ is a homomorphism if and only if $f(x_1 \circ x_2) = f(x_1) \circ f(x_2)$ for any two elements x_1, x_2 of A. Two common operations, for example, are addition and multiplication. Here f is a homomorphism if and only if $f(x_1 x_2) = f(x_1)f(x_2)$ and $f(x_1 + x_2) = f(x_1) + f(x_2)$, that is, if multiplication and addition are preserved. The logarithmic transformation is not a homomorphism since it does not preserve addition or multiplication.

Definition 1.20. Consider set A, consisting of specified empirical objects or properties (attributes), together with m empirical relations R_1, $R_2, \ldots, R_m$ and the set R of real numbers with relations S_1, $S_2, \ldots, S_m$. Then a real-valued function ϕ on A is a homomorphism if it takes each R_i from the empirical relational system into S_i, the numerical relational system, $i = 1, 2, \ldots, m$.

More generally we can have n sets A_1, $A_2, \ldots, A_n$, m relations R_1, $R_2, \ldots, R_m$, defined on the Cartesian product $A_1 \times A_2 \times \cdots A_n$ and a vector-valued homomorphism ϕ with components ϕ_1, $\phi_2, \ldots, \phi_n$ such that ϕ takes each R_i into S_i, each ϕ_i being defined on A_i, $i = 1, 2, \ldots, n$.

Homomorphisms are important in measurement theory since measurement can be considered as the construction of homomorphisms from empirical relational structures into numerical relational systems. The mapping must of course be a homomorphism since arithmetic operations defined between measurements must preserve equivalent empirical relations. A set of measurements is also known as a measurement scale. A scale can be either univariate or multivariate.

Definition 1.21. Consider a set A together with m empirical relations R_1, $R_2, \ldots, R_m$ and the r-dimensional plane R^r with relations S_1, $S_2, \ldots, S_m$. Then an r-dimensional scale is a homomorphism of the empirical relational system into the r-dimensional numerical relational system.

These definitions determine in a general sense what a measurement scale is, but they do not specify how to construct such scales. The actual scale construction is largely a practical issue which depends on the objective(s) of the analysis, the type of phenomenon under consideration, and the extent and quality of information and data available. There is no single method for constructing measurements, and scales are therefore not unique since an infinite number of scales can exist which will map homomorphically any given empirical relational system into a numerical relational system. Thus if no single criterion is available to select a scale from members of its equivalence class, we are left with the additional task of having to chose that particular scale which will meet the objectives of the analysis. Many different types of scales exist, and over the years they have been classified into six main groupings depending on the relative uniqueness of the set of permissible mappings.

1.5.2   Admissible Transformations and the Classification of Scales

Since a given measurement scale is not uniquely determined, the question arises as to the degrees of freedom available when specifying a particular scale. Thus for a given representation characterizing the relationship of an empirical relational system, a scale may be required to preserve the ordering of elements, the ordering of differences between values assigned to elements, and so forth. If a given scale represents an empirical relational system adequately, any other scale is also permitted (and is referred to as an admissible transformation) if it preserves the representational relation between empirical and numerical entities. There are six major measurement scales which are widely employed in the natural and social sciences, and which are based on the positive affine linear mapping:

$$y = a + bx \qquad (b > 0) \tag{1.33}$$

Equation (1.33) preserves both order and interval, a property which is of crucial importance to scientific measurement. In what follows we therefore consider only unidimensional and linear scales, since multidimensional and nonlinear scales can be obtained by extension.

Absolute Scale

The only unique scale that can be defined is the absolute scale, which possesses a unique unit of measure. Thus if x is an absolute scale, the only admissible transformation is $y = x$. The best known example of an absolute scale is counting the number of elements in a set (e.g., the binomial random variable x). Data of this type are usually referred to as count data (Chapter 8).

Ratio Scales

A more familiar type of scale is the so-called ratio scale, which conforms more closely to our intuitive motion of what constitutes a measurement. It is mainly employed for properties that correspond to continuous quantities such as length, intervals of time, heat, mass, age, money income and so forth. Such scales are also known as quantitative scales since they possess a natural zero origin, order, and differential gradation. Here the empirical relational system is mapped homomorphically into a numerical relational system (the scale) which possesses the following properties:

1. Existence of a natural zero, so that complete absence of a quantity corresponds to the number zero on the scale, and vice versa. For example, a

zero reading on a chronometer only makes sense when no time has elapsed, and zero on the Kelvin thermometer implies a complete (theoretical) absence of heat. The Centigrade or Fahrenheit scales, on the other hand, do not represent a ratio scale for measuring heat since the choice of origin (the zero point) is arbitrary.

2. Ordinal property: let x, y be readings on a ratio scale. Then $x < y$ only when x denotes a lesser quantity than y.

3. Distance property: let x_1, y_1 and x_2, y_2 represent readings on a ratio scale such that $x_1 > y_1$ and $x_2 > y_2$, and let $(x_1 - y_1) = d_1$ and $(x_2 - y_2) = d_2$. Then $d_1 = d_2$ implies equal differences in the magnitudes of the readings. For example, when measuring length the difference d_1 between two lengths x_1, y_1 must be the same as the difference d_2 between two other lengths x_2, y_2 in order that $x_1 = x_2$, $y_1 = y_2$.

Clearly a ratio scale is not unique since it can be dilated (or compressed) by a constant factor of proportionality. A common example is converting meters into yards or yards into meters. Thus if x is a ratio scale, then

$$y = bx \qquad (b > 0) \qquad (1.34)$$

is also a ratio scale and Eq. (1.34) is the only admissible transformation for this type of measurement. Any other transformation will destroy properties 1–3 given above. Note that ratios of the form x/y are constant (invariant) and that ratios of the scale are also ratio scales. Clearly Eq. (1.34) can be easily extended to nonlinear relations, such as polynomials, by suitable transformations on x, y. Ratio scales are used to represent physical quantities and in this sense express the cardinal aspect of a number. In the scientific and engineering literature ratio scales are also known as physical variables.

Difference Scale

The origin of a ratio scale is unique and can only be represented by the number zero. If we remove this restriction we have what is known as a difference scale. Difference scales are thus not unique since their origins can assume any real number. For any different scale x, the linear transformation

$$y = a + x \qquad (1.35)$$

is also a difference scale. Thus although ratio scales must possess the same origin (but may differ in the unit of measure b), difference scales must possess the same unit of measure ($b = 1$) but may only differ in the origin. Difference scales can therefore be used in place of ratio scales when the true zero origin is either unknown or does not exist. Such scales are not often employed, but an example would be the measurement of historic time based on a religious calendar. Although the unit of measure is the same (the yearly cycle of the earth revolving around the sun), the origin is arbitrary since the starting times of the calendars coincide with different events.

Interval Scales

A more general type of scale, of which the difference scale is a special case, is the interval scale. Interval scales derive their name from the fact that the only meaningful comparison between readings on such scales is between intervals. Thus if x is any interval scale,

$$y = a + bx \qquad (b > 0) \qquad (1.36)$$

is also an interval scale and Eq. (1.36) is the only admissible transformation for such scales. Many examples exist of linear interval scales, the best known being the Centigrade and Fahrenheit thermometers. Both are related by the equation $C = (5/9)(F\text{-}32)$ and neither possesses a natural origin since $C = 0$ and $F = 0$ do not imply total absence of heat. The ordinal property on the other hand is preserved since $90°C > 80°C$, for example, holds only when the first reading corresponds to a greater heat content than the second. The distance property also holds since the difference $(90°C - 80°C) = 10°C$ is the same as $(50°C - 40°C) = 10°C$. Other examples of an interval scale include geological measurements such as the Wadell measure of sphericity (round-ness), grain size, and isotopic abundance. A closely related scale is the log-interval scale whose admissible transformation is

$$y = ax^{b} \qquad (a, b > 0) \qquad (1.37)$$

It derives its name from the fact that a logarithmic transformation yields an interval scale.

Ordinal Scales

An ordinal scale possesses no natural origin, and distance between points of the scale is undefined. It simply preserves ranks of the elements of an empirical set. It is usual to use the positive integers to denote "first," "second," ... , "nth" position in the ranking although this is not essential. If x is an ordinal scale and $f(x)$ is a strictly monotonically increasing continuous function of x, then $y = f(x)$ is also an ordinal scale. Of course in practice such scales tend to be discrete rather than continuous. Ordinal scales are used, for example, to represent preferences and judgments, ordering winners of tournaments, classifying (ranking) objects, and generally as indicators of relative magnitude such as Moh's hardness scale in geology. They can also be used in place of ratio scales when we have excessive measurement error, that is, when exact values are not known. Clearly ordinal scales represent orders of magnitude or the ordinal aspect of numbers.

Nominal Scales

Finally, when the ordinal property is removed we obtain the so-called nominal scale. A nominal scale is discrete and simply indicates in which group a given object is found. Since nominal scales possesses none of the three "quantitative" properties of a ratio scale, they also are known as qualitative or categorical scales, and are simply characterized by the notion of equivalence (Definition 1.16). Some authors do not consider nominal scales to be measurements at all, since they impart no quantitative information but represent pure qualities. They do however fall within the realm of definitions of measurement theory and are frequently encountered in practice. It was seen in Theorem 1.1 that the set of equivalence classes forms a partition of a set, that is, an equivalence relation partitions a set A into mutually exclusive and exhaustive subsets. Such a partition is equivalent to the forming of mutually exclusive and exhaustive categories, and this is sufficient to constitute a nominal measure. Thus a nominal scale is a one-to-one homomorphism of equivalence classes into the real numbers, where the equivalence relation is "equal to." The permissible transformation for a nominal scale is any one-to-one mapping from the real numbers R into themselves. For example, consider four different types of pesticides labeled A, B, C, and D. A nominal scale can be constructed from the four mutually exclusive and exhaustive categories by replacing them by the integers 1, 2, 3, and 4. These integers however are arbitrary, since any real numbers can be used in their place so long as no two are the same. They impart no numerical information and are merely used as convenient labels. Other examples of nominal scales include the male–female classification, type of occupation, presence or absence of a treatment in an experiment, and so forth. In Chapter 8 we see that since nominal scales cannot be manipulated using standard arithmetic operations they are usually replaced by the so-called "dummy" variables, taking on values of 0, 1, or any other suitable integers. Nominal scales are also closely related to the absolute scale since the only numerical information they impart (indirectly) is a count of the number of elements contained in a category or a set. This also includes relative frequencies such as probabilities and percentages associated with discrete events.

1.5.3   Scale Classification and Meaningful Statistics

The particular scale classification described in the previous section was first considered by Stevens (1946; see also Anderson, 1961; Pfanzagl, 1968), although the essential notions implicit in the scales are not new. Stevens' classification and terminology were initially adopted by the psychological and social sciences, but have since spread into other disciplines and are by no means unknown in the statistical sciences as well (Thomas, 1985). It is not the only system available or possible and Coombs et al. (1954), for example, have argued for a still finer classification. Stevens' classification however is widely used today and has become a convenient peg on which to hang a discussion of the application of stochastic models to data analysis. Some authors have even gone overboard with attempts to superimpose the concepts of measurement onto statistical theory and practice, without the apparent realization that the appropriate conceptualization for statistical measurement is accomplished by a more general theoretical structure—that of mathematical probability theory. The end result of such notions has often been a restrictively rigid and subjective view of what is thought to be "meaningful" in statistics (however, see Gaito, 1980). Thus a given statistical model, so the argument goes, can only be used with an "appropriate" type of measurement scale (together with its admissible transformations); for example, ordinal scales can only be used in conjunction with a "nonparametric" model, and so forth. Although this is relevant for many empirical applications, it is not valid in any universal sense, since the more fundamental distinction in statistics is that between a continuous and a discrete random variable.

As we have noted, the basic purpose of a scale is to convey, in numerical form, empirical information in the most convenient and direct way possible. This in turn facilitates computation and interpretation of the results. Also, since scales relate an empirical relational system to a numerical relational system, their construction must be guided by mathematical as well as physical considerations. The type of scale finally chosen therefore depends on the object or phenomena under study, the scope and purpose of the analysis, and the quality of data available. The situation can also be aggravated by the presence of measurement error, missing data, and unobservable or difficult-to-observe variables. For example, when variables are not observable directly, they must be estimated from those that are. Here the type of scale required for a population is not necessarily the same as that which is available in a sample. A question then arises as to the applicability of a particular statistical model for a given measurement scale and vice versa. Two misconceptions are common in the nonstatistical literature. Individual features of certain scales are at times ignored, and this leads to nonsense results—for example, when employing an incorrect dummy-variable coding in a regression ANOVA model (design) when estimating factor effects. On the other end of the spectrum, at times attempts are made to place overly rigid restrictions on the type of computation that can (or should) be carried out for a given scale. For example, arithmetic operations, it is at times claimed, can only be applied to ratio or interval scales if we are to obtain meaningful statistics (e.g., see Marcus-Roberts and Roberts, 1987). Also factor and regression analyses, we are cautioned, are not to be used with discrete (ordinal or nominal) data if we are to avoid "meaningless statistics" (Katzner, 1983). Not only is such a view contrary to much of statistical practice, it also ignores the fact that meaningfulness is not merely a function of scale–arithmetic interrelationships but depends on the a priori objectives of a statistical model, the mechanics of the actual computations, and the interpretability of the final result(s). The rationale for using a particular scale is therefore intimately bound with a particular statistical analysis or model and the scientific hypothesis to be tested or physical mechanism to be revealed. The use of a particular scale is, in the final analysis, valid because statistical tests (estimators) concern, in the first instance, numbers and not the particular type of scale used. This is because arithmetic operations do not involve units of measurement, and because scale values can always be redefined to yield interpretable results. The numbers themselves clearly cannot distinguish from what type of scale they have originated, although this must always be kept in mind by the analyst. Also, estimation is normally carried out in samples, not in populations, so that although observations may consist of ordinal or nominal scales this need not be the case for population values. Inherent measurement error and unobservability of population values frequently makes it impossible to measure sample observations on the same scale as the population. To reiterate, as long as understandable and useful interpretation is possible, any type of scale can be used in conjunction with any type of statistical model. For further discussion the reader is referred to Coombs et al. (1954), Adams et al. (1965), and Anderson et al. (1983a). A comprehensive and extensive treatment of the algebraic theory of measurement has also been provided by Roberts (1979).

1.5.4 Units of Measure and Dimensional Analysis for Ratio Scales

Generally speaking a measurement can be considered as a relationship between a real number and some phenomenon or object(s) under study. A more traditional and restrictive view common in the physical sciences is to consider measurement as a process of quantification, the objective being to determine the degree of "quantity" that can be associated with some object, material or otherwise. Here the concept of measurement is inherently bound up with relating quantities to ratio scales, usually by the use of a suitable apparatus which registers the readings. The particular advantage of using ratio scales, apart from purposes of recording and classifying data, is to enable us to relate empirical information to the entire (positive) real number system, which in turn allows the use of mathematical expressions to process or manipulate data. The end result, it is hoped, is a mathematical equation that describes or reflects law-like behavior of some specific aspect of the natural world, which can be used to predict future events. A question then arises as to which physical variables (units of measure) can be used together consistently within a single equation. Clearly, if we are to avoid "mixing apples and oranges," the units of measure must also conform with respect to the arithmetic operations implied by the equation. In other words, an equation should not only hold numerically but must also balance in terms of the symbols and concepts assigned to the physical variables. Note that the principle can also be used to assign units to the constants or parameters of an equation. Thus the actual symbols themselves are immaterial—all that is required is that they be employed in a consistent manner. The concept is known as the principle of dimensionality and was first introduced into physics by Fourier in 1822 in his well-known book *Théorie Analytique de la Chaleur*, where the comment is made that every physical variable or constant in a physical equation has its own "dimension," and that terms of the same equation cannot be compared if they do not have the same apparent dimension. The principle has since become a prominent feature of physics and engineering (Ipsen, 1960).

Example 1.6. Consider the familiar expression for an accelerating object. We have

$$\text{acceleration} = \text{velocity}/\text{time}$$
$$= \frac{\text{distance}/\text{time}}{\text{time}}$$
$$= \text{distance}/\text{time}^2$$

a quadratic function of time.

The reason for introducing rules of functional manipulation to units of measure is to be able to establish fundamental and consistent relationships within the four physical coordinate axes. In more empirical work however this is not always possible, for example, when using linear expressions of the form

$$y = b_0 + b_1 x_1 + b_2 x_2 + \cdots + b_p x_p$$

to study and/or predict the empirical behavior of some variable y based on a knowledge of p independent variables. Here disparate units are frequently used for the variables, particularly in exploratory research or in areas such as the social sciences, where it is difficult to use dimensional consistency (see, however, De Jong, 1967). As well, variables may be unit-free index numbers or ratios measured with respect to some common base. The principle, however, can be used to identify the constants as rates of change since

$$b_i = \frac{\text{no. of units of } y}{\text{unit of } x_i} \tag{1.38}$$

$\square$

1.6 STATISTICAL ENTROPY

We sometimes wish to compare the degree of concentration of two or more sets of probabilities. Three well-known indices of the form $\Sigma_i x_i p_i$ are available, where p_i is the probability measuring the share of the ith element and x_i is the weight attached to the probability. The distinction is arbitrary since we can also think of p_i as weights, for example, as in the case of mathematical expectation. The importance of such concentration indices lies in their use as measures of the degree of redundancy (information) or entropy of a system. The best known of such measures is the entropy index, which is derived from the negative binomial distribution. Consider an experiment with two equally likely outcomes, and let x represent the trial number on which we obtain a "success" outcome. Then the probability $p = f(x)$ of obtaining a success on the xth trial is given by

$$f(x) = p = (1/2)^x \tag{1.39}$$

and taking logarithms (base 2) we have

$$x = -\log_2 p \qquad x = 1, 2, \ldots \tag{1.40}$$

Equation (1.40) is also known as the information content of an event or a message. Note that here the more probable an event the less "information" it contains, and in the limit as $p \to 1$ we have $x \to 0$. Thus the term "information" here refers to empirical information, that is, to the occurrence of an event and not to its truth (or falsehood) or psychological content. An analogy which can be used in conjunction with Eq. (1.40) is that of a chess board, where a player makes a mental note of some particular square, the location of which the other player is to guess. Without prior information the optimal strategy is to divide the board into two equal areas, and under the binary response of "yes" or "no" to successively subdivide the board until the unknown square is found. Of course in this analogy the process is finite; generally the process is infinite. Using Eq. (1.40) the total entropy or information of the system can be defined as

$$
\begin{aligned}
I &= \sum_x f(x)x \\
&= -\sum_{i=1}^n p_i \log_2 p_i \\
&= \sum_{i=1}^n p_i \log_2 \frac{1}{p_i}
\end{aligned}
\tag{1.41}
$$

which is the expected (average) information content of an experiment consisting of n trials. Equation (1.41) is also known as the Shannon information index. Using $x = \log_2 p$ we may think of the expression as a measure of concentration of the probabilities. Two other measures that can also be used are the Herfindahl index,

$$c = \sum_{i=1}^n p_i p_i \tag{1.42}$$

where each probability is weighted by itself, and the Hall–Tideman index,

$$c = \sum_{i=1}^{n} i p_i \tag{1.43}$$

where the weights are taken as the rank order i of each probability in the series. The entropy or information measure Eq. (1.41) however is more commonly used since it has the advantage of decomposability; that is, if a series of probabilities are split into subgroups, the concentration within the groups, plus that between the groups, will equal the concentration of the overall group. This has an analogy in the decomposition of variance common in statistical analysis. The Shannon information index is maximum over all probability laws at the uniform distribution (Exercise 1.6) and minimum at the normal distribution. Minimizing Eq. (1.41) therefore can serve as a criterion for normality (Example 3.1).

1.7 COMPLEX RANDOM VARIABLES

The classification of measurement scales considered in Section 1.5 assumes that the admissible transformations are defined in the real field. Evidently this is almost always the case. Not all variables employed in statistical analysis however are real. For example, in the study of multivariate measurements distributed over time (or physical space) it is not uncommon to first transform the data by means of a bivariable (univariable) Fourier transform (Chapter 7) to restore independence to the observations. The end result is a set of power cross-spectra which depend on the imaginary number $\sqrt{-1} = i$.

Consider any two real numbers x, y. A complex number z is then defined as

$$z = x + iy \tag{1.44}$$

Geometrically, a complex number can be considered as a point in the two-dimensional plane (Fig. 1.9). Let z_1, z_2 be any two complex numbers. Then their sum and product are defined as

$$\begin{aligned} z_1 + z_2 &= (x_1 + iy_1) + (x_2 + iy_2) \\ &= (x_2 + x_2) + i(y_1 + y_2) \end{aligned} \tag{1.45}$$

and

$$\begin{aligned} z_1 z_2 &= (x_1 + iy_1)(x_2 + iy_2) \\ &= x_1 x_2 + i x_1 y_2 + i y_1 x_2 + i^2 y_1 y_2 \\ &= (x_1 x_2 - y_1 y_2) + i(x_1 y_2 + x_2 y_1) \end{aligned} \tag{1.46}$$

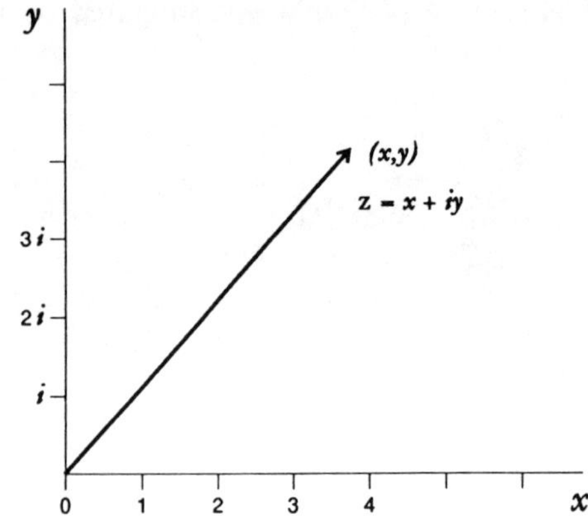

Figure 1.9   A geometric representation of a complex number Z in the two-dimensional plane.

Let

$$\bar{z} = x - iy \tag{1.47}$$

the complex conjugate of z. Then a special case of Eq. (1.46) is the product

$$
\begin{aligned}
z_1 \bar{z}_2 &= (x_1 + iy_1)(x_2 - iy_2) \\
&= (x_1 x_2 + y_1 y_2) + i(y_1 x_2 - x_1 y_2)
\end{aligned}
\tag{1.48}
$$

Equations (1.45)–(1.47) are complex, but to define the familiar concepts of length, distance, and angle we require real-valued functions. Using Eq. (1.47) we have

$$
\begin{aligned}
z\bar{z} &= (x + iy)(x - iy) \\
&= x^2 + y^2 \\
&= |z|^2
\end{aligned}
\tag{1.49}
$$

the squared magnitude (length) of a complex variable, which can be used to define real-valued functions of complex expressions.

Of particular interest is the case when z is a random variable. Since $i = \sqrt{-1}$ is fixed, z can only vary randomly when its real and imaginary parts x, y are random. This leads us to the following definition of a complex random variable.

Definition 1.22.   The (complex) expression Eq. (1.44) is said to be a complex random variable if and only if its real and imaginary parts are distributed as a (real) bivariate probability function. The definition of a complex random variable is easily extended to that of a complex multi-variate distribution.

Definition 1.23. Let Z_1, $Z_2, \ldots, Z_p$ be a set of complex random variables. Then $g(Z_1, Z_2, \ldots, Z_p)$ is a complex probability function if and only if the real and imaginary parts are distributed as $f(X_1,\ Y_1,\ X_2,\ Y_2, \ldots, X_p,\ Y_p)$, a $2p$-variate probability distribution.

For the first two moments of a complex random variable we have

$$\begin{aligned}
E(Z) &= E(X + iY) \\
&= E(X) + iE(Y) \\
&= \mu_x + i\mu_y
\end{aligned} \qquad (1.50)$$

and

$$\begin{aligned}
\operatorname{cov}(Z_1, Z_2) &= E\{[Z_1 - E(Z_1)][\overline{Z_2 - E(Z_2)}]\} \\
&= [\operatorname{cov}(x_1, x_2) + \operatorname{cov}(y_1, y_2)] + i[\operatorname{cov}(y_1, x_2) - \operatorname{cov}(x_1, y_2)]
\end{aligned}$$
$$(1.51)$$

using Eq. (1.48). Letting $Z_1 = Z_2$ it can then be shown that

$$\operatorname{var}(Z) = \sigma_x^2 + \sigma_y^2 \qquad (1.52)$$

where $\sigma_x^2 = \operatorname{var}(x)$, $\sigma_y^2 = \operatorname{var}(y)$.

EXERCISES

1.1 Prove Eq. (1.17).

1.2 Prove relationships in Eq. (1.8) using Eqs. (1.6) for the expectation.

1.3 Using Eq. (1.3) (or 1.4) prove the three results of Eq. (1.5).

1.4 Describe which scale you would use to measure the following random variables. State reasons for your choice and whether sample scale values correspond to population values.
   (a) Gold, silver, and bronze medals in an Olympic competition.
   (b) Letter grades of an examination in statistics.
   (c) Iron content in an ore sample
   (d) Artifacts found in a series of archaeological closed assemblages (e.g., graves).

    (i) Types of artifacts.

    (ii) Quantity of each type of artifact.

(e) Body length of a 3-year female sperm whale.

(f) Cross-classification of hair color versus eye color of a sample of school children.

1.5 Prove that for the entropy of a probabilistic system we have

$$0 \le \sum_{i=1}^{n} p_i \ln \left(\frac{1}{p_i} \right) \le \ln n$$

where ln represents the natural logarithm (e.g., see Shier, 1988). The entropy is thus maximized when all outcomes are equally probable (uniform distribution).

1.6 Prove the relation in Eq. (1.52), where Z is any complex random variable.

CHAPTER 2

Matrices, Vector Spaces

2.1 INTRODUCTION

Matrices together with the concept of a vector space play a leading role in the definition and interpretation of factor analysis and related multivariate models. Besides providing a unified approach to the subject, they also result in straightforward proofs and derivations of the main results, and make it possible to compare different factor model specifications without introducing unnecessary detail. In this chapter we describe, in a summary fashion, matrix topics and multivariate concepts that feature prominently in factor analysis and which the reader will find a handy reference. A more detailed account however is available in Graybill (1983), Searle (1982), and Basilevsky (1983). The main purpose of utilizing matrices in statistics is to be able to handle two-way numerical arrays of observations arranged in tables, independently of the size and complexity of such tables. Inherent in such a treatment is the concept of "dimension" for which the reader is referred to Hurewicz and Wallman (1974). The simplest example is that of a linear array of real numbers:

$$\mathbf{V} = [v_1, \ldots, v_2] \tag{2.1}$$

known as a (row) vector. The components v_i of the vector may be negative, positive, or zero and are referred to as scalars. When all elements of $\mathbf{V}$ are zero, $\mathbf{V}$ is known as a zero vector, and when $v_i = 1(i = 1, 2, \ldots, n)$ it is known as a unity vector to distinguish it from the unit vector consisting of a single unit entry with remaining elements identically zero. More generally a unit vector is any vector with unit length.

A vector may also be written as the column array

$$\mathbf{V} = \begin{bmatrix} v_1 \\ v_2 \\ \vdots \\ v_n \end{bmatrix} \tag{2.2}$$

the choice between Eqs. (2.1) and (2.2) being essentially arbitrary. In what follows we assume vectors to be column arrays of the form shown in Eq. (2.2) unless specified otherwise. Also, vectors are denoted by boldface uppercase letters and their elements as italic lowercase letters. To conserve space, in what follows Eq. (2.2) is represented in the form $\mathbf{V} = [v_1, v_2, \ldots, v_n]^T$ where superscript T denotes the operation of transposition—an operation that transforms row (column) vectors into column (row) vectors.

A matrix is a two-dimensional array of numbers

$$\mathbf{A} = \begin{bmatrix} a_{11} & a_{12} & \cdots & a_{1k} \\ a_{21} & a_{22} & \cdots & a_{2k} \\ \vdots & \vdots & & \vdots \\ a_{n1} & a_{n2} & \cdots & a_{nk} \end{bmatrix} \tag{2.3}$$

said to be of order $(n \times k)$ with typical element a_{ij}. Matrix $\mathbf{A}$ is then said to be of rectangular form. A matrix can be considered as a generalization of a vector since it is composed of both column and row vectors. Alternatively, a vector can be viewed as a reduced column or row matrix.

The transposed matrix $\mathbf{A}^T$ is one whose rows and columns have been interchanged. Clearly if $\mathbf{A}$ is $(n \times k)$, $\mathbf{A}^T$ is $(k \times n)$, that is, with k rows, n columns. A matrix can be multiplied by a scalar number, a vector array, or another matrix. When $n = k$ it can also be inverted, that is, the unique matrix inverse $\mathbf{A}^{-1}$ may exist such that $\mathbf{A}\mathbf{A}^{-1} = \mathbf{A}^{-1}\mathbf{A} = \mathbf{I}$ where $\mathbf{I}$ denotes the square identity matrix with unities on the main diagonal and zeroes elsewhere. A matrix that possesses inverse $\mathbf{A}^{-1}$ is then said to be nonsingular, otherwise $\mathbf{A}$ is a singular matrix.

Although matrices represent numerical arrays or tables, their elements can be used to define scalar functions. Two such functions are the determinant and the trace, both defined for square matrices only. The determinant of a square $(n \times n)$ matrix $\mathbf{A}$, written as $|\mathbf{A}|$, is a scalar number which can be used to obtain volume in multidimensional space. When applied to certain matrices the determinant provides a generalization of the statistical concept of (univariate) variance. The trace of a square matrix, written $\mathrm{tr}(\mathbf{A})$, is the sum of the diagonal elements of $\mathbf{A}$ and can also be used to measure the total variance contained in a set of random variables.

2.2 LINEAR, QUADRATIC FORMS

Besides being useful for manipulating large tables of data, matrices also lend themselves to the handling of large systems of linear and quadratic equations. Let $\mathbf{C}$ and $\mathbf{X}$ represent column vectors with typical elements c_i, x_i respectively. Then a linear equation can be expressed as the product

$$y = \mathbf{C}^T\mathbf{X}$$

$$= [c_1, c_2, \ldots, c_k] \begin{bmatrix} x_1 \\ x_2 \\ \vdots \\ x_k \end{bmatrix}$$

$$= c_1 x_1 + c_2 x_2 + \cdots + c_k x_k \tag{2.4}$$

by the definition of a vector product. A system of linear equations can then be expressed as

$$y_1 = a_{11}x_1 + a_{12}x_2 + \cdots + a_{1k}x_k$$
$$y_2 = a_{21}x_1 + a_{22}x_2 + \cdots + a_{2k}x_k$$
$$y_n = a_{n1}x_1 + a_{n2}x_2 + \cdots + a_{nk}x_k$$

or

$$\begin{bmatrix} y_1 \\ y_2 \\ \vdots \\ y_n \end{bmatrix} = \begin{bmatrix} a_{11} & a_{12} & \cdots & a_{1k} \\ a_{21} & a_{22} & \cdots & a_{2k} \\ \vdots & \vdots & & \vdots \\ a_{n1} & a_{n2} & \cdots & a_{nk} \end{bmatrix} \begin{bmatrix} x_1 \\ x_2 \\ \vdots \\ x_n \end{bmatrix}$$

$$\mathbf{Y} = \mathbf{AX} \tag{2.5}$$

where $\mathbf{Y}$ and $\mathbf{X}$ are $(n \times 1)$ vectors and $\mathbf{A}$ is the $(n \times k)$ matrix of coefficients. When $\mathbf{Y}$ is a vector of known coefficients and $n = k$ the system (Eq. 2.5) can be solved uniquely by computing the inverse $\mathbf{A}^{-1}$ (if it exists); that is, we can write

$$\mathbf{X} = \mathbf{A}^{-1}\mathbf{Y} \tag{2.6}$$

Quadratic forms also can be expressed in matrix notation. The general equation of second degree in the variables $x_1, x_2, \ldots, x_n$ is given by the scalar

$$y = \sum_{i=1}^{n} \sum_{j=1}^{n} a_{ij} x_i x_j \tag{2.7}$$

where it is convenient to assume $a_{ij} = a_{ji}$. Then Eq. (2.7) can be expressed uniquely in matrix form as

$$y = (x_1, x_2, \ldots, x_n) \begin{bmatrix} a_{11} & a_{21} & \cdots & a_{n1} \\ a_{21} & a_{22} & \cdots & a_{n2} \\ \vdots & \vdots & & \vdots \\ a_{n1} & a_{n2} & & a_{nn} \end{bmatrix} \begin{bmatrix} x_1 \\ x_2 \\ \vdots \\ x_n \end{bmatrix}$$

$$= \mathbf{X}^T\mathbf{AX} \tag{2.8}$$

THEOREM 2.1. The quadratic form (Eq. 2.7) can always be expressed uniquely, with respect to a given coordinate system, as $y = \mathbf{X}^\mathrm{T}\mathbf{A}\mathbf{X}$ where $\mathbf{X}$ is a vector of variables and $\mathbf{A}$ is a symmetric matrix of known coefficients.

The representational uniqueness of a quadratic form is conditional on the symmetry of $\mathbf{A}$, since a quadratic form can also be written in terms of a nonsymmetric matrix.

Definition 2.1. Let $\mathbf{X}$ and $\mathbf{Y}$ be two $(n \times 1)$ vectors. Then the inner product $\mathbf{X} \cdot \mathbf{Y}$ between $\mathbf{X}$ and $\mathbf{Y}$ is the sum of products of their components.

An inner product of two vectors can also be expressed as a matrix product, since

$$\mathbf{X} \cdot \mathbf{Y} = x_1 y_1 + x_2 y_2 + \cdots + x_n y_n$$

$$= [x_1, x_2, \ldots, x_n]\begin{bmatrix} y_1 \\ y_2 \\ \vdots \\ y_n \end{bmatrix}$$

$$= \mathbf{X}^\mathrm{T}\mathbf{Y}$$

When $\mathbf{X} = \mathbf{Y}$ we obtain

$$\mathbf{X}^\mathrm{T}\mathbf{X} = \sum_{i=1}^{n} x_i^2$$

the sum of squares of the components. This is also a special case of Eq. (2.8) when $\mathbf{A} = \mathbf{I}$, the identity matrix. Quadratic forms are not independent of the positioning of the coordinate axes. They may however be easily transformed from one set of coordinates to another by the use of the following theorem.

THEOREM 2.2. Two symmetric matrices $\mathbf{A}$ and $\mathbf{B}$ represent the same quadratic form if and only if $\mathbf{B} = \mathbf{P}^\mathrm{T}\mathbf{A}\mathbf{P}$ where $\mathbf{P}$ is nonsingular.

Quadratic forms, together with their associated symmetric matrices, can be classified into five major categories.

1. *Positive Definite*. A quadratic form $y = \mathbf{X}^\mathrm{T}\mathbf{A}\mathbf{X}$ is said to be positive definite if and only if $y = \mathbf{X}^\mathrm{T}\mathbf{A}\mathbf{X} > 0$ for all $\mathbf{X} \neq \mathbf{0}$. Here $\mathbf{A}$ is referred to as a symmetric positive definite matrix, the so-called Grammian matrix. Positive definite matrices always possess positive determinants.

Note however that a non-symmetric matrix can also have a positive determinant.

2. *Positive Semidefinite* (Nonnegative Definite). A quadratic form is said to be positive semidefinite (or nonnegative definite) if and only if $y = \mathbf{X}^T\mathbf{AX} > 0$ for all $\mathbf{X} \neq \mathbf{0}$. Here $|\mathbf{A}| > 0$ and $\mathbf{A}$ is said to be positive semidefinite.

3. *Negative Definite.* A quadratic form is said to be negative definite if and only if $y = \mathbf{X}^T\mathbf{AX} > 0$ for all $\mathbf{X} \neq \mathbf{0}$. Thus $\mathbf{A}$ is negative definite if and only if $-\mathbf{A}$ is positive definite.

4. *Negative Semidefinite* (Nonpositive definite). A quadratic form is said to be negative semidefinite (or nonpositive definite) if and only if $y = \mathbf{X}^T\mathbf{AX} \leq 0$.

5. *Indefinite.* Quadratic forms and their associated symmetric matrices need not be definite or semidefinite in any of the senses described above. In this case the quadratic form can be negative, zero, or positive depending on the values of $\mathbf{X}$.

Positive (semi) definite Grammian matrices possess the following properties.

THEOREM 2.3. Let $y = \mathbf{X}^T\mathbf{AX}$ be a positive definite quadratic form. Then:

1. The positive definiteness of y (and of $\mathbf{A}$) is preserved under nonsingular linear transformations of $\mathbf{X}$.
2. If $\mathbf{A}$ is positive definite then $\mathbf{B}^T\mathbf{AB}$ is also positive definite, where $\mathbf{B}$ is a nonsingular matrix.
3. Let $\mathbf{A}$ be symmetric. Then $\mathbf{A}$ is also positive definite if and only if it can be factored as $\mathbf{A} = \mathbf{P}^T\mathbf{P}$ where $\mathbf{P}$ is nonsingular.
4. Let $\mathbf{A}$ be positive definite. Then for any two vectors $\mathbf{X}, \mathbf{Y}$ we have

$$(\mathbf{X}^T\mathbf{Y})^2 \leq (\mathbf{X}^T\mathbf{AX})(\mathbf{Y}^T\mathbf{A}^{-1}\mathbf{Y}) \tag{2.9}$$

When $\mathbf{A}$ is large it is usually difficult to obtain the value of its determinant $|\mathbf{A}|$. At times only upper (lower) bounds of $|\mathbf{A}|$ are required. When $\mathbf{A}$ is positive definite these bounds are particularly easy to obtain.

THEOREM 2.4. Let $\mathbf{A}$ be a positive definite matrix. Then we have the following bounds for $|\mathbf{A}|$.

1.
$$0 < |\mathbf{A}| \leq a_{11}a_{22}\ldots a_{nn}$$

with equality holding when $\mathbf{A}$ is either diagonal or triangular.

2.
$$0 < |\mathbf{A}| \leq \prod_{j=1}^{n} (a_{1j}^2 + a_{2j}^2 + \cdots + a_{nj}^2)$$

3. Let **B** be positive definite. Then

$$|\mathbf{A} + \mathbf{B}|^{1/n} \geq |\mathbf{A}|^{1/n} + |\mathbf{B}|^{1/n}$$

the Minkowski inequality for determinants, with equality holding when one matrix is a scalar product of the other.

Finally, we can define polynomials of the form

$$\mathbf{X}^{\mathrm{T}}\mathbf{A}\mathbf{Y} = \sum_{i=1}^{n} \sum_{j=1}^{k} a_{ij} x_i y_j \tag{2.10}$$

which depend on two sets of variables x_i, y_i. Expressions such as Eq. (2.10) are known as bilinear forms. Here **A** is no longer necessarily a square matrix.

2.3 MULTIVARIATE DIFFERENTIATION

The vector analog of the partial derivative $\partial y / \partial x$ is the column vector of partial derivatives

$$\frac{\partial y}{\partial \mathbf{X}} = \begin{bmatrix} \dfrac{\partial}{\partial x_1} \\[2ex] \dfrac{\partial}{\partial x_2} \\[1ex] \vdots \\[1ex] \dfrac{\partial}{\partial x_n} \end{bmatrix} \qquad y = \begin{bmatrix} \dfrac{\partial y}{\partial x_1} \\[2ex] \dfrac{\partial y}{\partial x_2} \\[1ex] \vdots \\[1ex] \dfrac{\partial y}{\partial x_n} \end{bmatrix} \tag{2.11}$$

where $y = f(\mathbf{X})$; **X** is an $(n \times 1)$ column vector and y a scalar denoting the values of the multivariate function. Multivariate derivatives are useful in factor analysis since they can be used to derive normal equations of a model as well as multivariate estimators such as maximum likelihood estimators (see Dwyer and MacPhail, 1948; Dwyer, 1967; Mardia et al., 1979; Rayner, 1985). Multivariate derivatives may be conveniently classified into two main types depending on whether they result in a vector or in a matrix.

2.3.1 Derivative Vectors

1. $y = f(x) = $ constant. Then

$$\frac{\partial y}{\partial \mathbf{X}} = \begin{bmatrix} 0 \\ 0 \\ \vdots \\ 0 \end{bmatrix}$$

2. $y = f(x) = $ linear. Then

$$y = a_1 x_1 + a_2 x_2 + \cdots + a_n x_n$$

$$= \mathbf{A}^\mathrm{T}\mathbf{X}$$

and

$$\frac{\partial y}{\partial \mathbf{X}} = \begin{bmatrix} \dfrac{\partial y}{\partial x_1} \\ \dfrac{\partial y}{\partial x_2} \\ \vdots \\ \dfrac{\partial y}{\partial x_n} \end{bmatrix} = \begin{bmatrix} a_1 \\ a_2 \\ \vdots \\ a_n \end{bmatrix} = \mathbf{A} \tag{2.12}$$

3. $y = f(x) = $ quadratic. Here $y = \mathbf{X}^\mathrm{T}\mathbf{A}\mathbf{X}$ and the column vector of derivatives is easily seen to be

$$\frac{\partial y}{\partial \mathbf{X}} = \begin{bmatrix} \dfrac{\partial y}{\partial x_1} \\ \dfrac{\partial y}{\partial x_2} \\ \vdots \\ \dfrac{\partial y}{\partial x_n} \end{bmatrix} = 2 \begin{bmatrix} a_{11}x_1 + a_{12}x_2 + \cdots + a_{1n}x_n \\ a_{21}x_1 + a_{22}x_2 + \cdots + a_{2n}x_n \\ a_{n1}x_1 + a_{n2}x_x + \cdots + a_{nn}x_n \end{bmatrix}$$

$$= 2 \begin{bmatrix} a_{11} & a_{12} & \cdots & a_{1n} \\ a_{21} & a_{22} & \cdots & a_{2n} \\ \vdots & \vdots & & \vdots \\ a_{n1} & a_{n2} & & a_{nn} \end{bmatrix} \begin{bmatrix} x_1 \\ x_2 \\ \vdots \\ x_n \end{bmatrix} \tag{2.13}$$

$$= 2\mathbf{A}\mathbf{X}$$

where $\mathbf{A}$ is symmetric. For the special case $\mathbf{A} = \mathbf{I}$ we have

$$\frac{\partial y}{\partial \mathbf{X}} = 2\mathbf{X} \tag{2.14}$$

Example 2.1. Consider the quadratic equation $y = 9x_1^2 + 12x_1x_2 + 4x_2$ in the $n = 2$ independent variables x_1, x_2. We have

$$y = 9x_1^2 + 12x_1x_2 + 4x_2^2$$

$$= [x_1, x_2]\begin{bmatrix} 9 & 6 \\ 6 & 4 \end{bmatrix}\begin{bmatrix} x_1 \\ x_2 \end{bmatrix}$$

$$= \mathbf{X}^\mathrm{T}\mathbf{A}\mathbf{X}$$

and differentiating with respect to x_1, x_2 yields

$$\frac{\partial y}{\partial x_1} = 18x_1 + 12x_2 = 2(9x_1 + 6x_2)$$

$$\frac{\partial y}{\partial x_1} = 12x_1 + 8x_2 = 2(6x_1 + 4x_2)$$

or

$$\frac{\partial y}{\partial \mathbf{X}} = \begin{bmatrix} \dfrac{\partial y}{\partial x_1} \\[2mm] \dfrac{\partial y}{\partial x_2} \end{bmatrix} = 2 \begin{bmatrix} 9 & 6 \\ 6 & 4 \end{bmatrix} \begin{bmatrix} x_1 \\ x_2 \end{bmatrix} \qquad \square$$

2.3.2 Derivative Matrices

More generally we can define a vector derivative with respect to a vector of unknowns. Let

$$\mathbf{Y} = \begin{bmatrix} y_1 \\ y_2 \\ \vdots \\ y_n \end{bmatrix}, \qquad \mathbf{X} = \begin{bmatrix} x_1 \\ x_2 \\ \vdots \\ x_n \end{bmatrix}$$

Then the derivative of y_i with respect to every x_j can be expressed as

$$\frac{\partial \mathbf{Y}}{\partial \mathbf{X}} = \begin{bmatrix} \dfrac{\partial y_1}{\partial x_1} & \dfrac{\partial y_2}{\partial x_1} & \cdots & \dfrac{\partial y_n}{\partial x_1} \\[2mm] \dfrac{\partial y_1}{\partial x_2} & \dfrac{\partial y_2}{\partial x_2} & \cdots & \dfrac{\partial y_n}{\partial x_2} \\[2mm] \vdots & \vdots & & \vdots \\[2mm] \dfrac{\partial y_1}{\partial x_n} & \dfrac{\partial y_2}{\partial x_n} & \cdots & \dfrac{\partial y_n}{\partial x_n} \end{bmatrix} \qquad (2.15)$$

The following derivatives play an important role in multivariate analysis.

1. *Systems of Linear Equations.* Let $\mathbf{Y} = \mathbf{AX}$ be a system of linear equations in n unknowns. Then

$$\frac{\partial \mathbf{Y}}{\partial \mathbf{X}} = \begin{bmatrix} a_{11} & a_{21} & \cdots & a_{n1} \\ a_{12} & a_{22} & \cdots & a_{n2} \\ \vdots & \vdots & & \vdots \\ a_{1n} & a_{2n} & \cdots & a_{nn} \end{bmatrix}$$

$$= \mathbf{A}^{\mathrm{T}} \qquad (2.16)$$

2. **With Respect to Its Own Elements.** The derivative of a $(n \times r)$ martix $\mathbf{X}$ with respect to any of its element x_{ij} is given by

$$\frac{\partial \mathbf{X}}{\partial x_{ij}} = \mathbf{J}_{ij} \tag{2.17a}$$

the $(n \times r)$ matrix with unity in its (i, j)th position and zeros elsewhere. If $\mathbf{X}$ is symmetric, then $x_{ij} = x_{ji}$ and

$$\frac{\partial x}{\partial x_{ij}} = \begin{cases} \mathbf{J}_{ij} + \mathbf{J}_{ji} & i \neq j \\ \mathbf{J}_{ii} & i = j \end{cases} \tag{2.17b}$$

3. *Matrix Product.* Suppose that two matrices $\mathbf{X}$, $\mathbf{Y}$ are conformable for multiplication and that both are functions of the variable z. Then

$$\frac{\partial \, \mathbf{XY}}{\partial z} = \frac{\mathbf{Y} \, \partial \mathbf{X}}{\partial z} + \frac{\mathbf{X} \, \partial \mathbf{Y}}{\partial z} \tag{2.18}$$

4. *Determinants.* Let $y = |\mathbf{X}|$ and let $|\mathbf{X}_{ij}|$ be the cofactor of element x_{ij}. Then we have $|\mathbf{X}| = x_{i1}|\mathbf{X}_{i1}| + x_{i2}|\mathbf{X}_{i2}| + \cdots + x_{in}|\mathbf{X}_{in}|$ so that

$$\frac{\partial |\mathbf{X}|}{\partial x_{ij}} = \frac{\partial}{\partial x_{ij}} \left(x_{i1}|\mathbf{X}_{i1}| + x_{i2}|\mathbf{X}_{i2}| + \cdots + x_{in}|\mathbf{X}_{in}| \right)$$

$$= |\mathbf{X}_{ij}| \tag{2.19}$$

for nonsymmetric $\mathbf{X}$. When $\mathbf{X}$ is symmetric we have

$$\frac{\partial |\mathbf{X}_{ij}|}{\partial x_{ij}} = \begin{cases} 2|\mathbf{X}_{ij}| & i \neq j \\ |\mathbf{X}_{ij}| & i = j \end{cases} \tag{2.20}$$

Furthermore if $\mathbf{X}$ is nonsingular,

$$\frac{\partial}{\partial x_{ij}} (\log_e |\mathbf{X}|) = \frac{1}{|\mathbf{X}|} \frac{\partial |\mathbf{X}|}{\partial x_{ij}}$$

$$= \frac{|\mathbf{X}_{ij}|}{|\mathbf{X}|} \tag{2.21}$$

$$= (\mathbf{X}^{-1})^{\mathrm{T}}$$

and when $\mathbf{X}$ is symmetric we have

$$\frac{\partial}{\partial x_{ij}} \log_e |\mathbf{X}| = \begin{cases} 2\mathbf{X}^{-1} & i \neq j \\ \mathbf{X}^{-1} & i = j \end{cases} \tag{2.22}$$

If the elements of $\mathbf{X}$ depend on some variable y, then

$$\frac{\partial}{\partial y} \log_e |\mathbf{X}| = \frac{1}{|\mathbf{X}|} \frac{\partial |\mathbf{X}|}{\partial y}$$

$$= \frac{1}{|\mathbf{X}|} \sum_i \sum_j \frac{\partial |\mathbf{X}|}{\partial x_{ij}} \frac{\partial x_{ij}}{\partial y} \qquad (2.23)$$

$$= \text{tr}\left(\mathbf{X}^{-1} \frac{\partial \mathbf{X}}{\partial y} \right)$$

which applies to any nonsingular matrix, symmetric or otherwise.

5. *Trace Function.* Let $y = \text{tr}(\mathbf{X}) = \text{tr}(\mathbf{X}^{\text{T}})$. Then

$$\frac{\partial y}{\partial x_{ij}} = \frac{\partial(x_{11} + x_{22} + \cdots + x_{nn})}{\partial x_{ij}}$$

$$= \begin{cases} 0 & i \neq j \\ 1 & i = j \end{cases} \qquad (2.24)$$

so that the matrix of derivatives is given by $\partial y/\partial x = \mathbf{I}$.

Let $\mathbf{A}$ be another square matrix and $\mathbf{X}$ be nonsingular. Then

$$\frac{\partial}{\partial x_{ij}} \text{tr}(\mathbf{A}\mathbf{X}^{-1}) = \text{tr}\left[\mathbf{A} \frac{\partial \mathbf{X}^{-1}}{\partial x_{ij}} \right]$$

$$= \text{tr}\left[-\mathbf{A}\mathbf{X}^{-1} \left(\frac{\partial \mathbf{X}}{\partial x_{ij}} \right) \mathbf{X}^{-1} \right]$$

$$= \text{tr}\left[\mathbf{X}^{-1}\mathbf{A}\mathbf{X}^{-1} \left(\frac{\partial \mathbf{X}}{\partial x_{ij}} \right) \right]$$

$$= -(\mathbf{X}^{-1}\mathbf{A}\mathbf{X}^{-1})_{ji}$$

the (j, i)th element. Thus

$$\frac{\partial}{\partial \mathbf{X}} \text{tr}(\mathbf{A}\mathbf{X}^{-1}) = -(\mathbf{X}^{-1}\mathbf{A}\mathbf{X}^{-1})^{\text{T}} \qquad (2.25)$$

When both $\mathbf{A}$ and $\mathbf{X}$ are symmetric we have

$$\frac{\partial}{\partial x} \text{tr}(\mathbf{A}\mathbf{X}^{-1}) = \begin{cases} -2(\mathbf{X}^{-1}\mathbf{A}\mathbf{X}^{-1})^{\text{T}} & i \neq j \\ -(\mathbf{X}^{-1}\mathbf{A}\mathbf{X}^{-1})^{\text{T}} & i = j \end{cases} \qquad (2.26)$$

6. *Matrix Inverse.* We have the product $\mathbf{I} = \mathbf{X}\mathbf{X}^{-1}$ and using Eq. (2.18),

$$\frac{\partial \mathbf{I}}{\partial x_{ij}} = 0 = \mathbf{J}_{ij}\mathbf{X}^{-1} + \frac{\partial \mathbf{X}^{-1}}{\partial x_{ij}}$$

or   (2.27)

$$\frac{\partial \mathbf{X}^{-1}}{\partial x_{ij}} = -\mathbf{X}^{-1}\mathbf{J}_{ij}\mathbf{X}^{-1}$$

7. *Total Derivatives.* Let $y = f(\mathbf{X})$ where y is a scalar and $\mathbf{X}$ some matrix. Then if $d\mathbf{X}$ is the matrix of total derivatives we have

$$dy = \operatorname{tr}\frac{(dy)}{d\mathbf{X}}\,d\mathbf{X}^{\mathrm{T}}$$

and hence if

$$dy = \operatorname{tr}(\mathbf{C}\,d\mathbf{X}^{\mathrm{T}})$$

$$= \operatorname{tr}(\mathbf{C}^{\mathrm{T}}\,d\mathbf{X})$$

where vector $\mathbf{C}$ may depend on $\mathbf{X}$ (but not on $d\mathbf{X}$), then

$$\frac{\partial y}{\partial \mathbf{X}} = \mathbf{C} \tag{2.28}$$

Also,

$$d(\mathbf{YX}) = (d\mathbf{Y})\mathbf{X} + \mathbf{Y}(d\mathbf{X}) \tag{2.29}$$

$$d\mathbf{X}^{-1} = -\mathbf{X}^{-1}\,d\mathbf{X}\mathbf{X}^{-1} \tag{2.30}$$

$$d(\operatorname{tr}\mathbf{X}) = \operatorname{tr}(d\mathbf{X}) \tag{2.31}$$

$$d\log_e|\mathbf{X}| = \operatorname{tr}(\mathbf{X}^{-1}\,d\mathbf{X}) \tag{2.23}$$

2.4 GRAMMIAN ASSOCIATION MATRICES

The moments considered in Chapter 1 are all defined in terms of population probability functions. Two major difficulties exist which tend to render these definitions less effective in a multivariate sample. First, the population distributions required for the univariate and bivariate moments are rarely known in practice, so that sample estimates of the true moments must be used instead. Second, these product moments are at most binary in nature, that is, they are only defined for at most two random variables and when more than two are available we are left without a measure of association. One way out of the difficulty is to define the notion of a (sample) association matrix.

Consider a table of data observed for p random variables and n sample points, for example, as in Table 2.2 where $p = 7$ and $n = 24$. Viewed as a matrix the table can generally consist of any one, or a mixture, of measurement scales considered in Section 1.5. We have

$$\mathbf{Y} = \begin{bmatrix} y_{11} & y_{12} & \cdots & y_{1p} \\ y_{21} & y_{22} & \cdots & y_{2p} \\ \vdots & \vdots & & \vdots \\ y_{n1} & y_{n2} & \cdots & y_{np} \end{bmatrix} \tag{2.33}$$

a matrix consisting of n rows and p columns, where y_{ij} represents an observation for the jth variable and ith sample point. The p column vectors thus represent random variables and the n rows the sample points. Usually interest centers on an analysis of the p random variables and in this case $n > p$, although this is by no means the only situation possible. One of the chief objectives of factor analysis is to be able to discern patterns of interrelationships which may exist between the p random variables and/or the n sample points. To achieve this objective we require a multivariate measure of association, or an association matrix. We first consider association moments for the p random variables.

Definition 2.2. Let $\mathbf{Y}$ denote a $(n \times p)$ data matrix. Then any symmetric $(p \times p)$ matrix $\mathbf{S}$ whose (l, h)th element measures the degree of association between variables $\mathbf{Y}_l$ and $\mathbf{Y}_h$ is known as an association matrix.

Association matrices can be either Grammian or non-Grammian depending on the binary measure of association used. An example of a non-Grammian association matrix is the (symmetric) distance matrix $\mathbf{D}$ whose (l, h)th element is the distance (Euclidian or otherwise) between any two random variables. Although $\mathbf{D}$ is symmetric, it is not positive definite owing to the zero diagonal entries, and thus cannot be Grammian. Although factor analysis is usually restricted to Grammian association matrices, it is possible to extend the method to distance matrices as well.

Four broad categories of Grammian matrices are possible for summarizing the pattern of associations between observed random variables, as indicated in Table 2.1.

Table 2.1 Four Types of Grammian Association Matrices Which are Possible Depending on Parameters of Magnitude and Location

	About Origin	About Mean
Unstandardized	Inner product matrix	Covariance matrix
Standardized to unit variance	Cosine matrix	Correlation matrix

2.4.1 The Inner Product Matrix

The simplest measure of association between any two column vectors (variables) of a data matrix is the inner product, normally considered in linear algebra. The inner product is also refered to as the scalar or the dot product. Let $\mathbf{Y}_l$ and $\mathbf{Y}_h$ be n-component column vectors of $\mathbf{Y}$. Then the inner product between $\mathbf{Y}_l$ and $\mathbf{Y}_h$ is defined as the sum of products of components, that is,

$$\mathbf{Y}_l \cdot \mathbf{Y}_h = \sum_{i=1}^{n} y_{il} y_{ih} \qquad (2.34)$$

The inner product can also be expressed as

$$\mathbf{Y}_l \cdot \mathbf{Y}_h = \|\mathbf{Y}_l\| \, \|\mathbf{Y}_h\| \cos \theta \qquad (2.35)$$

where $\|\mathbf{Y}_l\| = (\Sigma_{i=1}^{n} y_{il}^2)^{1/2}$ and $\|\mathbf{Y}_h\| = (\Sigma_{i=1}^{n} y_{ih}^2)^{1/2}$ are the lengths of $\mathbf{Y}_l$, $\mathbf{Y}_h$ respectively and θ is the angle between them. This is portrayed in Figure 2.1. Although vectors are often drawn symbolically as arrows, it should be kept in mind that they represent points in multidimensional vector space.

The following three properties of the inner product can be deduced using Eq. (2.35).

1. The inner product is not independent of the number of vector components, that is, it depends on the sample size n. Thus increasing (or decreasing) the sample size will alter the value of $\mathbf{Y}_l \cdot \mathbf{Y}_h$ without necessarily altering θ.

2. The inner product depends on the magnitudes of the two vectors, so that a simple proportional change in the unit of measure will alter its value.

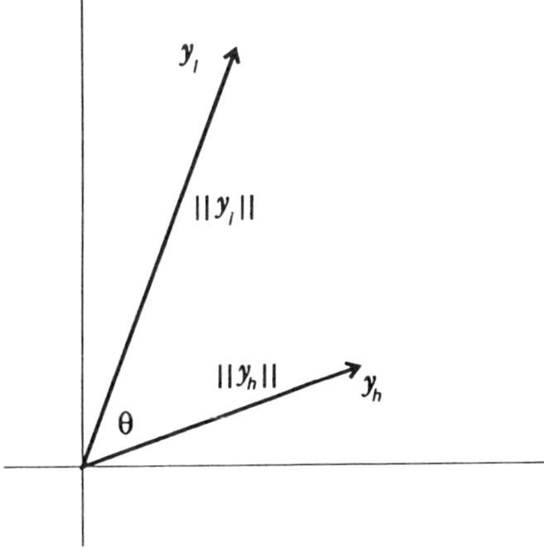

Figure 2.1 Angle θ between two vectors $\mathbf{Y}_l$ and $\mathbf{Y}_k$ with lengths $\|\mathbf{Y}_l\|$ and $\|\mathbf{Y}_k\|$, respectively.

3. It is not independent of the origin or the means of the random variables, and can be altered by a simple translation of the axis produced by an addition or subtraction of a constant.

When 2 and 3 are not considered to be objectionable, or are indeed essential to the measuring process, the inner product can be modified by defining an adjusted or "average" value:

$$\frac{1}{n}(\mathbf{Y}_l \cdot \mathbf{Y}_h) = \frac{1}{n}\|\mathbf{Y}_l\| \|\mathbf{Y}_h\| \cos\theta \qquad (2.36)$$

which is independent of the sample size n. This leads to the following definition of an inner product matrix.

Definition 2.3. A matrix of (sample) inner products is the matrix $\mathbf{S}_I$ whose (l, h)th element is the average inner product between the lth and hth columns $\mathbf{Y}$, that is,

$$\mathbf{S}_I = \frac{1}{n}\mathbf{Y}^T\mathbf{Y} \qquad (2.37)$$

where

$$\mathbf{Y}_l{}^T\mathbf{Y}_h = \begin{cases} \|\mathbf{Y}_l\| \|\mathbf{Y}_h\| \cos\theta & (l \neq h) \\ \sum_{i=1}^{n} y_{il}^2 & (l = h) \end{cases} \qquad (2.38)$$

2.4.2 The Cosine Matrix

The dependence of the inner product on the lengths of the vectors is at times an undesirable property, particularly when considering variables with uncomparable units of measure. Also, even for comparable units of measure a change from inches to centimeters, for example, will alter the inner product even though no intrinsic change has occurred in the degree of relationship between $\mathbf{Y}_l$ and $\mathbf{Y}_h$. Thus when a unit-free measure of association is required the inner product can be standardized to yield a magnitude-free coefficient which does not change when a variable is multiplied by a constant, such as occurs when inches are transformed to centimeters. Dividing both sides of Eq. (2.36) by the magnitudes of the two vectors yields

$$\cos\theta = \frac{(\mathbf{Y}_l \cdot \mathbf{Y}_h)}{\|\mathbf{Y}_l\| \|\mathbf{Y}_h\|} \qquad (2.39)$$

the cosine of the angle between $\mathbf{Y}_l$, $\mathbf{Y}_h$. When $\theta = 0$ both $\mathbf{Y}_l$ and $\mathbf{Y}_h$ lie on the same straight line and are thus linearly dependent. For $\theta = 90°$ the random variables are orthogonal, and in general $-1 \le \cos\theta \le 1$. It is also easy to verify that Eq. (2.39) is a unit-free measure of linear association.

Definition 2.4. The cosine matrix is the matrix $\mathbf{C}$ whose (l, h)th element is the cosine of the angle between $\mathbf{Y}_l$, $\mathbf{Y}_h$, that is,

$$\mathbf{C} = \mathbf{M}^{-1/2}(\mathbf{Y}^T\mathbf{Y})\mathbf{M}^{-1/2} \tag{2.40}$$

where $\mathbf{M}$ is the diagonal matrix of vector magnitudes.

A typical element c_{lh} of $\mathbf{C}$ is given by

$$c_{lh} = \begin{cases} \cos\theta & l \ne h \\ 1 & l = h \end{cases} \tag{2.41}$$

In terms of vector components $\cos\theta$ can be expressed as

$$\cos\theta = \frac{\sum\limits_{i=1}^{n} y_{il}y_{ih}}{\left(\sum\limits_{i=1}^{n} y_{il}^2\right)^{1/2}\left(\sum\limits_{i=1}^{n} y_{ih}^2\right)^{1/2}} \tag{2.42}$$

2.4.3 The Covariance Matrix

Both the inner product and the cosine are dependent on the positions of the coordinate axes. Since variances and covariance are defined as second moments about the mean, however, they are independent of average or general levels of a variable; that is, they are invariant with respect to shifts of the axes effected by addition or subtraction of constants.

Definition 2.5 The sample covariance matrix of a set of p random variables is a matrix $\mathbf{S}$ whose (l, h)th element is the covariance between the lth, hth columns of a data matrix $\mathbf{Y}$, that is, $\mathbf{S}$ has the typical element

$$s_{lh} = \sum_{i=1}^{n} \frac{(y_{il} - \bar{y}_l)(y_{ih} - \bar{y}_h)}{n - 1}$$

$$= \frac{1}{n-1}\left[\sum_{i=1}^{n} y_{il}y_{ih} - n\bar{y}_l\bar{y}_h\right] \tag{2.43}$$

When $\mathbf{Y}_l = \mathbf{Y}_h$, Eq. (2.43) yields the sample variance. A sample covariance has the following properties:

$$s_{(y_l+c),(y_h+k)} = s_{lh}$$

$$s_{cy_l, ky_k} = cks_{lh} \tag{2.44}$$

Adding (subtracting) constant numbers to random variables does not alter the magnitude of the covariance, but multiplying by constants does.

The covariance martix can also be computed in terms of matrix operations. Let

$$
\bar{\mathbf{Y}} =
\begin{bmatrix}
\bar{y}_1 & \bar{y}_2 & \cdots & \bar{y}_p \\
\bar{y}_1 & \bar{y}_2 & \cdots & \bar{y}_p \\
\vdots & \vdots & & \vdots \\
\bar{y}_1 & \bar{y}_2 & \cdots & \bar{y}_p
\end{bmatrix}
$$

by the mean value matrix whose columns are the sample means of $\mathbf{Y}_1, \mathbf{Y}_2, \ldots, \mathbf{Y}_p$. Then the covariance matrix is given by

$$
\mathbf{S} = \frac{1}{n-1} (\mathbf{Y} - \bar{\mathbf{Y}})^{\mathrm{T}} (\mathbf{Y} - \bar{\mathbf{Y}})
$$

$$
= \frac{1}{n-1} \mathbf{X}^{\mathrm{T}} \mathbf{X}
$$

where $\mathbf{X} = \mathbf{Y} - \bar{\mathbf{Y}}$ is the deviations-about-means matrix. The matrix equivalent of Eq. (2.43) can then be expressed as

$$
\mathbf{S} = \frac{1}{n-1} (\mathbf{Y} - \bar{\mathbf{Y}})^{\mathrm{T}} (\mathbf{Y} - \bar{\mathbf{Y}})
$$

$$
= \frac{1}{n-1} (\mathbf{Y}^{\mathrm{T}}\mathbf{Y} - \mathbf{Y}^{\mathrm{T}}\bar{\mathbf{Y}} - \bar{\mathbf{Y}}^{\mathrm{T}}\mathbf{Y} + \bar{\mathbf{Y}}^{\mathrm{T}}\bar{\mathbf{Y}})
$$

$$
= \frac{1}{n-1} (\mathbf{Y}^{\mathrm{T}}\mathbf{Y} - n\bar{\mathbf{Y}}^{\mathrm{T}}\bar{\mathbf{Y}}) \tag{2.45}
$$

(Exercise 2.17). The symmetric matrix $\bar{\mathbf{Y}}^{\mathrm{T}}\bar{\mathbf{Y}}$ contains terms of the form $\bar{y}_l \bar{y}_h$ for $l \neq h$ and $\bar{y}_l^2$ for $l = h$. Equation (2.45) is the multivariate equivalent of the familiar computing formula for the sample variance (see also Exercise 2.2). Since in most applications the levels of random variables should not influence their degree of closeness or association, the covariance matrix is more widely employed than the inner product or the cosine matrix, particularly when the random variables are continuous. Covariances are also independent of the sample size since the numerator of Eq. (2.43) is divided by $n - 1$.

2.4.4 The Correlation Matrix

Since the covariance between any two random variables depends on the variances (magnitudes) of the variables, it can only be used for random variables with the same units of measure. It can however be adjusted to yield a new coefficient of association that is scale free. Such a coefficient is known as the correlation coefficient and can be obtained from a covariance in much the same way as the cosine is obtained from the inner product. Let $|s_{lh}|$ denote the absolute value of the sample covariance between variables $\mathbf{Y}_l$, $\mathbf{Y}_h$. Then it follows from the Cauchy–Schwartz inequality that $|s_{lh}| \leq s_l s_h$, that is, $-s_l s_h \leq s_{lh} \leq s_l s_h$ where s_l and s_h are sample standard deviations. Dividing by $s_l s_h$ we have

$$-1 \leq \frac{s_{lh}}{s_l s_h} \leq 1 \tag{2.46}$$

where

$$r_{lh} = \frac{s_{lh}}{s_l s_h}$$

$$= \sum_{i=1}^{n} \frac{(y_{il} - \bar{y}_l)(y_{ih} - \bar{y}_h)}{\left[\sum_{i=1}^{n} (y_{il} - \bar{y}_l)^2\right]^{1/2}\left[\sum_{i=1}^{n} (y_{ih} - \bar{y}_h)^2\right]^{1/2}} \tag{2.47}$$

is the sample correlation coefficient between $\mathbf{Y}_l$ and $\mathbf{Y}_h$. It can be shown that

$$r_{(y_l+c),\,(y_h+k)} = r_{lh}$$

$$r_{cy_l,\,ky_h} = r_{lh} \tag{2.48}$$

so that a correlation coefficient is independent of addition and multiplication by constants. It is also independent of the sample size. In this sense Eq. (2.47) represents a pure, unitless measure of linear association between two observed random variables.

Definition 2.6 A correlation matrix is a matrix of correlation coefficients $\mathbf{R}$ whose off-diagonal elements consist of terms of the form given by Eq. (2.47) and whose diagonal elements are identically equal to unity.

The correlation matrix can also be computed by matrix operations. Let $\mathbf{S}_d$ be a diagonal matrix whose diagonal elements are the sum of squares $(\mathbf{Y}_1 - \bar{\mathbf{Y}}_1)^T(\mathbf{Y}_1 - \bar{\mathbf{Y}}_1), (\mathbf{Y}_2 - \bar{\mathbf{Y}}_2)^T(\mathbf{Y}_2 - \bar{\mathbf{Y}}_2), \ldots, (\mathbf{Y}_p - \bar{\mathbf{Y}}_p)^T(\mathbf{Y}_p - \bar{\mathbf{Y}}_p)$. Then R is given by

$$\mathbf{R} = \mathbf{S}_d^{-1/2}(\mathbf{Y} - \bar{\mathbf{Y}})^T(\mathbf{Y} - \bar{\mathbf{Y}})\mathbf{S}_d^{-1/2}$$

$$= \mathbf{S}_d^{-1/2}\mathbf{X}^T\mathbf{X}\mathbf{S}_d^{-1/2}$$

$$= \mathbf{Z}^T\mathbf{Z} \tag{2.49}$$

where $\mathbf{Z} = (\mathbf{Y} - \bar{\mathbf{Y}})\mathbf{S}_d^{-1/2}$ is the $(n \times p)$ matrix of standardized variables, that is, columns of $\mathbf{Y}$ adjusted to zero mean, unit variance. From Eq. (2.49) it follows that a sample correlation coefficient between random variables can also be viewed as the inner product between two vectors standardized to zero mean and unit length. When common units of measure exist the covariance matrix generally should be used since a difference in variances imparts important information to the estimation process. At times this is not possible and a typical data base will contain diverse units of measure. In this situation variances are no longer comparable and yield no useful information, and random variables should be standardized to equal (unit) variance. Although standardization can introduce an element of artificiality into the analysis, there does not seem to be much choice in the matter once different units of measure are present.

Example 2.2.   Table 2.2 presents a data matrix **Y** which consists of seven

Table 2.2   Seven Socioeconomic Random Variables Observed for $n = 29$ London Boroughs

Boroughs	Y_1	Y_2	Y_3	Y_4	Y_5	Y_6	Y_7
1. City	−44.3	182.4	8.1	0.647	0.0304	65.58	0.00
2. Battersea	−31.5	101.4	15.19	1.957	0.0174	83.75	1.62
3. Bermondsey	−50.3	97.2	16.02	1.697	0.0170	93.69	1.13
4. Bethnal Green	−49.8	102.6	15.26	1.763	0.0080	95.74	2.03
5. Camberwell	−32.5	99.8	15.91	1.913	0.0190	86.85	2.23
6. Chelsea	−12.1	134.0	11.54	1.363	0.0343	46.87	2.23
7. Deptford	−35.1	100.4	15.46	2.073	0.0156	90.10	2.24
8. Finsbury	50.7	101.5	15.70	1.850	0.0166	89.48	1.99
9. Fulham	−23.6	104.9	13.52	1.693	0.0231	77.77	1.72
10 Greenwich	−16.5	101.1	16.22	1.501	0.0174	80.74	1.26
11. Hackney	−26.4	104.1	14.92	2.193	0.0174	90.43	2.30
12. Hammersmith	−18.0	95.4	14.67	2.113	0.0167	78.94	2.72
13. Hampstead	4.7	149.8	11.27	1.767	0.0300	46.77	2.09
14. Holborn	−31.6	173.3	9.91	0.983	0.0410	63.12	2.28
15. Islington	−32.6	109.3	15.04	2.427	0.0146	87.62	2.94
16. Kensington	−8.9	149.5	11.59	1.920	0.0297	46.36	3.84
17. Lambeth	−26.1	101.9	15.48	2.250	0.0212	83.78	2.75
18. Lewisham	−4.3	102.3	15.34	1.700	0.0167	80.39	1.21
19. Paddington	−17.6	120.5	12.17	2.260	0.0287	66.35	2.80
20. Poplar	−58.1	94.1	17.07	1.993	0.0166	95.36	2.95
21. St. Marylebone	−18.7	167.6	8.66	1.107	0.0311	50.22	2.28
22. St. Pancras	−31.9	115.6	13.43	1.910	0.0246	77.32	3.04
23. Shoreditch	−57.6	94.0	17.45	1.650	0.0106	96.09	3.18
24. Shouthwark	−46.0	100.4	16.40	1.907	0.0245	93.58	2.47
25. Stepney	−59.1	104.0	15.97	1.977	0.0200	93.21	3.38
26. Stoke Newington	−11.0	99.6	15.93	2.383	0.0166	84.88	1.66
27. Wandsworth	−8.9	100.7	14.61	1.723	0.0182	74.79	1.83
28. Westminster	−26.2	129.0	9.86	1.170	0.0315	54.73	1.96
29. Woolich	−6.5	94.3	16.63	1.520	0.0109	80.87	1.28

Source: Wallis and Maliphant, 1967.

random variables observed from 29 London Boroughs. The variables are defined as follows:

Y_1 = Percentage of population change, 1931–1951.

Y_2 = Female/male ratio (percentage) of 15- to 20-year olds.

Y_3 = Percentage of population that is male, under 21 years of age.

Y_4 = The demographic fertility rate.

Y_5 = Suicide rate during 1960–1962; 14-year olds and older (percentage of population).

Y_6 = Percent of population whose terminal education age is 15 years or less.

Y_7 = Percentage of labor force that is unemployed. $\qquad\qquad\square$

Using Eq. (2.34) the inner product matrix (Eq. 2.37) is given by

$$S_1 = \begin{bmatrix} 740.0 & & & & & & \\ -1582.9 & 30786.7 & & & & & \\ -709.7 & 146.1 & 381.8 & & & & \\ -36.8 & 396.2 & 43.6 & 7.9 & & & \\ .03 & 7.0 & -5.0 & -0.01 & .002 & & \\ 1764.2 & -46.0 & 2059.6 & 225.2 & -1.14 & 12936.5 & \\ 23.9 & 185.9 & 45.1 & 8.6 & 1.13 & 176.6 & 21.9 \end{bmatrix} \begin{matrix} Y_1 \\ Y_2 \\ Y_3 \\ Y_4 \\ Y_5 \\ Y_6 \\ Y_7 \end{matrix}$$

with columns $Y_1 \; Y_2 \; Y_3 \; Y_4 \; Y_5 \; Y_6 \; Y_7$.

Since S_1 is symmetric, the upper half of the matrix is omitted. Positive inner products imply that variables are related directly, whereas negative values indicate an inverse relationship.

The inner product depends on the mean values of the random variables. To see the extent of this dependence we compute the covariance matrix S. The mean values derived from Table 2.2 are $\bar{y}_1 = -25.1655$, $\bar{y}_2 = 114.8517$, $\bar{y}_3 = 14.1145$, $\bar{y}_4 = 1.7728$, $\bar{y}_5 = 0.0214$, $\bar{y}_6 = 77.7717$, $\bar{y}_7 = 2.1866$ so that matrix $\bar{Y}$ contains $n = 29$ equal row vectors and $r = 7$ columns consisting of the means $\bar{y}_1, \bar{y}_2, \ldots, \bar{y}_7$. Using Eq. (2.43) then yields the (7×7) covariance matrix

$$S = \begin{bmatrix} 14224.9 & & & & & & \\ 1354.1 & 18224.4 & & & & & \\ -245.3 & -1527.1 & 189.1 & & & & \\ 8.1 & -199.5 & 19.3 & 4.9 & & & \\ .59 & 4.7 & -.49 & -.05 & .0017 & & \\ 3854.2 & -8977.9 & 996.3 & 90.4 & -2.9 & 7182.7 & \\ 81.7 & -67.6 & 8.5 & 4.7 & .008 & 2.6 & 17.7 \end{bmatrix}$$

We observe large changes in the corresponding entries of the two matrices, particularly for variables with large means. Indeed, the change in some cases is so acute as to reverse the signs of the inner product. The reason for this lies mainly in the large differences in the means. Covariances therefore measure linear association when random variables are standardized to the same (zero) mean, and in this sense variables are compared on an equal footing. This is generally a desirable property for continuous variables since a relationship should depend on the degree of covariation between the random variables rather than on the levels of the measurement scales.

Although the covariance matrix is adjusted for means, it still depends on the diagonal variance terms. When these reveal large differences, particularly when units of measurement are not comparable, it makes little sense to allow the variances to influence the analysis. For example, here $s_{21} = 1354.1$ and $s_{52} = 4.7$, which would indicate that Y_2 is more closely related to Y_1 than to Y_5. However, Y_5 tends not to vary to any great extent and a very small (but significant) covariation of Y_2 and Y_5 may go undetected if the large difference in the variances is not taken into account. To remove the effects of the variances we therefore compute the correlation matrix (Eq. 2.49):

$$
\mathbf{R} = \begin{bmatrix}
1.0000 & & & & & & \\
.0841 & 1.0000 & & & & & \\
-.1496 & -.9312 & 1.0000 & & & & \\
.0308 & -.6680 & .6348 & 1.0000 & & & \\
.1193 & .8310 & -.8481 & -.5132 & 1.000 & & \\
-.3813 & -.7847 & .8550 & .4819 & -.8088 & 1.0000 & \\
-.1628 & -.1190 & .1462 & .5039 & .0473 & .0073 & 1.000
\end{bmatrix}
$$

where the diagonal matrix $\mathbf{S}_\mathrm{d}^{-1/2}$ is given by

$$
\mathbf{S}_\mathrm{d}^{-1/2} = \begin{bmatrix}
1/119.3 & & & & & & \\
& 1/135.0 & & & & & \\
& & 1/13.8 & & & & \\
& & & 1/2.2 & & & \\
& & & & 1/.04 & & \\
& & & & & 1/84.8 & \\
& & & & & & 1/4.2
\end{bmatrix}
$$

We now observe Y_2 to be more closely related to Y_5 than to Y_1.

2.5 TRANSFORMATION OF COORDINATES

There exists a basic source of indeterminacy in representing vectors as points in multidimensional space, since the origin together with the position of the coordinate axes essentially is arbitrary. This was seen in Section 2.4, where transformations of the variates lead to alternative measures of association. Any movement of the axes results in a new set of coordinates, and thus in different numerical values of the observations, although in a basic sense the variates themselves remain unchanged. Thus expressing the original variables as differences from their means, for example, is simply equivalent to a parallel shift of the coordinates axes where the origin is placed at the mean point of the sample observations. Also, transforming the covariance to a correlation coefficient may be viewed as the shrinking (extension) of the coordinate axes to unit length.

A more fundamental procedure of transforming the coordinates however is to alter their orientation by means of a rotation, clockwise or anticlockwise. When the angle is maintained at $90°$ the rotation is said to be orthogonal, whereas for alternative angles we say the rotation is oblique. In a typical statistical analysis the axes are left unrotated, but in factor analysis such rotations of the axes form a part of the methodology.

2.5.1 Orthogonal Rotation

Let $\mathbf{V} = (x, y)^\mathrm{T}$ be a vector in two-dimensional space with respect to a set of coordinate axes. Assume the coordinate axes are rotated clockwise through an angle θ. This is portrayed in Fig. 2.2. If $\mathbf{V}^* = (x^*, y^*)^\mathrm{T}$ represents the same vector with respect to a new position of the axes we have

$$x = r \cos \alpha, \qquad y = r \sin \alpha \tag{2.50}$$

and

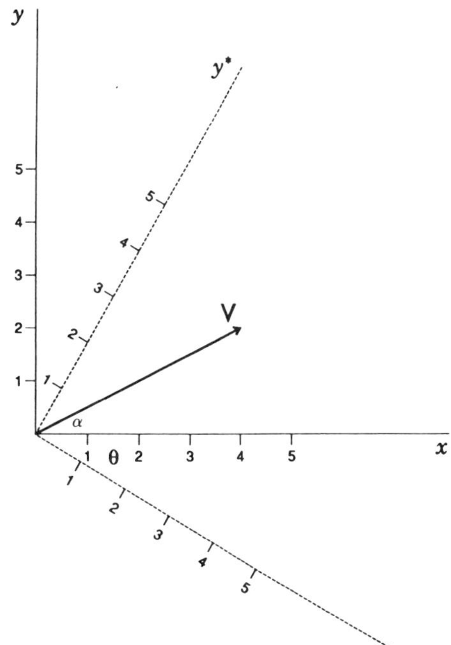

Figure 2.2 An orthogonal clockwise rotation of coordinate axes x and y through the angle $\theta = 30°$ to the new position x^*, y^*.

$$x^* = r \cos(\alpha + \theta) = r \cos \alpha \cos \theta - r \sin \alpha \sin \theta$$

$$y^* = r \sin(\alpha + \theta) = r \sin \alpha \cos \theta + r \cos \alpha \sin \theta$$

where $r = \|\mathbf{V}\|$. Using Eq. (2.50) we have

$$x^* = x \cos \theta - y \sin \theta$$

$$y^* = x \sin \theta + y \cos \theta \tag{2.51}$$

or

$$\begin{bmatrix} x^* \\ y^* \end{bmatrix} = \begin{bmatrix} \cos \theta & -\sin \theta \\ \sin \theta & \cos \theta \end{bmatrix} \begin{bmatrix} x \\ y \end{bmatrix}$$

that is,

$$\mathbf{V}^* = \mathbf{PV} \tag{2.52}$$

where $\mathbf{P}$ is an orthogonal matrix. Since $\sin \theta = \cos(90 - \theta)$ and $\sin^2\theta + \cos^2\theta = 1$, the columns of $\mathbf{P}$ are direction cosines of the new axes x^*, y^*. Also, the counterclockwise rotation is given by $\mathbf{V}^* = \mathbf{P}^{\mathrm{T}}\mathbf{V}$ where $\mathbf{P}^{\mathrm{T}} = \mathbf{P}^{-1}$.

Example 2.3. Let $\mathbf{V} = (x, y)^{\mathrm{T}} = (2, 4)^{\mathrm{T}}$. Rotating x, y clockwise through $30°$ results in the new system $(x^*, y^*)^{\mathrm{T}}$, with respect to which the coordinates of $\mathbf{V}$ are

$$\begin{bmatrix} x^* \\ y^* \end{bmatrix} = \begin{bmatrix} \cos \theta & -\sin \theta \\ \sin \theta & \cos \theta \end{bmatrix} \begin{bmatrix} x \\ y \end{bmatrix}$$

$$= \begin{bmatrix} .8660 & -.5000 \\ .5000 & .8660 \end{bmatrix} \begin{bmatrix} 4 \\ 2 \end{bmatrix}$$

$$= \begin{bmatrix} 2.464 \\ 3.173 \end{bmatrix}$$

The direction cosine of axes $(x^*, y^*)^{\mathrm{T}}$ referred to the original axes $(x, y)^{\mathrm{T}}$ are given by the vectors

$$\mathbf{P}_1 = \begin{bmatrix} \cos 30 \\ \sin 30 \end{bmatrix} = \begin{bmatrix} .8660 \\ .5000 \end{bmatrix}$$

$$\mathbf{P}_2 = \begin{bmatrix} -\sin 30 \\ \cos 30 \end{bmatrix} = \begin{bmatrix} -.5000 \\ .8660 \end{bmatrix}$$

that is, by the columns of $\mathbf{P}$. For an anticlockwise rotation the matrix $\mathbf{P}$ is replaced by $\mathbf{P}^{\mathrm{T}}$.

When more than two coordinate axes are present the situation becomes less straightforward because any pair of axes (or more) can be rotated independently of the remaining set. It consequently becomes difficult to give a general expression for the orthogonal transformation matrix $\mathbf{P}$. In practice however axes rotations can be carried out by taking two axes at a time until a satisfactory position has been arrived at (although Euler's method allows the simultaneous rotation of up to three axes). Also, axes can be rotated in the presence of more than a single vector. This is illustrated in the following example using five vectors in three-dimensional vector space.   □

Example 2.4.   Consider three coordinate axes x, y, z and vectors $\mathbf{V}_1 = (1, 2, 3)^T$, $\mathbf{V}_2 = (2, 4, 1)^T$, $\mathbf{V}_3 = (0, 3, 5)^T$, $\mathbf{V}_4 = (-2, 1, 4)^T$ and $\mathbf{V}_5 = (1, -3, 4)^T$. Arranging the five vectors as columns of a matrix $\mathbf{X}$, the first rotation is given by rotating the xz plane (clockwise) through some angle θ_1. Let $\theta_1 = 20°$. Then

$$
\mathbf{P}_1\mathbf{X} = \begin{bmatrix} \cos 20° & 0 & -\sin 20° \\ 0 & 1 & 0 \\ \sin 20° & 0 & \cos 20° \end{bmatrix} \begin{bmatrix} 1 & 2 & 0 & -2 & 1 \\ 2 & 4 & 3 & 1 & -3 \\ 3 & 1 & 5 & 4 & 4 \end{bmatrix}
$$

$$
= \begin{bmatrix} -.0864 & 1.5373 & -1.7101 & -3.2475 & -.4284 \\ 2 & 4 & 3 & 1 & -3 \\ 3.1611 & 1.6237 & 4.6985 & 3.0747 & 4.1008 \end{bmatrix}
$$

$$
= \mathbf{X}_1
$$

Next we rotate axes x and y counterclockwise, through $\theta_2 = 35°$. The matrix of rotation is

$$
\mathbf{P}_2 = \begin{bmatrix} \cos 35° & \sin 35° & 0 \\ -\sin 35° & \cos 35° & 0 \\ 0 & 0 & 1 \end{bmatrix}
$$

so that the second stage rotation is given by

$$
\mathbf{P}_2\mathbf{X}_1 = \begin{bmatrix} \cos 35° & \sin 35° & 0 \\ -\sin 35° & \cos 35° & 0 \\ 0 & 0 & 1 \end{bmatrix}
$$

$$
\times \begin{bmatrix} -.0864 & 1.537 & -1.7101 & -3.2475 & -.4284 \\ 2 & 4 & 3 & 1 & -3 \\ 3.1611 & 1.6237 & 4.6985 & 3.0747 & 4.1018 \end{bmatrix}
$$

$$
= \begin{bmatrix} 1.0763 & 3.5533 & .3199 & -2.0866 & -2.0716 \\ 1.6879 & 2.3950 & 3.4383 & 2.6818 & -2.2118 \\ 3.1611 & 1.6237 & 4.6985 & 3.0747 & 4.1018 \end{bmatrix}
$$

Last, we rotate the yz axis clockwise through $\theta_3 = 60°$:

$$\mathbf{P}_3\mathbf{X}_2 = \begin{bmatrix} 1 & 0 & 0 \\ 0 & \cos 60^\circ & -\sin 60^\circ \\ 0 & \sin 60^\circ & \cos 60^\circ \end{bmatrix}$$

$$\times \begin{bmatrix} 1.0763 & 3.5533 & .3199 & -2.0866 & -2.0716 \\ 1.6879 & 2.3950 & 3.4383 & 2.6818 & -2.2118 \\ 3.1611 & 1.6237 & 4.6985 & 3.0747 & 4.1018 \end{bmatrix}$$

$$= \begin{bmatrix} 1.0763 & 3.5533 & .3199 & -2.0866 & -2.0716 \\ -1.8936 & -.2086 & -2.3498 & -1.3219 & -4.6581 \\ 3.0423 & 2.8859 & 5.3269 & 3.8597 & -3.9663 \end{bmatrix}$$

The final rotation matrix $\mathbf{P}$ is then obtained as the product of the pairwise rotation matrices, that is,

$$\mathbf{P} = \mathbf{P}_3\mathbf{P}_2\mathbf{P}_1$$

Since $\mathbf{P}_1$, $\mathbf{P}_2$, and $\mathbf{P}_3$ are orthogonal, $\mathbf{P}$ must also be an orthogonal matrix. $\quad\square$

2.5.2 Oblique Rotations

A more general rotation of axes is the oblique rotation where axes are allowed to intersect at arbitrary angles. Consider an orthonormal coordinate system $\mathbf{E}_1 = (1, 0, 0, \ldots, 0)^\mathrm{T}, \mathbf{E}_2 = (0, 1, 0, \ldots, 0)^\mathrm{T}, \ldots, \mathbf{E}_n = (0, 0, \ldots, 1)^\mathrm{T}$ and the set of linearly independent unit vectors $\mathbf{F}_1 = (a_{11}, a_{12}, \ldots, a_{1n})^\mathrm{T}$, $\mathbf{F}_2 = (a_{21}, a_{22}, \ldots, a_{2n})^\mathrm{T}, \ldots, \mathbf{F}_n = (a_{n1}, a_{n2}, \ldots, a_{nn})^\mathrm{T}$, where coordinates a_{ij} are measured with respect to the base $\mathbf{E}_i$ $(i = 1, 2, \ldots, n)$. Vectors $\mathbf{F}_1$, $\mathbf{F}_2, \ldots, \mathbf{F}_n$ can be expressed as

$$\begin{aligned} \mathbf{F}_1 &= a_{11}\mathbf{E}_1 + a_{12}\mathbf{E}_2 + \cdots + a_{1n}\mathbf{E}_n \\ \mathbf{F}_2 &= a_{21}\mathbf{E}_1 + a_{22}\mathbf{E}_2 + \cdots + a_{2n}\mathbf{E}_n \\ \mathbf{F}_n &= a_{n1}\mathbf{E}_1 + a_{n2}\mathbf{E}_2 + \cdots + a_{nn}\mathbf{E}_n \end{aligned} \tag{2.53}$$

Consider any other vector $\mathbf{V} = (v_1, v_2, \ldots, v_n)^\mathrm{T}$ with respect to the orthonormal basis. Then

$$\mathbf{V} = v_1\mathbf{E}_1 + v_2\mathbf{E}_2 + \cdots + v_n\mathbf{E}_n \tag{2.54}$$

Also since $\mathbf{F}_1, \mathbf{F}_2, \ldots, \mathbf{F}_n$ are linearly independent they form a basis of the vector space, and

$$\mathbf{V} = v_1^*\mathbf{F}_1 + v_2^*\mathbf{F}_2 + \cdots + v_n^*\mathbf{F}_n \tag{2.55}$$

where the v_i^* denote coordinates of $\mathbf{V}$ with respect to $\mathbf{F}_1, \mathbf{F}_2, \ldots, \mathbf{F}_n$. Substituting Eq. (2.53) into Eq. (2.55) yields

$$\mathbf{V} = v_1^*(a_{11}\mathbf{E}_1 + a_{12}\mathbf{E}_2 + \cdots + a_{1n}\mathbf{E}_n) + v_2^*(a_{21}\mathbf{E}_1 + a_{22}\mathbf{E}_2 + \cdots + a_{2n}\mathbf{E}_n)$$

$$+ \cdots + v_n^*(a_{n1}\mathbf{E}_1 + a_{n2}\mathbf{E}_2 + \cdots + a_{nn}\mathbf{E}_n)$$

$$= (v_1^* a_{11} + v_2^* a_{21} + \cdots + v_n^* a_{n1})\mathbf{E}_1 + (v_1^* a_{12} + v_2^* a_{22} + \cdots + v_n^* a_{n2})\mathbf{E}_2$$

$$+ \cdots + (v_1^* a_{1n} + v_2^* a_{2n} + \cdots + v_n^* a_{nn})\mathbf{E}_n$$

and equating coefficients with those of Eq. (2.54) we have

$$\begin{aligned}
v_1 &= a_{11} v_1^* + a_{21} v_2^* + \cdots + a_{n1} v_n^* \\
v_2 &= a_{12} v_1^* + a_{22} v_2^* + \cdots + a_{n2} v_n^* \\
\hline
v_n &= a_{1n} v_1^* + a_{2n} v_2^* + \cdots + a_{nn} v_n^*
\end{aligned} \tag{2.56}$$

The system of n equations (Eq. 2.56), in terms of the unknowns v_1^*, $v_2^*, \ldots, v_n^*$, possesses a unique solution when the $\mathbf{F}_1, \mathbf{F}_2, \ldots, \mathbf{F}_n$ are linearly independent, yielding oblique coordinates in terms of the original coordinates.

Distance and angle can also be expressed in terms of oblique coordinates. Let d be the length of a vector $\mathbf{V}$ whose coordinates refer to an oblique system (Fig. 2.3) such that $0 \le \theta_{12} \le 180°$. Using the Law of Cosines (e.g., see Basilevsky, 1983) the squared distance between $\mathbf{V}$ and the origin is given by

$$d^2 = x_1^2 + x_2^2 - 2x_1 x_2 \cos(180° - \theta_{12})$$

$$= x_1^2 + x_2^2 + 2x_1 x_2 \cos \theta_{12}$$

$$= \sum_{i=1}^{2} \sum_{j=1}^{2} x_i x_j \cos \theta_{ij} \tag{2.57}$$

Equation (2.57) represents a quadratic form in x_1 and x_2. More generally,

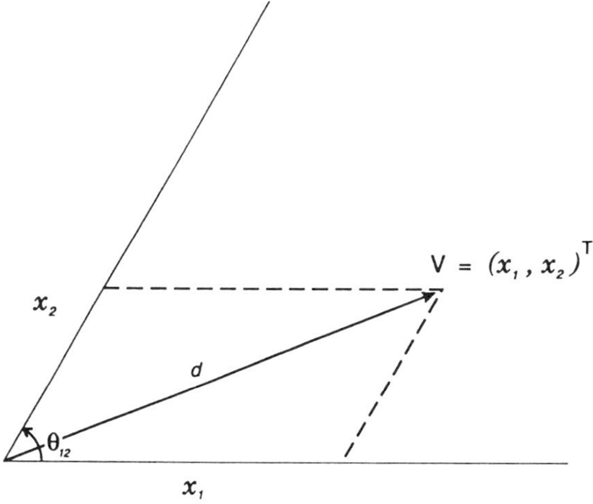

Figure 2.3 Euclidian distance in terms of oblique coordinates.

given a k-dimensional space, Eq. (2.57) can be generalized to the quadratic form

$$d^2 = \sum_{i=1}^{k} \sum_{j=1}^{k} x_i x_j \cos \theta_{ij}$$

$$= [x_1, x_2, \ldots, x_k] \begin{bmatrix} \cos \theta_{11} & \cos \theta_{12} & \cdots & \cos \theta_{1k} \\ \cos \theta_{21} & \cos \theta_{22} & \cdots & \cos \theta_{2k} \\ \vdots & \vdots & & \vdots \\ \cos \theta_{k1} & \cos \theta_{k2} & \cdots & \cos \theta_{kk} \end{bmatrix} \begin{bmatrix} x_1 \\ x_2 \\ \vdots \\ x_k \end{bmatrix}$$

$$= \mathbf{X}^T \mathbf{C} \mathbf{X} \tag{2.58}$$

where $\mathbf{C}$ is a symmetric cosine matrix. Similarly the Euclidian distance between two vectors $\mathbf{X}_1$, $\mathbf{X}_2$ is given by

$$d_2 = (\mathbf{X}_1 - \mathbf{X}_2)^T \mathbf{C} (\mathbf{X}_1 - \mathbf{X}_2)$$

$$= \sum_{i=1}^{k} \sum_{j=1}^{k} (x_{1i} - x_{2i})(x_{1j} - x_{2j}) \cos \theta_{ij} \tag{2.59}$$

Let α_i, β_j represent direction cosines and let θ be an angle lying between two vectors expressed in terms of k oblique coordinates. Then the cosine of θ is

$$\cos \theta = \sum_{i=1}^{k} \sum_{j=1}^{k} \alpha_i \beta_j \cos \theta_{ij} \tag{2.60}$$

2.6   LATENT ROOTS AND VECTORS OF GRAMMIAN MATRICES

To a large extent the mathematical core of factor analysis consists of computing latent roots and latent vectors of symmetric, positive definite matrices. Other names include eigenroot, eigenvalue (eigenvector), characteristic root, and characteristic value (characteristic vector). In certain areas of engineering and applied mathematics latent roots are also known as singular values. In classical factor analysis interest is restricted to (semi) definite Grammian matrices such as the covariance or correlation matrix. Latent roots and latent vectors then estimate latent or underlying tendencies which may exist within an intercorrelated set of random variables. In this section we consider principal properties of latent roots and latent vectors of Grammian matrices since these play a prominent role in subsequent chapters.

Definition 2.7.   A nonzero latent vector $\mathbf{P}_i$ of a $(k \times k)$ matrix $\mathbf{A}$ is any $(k \times 1)$ vector satisfying the equation

$$\mathbf{AP}_i = \lambda_i \mathbf{P}_i \qquad (i = 1, 2, \ldots, k)$$

where λ_i is the corresponding latent root associated with $\mathbf{P}_i$.

The latent roots and vectors can also be combined into matrix form, and we write

$$\mathbf{P}^{-1}\mathbf{AP} = \Lambda \tag{2.61}$$

where the columns of $\mathbf{P}$ are linearly independent latent vectors $\mathbf{P}_1$, $\mathbf{P}_2, \ldots, \mathbf{P}_k$ and Λ is the diagonal matrix

$$\Lambda = \begin{bmatrix} \lambda_1 & & & \\ & \lambda_2 & & \mathbf{0} \\ & & \ddots & \\ \mathbf{0} & & & \lambda_r \end{bmatrix} \tag{2.62}$$

For an arbitrary matrix $\mathbf{A}$, Λ, and $\mathbf{P}$ contain complex numbers. The set of all latent vectors of $\mathbf{A}$ which correspond to a single latent root λ form a vector space of dimension $(k - r)$ where r is the rank of $(\mathbf{A} - \lambda\mathbf{I})$. Actually $\mathbf{A}$ always possesses k linearly independent latent vectors $\mathbf{P}_1$, $\mathbf{P}_2, \ldots, \mathbf{P}_k$ if they correspond to distinct latent roots λ_1, $\lambda_2, \ldots, \lambda_k$. The converse is not necessarily true, and a matrix can possess linearly independent latent vectors even though roots are not distinct. Equation (2.61) is known as a similarity transformation, and we say that the diagonal matrix Λ is similar to the matrix $\mathbf{A}$.

The similarity transformation (Eq. 2.61) preserves many important properties of a (square) matrix. When $\mathbf{A}$ is positive (semi) definite, Λ is also positive (semi) definite and when $\mathbf{A}$ is nonsingular, so is Λ, that is $\mathbf{A}$ is nonsingular if and only if all latent roots are nonzero. However, in general the rank of $\mathbf{A}$ does not equal the number of nonzero latent roots, although for a Grammian matrix this is always true. The principal properties which are of interest in factor analysis can be summarized by the following theorem.

THEOREM 2.5. Let $\mathbf{A}$ be a real $(k \times k)$ matrix with distinct latent roots $\lambda_1, \lambda_2, \ldots, \lambda_k$. Then
(1) $|\mathbf{A}| = |\Lambda| = \lambda_1, \lambda_2, \ldots, \lambda_k$.
(2) $tr(A) = tr(\Lambda) = \lambda_1 + \lambda_2 + \cdots + \lambda_k$.
(3) When $\mathbf{A}$ is similar to Λ, $\mathbf{A}^{-1}$ is similar to Λ^{-1}.
(4) $\mathbf{A}$, $\mathbf{A}^{\mathrm{T}}$ possess the same latent roots but different latent vectors.

When $\mathbf{A}$ is symmetric, latent roots and vectors assume a particular and well-known structure which has far-reaching consequences for applied work.

THEOREM 2.6.   Let $\mathbf{A}$ be any symmetric matrix. Then

   (1) Latent roots and latent vectors of $\mathbf{A}$ are real.
   (2) Any two latent vectors $\mathbf{P}_i$, $\mathbf{P}_j$ which correspond to latent roots λ_i, λ_j $(i, j = 1, 2, \ldots, k)$ respectively are orthogonal; that is, there always exist a matrix $\mathbf{P}$ such that

$$\mathbf{P}^T \mathbf{A} \mathbf{P} = \Lambda \tag{2.63}$$

where $\mathbf{P}^T \mathbf{P} = \mathbf{P} \mathbf{P}^T = \mathbf{I}$ and diagonal elements of Λ are not necessarily distinct.

   (3) The number of nonzero latent roots (latent vectors) of $\mathbf{A}$ is equal to the rank of $\mathbf{A}$.

Thus given a symmetric matrix $\mathbf{A}$ we can always define a linear decomposition

$$\mathbf{A} = \mathbf{P}\Lambda\mathbf{P}^T$$

$$= \lambda_1 \mathbf{P}_1 \mathbf{P}_1^T + \lambda_2 \mathbf{P}_2 \mathbf{P}_2^T + \cdots + \lambda_k \mathbf{P}_k \mathbf{P}_k^T \tag{2.64}$$

where $\mathbf{P}^T = \mathbf{P}^{-1}$, $\mathbf{P}_i \mathbf{P}_i^T$ are unit rank $(r \times r)$ matrices, and

$$\mathbf{P}_1 \mathbf{P}_1^T + \mathbf{P}_2 \mathbf{P}_2^T + \cdots + \mathbf{P}_k \mathbf{P}_k^T = \mathbf{I} \tag{2.65}$$

The decomposition (Eq. 2.64) is also known as the spectral or the singular value decomposition of $\mathbf{A}$. When all elements of a symmetric matrix are strictly positive, using the well-known Rayleigh quotient it can be shown that (1) the largest latent root λ_1 is strictly positive, and (2) its associated latent vector $\mathbf{P}_1$ can always be chosen to have strictly positive elements (e.g., see Barnett, 1978, Appendix). Alternatively, the result may be derived as a special case of Peron's theorem (Basilevsky, 1983; Exercise 2.16).

When $\mathbf{A}$ is further specified to be positive (semi) definite, sharper results can be obtained concerning the spectrum of $\mathbf{A}$. The following theorem is a special case of Theorem 2.6.

THEOREM 2.7.   Let $\mathbf{A}$ be a symmetric, positive (semi) definite matrix. Then

   (1) The λ_i are all strictly positive when $\mathbf{A}$ is positive definite.
   (2) The λ_i are nonnegative when $\mathbf{A}$ is positive semidefinite.
   (3) Elements of $\mathbf{P}$ are real (negative, positive, zero)

Because of the preservation of Grammian properties of a matrix under similarity transformations, quadratic forms can also be classified in terms of their latent roots (Section 2.2). Let $\mathbf{A}$ be a $(k \times k)$ symmetric matrix. Then a quadratic form $y = \mathbf{X}^T \mathbf{A} \mathbf{X}$ is:

(1) Positive definite if and only if $\lambda_i > 0$, $i = 1, 2, \ldots, k$.

(2) Positive semidefinite if and only if $\lambda_i \geq 0$.

(3) Negative definite if and only if $\lambda_i < 0$, $i = 1, 2, \ldots, k$.

(4) Negative semidefinite if and only if $\lambda_i \leq 0$.

(5) Indefinite if and only if some latent roots are positive and others are negative. When a latent root is zero, the indefinite form is singular.

The task of actually computing latent roots and latent vectors is carried out by specialized numerical algorithms (e.g., see Hammarling, 1970). The algebraic basis of a solution, however, can be described as follows. From Definition 2.6 we have $\mathbf{A}\mathbf{P}_i = \lambda_i \mathbf{P}_i$ or

$$(\mathbf{A} - \lambda_i \mathbf{I})\mathbf{P}_i = \mathbf{0} \qquad (i = 1, 2, \ldots, k) \tag{2.66}$$

a system of k homogenous linear equations in $k + 1$ unknowns; k unknown components of $\mathbf{P}_i$ plus the unknown latent root λ_i. From the general theory of linear equations we know a nonzero solution vector exists if the matrix of coefficients $\mathbf{B} = (\mathbf{A} - \lambda \mathbf{I})$ is singular for some root $\lambda_i = \lambda$; that is, if

$$|\mathbf{B}| = |\mathbf{A} - \lambda \mathbf{I}| = 0 \tag{2.67}$$

or

$$\phi(\lambda) = \begin{bmatrix} (a_{11} - \lambda) & a_{12} & \cdots & a_{1k} \\ a_{12} & (a_{22} - \lambda) & \cdots & a_{2k} \\ \vdots & \vdots & & \vdots \\ a_{1k} & a_{2k} & \cdots & (a_{kk} - \lambda) \end{bmatrix} \tag{2.68}$$

$$= (-1)^k \lambda^k + (-1)^{k-1} q_1 \lambda^{k-1} + (-1)^{k-2} q_2 \lambda^{k-2} + \cdots + (-1)q_{k-1}\lambda + q$$

$$= 0$$

a kth-order polynomial in λ. The roots $\lambda_1, \lambda_2, \ldots, \lambda_k$ are conventionally ranked as $\lambda_1 \geq \lambda_2 \geq \cdots \geq \lambda_k$. Once the values of λ_i are known, corresponding latent vector $\mathbf{P}_i$ can be found using Eq. (2.66). For example, when $i = 1$ we have the following system of k linear equations:

$$\begin{bmatrix} (a_{11} - \lambda_1) & a_{12} & \cdots & a_{1k} \\ a_{12} & (a_{22} - \lambda_1) & \cdots & a_{2k} \\ \vdots & \vdots & \ddots & \vdots \\ a_{1k} & a_{2k} & \cdots & (a_{kk} - \lambda_1) \end{bmatrix} \begin{bmatrix} p_{11} \\ p_{21} \\ \vdots \\ p_{k1} \end{bmatrix} = \begin{bmatrix} 0 \\ 0 \\ \vdots \\ 0 \end{bmatrix} \tag{2.69}$$

in k unknowns. The system does not have a unique solution, however, owing to the arbitrary magnitudes of the latent vectors. Since we can choose any magnitude for $\mathbf{P}_1$, in practice $\mathbf{P}_1$ is given unit length for convenience. This introduces the additional equation

$$\mathbf{P}_1{}^T\mathbf{P}_1 = p_{11}^2 + p_{21}^2 + \cdots + p_{ki}^2 = 1 \tag{2.70}$$

which together with Eq. (2.69) yields a unique solution. The process is repeated for $i = 2, 3, \ldots, k$ until all the latent roots λ_i and their corresponding latent vectors $\mathbf{P}_i$ are known. Since $\mathbf{A}$ is Grammian, all roots are real and nonnegative and the real latent vectors form an orthonormal set. Note however that if $\mathbf{P}_i$ is a latent vector, so is $-\mathbf{P}_i$ even though $\mathbf{P}_i^T\mathbf{P}_i = 1$.

Example 2.5. We have the Grammian matrix

$$\mathbf{A} = \begin{bmatrix} 11 & 5 \\ 5 & 5 \end{bmatrix}$$

where

$$|\mathbf{A} - \lambda\mathbf{I}| = \begin{vmatrix} 11 - \lambda & 5 \\ 5 & 5 - \lambda \end{vmatrix}$$
$$= \lambda^2 - 16\lambda + 30$$

a second-degree polynomial with solutions $\lambda_1 = 13.83$, $\lambda_2 = 2.17$. Substituting $\lambda_1 = 13.83$ in Eq. (2.69) we have

$$\begin{bmatrix} 11 - \lambda_1 & 5 \\ 5 & 5 - \lambda_1 \end{bmatrix}\begin{bmatrix} p_{11} \\ p_{21} \end{bmatrix} = \begin{bmatrix} 0 \\ 0 \end{bmatrix}$$

or

$$\begin{bmatrix} -2.83 & 5 \\ 5 & -8.83 \end{bmatrix}\begin{bmatrix} p_{11} \\ p_{21} \end{bmatrix} = \begin{bmatrix} 0 \\ 0 \end{bmatrix}$$

so that

$$-2.83p_{11} + 5p_{21} = 0$$
$$5p_{11} - 8.83p_{21} = 0$$
$$p_{11}^2 + p_{21}^2 = 1$$

with solution $\mathbf{P}_1 = (.8700, .4925)^T$. Similarly when $\lambda_2 = 2.17$ we have $\mathbf{P}_2 = (-.4925, .8700)^T$. The latent vectors can be arranged as columns of the matrix

$$\mathbf{P} = \begin{bmatrix} .8700 & -.4925 \\ .4925 & .8700 \end{bmatrix}$$

where it is easy to verify that $\mathbf{P}^T\mathbf{P} = \mathbf{P}\mathbf{P}^T = \mathbf{I}$. $\square$

Example 2.6. Consider the correlation matrix **R** of Example 2.2. The latent roots and vectors are found to be

$$\Lambda = \begin{bmatrix} 4.0437 & & & & & & \\ & 1.2717 & & & & 0 & \\ & & 1.0896 & & & & \\ & & & .2362 & & & \\ & & & & .1644 & & \\ & 0 & & & & .1427 & \\ & & & & & & .0516 \end{bmatrix}$$

$$\begin{array}{ccccccc} \mathbf{P}_1 & \mathbf{P}_2 & \mathbf{P}_3 & \mathbf{P}_4 & \mathbf{P}_5 & \mathbf{P}_6 & \mathbf{P}_7 \end{array}$$

$$\mathbf{P} = \begin{bmatrix} .1115 & .0005 & .9219 & -.2196 & .0168 & .2896 & -.0720 \\ .4694 & .0495 & -.1389 & .1069 & -.4257 & .5056 & .5565 \\ -.4787 & -.0590 & .0550 & -.2644 & .3449 & -.0230 & .7583 \\ -.3639 & .4561 & .2336 & .7656 & -.0945 & .0068 & .0990 \\ .4440 & .2386 & -.0804 & .1921 & .8177 & .1844 & .0053 \\ -.4453 & -.2084 & -.2122 & .0213 & .1019 & .7837 & -.2971 \\ -.1017 & .8281 & -.1472 & -.4968 & -.1078 & .1083 & -.1101 \end{bmatrix}$$

and can be used to verify Theorems 2.5–2.7 (Exercise 2.4). □

2.7 ROTATION OF QUADRATIC FORMS

The latent roots and vectors of a symmetric matrix can be used to rotate quadratic forms to a more simple structure. Let $y = \mathbf{X}^T\mathbf{A}\mathbf{X}$, the equation of a kth dimensional ellipse with center at the origin, $\mathbf{X} = (x_1, x_2, \ldots, x_k)^T$ and $y = q > 0$, an arbitrary constant. Since **A** need not be diagonal, the quadratic form generally contains product terms of the form $x_i x_j$ $(i \neq j)$ in addition to the squared expressions x_i^2 $(i = 1, 2, \ldots, k;$ Section 2.2). The principal axes of the ellipse do not coincide with the k coordinate axes and this gives the quadratic expression a cumbersome "interactive" form. The $\partial y / \partial x_i$ contain variables other than x_i, which magnifies the problem of locating extremum points. Transforming **A** to diagonal form simplifies the equation, and as shown in Chapter 3 allows for a clearer understanding of data.

Let **X** be a radius vector with squared length

$$r^2 = \mathbf{X}^T\mathbf{X} = x_1^2 + x_2^2 + \cdots + x_k^2 \tag{2.71}$$

where $\mathbf{X} = (x_1, x_2, \ldots, x_k)^T$ is any point on the ellipse. Let r_i denote the lengths of **X** when the radius vector (Eq. 2.71) coincides with the ith principal axis of the ellipse, that is, **X** has length r_i when its vertex lies on the point of intersection of the ellipse and a sphere with center at the origin and radius r_i. Let the points of intersection be $\mathbf{X}_i = \mathbf{Q}_i = (q_{1i}, q_{2i}, \ldots, q_{ki})^T$

so that $\mathbf{Q}_i^{\mathrm{T}}\mathbf{Q}_i = r_i$ $(i = 1, 2, \ldots, k)$. Using the theory of Lagrange multipliers we know that constrained extremum points $\mathbf{Q}_i$ are also stationary values of the function

$$\phi = \mathbf{Q}_i^{\mathrm{T}}\mathbf{A}\mathbf{Q}_i - \lambda_i(\mathbf{Q}_i^{\mathrm{T}}\mathbf{Q}_i - r_i^2) \tag{2.72}$$

Differentiating Eq. (2.72) with respect to $\mathbf{Q}_i$ and setting to zero yields the system of linear equations

$$\frac{\partial \phi}{\partial \mathbf{Q}_i} = 2\mathbf{A}\mathbf{Q}_i - 2\lambda_i\mathbf{Q}_i = \mathbf{0} \tag{2.73}$$

or

$$(\mathbf{A} - \lambda_i\mathbf{I})\mathbf{Q}_i = \mathbf{0} \qquad (i = 1, 2, \ldots, k) \tag{2.74}$$

It follows from Eq. (2.66) that points $\mathbf{Q}_1, \mathbf{Q}_2, \ldots, \mathbf{Q}_k$ are latent vectors of the symmetric, positive definite matrix $\mathbf{A}$ (scaled to lengths r_i) and λ_1, $\lambda_2, \ldots, \lambda_k$ are k latent roots. Once λ_i and $\mathbf{Q}_i$ are known, they may be arranged in matrix form and Eq. (2.74) is equivalent to

$$\mathbf{A}\mathbf{Q} = \mathbf{Q}\mathbf{\Lambda} \tag{2.75}$$

where $\mathbf{\Lambda}$ is diagonal and $\mathbf{Q}$ is orthogonal. Also, rewriting Eq. (2.74) as $\mathbf{A}\mathbf{Q}_i = \lambda_i\mathbf{Q}_i$ we have

$$\mathbf{Q}_i^{\mathrm{T}}\mathbf{A}\mathbf{Q}_i = \lambda_i\mathbf{Q}_i^{\mathrm{T}}\mathbf{Q}_i$$

$$= \lambda_i r_i^2$$

$$= q \tag{2.76}$$

so that

$$r_i = \frac{q^{1/2}}{\lambda_i} \tag{2.77}$$

is the half-length of the principal axes of the ellipse, where $\lambda_i > 0$ and $\mathbf{Q}^{\mathrm{T}}\mathbf{Q} = \mathbf{R}$ a diagonal matrix with elements r_i $(i = 1, 2, \ldots, k)$. The column vectors of $\mathbf{Q}$ can be scaled to unit length so that columns of $\mathbf{P} = \mathbf{Q}\mathbf{R}^{-1/2}$ are unit latent vectors. The orthogonal transformation

$$\mathbf{Z} = \mathbf{P}^{\mathrm{T}}\mathbf{Z} \tag{2.78}$$

then defines an orthogonal rotation of axes (orthogonal rotation of the k-dimensional ellipse) such that axes of the ellipse coincide with new coordinates axes $\mathbf{Z}_1, \mathbf{Z}_2, \ldots, \mathbf{Z}_k$. The ellipse can now be expressed as

$$y = \mathbf{X}^T \mathbf{A} \mathbf{X}$$

$$= \mathbf{X}^T (\mathbf{P} \mathbf{\Lambda} \mathbf{P}^T) \mathbf{X}$$

$$= \mathbf{Z}^T \mathbf{\Lambda} \mathbf{Z}$$

$$= \lambda_1 z_1^2 + \lambda_2 z_2^2 + \cdots + \lambda_k z_k^2 \tag{2.79}$$

which is free of cross-product terms $z_i z_j$ $(i \neq j)$. As demonstrated in Chapter 3, this is the geometric basis for a principal components decomposition of a covariance (correlation) matrix. A more general result concerning the equivalence of two quadratic forms is as follows.

THEOREM 2.8. Two symmetric matrices $\mathbf{A}$, $\mathbf{B}$ represent the same quadratic form if and only if $\mathbf{B} = \mathbf{P}^T \mathbf{A} \mathbf{P}$ where $\mathbf{P}$ is nonsingular.

Example 2.7. Consider the quadratic form

$$y = 11x_1^2 + 10x_1 x_2 + 5x_2^2$$

$$= [x_1, x_2] \begin{bmatrix} 11 & 5 \\ 5 & 5 \end{bmatrix} \begin{bmatrix} x_1 \\ x_2 \end{bmatrix}$$

$$= \mathbf{X}^T \mathbf{A} \mathbf{X}$$

which contains the cross-product term $10x_1 x_2$. From Example 2.5 we have

$$\mathbf{\Lambda} = \begin{bmatrix} 13.83 & 0 \\ 0 & 2.17 \end{bmatrix}, \qquad \mathbf{P} = \begin{bmatrix} .8700 & -.4925 \\ .4925 & .8700 \end{bmatrix}$$

and the two half-lengths of the ellipse are, for $y = q = 1$, $r_1 = (1/13.83)^{1/2} = .2689$, $r_2 = (1/2.17)^{1/2} = .6788$. Then from Eq. (2.79) it follows that the quadratic, with respect to the new axes, is

$$y = 13.83 z_1^2 + 2.17 z_2^2 \qquad \qquad \square$$

2.8 ELEMENTS OF MULTIVARIATE NORMAL THEORY

Latent roots and vectors of Grammian matrices play a crucial role in many areas of multivariate analysis, particularly when dealing with the multivariate normal distribution. Recall that univariate t, F, and χ^2 distributions all assume that the underlying random variable is distributed normally (Section 1.2). In multivariate analysis the equivalent assumption is that of multivariate normality, from which multivariate versions of the t, F, and χ^2 distributions can be derived. The multivariate normal distribution plays a central role in multivariate tests of significance and in deriving likelihood ratio tests for principal components analysis and maximum likelihood factor analysis.

2.8.1　The Multivariate Normal Distribution

Consider a set of k random variables $X_1, X_2, \ldots, X_k$, not necessarily independent, where $\mathbf{X} = (x_1, x_2, \ldots, x_k)^{\mathrm{T}}$ represents a vector of values of the random variables. Then $\mathbf{X}$ follows a multivariate normal distribution if and only if

$$f(\mathbf{X}) = \frac{1}{(2\pi)^{k/2}|\mathbf{\Sigma}|^{1/2}} \exp\left\{-\frac{1}{2}\left[(\mathbf{X} - \mathbf{\mu})^{\mathrm{T}} \mathbf{\Sigma}^{-1}(\mathbf{X} - \mathbf{\mu})\right]\right\} \qquad (2.80)$$

Here $\mathbf{\Sigma}$ and $\mathbf{\mu}$ are the population covariance matrix and mean vector, respectively, where

$$\mathbf{\Sigma} = \begin{bmatrix} \sigma_1^2 & \sigma_{12} & \cdots & \sigma_{1k} \\ \sigma_{12} & \sigma_2^2 & \cdots & \sigma_{2k} \\ \vdots & \vdots & & \vdots \\ \sigma_{1k} & \sigma_{2k} & \cdots & \sigma_k^2 \end{bmatrix} \qquad (2.81)$$

and $\mathbf{\mu} = (\mu_1, \mu_2, \ldots, \mu_k)^{\mathrm{T}}$. Since $\mathbf{\Sigma}^{-1}$ is assumed to exist we have $|\mathbf{\Sigma}| \neq 0$ and the multivariate normal (Eq. 2.80) is said to be nonsingular. The covariance matrix can also be expressed as

$$\mathbf{\Sigma} = \begin{bmatrix} \sigma_1^2 & \rho_{12}\sigma_1\sigma_2 & \cdots & \rho_{1k}\sigma_1\sigma_k \\ \rho_{12}\sigma_1\sigma_2 & \sigma_2^2 & & \rho_{2k}\sigma_2\sigma_k \\ \vdots & \vdots & \ddots & \vdots \\ \rho_{1k}\sigma_1\sigma_k & \rho_{2k}\sigma_2\sigma_k & \cdots & \sigma_k^2 \end{bmatrix} \qquad (2.82)$$

where ρ_{ij} is the population correlation between X_i, X_j. Since the multivariate normal is completely determined by $\mathbf{\mu}$ and $\mathbf{\Sigma}$ it is usually written as $N(\mathbf{\mu}, \mathbf{\Sigma})$.

The exponent of Eq. (2.80) represents a quadratic form with cross product terms unless $\mathbf{\Sigma} = \mathbf{D}$, a diagonal matrix. Consequently it follows from Section 2.7 that

$$(\mathbf{X} - \mathbf{\mu})^{\mathrm{T}} \mathbf{\Sigma}^{-1}(\mathbf{X} - \mathbf{\mu}) = c \qquad (2.83)$$

is a k-dimensional ellipsoid with center at $\mathbf{\mu} = (\mu_1, \mu_2, \ldots, \mu_k)^{\mathrm{T}}$ and shape parameters $\mathbf{\Sigma}$ (Fig. 2.4). Also, Eq. (2.83) is distributed as chi-squared with k degrees of freedom (Section 1.2.1). Since $\mathbf{\Sigma}^{-1}$ is symmetric and positive definite, the quadratic form (Eq. 2.83) is positive definite and can (see

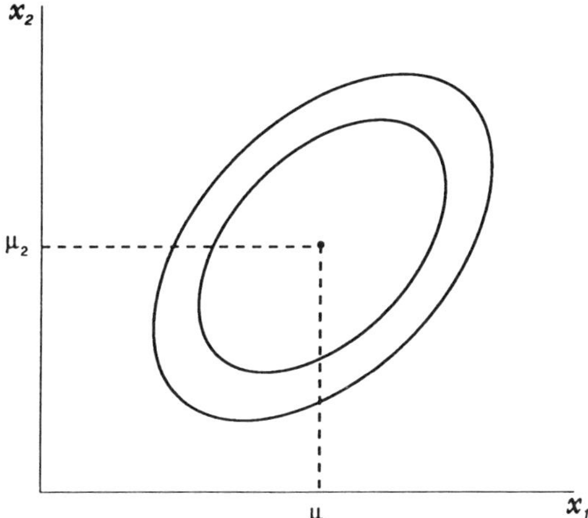

Figure 2.4 A finite representation of elliptic bases of two bivariate normal distributions with common $\boldsymbol{\Sigma}$ and $\boldsymbol{\mu}$ but different values of c.

Section 2.6) be rotated to diagonal form using latent roots and vectors of $\boldsymbol{\Sigma}$. Maximizing Eq. (2.83) subject to constraint

$$(\mathbf{X} - \boldsymbol{\mu})^{\mathrm{T}}(\mathbf{X} - \boldsymbol{\mu}) = r \tag{2.84}$$

leads to the Lagrange expression

$$\phi = (\mathbf{X} - \boldsymbol{\mu})^{\mathrm{T}}\boldsymbol{\Sigma}^{-1}(\mathbf{X} - \boldsymbol{\mu}) - \lambda[(\mathbf{X} - \boldsymbol{\mu})^{\mathrm{T}}(\mathbf{X} - \boldsymbol{\mu}) - r] \tag{2.85}$$

and differentiating with respect to vector $\mathbf{X}$ (Section 2.3) leads to

$$\frac{\partial \phi}{\partial \mathbf{X}} = 2\boldsymbol{\Sigma}^{-1}(\mathbf{X} - \boldsymbol{\mu}) - 2\lambda(\mathbf{X} - \boldsymbol{\mu}) = \mathbf{0}$$

or

$$(\boldsymbol{\Sigma}^{-1} - \lambda \mathbf{I})(\mathbf{X} - \boldsymbol{\mu}) = \mathbf{0}$$

For part (3) of Theorem 2.5 we know that $\boldsymbol{\Sigma}, \boldsymbol{\Sigma}^{-1}$ possess identical latent vectors, but reciprocal latent roots. Since the ordering of latent roots is arbitrary, $\boldsymbol{\Sigma}^{-1}$ can be replaced by $\boldsymbol{\Sigma}$ where

$$(\boldsymbol{\Sigma} - \delta \mathbf{I})(\mathbf{X} - \boldsymbol{\mu}) = \mathbf{0} \tag{2.86}$$

and $\delta = 1/\lambda$. For a repeated root of multiplicity m the k-dimensional ellipsoid is hyperspherical of dimension $m < k$. Random variables associated with repeated roots are said to possess isotropic variation in that subspace. The variables X_i can also be transformed to independent (uncorrelated) form by the following theorem.

THEOREM 2.9.   A matrix Σ is Grammian if and only if it can be expressed as $\Sigma = \mathbf{P}^\mathrm{T}\mathbf{P}$ where $\mathbf{P}$ is nonsingular.

From Eq. (2.86) we have

$$(\Sigma - \delta\mathbf{I})(\mathbf{X} - \mu) = (\mathbf{P}^\mathrm{T}\mathbf{P} - \delta\mathbf{I})(\mathbf{X} - \mu)$$

$$= \mathbf{0}$$

or

$$\mathbf{P}^\mathrm{T}\mathbf{P}(\mathbf{X} - \mu) = (\mathbf{X} - \mu)\delta\mathbf{I}$$

so that

$$(\mathbf{X} - \mu)^\mathrm{T}\mathbf{P}^\mathrm{T}\mathbf{P}(\mathbf{X} - \mu) = r\delta\mathbf{I} \tag{2.87}$$

using Eq. (2.84). The transformed set of variables is then given by

$$\mathbf{Z} = \mathbf{P}(\mathbf{X} - \mu) \tag{2.88}$$

which represents a new set of orthogonal normal random variables. The columns of $\mathbf{P}$ are normalized latent vectors and the new variables, $\mathbf{Z} = (z_1, z_2, \ldots, z_k)^\mathrm{T}$ are orthonormal, that is,

$$z_i{}^\mathrm{T}z_j = \begin{cases} 1 & (i = j) \\ 0 & (i \neq j) \end{cases} \tag{2.89}$$

Also, it is easy to see that $\mathrm{E}(\mathbf{Z}) = \mathbf{0}$. More generally,

$$\mathrm{var}(z_i) = \mathbf{P}_i{}^\mathrm{T}\Sigma\mathbf{P}_i$$

$$= \delta_i \tag{2.90}$$

Thus from the theory of quadratic forms it follows that a set of correlated multivariate normal variables can always be rotated to a new uncorrelated univariate set.

Example 2.8.   Consider the bivariate normal distribution where we let $x_1 = x$ and $x_2 = y$. Equation (2.80) becomes

$$f(x, y) = \frac{1}{2\pi\sigma_x\sigma_y\sqrt{1 - \rho^2}}$$

$$\times \exp\left\{ -\frac{1}{2(1 - \rho^2)} \left[\left(\frac{x - \mu_x}{\sigma_x}\right)^2 - 2\rho\frac{(x - \mu_x)}{\sigma_x}\frac{(y - \mu_y)}{\sigma_y} + \left(\frac{y - \mu_y}{\sigma_y}\right)^2 \right] \right\} \tag{2.91}$$

where $(2\pi)^{k/2} = 2\pi$ for $k = 2$ and

$$|\mathbf{\Sigma}|^{1/2} = \begin{vmatrix} \sigma_x^2 & \rho_{xy}\sigma_x\sigma_y \\ \rho_{xy}\sigma_x\sigma_y & \sigma_y^2 \end{vmatrix}^{1/2}$$

$$= \sigma_x\sigma_y\sqrt{1-\rho^2} \qquad (2.92)$$

Note that in order to avoid singularity, $-1 < \rho < 1$. Also, for the exponent,

$$-\frac{1}{2}[(\mathbf{X}-\boldsymbol{\mu})^{\mathrm{T}}\mathbf{\Sigma}^{-1}(\mathbf{X}-\boldsymbol{\mu})] = -\frac{1}{2}[(x-\mu_x),\,(y-\mu_y)]$$

$$\times \begin{bmatrix} \dfrac{1}{\sigma_x^2(1-\rho^2)} & \dfrac{-\rho}{\sigma_x\sigma_y(1-\rho^2)} \\ \dfrac{-\rho}{\sigma_x\sigma_y(1-\rho^2)} & \dfrac{1}{\sigma_y^2(1-\rho^2)} \end{bmatrix}\begin{bmatrix} (x-\mu_y) \\ (y-\mu_y) \end{bmatrix} \qquad (2.93)$$

$$= -\frac{1}{2(1-\rho^2)}\left[\frac{(x-\mu_x)^2}{\sigma_x^2} + \frac{2\rho(x-\mu_x)(y-\mu_y)}{\sigma_x\sigma_y} + \frac{(y-\mu_y)^2}{\sigma_y^2}\right]$$

a quadratic form with the cross-product term $(x-\mu_x)(y-\mu_y)$.

The multivariate normal can also be written in a more compact form. Let

$$z_1 = \frac{x_1-\mu_1}{\sigma_1}, \quad z_2 = \frac{x_2-\mu_2}{\sigma_2}, \ldots, z_k = \frac{x_k-\mu_k}{\sigma_k}$$

Then Eq. (2.80) can be expressed as

$$f(z_1, z_2, \ldots, z_k) = f(\mathbf{Z}) = \frac{1}{(2\pi)^{k/2}|\mathbf{R}|^{1/2}}\exp\left\{-\frac{1}{2}[\mathbf{Z}^{\mathrm{T}}\mathbf{R}^{-1}\mathbf{Z}]\right\} \quad (2.94)$$

where

$$\mathbf{R} = \begin{bmatrix} 1 & \rho_{21} & \cdots & \rho_{k1} \\ \rho_{21} & 1 & \cdots & \rho_{k2} \\ \vdots & \vdots & \ddots & \vdots \\ \rho_{k1} & \rho_{k2} & \cdots & 1 \end{bmatrix} \qquad (2.95)$$

is the correlation matrix. The $z_1,\, z_2, \ldots, z_k$ are standardized normal variates with zero mean and unit variance. Equation (2.94) is easier to manipulate because of its simplified structure. For the bivariate normal we have

$$f(u, v) = \frac{1}{2\pi\sqrt{1-\rho^2}}\exp\left\{-\frac{1}{2(1-\rho^2)}[u^2 - 2\rho uv + v^2]\right\} \qquad (2.96)$$

where $z_1 = u$, $z_2 = v$.

Let $\mathbf{S} = (s_1, s_2, \ldots, s_k)^{\mathrm{T}}$ be a set of arbitrary variables. Then the multivariate normal moment generating function is given by

$$\mathbf{M}_{\mathbf{X}}(\mathbf{S}) = \mathbf{E}(e^{\mathbf{S}^{\mathrm{T}}\mathbf{X}})$$

$$= \exp\left(\mathbf{S}^{\mathrm{T}}\boldsymbol{\mu} - \frac{1}{2}\mathbf{S}^{\mathrm{T}}\boldsymbol{\Sigma}\mathbf{S}\right) \tag{2.97}$$

and for standardized variables we have

$$\mathbf{M}_{\mathbf{Z}}(\mathbf{S}) = \mathbf{E}[\exp(\mathbf{S}^{\mathrm{T}}\mathbf{Z})]$$

$$= \exp\left(-\frac{1}{2}\mathbf{S}^{\mathrm{T}}\mathbf{R}\mathbf{S}\right)$$

$$= \exp\left(-\frac{1}{2}\sum_{i=1}^{k}\sum_{j=1}^{k}\rho_{ij}s_is_j\right) \tag{2.98}$$

Also for $k = 2$,

$$\mathbf{M}_{\mathbf{Z}}(\mathbf{S}) = \exp\left(-\frac{1}{2}\sum_{i=1}^{2}\sum_{j=1}^{2}\rho_{ij}s_is_j\right)$$

$$= \exp\left[-\frac{1}{2}(s_1^2 + s_2^2 + 2\rho_{12}s_1s_2)\right] \tag{2.99}$$

The multivariate normal provides a convenient starting point for factor analysis owing to the equivalence between zero correlation and independence, as indicated by the following theorems. $\square$

THEOREM 2.10.   A set of multivariate normal variates $X_1, X_2, \ldots, X_k$ are independent if and only if $\boldsymbol{\Sigma}$ is diagonal.

PROOF.   Independence implies zero correlation for any two normal random variables X and Y since

$$\mathrm{cov}(X, Y) = E(xy) - E(x)E(y)$$

$$= E(x)E(y) - E(x)E(y)$$

$$= 0 \tag{2.100}$$

using Eq. (1.28). Conversely, when $\mathrm{cov}(X, Y) = 0$, $\boldsymbol{\Sigma}$ is diagonal and the multivariate normal can be factored into a product of univariate normals.

Example 2.9.   Under zero correlation the bivariate normal (Eq. 2.91) can be expressed as

$$f(x, y) = \frac{1}{2\pi\sigma_x\sigma_y} \exp\left[-\frac{1}{2}\left(\frac{x - \mu_x}{\sigma_x}\right)^2 + \left(\frac{y - \mu_y}{\sigma_y}\right)^2\right]$$

$$= \frac{1}{\sqrt{2\pi}\sigma_x} \exp\left[-\frac{1}{2}\left(\frac{x - \mu_x}{\sigma_x}\right)^2\right] \frac{1}{\sqrt{2\pi}\sigma_y} \exp\left[-\frac{1}{2}\left(\frac{y - \mu_y}{\sigma_y}\right)^2\right]$$

$$= f(x)f(y)$$

a product of univariate normals. Two points must be kept in mind concerning independence, zero correlation, and normality. First, it is insufficient that a multivariate distribution be factored into a product of functions, each involving only a single variable, to establish distributional independence—each term in the product must be a univariate probability function. Second, although multivariate normality implies marginal normality, the converse does not hold, since it is easy to provide examples of nonnormal bivariate distributions whose marginal distributions are normal (Anderson 1984, p. 47; Broffitt, 1986). Thus independence and zero correlation are equivalent concepts only under joint (multivariate) normality, not simply marginal normality. It is possible, therefore, to have univariate normal variates which are uncorrelated but dependent (e.g., see Behboodian, 1990). □

THEOREM 2.11. Let $\mathbf{X} \sim N(\boldsymbol{\mu}, \boldsymbol{\Sigma})$. Then any subvector of $\mathbf{X}$ is also multivariate normal, with the correspondingly reduced mean vector and covariance matrix.

Theorem 2.11 states that multivariate normality is preserved in any subspace so that multivariate normality implies normality of all marginal distributions.

Multivariate independence can also be generalized as follows. Consider the two subvectors of $\mathbf{X} = (\mathbf{X}_1, \mathbf{X}_2)^T$ which partition the covariance matrix of $\mathbf{X}$ as

$$\boldsymbol{\Sigma} = \begin{bmatrix} \boldsymbol{\Sigma}_{11} & \boldsymbol{\Sigma}_{12} \\ \boldsymbol{\Sigma}_{21} & \boldsymbol{\Sigma}_{22} \end{bmatrix} \tag{2.101}$$

where $\boldsymbol{\Sigma}_{21} = \boldsymbol{\Sigma}_{12}^T$. Then the following theorem holds.

THEOREM 2.12. The multivariate normal subvectors $\mathbf{X}_1$, $\mathbf{X}_2$ are distributed independently if and only if $\boldsymbol{\Sigma}_{12} = \boldsymbol{\Sigma}_{21}^T = 0$.

PROOF. When $\boldsymbol{\Sigma}_{12} = 0$ the second part of the exponent in the moment generating function (Eq. 2.97) can be expressed in the partitioned form (Eq. 2.101) as

$$\mathbf{S}^T \boldsymbol{\Sigma} \mathbf{S} = \mathbf{S}_1^T \boldsymbol{\Sigma}_{11} \mathbf{S}_1 + \mathbf{S}_2^T \boldsymbol{\Sigma}_{22} \mathbf{S}_2$$

We have

$$\mathbf{M}_\mathbf{X}(\mathbf{S}) = \exp\!\left(\mathbf{S}_1^T \boldsymbol{\mu}_1 - \frac{1}{2}\mathbf{S}_1^T \boldsymbol{\Sigma}_{11}\mathbf{S}_1\right)\exp\!\left(\mathbf{S}_2^T \boldsymbol{\mu}_2 - \frac{1}{2}\mathbf{S}_2^T \boldsymbol{\Sigma}_{22}\mathbf{S}_2\right)$$

implying $f(\mathbf{X}) = f_1(\mathbf{X}_1)f_2(\mathbf{X}_2)$, where $f(\mathbf{X}_1) = \mathbf{N}(\boldsymbol{\mu}_1, \boldsymbol{\Sigma}_{11})$ and $f(\mathbf{X}_2) = \mathbf{N}(\boldsymbol{\mu}_2, \boldsymbol{\Sigma}_{22})$. It follows that the subvectors are distributed independently. The converse of the statement is also true.

THEOREM 2.13. The multivariate normal random variables X_1, $X_2, \ldots, X_k$ are mutually independent if and only if they are pairwise uncorrelated.

The proof consists of invoking Theorem 2.10 and using the moment generating function (Eq. 2.97) (see Exercise 2.1).

It can be concluded from these theorems that independence of normal random variables depends on whether $\boldsymbol{\Sigma}$ is diagonal. Since any Grammian matrix can be diagonalized by a similarity transformation (Section 2.6) it follows that a set of multivariate normal variates can always be transformed to an independent set by a rotation of the associated quadratic form (Theorem 2.10). Of course a correlated set of nonnormal variables can also be rotated to an uncorrelated form, but this does not necessarily achieve independence.

Example 2.10. For the bivariate normal the correlation matrix is

$$\mathbf{R} = \begin{bmatrix} 1 & \rho \\ \rho & 1 \end{bmatrix}$$

and to reduce $\mathbf{R}$ to diagonal form we solve $(\mathbf{R} - \lambda_i \mathbf{I})\mathbf{P}_i = 0$ where λ_i, $\mathbf{P}_i$ are latent roots—vectors. We have

$$\begin{vmatrix} 1 - \lambda & \rho \\ \rho & 1 - \lambda \end{vmatrix} = 0$$

or

$$\lambda^2 - 2\lambda + (1 - \rho^2) = \lambda^2 - 2\lambda + (1 - \rho)(1 + \rho)$$
$$= [\lambda(1 - \rho)][\lambda - (1 + \rho)]$$
$$= 0$$

with solutions $\lambda = \lambda_1 = 1 + \rho$, $\lambda = \lambda_2 + 1 - \rho$. When $\rho = 1$ the correlation matrix is singular, $\lambda_1 = 2$, $\lambda_2 = 0$, and the bivariate normal degenerates to

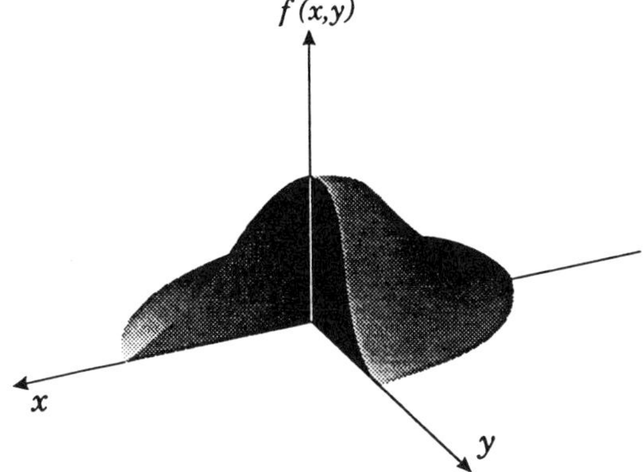

Figure 2.5 A bivariate normal distribution rotated to independent form.

the univariate case. Also when $\rho = 0$, $\lambda_1 = \lambda_2$ and the ellipse (Fig. 2.5) assumes the isotropic circular form. For $0 < \rho < 1$ we have $\lambda_1 > \lambda_2$, and for $-1 < \rho < 0$ we have $\lambda_1 < \lambda_2$. To obtain the latent vectors we solve the normal equations (Eq. 2.86), that is,

$$\rho p_{11} - \rho p_{12} = 0$$
$$\rho p_{11} - \rho p_{12} = 0$$
$$p_{11}^2 + p_{21}^2 = 1$$

for $\lambda_1 = 1 + \rho$ and

$$\rho p_{21} - \rho p_{22} = 0$$
$$\rho p_{21} - \rho p_{22} = 0$$
$$p_{21}^2 + p_{22}^2 = 1$$

for $\lambda_2 = 1 - \rho$. The solutions are then given by $\mathbf{P}_1 = (\sqrt{1/2}, \sqrt{1/2})^T$, $\mathbf{P}_2 = (\sqrt{1/2}, -\sqrt{1/2})^T$. Note that the latent vectors do not depend on ρ. This is considered further in Chapter 3 when we examine the case of equally correlated variates. Continuing with the example, the exponent of the bivariate normal [omitting the proportionality constant $-(1/2)(1 - \rho^2)$] can be expressed as

$$c = u^2 - 2\rho uv + v^2 = [u, v] \begin{bmatrix} 1 & \rho \\ \rho & 1 \end{bmatrix} \begin{bmatrix} u \\ v \end{bmatrix}$$

$$= [z_1, z_2] \begin{bmatrix} (1 + \rho) & 0 \\ 0 & (1 - \rho) \end{bmatrix} \begin{bmatrix} z_1 \\ z_2 \end{bmatrix}$$

$$= (1 + \rho)z_1^2 + (1 - \rho)z_2^2$$

where z_1, z_2 are new independent normal variables.

The multivariate normal distribution has two additional useful properties apart from those considered in Theorems 2.9–2.13; (1) all conditional distributions are normal, and (2) the conditional expectation is linear in the conditioning variables. $\qquad\square$

THEOREM 2.14. Let $\mathbf{X} = (\mathbf{X}_1, \mathbf{X}_2)^\mathrm{T}$ where $\mathbf{X}_1, \mathbf{X}_2$ are partitions of a multivariate normal vector $\mathbf{X}$. Then the conditional distribution of $\mathbf{X}_1$, given $\mathbf{X}_2$, is multivariate normal with expectation

$$\boldsymbol{\mu}_1 + \boldsymbol{\Sigma}_{12}\boldsymbol{\Sigma}_{22}^{-1}(\mathbf{X}_2 - \boldsymbol{\mu}_2)$$

and covariance matrix

$$\boldsymbol{\Sigma}_{11} - \boldsymbol{\Sigma}_{12}\boldsymbol{\Sigma}_{22}^{-1}\boldsymbol{\Sigma}_{12}^\mathrm{T}$$

$\boldsymbol{\mu}_i$ and $\boldsymbol{\Sigma}_{ij}$ $(i, j = 1, 2)$ are suitable partitions of the mean vector and covariance matrix respectively. Note that the expected values of $\mathbf{X}_1$ are linear functions of the conditioning random variables $\mathbf{X}_2$. In this sense, it may be said that linearity characterizes the normal distribution and vice versa. Several other characterizations involving linearity can also be established.

THEOREM 2.15. Let $X_1, X_2, \ldots, X_k$ be independent random variables each distributed as the univariate normal $\mathrm{N}(\mu_i, \sigma_i^2)$, $i = 1, 2, \ldots, k$. Then any linear combination $y = a_1 x_1 + a_2 x_2 + \cdots + a_k x_k = \mathbf{a}^\mathrm{T}\mathbf{X}$ is also univariate normal with the mean $\mathbf{a}^\mathrm{T}\boldsymbol{\mu}$ and variance $\mathbf{a}^\mathrm{T}\boldsymbol{\Sigma}\mathbf{a}$, where $\boldsymbol{\Sigma} = \mathrm{diag}(\sigma_i^2)$.

PROOF. The moment generating function of X_i is

$$M_{x_i}(s) = \exp\left(\mu_i s - \frac{1}{2}\sigma_i^2 s^2\right) \qquad (i = 1, 2, \ldots, k)$$

and since X_i are independent normal they must also be uncorrelated (Theorem 2.13). The moment generating function of a linear combination y is then

$$M_y(s) = \prod_{i=1}^{k} M_{x_i}(a_i s)$$

$$= \prod_{i=1}^{k} \exp\left(\mu_i a_i s - \frac{1}{2}\sigma_i^2 a_i^2 s^2\right)$$

$$= \exp\left(s\sum_{i=1}^{k}\mu_i a_i - \frac{1}{2}s^2\sum_{i=1}^{k}a_i^2\sigma_i^2\right) \tag{2.102}$$

the moment generating function of a univariate normal variable with mean

$$\mu_y = \sum_{i=1}^{k} a_i \mu_i = \mathbf{a}^{\mathrm{T}} \boldsymbol{\mu}$$

and variance

$$\sigma_y^2 = \sum_{i=1}^{k} a_i^2 \sigma_i^2$$

$$= \mathbf{a}^{\mathrm{T}} \boldsymbol{\Sigma} \mathbf{a}$$

where $\boldsymbol{\Sigma}$ is diagonal owing to the independence of X_i, $X_2, \ldots, X_k$.

It is incorrect however to conclude that linear combinations of univariate normal variables are necessarily normal. The correct condition is that the random variables be multivariate normal. Also, using the Central Limit Theorem, Theorem 2.15 can be extended to sums of nonnormal independent random variables with finite means and variances. The resultant distribution however is only asymptotically normal. For further results along these lines see Muirhead (1982). Exact normality of linear combinations of random variables can be obtained under conditions of the following theorem.

THEOREM 2.16. Let $\mathbf{X}$ be distributed as the multivariate normal $N(\boldsymbol{\mu}, \boldsymbol{\Sigma})$. Then $r \leq k$ linear combinations $\mathbf{Y} = \mathbf{A}\mathbf{X}$, where $\mathbf{A}$ is $(r \times k)$, are distributed as the multivariate normal $N(\mathbf{A}\boldsymbol{\mu}, \mathbf{A}^{\mathrm{T}}\boldsymbol{\Sigma}\mathbf{A})$.

PROOF. The moment generating function of the vector $\mathbf{Y}$ is

$$E(e^{\mathbf{S}^{\mathrm{T}}\mathbf{Y}}) = E(e^{\mathbf{S}^{\mathrm{T}}\mathbf{A}\mathbf{X}})$$

$$= E[e^{(\mathbf{A}^{\mathrm{T}}\mathbf{S})^{\mathrm{T}}\mathbf{X}}]$$

$$= \exp\left[(\mathbf{A}^{\mathrm{T}}\mathbf{S})^{\mathrm{T}}\boldsymbol{\mu} - \frac{1}{2}(\mathbf{A}^{\mathrm{T}}\mathbf{S})^{\mathrm{T}}\boldsymbol{\Sigma}(\mathbf{A}^{\mathrm{T}}\mathbf{S}) \right]$$

$$= \exp\left[\mathbf{S}^{\mathrm{T}}(\mathbf{A}\boldsymbol{\mu}) - \frac{1}{2}\mathbf{S}^{\mathrm{T}}(\mathbf{A}\boldsymbol{\Sigma}\mathbf{A}^{\mathrm{T}})\mathbf{S} \right]$$

which is the moment generating function of the multivariate normal $N(\mathbf{A}\boldsymbol{\mu}, \mathbf{A}\boldsymbol{\Sigma}\mathbf{A}^{\mathrm{T}})$.

The multivariate normal also possess a unique linearity property in that every linear combination of its variates is also normal.

THEOREM 2.17. (Fréchet, 1951). Let $\mathbf{X}$ possess a multivariate distribution with mean vector $\boldsymbol{\mu}$ and covariance matrix $\boldsymbol{\Sigma}$. Let $y = \mathbf{a}^{\mathrm{T}}\mathbf{X}$ be a linear combination such that y is univariate normal for any $\mathbf{a} \neq 0$. Then $\mathbf{X}$ is distributed as the multivariate normal $N(\boldsymbol{\mu}, \boldsymbol{\Sigma})$.

Proof. The moment generating function of y is

$$E(e^{sa^\mathrm{T}X}) = \exp\left(sa^\mathrm{T}\mu - \frac{1}{2}s^2 a^\mathrm{T}\Sigma a\right)$$

for every value of the scalar s. Letting $s = 1$ we have

$$\exp\left(a^\mathrm{T}\mu - \frac{1}{2}a^\mathrm{T}\Sigma a\right)$$

which is the moment generating function of the multivariate normal $N(\mu, \Sigma)$.

The converse question to ask is how many normal linear combinations are required to establish multivariate normality of the random variables.

THEOREM 2.18. If every linear combination of X_1, $X_2, \ldots, X_k$ is normally distributed, then $X = (X_1, X_2, \ldots, X_k)^\mathrm{T}$ is multivariate normal.

The stress in Theorem 2.18 is on the word every, since *all* linear combinations must be normal to ensure multivariate normality of X. If only $r < k$ linear combinations are normal, this is not sufficient to establish multivariate normality of X. A counterexample is given by Melnick and Tenenbien (1982) for $k = 2$.

THEOREM 2.19. (Generalization of Theorem 2.15).
Let $X_1, X_2, \ldots, X_k$ be $(k \times 1)$ vectors, each distributed independently as $N(\mu_i, \Sigma_i)$ $(i = 1, 2, \ldots, k)$. Let $Y = A_1 X_1 + A_2 X_2 + \cdots + A_k X_k$ where A_i are fixed $(m \times k)$ matrices such that $m \leq k$. Then $Y = \Sigma_{i=1}^{k} A_i X_i$ is distributed as the m-dimensional multivariate normal with mean vector $\Sigma_{i=1}^{k} A_i \mu_i$ and covariance matrix $\Sigma_{i=1}^{k} A_i \Sigma_i A_i^\mathrm{T}$.

This theorem possesses a number of special cases of interest. For example, when $A_i = I$ the sum of multivariate vectors $X_1 + X_2 + \cdots + X_k$ is distributed as the multivariate normal with mean vector $\Sigma_{i=1}^{k} \mu_i$ and covariance matrix $\Sigma_{i=1}^{k} \Sigma_i$. The converse, a multivariate generalization of Cramér's (1937) theorem, is also true.

THEOREM 2.20. (Multivariate generalization of Cramér Theorem). Let X_1, $X_2, \ldots, X_k$ be independent random vectors. Then when the sum $Y = X_1 + X_2 + \cdots + X_k$ is multivariate normal, each X_i $(i = 1, 2, \ldots, k)$ is also multivariate normal.

For a proof the reader is referred to Srivastava and Khatri (1979). Cramér's (1937) original theorem states that if the sum of k independent random variables is normally distributed, each random variable is also normal. When the random variables are not independent, the result is not necessarily true.

The main properties of the normal distribution can be summarized as follows.

1. Let $\mathbf{X}$ be a k-dimensional random variable such that $\mathbf{X} \sim N(\boldsymbol{\mu}, \boldsymbol{\Sigma})$. Then each component $\mathbf{X}_i$ of $\mathbf{X}$ is marginally univariate normal. Also, each subvector $\mathbf{X}^*$ of $\mathbf{X}$ is marginally multivariate normal $N(\boldsymbol{\mu}^*, \boldsymbol{\Sigma}^*)$ where $\boldsymbol{\mu}^*$ and $\boldsymbol{\Sigma}^*$ are conformable partitions of $\boldsymbol{\mu}$ and $\boldsymbol{\Sigma}$. The converse, however, is not true since a set of k normal random variables need not possess a joint multivariate normal distribution. It follows that if random variables are not marginally normal, they cannot possess a multivariate normal distribution.

2. Linear combinations of independent normal random variables are normally distributed (Theorem 2.15). Also, linear combinations of multivariate normal variates are normal (Theorem 2.16). A linear combination of correlated normal random variables however is not necessarily normal. This demonstrates the looseness of the statement "linear combinations of normally distributed random variables are normally distributed." For a bivariate illustration see Rosenberg (1965).

3. We have from Theorem 2.19 that a sum of independent normal variables is normal. Also from Theorem 2.20 we have the converse that if k vectors $\mathbf{X}_1, \mathbf{X}_2, \ldots, \mathbf{X}_k$ are distributed independently, and if in addition their sum $\mathbf{Y}$ is normal, then each $\mathbf{X}_i$ $(i = 1, 2, \ldots, k)$ must be normal. When $\mathbf{X}_i$ are dependent, the result does not necessarily follow, since one of the following may hold.

(i) The $\mathbf{X}_i$ are marginally normal, but the sum $\mathbf{Y} = \Sigma_{i=1}^{k} \mathbf{X}_i$ is not normal.

(ii) The $\mathbf{X}_i$ are marginally not normal, but the sum $\mathbf{Y} = \Sigma_{i=1}^{k} \mathbf{X}_i$ is normal.

For examples of these two cases see Kale (1970).

4. Finally, of major importance in multivariate analysis is the relationship between independence and uncorrelatedness (Theorem 2.10), namely, that multivariate normal variates are mutually independent if and only if they are mutually uncorrelated. The essential condition is that the variates be multivariate normal, not merely marginally normal, since the latter condition is not enough to establish uncorrelatedness as a necessary and sufficient condition for independence.

The following example (Behboodian, 1972, 1990) illustrates the looseness of the statement "two normal random variables are independent if and only if they are uncorrelated."

Example 2.11. Consider two k-dimensional normal distributions $f_1(x_1, x_2, \ldots, x_k)$, $f_2(x_1, x_2, \ldots, x_k)$ together with the linear combination

$$f(x_1, x_2, \ldots, x_k) = \sum_{i=1}^{2} p_i f_i(x_1, x_2, \ldots, x_k) \qquad (2.103)$$

where $p_1 + p_2 = 1$ and $0 < p_i < 1$ for $i = 1, 2$. Equation (2.103) is a probability density function known as a mixture of (two) normal distributions f_1, f_2. Assume both normal distributions are standardized to zero mean and unit variance. Let $\mathbf{R}_1 = (\rho_{ij1})$ and $\mathbf{R}_2 = (\rho_{ij2})$ be the two correlation matrices for f_1, f_2 respectively. Then the moment generating functions of f_1, f_2 are given by Eq. (2.98) and that of the mixture by

$$M(s_1, s_2, \ldots, s_k) = \sum_{i=1}^{2} p_i M_i(s_1, s_2, \ldots, s_k)$$

$$= p_1 \exp\left(-\frac{1}{2} \sum_{i=1}^{k} \sum_{j=1}^{k} \rho_{ij1} s_i s_j\right)$$

$$+ p_2 \exp\left(-\frac{1}{2} \sum_{i=1}^{k} \sum_{j=1}^{k} \rho_{ij2} s_i s_j\right) \qquad (2.104)$$

The marginal distributions of the x_i are univariate normal, whereas their joint density is a mixture of two bivariate normal densities and thus cannot be normal. Also, from Eq. (2.104) it follows that the correlation coefficient of any two normal variates x_i and x_j (in the mixture) is $p_1 \rho_{ij1} + p_2 \rho_{ij2}$. Now let $p_1 = p_2$ and $\rho_{ij1} = \rho_{ij2}$. Then although x_i and x_j are not independent in f_1 and f_2, since $\mathbf{R}_1$, $\mathbf{R}_2$ are not diagonal, they are nevertheless uncorrelated in the mixture Eq. (2.103).

To illustrate further the relationship between normality and independence consider a mixture or two trivariate normal distributions with correlation matrices

$$\mathbf{R}_1 = \begin{bmatrix} 1 & 0 & -\rho \\ 0 & 1 & \rho \\ -\rho & \rho & 1 \end{bmatrix} \qquad \mathbf{R}_2 = \begin{bmatrix} 1 & -\rho & \rho \\ -\rho & 1 & 0 \\ \rho & 0 & 1 \end{bmatrix}$$

where $\rho < 1/\sqrt{2}$ to ensure positive definiteness (Behboodian, 1972). The moment generating function of the mixture is

$$M(s_1, s_2, s_3) = p_1 \exp\left[-\frac{1}{2}(s_1^2 + s_2^2 + s_3^2 - 2\rho s_1 s_3 + 2\rho s_2 s_3)\right]$$

$$+ p_2 \exp\left[-\frac{1}{2}(s_1^2 + s_2^2 + s_3^2 - 2\rho s_1 s_2 + 2\rho s_1 s_3)\right]$$

$$(2.105)$$

and letting $p_1 = p_2 = 1/2$ we observe that although the normal variates x_1 and x_3 are uncorrelated, they are not independent. Also, the linear combination $y_1 = x_1 + x_2 + x_3$ is normal but $y_2 = x_1 - x_2 + x_3$ is not.

2.8.2 Sampling From the Multivariate Normal

The multivariate normal distribution is completely determined by its mean vector $\boldsymbol{\mu}$ and covariance matrix $\boldsymbol{\Sigma}$. In most instance, however, when dealing with empirical observations the values of $\boldsymbol{\mu}$ and $\boldsymbol{\Sigma}$ are not known. Thus given a multivariate normal population, and given a finite random sample from the population, there remains the additional question of how to estimate $\boldsymbol{\mu}$ and $\boldsymbol{\Sigma}$. A commonly used technique is that of maximum likelihood (ML), which possesses certain optimality properties. Analogously to Eq. (1.15), the multivariate normal likelihood function can be obtained as follows. Given the ith observation vector we have

$$f(y_{i1}, y_{i2}, \ldots, y_{ik}) = \frac{1}{(2\pi)^{k/2}|\boldsymbol{\Sigma}|^{1/2}} \exp\left\{-\frac{1}{2}[(\mathbf{y}_i - \boldsymbol{\mu})^{\mathrm{T}}\boldsymbol{\Sigma}^{-1}(\mathbf{y}_i - \boldsymbol{\mu})]\right\}$$

$$(i = 1, 2, \ldots, n)$$

where $\mathbf{y}_i$ is the ith row of $\mathbf{Y}$. For an independent sample the multivariate likelihood function becomes

$$L(\boldsymbol{\mu}, \boldsymbol{\Sigma}) = f(\mathbf{y}_1^{\mathrm{T}})f(\mathbf{y}_2^{\mathrm{T}}) \cdots f(\mathbf{y}_n^{\mathrm{T}})$$

$$= \frac{1}{(2\pi)^{nk/2}|\boldsymbol{\Sigma}|^{n/2}} \exp\left\{-\frac{1}{2}\sum_{i=1}^{n}(\mathbf{y}_i - \boldsymbol{\mu})^{\mathrm{T}}\boldsymbol{\Sigma}^{-1}(\mathbf{y}_i - \boldsymbol{\mu})\right\}$$

$$(2.106)$$

and taking natural logarithms we obtain

$$L^*(\boldsymbol{\mu}, \boldsymbol{\Sigma}) = -\frac{1}{2}nk\ln(2\pi) - \frac{1}{2}n\ln|\boldsymbol{\Sigma}| - \frac{1}{2}\sum_{i=1}^{n}(\mathbf{y}_i - \boldsymbol{\mu})^{\mathrm{T}}\boldsymbol{\Sigma}^{-1}(\mathbf{y}_i - \boldsymbol{\mu})$$

$$(2.107)$$

where ML solutions are given by

$$\frac{\partial L}{\partial \boldsymbol{\mu}} = 0, \qquad \frac{\partial L}{\partial \boldsymbol{\Sigma}} = 0$$

The ML estimator of the mean vector $\boldsymbol{\mu}$ is thus given by

$$\frac{\partial L}{\partial \boldsymbol{\mu}} = -\boldsymbol{\Sigma}^{-1}(\mathbf{y}_1 - \boldsymbol{\mu}) - \boldsymbol{\Sigma}^{-1}(\mathbf{y}_2 - \boldsymbol{\mu}) - \cdots - \boldsymbol{\Sigma}^{-1}(\mathbf{y}_n - \boldsymbol{\mu})$$

$$= -\boldsymbol{\Sigma}^{-1}\sum_{i=1}^{n}(\mathbf{y}_i - \boldsymbol{\mu})$$

$$= \mathbf{0}$$

or

$$\hat{\boldsymbol{\mu}} = (\hat{\boldsymbol{\mu}}_1, \hat{\boldsymbol{\mu}}_2, \ldots, \hat{\boldsymbol{\mu}}_k)^{\mathrm{T}} = (\bar{y}_1, \bar{y}_2, \ldots, \bar{y}_k)^{\mathrm{T}}$$

$$= \bar{\mathbf{Y}} \tag{2.108}$$

The ML estimator of $\boldsymbol{\mu}$ is thus simply $\bar{\mathbf{Y}}$, whose components are univariate sample means of the k random variables. The ML estimator for $\boldsymbol{\Sigma}$ can be derived in at least one of three ways (see Anderson, 1984; Rao, 1965; Morrison, 1967). In what follows we outline the proof presented by Morrison, which is somewhat less involved. Using matrix derivatives (Section 2.2) and Eq. (2.107) we can write

$$\frac{\partial L}{\partial \hat{\sigma}_{ij}} = n\hat{\sigma}^{ij} + \frac{1}{2} \sum_{i=1}^{n} (\mathbf{y}_i - \boldsymbol{\mu})^{\mathrm{T}} \hat{\boldsymbol{\Sigma}}^{-1} (\mathbf{J}_{ij} + \mathbf{J}_{ji}) \hat{\boldsymbol{\Sigma}}^{-1} (\mathbf{y}_i - \boldsymbol{\mu}) = 0 \tag{2.109}$$

where $\hat{\sigma}^{ij}$ is the (i, j)th element of $\hat{\boldsymbol{\Sigma}}^{-1}$ and matrices $\mathbf{J}_{ij}$ are defined in Section 2.3. Since the expression under the summation is scalar, it equals its own trace and Eq. (2.109) becomes

$$\sum_{i=1}^{n} (\mathbf{y}_i - \boldsymbol{\mu})^{\mathrm{T}} \boldsymbol{\Sigma}^{-1} (\mathbf{J}_{ij} + \mathbf{J}_{ji}) \boldsymbol{\Sigma}^{-1} (\mathbf{y}_i - \boldsymbol{\mu})$$

$$= \mathrm{tr} \sum_{i=1}^{n} (\mathbf{y}_i - \boldsymbol{\mu})(\mathbf{y}_i - \boldsymbol{\mu})^{\mathrm{T}} \boldsymbol{\Sigma}^{-1} (\mathbf{J}_{ij} + \mathbf{J}_{ji}) \boldsymbol{\Sigma}^{-1}$$

a well-known property of the trace function. The $k(k+1)/2$ normal equations are then given by

$$-n\hat{\boldsymbol{\Sigma}}^{-1} + \hat{\boldsymbol{\Sigma}}^{-1} \sum_{i=1}^{n} (\mathbf{y}_i - \bar{\mathbf{Y}})(\mathbf{y}_i - \bar{\mathbf{Y}})^{\mathrm{T}} \hat{\boldsymbol{\Sigma}}^{-1} = 0 \tag{2.110}$$

where $\boldsymbol{\mu}$ has been replaced by the ML estimator Eq. (2.108). Solving Eq. (2.110) yields

$$\hat{\boldsymbol{\Sigma}} = \frac{1}{n} \sum_{i=1}^{n} (\mathbf{y}_i - \bar{\mathbf{Y}})(\mathbf{y}_i - \bar{\mathbf{Y}})^{\mathrm{T}} = \frac{1}{n} \sum_{i=1}^{n} \mathbf{x}_i \mathbf{x}_i^{\mathrm{T}}$$

$$= \mathbf{S} \tag{2.111}$$

the (biased) ML estimator of $\boldsymbol{\Sigma}$. An unbiased estimator is obtained by replacing n with $n - 1$.

Results that are similar to univariate sampling theory can also be derived for the multivariate case.

THEOREM 2.21. Let $\mathbf{Y}$ denote a $(n \times k)$ sample matrix taken from $N(\boldsymbol{\mu}, \boldsymbol{\Sigma})$. Then for an independent and identically distributed sample

(1) $\bar{\mathbf{Y}}$ is distributed as $N\left(\boldsymbol{\mu}, \dfrac{1}{n} \boldsymbol{\Sigma}\right)$.

(2) Let $\Sigma_{i=1}^{n} \mathbf{x}_i \mathbf{x}_i^T = \mathbf{X}^T\mathbf{X}$ where $\mathbf{X} = \mathbf{Y} - \bar{\mathbf{Y}}$ and $\mathbf{x}_i$ denotes the ith row of $\mathbf{X}$. Then $\mathbf{X}^T\mathbf{X}$ is distributed as the Wishart distribution.

$$f(\mathbf{X}^T\mathbf{X}) = c|\mathbf{\Sigma}|^{-n/2}|\mathbf{X}^T\mathbf{X}|^{(n-k-1)/2} \exp\left(-\frac{n}{2}\operatorname{tr}\mathbf{S}\mathbf{\Sigma}^{-1}\right) \qquad (2.112)$$

where the constant c is given by

$$\frac{1}{c} = 2^{nk/2} \pi^{1/4k(k-1)} \prod_{i=1}^{k} \Gamma\left[\frac{1}{2}(n+1-i)\right] \qquad (2.113)$$

When $k = 1$ the Wishart assumes its special univariate case, the chi-squared distribution. The Wishart distribution plays an important role in maximum likelihood factor analysis and is discussed further in Chapters 4 and 6.

The normal maximum likelihood function can be expressed in several alternative forms, all of which will be found useful in subsequent chapters. We have, for the exponent part of Eq. (2.106),

$$-\frac{1}{2}\sum_{i=1}^{n}(\mathbf{y}_i - \boldsymbol{\mu})^T\mathbf{\Sigma}^{-1}(\mathbf{y}_i - \boldsymbol{\mu}) = -\frac{1}{2}\operatorname{tr}\sum_{i=1}^{n}(\mathbf{y}_i - \boldsymbol{\mu})^T\mathbf{\Sigma}^{-1}(\mathbf{y}_i - \boldsymbol{\mu})$$

$$= -\frac{1}{2}\sum_{i=1}^{n}\operatorname{tr}(\mathbf{y}_i - \boldsymbol{\mu})^T\mathbf{\Sigma}^{-1}(\mathbf{y}_i - \boldsymbol{\mu})$$

$$= -\frac{1}{2}\sum_{i=1}^{n}\operatorname{tr}\mathbf{\Sigma}^{-1}(\mathbf{y}_i - \boldsymbol{\mu})(\mathbf{y}_i - \boldsymbol{\mu})^T$$

$$= -\frac{1}{2}\operatorname{tr}\left[\mathbf{\Sigma}^{-1}\sum_{i=1}^{n}(\mathbf{y}_i - \boldsymbol{\mu})(\mathbf{y}_i - \boldsymbol{\mu})^T\right]$$

$$= -\frac{1}{2}\operatorname{tr}[\mathbf{\Sigma}^{-1}\mathbf{X}^T\mathbf{X}]$$

where $\mathbf{x}_i = \mathbf{y}_i - \boldsymbol{\mu}$. It follows that the likelihood (Eq. 2.106) can also be written as

$$f(\mathbf{X}) = \frac{|\mathbf{\Sigma}|^{-n/2}}{(2\pi)^{nk/2}} \exp\left[-\frac{1}{2}\operatorname{tr}(\mathbf{\Sigma}^{-1}\mathbf{X}^T\mathbf{X})\right] \qquad (2.114)$$

Also, for the likelihood function, $\mathbf{y}_i$ is the ith row of the matrix

$$\mathbf{Y} = \begin{bmatrix} y_{11} & y_{12} & \cdots & y_{1k} \\ y_{21} & y_{22} & \cdots & y_{2k} \\ \vdots & \vdots & & \vdots \\ y_{n1} & y_{n2} & \cdots & y_{nk} \end{bmatrix}$$

and $\boldsymbol{\mu}$ is the vector of population means. Let $\mathbf{1} = (1, 1, \ldots, 1)^T$, the unity vector. Then the likelihood (Eq. 2.106) can also be expressed as

$$f(\mathbf{Y}) = c \exp\left[-\frac{1}{2}(\mathbf{Y} - \mathbf{1}\boldsymbol{\mu})(\mathbf{Y} - \mathbf{1}\boldsymbol{\mu})^{\mathrm{T}} \right] \qquad (2.115)$$

where c is the constant of proportionality. When $\boldsymbol{\mu}$ is not known it can be replaced by $\bar{\mathbf{Y}}$, and another form of the likelihood is

$$f(\mathbf{Y}) = c \exp\left[-\frac{1}{2}\operatorname{tr}(\boldsymbol{\Sigma}^{-1})(\mathbf{Y}^{\mathrm{T}}\mathbf{Y} - n\bar{\mathbf{Y}}^{\mathrm{T}}\bar{\mathbf{Y}}) \right] \qquad (2.116)$$

since $\mathbf{X}^{\mathrm{T}}\mathbf{X} = \mathbf{Y}^{\mathrm{T}}\mathbf{Y} - n\bar{\mathbf{Y}}^{\mathrm{T}}\bar{\mathbf{Y}}$. For further information the reader is referred to Anderson (1984), Srivastava and Carter (1983), Muirhead (1982), and Morrison (1967).

2.9 THE KRONECKER PRODUCT

Let $\mathbf{A}$ and $\mathbf{B}$ be $(n \times k)$ and $(m \times r)$ matrices respectively. Then the product

$$\mathbf{C} = \mathbf{A} \otimes \mathbf{B} = \begin{bmatrix} a_{11}\mathbf{B} & a_{12}\mathbf{B} & \cdots & a_{1k}\mathbf{B} \\ a_{21}\mathbf{B} & a_{22}\mathbf{B} & \cdots & a_{2k}\mathbf{B} \\ \vdots & \vdots & & \vdots \\ a_{n1}\mathbf{B} & a_{n2}\mathbf{B} & & a_{nk}\mathbf{B} \end{bmatrix}$$

is known as the Kronecker product. Other names include direct product and tensor product. Since $\mathbf{C}$ is $(nm \times kr)$, $\mathbf{A}$ and $\mathbf{B}$ are always conformable for multiplication. The Kronecker product has the following properties (see Exercise 2.12).

1. Anticommutative law: $\mathbf{A} \otimes \mathbf{B} \neq \mathbf{B} \otimes \mathbf{A}$.
2. Let $\mathbf{A}$, $\mathbf{B}$, $\mathbf{C}$, and $\mathbf{D}$ be matrices for which the usual matrix products $\mathbf{AC}$ and $\mathbf{BD}$ are defined. Then the matrix product of Kronecker products is given by $(\mathbf{A} \otimes \mathbf{B})(\mathbf{C} \otimes \mathbf{D}) = \mathbf{AC} \otimes \mathbf{BD}$. This generalized to any number of products.
3. Associative law: $\mathbf{A} \otimes (\mathbf{B} \otimes \mathbf{C}) = (\mathbf{A} \otimes \mathbf{B}) \otimes \mathbf{C}$.
4. Transpose of a Product: $(\mathbf{A} \otimes \mathbf{B})^{\mathrm{T}} = \mathbf{A}^{\mathrm{T}} \otimes \mathbf{B}^{\mathrm{T}}$.
5. Let $\mathbf{A}$, $\mathbf{B}$, $\mathbf{C}$, and $\mathbf{D}$ be conformable for matrix addition. Then $(\mathbf{A} + \mathbf{B}) \otimes (\mathbf{C} + \mathbf{D}) = (\mathbf{A} \otimes \mathbf{C}) + (\mathbf{A} \otimes \mathbf{D}) + (\mathbf{B} \otimes \mathbf{C}) + (\mathbf{B} \otimes \mathbf{D})$.
6. Let $\mathbf{A}$ be an $(n \times k)$ matrix. The Kronecker power is defined as $\mathbf{A}^{(r+1)} = \mathbf{A} \otimes \mathbf{A}^{(r)}$ where $r = 1, 2, \ldots, n$.
7. The trace of the Kronecker product is the product of traces, that is $\operatorname{tr}(\mathbf{A} \otimes \mathbf{B}) = \operatorname{tr}(\mathbf{A}) \operatorname{tr}(\mathbf{B})$.
8. The rank of a Kronecker product equals the product of ranks, that is $\rho(\mathbf{A} \otimes \mathbf{B}) = \rho(\mathbf{A}) \otimes \rho(\mathbf{B})$.

9. Let $\mathbf{A}$ and $\mathbf{B}$ be $(n \times n)$ and $(m \times m)$ nonsingular matrices. Then $(\mathbf{A} \otimes \mathbf{B})^{-1} = \mathbf{A}^{-1} \otimes \mathbf{B}^{-1}$.

10. Let $\mathbf{A}$ and $\mathbf{B}$ be $(n \times n)$ positive definite (semidefinite) matrices. Then $\mathbf{A} \otimes \mathbf{B}$ is positive definite (semidefinite).

11. Let $\mathbf{A}$ and $\mathbf{B}$ be $(n \times n)$ and $(m \times m)$ nonsingular matrices. Then $|\mathbf{A} \otimes \mathbf{B}| = |\mathbf{A}|^{n}|\mathbf{B}|^{m}$.

12. Let $\mathbf{A}^{-}$ be a generalized inverses of $\mathbf{A}$, that is $\mathbf{A}\mathbf{A}^{-}\mathbf{A} = \mathbf{A}$. Then for any two matrices $\mathbf{A}$ and $\mathbf{B}$ we have $(\mathbf{A} \otimes \mathbf{B})^{-} = \mathbf{A}^{-} \otimes \mathbf{B}^{-}$.

13. Let $\mathbf{A}^{+}$ and $\mathbf{B}^{+}$ be left inverse of any two $(n \times k)$ and $(m \times r)$ matrices, that is $\mathbf{A}^{+}\mathbf{A} = \mathbf{I}_{k}$ and $\mathbf{B}^{+}\mathbf{B} = \mathbf{I}_{r}$, where $\mathbf{I}_{k}$ and $\mathbf{I}_{r}$ are $(k \times k)$ and $(r \times r)$ unit matrices. Then $(\mathbf{A}^{+} \otimes \mathbf{B}^{+})(\mathbf{A} \otimes \mathbf{B}) = \mathbf{I}_{kr}$, where $\mathbf{I}_{kr}$ is the $(kr \times kr)$ unit matrix.

14. Let $\mathbf{A} = \mathbf{PLP}^{\mathrm{T}}$ and $\mathbf{B} = \mathbf{QMQ}^{\mathrm{T}}$ where $\mathbf{A}$ and $\mathbf{B}$ are square matrices of order n, $\mathbf{P}$ and $\mathbf{Q}$ are orthogonal, and $\mathbf{L}$ and $\mathbf{M}$ are diagonal. Also let $\mathbf{A} \otimes \mathbf{B} = \mathbf{C}$. Then $\mathbf{C}$ has the spectral representation $\mathbf{C} = \mathbf{RNR}^{\mathrm{T}}$ where $\mathbf{R} = \mathbf{P} \otimes \mathbf{Q}$ is orthogonal and $\mathbf{N}$ a diagonal matrix.

PROOF. We have

$$\mathbf{C} = \mathbf{A} \otimes \mathbf{B} = (\mathbf{PLP}^{\mathrm{T}}) \otimes (\mathbf{QMQ}^{\mathrm{T}})$$

$$= (\mathbf{P} \otimes \mathbf{Q})(\mathbf{L} \otimes \mathbf{M})(\mathbf{P}^{\mathrm{T}} \otimes \mathbf{Q}^{\mathrm{T}})$$

$$= (\mathbf{P} \otimes \mathbf{Q})(\mathbf{L} \otimes \mathbf{M})(\mathbf{P} \otimes \mathbf{Q})^{\mathrm{T}}$$

where

$$(\mathbf{P} \otimes \mathbf{Q})^{\mathrm{T}}(\mathbf{P} \otimes \mathbf{Q}) = (\mathbf{P}^{\mathrm{T}} \otimes \mathbf{Q}^{\mathrm{T}})(\mathbf{P} \otimes \mathbf{Q})$$

$$= \mathbf{P}^{\mathrm{T}}\mathbf{P} \otimes \mathbf{Q}^{\mathrm{T}}\mathbf{Q}$$

$$= \mathbf{I}_{k} \otimes \mathbf{I}_{r}$$

$$= \mathbf{I}_{kr}$$

so that $\mathbf{R}$ is orthogonal. Also $\mathbf{N} = \mathbf{L} \otimes \mathbf{M}$ is diagonal so that $\mathbf{C} = \mathbf{RNR}^{\mathrm{T}}$ is the spectral representation of $\mathbf{C}$.

15. More generally, let $\mathbf{C} = (\mathbf{A}_1 \otimes \mathbf{B}_1) + (\mathbf{A}_2 \otimes \mathbf{B}_2) + \cdots + (\mathbf{A}_k \otimes \mathbf{B}_k)$ where $\mathbf{A}_i$ and $\mathbf{B}_i$ are square matrices of order n. Then if $\mathbf{y}$ is a latent vector of all the matrices $\mathbf{B}_1, \mathbf{B}_2, \ldots, \mathbf{B}_k$ so that $\mathbf{B}_i \mathbf{y} = \lambda_i \mathbf{y}$ $(i = 1, 2, \ldots, k)$ and if $\mathbf{x}$ is a vector such that $(\lambda_1 \mathbf{A}_1 + \lambda_2 \mathbf{A}_2 + \cdots + \lambda_k \mathbf{A}_k)\mathbf{x} = \mu \mathbf{x}$ then μ is a latent root of $\mathbf{C}$ and

$$C(\mathbf{x} \otimes \mathbf{y}) = \sum_{i=1}^{k} (\mathbf{A}_i \otimes \mathbf{B}_i)\mathbf{x} \otimes \mathbf{y}$$

$$= \sum_{i=1}^{k} (\mathbf{A}_i\mathbf{x}) \otimes (\mathbf{B}_i\mathbf{y})$$

$$= \sum_{i=1}^{k} (\lambda_i\mathbf{A}_i\mathbf{x}) \otimes \mathbf{y}$$

$$= \mu(\mathbf{x} \otimes \mathbf{y})$$

Further properties and applications are given by Graham (1981).

2.10 SIMULTANEOUS DECOMPOSITION OF TWO GRAMMIAN MATRICES

The spectral decomposition of a Grammian matrix (Section 2.5) can be generalized to that of a decomposition of a Grammian matrix $\mathbf{A}$ in the metric of $\mathbf{B}$, where $\mathbf{B}$ is any nonsingular matrix. This yields a simultaneous decomposition of the two matrices $\mathbf{A}$ and $\mathbf{B}$. When $\mathbf{B}$ is positive definite we have the following important special case.

THEOREM 2.22. Let $\mathbf{A}$ and $\mathbf{B}$ be symmetric matrices such that $\mathbf{B}$ is positive definite, and let

$$(\mathbf{A} - \lambda_i\mathbf{B})\mathbf{P}_i = \mathbf{0} \tag{2.117}$$

Then the latent roots λ_i and latent vectors $\mathbf{P}_i$ of $\mathbf{A}$, said to be in the metric of $\mathbf{B}$, have the following properties:

(1) There exists a real diagonal matrix $\mathbf{\Lambda} = \mathrm{diag}(\lambda_i) \geq \mathbf{0}$, and a real matrix $\mathbf{P}$, such that

$$\mathbf{P}^{\mathrm{T}}\mathbf{A}\mathbf{P} = \mathbf{\Lambda} \tag{2.118}$$

$$\mathbf{P}^{\mathrm{T}}\mathbf{B}\mathbf{P} = \mathbf{I} \tag{2.119}$$

(2) $\mathbf{A}$ and $\mathbf{B}$ can be decomposed simultaneously as

$$\mathbf{A} = (\mathbf{P}^{-1})^{\mathrm{T}}\mathbf{\Lambda}\mathbf{P}^{-1}$$

$$= \lambda_1\mathbf{S}_1\mathbf{S}_1^{\mathrm{T}} + \lambda_2\mathbf{S}_2\mathbf{S}_2^{\mathrm{T}} + \cdots \lambda_k\mathbf{S}_k\mathbf{S}_k^{\mathrm{T}} \tag{2.120}$$

$$\mathbf{B} = (\mathbf{P}^{-1})^{\mathrm{T}}\mathbf{P}^{-1}$$

$$= \mathbf{S}_1\mathbf{S}_1^{\mathrm{T}} + \mathbf{S}_2\mathbf{S}_2^{\mathrm{T}} + \cdots + \mathbf{S}_k\mathbf{S}_k^{\mathrm{T}} \tag{2.121}$$

where $\mathbf{S}_i$ is the ith column of $(\mathbf{P}^{-1})^{\mathrm{T}}$.

(3) The roots λ_i are invariant with respect to coordinate transformation, that is, for a nonsingular matrix $\mathbf{C}$ we have

$$|\mathbf{C}^{\mathrm{T}}\mathbf{A}\mathbf{C} - \lambda(\mathbf{C}^{\mathrm{T}}\mathbf{B}\mathbf{C})| = |\mathbf{C}^{\mathrm{T}}(\mathbf{A} - \lambda\mathbf{B})\mathbf{C}|$$

$$= |\mathbf{C}^{\mathrm{T}}| \ |\mathbf{A} - \lambda\mathbf{B}| \ |\mathbf{C}|$$

and setting to zero yields

$$|\mathbf{C}^{\mathrm{T}}\mathbf{A}\mathbf{C} - \lambda(\mathbf{C}^{\mathrm{T}}\mathbf{B}\mathbf{C})| = |\mathbf{A} - \lambda\mathbf{B}| = 0$$

the characteristic equation of Eq. (2.117).

(4) Equation (2.117) is equivalent to

$$(\mathbf{B}^{-1}\mathbf{A} - \lambda\mathbf{I})\mathbf{P} = 0 \tag{2.122}$$

where $\mathbf{B}^{-1}\mathbf{A}$ is not necessarily symmetric and the latent vectors need not represent an orthogonal coordinate system, that is, $\mathbf{P}^{\mathrm{T}}\mathbf{P} \neq \mathbf{I}$.

(5) An alternative way to express Eq. (2.117) which may be more convenient for statistical computation is as follows. Since $\mathbf{B}$ is Grammian, by assumption we can write $\mathbf{B} = \mathbf{W}^{\mathrm{T}}\mathbf{W}$ where $\mathbf{W}$ is nonsingular (Theorem 2.9). Let $\mathbf{Q}_i = \mathbf{W}\mathbf{P}_i$. Then Eq. (2.117) can be expressed as

$$\mathbf{A}\mathbf{W}^{-1}\mathbf{Q}_i = \lambda_i \mathbf{W}^{\mathrm{T}} \mathbf{W}\mathbf{P}_i$$

$$= \lambda_i \mathbf{W}^{\mathrm{T}} \mathbf{Q}_i$$

that is,

$$[(\mathbf{W}^{\mathrm{T}})^{-1}\mathbf{A}\mathbf{W}^{-1} - \lambda_i\mathbf{I}]\mathbf{Q}_i = 0 \tag{2.123}$$

$$(i = 1, 2, \ldots, k)$$

The advantage of Eq. (2.123) is that matrix $(\mathbf{W}^{\mathrm{T}})^{-1}\mathbf{A}\mathbf{W}^{-1}$ is symmetric, unlike $\mathbf{B}^{-1}\mathbf{A}$, so that the latent vectors $\mathbf{Q}_i$ can be chosen orthonormal. Also it can be shown that Eqs., (2.118) and (2.119) still hold in Eq. (2.123) (Exercise 2.18).

Finally we note that Eq. (2.117) can be derived by maximizing the quadratic form (Eq. 2.118) subject to a constraint (Eq. 2.123). Let λ_i be a Lagrange multiplier. Thus maximizing

$$\phi = \mathbf{P}_i^{\mathrm{T}}\mathbf{A}\mathbf{P}_i - \lambda_i(\mathbf{P}_i^{\mathrm{T}}\mathbf{B}\mathbf{P}_i - \mathbf{I}) \tag{2.123a}$$

we have

$$\frac{\partial \phi}{\partial \mathbf{P}_i} = \mathbf{AP}_i - \lambda_i \mathbf{BP}_i = 0$$

or

$$(\mathbf{A} - \lambda_i \mathbf{B})\mathbf{P}_i = 0 \qquad (1, 2, \dots, k) \tag{2.124}$$

2.11 THE COMPLEX MULTIVARIATE NORMAL DISTRIBUTION

Although most statistical analyses are conducted in terms of real numbers, it is at times necessary to consider the more general complex plane. Complex random variables were already encountered in Section 1.7 together with the notion of a complex multivariate distribution. In this section we examine the complex multivariate normal distribution together with the Hermitian complex matrix and its spectral representation.

2.11.1 Complex Matrices, Hermitian Forms

Definition 2.8. A complex matrix $\mathbf{C}$ is a matrix that contains at least one complex entry. The complex conjugate $\bar{\mathbf{C}}$ is the matrix whose entries consist of the complex conjugates $\bar{c}_{ij}$, where c_{ij} is the (i, j)th element of $\mathbf{C}$.

Of special interest are the so-called Hermitian matrices which, in a sense, are complex equivalents of real symmetric matrices, and which play an important role in the complex multivariate normal distribution and its uses, e.g. in multiple time series analysis.

Definition 2.9. The Hermitian transpose of a complex matrix $\mathbf{C}$ is the matrix $\mathbf{C}^H = \bar{\mathbf{C}}^T$. A matrix is said to be Hermitian if and only if $\mathbf{C}^H = \mathbf{C}$.

THEOREM 2.23. Let $\mathbf{A}$ and $\mathbf{B}$ be any two complex matrices. Then
(1) $\overline{(\mathbf{AB})} = \bar{\mathbf{A}}\bar{\mathbf{B}}$
(2) $(\bar{\mathbf{A}})^T = \overline{(\mathbf{A}^T)}$
(3) $\mathbf{A}^H \mathbf{A}$ is real
(4) $(\mathbf{A}^H)^H = \mathbf{A}$
(5) $(\mathbf{AB})^H = \mathbf{B}^H \mathbf{A}^H$

The proof of the theorem is left to Exercise 2.19.

Definition 2.10. The matrix $\mathbf{P}$ is said to be unitary if and only if $\mathbf{P}^H \mathbf{P} = \mathbf{P}\mathbf{P}^H = \mathbf{I}$, the identity matrix.

A unitary matrix is the complex counterpart of a real orthogonal matrix.

Definition 2.11. The matrix $\mathbf{A}$ is said to be normal if and only if $\mathbf{A}^H \mathbf{A} = \mathbf{A} \mathbf{A}^H$.

Examples of normal matrices are unitary and Hermitian matrices (Exercise 2.15).

Definition 2.12. Let $\mathbf{X}$ and $\mathbf{Y}$ be two $(n \times 1)$ complex vectors. The complex inner product is the scalar $\mathbf{X} \cdot \mathbf{Y} = \bar{x}_1 y_1 + \bar{x}_2 y_2 + \cdots + \bar{x}_n y_n$. The norm (length) of a complex vector $\mathbf{X}$ is then given by $(\mathbf{X} \cdot \mathbf{X})^{1/2}$.

Note that by Definition 2.12 $(\mathbf{X} \cdot \mathbf{X}) > 0$ if $\mathbf{X} \neq 0$. Also $(\mathbf{X} \cdot \mathbf{Y}) = \overline{(\mathbf{Y} \cdot \mathbf{X})}$, the property of Hermitian symmetry. In matrix notation we can write the inner product as $\bar{\mathbf{X}}^T \mathbf{Y}$ so that $\bar{\mathbf{X}}^T \mathbf{Y} = \overline{\bar{\mathbf{Y}}^T \mathbf{X}}$. A quadratic form can then be expressed as $\bar{\mathbf{X}}^T \mathbf{A} \mathbf{Y}$ and the question arises as to whether a complex form can be rotated to independence using the spectral properties of $\mathbf{A}$.

THEOREM 2.23a.   Let $\mathbf{C}$ be a $(k \times k)$ Hermitian matrix. Then
   (1) The latent roots of $\mathbf{C}$ are real.
   (2) $\mathbf{C}$ possesses k orthonormal latent vectors.

PROOF
   (1) Let $\mathbf{P}_i$ be a latent vector which corresponds to the latent root λ_i $(i = 1, 2, \ldots, k)$. Then $\mathbf{C} \mathbf{P}_i = \lambda_i \mathbf{P}_i$ and premultiplying by $\bar{\mathbf{P}}_i^T$ we have

$$\bar{\mathbf{P}}_i^T \mathbf{C} \mathbf{P}_i = \lambda_i \bar{\mathbf{P}}_i^T \mathbf{P}_i \tag{2.125}$$

Also,

$$(\mathbf{C} \mathbf{P}_i)^H = (\lambda_i \mathbf{P}_i)^H$$

or

$$\mathbf{P}_i^H \mathbf{C}^H = \mathbf{P}_i^H \lambda_i^H$$

or

$$\bar{\mathbf{P}}_i^T \bar{\mathbf{C}}^T = \bar{\lambda}_i \bar{\mathbf{P}}_i^T \tag{2.126}$$

since $\mathbf{C}$ is Hermitian. Post-multiplying (2.126) by $\mathbf{P}_i$, subtracting Eqs. (2.125) and (2.126) we have

$$\bar{\mathbf{P}}_i^T \mathbf{P}_i \lambda_i - \bar{\mathbf{P}}_i^T P_i \bar{\lambda}_i = \bar{\mathbf{P}}_i^T \mathbf{P}_i (\lambda_i - \bar{\lambda}_i) = 0$$

and since $\bar{\mathbf{P}}_i^T \mathbf{P}_i \neq 0$ for $\mathbf{P}_i \neq 0$ we conclude $(\lambda_i - \bar{\lambda}_i) = 0$, that is, $\lambda_i = \bar{\lambda}_i$ which is only possible when λ_i is real.

(2) We have $\mathbf{CP}_i = \lambda_i \mathbf{P}_i$, $\mathbf{CP}_j = \lambda_j \mathbf{P}_j$ for any two real, distinct latent roots of $\mathbf{C}$. Premultiplying by $\bar{\mathbf{P}}_j$ and $\bar{\mathbf{P}}_i$ respectively we have

$$\bar{\mathbf{P}}_j^{\mathrm{T}} \mathbf{CP}_i = \lambda_i \bar{\mathbf{P}}_j^{\mathrm{T}} \mathbf{P}_i \, , \qquad \bar{\mathbf{P}}_i^{\mathrm{T}} \mathbf{CP}_j = \lambda_j \bar{\mathbf{P}}_i^{\mathrm{T}} \mathbf{P}_j$$

Subtracting the two expressions yields

$$(\bar{\mathbf{P}}_j^{\mathrm{T}} \mathbf{CP}_i - \bar{\mathbf{P}}_i^{\mathrm{T}} \mathbf{CP}_j) = \bar{\mathbf{P}}_j^{\mathrm{T}} \mathbf{P}_i (\lambda_i - \lambda_j) \qquad (2.127)$$

Since both terms on the left-hand side are scalar quantities, each must equal its Hermitian transpose, and we have

$$\bar{\mathbf{P}}_j^{\mathrm{T}} \mathbf{CP}_i = (\bar{\mathbf{P}}_j^{\mathrm{T}} \mathbf{CP}_i)^{\mathrm{H}}$$

$$= \mathbf{P}_i^{\mathrm{H}} \mathbf{C}^{\mathrm{H}} (\bar{\mathbf{P}}_j^{\mathrm{T}})^{\mathrm{H}}$$

$$= \bar{\mathbf{P}}_i^{\mathrm{T}} \mathbf{CP}_j$$

Since $\mathbf{C}$ is Hermitian. It follows that the left-hand side of Eq. (2.127) is zero and

$$\bar{\mathbf{P}}_j^{\mathrm{T}} \mathbf{P}_i (\lambda_i - \lambda_j) = 0$$

Since $\lambda_i \neq \lambda_j$ by assumption it follows that $\bar{\mathbf{P}}_j^{\mathrm{T}} \mathbf{P}_i = 0$, implying that $\mathbf{P}_i$ and $\mathbf{P}_j$ are orthogonal vectors.

When all k latent roots are not distinct it is still possible to find a set of k mutually orthogonal latent vectors, analogously to the real case. Since a real symmetric matrix is a special case of a Hermitian matrix, we have also proved that a real symmetric matrix must possess real latent roots and mutually orthogonal latent vectors. When the symmetric matrix is positive (semi) definite the latent roots are also positive (nonnegative). Of interest in statistical analysis is the complex generalization of a real quadratic form, known as a Hermitian form.

Definition 2.13.   A Hermitian form in the complex-valued variables x_1, $x_2, \ldots, x_k$ is a function of the type

$$h = \sum_{i=1}^{k} \sum_{j=1}^{k} c_{ij} \bar{x}_i x_j$$

$$= \bar{\mathbf{X}}^{\mathrm{T}} \mathbf{CX}$$

where $\mathbf{C}$ is a Hermitian matrix.

The matrix $\mathbf{C}$ can be real or complex. When $\mathbf{C}$ is real h is known as a real Hermitian form. When the variables $\mathbf{X}$ are also real-valued, then a Hermitian form becomes a (real) quadratic form (Section 2.7). Actually the main results concerning Hermitian forms parallel closely those obtained for quadratic forms. Thus a Hermitian form can be rotated to independent form, analogously to a real quadratic form. We have the following theorem, which may be considered as the basis of a principal component analysis in the complex plane (see Section 5.7).

THEOREM 2.24. Let $\mathbf{C}$ be a $(k \times k)$ Hermitian matrix and $\mathbf{P}$ a (column) latent vector (unitary) matrix. Let $\mathbf{X}$ be a $(k \times 1)$ vector of complex valued variables such that $\mathbf{X} = \mathbf{PZ}$. Then

$$\bar{\mathbf{X}}^{\mathrm{T}}\mathbf{C}\mathbf{X} = \lambda_1\bar{\mathbf{Z}}_1\mathbf{Z}_1 + \lambda_2\bar{\mathbf{Z}}_2\mathbf{Z}_2 + \cdots + \lambda_k\bar{\mathbf{Z}}_k\mathbf{Z}_k \qquad (2.128)$$

PROOF. We have

$$\bar{\mathbf{X}}^{\mathrm{T}}\mathbf{C}\mathbf{X} = \bar{\mathbf{Z}}^{\mathrm{T}}\bar{\mathbf{P}}^{\mathrm{T}}\mathbf{C}\mathbf{P}\mathbf{Z}$$

$$= \bar{\mathbf{Z}}^{\mathrm{T}}\mathbf{\Lambda}\mathbf{Z}$$

$$= \lambda_1\bar{\mathbf{Z}}_1\mathbf{Z}_1 + \lambda_2\bar{\mathbf{Z}}_2\mathbf{Z}_2 + \cdots + \lambda_k\bar{\mathbf{Z}}_k\mathbf{Z}_k$$

2.11.2 The Complex Multivariate Normal

Let

$$\mathbf{Z} = [z_1, z_2, \ldots, z_p]^{\mathrm{T}}$$

$$= [x_1 + iy_1, x_2 + iy_2, \ldots, x_p + iy_p] \qquad (2.129)$$

a complex vector of random variables (Section 1.7) where $\mathbf{V} = [x_1,\ y_1,\ x_2,\ y_2, \ldots, x_p,\ y_p]^{\mathrm{T}}$ is $2p$-variate normal. For the sake of simplicity we assume $\mathbf{E}(\mathbf{V}) = 0$. Then $\mathbf{Z}$ is said to follow a p-variate complex distribution. A question arises as to whether $\mathbf{Z}$ can be expressed in a functional form equivalent to Eq. (2.80). It turns out that this is only possible in the special case where

$$\mathrm{var}(x_j) = \mathrm{var}(y_i) = \frac{1}{2}\sigma_j^2$$

$$\mathrm{cov}(x_j, y_j) = 0$$

for the same complex variate, and

$$\mathrm{cov}(x_j, x_k) = \mathrm{cov}(y_j, y_k) = \frac{1}{2}\alpha_{jk}\sigma_j\sigma_k$$

$$\mathrm{cov}(x_j, y_k) = -\mathrm{cov}(x_k, y_j) = \frac{1}{2}\beta_{jk}\sigma_j\sigma_k$$

for different complex variates (Wooding, 1956; Goodman, 1963). Here α_{jk}

and β_{jk} are arbitrary constants. Thus the $(2p \times 2p)$ covariance matrix $E(\mathbf{V}\bar{\mathbf{V}}^{\mathrm{T}}) = \boldsymbol{\Sigma}$ consists of (2×2) submatrices of the form

$$\begin{bmatrix} E(x_j x_k) & E(x_j y_k) \\ E(y_j x_k) & E(y_j y_k) \end{bmatrix} = \begin{cases} \dfrac{1}{2}\begin{bmatrix} 1 & 0 \\ 0 & 1 \end{bmatrix}\sigma_j^2 & j = k \\ \dfrac{1}{2}\begin{bmatrix} \alpha_{jk} & -\beta_{jk} \\ \beta_{jk} & \alpha_{jk} \end{bmatrix}\sigma_j\sigma_k & j \neq k \end{cases} \quad (2.130)$$

The corresponding zero mean p-variate complex normal random variables $\mathbf{Z}$ then have their distributions specified by $(p \times p)$ Hermitian covariance matrices

$$\boldsymbol{\Sigma} = \mathbf{E}[\mathbf{Z}\mathbf{Z}^{\mathrm{H}}]$$

with typical element

$$E(\mathbf{Z}_j \bar{\mathbf{Z}}_k) = (\sigma_{jk})$$

where

$$\sigma_{jk} = \begin{cases} \sigma_k^2 & (j = k) \\ (\alpha_{jk} + i\beta_{jk})\sigma_j\sigma_k & (j \neq k) \end{cases} \quad (2.131)$$

and can be expressed in the form

$$f(\mathbf{Z}) = \frac{1}{\pi^{\mathrm{P}}|\boldsymbol{\Sigma}|} \exp(\mathbf{Z}^{\mathrm{H}}\boldsymbol{\Sigma}^{-1}\mathbf{Z}) \quad (2.132)$$

The term "complex multivariate normal distribution" is then usually restricted to the special case for which the restrictions of Eq. (2.130) hold. If $\mathbf{Z}_1, \mathbf{Z}_2, \ldots, \mathbf{Z}_n$ is a sample of n complex-valued vectors from Eq. (2.132), then the sample Hermitian covariance matrix

$$\hat{\boldsymbol{\Sigma}} = \frac{1}{n}\sum_{j=1}^{n} \mathbf{Z}_j\mathbf{Z}_j^{\mathrm{H}} \quad (2.133)$$

is the (sufficient) maximum likelihood estimator of $\boldsymbol{\Sigma}$. Let $\mathbf{A} = n\hat{\boldsymbol{\Sigma}}$. Then the joint distribution of the (distinct) elements of $\mathbf{A}$ is the complex form of the Wishart distribution given by

$$f(\mathbf{A}) = \frac{|\mathbf{A}|^{n-p}}{I(\boldsymbol{\Sigma})} \exp[-\mathrm{tr}(\boldsymbol{\Sigma}^{-1}\mathbf{A})] \quad (2.134)$$

where $I(\boldsymbol{\Sigma}) = \pi^{1/2p(p-1)}\Gamma(n)\Gamma(n-1)\ldots\Gamma(n-p+1)|\boldsymbol{\Sigma}|^n$ (Goodman, 1963; Srivastava and Khatri, 1979).

EXERCISES

2.1 Prove Theorem 2.13.

2.2 Let $\mathbf{Y}$ be a $(n \times p)$ data matrix and $\mathbf{J}$ the $(n \times n)$ matrix consisting of unities. Show that the sample covariance matrix can be expressed as

$$\mathbf{S} = \frac{1}{n-1}\left[\mathbf{Y}^{\mathrm{T}}\left(\mathbf{I} - \frac{1}{n}\mathbf{J}\right)\mathbf{Y}\right]$$

where $\mathbf{I}$ is the identity matrix.

2.3 (Albert and Tittle, 1967). For the probability function

$$f(x_1, x_2) = \begin{cases} \dfrac{1}{\pi}\exp\left[-\dfrac{1}{2}(x_1^2 + x_2^2)\right] & x_1 \geq x_2 \geq 0; \, x_1 \leq 0, \, x_2 \leq 0 \\ \\ 0 & x_1 \geq 0, \, x_2 \leq 0; \, x_1 \leq 0, \, x \geq 0 \end{cases}$$

(a) show that it is zero in the second and fourth quadrants of the (x_1, x_2) plane and is consequently not bivariate normal.
(b) Show that the marginal normal densities $f_1(x_1)$ and $f_2(x_2)$ of $f(x_1, x_2)$ are standard normal.

2.4 Verify Theorems 2.5–2.7 using matrices of Example 2.6.

2.5 Prove Theorem 2.22.

2.6 Let $\mathbf{X} = (x_{ij})$ be a $(n \times p)$ matrix. Show that

$$\sum_{i=1}^{n}\sum_{j=1}^{p} x_{ij}^2 = \mathrm{tr}(\mathbf{X}\mathbf{X}^{\mathrm{T}}).$$

2.7 Prove Eq. (2.35) for the inner product, and verify the three properties of the inner product.

2.8 Verify Eqs. (2.44) and (2.48) for the covariance and correlation coefficients respectively.

2.9 Using Example 2.2 verify (within rounding errors) Theorems 2.5–2.7.

2.10 Show, for Example 2.10, that

$$\rho = \frac{\lambda_1 - \lambda_2}{\lambda_1 + \lambda_2}$$

$$= 1/2(\lambda_1 - \lambda_2)$$

2.11 Let $\mathbf{X} = (x_1, x_2, \ldots, x_k)^T$, a $(k \times 1)$ column vector. Show that the unit rank matrix $\mathbf{X}\mathbf{X}^T$ possesses the unique nonzero latent root $\lambda = \mathbf{X}^T\mathbf{X}$.

2.12 Verify the Kronecker product properties of Section 2.9.

2.13 Prove that a set of k random variables which do not have marginal normal distributions cannot be distributed as a multivariate normal distribution.

2.14 Prove that unitary and Hermitian matrices are normal.

2.15 Prove that for a symmetric matrix with strictly positive elements, (1) there exists a largest, unique latent root $\lambda_1 > 0$, and (2) the latent vector $\mathbf{P}_1$ associated with λ_1 can always be chosen to have strictly positive elements.

2.16 Prove Eq. (2.45).

2.17 Show that Eq. (2.123) preserves Eqs. (2.118) and (2.119).

CHAPTER 3

The Ordinary Principal Components Model

3.1 INTRODUCTION

Let $X_1, X_2, \ldots, X_p$ represent a set of random variables distributed as some multivariate probability function $f(x_1, x_2, \ldots, x_p)$. The p variates need not be independent, and this gives rise to the essential distinction between univariate and multivariate distributions (Definition 1.10). An important special case arises when $f(x_1, x_2, \ldots, x_p)$ denotes a multivariate normal distribution. As noted in Section 2.8, two further properties follow which make the multivariate normal of practical importance: (1) conditional means assume a linear structure, and (2) zero correlations become a necessary and sufficient condition for complete independence. Thus for a multivariate normal sample a significant lack of pairwise correlation infers population independence. This is the simplest situation for detecting independence between continuous random variables. Frequently random variables will be independent, for example, in experimental set-ups where treatments are manipulated independently of each other or other covarying influences. In both experimental as well as nonexperimental research, however, independence is a relatively rare phenomenon in situations where variables cannot be collected in isolation from each other. This is because for a given sample (population) variables are usually collected within fairly narrow and well-defined areas of interest, and this results in an intercorrelated set. There often exist unobservable or "latent" variables which have given rise to the correlation among the observed set, for example, responses of individuals to an opinion poll such as a market research questionnaire, which usually depend on unobserved traits such as taste and attitude. Also, variables may represent measurements of the same object, for example, body parts of a biological organism where dimensions of the constituent parts will depend on latent traits such as overall body "size" and body "shape". Alternatively, in the absence of latent variables, correlation may result from the existence of distinct groups or clusters of variables which possess a common property not shared by the remaining set(s), and here we may wish to extend the concept of correlation to apply to more than two variables. Furthermore, when an experiment possesses more than a single dependent (experimental) variable, we may observe significant correlation amongst these variables, even though the treatments are orthogonal. The class of models known as factor analysis then attempts to answer what is perhaps the most elementary and fundamental question in multivariate analysis: given a set of intercorrelated random variables, how can we reduce (estimate) the dimensionality of the vector space containing these variables, and at the same time (possibly) identify the source of the observed correlations? Differently put, how can we explain the systematic behavior of the observed variables by means of a smaller set of computed but unobserved latent random variables? From the perspective of data analysis these objectives are equivalent to the attempt of representing, with as little loss as possible, a large set of data by means of a parsimonious set of linear relations, which in turn can be considered as newly created random variables.

The most straightforward model that seeks to achieve this objective is that of principal components analysis (PCA). The situation here is somewhat different from the least squares regression model since neither of the observed variables is assumed to be the dependent or the independent variable. Rather we suspect that the observed variates depend on a smaller number of unobserved dimensions or "variables" which may account for the systematic or true variance/covariance structure of the observations (Fig. 3.1). The purpose for a dimensional reduction may then lie in the convenience of working within a smaller space, or else we may wish to estimate and identify what are felt to be real (but unobserved) influences or random variables. The former task is at times ascribed to principal components (PCs), whereas the latter is conventionally viewed as the objective of a "proper" factor analysis model, such as the maximum likelihood factor model (Chapter 6). A sharp distinction between the two types of models however is not warranted since the mathematical and statistical structures of both are not unrelated. Certainly in many applications PCs may be treated as if they were random variables capable of explaining observed behavior. The most elementary framework however in which to consider the ordinary PC model is in terms of an extension of the usual binary (bivariate) correlation coefficient to the multivariate case. The model has a wide range of more specific applications which include the following:

1. Given a set of random variables, we may wish to examine the correlational pattern(s) that can exist amongst the variables. This may also be done interactively (Dumitriu et al., 1980). A frequent result is a nonhierarchical cluster analysis or a grouping (aggregation) of the variables. Conversely, we may be given samples of chemical or physical mixtures with the task of uncovering or "unmixing" their constituent parts.

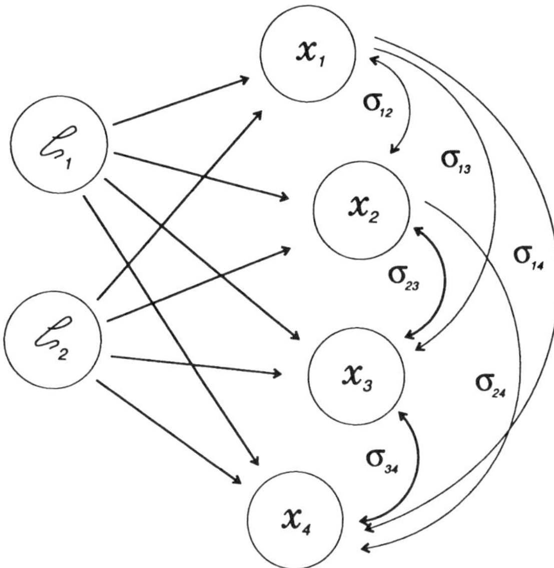

Figure 3.1 A set of $p = 4$ observed random variables influenced by $r = 2$ latent factors.

2. Uncovering unobserved explanatory variables which in a regression sense account for the variation in the observed set. In this context it is usual to attempt to identify the components (factors) in terms of real phenomenon. A related problem is to identify which factors are to be retained and which are to be discarded. Alternatively, for multivariate normal data we may wish to rotate the corresponding ellipse to its major (minor) axes. This results in a set of orthogonal (independent) normal variates, and is in fact the original context in which PC-type orthogonal rotations were considered by Bravais (1846).

3. Rather than reduce the dimensionality of an observed set by discarding needless components, we may wish to discard unnecessary variables from the observed set.

4. Given a set of homogeneous time series, for example, commodity prices, the aim may be to construct index numbers which represent the observed variation over a period of time.

5. Given a stochastic process with at least a single realization, we may wish to perform a spectral decomposition of the observed process. The objective may be to filter random error or noise from a time series or a two-dimensional image, or to recognize a pattern in some configuration.

6. Regression-type curve fitting. This may be carried out in at least four distinct fashions depending on the objective(s). (i) Given a least squares regression equation with highly multicollinear and/or error-prone explanatory variables, we are to estimate the original equation while at the same

time correcting for the effects of multicollinearity and/or errors in the independent variables. Here the factors or components are used as instrumental variables and need not be identified with any real behavior. A byproduct of the exercise is an increase in the number of degrees of freedom, resulting from reduced dimensionality of the regressor subspace. (ii) Independent regression variables are decomposed into a smaller number of components-factors and these are used in place of the original explanatory variables. This also provides an effective solution of multicollinearity and/or errors in variables and increased degrees of freedom. However the component-factors must actually be identified with some "real" behavior in order to achieve a sensible interpretation of the regression equation. (iii) Estimate a regression of the form $f(x_1, x_2, \ldots, x_p) = 0$ where all variables are treated symmetrically, that is, all are assumed to possess error terms. This is the original use of "principal components" in a regression setting, from Pearson (1901), who termed the procedure as the method of "planes of closest fit." As in other regression models, the variables should not be highly collinear. (iv) Estimate functional forms and growth curves.

7. Identification of outlier observations; optimal matrix approximations.

8. Classifying (scaling) individuals (sample points). Rather than analyze intercorrelations between variables, we may wish to study the interelationships between the individual sample points. More generally we may wish to plot both variables as well as sample points in a joint space in order to reveal particular affinities of certain sample points to particular variables.

9. Hierarchical clustering. Here the purpose is to construct a tree-like graph or a dendogram for classification purposes of sample points, given a set of variables or characteristics. When many variables are present it can be more convenient to use a reduced number of PCs to construct the dendogram rather than the original variables.

10. For discrete (ordinal, nominal) data we may wish to investigate the existence of optimal continuous scales which maximize the Cramer correlation within a $(p \times q)$ contingency table. Boolean $(0, 1)$ matrices can also be analyzed, either within a network-type graph or an ordination (seriation) situation.

11. Finally, graphical techniques such as biplots and others can be used in conjunction with principal components.

More general decompositions, for example in the complex field, are also common and some of these are considered in subsequent chapters. Owing to the existence of statistical electronic computer programs, relatively large sets of data can be analyzed without great difficulty. For a brief description of factor analysis as a data analytic tool the reader is referred to Frane and Hill (1976).

3.2 PRINCIPAL COMPONENTS IN THE POPULATION

From a purely mathematical viewpoint the purpose of a population principal component (PC) model is to transform p correlated random variables to an orthogonal set which reproduces the original variance/covariance structure. This is equivalent to rotating a pth dimensional quadratic form to achieve independence between the variables (Section 2.7). A number of methods exist which can be used to orthogonalize a set of random variables. A PCA employs latent roots and latent vectors of a Grammian matrix to achieve this aim, initiated in statistics by Bravais (1846), Pearson (1901), Frisch (1929), Hotelling (1933), and Girschick (1936, 1939). The origins of the method probably go back to Bravais (1846) in the form of rotating an ellipse to "axes principaux" in order to achieve independence in a multivariate normal distribution (Sections 2.7, 2.8).

Let $\mathbf{X} = (X_1, X_2, \ldots, X_p)^T$ be a $(p \times 1)$ vector of continuous random variables with zero mean and covariance matrix $E(\mathbf{XX}^T) = \mathbf{\Sigma}$, and consider the linear transformation

$$
\begin{aligned}
\zeta_1 &= \pi_{11}x_1 + \pi_{21}x_2 + \cdots + \pi_{p1}x_p = \mathbf{\Pi}_1^T \mathbf{X} \\
\zeta_2 &= \pi_{12}x_1 + \pi_{22}x_2 + \cdots + \pi_{p2}x_p = \mathbf{\Pi}_2^T \mathbf{X} \\
\hline
\zeta_p &= \pi_{1p}x_1 + \pi_{2p}x_2 + \cdots + \pi_{pp}x_p = \mathbf{\Pi}_p^T \mathbf{X}
\end{aligned}
\tag{3.1}
$$

where

$$
\begin{aligned}
\operatorname{var}(\zeta_i) = E(\zeta_i^2) &= E[(\mathbf{\Pi}_i^T \mathbf{X})(\mathbf{\Pi}_i^T \mathbf{X})^T] \\
&= \mathbf{\Pi}_i^T E(\mathbf{XX}^T)\mathbf{\Pi}_i \\
&= \mathbf{\Pi}_i^T \mathbf{\Sigma} \mathbf{\Pi}_i \qquad i = 1, 2, \ldots, p
\end{aligned}
\tag{3.2}
$$

and

$$
\begin{aligned}
\operatorname{cov}(\zeta_i, \zeta_j) = E(\zeta_i, \zeta_j) &= E[(\mathbf{\Pi}_i^T \mathbf{X})(\mathbf{\Pi}_j^T \mathbf{X})^T] \\
&= \mathbf{\Pi}_i^T E(\mathbf{XX}^T)\mathbf{\Pi}_j \\
&= \mathbf{\Pi}_i^T \mathbf{\Sigma} \mathbf{\Pi}_j \\
&= 0 \qquad\qquad i \neq j
\end{aligned}
\tag{3.3}
$$

In matrix form, $\boldsymbol{\zeta} = \mathbf{\Pi}^T \mathbf{X}$ where $\boldsymbol{\zeta} = (\zeta_1, \zeta_2, \ldots, \zeta_p)^T$ are the new random variables and $\mathbf{\Pi}$ is a $(p \times p)$ matrix of fixed, nonrandom coefficients. For the special case when $\mathbf{\Sigma}$ is diagonal, X_i are uncorrelated random variables and, hopefully, each can be studied more simply by univariate methods. Here orthogonalization is clearly unnecessary since the new variables ζ_i are simply proportional to the original X_i. As the covariances between the random variables $X_1, X_2, \ldots, X_p$ increase, however, they tend to contain more and more redundant information and may be replaced by a smaller number r of orthogonal variables which account for a smaller but sufficiently higher percentage of the original variance. In the limit when all observed variables become highly intercorrelated (linearly interdependent), a single dimension suffices to account for the total variance. The principal components model therefore represents an extension of the notion of (bivariate) correlation to three or more random variables in terms of their mutual correlation to a set of common variables or "factors." The creation of such mathematical variables or dimensions is necessitated by the inherent restrictiveness of the usual correlation coefficient which is, by definition, a binary function. Although initially such factors are viewed as mathematical constructs, should it prove possible to identify them in terms of real behavior they may be regarded as genuine explanatory dimensions or variables. This is illustrated in Figure 3.1 for $p = 4$ and $r = 2$. The new random variables ζ_i are known as principal components (PCs) and the coefficients π_{ij} are known as loadings. Clearly if parsimony is to be at a maximum the ζ_i should be orthogonal, but may be transformed to an oblique set (Section 2.5.2) if required. Also, the number of orthogonal components that account for a given percentage of variance should be kept to a minimum. The former condition of orthogonality is satisfied by Theorem 2.6, whereas the latter is met by ranking components in a decreasing order of explanatory power or variance. The expansion (Eq. 3.1) may be terminated at any stage if it is felt a satisfactory percentage of variance of the original variables has been accounted for by the ζ_i $(i = 1, 2, \ldots, r)$ for some r. Using the results of Section 2.6, the number of (nonzero) principal components is then equal to the rank of $\boldsymbol{\Sigma}$. The latent vectors are also standardized to unit length (variance) so that $\boldsymbol{\Pi}_i^{\mathrm{T}}\boldsymbol{\Pi}_i = 1$ $(i = 1, 2, \ldots, p)$. Alternative standardization can also be used depending on the application.

In mathematical terms the objective is to compute linear combinations given by Eq. (3.1) which maximize the overall variance of $\mathbf{X}_1, \mathbf{X}_2, \ldots, \mathbf{X}_p$, subject to the constraint that $\boldsymbol{\Pi}_i$ are unit vectors. This is equivalent to maximizing the Lagrangean expression (see, eg., Tracy and Dwyer, 1969)

$$\phi_i = \boldsymbol{\Pi}_i^{\mathrm{T}}\boldsymbol{\Sigma}\boldsymbol{\Pi}_i - \lambda_i(\boldsymbol{\Pi}_i^{\mathrm{T}}\boldsymbol{\Pi}_i - 1) \qquad (i = 1, 2, \ldots, p) \qquad (3.4)$$

with respect to the $\boldsymbol{\Pi}_i$, that is, finding the latent roots and vectors of a quadratic form (Section 2.7). Setting partial derivatives to zero then yields

$$\frac{\partial \phi_i}{\partial \boldsymbol{\Pi}_i} = 2\boldsymbol{\Sigma}\boldsymbol{\Pi}_i - 2\lambda_i\boldsymbol{\Pi}_i = \mathbf{0}$$

or

$$(\boldsymbol{\Sigma} - \lambda_i\mathbf{I})\boldsymbol{\Pi}_i = \mathbf{0} \qquad (i = 1, 2, \ldots, p) \qquad (3.5)$$

Equation (3.5) is maximized by that latent vector which corresponds to the largest latent root, and ranking the roots in the decreasing order $\lambda_1 \geq \lambda_2 \geq \cdots \geq \lambda_p$ maximizes the quadratic form (Eq. 3.5) in a sequential manner. The latent roots and vectors can also be derived simultaneously by maximizing

$$\phi = \Pi^T \Sigma \Pi - \lambda(\Pi^T \Pi - I) \tag{3.6}$$

that is, by solving

$$\frac{\partial \phi}{\partial \Pi} = 2\Sigma\Pi - 2\lambda\Pi = 0$$

or

$$(\Sigma - \lambda I)\Pi = 0 \tag{3.7}$$

From Section 2.6 the terms $\lambda_1 \geq \lambda_2 \geq \cdots \geq \lambda_p$ are real, nonnegative, and roots of the determinantal polynomial

$$|\Sigma - \lambda I| = 0 \tag{3.8}$$

Once λ_i are known, the latent vectors are obtained by solving Eq. (3.7). The variance/covariance structure of the PCs can then be derived as follows.

THEOREM 3.1. Let $\zeta_i = \Pi_i^T X$, and $\zeta_j = \Pi_j^T X$ be any two linear combinations of p random variables X such that Π_i and Π_j are latent vectors of $E(XX^T) = \Sigma$. Then

(1) Π_i and Π_j are orthogonal vectors when they correspond to unequal latent roots.

(2) ζ_i and ζ_j are uncorrelated random variables such that $\text{var}(\zeta_i) = \lambda_i$ $(i = 1, 2, \ldots, p)$.

PROOF

(1) From Eq. (3.5) we have $\Sigma\Pi_i = \lambda_i\Pi_i$, $\Sigma\Pi_j = \lambda_j\Pi_j$ and premultiplying by Π_j^T and Π_i^T respectively yields $\Pi_j^T\Sigma\Pi_i = \lambda_i\Pi_j^T\Pi_i$ and $\lambda_i\Pi_i^T\Sigma\Pi_j = \lambda_j\Pi_i^T\Pi_j$. Subtracting we then have

$$\Pi_j^T\Sigma\Pi_i - \Pi_i^T\Sigma\Pi_j = \lambda_i\Pi_j^T\Pi_i - \lambda_j\Pi_i^T\Pi_j$$

$$= (\lambda_i - \lambda_j)\Pi_i^T\Pi_j$$

$$= 0$$

since $\Pi_i^T\Pi_j = \Pi_j^T\Pi_i$ and $\Pi_j^T\Sigma\Pi_i = \Pi_i^T\Sigma\Pi_j$. Thus for $\lambda_i \neq \lambda_j$ we conclude that $\Pi_i^T\Pi_j = 0$.

(2) We have

$$\text{cov}(\zeta_i, \zeta_j) = \text{cov}(\mathbf{\Pi}_i^T \mathbf{X}, \mathbf{\Pi}_j^T \mathbf{X})$$

$$= \mathbf{\Pi}_i^T \mathbf{\Sigma} \mathbf{\Pi}_j$$

$$= \mathbf{\Pi}_i^T (\lambda_j \mathbf{\Pi}_j)$$

$$= \lambda_j \mathbf{\Pi}_i^T \mathbf{\Pi}_j$$

$$= 0$$

since $\mathbf{\Pi}_i^T \mathbf{\Pi}_j = 0$ from part (1). The PCs are thus orthogonal whenever the latest vectors are. Their variances are given by

$$\text{var}(\zeta_i) = \mathbf{\Pi}_i^T \mathbf{\Sigma} \mathbf{\Pi}_i$$

$$= \lambda_i \qquad (i = 1, 2, \ldots, p)$$

using Eqs. (3.2) and (3.5).

THEOREM 3.2.   Let $\mathbf{\Sigma}$ possess latent roots $\mathbf{\Lambda}$ and latent vectors $\mathbf{\Pi}$. Then
(1) When $X_1, X_2, \ldots, X_p$ are uncorrelated,

$$\sigma_i^2 = \text{var}(X_i) = \lambda_i \qquad (i = 1, 2, \ldots, p).$$

(2) When $\lambda_1 = \lambda_2 = \cdots = \lambda_p$, $X_1, X_2, \ldots, X_p$ are homoscedastic and uncorrelated.

PROOF
(1) From Eq. (3.8) we have

$$\begin{vmatrix} \sigma_1^2 - \lambda_1 & & & \\ & \sigma_2^2 - \lambda_2 & & \bigcirc \\ & & \ddots & \\ & \bigcirc & & \sigma_p^2 - \lambda_p \end{vmatrix} = 0$$

or

$$(\sigma_1^2 - \lambda_1)(\sigma_2^2 - \lambda_2) \ldots (\sigma_p^2 - \lambda_p) = 0$$

so that

$$\text{var}(X_i) = \sigma_i^2 = \lambda_i \qquad (i = 1, 2, \ldots, p)$$

(2) The second part follows as a special case.

Theorem 3.2 demonstrates that when a covariance matrix is diagonal, there is no gain in performing a PCA. As correlations between the random variables increase, however, economy is achieved by replacing $X_1, X_2, \ldots, X_p$ with a reduced number of components. Also, Theorem 3.1 indicates that the latent vectors, together with their PCs, are always orthogonal whenever their corresponding latent roots are different. Actually, it can be shown that a symmetric matrix always possesses orthogonal (unit) latent vectors even when some (or all) roots are equal (Theorem 2.6). The following three cases can be distinguished in practice:

1. All latent roots distinct, $\lambda_1 > \lambda_2 > \cdots > \lambda_p$. Here $\mathbf{\Pi}$ is unique as are the components $\boldsymbol{\zeta} = \mathbf{\Pi}^T \mathbf{X}$ (up to sign changes).

(2) All roots equal, $\lambda_1 = \lambda_2 = \cdots = \lambda_p$. The quadratic form associated with Eq. (3.5) becomes a pth dimensional sphere and it no longer makes sense to maximize Eq. (3.4). The quadratic form is said to be isotropic, and although $\mathbf{\Pi}$ is still orthogonal, it is no longer unique.

(3) $r < p$ roots distinct, $\lambda_1 > \lambda_2 > \cdots > \lambda_r > \lambda_{r+1} = \cdots = \lambda_p$. Latent vectors corresponding to $\lambda_1 > \lambda_2 > \cdots > \lambda_r$ are unique (up to sign change), but those corresponding to $\lambda_{r+1} = \cdots = \lambda_p$ are not. We have both nonisotropic and isotropic variation in a p-dimensional space. The p-dimensional ellipse contains an embedded sphere of dimension $(p - r)$ with a constant-length radius vector.

The maximization of the quadratic form results in an important property of the PCs—at each stage they maximize the global variance of the random variables X_i $(i = 1, 2, \ldots, p)$. Actually, PCs possess several other desirable optimality properties which hinge on the following theorem.

THEOREM 3.3 (Rayleigh Quotient). Let $\mathbf{\Sigma}$ be a covariance matrix with latent roots $\lambda_1 \geq \lambda_2 \geq \cdots \geq \lambda_p$ and let

$$q = \frac{\mathbf{U}^T \mathbf{\Sigma} \mathbf{U}}{\mathbf{U}^T \mathbf{U}} \tag{3.9}$$

for some vector $\mathbf{U} \neq \mathbf{0}$. Then

    (i) $\lambda_p \leq q \leq \lambda_1$

    (ii) $\lambda_1 = \max \dfrac{\mathbf{U}^T \mathbf{\Sigma} \mathbf{U}}{\mathbf{U}^T \mathbf{U}}, \qquad \lambda_p = \min \dfrac{\mathbf{U}^T \mathbf{\Sigma} \mathbf{U}}{\mathbf{U}^T \mathbf{U}}$

if and only if $\mathbf{U}$ is, respectively, the latent vector of $\mathbf{\Sigma}$ corresponding to the first and last latent root.

PROOF.

(i) Since $\boldsymbol{\Sigma}$ is Grammian it must possess real, orthogonal, latent vectors $\boldsymbol{\Pi}_1, \boldsymbol{\Pi}_2, \ldots, \boldsymbol{\Pi}_p$ such that $\boldsymbol{\Sigma\Pi}_i = \boldsymbol{\Pi}_i \lambda$ $(i = 1, 2, \ldots, p)$. Also, the vector $\mathbf{U}$ can always be expanded as

$$\mathbf{U} = a_1\boldsymbol{\Pi}_1 + a_2\boldsymbol{\Pi}_2 + \cdots + a_p\boldsymbol{\Pi}_p$$

$$= \boldsymbol{\Pi}\mathbf{A}$$

for some coefficients $a_1, a_2, \ldots, a_p$. We have

$$q = \frac{\mathbf{U}^t\boldsymbol{\Sigma}\mathbf{U}}{\mathbf{U}^T\mathbf{U}}$$

$$= \frac{\mathbf{A}^T(\boldsymbol{\Pi}^T\boldsymbol{\Sigma}\boldsymbol{\Pi})\mathbf{A}}{\mathbf{A}^T\boldsymbol{\Pi}^T\boldsymbol{\Pi}\mathbf{A}}$$

$$= \frac{\mathbf{A}^T\boldsymbol{\Lambda}\mathbf{A}}{\mathbf{A}^T\mathbf{A}}$$

$$= \frac{a_1^2\lambda_1 + a_2^2\lambda_2 + \cdots + a_p^2\lambda_p}{a_1^2 + a_2^2 + \cdots + a_p^2}$$

so that

$$q - \lambda = \frac{a_1^2\lambda_1 + a_2^2\lambda_2 + \cdots + a_p^2\lambda_p}{a_1^2 + a_2^2 + \cdots + a_p^2} - \frac{\lambda_1(a_1^2 + a_2^2 + \cdots + a_p^2)}{a_1^2 + a_2^2 + \cdots + a_p^2}$$

$$= \frac{(\lambda_2 - \lambda_1)a_1^2 + (\lambda_3 - \lambda_1)a_2^2 + \cdots + (\lambda_p - \lambda_1)a_p^2}{a_1^2 + a_2^2 + \cdots + a_p^2}$$

Since the right-hand side of the equation is non-negative and since $\lambda_1 \geq \lambda_2 \geq \cdots \geq \lambda_p$ it follows that $q \leq \lambda_1$ and repeating the process using λ_p we obtain $q \geq \lambda_p$.

(ii) Let $\mathbf{U} = \boldsymbol{\Pi}_1$. Then

$$q - \lambda_1 = \frac{\boldsymbol{\Pi}_1^T\boldsymbol{\Sigma}\boldsymbol{\Pi}_1}{\boldsymbol{\Pi}_1^T\boldsymbol{\Pi}_1} - \lambda_1$$

$$= \lambda_1 - \lambda_1$$

$$= 0$$

Likewise $q - \lambda_p = 0$ so that

$$\lambda_1 = \max \frac{\boldsymbol{\Pi}_1^T\boldsymbol{\Sigma}\boldsymbol{\Pi}_1}{\boldsymbol{\Pi}_1^T\boldsymbol{\Pi}_1} \quad \text{and} \quad \lambda_p = \min \frac{\boldsymbol{\Pi}_p^T\boldsymbol{\Sigma}\boldsymbol{\Pi}_p}{\boldsymbol{\Pi}_p^T\boldsymbol{\Pi}_p}$$

THEOREM 3.4.   Let $\mathbf{\Sigma\Pi} = \mathbf{\Pi\Lambda}$, where $\mathbf{\Pi}$ is the orthogonal matrix of latent vectors and $\mathbf{\Lambda}$ is a diagonal matrix of latent roots such that $\lambda_1 \geq \lambda_2 \geq \cdots \geq \lambda_p$. Then

$$\lambda_i = \max(\boldsymbol{\alpha}^T\mathbf{\Sigma}\boldsymbol{\alpha}) = \mathbf{\Pi}_i^T\mathbf{\Sigma}\mathbf{\Pi}_i \qquad (i = 1, 2, \ldots, p)$$

where $\boldsymbol{\alpha}$ is any unit vector such that $\boldsymbol{\alpha}^T\mathbf{\Pi}_i = 0$.

PROOF.   Let $\boldsymbol{\alpha}^T\mathbf{X} = \theta$ be any liner combination of the random variables. Then

$$\begin{aligned}
\mathrm{var}(\theta) &= \boldsymbol{\alpha}^T\mathbf{\Sigma}\boldsymbol{\alpha} \\
&= \boldsymbol{\alpha}^T\mathbf{\Pi\Pi}^T\mathbf{\Sigma}\mathbf{\Pi\Pi}^T\boldsymbol{\alpha} \\
&= \boldsymbol{\alpha}^T\mathbf{\Pi\Lambda\Pi}^T\boldsymbol{\alpha} \\
&= \boldsymbol{\gamma}^T\mathbf{\Lambda}\boldsymbol{\gamma} \\
&= \sum_{i=1}^{p} \lambda_i\gamma_i^2
\end{aligned}$$

where $\boldsymbol{\gamma} = \mathbf{\Pi}^T\boldsymbol{\alpha}$. Also $\boldsymbol{\gamma}^T\boldsymbol{\gamma} = \boldsymbol{\alpha}^T\mathbf{\Pi\Pi}^T\boldsymbol{\alpha} = 1$ since $\boldsymbol{\alpha}^T\boldsymbol{\alpha} = 1$. We have

$$\begin{aligned}
\mathrm{var}(\theta) &= \boldsymbol{\alpha}^T\mathbf{\Sigma}\boldsymbol{\alpha} \\
&= \lambda_1\gamma_1^2 + \lambda_2\gamma_2^2 + \cdots + \lambda_p\gamma_p^2 \\
&\leq \lambda_1\gamma_1^2 + \lambda_1\gamma_2^2 + \cdots + \lambda_1\gamma_p^2 \\
&\leq \lambda_1
\end{aligned}$$

since $\lambda_1 \geq \lambda_2 \geq \cdots \geq \lambda_p$. Thus for any linear combination θ we have $\mathrm{var}(\theta) = \boldsymbol{\alpha}^T\mathbf{\Sigma}\boldsymbol{\alpha} \leq \lambda_1$. Since equality only holds when $\boldsymbol{\gamma} = (1, 0, \ldots, 0)^T$, that is, when $\boldsymbol{\alpha} = \mathbf{\Pi}_1$, it follows that

$$\lambda_1 = \max_{\boldsymbol{\alpha}^T\boldsymbol{\alpha}=1} (\boldsymbol{\alpha}^T\mathbf{\Sigma}\boldsymbol{\alpha}) = \lambda_1^T\mathbf{\Sigma}\mathbf{\Pi}_1$$

and λ_1 maximizes the variance of $\theta = \boldsymbol{\alpha}^T\mathbf{X}$

For the second root λ_2, condition $\boldsymbol{\alpha}^T\mathbf{\Pi}_1 = 0$ is replaced by the orthogonality condition $\boldsymbol{\gamma}^T\mathbf{\Pi}^T\mathbf{\Pi}_1 = 0$, so that $\gamma_1 = 0$. Since $\boldsymbol{\gamma}^T\boldsymbol{\gamma} = \boldsymbol{\alpha}^T\boldsymbol{\alpha} = 1$, by definition we have

$$\boldsymbol{\alpha}^T\mathbf{\Sigma}\boldsymbol{\alpha} = \sum_{i=1}^{p} \lambda_i\gamma_1^2 \leq \lambda_2(\gamma_2^2 + \gamma_3^2 + \cdots + \gamma_p^2) = \lambda_2$$

with equality holding only when $\boldsymbol{\gamma} = (0, 1, 0, \ldots, 0)^T$. This implies that

$\gamma = \Pi_2$, and

$$\lambda_2 = \max_{\alpha^T\alpha=1} (\alpha^T \Sigma \alpha) = \Pi_2^T \Sigma \Pi_2$$

where $\alpha^T \Pi_2 = 0$. The process continues until all p latent roots are extracted.

Theorem 3.4 demonstrates that of all linear combinations of some random variables $\mathbf{X} = (X_1, X_2, \ldots, X_p)^T$, variance is maximized (at each stage) only when coefficients of linear combinations are latent vectors of Σ such that $\lambda_1 \geq \lambda_2 \geq \cdots \geq \lambda_p$. The linear combinations are then the PCs $\zeta_i = \Pi_i^T \mathbf{X}$ $(i = 1, 2, \ldots, p)$, and this effectively solves the optimality problem introduced at the outset of the chapter: given a set of random variables $X_1, X_2, \ldots, X_p$, how can we find another set $\zeta_1, \zeta_2, \ldots, \zeta_p$ which maximize total variance of the X_i? Since each PC maximizes variance, we can always retain that number $r \leq p$ which accounts for some predetermined minimal percentage of variance. Alternatively, all PCs may be computed and the last $(p - r)$ omitted if they account for an insignificant (negligible) proportion of variance. Both approaches result in identical coefficients Π_i. The higher the correlation between the value of X_i, the smaller is the number of PCs required to account for a fixed percentage of variance. The principal properties of the PC model can be summarized by the following five theorems.

THEOREM 3.5. Let $\mathbf{X} = (X_1, X_2, \ldots, X_p)^T$ be a set of random variables where $E(\mathbf{X}) = 0$, $E(\mathbf{XX}^T) = \Sigma$, and $\Pi^T \Sigma \Pi = \Lambda$. Then

    (i) $E(\zeta) = 0$

   (ii) $E(\zeta\zeta^T) = \Lambda$

PROOF.

    (i) Since $\zeta = \Pi^T \mathbf{X}$ we have

$$E(\zeta_i) = \pi_{1i}E(X_1) + \pi_{2i}E(X_2) + \cdots + \pi_{pi}E(X_p)$$

$$= 0$$

When $E(\mathbf{X}) \neq 0$, however, population means of components ζ_i are linear functions of the expected values of the X_i.

   (ii)
$$E(\zeta\zeta^T) = E[(\Pi^T\mathbf{X})(\Pi^T\mathbf{X})^T]$$

$$= \Pi^T E(\mathbf{XX}^T)\Pi$$

$$= \Pi^T \Sigma \Pi$$

$$= \Lambda$$

so that

$$E(\zeta_i \zeta_j) = \begin{cases} \lambda_i & i = j \\ 0 & i \neq j \end{cases}$$

Once latent vectors are known Eq. (3.1) can be inverted to express the random variables in terms of the PCs,

$$
\begin{aligned}
X_1 &= \pi_{11}\zeta_1 + \pi_{12}\zeta_2 + \cdots + \pi_{1p}\zeta_p = \mathbf{\Pi}_{(1)}^{\mathrm{T}}\zeta \\
X_2 &= \pi_{21}\zeta_1 + \pi_{22}\zeta_2 + \cdots + \pi_{2p}\zeta_p = \mathbf{\Pi}_{(2)}^{(\mathrm{T})}\zeta \\
\hline
X_p &= \pi_{p1}\zeta_1 + \pi_{p2}\zeta_2 + \cdots + \pi_{pp}\zeta_p = \mathbf{\Pi}_{(p)}^{\mathrm{T}}\zeta
\end{aligned}
\tag{3.10}
$$

where $\mathbf{\Pi}_{(i)}$ is the ith row vector of $\mathbf{\Pi}$. Note that $\mathbf{\Pi}_{(i)}^{\mathrm{T}}\mathbf{\Pi}_{(j)} \neq 0$. In matrix form,

$$\mathbf{X} = \mathbf{\Pi}\zeta$$

THEOREM 3.6.   The PCs $\zeta_1, \zeta_2, \ldots, \zeta_p$ reproduce both the variances as well as the covariances of the random variables $X_1, X_2, \ldots, X_p$.

PROOF

$$
\begin{aligned}
\mathbf{\Sigma} &= E(\mathbf{X}\mathbf{X}^{\mathrm{T}}) \\
&= E[(\mathbf{\Pi}\zeta(\mathbf{\Pi}\zeta)^{\mathrm{T}}] \\
&= \mathbf{\Pi}E(\zeta\zeta^{\mathrm{T}})\mathbf{\Pi}^{\mathrm{T}} \\
&= \mathbf{\Pi}\mathbf{\Lambda}\mathbf{\Pi}^{\mathrm{T}} \\
&= \lambda_1\mathbf{\Pi}_1\mathbf{\Pi}_1^{\mathrm{T}} + \lambda_2\mathbf{\Pi}_2\mathbf{\Pi}_2^{\mathrm{T}} + \cdots + \lambda_p\mathbf{\Pi}_p\mathbf{\Pi}_p^{\mathrm{T}}
\end{aligned}
\tag{3.11}
$$

from Theorem 3.5.

The PCs are not invariant with respect to scale change (linear or affine transformation) and poststandardization of the variables will not yield the same latent roots and latent vectors as prestandardization.

THEOREM 3.7.   Let $\mathbf{\Lambda}$ and $\mathbf{\Pi}$ be the latent roots and latent vectors of $\mathbf{\Sigma}$. Then covariances between X_i and ζ_j are given by the ith and jth elements of $\mathbf{\Pi}\mathbf{\Lambda}$.

PROOF. We have

$$\mathrm{cov}(\mathbf{X}, \boldsymbol{\zeta}) = E(\mathbf{X}\boldsymbol{\zeta}^{\mathrm{T}})$$

$$= E(\boldsymbol{\Pi}\boldsymbol{\zeta}\boldsymbol{\zeta}^{\mathrm{T}})$$

$$= \boldsymbol{\Pi}E(\boldsymbol{\zeta}\boldsymbol{\zeta}^{\mathrm{T}})$$

$$= \boldsymbol{\Pi}\boldsymbol{\Lambda} \tag{3.12}$$

so that

$$
E(\mathbf{X}\boldsymbol{\zeta}^{\mathrm{T}}) =
\begin{bmatrix}
E(X_1\zeta_1) & E(X_1\zeta_2) & \cdots & E(X_1\zeta_p) \\
E(X_2\zeta_1) & E(X_2\zeta_2) & \cdots & E(X_2\zeta_p) \\
\vdots & \vdots & & \vdots \\
E(X_p\zeta_1) & E(X_p\zeta_2) & \cdots & E(X_p\zeta_p)
\end{bmatrix}
$$

$$
=
\begin{bmatrix}
\pi_{11}\lambda_1 & \pi_{12}\lambda_2 & \cdots & \pi_{1p}\lambda_p \\
\pi_{21}\lambda_1 & \pi_{22}\lambda_2 & \cdots & \pi_{2p}\lambda_p \\
\vdots & \vdots & & \vdots \\
\pi_{p1}\lambda_1 & \pi_{p2}\lambda_2 & \cdots & \pi_{pp}\lambda_p
\end{bmatrix}
$$

or

$$E(X_i\zeta_j) = \lambda_j\pi_{ij}. \qquad (i, j = 1, 2, \ldots, p). \tag{3.13}$$

Equation (3.13) assumes unstandardized $\boldsymbol{\zeta}$. Standardizing, we obtain

$$\boldsymbol{\zeta}^* = \boldsymbol{\Lambda}^{-1/2}\boldsymbol{\zeta} \tag{3.14}$$

so that columns of $\boldsymbol{\zeta}^*$ become orthonormal random variables. The covariances between $\mathbf{X}$ and $\boldsymbol{\zeta}^*$ are then given by

$$\mathrm{cov}(\mathbf{X}, \boldsymbol{\zeta}^*) = E(\mathbf{X}\boldsymbol{\zeta}^{*\mathrm{T}})$$

$$= E[\mathbf{X}(\boldsymbol{\Lambda}^{-1/2}\boldsymbol{\zeta})^{\mathrm{T}}]$$

$$= E[\mathbf{X}\boldsymbol{\zeta}^{\mathrm{T}}\boldsymbol{\Lambda}^{-1/2}]$$

$$= \boldsymbol{\Pi}\boldsymbol{\Lambda}\boldsymbol{\Lambda}^{-1/2}$$

$$= \boldsymbol{\Pi}\boldsymbol{\Lambda}^{1/2}$$

so that

$$\mathrm{cov}(X_i, \zeta_j^*) = \lambda_j^{1/2}\pi_{ij} \tag{3.15}$$

Equation (3.15) still depends on σ_i^2. To free coefficients (covariances) from the effects of the variances σ_i^2 of the random variables they can further be standardized to the form

$$\alpha_{ij} = \frac{\lambda_j^{1/2} \pi_{ij}}{\sigma_i} = \frac{\text{cov}(X_i, \zeta_i)}{\lambda_j^{1/2} \sigma_i} \qquad (i, j = 1, 2, \ldots, p) \qquad (3.16)$$

using Eq. (3.13). The value of α_{ij} are the correlations or the loadings between the ith variate and jth component. In what follows they are referred to as correlation loadings to distinguish them from the covariance loadings (Eq. 3.15). The correlation loadings are thus latent vectors which have been normalized such that

$$\boldsymbol{\alpha}^T \boldsymbol{\alpha} = \boldsymbol{\Lambda}, \qquad \boldsymbol{\alpha}\boldsymbol{\alpha}^T = \mathbf{P} \qquad (3.17)$$

where $\boldsymbol{\Lambda}$ consists of latent roots of $\mathbf{P}$, the correlation matrix of the random variables. Equations (3.10) and (3.11) can also be expressed in terms of the $\boldsymbol{\alpha}$ coefficients if the latent vectors are scaled such that both random variables and PCs possess unit variance (Example 3.1). We have

$$\boldsymbol{\Pi}^T \boldsymbol{\Sigma} \boldsymbol{\Pi} = \boldsymbol{\Pi}^T E(\mathbf{X}\mathbf{X}^T)\boldsymbol{\Pi}$$

$$= E(\boldsymbol{\Pi}^T \mathbf{X}\mathbf{X}^T \boldsymbol{\Pi})$$

$$= \boldsymbol{\Lambda}$$

and pre- and postmultiplying by $\boldsymbol{\Lambda}^{-1/2}$ yields

$$E(\boldsymbol{\Lambda}^{-1/2} \boldsymbol{\Pi}^T \mathbf{X}\mathbf{X}^T \boldsymbol{\Pi}\boldsymbol{\Lambda}^{-1/2}) = \mathbf{I}$$

or

$$\boldsymbol{\zeta}^* \mathbf{X}\mathbf{X}^T \boldsymbol{\zeta}^{*T} = \mathbf{I} \qquad (3.18)$$

where

$$\boldsymbol{\zeta}^* = \boldsymbol{\Lambda}^{1/2} \boldsymbol{\Pi}^T \mathbf{X} \qquad (3.19)$$

Equation (3.19) implies that

$$\mathbf{X} = (\boldsymbol{\Pi}^T \boldsymbol{\Lambda}^{-1/2})^{-1} \boldsymbol{\zeta}^*$$

$$= \boldsymbol{\Pi}\boldsymbol{\Lambda}^{1/2} \boldsymbol{\zeta}^* \qquad (3.20)$$

and to convert covariance loadings into correlation loadings we have

$$\mathbf{X}^* = \boldsymbol{\Delta}^{-1} \mathbf{X} = \boldsymbol{\Delta}^{-1} \boldsymbol{\Pi}\boldsymbol{\Lambda}^{1/2} \boldsymbol{\zeta}^* \qquad (3.21)$$

where $\boldsymbol{\Delta}^2$ is the diagonal variance matrix of $\mathbf{X}$ and $\boldsymbol{\alpha} = \boldsymbol{\Delta}^{-1} \boldsymbol{\Pi}\boldsymbol{\Lambda}^{1/2}$. When $E(\mathbf{X}\mathbf{X}^T) = \mathbf{P}$, the correlation matrix, the loadings are given by $\boldsymbol{\alpha} = \boldsymbol{\Pi}\boldsymbol{\Lambda}^{1/2}$

since $\Delta = I$. In element form,

$$
\begin{aligned}
X_1^* &= \alpha_{11}\zeta_1^* + \alpha_{12}\zeta_2^* + \cdots + \alpha_{1p}\zeta_p^* \\
X_2^* &= \alpha_{21}\zeta_1^* + \alpha_{22}\zeta_2^* + \cdots + \alpha_{2p}\zeta_p^* \\
\hline
X_p^* &= \alpha_{p1}\zeta_1^* + \alpha_{p2}\zeta_2^* + \cdots + \alpha_{pp}\zeta_p^*
\end{aligned}
\tag{3.22}
$$

where both variables and PCs are standardized to unit length. The standardized loadings α obtained from the correlation matrix however are not equal to the standardized loadings obtained from the covariance matrix. The point is pursued further in Section 3.4. The standardized loadings are usually displayed in a table such as

	ζ_1^*	ζ_2^*	$\cdots$	ζ_p^*
X_1^*	α_{11}	α_{12}	$\cdots$	α_{1p}
X_2^*	α_{21}	α_{22}	$\cdots$	α_{2p}
$\cdot$	$\cdot$	$\cdot$		$\cdot$
$\cdot$	$\cdot$	$\cdot$		$\cdot$
$\cdot$	$\cdot$	$\cdot$		$\cdot$
X_p^*	α_{p1}	α_{p2}	$\cdots$	α_{pp}

which can be of aid when deciding which components are to be retained (deleted).

The following two theorems develop further the properties of a PCA.

THEOREM 3.8. Let Σ be a $(p \times p)$ covariance matrix with latent roots Λ and latent vectors Π. Then Σ^{-1} possesses latent roots Λ^{-1} and latent vectors Π.

PROOF. From Eq. (3.11) we have

$$\Sigma^{-1} = (\Pi\Lambda\Pi^T)^{-1}$$

$$= (\Pi^T)^{-1}\Lambda^{-1}\Pi^{-1}$$

$$= \Pi\Lambda^{-1}\Pi^T$$

$$= \frac{1}{\lambda_1}\Pi_1\Pi_1^T + \frac{1}{\lambda_2}\Pi_2\Pi_2^T + \cdots + \frac{1}{\lambda_p}\Pi_p\Pi_p^T \tag{3.23}$$

for $\lambda_i > 0$ $(i = 1, 2, \ldots, p)$.

Thus Σ is nonsingular if and only if $\lambda_i > 0$ for all i. When one (or more) of the latent roots are small, elements of $\mathbf{\Sigma}^{-1}$ become large and $\mathbf{\Sigma}$ is said to be ill-conditioned.

An important outcome of Theorem 3.3 is that the PC model possesses a number of important optimality properties, which make it an attractive model in many situations where optimization is of importance.

THEOREM 3.9.   Let $\mathbf{X} = (X_1, X_2, \ldots, X_p)^{\mathrm{T}}$ be a set of random variables as in Theorem 3.5. Then the PC model possesses the following optimality properties:

> (i) The PCs maximize the total trace (univariate variance) of the X_i.
>
> (ii) The PCs maximize the generalized variance of the X_i.
>
> (iii) The PCs minimize total entropy, that is, they maximize the information content of the variables.
>
> (iv) The PCs maximize Euclidian distance.
>
> (v) The PCs minimize the mean squared error criterion.

PROOF

> (i) From Theorem 3.6 we have
>
> $$\mathrm{tr}(\mathbf{\Sigma}) = \mathrm{tr}(\mathbf{\Pi}\mathbf{\Lambda}\mathbf{\Pi}^{\mathrm{T}})$$
> $$= \mathrm{tr}(\mathbf{\Lambda}\mathbf{\Pi}\mathbf{\Pi}^{\mathrm{T}})$$
> $$= \mathrm{tr}(\mathbf{\Lambda})$$
> $$= \lambda_1 + \lambda_2 + \cdots + \lambda_p \qquad (3.24)$$
>
> where $\lambda_1 \geq \lambda_2 \geq \cdots \geq \lambda_p$ are the variances of the PCs. Since $\mathrm{tr}(\mathbf{\Sigma})$ is the total univariate variance, it follows that at each stage the first PCs maximize the overall univariate variance of $\mathbf{X}$. Conversely, trace is minimized by the last set of PC's.
>
> (ii) Since the generalized variance is equivalent to (squared) volume in p-dimensional space we have
>
> $$|\mathbf{\Sigma}| = |\mathbf{\Pi}\mathbf{\Lambda}\mathbf{\Pi}^{\mathrm{T}}|$$
> $$= |\mathbf{\Pi}|\,|\mathbf{\Lambda}|\,|\mathbf{\Pi}^{\mathrm{T}}|$$
> $$= |\mathbf{\Lambda}|\,|\mathbf{\Pi}|\,|\mathbf{\Pi}^{\mathrm{T}}|$$
> $$= |\mathbf{\Lambda}|$$
> $$= \lambda_1 \lambda_2 \ldots \lambda_p \qquad (3.25)$$
>
> so that highest order PCs maximize the generalized variance or the determinant of the covariance matrix.
>
> (iii) From Section 1.6 the total information content of the variables

can be defined as

$$I = \sum_{i=1}^{p} \lambda_i \ln \lambda_i \tag{3.26}$$

where $\lambda_1 \geq \lambda_2 \geq \cdots \geq \lambda_p$. Since the logarithmic function is monotonic (order preserving), it follows that Eq. (3.26) is maximized for each PC.

(iv) Let d_{kh} denote the Euclidian distance between X_k and X_h. Using Theorem 3.5 we have

$$d_{kh}^2 = E(X_k - X_h)^2$$

$$= E(X_k^2) + E(X_h^2) - 2E(X_k X_h)$$

$$= \sum_{i=1}^{p} \pi_{ki}^2 \lambda_i + \sum_{i=1}^{p} \pi_{hi}^2 \lambda_i - 2 \sum_{i=1}^{p} \pi_{ki} \pi_{hi} \lambda_i$$

$$= \sum_{i=1}^{p} (\pi_{ki} - \pi_{hi})^2 \lambda_i \tag{3.27}$$

for $k \neq h$. Maximum distances between random variables are therefore associated with the largest latent roots.

(v) The proof is a consequence of the fifth part of Theorem 3.17 (see also Fukunaga, 1990).

It follows from part i of the theorem that for any other set of orthonormal vectors $\mathbf{Q}_1, \mathbf{Q}_2, \ldots, \mathbf{Q}_r$ we have

$$\sum_{i=1}^{r} \mathbf{Q}_i^T \mathbf{\Sigma} \mathbf{Q}_i \leq \sum_{i=1}^{r} \mathbf{\Pi}_i^T \mathbf{\Sigma} \mathbf{\Pi}_i = \lambda_1 + \lambda_2 + \cdots + \lambda_r \tag{3.27a}$$

for $1 \leq r \leq p$. Further optimality properties are considered in Section 5.8.1 (See also Jolliffe, 1986; Fukunaga, 1990; Obenchain, 1972; Okamoto, 1969).

Theorem 3.9 demonstrates that the PCs provide an optimal orthogonal coordinate system with respect to the five criteria considered. When the PCs are standardized to unit variance we also have, for expression 3.27,

$$d_{kh}^2 = \sum_{i=1}^{p} (\alpha_{ki} - \alpha_{hi})^2 \tag{3.28}$$

where α_{ki} and α_{hi} are defined by Eq. (3.13). For the correlation matrix we have $d_{kh}^2 = 2(1 - \rho_{kh})$ where ρ_{kh} is the correlation between X_k and X_h. Also note that entropy (Eq. 3.26) is maximized when $\lambda_1 = \lambda_2 = \cdots = \lambda_p$, that is, in the uniform or isotropic case where a PCA does not normally make sense (see Section 3.3). On the other extreme the minimum entropy occurs when random variables are multivariate normal, as illustrated by the following example.

Example 3.1. For the multivariate normal distribution $f(\mathbf{X})$ the information content is maximized by maximizing

$$\mathbf{I} = -E[\ln f(\mathbf{X})]$$

$$= \int (2\pi)^{-p/2}|\mathbf{\Sigma}| \exp[-\tfrac{1}{2}\mathbf{X}^T\mathbf{\Sigma}^{-1}\mathbf{X}]\{-\tfrac{1}{2}\mathbf{X}^T\mathbf{\Sigma}^{-1}\mathbf{X} - \tfrac{1}{2}\ln|\mathbf{\Sigma}|$$

$$- p/2\ln(2\pi)\}\, d\mathbf{X}$$

$$= \frac{p}{2} + \tfrac{1}{2}\ln|\mathbf{\Sigma}| + \frac{p}{2}\ln(2\pi)$$

$$= \tfrac{1}{2}\sum_{i=1}^{p}[1 + \ln \lambda_i + \ln(2\pi)]$$

which, as was seen above, is maximized by high-order latent roots since I is simply a function of $|\mathbf{\Sigma}|$. It is also possible to define the so-called cross entropy, as well as the conditional entropy of a random variable, and to express the latter in terms of the latent roots of the covariance matrix (see Parzen, 1983).   □

The main outcome of Theorems 3.3–3.9 is to establish that the principal objective of a PCA is to retain a smaller number $r < p$ of PCs, which reproduce a sufficiently high portion of the selected criterion, but in a lower dimensional space. Alternatively, optimality can be achieved by selecting the smallest latent roots, for example when minimizing the M.S. error criterion. Normally in practice the criterion chosen is that of univariate variance, that is, $\mathrm{tr}(\mathbf{\Sigma})$, although this owes more to tradition than to any intrinsically superior property of the function. The trace of a covariance matrix however does provide a general benchmark, since PCs that maximize Eq. (3.24) are likely also to be in the optimal region of the remaining criteria. The PC model does not require distributional assumptions, but when $\mathbf{X}$ is $N(\mathbf{0}, \mathbf{\Sigma})$, the model is identical to rotating a p-dimensional ellipse (Section 2.7). Here correlated normal variates $\mathbf{X} = (X_1, X_2, \ldots, X_p)^T$ are transformed to an uncorrelated set of normal variates $\boldsymbol{\zeta} = (\zeta_1, \zeta_2, \ldots, \zeta_p)^T$ such that $\boldsymbol{\zeta}$ is $N(\mathbf{0}, \mathbf{\Lambda})$, where $\mathbf{\Lambda}$ is diagonal. Consequently normal PCA can be viewed as a statistical procedure for converting a multivariate normal distribution into a set of independent univariate normal distributions, since here zero correlation implies distributional independence (Section 2.8). The advantage of the transformation, particularly when p is large, lies in the relative simplicity of dealing with optimal univariate variables as opposed to an interdependent multivariate set. As shown in Chapter 4, multivariate normality is also essential for parametric statistical significance testing using the chi-squared and other sampling distributions.

Example 3.2. The starting point for a PCA of an infinite population is a dispersion matrix of the variates, usually Σ or $\mathbf{P}$. Consider four random variables X_1, X_2, X_3, and X_4 with covariance matrix

$$\Sigma = \begin{bmatrix} 471.51 & 324.71 & 73.24 & 4.35 \\ 324.71 & 224.84 & 50.72 & 2.81 \\ 73.24 & 50.72 & 11.99 & 1.23 \\ 4.35 & 2.81 & 1.23 & .98 \end{bmatrix}$$

which can be converted to the correlation matrix

$$\mathbf{P} = \begin{bmatrix} 10000 & .9973 & .9741 & .2024 \\ .9973 & 1.0000 & .9768 & .1893 \\ .9741 & .9768 & 1.0000 & .3588 \\ .2024 & .1893 & .3588 & 1.0000 \end{bmatrix}$$

The latent roots and vectors of Σ are given by

$$\Lambda = \begin{bmatrix} 706.97940 & & & 0 \\ & 1.34915 & & \\ & & .89430 & \\ 0 & & & .09715 \end{bmatrix}$$

$$\Pi = \begin{bmatrix} .8164 & -.0248 & .5707 & .0844 \\ .5633 & -.1059 & -.7662 & -.2917 \\ .1272 & .5677 & -.2747 & .7652 \\ .0075 & .8160 & .1081 & -.5676 \end{bmatrix}$$

and using Eq. (3.10) the random variables can be expressed as

$$X_1 = .8164\zeta_1 - .0248\zeta_2 + .5707\zeta_3 + .0844\zeta_4$$

$$X_2 = .5633\zeta_1 - .1059\zeta_2 - .7662\zeta_3 - .2917\zeta_4$$

$$X_3 = .1272\zeta_1 + .5677\zeta_2 - .2747\zeta_3 + .7652\zeta_4$$

$$X_4 = .0075\zeta_1 + .8160\zeta_2 + .1081\zeta_3 - .5676\zeta_4 \qquad \square$$

Standardizing the random variables, together with the principal components, to unit length using Eq. (3.16) then yields the expansion

$$X_1^* = .9997\zeta_1^* - .0013\zeta_2^* + .0249\zeta_3^* + .0266\zeta_4^*$$

$$X_2^* = .9989\zeta_1^* - .0082\zeta_2^* - .0483\zeta_3^* - .0061\zeta_4^*$$

$$X_3^* = .9767\zeta_1^* + .1904\zeta_2^* - .0750\zeta_3^* + .0688\zeta_4^*$$

$$X_4^* = .2051\zeta_1^* + .9575\zeta_2^* + .1033\zeta_3^* - .1787\zeta_4^*$$

Now consider the latent roots and vectors of the correlation matrix $\mathbf{P}$,

$$\Lambda = \begin{bmatrix} 3.0568 & & & \bigcirc \\ & .9283 & & \\ & & .0130 & \\ \bigcirc & & & .0018 \end{bmatrix}$$

and

$$\Pi = \begin{bmatrix} .5627 & .1712 & .5685 & .5750 \\ .5623 & .1849 & .1820 & -.7851 \\ .5696 & -.0008 & -.7906 & .2241 \\ .2065 & -.9677 & .1360 & -.0484 \end{bmatrix}$$

Since the random variables have been prestandardized, we only have to standardize the components. Multiplying the columns of Π by the square roots of corresponding latent roots (Eq. 3.15), we obtain the expansion

$$X_1^* = .9838\zeta_1^* + .1650\zeta_2^* + .0648\zeta_3^* + .0244\zeta_4^*$$

$$X_2^* = .9831\zeta_1^* + .1782\zeta_2^* + .0207\zeta_3^* - .0333\zeta_4^*$$

$$X_3^* = .9959\zeta_1^* - .0008\zeta_2^* - .0901\zeta_3^* + .0095\zeta_4^*$$

$$X_4^* = .3610\zeta_1^* - .9324\zeta_2^* + .0155\zeta_3^* - .0026\zeta_4^*$$

Although the covariance loadings are similar to the correlation loadings, they are clearly not the same. Even signs can be reversed—although α_{42} is positive as a covariance loading, it becomes highly negative as a correlation loadings. This illustrates the importance of deciding initially which dispersion matrix to use.

Since loadings are correlation coefficients, they indicate the relative influence of a component (variate) on the variates (components). Using the covariance matrix we see that X_1, X_2, and X_3 are very highly intercorrelated, their joint effect being picked up by the first PC. However, X_4 is largely independent of the first three variables and is mainly represented by the second component. The interpretation for the correlation matrix is the same except that the second component ζ_2^* is negatively correlated to X_4. Actually the two sets of results can be made to agree by noting that the covariance loading $\alpha_{42} = .9575$, together with its components ζ_2^*, can be multiplied by -1, which converts α_{42} to a negative quantity. Alternatively, the same operation can be applied to the correlation loadings. This illustrates the point that the latent vectors are unique up to sign changes only. The interpretation of ζ_2^* is then simply applied to $(-\zeta_2^*)$ with X_4^* retaining its sign. Both sets of loadings indicate that the variables can be well estimated within a two-dimensional vector space. For example, using the covariance matrix we have

$$X_1^* = .9997\zeta_1^*$$

$$X_2^* = .9989\zeta_1^*$$

$$X_3^* = .9767\zeta_1^*$$

$$X_4^* = .2015\zeta_1^* + .9575\zeta_2^*$$

where omitted components may be considered as residual terms which represent natural variation in the population. The four variables are plotted in Figure 3.2 in the plane defined by ζ_1^* and ζ_2^*. Clearly ζ_1^* can be identified with cluster A, whereas ζ_2^* represents cluster B, containing the single variable X_4. Given the high tendency for the first three variables to cluster, we may suspect that once residual variation is removed the variables measure the same underlying influence or "factor." Thus any two of these variables can be removed without excessive loss of information. A PCA thus provides a cross-sectional view of the pattern of intercorrelations within a set of random variables.

Finally, using latent roots it is possible to compute the percentage of variance accounted for by the retained components ζ_1^* and ζ_2^*. Using the

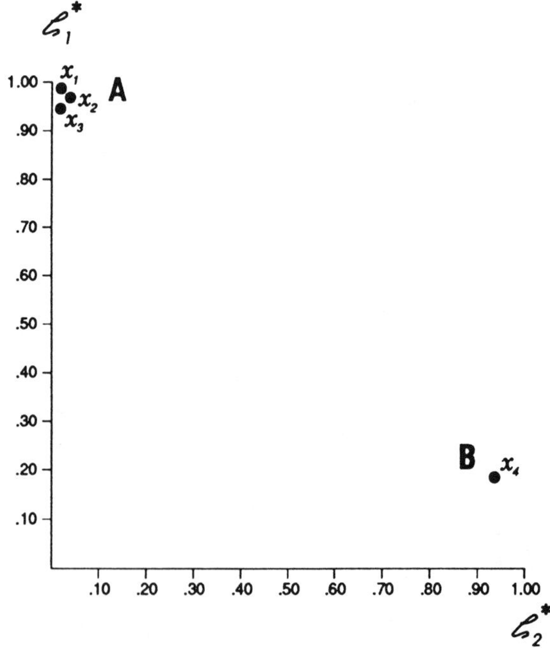

Figure 3.2 Random variables X_1, X_2, X_3, X_4 (Example 3.2) plotted in the reduced space defined by the first two principal components ζ_1^* and ζ_2^*.

covariance matrix we have

$$R^2 = \frac{\sum\limits_{i=1}^{2} \lambda_i}{\sum\limits_{i=1}^{4} \lambda_i} = \frac{706.9794 + 1.3492}{\mathrm{tr}(\Sigma)}$$

$$= \frac{708.33}{709.32}$$

$$= .9986$$

or 99.86% of the total (univariate) variance.

The main computational burden in carrying out a PCA lies in determining the latent roots Λ and latent vectors Π of Σ (or $\mathbf{P}$). Several well-known iterative numerical methods exist to achieve this, for which the reader is referred to Hotelling (1936a), Anderson (1958), and Hammarling (1970).

3.3. ISOTROPIC VARIATION

Principal components are, in a sense, mathematical artifacts created by the requirements of the correlational analysis, which usually cannot be observed directly unlike the original variables which represent the primary measurement scales. The PCs are linear combinations of the original variables, and as such cannot be used for explanatory purposes unless they too are identified in terms of real behavior. This is usually achieved by using the correlation loadings to tell us which components are most important, in the sense of being highly correlated with some (or all) of the variables. The exact method(s) of identification vary from application to application, and these are discussed more fully in the following chapters. One special case however is particularly easy to deal with and makes interpretation straightforward—when we have a single nonisotropic dimension or component explaining the behavior of all of the variates.

Consider p equally correlated random variables $\mathbf{X} = (X_1, X_2, \ldots, X_p)^{\mathrm{T}}$ with common correlation coefficient ρ and equal variance σ^2. Using Eq. (2.82) we see that the covariance matrix has a particularly straightforward structure (see also Eq. 4.14),

$$\Sigma = \begin{bmatrix} \sigma^2 & \sigma\rho^2 & \cdots & \sigma^2\rho \\ \sigma^2\rho & \sigma^2 & \cdots & \sigma^2\rho \\ \sigma^2\rho & \dfrac{\sigma^2\rho}{\sigma^2} & \cdot & \cdot \\ \vdots & \vdots & \ddots & \vdots \\ \sigma^2\rho & \sigma^2\rho & \cdots & \sigma^2 \end{bmatrix}$$

$$= \sigma^2 \begin{bmatrix} 1 & \rho & \rho & \rho & \cdots & \rho \\ \rho & 1 & \rho & \rho & \cdots & \rho \\ \rho & \rho & 1 & \rho & \cdots & \rho \\ \cdot & \cdot & \cdot & \cdot & \cdot & \cdot \\ & & & & & \\ \rho & \rho & \cdot & \cdot & \cdot & 1 \end{bmatrix}$$

$$= \sigma^2 \mathbf{P} \tag{3.29}$$

where $\mathrm{var}(X_i) = \sigma^2$, $(i = 1, 2, \ldots, p)$. Since the latent roots of Eq. (3.29) are proportional to those of ρ, there is no loss of generality in considering latent roots and vectors of the latter matrix. Besides having theoretical interest the equicorrelation matrix $\mathbf{P}$ finds application in a number of diverse areas. It is also particularly easy to work with using analytic means and provides an easy-to-understand illustration of a population PCA.

Consider a $(p \times p)$ matrix $\mathbf{A}$ with a typical element

$$a_{ij} = \begin{cases} x & i = j \\ y & i \neq j \end{cases} \tag{3.30}$$

By performing row/column operations, $\mathbf{A}$ can be reduced to diagonal form, and consequently its determinant can be expressed as

$$|\mathbf{A}| = |x + (p-1)y|(x-y)^{(p-1)} \tag{3.31}$$

(Exercise 3.6). From Eq. (3.31) the latent roots of $\mathbf{P}$ are solutions of

$$|\mathbf{P} - \lambda \mathbf{I}| = \begin{bmatrix} 1-\lambda & \rho & \cdots & \rho \\ \rho & 1-\lambda & \cdots & \rho \\ \vdots & \vdots & \ddots & \vdots \\ \rho & \rho & \cdots & 1-\lambda \end{bmatrix}$$

$$= [(1-\lambda) + (p-1)\rho](1-\lambda-\rho)^{p-1}$$

$$= 0$$

so that the largest root is given by

$$\lambda_1 = 1 + (p-1)\rho \tag{3.32}$$

and the remaining $(p-1)$ isotropic roots are solutions of $(1-\lambda-\rho)^{p-1} = 0$, that is,

$$\lambda_2 = \lambda_3 = \cdots = \lambda_p = 1 - \rho \tag{3.33}$$

Solving Eq. (3.32) then yields

$$\rho = \frac{\lambda_1 - 1}{p - 1}$$

λ_1 is also the largest root equal to the constant row (column) sum of $(\mathbf{P} - \lambda \mathbf{I})$. When $\rho = 1$, all last $p - 1$ roots are identically zero. It can be shown that

$$\mathbf{\Pi}_1 = \left(\frac{1}{\sqrt{p}}, \frac{1}{\sqrt{p}}, \cdots \frac{1}{\sqrt{p}}\right)^{\mathrm{T}} \tag{3.34}$$

the unit latent vector corresponding to λ_1. Since remaining $p - 1$ roots are equal, coordinate axes in the $p - 1$ dimensional subspace can be rotated to any arbitrary position without altering variance. Thus any set of $p - 1$ orthogonal vectors can be taken for these latent vectors. It is convenient to select the $p - 1$ latent vectors from the last $p - 1$ columns of the orthogonal Helmert matrix as

$$\mathbf{\Pi}_2 = \left(\frac{1}{\sqrt{1 \cdot 2}}, -\frac{1}{\sqrt{1 \cdot 2}}, 0, \ldots, 0\right)^{\mathrm{T}}$$

$$\mathbf{\Pi}_3 = \left(\frac{1}{\sqrt{2 \cdot 3}}, \frac{1}{\sqrt{2 \cdot 3}}, -\frac{2}{\sqrt{2 \cdot 3}}, 0, \ldots, 0\right)^{\mathrm{T}}$$

$$\mathbf{\Pi}_p = \left(\frac{1}{\sqrt{p(p-1)}}, \frac{1}{\sqrt{p(p-1)}}, \ldots, \frac{1}{\sqrt{p(p-1)}}, -\frac{p-1}{\sqrt{p(p-1)}}\right)^{\mathrm{T}} \tag{3.35}$$

Note that the (2×2) correlation matrix of Example 2.10 is necessarily both a special case of the isotropic model (Eq. 3.32) as well as a special case of a PCA of an arbitrary correlation matrix.

The first PC that corresponds to the nonisotropic latent root λ_1 is given by

$$\zeta_1 = \mathbf{\Pi}_1^{\mathrm{T}} \mathbf{X}$$

$$= \left(\frac{1}{\sqrt{p}}, \frac{1}{\sqrt{p}}, \ldots, \frac{1}{\sqrt{p}}\right) \begin{bmatrix} x_1 \\ x_2 \\ \vdots \\ x_p \end{bmatrix}$$

$$= \frac{1}{\sqrt{p}} \sum_{i=1}^{p} x_i \tag{3.36}$$

$$= \sqrt{p}\bar{x}$$

where $\bar{x}$ is the arithmetic mean of the p random variables. The variance of ζ_1

is given by

$$\operatorname{var}(\zeta_1) = \frac{1}{p}\operatorname{var}(x_1 + x_2 + \cdots + x_p)$$

$$= \frac{1}{p}\left[\operatorname{var}(x_1) + \operatorname{var}(x_2) + \cdots + \operatorname{var}(x_p) + (\text{sums of crossproducts})\right]$$

$$= \frac{1}{p}\left[\text{sum of all elements of matrix } \mathbf{P}\right]$$

$$= \frac{1}{p}\left\{[1 + (1-p)\rho] + [1 + (1-p)\rho] + \cdots + [1 + (1-p)\rho]\right\}$$

$$= 1 + (1-p)\rho$$

$$= \lambda_1$$

and correlation loadings between ζ_1 and the X_i are given by $\lambda_1^{1/2}\mathbf{\Pi}_1$ or

$$\alpha = \left[\frac{1}{p} + \frac{(p-1)}{p}\rho\right]^{1/2}$$

$$= \left[\rho + \frac{(1-\rho)}{p}\right]^{1/2} \tag{3.37}$$

The dominant component ζ_1 is thus equally correlated to all of the p random variables. The proportion of variance explained by ζ_1 is

$$\frac{\lambda_1}{p} = \frac{1 + (p-1)\rho}{p}$$

$$= \rho + \frac{(1-p)}{p}$$

$$= \alpha^2 \tag{3.38}$$

The structure of the latent roots for equicorrelated random variables is portrayed in Figure 3.3 for $\rho = .50$ and $p = 7$.

These results can be summarized by the following theorem.

THEOREM 3.10.   Let $\mathbf{\Sigma}$ be a covariance matrix such that $\lambda_1 > \lambda_2 = \lambda_3 = \cdots = \lambda_p = \delta$. Then

$$\mathbf{\Sigma} = \mathbf{\alpha}_1 \mathbf{\alpha}_1^{\mathrm{T}} + \delta\mathbf{I} \tag{3.39}$$

where $\mathbf{\alpha}_1$ is the vector of (equal) loadings for the first nonisotropic PC.

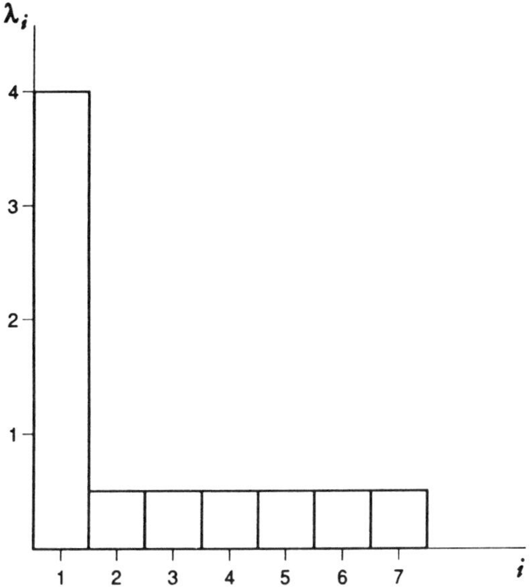

Figure 3.3   Latent roots for equicorrelated random variables where $\rho = .50$ and $p = 7$.

PROOF.   Using Eq. (3.11) we have

$$\Sigma = \lambda_1 \Pi_1 \Pi_1^T + \delta \Pi_2 \Pi_2^T + \cdots + \delta \Pi_p \Pi_p^T + \delta \Pi_1 \Pi_1^T - \delta \Pi_1 \Pi_1^T$$

$$= (\lambda_1 - \delta) \Pi_1 \Pi_1^T + \delta(\Pi_1 \Pi_1^T + \Pi_2 \Pi_2^T + \cdots + \Pi_p \Pi_p^T)$$

$$= (\lambda_1 - \delta) \Pi_1 \Pi_1^T + \delta I$$

Let $\lambda = \lambda_1 - \delta$. Then

$$\Sigma = \lambda \Pi_1 \Pi_1^T + \delta I$$

$$= \lambda^{1/2} \Pi_1 \lambda^{1/2} \Pi_1^T + \delta I$$

$$= \alpha_1 \alpha_1^T + \delta I$$

where $\alpha_1 = \lambda^{1/2} \Pi_1 = (\lambda_1 - \sigma)^{1/2} \Pi_1$ is the column vector of (equal) loadings for the first PC.

Note that Π_1 is weighted by $(\lambda_1 - \delta)^{1/2}$, not by $\lambda^{1/2} 1$. Also Eq. (3.39) implies that we can write

$$X = \alpha_1 \zeta_1^* + \epsilon \qquad\qquad (3.40)$$

where $E(\epsilon \epsilon^T) = \delta I$ such that $\zeta_1^* \epsilon^T = 0$. An important point to note is the residual error term ϵ is homoscedastic and uncorrelated. Equations (3.39) and (3.40) represent the simplest form of a factor analysis model (see Chapter 6).

Example 3.3. Equicorrelated random variables can occur in the context of biological allometry, which is the study of differences in shapes of biological organisms that can be associated with size. Consider the growth of an organism with physical attributes measured by five variables $Y_1, Y_2, \ldots, Y_5$. Also, assume growth occurs in such a way as to result in all body attributes of the organism to be equally correlated, say .80. Then the correlation matrix takes the form

$$\mathbf{P} = \begin{bmatrix} 1.00 & & & & \\ .80 & 1.00 & & & \\ .80 & .80 & 1.00 & & \\ .80 & .80 & .80 & 1.00 & \\ .80 & .80 & .80 & .80 & 1.00 \end{bmatrix}$$

Using Eqs. (3.32) and (3.33) the latent roots are $\lambda_1 = 1 + (5-1).80 = 4.2$, $\lambda_2 = \cdots = \lambda_5 = 1 - p = .20$. Also using Eq. (3.34) the latent vector corresponding to $\lambda_1 = 4.2$ is

$$\mathbf{\Pi}_1 = (1/\sqrt{5}, 1/\sqrt{5}, \ldots, 1/\sqrt{5})^\mathrm{T}$$

$$= (.4472, .4472, \ldots, .4472)^\mathrm{T}$$

the remaining latent vectors being given by Eq. (3.35). The correlation loadings for $\mathbf{\Pi}_1$ can also be computed from Eq. (3.37) as

$$\left[p + \frac{(1-p)}{p} \right]^{1/2} = \left[.80 + \frac{.20}{5} \right]^{1/2}$$

$$= .91655$$

and ζ_1^* represents the general size component of the organism, accounting for $\lambda_1/p = 4.2/5 = 84\%$ of the total trace of $\mathbf{P}$. The remaining 16% of the variance is isotropic and cannot be identified with systematic biological phenomenon such as shape—it is simply due to the unique properties of the five attributes or variables, much in the same way as for an ordinary least squares equation. For a theoretical review of allometry see Sprent (1972). The problem is considered further in Section 6.3.

Example 3.4. To see how the equicorrelation model works with real data we consider a well-known example from Jolicoeur and Mosimann (1960) which deals with the size and shape of the Painted Turtle, *chrysemys picta marginata*. A total of 84 specimens were collected in the St. Lawrence valley 35 miles southwest of Montreal, Quebec, of which 24 males and an equal number of females were selected and their carapace dimensions measured in three perpendicular directions. Note that for a sample PCA the starting point is a data matrix $\mathbf{Y}$, not a dispersion matrix as in the case of an infinite population (Examples 3.2, 3.3) The data are reproduced in Table 3.1. The mean vector and covariance matrix for the females is $\bar{\mathbf{Y}} = (136.00, 102.58, 51.96)^T$ and

$$\mathbf{S} = \begin{bmatrix} 451.39 & 271.17 & 168.70 \\ 271.17 & 171.73 & 103.29 \\ 168.70 & 103.29 & 66.65 \end{bmatrix}$$

A comparison of the female and male covariance matrices is carried out in Example 4.5. Decomposing $\mathbf{S}$ yields the following latent roots and latent

Table 3.1 Carapace Measurements (mm) of the Painted Turtle

Males			Females		
Length $(\mathbf{Y}_1)$	Width $(\mathbf{Y}_2)$	Height $(\mathbf{Y}_3)$	Length $(\mathbf{Y}_1)$	Width $(\mathbf{Y}_2)$	Height $(\mathbf{Y}_3)$
93	74	37	98	81	38
94	78	35	103	84	38
96	80	35	103	86	42
101	84	39	105	86	40
102	85	38	109	88	44
103	81	37	123	92	50
104	83	39	123	95	46
106	83	39	133	99	51
107	82	38	133	102	51
112	89	40	133	102	51
113	88	40	134	100	48
114	86	40	136	102	49
116	90	43	137	98	51
117	90	41	138	99	51
117	91	41	141	105	53
119	93	41	147	108	57
120	89	40	149	107	55
120	93	44	153	107	56
121	95	42	155	115	63
125	93	45	155	117	60
127	96	45	158	115	62
128	95	45	159	118	63
131	95	46	162	124	61
135	106	47	177	132	67

Source: Jolicoeur and Mosimann, 1960; reproduced with permission.

vectors:

$$\mathbf{L} = \begin{bmatrix} 680.40 & & 0 \\ & 6.50 & \\ 0 & & 2.86 \end{bmatrix}, \qquad \mathbf{P} = \begin{bmatrix} .8126 & .4955 & .3068 \\ -.5454 & .8321 & .1006 \\ -.2054 & -.2491 & .9465 \end{bmatrix}$$

where Latin letters indicate matrices of sample values. The covariance matrix $\mathbf{S}$ does not exhibit the homoscedastic, equicovariance structure of Eq. (3.29). However, converting the latent vectors into correlation loadings reveals the following. Let

$$\mathbf{S}^{-1/2} = \begin{bmatrix} .047067 & & 0 \\ & .07631 & \\ 0 & & .12249 \end{bmatrix}$$

be a diagonal matrix whose nonzero elements are reciprocal square roots of the variances. Then correlation loadings are given by

$$\mathbf{A}^{\mathrm{T}} = \mathbf{S}^{-1/2}\mathbf{P}^{\mathrm{T}}\mathbf{L}^{1/2}$$

$$= \begin{bmatrix} .04707 & 0 & 0 \\ 0 & .07631 & 0 \\ 0 & 0 & .12249 \end{bmatrix}\begin{bmatrix} .8126 & -.5454 & -.2054 \\ .4955 & .8321 & -.2491 \\ .3068 & .1006 & .9465 \end{bmatrix}$$

$$\times \begin{bmatrix} 26.0845 & 0 & 0 \\ 0 & 2.5495 & 0 \\ 0 & 0 & 1.6912 \end{bmatrix}$$

$$= \begin{bmatrix} .9976 & .0654 & .0164 \\ .9863 & .1619 & .0321 \\ .9803 & .0314 & .1961 \end{bmatrix}$$

The main results can be conveniently summarized in a table of the form (see Section 5, 3.2 and 3.4)

	$\mathbf{Z}_1^*$	$\mathbf{Z}_2^*$	$\mathbf{Z}_3^*$
Length: $\mathbf{X}_1$	.9976	.0654	.0164
Width: $\mathbf{X}_2$	.9863	.1619	.0321
Height: $\mathbf{X}_3$	.9803	.0314	.1961
Latent roots	680.40	6.50	2.86
Percent of trace	98.64	.94	.41

where the $\mathbf{Z}_i^*$ represent standardized PCs. Only a single size component is required to account for most of the variation, indicating a high degree of "redundancy" in the measurements, the last two representing random variation. The concepts have also been applied to measurements of the human body (Relethford et al., 1978).

3.4 PRINCIPAL COMPONENTS IN THE SAMPLE

3.4.1 Introduction

The preceding section deals with the theory of PCA as it applies to an infinite population of measurements, where the input consists of a $(p \times p)$ population dispersion matrix. In practice populations are rarely observed directly and instead a finite sample of n observations is taken for p random variables (Example 3.3). The starting point here is a $(n \times p)$ data matrix $\mathbf{X} = \mathbf{Y} - \bar{\mathbf{Y}}$, which can then be converted to any one of the four Grammian forms (Section 2.4).

A data matrix can be represented in a finite dimensional vector space. This may be done in two ways. First, we can define a vector space of dimension p, the so-called variable space where each sample point (row of $\mathbf{X}$) is represented as a point in a p-dimensional vector. This is of course conditional on the linear independence of the columns of $\mathbf{X}$ since otherwise the points will be contained in a subspace of smaller dimension. Thus, generally, the dimension of the variable space defined by the columns of $\mathbf{X}$ is equal to the rank $r \leq p \leq n$ of $\mathbf{X}$, that is, the number of linearly independent columns of $\mathbf{X}$. Alternatively, we can define an n-dimensional sample vector space which contains the p columns of $\mathbf{X}$. Assuming for the moment that $p \leq n$, the r-dimensional column space of $\mathbf{X}$ is a subspace of the n-dimensional sample space. The purpose of PCA in a random independent sample taken for a set of continuous (ratio, difference) random variables is to locate the subspace which, in a sense, represents the random variables. The initial objective for a sample is to discard those PCs that represent sampling variation, measurement error, and natural individual variation of the population. Using the optimality theorem (Theorem 3.9) these will be the last PCs. The end result is an overall enhancement of the data. If too many PCs are retained the enhancement will be insufficient, whereas if too many are discarded the result will be a distortion of the data (the topic of testing for residual PCs is pursued further in Chapter 4). This is accomplished by computing latent roots and vectors of a suitable sample Grammian matrix and then examining for isotropic variation (Section 3.3). Again either the covariance or correlation matrix can be used, depending on the unit(s) of measure and the extent of variation of the random variables. For the sake of convenience we shall work with a Grammian matrix of the general form $\mathbf{X}^{\mathrm{T}}\mathbf{X}$, the specific scaling of columns of $\mathbf{X}$ being evident within the context of discussion. The following theorem is fundamental to a PCA of a rectangular data matrix $\mathbf{X}$.

THEOREM 3.11. Let $\mathbf{X}$ be a $(n \times p)$ matrix. Then the vector space generated by the columns of $\mathbf{X}$ is identical to that generated by the columns of $\mathbf{X}\mathbf{X}^{\mathrm{T}}$.

PROOF. Let $\mathbf{V}$ be a $(\mathbf{n} \times 1)$ vector such that $\mathbf{V}^T\mathbf{X} = \mathbf{0}$. Hence $\mathbf{V}^T\mathbf{X}\mathbf{X}^T = \mathbf{0}$. Conversely, let $\mathbf{V}^T\mathbf{X}\mathbf{X}^T = \mathbf{0}$ implying that $\mathbf{V}^T\mathbf{X}\mathbf{X}^T\mathbf{V} = \mathbf{0}$ so that $\mathbf{V}^T\mathbf{X} = \mathbf{0}$. It follows that every vector which is orthogonal to $\mathbf{X}$ is also orthogonal to $\mathbf{X}\mathbf{X}^T$ and vice versa, so that columns of $\mathbf{X}$ and $\mathbf{X}\mathbf{X}^T$ generate the same vector space.

It can also be shown that columns of $\mathbf{X}$ generate the same vector space as columns (rows) of $\mathbf{X}^T\mathbf{X}$. A rectangular data matrix $\mathbf{X}$ can therefore be analyzed by using the spectral decomposition of either $\mathbf{X}^T\mathbf{X}$ or $\mathbf{X}\mathbf{X}^T$ which are both square matrices (Theorem 3.17). An analysis using $\mathbf{X}^T\mathbf{X}$ is at times known as R-mode analysis, while that based on $\mathbf{X}\mathbf{X}^T$ as Q-mode analysis. As will be seen below, the two types of PC analyses are duals of each other. However, an $\mathbf{R}$ and a $\mathbf{Q}$ analysis may have different substantive objectives in mind. When the purpose is to analyze the random variables, normally $p < n$, and an R-mode decomposition is more handy, but the situation can be reversed when interelationships between the sample points are of interest (Section 5.4). Since rows of $\mathbf{X}^T\mathbf{X}$ generate the same vector space as rows of $\mathbf{X}$, we have $\rho(\mathbf{X}^T\mathbf{X}) = \rho(\mathbf{X}) = r \leq p$. Similarly, columns of both $\mathbf{X}\mathbf{X}^T$ and $\mathbf{X}$ generate the same space so that $\rho(\mathbf{X}^T\mathbf{X}) = \rho(\mathbf{X}\mathbf{X}^T) = \rho(\mathbf{X}) = r \leq p$.

3.4.2 The General Model

Consider n observations on p random variables represented by the vectors $\mathbf{X}_1, \mathbf{X}_2, \ldots, \mathbf{X}_p$, where $\bar{\mathbf{X}}_1 = \bar{\mathbf{X}}_2 = \cdots = \bar{\mathbf{X}}_p = 0$ and $\mathbf{S} = (\frac{1}{n-1})\mathbf{X}^T\mathbf{X}$. Since degrees of freedom represent a scalar constant they are usually omitted, and the analysis is based on $\mathbf{X}^T\mathbf{X}$ rather than on the sample covariance matrix $\mathbf{S}$. Let $\mathbf{P}$ denote a $(p \times p)$ matrix of unknown coefficients such that the quadratic form $\mathbf{P}^T\mathbf{X}^T\mathbf{X}\mathbf{P}$ is maximized subject to the constraint $\mathbf{P}^T\mathbf{P} = \mathbf{I}$. This is equivalent to maximizing the Lagrangean expression

$$\Phi = \mathbf{P}^T\mathbf{X}^T\mathbf{X}\mathbf{P} - l\mathbf{I}(\mathbf{P}^T\mathbf{P} - \mathbf{I}) \tag{3.41}$$

where $l\mathbf{I}$ is a diagonal matrix of Lagrange multipliers. Differentiating with respect to $\mathbf{P}$ and setting to zero we have

$$\frac{\partial \Phi}{\partial \mathbf{P}} = 2\mathbf{X}^T\mathbf{X} - 2l\mathbf{P} = \mathbf{0}$$

or

$$(\mathbf{X}^T\mathbf{X} - l\mathbf{I})\mathbf{P} = \mathbf{0} \tag{3.42}$$

The normal equations (Eq. 3.42) yield estimates $l\mathbf{I}$ and $\mathbf{P}$ of population latent roots and vectors $\lambda\mathbf{I}$ and Π of Section 3.2. When $\mathbf{X}^T\mathbf{X}$ is nonsingular, all latent roots are strictly positive. However in a sample, unlike a population, strict equality of the sample latent roots is precluded, as indicated in the following theorem.

THEOREM 3.12. Let $\mathbf{X}$ be a $(n \times p)$ data matrix, such that the joint distribution of the p variates is absolutely continuous with respect to a np-dimensional Lebesgue measure. Let $\mathbf{X}^{\mathrm{T}}\mathbf{A}\mathbf{X} = \mathbf{S}$ where $\mathbf{A}$ is a real $(n \times n)$ symmetric matrix of rank r. Let $\rho(\cdot)$ denote the rank of a matrix. Then

(i) $\rho(\mathbf{S}) = \min(p, r)$

(ii) Nonzero latent roots of $\mathbf{S}$ are distinct with probability one.

For a proof of the theorem see Okamoto (1973). Theorem 3.12 implies that when $|\mathbf{\Sigma}| \neq 0$ the latent roots $\mathbf{X}^{\mathrm{T}}\mathbf{X}$ can be ordered in a strictly decreasing sequence $l_1 > l_2 > \cdots > l_p$, which are solutions of the characteristic polynomial

$$|\mathbf{X}^{\mathrm{T}}\mathbf{X} - l\mathbf{I}| = 0 \tag{3.43}$$

Once the latent roots and vectors are known, Eq. (3.42) can be written in matrix form as

$$\mathbf{P}^{\mathrm{T}}\mathbf{X}^{\mathrm{T}}\mathbf{X}\mathbf{P} = \mathbf{L} \tag{3.44}$$

where

$$\mathbf{L} = \begin{bmatrix} l_1 & & & \bigcirc \\ & l_2 & & \\ & & \ddots & \\ \bigcirc & & & l_p \end{bmatrix} \tag{3.45}$$

and columns of $\mathbf{P}$ are latent vectors of $\mathbf{X}^{\mathrm{T}}\mathbf{X}$ such that $\mathbf{P}^{\mathrm{T}}\mathbf{P} = \mathbf{I}$. Since latent vectors correspond to distinct sample latent roots they are unique up to sign changes, and in a sample we need not concern ourselves with the problem of equal latent, roots, at least theoretically speaking.

Sample PCs are obtained in a similar manner to population values. Since $\mathbf{X}$ represents a $(n \times p)$ data matrix (rather than a vector of random variables), the sample PCs are $(n \times 1)$ vectors, say $\mathbf{Z}_1, \mathbf{Z}_2, \ldots, \mathbf{Z}_p$. Once latent roots and vectors are known (Eq. 3.44) can be rewritten as $\mathbf{Z}^{\mathrm{T}}\mathbf{Z} = \mathbf{L}$, where

$$\mathbf{Z} = \mathbf{X}\mathbf{P} \tag{3.46}$$

is the $(n \times p)$ matrix whose columns consist of unstandardized PCs. Standardizing to unit length, we have

$$\mathbf{Z}^* = \mathbf{Z}\mathbf{L}^{-1/2} = \mathbf{X}\mathbf{P}\mathbf{L}^{-1/2} \tag{3.47}$$

where it is easy to show that $\mathbf{Z}^{*\mathrm{T}}\mathbf{Z}^* = \mathbf{I}$.

In what follows we assume that the PCs are standardized so that $\mathbf{Z}$ denotes a matrix of p standardized PCs (unless stated otherwise). For alternative scaling criteria see Section 3.8. The data matrix can then be expressed as

$$\mathbf{X} = \mathbf{Z}(\mathbf{PL}^{-1/2})^{-1}$$

$$= \mathbf{Z}\mathbf{L}^{1/2}\mathbf{P}^{\mathrm{T}}$$

$$= \mathbf{Z}\mathbf{A}^{\mathrm{T}} \tag{3.48}$$

where $\mathbf{A}$ is a $(p \times p)$ matrix of loading coefficients. We have

$$\begin{bmatrix} x_{11} & x_{12} & \cdots & x_{1p} \\ x_{21} & x_{22} & \cdots & x_{2p} \\ \vdots & \vdots & & \vdots \\ x_{n1} & x_{n2} & \cdots & x_{np} \end{bmatrix} = \begin{bmatrix} z_{11} & z_{12} & \cdots & z_{1p} \\ z_{21} & z_{22} & \cdots & z_{2p} \\ \vdots & \vdots & & \vdots \\ z_{n1} & z_{n2} & \cdots & z_{np} \end{bmatrix} \begin{bmatrix} a_{11} & a_{21} & \cdots & a_{p1} \\ a_{12} & a_{22} & \cdots & a_{p2} \\ \vdots & \vdots & & \vdots \\ a_{1p} & a_{2p} & \cdots & a_{pp} \end{bmatrix}$$

or

$$x_{ij} = \sum_{h=1}^{P} z_{ih} a_{jh}$$

$$= \sum_{h=1}^{P} z_{ih} l^{1/2} p_{jh} \tag{3.49}$$

Equation (3.48) implies that the decomposition of the p columns of $\mathbf{X}$ is

$$\begin{aligned} \mathbf{X}_1 &= a_{11}\mathbf{Z}_1 + a_{12}\mathbf{Z}_2 + \cdots + a_{1p}\mathbf{Z}_p \\ \mathbf{X}_2 &= a_{21}\mathbf{Z}_1 + a_{22}\mathbf{Z}_2 + \cdots + a_{2p}\mathbf{Z}_p \\ \hline \mathbf{X}_p &= a_{p1}\mathbf{Z}_1 + a_{p2}\mathbf{Z}_2 + \cdots + a_{pp}\mathbf{Z}_p \end{aligned} \tag{3.50}$$

where $-1 \le a_{ij} \le 1$. Note that notationally the matrix of loading coefficients is defined as $\mathbf{A}^{\mathrm{T}}$ rather than $\mathbf{A}$, since we initially assume that interest lies in "explaining" the correlational (covariance) structure of the observed variates in terms of PCs as in Eq. (3.50). The coefficients

$$\mathbf{A} = \mathbf{PL}^{1/2} \tag{3.51}$$

have a somewhat different interpretation, and are considered further in Section 3.8.2. Note also that $\mathbf{A}^{\mathrm{T}} \ne \mathbf{A}^{-1}$. When both variables and components are standardized to unit length the coefficients a_{ij} are correlations between the variables and PCs (correlation loadings), and the elements of $\mathbf{Z}$ are referred to as component scores. For unstandardized variates and/or PCs, a_{ij} represent covariances (covariance loadings). Equation (3.50) forms the basis of a PCA of a set of random variables. Since components account for a decreasing percentage of the observed variables the idea is to retain a smaller number r of PCs which account for a sufficiently high percentage of the total variance/covariance structure and which can be interpreted meaningfully within a given framework. The remaining $p - r$ components can then be grouped into residual error terms, one for each random variable. Generally speaking the first r components describe (in a decreasing order) the common effects or "redundancies" (if they exist) whereas the $p - r$ residual components represent specific or unique residual effects. Also, when variables form distinct clusters the PCs can be viewed as estimates of these clusters. Alternatively, the last $(p - r)$ PCs may be of main interest in which case the linear combinations account for an increasing percentage of variance.

Example 3.5. To set ideas, consider the sample correlation matrix

$$\mathbf{R} = \begin{bmatrix} 1.00 & -.3649 & .5883 \\ -.3649 & 1.00 & -.0384 \\ .5883 & -.0384 & 1.00 \end{bmatrix}$$

with latent roots and latent vectors given by

$$\mathbf{L} = \begin{bmatrix} 1.7099 & & 0 \\ & .9656 & \\ 0 & & .3245 \end{bmatrix} \quad \mathbf{P} = \begin{bmatrix} .6981 & .0248 & .7156 \\ -.3912 & .8502 & .3522 \\ .6000 & .5258 & -.6032 \end{bmatrix}$$

respectively where the observed variates are assumed to be standardized to unit length for the sake of simplicity (for unstandardized variates see Section 3.4.3). Unstandardized PCs (the "scores") are then given by

$$\mathbf{Z}_1 = \mathbf{XP}_1 = (\mathbf{X}_1, \mathbf{X}_2, \mathbf{X}_3) \begin{bmatrix} .6981 \\ -.3912 \\ .6000 \end{bmatrix} = .6981\mathbf{X}_1 - .3912\mathbf{X}_2 + .6000\mathbf{X}_3$$

$$\mathbf{Z}_2 = \mathbf{XP}_2 = (\mathbf{X}_1, \mathbf{X}_2, \mathbf{X}_3) \begin{bmatrix} .0248 \\ .8502 \\ .5258 \end{bmatrix} = .0248\mathbf{X}_1 + .8502\mathbf{X}_2 + .5258\mathbf{X}_3$$

$$\mathbf{Z}_3 = \mathbf{XP}_3 = (\mathbf{X}_1, \mathbf{X}_2, \mathbf{X}_3) \begin{bmatrix} .7156 \\ .3522 \\ -.6032 \end{bmatrix} = .7156\mathbf{X}_1 + .3522\mathbf{X}_2 - .6032\mathbf{X}_3$$

where the $\mathbf{X}_i$, $\mathbf{Z}_j$ $(i, j = 1, 2, 3)$ are $(n \times 1)$ sample vectors. Using Eq. (3.48) we can then express the variates in terms of standardized PCs as

$$(\mathbf{X}_1, \mathbf{X}_2, \mathbf{X}_3) = (\mathbf{Z}_1, \mathbf{Z}_2, \mathbf{Z}_3)$$

$$\times \begin{bmatrix} 1.3076 & & 0 \\ & .9827 & \\ 0 & & .5696 \end{bmatrix} \begin{bmatrix} .6981 & -.3912 & .6000 \\ .0248 & .8502 & .5258 \\ .7156 & .3522 & -.6032 \end{bmatrix}$$

or

$$\mathbf{X}_1 = (.6981)(1.3076)\mathbf{Z}_1 + (.0248)(.9827)\mathbf{Z}_2 + (.7156)(.5696)\mathbf{Z}_3$$

$$= .9128\mathbf{Z}_1 + .0243\mathbf{Z}_2 + .4076\mathbf{Z}_3$$

$$\mathbf{X}_2 = (-.3912)(1.3076)\mathbf{Z}_1 + (.8502)(.9827)\mathbf{Z}_2 + (.3522)(.5696)\mathbf{Z}_3$$

$$= -.5116\mathbf{Z}_1 + .8355\mathbf{Z}_2 + .2006\mathbf{Z}_3$$

$$\mathbf{X}_3 = (.6000)(1.3076)\mathbf{Z}_1 + (.5258)(.9827)\mathbf{Z}_2 - (.6032)(.5696)\mathbf{Z}_3$$

$$= .7842\mathbf{Z}_1 + .5167\mathbf{Z}_2 - .3436\mathbf{Z}_3$$

that is,

$$\mathbf{A}^{\mathrm{T}} = \begin{array}{c} \\ \mathbf{X}_1 \\ \mathbf{X}_2 \\ \mathbf{X}_3 \end{array} \begin{array}{ccc} \mathbf{Z}_1 & \mathbf{Z}_2 & \mathbf{Z}_3 \\ \left[\begin{array}{ccc} .9128 & .0243 & .4076 \\ -.5116 & .8355 & .2006 \\ .7842 & .5167 & -.3436 \end{array} \right] \end{array}$$

The statistical properties of sample PCs are summarized in the following theorems.

THEOREM 3.13. Let $\mathbf{X}$ represent a $(n \times p)$ data matrix such that $\bar{\mathbf{X}}_1 = \bar{\mathbf{X}}_2 = \cdots = \bar{\mathbf{X}}_p = 0$ and let $\mathbf{P}$ and $\mathbf{L}$ be the latent vectors and latent roots of $\mathbf{X}^{\mathrm{T}}\mathbf{X}$ respectively. Then

(i) The columns of $\mathbf{Z}$ have zero means, that is, $\bar{\mathbf{Z}}_1 = \bar{\mathbf{Z}}_2 = \cdots = \bar{\mathbf{Z}}_p = 0$.

(ii) The i, jth element $\mathbf{X}_i^{\mathrm{T}}\mathbf{X}_j$ of $\mathbf{X}^{\mathrm{T}}\mathbf{X}$ can be expressed as

$$\mathbf{X}_i^{\mathrm{T}}\mathbf{X}_j = a_{i1}a_{j1} + a_{i2}a_{j2} + \cdots + a_{ip}a_{jp} \tag{3.52}$$

that is, $\mathbf{X}^{\mathrm{T}}\mathbf{X} = \mathbf{A}\mathbf{A}^{\mathrm{T}}$.

(iii) $\mathbf{A}^{\mathrm{T}}\mathbf{A} = \mathbf{L}$ $\tag{3.53}$

PROOF

(i) From Eq. (3.46) we have

$$\mathbf{Z}_j = p_{1j}\mathbf{X}_1 + p_{2j}\mathbf{X}_2 + \cdots + p_{pj}\mathbf{X}_p$$

and summing over vector elements yields

$$\sum_{i=1}^{n} z_{ij} = p_{1j} \sum_{i=1}^{n} x_{i1} + p_{2j} \sum_{i=1}^{n} x_{i2} + \cdots + p_{pj} \sum_{i=1}^{n} x_{ip}$$

$$= 0$$

When the observed variables are not expressed as differences about means, the sample means of the principal components are linear combinations of the variate means.

(ii) Equation (3.44) implies that $\mathbf{X}^T\mathbf{X} = \mathbf{PLP}^T$, and for the (i, j)th element of $\mathbf{X}^T\mathbf{X}$ we have

$$\mathbf{X}_i^T \mathbf{X}_j = \mathbf{P}_i \mathbf{L} \mathbf{P}_j^T$$

$$= l_1 p_{i1} p_{j1} + l_2 p_{i2} p_{j2} + \cdots + l_p p_{ip} p_{jp}$$

$$= (l_1^{1/2} p_{i1})(l_1^{1/2} p_{j1}) + (l_2^{1/2} p_{i2})(l_2^{1/2} p_{j2}) + \cdots + (l_p^{1/2} p_{ip})(l_p^{1/2} p_{jp})$$

$$= a_{i1} a_{j1} + a_{i2} a_{j2} + \cdots + a_{ip} a_{jp}$$

$$= (\mathbf{AA}^T)_{ij} \tag{3.54}$$

the (i, j)th element of $\mathbf{AA}^T$ so that $\mathbf{X}^T\mathbf{X} = \mathbf{AA}^T$.

(iii) From Eq. (3.51) we have

$$\mathbf{A}^T\mathbf{A} = \mathbf{L}^{1/2}\mathbf{P}^T\mathbf{PL}^{1/2}$$

$$= \mathbf{L}$$

which establishes Eq. (3.53). Also

$$\mathbf{A}_i^T \mathbf{A}_j = \begin{cases} l_i & i = j \\ 0 & i \neq j \end{cases} \tag{3.55}$$

THEOREM 3.14.   Let $\mathbf{X}$ be a $(n \times p)$ data matrix such that $\bar{\mathbf{X}} = 0$, and let $\mathbf{Z}$ represent the matrix of unstandardized principal components. Then the sample covariance loadings are given by

(i)

$$\mathrm{cov}(\mathbf{Z}_j, \mathbf{X}_i) = \left(\frac{1}{n-1}\right)\mathbf{LP}^T \qquad (i, j = 1, 2, \ldots, p) \tag{3.56}$$

When the variables are standardized then

(ii) The loadings $\mathbf{A}$ are correlation coefficients between the $\mathbf{Z}_j$ and $\mathbf{X}_i$ $(i, j = 1, 2, \ldots, p)$.

PROOF

(i) We have, in matrix form,

$$\text{cov}(\mathbf{Z}, \mathbf{X}) = \left(\frac{1}{n-1}\right) \mathbf{Z}^T \mathbf{X}$$

$$= \left(\frac{1}{n-1}\right) (\mathbf{XP})^T \mathbf{X}$$

$$= \left(\frac{1}{n-1}\right) \mathbf{P}^T \mathbf{X}^T \mathbf{X}$$

$$= \left(\frac{1}{n-1}\right) \mathbf{LP}^T$$

using Eqs. (3.51) and (3.49).

(ii) Let $\mathbf{S} = (1/n-1)\mathbf{X}^T\mathbf{X}$ be the sample covariance matrix. The correlation matrix between unstandardized variates $\mathbf{X}$ and unstandardized components $\mathbf{Z}$ is given by

$$\mathbf{R} = \mathbf{L}^{-1/2}\mathbf{Z}^T\mathbf{X}\mathbf{S}^{1/2}$$

$$= \mathbf{L}^{-1/2}\mathbf{P}^T\mathbf{X}^{*T}\mathbf{X}^*$$

$$= \mathbf{L}^{-1/2}\mathbf{LP}^T$$

$$= \mathbf{A}^T \tag{3.57}$$

where $\mathbf{X}^*$ denotes the standardized version of $\mathbf{X}$. Thus when variates and components are standardized their intercorrelations are given by

$$\mathbf{A} = \mathbf{X}^{*T}\mathbf{Z}^* \tag{3.58}$$

In what follows we assume that the variates and components are unit vectors so that $\mathbf{A}$ consists of correlation loading coefficients unless stated otherwise. When unstandardized variables are used, however, the elements of $\mathbf{A}$ must be further adjusted by diagonal elements of $(\mathbf{X}^T\mathbf{X})^{-1/2}$, assuming the PCs are unit vectors. In the most general case, when neither the variates nor the components are standardized, correlation loadings in element form are given by

$$a_{ij} = \frac{l_j^{1/2} p_{ij}}{(\mathbf{X}_i^T\mathbf{X}_i)^{1/2}} \tag{3.59}$$

for $i, j = 1, 2, \ldots, p$ where p_{ij} is the (i, j)th element of $\mathbf{P}$. In matrix notation we have as in Eq. 3.57 $\mathbf{A}^T = \mathbf{S}^{-1/2}\mathbf{P}^T\mathbf{L}^{1/2}$ where $\mathbf{S}^{-1/2}$ is the diagonal matrix of reciprocal values of the sum-of-squares. It is clear from the preceding theorems that sample corelation or covariance loadings can be obtained in two distinct ways (as in the case of the population model—see Sec. 3.2). The variates, the components, or both can be either prestandardized or post-standardized, but this does not yield the same loading coefficients. Clearly the latent roots and latent vectors also differ for the correlation and covariance matrices (Section 3.4.3).

At times we require latent roots and vectors of certain matrix functions. The following theorem is a special case of the well-known Caley–Hamilton theorem.

THEOREM 3.15. Let $\mathbf{X}$ be a $(n \times p)$ data matrix such that $\bar{\mathbf{X}}_1 = \bar{\mathbf{X}}_2 = \cdots = \bar{\mathbf{X}}_p = 0$, and where $\mathbf{X}^T\mathbf{X}$ has latent roots $\mathbf{L}$ and latent vectors $\mathbf{P}$. Then for some scalar c,

(i) $c\mathbf{X}^T\mathbf{X}$ has latent roots $c\mathbf{L}$ and latent vectors $\mathbf{P}$.

(ii) $\mathbf{X}^T\mathbf{X} + c\mathbf{I}$ has latent roots $\mathbf{L} + c\mathbf{I}$ and latent vectors $\mathbf{P}$.

(iii) $(\mathbf{X}^T\mathbf{X})^c$ has latent roots $\mathbf{L}^c$ and latent vectors $\mathbf{P}$.

PROOF

(i) Diagonalization of $c\mathbf{X}^T\mathbf{X}$ yields

$$\mathbf{P}^T(c\mathbf{X}^T\mathbf{X})\mathbf{P}^T = c\mathbf{P}^T\mathbf{X}^T\mathbf{X}\mathbf{P}$$

$$= c\mathbf{L} \tag{3.60}$$

using Eq. (3.44).

(ii) We have

$$\mathbf{P}^T(\mathbf{X}^T\mathbf{X} + c\mathbf{I})\mathbf{P} = \mathbf{P}^T\mathbf{X}^T\mathbf{X}\mathbf{P} + \mathbf{P}^Tc\mathbf{I}\mathbf{P}$$

$$= \mathbf{L} + c\mathbf{P}^T\mathbf{P}$$

$$= \mathbf{L} + c\mathbf{I} \tag{3.61}$$

(iii) We have $\mathbf{X}^T\mathbf{X}\mathbf{P} = \mathbf{P}\mathbf{L}$, and premultiplying by $(\mathbf{X}^T\mathbf{X})$ yields

$$(\mathbf{X}^T\mathbf{X})^2\mathbf{P} = \mathbf{X}^T\mathbf{X}\mathbf{P}\mathbf{L}$$

$$= \mathbf{P}\mathbf{L}^2$$

By repeated multiplications and induction we conclude that

$$(\mathbf{X}^T\mathbf{X})^c\mathbf{P} = \mathbf{P}\mathbf{L}^c \tag{3.62}$$

so that $(\mathbf{X}^T\mathbf{X})^c$ possesses latent roots $\mathbf{L}^c$ and latent vectors $\mathbf{P}$.

Property i of Theorem 3.15 permits the use of $\mathbf{X}^T\mathbf{X}$ in place of the sample covariance matrix $\mathbf{S}$ since neither correlation loadings or standardized component scores are affected by the scalar $1/(n-1)$ (but latent roots are). Note that unstandardized component scores are not the same for the two matrices. In practice, however, for large n the matrix $\mathbf{X}^T\mathbf{X}$ can be awkward to work with because of large entries. To fix ideas we consider two examples taken from sociology/demography and geology.

Example 3.6. Consider the latent roots and vectors of the sociodemographic variables of Example 2.2. For the correlation matrix the first three PCs account for

$$\frac{\sum_{i=1}^{3} l_i}{\sum_{i=1}^{7} l_i} \times 100 = \frac{6.4051}{7.00} \times 100$$

$$= 91.5$$

percent of the trace (univariate variance) of the variables (see also Exercise 3.21). Using Table 3.4 and Eq. (3.59) we see that the PCs also account for a high percentage of the correlation amongst the seven variables. Table 3.4 also indicates that only three components are required in order to explain most of the variance of the seven random variables. Also since $\Pi_{i=1}^{3} l_i = 5.6033$ the first three components account for a large portion of the generalized variance (Theorem 3.9). Since $\mathbf{L}$ and $\mathbf{P}$ are known the standardized component scores can be computed using Eq. (3.51), where

$$\mathbf{PL}^{-1/2} = \begin{bmatrix} .111 & .001 & .922 & -.220 & .017 & .290 & -.072 \\ .469 & .050 & -.139 & .107 & -.426 & .506 & .556 \\ -.479 & -.059 & .055 & -.264 & .345 & -.023 & .758 \\ -.364 & .456 & .234 & .766 & -.095 & .077 & .099 \\ .444 & .239 & -.080 & .192 & .818 & .184 & .005 \\ -.445 & -.208 & -.212 & .021 & .102 & .784 & -.297 \\ -.102 & .828 & -.147 & -.497 & -.108 & .108 & -.110 \end{bmatrix} \begin{bmatrix} 1/2.011 & & & & & & \\ & 1/1.128 & & & & & \\ & & 1/1.044 & & & & \\ & & & 1/.486 & & & \\ & & & & 1/.405 & & \\ & & & & & 1/.378 & \\ & & & & & & 1/.227 \end{bmatrix}$$

$$= \begin{bmatrix} .05542 & .00045 & .88322 & -.45216 & .04142 & .76670 & -.31677 \\ .23342 & .04393 & -.13308 & .22001 & -1.04983 & 1.33850 & 2.44888 \\ -.23803 & -.05228 & .05269 & -.54411 & .85066 & -.06085 & 3.33690 \\ -.18097 & .40447 & .22375 & 1.57529 & -.23315 & .01806 & .43577 \\ .22079 & .21155 & -.07704 & .39519 & 2.01659 & .48807 & .02319 \\ -.22142 & -.18477 & -.20332 & .04375 & .25132 & 2.07474 & -1.30726 \\ -.05059 & .73427 & -.14104 & -1.02221 & -.26598 & .28664 & -.48442 \end{bmatrix}$$

and $\mathbf{X}$ and $\mathbf{Z}^*$ are given in Tables 3.2 and 3.3 respectively. The individual scores can also be obtained from

$$\mathbf{Z}_i^* = \mathbf{X}\mathbf{P}_i l_i^{-1/2} \qquad (i = 1, 2, \ldots, p) \tag{3.63}$$

where $\mathbf{P}_i$ is the ith column of $\mathbf{P}$. Since here $\mathbf{X}$ contains standardized random variables, the matrix of correlation loadings is given by Eq. (3.56) where every column of $\mathbf{P}$ is multiplied by the square root of the corresponding latent root. The elements of $\mathbf{A}^T$ are exhibited in Table 3.4, where by convention rows correspond to variables and columns to the PCs. Evidently,

Table 3.2 Standardized Sociodemographic Random Variable Matrix X, Obtained by Standardizing Variables of Table 2.2

Borough	X_1	X_2	X_3	X_4	X_5	X_6	X_7
City	−.1604	.5004	−.4374	−.5088	.2173	−.1439	−.5195
Battersea	−.0531	−.0996	.0782	.0833	−.0951	.0705	−.1346
Bermondsey	−.2107	−.1308	.1386	−.0342	−.1047	.1878	−.2510
Bethnal Green	−.2065	−.0908	.0833	−.0044	−.3210	.2120	−.0372
Camberwell	−.0615	−.1115	.1306	.0634	−.0567	.1017	.0103
Chelsea	.1095	.1418	−.1872	−.1852	.3110	−.3646	.0103
Deptford	−.0833	−.1071	.0979	.1357	−.1384	.1455	.0127
Finnsbury	.6361	−.0989	.1153	.0349	−.1144	.1381	−.0467
Fulham	.0131	−.0737	−.0432	−.0361	.0418	0	−.1108
Greenwich	.0727	−.1019	.1531	−.1228	−.0951	.0350	−.2201
Hackney	−.0104	−.0796	.0586	.1899	−.0951	.1494	.0270
Hammersmith	.0601	−.1441	.0404	.1538	−.1120	.0138	.1267
Hampstead	.2504	.2589	−.2069	−.0026	.2077	−.3658	−.0229
Holborn	−.0539	.4330	−.3058	−.3570	.4720	−.1729	.0222
Islington	−.0623	−.0411	.0673	.2957	−.1624	.1162	.1790
Kensington	.1364	.2567	−.1836	.0666	.2005	−.3706	.3928
Lambeth	−.0078	−.0959	.0993	.2157	−.0038	.0709	.1339
Lewisham	.1749	−.0930	.0891	−.0329	−.1120	.0309	−.2320
Paddington	.0634	.0418	−.1414	.2202	.1760	−.1348	.1457
Poplar	−.2761	−.1537	.2149	.0995	−.1144	.2075	.1814
St. Marylebone	.0542	.3907	−.3967	−.3009	.2341	−.3251	.0222
St. Pancras	−.0565	.0055	−.0498	.0620	.0779	−.0053	.2028
Shoreditch	−.2719	−.1545	.2426	−.0555	−.2586	.2161	.2360
Southwark	−.1747	−.1071	.1662	.0607	.0755	.1865	.0673
Stepney	−.2845	−.0804	.1349	.0923	−.0327	.1822	.2835
Stoke-Newington	.1188	−.1130	.1320	.2758	−.1144	.0839	−.1251
Wandsworth	.1364	−.1048	.0360	−.0225	−.0759	−.0352	−.0847
Westminster	−.0087	.1048	−.3094	−.2724	.2437	−.2719	−.0538
Woolich	.1565	−.1522	.1829	−.1142	−.2513	.0366	−.2154

Z_3, picks up the influence of X_1 (population change 1931–1951), whereas Z_2 correlates with X_7 (percentage unemployed) and to a lesser extent with X_4 (fertility rate). Most of the variance, however, is explained by Z_1, which shows a contrast between X_3 and X_6 (young male population, low terminal education age) and X_2 and X_6 (high young female/male ratio, suicide rate). Such components are also known as polar components. The contrast indicates low education age, which is found in boroughs high in young males and low in young females, which in turn is negatively correlated with the suicide rate (X_5). Variable X_4 (fertility rate) is also negatively correlated with X_2 and X_5 and we conclude that high suicide rates appear to be found in boroughs containing a high young females/young male ratio, for reasons which are not apparent from Table 3.4.

The split between boroughs high in young females on the one hand and

Table 3.3 Standardized Matrix Z* of Principal Components Obtained from the Sociodemographic Variables of Table 3.2

Borough	Z_1^*	Z_2^*	Z_3^*	Z_4^*	Z_5^*	Z_6^*	Z_7^*
City	.4102	−.4699	−.2594	.2296	−.2452	.2229	.0395
Battersea	−.0897	−.1086	.0011	.1938	.0112	−.1160	.0409
Bermondsey	−.1201	−.2681	−.1638	.1606	.1572	−.0709	.0677
Bethnal Green	−.1676	−.1446	−.1801	−.0584	−.4256	−.0125	−.1473
Camberwell	−.1087	−.0103	−.0373	.0038	.1207	−.0057	.0635
Chelsea	.2661	.0819	.0753	−.1118	.2723	−.3198	.0860
Deptford	−.1409	−.0018	−.0445	.1133	−.0854	.0272	−.0496
Finnsbury	−.0751	−.0800	.5762	−.3085	.0367	.5663	−.2043
Fulham	.0152	−.0881	.0235	.0744	.1634	−.0980	−.2900
Greenwich	−.0516	−.2504	.0896	−.1431	.1444	−.1290	.2436
Hackney	−.1229	.0423	.0202	.1959	−.0727	.1565	−.1240
Hammersmith	−.1019	.1206	.0967	−.0118	−.1037	−.1362	−.2521
Hampstead	.2520	.1159	.2368	.1417	−.1038	−.1131	.3573
Holborn	.3768	.0387	−.2056	−.1200	.2689	.4284	.1276
Islington	−.1532	.1899	−.0623	.2063	−.3170	.1115	.0301
Kensington	.2056	.4472	.0906	−.1390	−.2289	−.0981	.3005
Lambeth	−.1088	.1622	.0263	.1330	.1091	.0470	.0353
Lewisham	−.0471	−.2217	.1993	−.0056	.0320	−.0534	.0694
Paddington	.0685	.2676	.0855	.3192	.0702	−.0345	−.1839
Poplar	−.2007	.0928	−.2588	−.0906	.0829	−.0020	.1099
St. Marylebone	.3656	.0421	−.0474	−.1411	−.2908	.0294	−.0955
St. Pancras	.0069	.1943	−.0729	−.0252	.0369	.0534	−.1971
Shoreditch	−.2157	.0366	−.2766	−.4644	−.1597	−.0411	.0903
Southwark	−.1133	.0420	−.1709	.0297	.4136	.1569	.0996
Stepney	−.1452	.1942	−.2873	−.1118	.0704	.1110	.0076
Stoke-Newington	−.1386	−.0318	.1980	.3705	−.0047	.0191	.1948
Wandsworth	−.0261	−.0873	.1562	−.0847	.0122	−.1727	−.1042
Westminster	.2637	−.0271	−.0548	−.0944	.1274	−.3130	−.5047
Woolich	−.1024	−.2805	.1849	−.2613	−.0918	−.2055	.1890

Table 3.4 The Matrix of Correlation Loadings A^T Representing Correlation Between PCs and Variates[a]

	Z_1^*	Z_2^*	Z_3^*	Z_4^*	Z_5^*	Z_6^*	Z_7^*	Total SS
X_1	.2241	.0006	.9623	−.1068	.0068	.1094	−.0163	1.0
X_2	.9439	.0559	−.1450	.05197	−.1726	.1910	.1264	1.0
X_3	−.9625	−.0665	.0574	−.1285	.1399	−.0087	.1722	1.0
X_4	−.7318	.5144	.2440	.3721	−.0383	.0026	.0225	1.0
X_5	.8928	.2690	−.0839	.0933	.3315	.0696	.0012	1.0
X_6	−.8954	−.2350	−.2215	.0103	.0413	.2961	−.0675	1.0
X_7	.2046	.9338	−.1537	−.2415	.0437	.0409	−.0250	1.0

[a] $l_1 = 4.043$; $l_2 = 1.271$; $l_3 = 1.089$; $l_4 = .2362$; $l_5 = .1644$; $l_6 = .1427$; $l_7 = .0516$.

young males on the other is probably due to structural aspects of the labor force in London. Since both $\mathbf{X}_2$ and $\mathbf{X}_5$ positively correlate to $\mathbf{Z}_1$ their high level is indicated by large and positive $\mathbf{Z}_1$ scores. Thus both high young female/male ratios, as well as high suicide rates, tend to be found in business-oriented districts such as City, Chelsea, Hampstead, Halborn, Kensington, St. Marylebone, and Westminster. These boroughs possess a higher young, single, female labor force mainly employed in government institutions and private service industries. Highly negative scores on the other hand are found for Poplar and Shoreditch, indicating a prevalence of young males and low suicide rates in the more traditional blue collar, working class districts. Note that signs of the loadings and scores, for a given PC, are relative in the sense that interchanging both will not alter interpretation. Correlation loadings for the first two components are plotted in Figure 3.4. The predicted (systematic) parts of the variables are given by

$$\hat{\mathbf{X}}_1 = .2241\mathbf{Z}_1 + .0006\mathbf{Z}_2 + .9623\mathbf{Z}_3$$
$$\hat{\mathbf{X}}_2 = .9439\mathbf{Z}_1 + .0559\mathbf{Z}_2 - .1450\mathbf{Z}_3$$
$$\overline{\hat{\mathbf{X}}_7 = .2046\mathbf{Z}_1 + .9338\mathbf{Z}_2 - .1537\mathbf{Z}_3}$$

Of course all three PCs need not be retained for all variates. Components

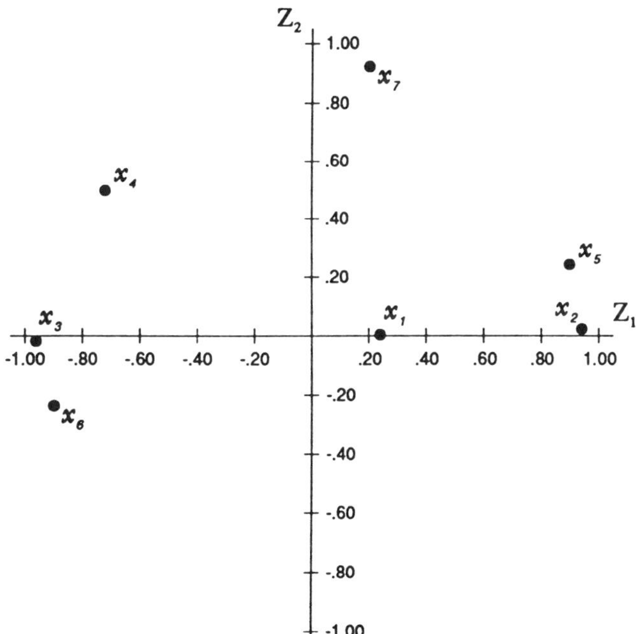

Figure 3.4 Correlation loading coefficients for the first two principal components $\mathbf{Z}_1$ and $\mathbf{Z}_2$ of Table 3.4.

exhibiting low correlation loadings may be omitted since not every PC is of equal importance for every variable, nor are all variables of equal importance for every component. For instance, although $\mathbf{Z}_1, \mathbf{Z}_2$, and $\mathbf{Z}_3$ explain $\mathbf{R}_1^2 = (.2241)^2 + (.0006)^2 + (.9623)^2 = .9762$ of the variance of $\mathbf{X}_1$, they account for only $\mathbf{R}_4^2 = .8600$ of the variance of $\mathbf{X}_4$.

Example 3.7. An interesting application of PCA in the natural sciences occurs in Geology. During the summer of 1983 sediment samples were taken at Lynn Lake, Manitoba, by the Geological Survey of Canada. Amongst other variables, the following $p = 16$ measurements of trace elements were taken for each sample location.

$$
\begin{aligned}
Y_1 &= \text{zinc, ppm} & Y_9 &= \text{molybdenum, ppm} \\
Y_2 &= \text{copper, ppm} & Y_{10} &= \text{iron, pct} \\
Y_3 &= \text{lead, ppm} & Y_{11} &= \text{mercury, ppb} \\
Y_4 &= \text{nickel, ppm} & Y_{12} &= \text{loss on ignition, pct} \\
Y_5 &= \text{cobalt, ppm} & Y_{13} &= \text{uranium, ppm} \\
Y_6 &= \text{silver, ppm} & Y_{14} &= \text{fluorine, ppm} \\
Y_7 &= \text{manganese, ppm} & Y_{15} &= \text{vanadium, ppm} \\
Y_8 &= \text{arsenic, ppm} & Y_{16} &= \text{cadmium, ppm}
\end{aligned}
$$

Sample means and standard deviations, together with the correlation matrix, are presented in Tables 3.5 and 3.6. Because of the effect of different units of measure (parts per million, parts per billion, percentages) the correlation

Table 3.5 Sample Means and Standard Deviations of Lake Sediment Data from Lynn Lake, Manitoba

Element	Means	Standard Deviations
$\mathbf{Y}_1$	81.68 ppm	43.74 ppm
$\mathbf{Y}_2$	15.68 ppm	6.85 ppm
$\mathbf{Y}_3$	2.08 ppm	3.46 ppm
$\mathbf{Y}_4$	14.63 ppm	8.21 ppm
$\mathbf{Y}_5$	7.88 ppm	5.68 ppm
$\mathbf{Y}_6$	.12 ppm	.19 ppm
$\mathbf{Y}_7$	975.54 ppm	2874.95 ppm
$\mathbf{Y}_8$	1.23 ppm	.49 ppm
$\mathbf{Y}_9$	2.56 ppm	2.63 ppm
$\mathbf{Y}_{10}$	4.42 pct	4.60 pct
$\mathbf{Y}_{11}$	64.40 ppb	48.96 ppm
$\mathbf{Y}_{12}$	35.79 pct	19.18 pct
$\mathbf{Y}_{13}$	4.12 ppm	3.09 ppm
$\mathbf{Y}_{14}$	242.34 ppm	116.19 ppm
$\mathbf{Y}_{15}$	31.35 ppm	50.92 ppm
$\mathbf{Y}_{16}$	.65 ppm	4.99 ppm

Table 3.6   Correlation Matrix for $p = 16$ Geological Measurements (Table 3.4)

Y_1	Y_2	Y_3	Y_4	Y_5	Y_6	Y_7	Y_8	Y_9	Y_{10}	Y_{11}	Y_{12}	Y_{13}	Y_{14}	Y_{15}	Y_{16}
1.000															
.505	1.000														
.031	.037	1.000													
.600	.557	.164	1.000												
.785	.513	.085	.820	1.000											
−.015	.005	.056	.152	.091	1.000										
.455	.289	.102	.745	.723	−.022	1.000									
.336	.292	.053	.415	.469	.057	.404	1.000								
.570	.229	.026	.441	.646	−.022	.482	.403	1.000							
.798	.298	−.031	.327	.598	−.061	.283	.280	.671	1.000						
.148	.049	.013	−.062	−.095	−.030	−.081	−.136	−.089	−.034	1.000					
−.084	−.047	−.039	−.210	−.283	.022	−.226	−.336	−.206	−.129	.423	1.000				
.333	.337	.050	.278	.257	−.076	.152	.101	.174	.236	−.092	−.143	1.000			
.100	.266	.046	.366	.280	−.072	.248	.351	.012	−.030	−.241	−.648	.087	1.000		
.427	.175	.050	.262	.284	−.032	.081	.054	.144	.411	−.052	−.010	.631	−.095	1.000	
.229	.019	.027	.119	.110	.009	.003	−.048	−.020	.165	−.002	.135	.580	−.2111	.929	1.000

Table 3.7 Correlation Loadings, Percentage Variance Explained, and Latent Roots for $p = 16$ Geological Variables (Table 3.5)

Element	Z_1	Z_2	Z_3	Z_4	Z_5	R^2	Latent Roots
Y_1	.839	.186	.296	−.102	−.089	.8445	5.10374
Y_2	.595	−.031	.092	.284	−.458	.6539	2.50188
Y_3	.096	−.065	.002	.512	.290	.3597	1.71054
Y_4	.823	−.160	.047	.377	−.029	.8481	1.23672
Y_5	.917	−.130	.148	.051	.064	.8864	1.08377
Y_6	.020	−.062	.094	.462	.606	.5938	.95891
Y_7	.701	−.269	.115	.153	.035	.6016	.77603
Y_8	.560	−.346	−.071	−.026	.067	.4435	.67026
Y_9	.690	−.093	.290	−.389	.254	.7847	.60995
Y_{10}	.698	.199	.262	−.462	.131	.8261	.47363
Y_{11}	−.121	.251	.603	.251	−.402	.6659	.25392
Y_{12}	−.344	.492	.600	.207	−.071	.7683	.24175
Y_{13}	.469	.507	−.435	.039	−.198	.7070	.16237
Y_{14}	.330	−.574	−.507	.113	−.313	.8062	.11937
Y_{15}	.471	.768	−.341	.030	.082	.9356	.07497
Y_{16}	.262	.834	−.352	.122	.104	.9138	.02219
Variance %	46.40	15.64	10.69	7.73	6.77		16.0000

rather than the covariance matrix is used to eliminate the effects of uneven variance (see Section 3.4.3).

Table 3.7 exhibits correlation loadings, "communalities" (R^2), latent roots, and percentage of variance accounted for by the first $r = 5$ PCs. Together, the five components explain 87.23% of the trace of the correlation matrix. The first component Z_1 accounts for almost half of the variance, and mainly correlates with Y_5, Y_1, Y_4, Y_7, Y_9, Y_{10}, Y_8, and Y_2. Here Z_1 represents a mixture (linear combination) of these elements. The component scores of Z_1 then indicate which sample points contain a particularly high (low) amount of such a metal combination (mineral type). The second component Z_2 indicates that Y_{16}, Y_{15}, Y_{13}, and Y_{12} form another mixture (linear combination) of elements, orthogonal to Z_1 and characterized by a low iron, a high cadmium, vanadium, and uranium content. It is also relatively high in flammable material (loss on ignition). The remaining dimensions have a similar interpretation, although they contribute less to the observations.

3.4.3 The Effect of Means and Variances on PCs

Grammian association matrices can be expressed in four different ways depending on measures of location and scale (Section 2.4). Both the loadings and the scores however are dependent on the mean values and variances of the variables, so that initially a choice must be made as to how

Table 3.8 Correlation Loading Coefficients for Standardized, Unstandardized Variates and PCs Obtained from the Correlation, Product-Moment Matrices

Variates	Principle Components	
	Standardized	Not Standardized
Standardized $\mathbf{R} = \mathbf{S}^{-1/2}\mathbf{X}^{\mathrm{T}}\mathbf{X}\mathbf{S}^{-1/2}$	$a_{ij} = p_{ij}$	$a_{ij} = l_j^{1/2} p_{ij}$
Not standardized $\mathbf{X}^{\mathrm{T}}\mathbf{X}$	$a_{ij} = (\mathbf{X}_i^{\mathrm{T}}\mathbf{X}_i)^{-1/2} p_{ij}$	$a_{ij} = \dfrac{l_j^{1/2} p_{ij}}{(\mathbf{X}_i^{\mathrm{T}}\mathbf{X}_i)^{1/2}}$

the variables are to be centered and scaled. The situation is portrayed in Table 3.8. First consider the effect of the mean vector. In most applications random variables are adjusted for differences in their general levels (distance from the origin). The affine transformation is applied in order to place the variables on an equal footing with respect to the origin. An added advantage is that the data vectors become points in a linear vector space. Using Eq. (3.46) where $\mathbf{X} = \mathbf{Y} - \bar{\mathbf{Y}}$ we have

$$\mathbf{X} = (\mathbf{Y} - \bar{\mathbf{Y}}) = \mathbf{Z}\mathbf{P}^{\mathrm{T}}$$

or

$$\mathbf{Y} = \bar{\mathbf{Y}} + \mathbf{Z}\mathbf{P}^{\mathrm{T}} \tag{3.64}$$

so that given PCs and latent vectors (loadings) of $\mathbf{X}^{\mathrm{T}}\mathbf{X}$ we can always reconstruct the original observations. Note however that latent roots and vectors of $\mathbf{Y}^{\mathrm{T}}\mathbf{Y}$ cannot be obtained in a straightforward way from those of $\mathbf{X}^{\mathrm{T}}\mathbf{X}$ since latent roots and vectors are not independent of affine (or linear) transformations. It was seen in Section 3.3 that the first PC tends to reflect the general correlational structure of the variables and is thus a reflection of general "size." This is even more true for the matrix $\mathbf{Y}^{\mathrm{T}}\mathbf{Y}$. Since the sums of squares and products tend to be dominated by the mean values, the first PC of $\mathbf{Y}^{\mathrm{T}}\mathbf{Y}$ simply reflects this mean influence. A PCA of $\mathbf{Y}$ therefore generally tends to yield trivial results, with large loadings for the first PC and much smaller ones for the remaining. The difficulty however is readily solved by replacing $\mathbf{Y}$ by the matrix $\mathbf{X}$.

Unequal variances also cause interpretational difficulties when variables either do not possess comparable units of measure or else are highly heteroscedastic (uneven variance) for the same unit of measure. This is because the lack of comparability between correlation and covariance matrices is caused by the (unequal) variances rather than the covariance terms (Exercise 3.23). In this case the difficulty is solved by standardizing $\mathbf{X}$ such that columns possess equal (unit) variance. Even when units of measure are the same, however, the correlation matrix may be more advantageous when there are large differences in the variance. This is because variables with relatively large sums of squares tend to dominate those PCs that correspond to large latent roots. The loadings for the first few PCs therefore are generally distorted by these differences in the variances. Thus it is precisely variables that could be most unstable which receive disproportionately larger weights and which tend to dominate the analysis and cause interpretational difficulties (e.g., see McGillivray, 1985). In this situation the high variance variables can be omitted from the analysis or the entire set standardized to unit variance. In certain applications where variables are related exponentially, for example, those dealing with constant rates of growth (Section 3.7), a logarithmic transformation will tend to reduce differences in the variances. More generally we may also wish to weight elements of $\mathbf{X}^T\mathbf{X}$ by numbers other than standard deviations. For example, not all variables may possess equal reliability, as when some data are missing (Section 4.7.3), and those more error prone may then be given smaller weight in the analysis. It must be stressed however that there is generally no straightforward relationship between a PCA of weighted and unweighted variables.

The following example illustrates the effect of the mean values of the variables.

Example 3.8. We consider the artificial data matrix given by Orloci (1967). We have

$$
\mathbf{Y} = \begin{bmatrix} 26 & 24 & 36 \\ 40 & 16 & 42 \\ 30 & 26 & 28 \\ 40 & 20 & 16 \end{bmatrix} \qquad \mathbf{X} = \begin{bmatrix} -8 & 2.5 & 5.5 \\ 6 & -5.5 & 11.5 \\ -4 & 4.5 & -2.5 \\ 6 & -1.5 & -14.5 \end{bmatrix}
$$

where $\mathbf{Y} = (34,\ 21.5,\ 30.5)^T$ and

$$
\mathbf{Y}^T\mathbf{Y} = \begin{bmatrix} 4776 & 2844 & 4096 \\ 2844 & 1908 & 2584 \\ 4098 & 2584 & 4100 \end{bmatrix} \qquad \mathbf{X}^T\mathbf{X} = \begin{bmatrix} 152 & -80 & -52 \\ -80 & 59 & -39 \\ -52 & -39 & 379 \end{bmatrix}
$$

The latent roots, vectors of the two matrices are

$$
\begin{bmatrix} 10306.050083 & & 0 \\ & 331.087990 & \\ 0 & & 146.861927 \end{bmatrix}, \quad \begin{bmatrix} .669641 & .631919 & .390204 \\ .416065 & .116015 & -.901904 \\ .615200 & -.766302 & .185231 \end{bmatrix}
$$

and

$$
\begin{bmatrix} 391.807285 & & 0 \\ & 194.811758 & \\ 0 & & 3.380958 \end{bmatrix}, \quad \begin{bmatrix} .189278 & .838846 & .510400 \\ .069281 & -.529910 & .845219 \\ -.979476 & .124620 & .158417 \end{bmatrix}
$$

respectively. The (unstandardized)PCs of $\mathbf{X}$ are then the columns of

$$\mathbf{Z} = \begin{bmatrix} -8 & 2.5 & 5.5 \\ 6 & -5.5 & 11.5 \\ -4 & 4.5 & -2.5 \\ 6 & -1.5 & -14.5 \end{bmatrix} \begin{bmatrix} .189278 & .838848 & .510400 \\ .069281 & -.529910 & .845219 \\ -.979476 & .124620 & .158417 \end{bmatrix}$$

$$= \begin{bmatrix} -6.72814 & -7.35015 & -1.09886 \\ -10.50935 & 9.38072 & .23550 \\ 2.00334 & -6.05154 & 1.36585 \\ 15.23415 & 4.02096 & -.50248 \end{bmatrix}$$

The original observations can be reproduced using Eq. (3.64). Thus for $\mathbf{Y}_1$ we have

$$\mathbf{Y}_1 = \bar{\mathbf{Y}}_1 + \mathbf{Z}\mathbf{P}'_1$$

$$= \begin{bmatrix} 34 \\ 34 \\ 34 \\ 34 \end{bmatrix} + \begin{bmatrix} -6.72814 & -7.35015 & -1.09886 \\ -10.50935 & 9.38072 & .23550 \\ 2.00334 & -6.05154 & 1.36585 \\ 15.23415 & 4.02096 & -.50248 \end{bmatrix} \begin{bmatrix} .189278 \\ .838848 \\ .510400 \end{bmatrix}$$

$$= \begin{bmatrix} 34 \\ 34 \\ 34 \\ 34 \end{bmatrix} + \begin{bmatrix} -8.00 \\ 6.00 \\ -4.00 \\ 6.00 \end{bmatrix}$$

$$= \begin{bmatrix} 26.00 \\ 40.00 \\ 30.00 \\ 40.00 \end{bmatrix}$$

where P'_1 is the first row of $\mathbf{P}'$, not to be confused with a transposed (row vector). The remaining columns can be determined in a similar fashion.

The mean vector makes itself felt through the first PC of the matrix $\mathbf{Y}^T\mathbf{Y}$. Standardized loadings for both $\mathbf{Y}^T\mathbf{Y}$ and $\mathbf{X}^T\mathbf{X}$ are given below. Since the columns of $\mathbf{Y}$ are not corrected for their means, the loadings for $\mathbf{Y}$ are direction cosines rather than correlations.

	$\mathbf{Z}_1$	$\mathbf{Z}_2$	$\mathbf{Z}_3$		$\mathbf{Z}_1$	$\mathbf{Z}_2$	$\mathbf{Z}_3$
$\mathbf{Y}_1$	.9837	.1664	.0684	$\mathbf{X}_1$	.3039	.9497	.0761
$\mathbf{Y}_2$	.9670	.0483	-.2502	$\mathbf{X}_2$	.1785	-.9629	.2023
$\mathbf{Y}_3$	.9754	-.2178	.0351	$\mathbf{X}_3$	-.9959	.0893	.0150

It is evident that the inclusion of the mean vector has made a substantial difference in the analysis. Although the mean vector dominates the analysis of $\mathbf{Y}$ through the first component $\mathbf{Z}_1$, once it is removed the loadings of $\mathbf{X} = \mathbf{Y} - \bar{\mathbf{Y}}$ reveal nontrivial linear combinations of the columns of $\mathbf{X}$.

3.5 PRINCIPAL COMPONENTS AND PROJECTIONS

Consider a set of p linearly independent variables $\mathbf{X}_1, \mathbf{X}_2, \ldots, \mathbf{X}_p$, each containing n observations. Let

$$\mathbf{P}_z = \mathbf{Z}(\mathbf{Z}^\mathrm{T}\mathbf{Z})^{-1}\mathbf{Z}^\mathrm{T} \tag{3.65}$$

be a $(n \times n)$ projection matrix which projects vectors orthogonally onto a subspace of dimension $p < n$. A well-known transformation which seeks to orthogonalize the observed variates is the Gram–Schmidt transformation

$$\mathbf{Z}_1 = \mathbf{X}_1$$

$$\mathbf{Z}_2 = (\mathbf{I} - \mathbf{P}_{z1})\mathbf{X}_2$$

$$\mathbf{Z}_3 = (\mathbf{I} - \mathbf{P}_{z1} - \mathbf{P}_{z2})\mathbf{X}_3$$

$$\overline{}$$

$$\mathbf{Z}_p = (\mathbf{I} - \mathbf{P}_{z1} - \mathbf{P}_{z2} - \cdots - \mathbf{P}_{z(p-1)})\mathbf{X}_p \tag{3.66}$$

where the $\mathbf{Z}_i$ form a new set of orthogonal variates (Exercise 3.2). The orthogonalization procedure is sequential. Beginning with a first arbitrary choice $\mathbf{X}_1 = \mathbf{Z}_1$, the second orthogonal vector $\mathbf{Z}_2$ is the difference between $\mathbf{X}_2$ and its orthogonal projection onto $\mathbf{X}_1$, $\mathbf{Z}_3$ is the difference between $\mathbf{X}_3$ and its orthogonal projection onto $\mathbf{Z}_1$ and $\mathbf{Z}_2$, and so forth until all p vectors are found.

The Gram–Schmidt projection can be viewed as a series of least squares regressions. Let $\mathbf{Y}$ be a random $(n \times 1)$ vector of observations which depends on p explanatory variables such that

$$\mathbf{Y} = \beta_1\mathbf{X}_1 + \beta_2\mathbf{X}_2 + \cdots + \beta_p\mathbf{X}_p + \boldsymbol{\epsilon}$$

$$= \mathbf{X}\boldsymbol{\beta} + \boldsymbol{\epsilon} \tag{3.67}$$

$\mathbf{X}$ is an $(n \times p)$ matrix, $\boldsymbol{\beta}$ is a $(p \times 1)$ vector of fixed coefficients, and $\boldsymbol{\epsilon}$ is the $(n \times 1)$ residual vector such that $E(\boldsymbol{\epsilon}) = \mathbf{0}$, $E(\mathbf{Y}) = \boldsymbol{\mu} = \mathbf{X}\boldsymbol{\beta}$. A simplifying assumption is that $\boldsymbol{\epsilon}$ possesses a diagonal covariance matrix with equal nonzero diagonal entries, say σ^2. The covariance matrix of $\mathbf{Y}$ can then be written as $\boldsymbol{\Sigma} = \sigma^2\mathbf{V}$ where $\mathbf{V}$ is symmetric, positive definite. Let Ω denote a vector space such that $\boldsymbol{\mu} \in \Omega$ and $\dim(\Omega) = p$. Then the range of $\mathbf{X}$ is $R(\mathbf{X}) = \{\boldsymbol{\mu}: \boldsymbol{\mu} = \mathbf{X}\boldsymbol{\beta}\}$ for some fixed $\boldsymbol{\beta}$, and an estimate $\mathbf{M} = \hat{\mathbf{Y}}$ of $\boldsymbol{\mu} = E(\mathbf{Y})$ can be found such that $\mathbf{M} \in \Omega$. Since $\mathbf{M}$ is not unique an additional condition of orthogonality is usually required so that $\mathbf{M}$ is orthogonal to the sample residual vector $\hat{\boldsymbol{\epsilon}} = \mathbf{Y} - \mathbf{M}$, that is $(\mathbf{Y} - \mathbf{M})\mathbf{M}^\mathrm{T} = 0$ for all $\mathbf{X} \in \Omega$. When

$\Omega = R(\mathbf{X})$, that is, $\mathbf{X}$ is of full rank, $\mathbf{M}$ is the orthogonal projection of $\mathbf{Y}$ onto $R(\mathbf{X})$, with projection matrix $\mathbf{P}_x = \mathbf{X}(\mathbf{X}^T\mathbf{X})^{-1}\mathbf{X}^T$ so that

$$\begin{aligned}
\mathbf{M} &= \mathbf{P}_x\mathbf{Y} \\
&= \mathbf{X}(\mathbf{X}^T\mathbf{X})^{-1}\mathbf{X}^T\mathbf{Y} \\
&= \mathbf{X}\hat{\boldsymbol{\beta}} \\
&= \hat{\mathbf{Y}}
\end{aligned} \tag{3.68}$$

The regression estimator $\hat{\boldsymbol{\beta}} = (\mathbf{X}^T\mathbf{X})^{-1}\mathbf{X}^T\mathbf{Y}$ possesses certain optimal properties which derive from the orthogonal nature of the projection. Thus $\hat{\boldsymbol{\epsilon}}$ is constrained to lie in the $n - p$ dimensional null space $N(\mathbf{X})$, perpendicular to $R(\mathbf{X})$.

Two points emerge from the geometry of regression, discussed above, which are relevant to factor analysis and which indicate the formal similarity of factor analysis models to least squares regression. First, least squares represent, in a sense, a dimension-reducing procedure since the n-dimensional sample space is reduced to a $p < n$ dimensional random variable space containing the systematic variation in $\mathbf{Y}$. The reduction results from the orthogonal projection of $\mathbf{Y}$ onto the column space of $\mathbf{X}$, which gives least squares regression optimal properties. It follows that prior information concerning the error structure of the variables is required, such as orthogonality. Second, regression requires prior information in the form of explanatory variables which are thought to account for a significant proportion of the variance in $\mathbf{Y}$. The dependence must be posited beforehand since statistical techniques, by themselves, cannot isolate the direction of causality if such causality exists.

Both PCA and least squares regression share common geometric properties and can be considered in a unified manner in terms of projections. Consider p intercorrelated random vectors $\mathbf{X}_1, \mathbf{X}_2, \ldots, \mathbf{X}_p$. Then computing $r \leq p < n$ PCs is equivalent to projecting vectors from a p-dimensional space $R(\mathbf{X})$ onto an r-dimensional subspace $S(\mathbf{Z})$ such that

$$\begin{aligned}
\mathbf{X} &= \mathbf{Z}_1\hat{\boldsymbol{\alpha}}_1^T + \mathbf{Z}_2\hat{\boldsymbol{\alpha}}_2^T + \cdots + \mathbf{Z}_r\hat{\boldsymbol{\alpha}}_r^T + \hat{\boldsymbol{\epsilon}}_{(r)} \\
&= \mathbf{Z}_{(r)}\hat{\boldsymbol{\alpha}}_{(r)}^T + \hat{\boldsymbol{\epsilon}}_{(r)}
\end{aligned} \tag{3.69}$$

$\mathbf{Z}_{(r)}$ is the $(n \times r)$ matrix of the first r PCs, $\hat{\boldsymbol{\alpha}}_{(r)}^T$ the $(r \times p)$ matrix of estimated coefficients, and $\boldsymbol{\epsilon}_{(r)}$ is the $(n \times p)$ matrix of residuals. Equation (3.69) is similar to least squares regression except that the PCs are unknown and must be estimated together with the coefficients. The first sample PC $\mathbf{Z}_1$ is that linear combination which accounts for the largest proportion of variance, where variance, where

$$\mathbf{X} = \mathbf{Z}_1 \hat{\boldsymbol{\alpha}}_1^T + \hat{\boldsymbol{\epsilon}}_{(1)} \tag{3.70}$$

and $S(\mathbf{Z}_1) = \{\boldsymbol{\Psi}_1 : \boldsymbol{\Psi}_1 = \mathbf{Z}_1 \boldsymbol{\alpha}_1^T\}$. An estimate of $\boldsymbol{\Psi}_1$, say $\mathbf{F}_1 = \mathbf{Z}_1 \hat{\boldsymbol{\alpha}}_1^T$, is found such that

$$(\mathbf{X} - \mathbf{F}_1)^T \mathbf{Z}_1 = \mathbf{0} \tag{3.71}$$

for all $\mathbf{Z}_1 \in S(\mathbf{Z}_1)$, that is, $\mathbf{Z}_1$ is orthogonal to the $(p-1)$ dimensional residual space $R(\mathbf{X} - \mathbf{F}_1)$. For any other estimate $\mathbf{F}_1^*$ the norm $|\mathbf{X} - \mathbf{F}_1^*|$ is minimized if and only if $\mathbf{F}_1^* = \mathbf{F}_1$. Thus $\mathbf{F}_1$ is the orthogonal projection of $\mathbf{X}$ onto $\mathbf{Z}_1$, and

$$\begin{aligned}
\mathbf{F}_1 &= \mathbf{P}_{z1} \mathbf{X} \\[4pt]
&= \mathbf{Z}_1 (\mathbf{Z}_1^T \mathbf{Z}_1)^{-1} \mathbf{Z}_1^T \mathbf{X} \\[4pt]
&= \mathbf{Z}_1 \mathbf{Z}_1^T \mathbf{X} \\[4pt]
&= \mathbf{Z}_1 \hat{\boldsymbol{\alpha}}_1^T
\end{aligned} \tag{3.72}$$

where $\hat{\boldsymbol{\alpha}}_1^T = \mathbf{Z}_1^T \mathbf{X}$ is the least squares regression estimator of $\boldsymbol{\alpha}_1^T$. Thus

$$\mathbf{X} = \mathbf{P}_{z1} \mathbf{X} + (\mathbf{I} - \mathbf{P}_{z1}) \mathbf{X} \tag{3.73}$$

where $\mathbf{P}_{z1} = \mathbf{Z}_1 (\mathbf{Z}_1^T \mathbf{Z}_1)^{-1} \mathbf{Z}_1^T$.

Equation (3.72) cannot be used as it stands since $\mathbf{Z}_1$ is unknown. The sum of squares of $\mathbf{X}$ due to $\mathbf{Z}_1$ is the squared norm

$$\begin{aligned}
\sum_{i=1}^{p} \|\mathbf{P}_{z1} \mathbf{X}_i\| &= \operatorname{tr}[(\mathbf{P}_{z1} \mathbf{X})^T (\mathbf{P}_{z1} \mathbf{X})] \\[4pt]
&= \operatorname{tr}[\mathbf{X}^T \mathbf{P}_{z1} \mathbf{X}] \\[4pt]
&= \operatorname{tr}[\mathbf{X}^T \mathbf{Z}_1 (\mathbf{Z}_1^T \mathbf{Z}_1)^{-1} \mathbf{Z}_1^T \mathbf{X}] \\[4pt]
&= \operatorname{tr}[\mathbf{X} \mathbf{X}^T \mathbf{Z}_1 (\mathbf{Z}_1^T \mathbf{Z}_1)^{-1} \mathbf{Z}_1^T] \\[4pt]
&= \operatorname{tr}[\mathbf{Z}_1^T \mathbf{X} \mathbf{X}^T \mathbf{Z}_1 (\mathbf{Z}_1^T \mathbf{Z}_1)^{-1}] \\[4pt]
&= \frac{\mathbf{Z}_1^T \mathbf{X} \mathbf{X}^T \mathbf{Z}_1}{\mathbf{Z}_1^T \mathbf{Z}_1}
\end{aligned} \tag{3.74}$$

using well-known properties of the trace function. To maximize Eq. (3.74) we maximize the Lagrangian expression

$$\phi = \boldsymbol{\zeta}_1^T \mathbf{X} \mathbf{X}^T \boldsymbol{\zeta}_1 - h_1 (\boldsymbol{\zeta}_1^T \boldsymbol{\zeta}_1 - 1) \tag{3.75}$$

where h_1 is the Lagrange multiplier. Differentiating with respect to ζ_1 and setting to zero we have

$$\frac{\partial \phi}{\partial \zeta_1} = \mathbf{X}\mathbf{X}^T\mathbf{Z}_1 - h_1\mathbf{Z}_1 = \mathbf{0}$$

or

$$(\mathbf{X}\mathbf{X}^T - h_1\mathbf{I})\mathbf{Z}_1 = \mathbf{0} \tag{3.76}$$

The vector of estimates $\mathbf{Z}_1$ of ζ_1 is therefore the latent vector of the $(n \times n)$ matrix $\mathbf{X}\mathbf{X}^T$ which corresponds to the largest root h_1. The vectors $\hat{\boldsymbol{\alpha}}_1$ and $\mathbf{Z}_1$ can therefore be determined jointly as the first latent vectors of $\mathbf{X}^T\mathbf{X}$ and $\mathbf{X}\mathbf{X}^T$, respectively. To minimize computational effort, however, $\mathbf{Z}_1$ is normally taken as the first vector of scores obtained from Eq. (3.63). Here the population PC ζ_1 can be regarded either as random or as fixed. (Section 6.8).

The second PC is obtained by estimating the coefficient vector $\boldsymbol{\alpha}_2$ such that

$$\mathbf{X} = \mathbf{Z}_1\boldsymbol{\alpha}_1^T + \mathbf{Z}_2\boldsymbol{\alpha}_2^T + \boldsymbol{\epsilon}_{(2)}$$

$$= \boldsymbol{\Psi}_1 + \boldsymbol{\Psi}_2 + \boldsymbol{\epsilon}_{(2)} \tag{3.77}$$

where $S(\mathbf{Z}_2) = (\boldsymbol{\Psi}_2 : \boldsymbol{\Psi}_2 = \mathbf{Z}_2\boldsymbol{\alpha}_2^T\}$ and

$$\mathbf{X} - \mathbf{Z}_1\boldsymbol{\alpha}_1^T = \mathbf{Z}_2\boldsymbol{\alpha}_2^T + \boldsymbol{\epsilon}_{(2)}$$

$$= \boldsymbol{\epsilon}_{(1)} \tag{3.78}$$

so that

$$\boldsymbol{\epsilon}_{(1)} = \mathbf{Z}_2\boldsymbol{\alpha}_2^T + \boldsymbol{\epsilon}_{(2)} \tag{3.79}$$

We seek an estimator of $\boldsymbol{\Psi}_2 = \mathbf{Z}_2\boldsymbol{\alpha}_2^T$, say $\mathbf{F}_2$, such that

$$(\boldsymbol{\epsilon}_{(1)} - \mathbf{F}_2)^T\mathbf{Z}_2 = \mathbf{0} \tag{3.80}$$

The distance is a minimum if and only if $\mathbf{F}_2$ is the orthogonal projection of the first stage residual matrix

$$\mathbf{H}_1 = \mathbf{X} - \mathbf{F}_1 \tag{3.81}$$

onto the second basis vector $\mathbf{Z}_2$, that is,

$$
\begin{aligned}
\mathbf{F}_2 &= \mathbf{P}_{z2}\mathbf{H}_1 \\
&= \mathbf{Z}_2(\mathbf{Z}_2^T\mathbf{Z}_2)^{-1}\mathbf{Z}_2^T\mathbf{H}_1 \\
&= \mathbf{Z}_2\mathbf{Z}_2^T(\mathbf{X} - \mathbf{F}_1) \\
&= \mathbf{Z}_2\mathbf{Z}_2^T\mathbf{X} \\
&= \mathbf{Z}_2\hat{\boldsymbol{\alpha}}_2^T
\end{aligned}
\tag{3.82}
$$

where $\hat{\boldsymbol{\alpha}}_2 = \mathbf{Z}_2^T\mathbf{X}$ and $\mathbf{Z}_2^T\mathbf{Z}_2 = 1$, $\mathbf{H}_1 = \mathbf{X} - \mathbf{Z}_1\hat{\boldsymbol{\alpha}}_1^T$, and $\mathbf{Z}_2^T\mathbf{F}_1 = 0$. An estimator $\mathbf{Z}_2$ of $\boldsymbol{\zeta}_2$ is obtained by taking the latent vector of $\mathbf{X}\mathbf{X}^T$ which corresponds to the largest latent root $h_2 = l_2$. The orthogonal projection induces the partition

$$
\mathbf{H}_1 = \mathbf{P}_{z2}\mathbf{H}_1 + (\mathbf{I} - \mathbf{P}_{z2})\mathbf{H}_1
\tag{3.83}
$$

and substituting for $\mathbf{H}_1$ yields

$$
\begin{aligned}
(\mathbf{X} - \mathbf{P}_{z1}\mathbf{X}) &= \mathbf{P}_{z2}(\mathbf{X} - \mathbf{P}_{z1}\mathbf{X}) + (\mathbf{I} - \mathbf{P}_{z2})(\mathbf{X} - \mathbf{P}_{z1}\mathbf{X}) \\
&= \mathbf{Z}_2\mathbf{Z}_2^T\mathbf{X} + \mathbf{X} - \mathbf{Z}_1\mathbf{Z}_1^T - \mathbf{Z}_2\mathbf{Z}_2^T\mathbf{X} \\
&= \mathbf{P}_{z2}\mathbf{X} + \mathbf{X} - \mathbf{P}_{z1}\mathbf{X} - \mathbf{P}_{z2}\mathbf{X}
\end{aligned}
$$

where $\mathbf{P}_{z1}\mathbf{P}_{z2} = 0$. The second stage decomposition of $\mathbf{X}$ is then

$$
\begin{aligned}
\mathbf{X} &= (\mathbf{P}_{z1} + \mathbf{P}_{z2})\mathbf{X} + [\mathbf{I} - (\mathbf{P}_{z1} + \mathbf{P}_{z2})]\mathbf{X} \\
&= \mathbf{P}_{(2)}\mathbf{X} + (\mathbf{I} - \mathbf{P}_{(2)})\mathbf{X}
\end{aligned}
\tag{3.84}
$$

We also have

$$
\begin{aligned}
\mathbf{X}^T\mathbf{X} &= (\mathbf{Z}_1\hat{\boldsymbol{\alpha}}_1^T + \mathbf{Z}_2\hat{\boldsymbol{\alpha}}_2^T + \mathbf{H}_2)^T(\mathbf{Z}_1\hat{\boldsymbol{\alpha}}_1^T + \mathbf{Z}_2\hat{\boldsymbol{\alpha}}_2^T + \mathbf{H}_2) \\
&= \hat{\boldsymbol{\alpha}}_1\hat{\boldsymbol{\alpha}}_1^T + \hat{\boldsymbol{\alpha}}_2\hat{\boldsymbol{\alpha}}_2^T + \mathbf{H}_2^T\mathbf{H}_2
\end{aligned}
\tag{3.85}
$$

so that the second-stage residual sums of squares matrix is given by

$$
\begin{aligned}
\mathbf{H}_2^T\mathbf{H}_2 &= \mathbf{X}^T\mathbf{X} - \hat{\boldsymbol{\alpha}}_1\hat{\boldsymbol{\alpha}}_1^T - \hat{\boldsymbol{\alpha}}_2\hat{\boldsymbol{\alpha}}_2^T \\
&= \mathbf{X}^T\mathbf{X} - \mathbf{X}^T\mathbf{P}_{z1}\mathbf{X} - \mathbf{X}^T\mathbf{P}_{z2}\mathbf{X} \\
&= \mathbf{X}^T(\mathbf{I} - \mathbf{P}_{z1} - \mathbf{P}_{z2})\mathbf{X}
\end{aligned}
\tag{3.86}
$$

where

$$
\mathbf{H}_2 = (\mathbf{I} - \mathbf{P}_{z1} - \mathbf{P}_{z2})
\tag{3.87}
$$

The process continues until $(\mathbf{I} - \mathbf{P}_{z1} - \mathbf{P}_{z2} - \cdots - \mathbf{P}_{zp})\mathbf{X} = \mathbf{0}$. Given $r \leq p$ we have

$$\mathbf{X} = (\mathbf{P}_{z1} + \mathbf{P}_{z2} + \cdots + \mathbf{P}_{zr})\mathbf{X} + [\mathbf{I} - (\mathbf{P}_{z1} + \mathbf{P}_{z2} + \cdots + \mathbf{P}_{zr})]\mathbf{X}$$

$$= \mathbf{Z}_{(r)}\hat{\boldsymbol{\alpha}}_{(r)}^{\mathrm{T}} + \hat{\boldsymbol{\epsilon}}_{(r)}$$

$$= \hat{\mathbf{X}}_{(r)} + \hat{\boldsymbol{\epsilon}}_{(r)} \tag{3.88}$$

where $\hat{\mathbf{X}}_{(r)}$ is the $(n \times p)$ matrix of predicted values, and $\hat{\boldsymbol{\alpha}}$ and $\mathbf{Z}_i$ are the latent vectors of $\mathbf{X}^{\mathrm{T}}\mathbf{X}$ and $\mathbf{X}\mathbf{X}^{\mathrm{T}}$ respectively. As is shown in Theorem 3.17, however, it is wasteful to decompose both matrices, and in practice $\mathbf{Z}_i$ are taken as PC scores of $\mathbf{X}^{\mathrm{T}}\mathbf{X}$. Evidently when $r = p$ decomposition (Eq. 3.88) reproduces $\mathbf{X}$ exactly. Note that Eq. (3.88) can also be viewed as representing a linear hypothesis of the type

$$\mathbf{H}_0: \quad \Sigma \text{ contains } r \text{ components (factors)}$$

$$\mathbf{H}_a: \quad \Sigma \text{ is arbitrary} \tag{3.89}$$

the testing of which is deferred to Chapter 4. When $\mathbf{H}_0$ is accepted, the first r components $\mathbf{Z}_{(r)}$ are viewed as estimators of the unobservable latent variables $\boldsymbol{\zeta}_{(r)}$, which explain the behavior of $\mathbf{X}$. Then $\hat{\boldsymbol{\epsilon}}_{(r)}$ is the $(n \times p)$ matrix of unexplained residual variation. The coordinate-free projection viewpoint of the PC model provides an analogy with the more familiar regression model. Also, the following two points emerge more clearly.

1. Since the residual sum-of-squares matrix $\hat{\boldsymbol{\epsilon}}_{(r)}^{\mathrm{T}}\hat{\boldsymbol{\epsilon}}_{(r)}$ is not diagonal, the residual terms are not assumed to be uncorrelated. The effect of residual variation is therefore assumed to be felt on both the diagonal as well as off-diagonal terms of $\mathbf{X}^{\mathrm{T}}\mathbf{X}$.

2. Equation (3.88) indicates the looseness of the statement "the PCA represents an exact decomposition of a dispersion matrix," since the residual matrix $\hat{\boldsymbol{\epsilon}}_{(r)}$ can clearly be endowed with stochastic characteristics which results in a stochastic decomposition.

Projective properties of PCA are summarized in the following theorem.

THEOREM 3.16. Let $\mathbf{X}$ be a $(n \times p)$ data matrix and let $\mathbf{P}_z = \mathbf{Z}(\mathbf{Z}^{\mathrm{T}}\mathbf{Z})^{-1}\mathbf{Z}^{\mathrm{T}} = \mathbf{Z}\mathbf{Z}^{\mathrm{T}}$ represent the $(n \times n)$ matrix which projects observations orthogonally onto the $r \leq p$ dimensional subspace spanned by $r \leq p$ PCs. Then

    (i) The predicted values of $\mathbf{X}$ are given by

$$\hat{\mathbf{X}} = \mathbf{P}_z\mathbf{X} \tag{3.90}$$

(ii) The predicted sums-of-squares are given by

$$\operatorname{tr}(\mathbf{X}^{\mathrm{T}}\mathbf{P}_z\mathbf{X}) = l_1 + l_2 + \cdots + l_r \tag{3.91}$$

(iii) The predicted sums of squares matrix is given by

$$\hat{\mathbf{X}}^{\mathrm{T}}\hat{\mathbf{X}} = \hat{\mathbf{a}}_{(r)}\hat{\boldsymbol{\alpha}}_{(r)}^{\mathrm{T}} \tag{3.92}$$

where $\hat{\boldsymbol{\alpha}}_{(r)} = \mathbf{P}_{(r)}\mathbf{L}_{(r)}^{1/2} = \mathbf{A}_{(r)}$ are sample correlation loading coefficients (Section 3.4.2) so that the predicted sample covariance matrix is

$$\hat{\mathbf{S}} = \frac{1}{n-1}\,\hat{\boldsymbol{\alpha}}_{(r)}\hat{\boldsymbol{\alpha}}_{(r)}^{\mathrm{T}} \tag{3.93}$$

PROOF

(i) We have

$$\begin{aligned}
\hat{\mathbf{X}} = \mathbf{P}_z\mathbf{X} &= (\mathbf{P}_{z1} + \mathbf{P}_{z2} + \cdots + \mathbf{P}_{zr})\mathbf{X} \\
&= \mathbf{P}_{z1}\mathbf{X} + \mathbf{P}_{z2}\mathbf{X} + \cdots + \mathbf{P}_{zr}\mathbf{X} \\
&= \mathbf{Z}_1\mathbf{Z}_1^{\mathrm{T}}\mathbf{X} + \mathbf{Z}_2\mathbf{Z}_2^{\mathrm{T}}\mathbf{X} + \cdots + \mathbf{Z}_r\mathbf{Z}_r^{\mathrm{T}}\mathbf{X} \\
&= \mathbf{Z}_1\hat{\boldsymbol{\alpha}}_1^{\mathrm{T}} + \mathbf{Z}_2\hat{\boldsymbol{\alpha}}_2^{\mathrm{T}} + \cdots + \mathbf{Z}_r\hat{\boldsymbol{\alpha}}_r^{\mathrm{T}}
\end{aligned} \tag{3.94}$$

the predicted values of X, using the first r PCs.

(ii) For the predicted sums of squares we have

$$\begin{aligned}
\operatorname{tr}(\mathbf{X}^{\mathrm{T}}\mathbf{P}_z\mathbf{X}) &= \operatorname{tr}(\mathbf{X}^{\mathrm{T}}\mathbf{Z}_{(r)}\mathbf{Z}_{(r)}^{\mathrm{T}}\mathbf{X}) \\
&= \operatorname{tr}(\hat{\boldsymbol{\alpha}}_{(r)}\hat{\boldsymbol{\alpha}}_{(r)}^{\mathrm{T}}) \\
&= \operatorname{tr}(\hat{\boldsymbol{\alpha}}_{(r)}^{\mathrm{T}}\hat{\boldsymbol{\alpha}}_{(r)}) \\
&= l_1 + l_2 + \cdots + l_r
\end{aligned}$$

(iii) We have

$$\begin{aligned}
\hat{\mathbf{X}}^{\mathrm{T}}\hat{\mathbf{X}} &= (\mathbf{P}_z\mathbf{X})^{\mathrm{T}}(\mathbf{P}_z\mathbf{X}) \\
&= \mathbf{X}^{\mathrm{T}}\mathbf{Z}_{(r)}\mathbf{Z}_{(r)}^{\mathrm{T}}\mathbf{X} \\
&= \hat{\boldsymbol{\alpha}}_{(r)}\hat{\boldsymbol{\alpha}}_{(r)}^{\mathrm{T}}
\end{aligned}$$

where $\hat{\boldsymbol{\alpha}}_{(r)} = \mathbf{P}_{(r)}\mathbf{L}_{(r)}^{1/2}$. Using Theorem 3.15 we have

$$\frac{1}{n-1}\hat{\boldsymbol{\alpha}}_{(r)}\hat{\boldsymbol{\alpha}}_{(r)}^{\mathrm{T}} = \frac{1}{n-1}\mathbf{P}_{(r)}\mathbf{L}_{(r)}^{1/2}\mathbf{L}_{(r)}^{1/2}\mathbf{P}_{(r)}^{\mathrm{T}}$$

$$= \frac{1}{n-1}\mathbf{P}_{(r)}\mathbf{L}_{(r)}\mathbf{P}_{(r)}^{\mathrm{T}}$$

$$= \frac{1}{n-1}\hat{\mathbf{X}}^{\mathrm{T}}\hat{\mathbf{X}}$$

$$= \hat{\mathbf{S}}$$

Deriving the PC model by orthogonal projections provides a geometric perspective of the model, akin to least squares theory since formally speaking we can consider the first r components as explanatory statistical variables, which happen to be unobserved. Also the PCs can be characterized by the spectral decompositions of $\mathbf{XX}^{\mathrm{T}}$, as well as $\mathbf{X}^{\mathrm{T}}\mathbf{X}$. Indeed there exists a dual relationship between the spectra of the two matrices, which makes it unnecessary to compute both sets of latent root and latent vectors.

THEOREM 3.17 (Singular Value Decomposition). Let $\mathbf{X}$ be a $(n \times p)$ data matrix such that $\rho(\mathbf{X}) = p \leq n$, $\bar{\mathbf{X}} = \mathbf{0}$, and let $\mathbf{X}^{\mathrm{T}}\mathbf{X}$ and $\mathbf{XX}^{\mathrm{T}}$ be $(p \times p)$ and $(n \times n)$ Grammian matrices respectively. Then

   (i) Nonzero latent vectors of $\mathbf{XX}^{\mathrm{T}}$ are standardized PCs of $\mathbf{X}^{\mathrm{T}}\mathbf{X}$.

   (ii) Nonzero latent roots of $\mathbf{X}^{\mathrm{T}}\mathbf{X}$ and $\mathbf{XX}^{\mathrm{T}}$ are equal.

   (iii) Any real $(n \times p)$ matrix $\mathbf{X}$ can be decomposed as

$$\mathbf{X} = l_1^{1/2}\mathbf{Q}_1\mathbf{P}_1^{\mathrm{T}} + l_2^{1/2}\mathbf{Q}_2\mathbf{P}_2^{\mathrm{T}} + \cdots + l_p^{1/2}\mathbf{Q}_p\mathbf{P}_p^{\mathrm{T}} \tag{3.95}$$

where $\mathbf{Q}_i$ is the ith latent vector of $\mathbf{XX}^{\mathrm{T}}$ and $\mathbf{P}_i$ is the ith latent vector of $\mathbf{X}^{\mathrm{T}}\mathbf{X}$.

   (iv) $\mathbf{XX}^{\mathrm{T}}$ can be decomposed as

$$\mathbf{XX}^{\mathrm{T}} = \mathbf{ZLZ}^{\mathrm{T}} = l_1\mathbf{Z}_1\mathbf{Z}_1^{\mathrm{T}} + l_2\mathbf{Z}_2\mathbf{Z}_2^{\mathrm{T}} + \cdots + l_p\mathbf{Z}_p\mathbf{Z}_p^{\mathrm{T}}$$

$$= l_1\mathbf{P}_{z1} + l_2\mathbf{P}_{z2} + \cdots + l_p\mathbf{P}_{zp} \tag{3.96}$$

where $\mathbf{P}_{zi} = \mathbf{Z}_i(\mathbf{Z}_i^{\mathrm{T}}\mathbf{Z}_i)^{-1}\mathbf{Z}_i^{\mathrm{T}} = \mathbf{Z}_i\mathbf{Z}_i^{\mathrm{T}} = \mathbf{Q}_i\mathbf{Q}_i^{\mathrm{T}}$ are matrices which project vectors orthogonally onto $\mathbf{Z}_i$ $(i = 1, 2, \ldots, p)$, and $\mathbf{Z} = \mathbf{Q}$.

   (v) Equation (3.95) minimizes the sum of squared errors

$$\mathbf{E} = \mathbf{X} - \mathbf{Q}_r\mathbf{L}_r^{1/2}\mathbf{P}_r^{\mathrm{T}}$$

PROOF

(i) Let

$$(\mathbf{X}^T\mathbf{X})\mathbf{P} = \mathbf{P}\mathbf{L} \tag{3.97a}$$

and

$$(\mathbf{X}\mathbf{X}^T)\mathbf{Q} = \mathbf{Q}\mathbf{M} \tag{3.97b}$$

where $\mathbf{L}$ and $\mathbf{M}$ are latent roots and $\mathbf{P}$ and $\mathbf{Q}$ latent vectors of $\mathbf{X}^T\mathbf{X}$ and $\mathbf{X}\mathbf{X}^T$ respectively. Since $\rho(\mathbf{X}) = \rho(\mathbf{X}^T\mathbf{X}) = \rho(\mathbf{X}\mathbf{X}^T) = p$, the $(n \times n)$ matrix $\mathbf{X}\mathbf{X}^T$ can only have p nonzero latent roots (Theorem 2.6). Premultiplying Eq. (3.97b) by $\mathbf{X}$ we have

$$(\mathbf{X}\mathbf{X}^T)\mathbf{X}\mathbf{P} = \mathbf{X}\mathbf{P}\mathbf{L} \tag{3.98}$$

where $\mathbf{Z} = \mathbf{X}\mathbf{P}$ are unstandardized latent vectors of $\mathbf{X}\mathbf{X}^T$ (PCs of $\mathbf{X}^T\mathbf{X}$). Standardizing, we have

$$\mathbf{Q} = \mathbf{X}\mathbf{P}\mathbf{L}^{-1/2} = \mathbf{Z}^* \tag{3.99}$$

the standardized matrix of PCs of $\mathbf{X}^T\mathbf{X}$.
From part i we have

(ii)

$$
\begin{aligned}
\mathbf{M} = \mathbf{Q}^T(\mathbf{X}\mathbf{X}^T)\mathbf{Q} &= (\mathbf{X}\mathbf{P})^T(\mathbf{X}\mathbf{X}^T)\mathbf{X}\mathbf{P} \\
&= \mathbf{P}^T\mathbf{X}^T(\mathbf{X}\mathbf{X}^T)\mathbf{X}\mathbf{P} \\
&= (\mathbf{Z}^T\mathbf{X})(\mathbf{X}^T\mathbf{Z}) \\
&= \mathbf{A}^T\mathbf{A} \\
&= \mathbf{L}
\end{aligned}
$$

from Theorem 3.13, so that

$$
\mathbf{M} = \begin{bmatrix} l_1 & & & & \\ & l_2 & & & \mathbf{0} \\ & & \ddots & & \\ & & & l_p & \\ & \mathbf{0} & & & \mathbf{0} \end{bmatrix} = \begin{bmatrix} \mathbf{L} & \mathbf{0} \\ \mathbf{0} & \mathbf{0} \end{bmatrix} \tag{3.100}
$$

and it follows that nonzero roots of $\mathbf{M}$ and $\mathbf{L}$ must be equal.

(iii) We have

$$\mathbf{X} = \mathbf{P}_z\mathbf{X} = (\mathbf{P}_{z1} + \mathbf{P}_{z2} + \cdots + \mathbf{P}_{zp})\mathbf{X}$$

$$= \mathbf{Z}_1\mathbf{Z}_1^{\mathrm{T}}\mathbf{X} + \mathbf{Z}_2\mathbf{Z}_2^{\mathrm{T}}\mathbf{X} + \cdots + \mathbf{Z}_p\mathbf{Z}_p^{\mathrm{T}}\mathbf{X}$$

$$= \mathbf{Z}_1\hat{\boldsymbol{\alpha}}_1^{\mathrm{T}} + \mathbf{Z}_2\hat{\boldsymbol{\alpha}}_2^{\mathrm{T}} + \cdots + \mathbf{Z}_p\hat{\boldsymbol{\alpha}}_p^{\mathrm{T}}$$

$$= l_1^{1/2}\mathbf{Q}_1\mathbf{P}_1^{\mathrm{T}} + l_2^{1/2}\mathbf{Q}_2\mathbf{P}_2^{\mathrm{T}} + \cdots + l_p^{1/2}\mathbf{Q}_p\mathbf{P}_p^{\mathrm{T}} \qquad (3.101)$$

since $\mathbf{Q} = \mathbf{Z}$ from part i of the proof, and $\hat{\boldsymbol{\alpha}}_i = l_i^{1/2}\mathbf{P}_i^{\mathrm{T}} = \mathbf{A}_i$

(iv) From Eq. (3.98) we have

$$\mathbf{X}\mathbf{X}^{\mathrm{T}} = \mathbf{Q}\mathbf{M}\mathbf{Q}^{\mathrm{T}} = \mathbf{Q}\mathbf{L}\mathbf{O}^{\mathrm{T}}$$

$$= l_1\mathbf{Q}_1\mathbf{Q}_1^{\mathrm{T}} + l_2\mathbf{Q}_2\mathbf{Q}_2^{\mathrm{T}} + \cdots + l_p\mathbf{Q}_p\mathbf{Q}_p^{\mathrm{T}}$$

$$= l_1\mathbf{Z}_1\mathbf{Z}_1^{\mathrm{T}} + l_2\mathbf{Z}_2\mathbf{Z}_2^{\mathrm{T}} + \cdots + l_p\mathbf{Z}_p\mathbf{Z}_p^{\mathrm{T}}$$

$$= l_1\mathbf{P}_{z1} + l_2\mathbf{P}_{z2} + \cdots + l_p\mathbf{P}_{zp}$$

(v) The sum of squared errors is given by, for $s = p - r$,

$$\mathbf{E} = \mathbf{X} - \mathbf{Q}_r\mathbf{L}_r^{1/2}\mathbf{P}_r^{\mathrm{T}} = \mathbf{Q}_s\mathbf{L}_s^{1/2}\mathbf{Q}_s\mathbf{L}_s^{1/2}\mathbf{P}_s^{\mathrm{T}} \qquad (3.102)$$

If $1 \le r < P$ components are retained. The sum of squared errors is then

$$\mathrm{tr}(\mathbf{E}^{\mathrm{T}}\mathbf{E}) = \mathrm{tr}(\mathbf{P}_s\mathbf{L}_s^{1/2}\mathbf{Q}_s^{\mathrm{T}}\mathbf{Q}_s\mathbf{L}_s^{1/2}\mathbf{P}_s^{\mathrm{T}})$$

$$= \mathrm{tr}(\mathbf{P}_s\mathbf{L}_s^{1/2}\mathbf{L}_s^{1/2}\mathbf{P}_s^{\mathrm{T}})$$

$$= \mathrm{tr}(\mathbf{P}_s\mathbf{L}_s\mathbf{P}_s^{\mathrm{T}})$$

$$= \mathrm{tr}(\mathbf{L}_s)$$

$$= l_s + l_{s+1} + \cdots + l_p \qquad (3.102a)$$

the sum of the last $s = p - r$ latent roots which, by Theorem 3.9, are minimized.

Equation (3.101) is known as the singular value decomposition of a rectangular matrix $\mathbf{X}$ where $\rho(\mathbf{X}) = p \le n$. When the columns of $\mathbf{X}$ are centered at zero, it is unnecessary to compute loadings and scores of $\mathbf{X}\mathbf{X}^{\mathrm{T}}$ since the loading coefficients of $\mathbf{X}^{\mathrm{T}}\mathbf{X}$ are suitably normalized scores of $\mathbf{X}\mathbf{X}^{\mathrm{T}}$, and vice versa. While loadings yield information concerning the variable space, scores provide information about the sample space. The point is made more clearly by the following. From the singular value Theorem 3.17

we have

$$Q^T X P = L^{1/2} \rightarrow X = Q L^{1/2} P^T \tag{3.103a}$$

and transposing yields

$$P^T X^T Q = L^{1/2} \rightarrow X^T = P L^{1/2} Q^T \tag{3.103b}$$

where $Q^T Q = I$, but $Q Q^T$ is a projection matrix. Also, $Q = Z$ are the left latent vectors of X and P are the right latent vectors of X. Whereas Eq. (3.103a) represents a decomposition of the column (variable) space of X, Eq. (3.103b) represents a decomposition of the row (sample) space of X. Clearly both spaces provide distinct but complementary information. It is easy to verify that pre (post) multiplying Eq. (3.103a) by Eq. (3.103b) yields the Grammian decompositions of Eqs. (3.97a) and (3.97b) respectively. This simply represents a restatement of Theorem 3.11 (see also Exercise 3.11). More generally it can be shown that for any two matrices (real or complex) A, B the nonzero latent roots of AB are the same as those of BA, assuming the products exist (Good, 1969). The latent vectors however are different. Note that the symmetry between the loadings and the scores does not hold for covariance matrices of the rows and columns since the centering of the row and column vectors is not the same, and the spaces are therefore not comparable.

Example 3.9.   Consider the data matrix from Example 3.8, where columns are adjusted to zero mean and unit length. The correlation matrix for the columns is

$$X^T X = \begin{bmatrix} 1.0000 & -.8448 & -.2160 \\ -.8448 & 1.0000 & -.2608 \\ -.2167 & -.2608 & 1.0000 \end{bmatrix}$$

and the sum of squares and products matrix between the rows

$$X X^T = \begin{bmatrix} .6068 & -.3819 & .3649 & -.5898 \\ -.3819 & 1.0985 & -.6532 & -.0633 \\ .3649 & -.6532 & .4650 & -.1766 \\ -.5898 & -.0633 & -.1767 & .8297 \end{bmatrix}$$

Note that this is not a correlation matrix since rows of Y are not unit vectors. The latent roots of $X X^T$ are given by

$$M = \begin{bmatrix} 1.84595 & & 0 \\ & 1.11718 & \\ 0 & & .03682 \end{bmatrix} = \begin{bmatrix} L & \vdots & 0 \\ \cdots & & \cdots \\ 0 & \vdots & 0 \end{bmatrix}$$

and those of $\mathbf{X}^T\mathbf{X}$ are

$$\mathbf{L} = \begin{bmatrix} 1.84595 & & \mathbf{0} \\ & 1.11718 & \\ \mathbf{0} & & .03682 \end{bmatrix}$$

The latent vectors of $\mathbf{X}^T\mathbf{X}$ and $\mathbf{X}\mathbf{X}^T$ are columns, respectively, of

$$\mathbf{P} = \begin{bmatrix} -.7009 & -.2626 & -.6632 \\ .7122 & -.2054 & -.6713 \\ -.0401 & .9428 & -.3309 \end{bmatrix}, \quad \mathbf{Q} = \begin{bmatrix} .4970 & .3500 & .6162 & 0 \\ -.6438 & .5451 & -.1965 & 0 \\ .4782 & -.1478 & -.7073 & 0 \\ -.3314 & -.7433 & .2852 & 0 \end{bmatrix}$$

and unstandardized PC scores of $\mathbf{X}$ are

$$\mathbf{Z} = \mathbf{X}\mathbf{P}$$

$$= \begin{bmatrix} -.6489 & .3255 & .2825 \\ .4867 & -.7160 & .5907 \\ -.3244 & .5859 & -.1284 \\ .4867 & -.1953 & -.7448 \end{bmatrix} \begin{bmatrix} -.7009 & -.2626 & -.6632 \\ .7122 & -.2054 & -.6713 \\ -.0401 & .9428 & -.3309 \end{bmatrix}$$

$$= \begin{bmatrix} .6753 & .3699 & .1184 \\ -.8748 & .5761 & -.0376 \\ .6498 & -.1562 & -.1356 \\ -.4504 & -.7899 & .0548 \end{bmatrix}$$

Standardizing columns to unit length then leads to

$$\mathbf{Z}^* = \mathbf{X}\mathbf{P}\mathbf{L}^{-1/2} = \mathbf{Z}\mathbf{L}^{-1/2}$$

$$= \begin{bmatrix} .6753 & .3699 & .1184 \\ -.8748 & .5761 & -.0376 \\ .6498 & -.1562 & -.1356 \\ -.4504 & -.7899 & .0548 \end{bmatrix} \begin{bmatrix} 1/1.35866 & & \mathbf{0} \\ & 1/1.05697 & \\ \mathbf{0} & & 1.19187 \end{bmatrix}$$

$$= \begin{bmatrix} .4970 & .3500 & .6162 \\ -.6438 & .5451 & -.1965 \\ .4782 & -.1478 & -.7033 \\ -.3314 & -.7473 & .2852 \end{bmatrix}$$

which is identical to the nonzero submatrix contained in $\mathbf{Q}$. Equivalent results also hold for the covariance matrix, the sum of products, and the cosine matrices described in Chapter 2.

Example 3.10. An immediate implication of Theorem 3.17 is that we may compare and relate points in the variable space to those in the sample space. Consider a horticultural experiment where bud-cut chrysanthemums are used to evaluate the effects of different levels of sucrose and 8-hydroxyquinoline citrate on the opening of immature flowers (Broschat, 1979). Experimental data representing means of three replications are given in Table 3.9 where

$$Y_1 = \text{initial fresh weight (gm)}$$
$$Y_2 = \text{final fresh weight (gm)}$$
$$Y_3 = \text{final dry weight (gm)}$$
$$Y_4 = \text{flower diameter (mm)}$$
$$Y_5 = \text{flower height (mm)}$$
$$Y_6 = \text{postharvest life (days)}$$

The treatments consist of nine combinations of % sucrose/ppm and 8-HQC, which are thought to affect flower-keeping quality. We thus have a MANOVA design resulting from the intercorrelations that exist between the dependent variables. A PCA can be carried out to determine which treatments influence which dependent variable(s). First the correlation loadings of $X^T X$ are computed for the first two PCs (Table 3.10, Fig. 3.5), where we perceive two distinct clusters. The first cluster consists of X_1, X_4 and X_5, which is most closely related to Z_1. Since the three variables measure initial flower size, component Z_1 can be identified with an underlying dimension of initial flower size or the overall flower quality. Consulting the scores of $X^T X$ (Table 3.11), we also see that the flower size component Z_1 is related positively to the last three treatments, all of which contain high sucrose concentrations. It scores negatively for treatments 1

Table 3.9   Effect of Sucrose and an Antimicrobal Agent, 8-Hydroxyquinoline Citrate (8-HQC), on the Opening and Keeping Quality of Bud-Cut Chrysanthemums

Treatment	Y_1	Y_2	Y_3	Y_4	Y_5	Y_6
0/0	52.9	52.6	12.1	54.6	32.7	5.7
0/200	53.5	62.3	11.6	56.3	32.4	9.8
0/400	54.5	65.9	12.2	57.8	34.1	8.2
2/0	52.4	57.4	13.4	54.4	32.9	4.2
2/200	52.8	74.1	15.9	52.5	32.7	14.8
2/400	53.1	78.4	15.4	54.9	33.0	9.8
4/0	59.2	64.1	15.0	61.4	35.6	3.7
4/200	54.6	85.5	21.6	55.2	33.3	18.8
4/400	58.7	88.2	20.1	60.1	33.9	12.0

Source: Broschat, 1979.

Table 3.10 Correlation Loadings for Six Experimental Variables and Two Principal Components, Using the Data of Table 3.9.

Variables	Z_1	Z_2
X_1 (fresh wt., initial)	.942	−.280
X_2 (fresh wt., final)	.626	.736
X_3 (dry wt., final)	.661	.689
X_4 (flower diameter)	.818	−.519
X_5 (flower height)	.797	−.501
X_6 (post harvest life)	.184	.939
Latent Roots	3.054	2.495
Variance (%)	50.90	41.58

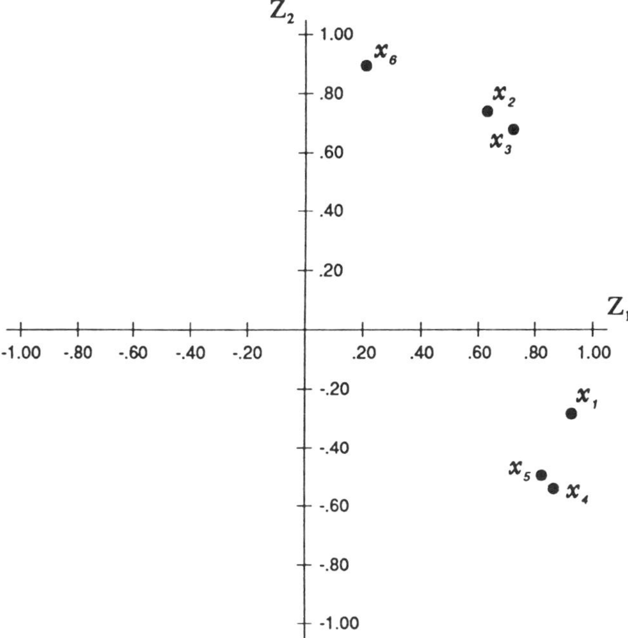

Figure 3.5 Correlation loadings of Table 3.10 representing $p = 6$ flower-keeping variables, in a two-dimensional subspace.

(controls) and 4, followed by 2 and 5, since these represent low combinations of sucrose and 8-HQC.

The second component Z_2 is positively correlated to X_2, X_3, and X_6 and represents flower-keeping quality. Again, consulting the scores (Table 3.11) we see that best results are obtained with 200 ppm of 8-HQC and 4% sucrose (treatment 8) followed by 2% sucrose (treatment 5). The worst results for keeping quality are produced by treatments 7, 3, 1, and 4, all of which lack either sucrose or 8-HQC. The best total effect of size and

Table 3.11 Principal Component Scores for the First Two Components. The Scores are Standardized so that Their Sum of Squares are Equal to the Degrees of Freedom $n - 1$

Treatment	Z_1	Z_2
0/0	-1.080	$-.612$
0/200	$-.749$	$-.199$
0/400	.031	$-.681$
2/0	$-.966$	$-.513$
2/200	$-.590$	1.037
2/400	$-.271$	.479
4/0	1.410	-1.613
4/200	.619	1.660
4/400	1.595	.441

keeping quality is therefore achieved in the vicinity of treatments 8 and 9, that is, combinations using high sucrose and/or high 8-HQC.

Although a PCA cannot be used directly for testing experimental hypotheses in a straightforward manner it can nevertheless be used as an exploratory procedure which may suggest hypotheses or reveal broad relationships between variables and treatments. The PCs can also be useful as a first step in a MANOVA of multivariate experimental data (Chapter 10.5). Since the first two PCs of Example 3.10 account for 92.5% of the variance, the remaining four can be taken to represent random residual error. The first two components therefore capture most of the experimental effect, and a MANOVA can be replaced by a standard ANOVA repeated separately for Z_1 and Z_2. Figure 3.5 also reveals that Z_1 and Z_2 can be rotated orthogonally to coincide with the two clusters, indicating that the treatments seem to have an independent effect on initial flower size and keeping quality.

3.6 PRINCIPAL COMPONENTS BY LEAST SQUARES

Referring to the last section, Eq. (3.69) assumes the existence of r explanatory PCs or factors which account for systematic variation in the observed variables. The error matrix $\hat{\epsilon}_{(r)}$ is then viewed as containing residual sampling variations and other random effects which are peculiar to each variable and sample point, but which are independent of the systematic variation $Z_{(r)}$. The explicit distinction between explained and residual variance, together with their orthogonal separation, leads to yet another development of PCA reminiscent of regression analysis, due to Whittle (1953). In this form, PCA is also one of the special cases of "proper" factor analysis which is considered further in Chapter 6.

Let ζ, α, represent fixed population parameters and let ϵ denote a random residual term. Then given a multivariate sample of n observations for p random variables we can write

$$\mathbf{X} = \zeta\alpha^{\mathrm{T}} + \epsilon \qquad (3.104)$$

where we wish to find estimators $\hat{\zeta}$ and $\hat{\alpha}$, of ζ and α which minimize the sums of squares of both $\epsilon^{\mathrm{T}}\epsilon$ and $\epsilon\epsilon^{\mathrm{T}}$. Assuming the residuals are homoscedastic and uncorrelated we have

$$\epsilon^{\mathrm{T}}\epsilon = \sigma^2\mathbf{I} = (\mathbf{X} - \zeta\alpha^{\mathrm{T}})^{\mathrm{T}}(\mathbf{X} - \xi\alpha^{\mathrm{T}})$$

$$= \mathbf{X}^{\mathrm{T}}\mathbf{X} - \mathbf{X}^{\mathrm{T}}\zeta\alpha^{\mathrm{T}} - \alpha\zeta^{\mathrm{T}}\mathbf{X} + \alpha\zeta^{\mathrm{T}}\zeta\alpha^{\mathrm{T}} \qquad (3.105)$$

and

$$\epsilon\epsilon^{\mathrm{T}} = \sigma^2\mathbf{I} = (\mathbf{X} - \zeta\alpha^{\mathrm{T}})(\mathbf{X} - \zeta\alpha^{\mathrm{T}})^{\mathrm{T}}$$

$$= \mathbf{X}\mathbf{X}^{\mathrm{T}} - \mathbf{X}\alpha\zeta^{\mathrm{T}} - \zeta\alpha^{\mathrm{T}}\mathbf{X} + \zeta\alpha^{\mathrm{T}}\alpha\zeta^{\mathrm{T}} \qquad (3.106)$$

This approach is somewhat different theoretically from the more standard PC model, where only loadings are fixed but the scores are assumed to be random. A similar distinction is also made by Okamoto (1976).

Because of the fixed nature of α and ζ, their components can be taken as parameters of the population, similar to the basic regression model. Differentiating Eq. (3.105) with respect to α and setting to zero we have

$$\frac{\partial(\epsilon^{\mathrm{T}}\epsilon)}{\partial\alpha} = -2\mathbf{X}^{\mathrm{T}}\hat{\zeta} + 2\hat{\alpha}\hat{\zeta}^{\mathrm{T}}\hat{\zeta} = 0 \quad \text{or} \quad \mathbf{X}^{\mathrm{T}}\hat{\zeta} = \hat{\alpha}\hat{\zeta}^{\mathrm{T}}\hat{\zeta} \qquad (3.107)$$

and premultiplying by $\mathbf{X}$ yields

$$\mathbf{X}\mathbf{X}^{\mathrm{T}}\hat{\zeta} = \mathbf{X}\hat{\alpha}\hat{\zeta}^{\mathrm{T}}\hat{\zeta}$$

$$= \hat{\zeta}\hat{\zeta}^{\mathrm{T}}\hat{\zeta}$$

since $\mathbf{X}\hat{\alpha} = \hat{\zeta}$. Thus

$$(\mathbf{X}\mathbf{X}^{\mathrm{T}})\hat{\zeta} = \hat{\zeta}\hat{\Lambda} \qquad (3.108)$$

where $\hat{\zeta}^{\mathrm{T}}\hat{\zeta} = \hat{\Lambda}$ is diagonal. The least squares estimators $\hat{\zeta}$ and $\hat{\Lambda}$ are thus latent vectors and latent roots of $\mathbf{X}\mathbf{X}^{\mathrm{T}}$. Also, differentiating Eq. (3.106) with respect to ζ and setting to zero yields

$$\frac{\partial(\epsilon\epsilon^{\mathrm{T}})}{\partial\zeta} = -2\mathbf{X}\hat{\alpha} + 2\hat{\zeta}\hat{\alpha}^{\mathrm{T}}\hat{\alpha} = 0 \quad \text{or} \quad \mathbf{X}\hat{\alpha} = \hat{\zeta}\hat{\alpha}^{\mathrm{T}}\hat{\alpha} \qquad (3.109)$$

and premultiplying by $\mathbf{X}^T$ and setting $\hat{\boldsymbol{\alpha}} = \mathbf{X}^T\hat{\mathbf{a}}$ we have

$$(\mathbf{X}^T\mathbf{X})\hat{\mathbf{a}} = \hat{\mathbf{a}}\hat{\boldsymbol{\Lambda}} \tag{3.110}$$

so that $\hat{\mathbf{a}} = \mathbf{P}$ and $\hat{\boldsymbol{\Lambda}} = \mathbf{L}$ are latent vectors and roots of $\mathbf{X}^T\mathbf{X}$ respectively (Section 3.4). Whittle's (1953) derivation thus leads to a PCA where $\hat{\boldsymbol{\alpha}}$ and $\hat{\boldsymbol{\zeta}}$ are latent vectors (loadings) and scores of $\mathbf{X}^T\mathbf{X}$ respectively (Theorem 3.17). Since the normal equations (Eqs. 3.107 and 3.108) assume that $\boldsymbol{\alpha}$ and $\boldsymbol{\zeta}$ are fixed population parameters, it may be of interest to find their maximum likelihood estimators (Section 1.3) under the assumption of multivariate normality. It has been shown however (Solari, 1969) that such estimators do not exist. The sole optimality property of Whittle's estimators is therefore that of least squares. This can be seen from Equations (3.107) and (3.109) by solving for $\hat{\boldsymbol{\alpha}}$ and $\hat{\boldsymbol{\zeta}}$.

3.7 NONLINEARITY IN THE VARIABLES

Both population as well as sample PC models represent linear decompositions in that (1) the components are linear combinations of the original variables and (2) the decompositions are linear in the loading coefficients. Conversely, once loadings and scores are known, the variables can be expressed as linear combinations of the PCs. However, much in the same way as for the regression model, linearity is not required for the original variables. The PCs, for instance, can represent linear combinations of nonlinear functions of the random variables. Two types of nonlinearities commonly occur in practice—those due to polynomial and exponential (constant-rate-of-growth) behavior. When these occur a set of highly related random variables can exhibit low correlation unless nonlinearity is taken into account. This however can be done in a straightforward manner, much the same as for the regression model.

As an example consider two variables X_1 and X_2 which are related by a quadratic polynomial. Clearly a correlation coefficient computed using X_1 and X_2 is somewhat meaningless since the interrelationship between the two variables is nonlinear. To reflect the quadratic nature of the interrelationship the correlation coefficient must therefore be computed using quadratic terms as well. A PC obtained from such quadratic correlation is then itself a quadratic function of the form

$$z_i = p_{1i}x_1 + p_{2i}x_2 + p_{3i}x_1^2 + p_{4i}(x_1x_2) + p_{5i}x_2^2 . \tag{3.111}$$

Letting $Y_1 = x_i$, $Y_2 = x_2, \ldots, Y_5 = x_2^2$, Eq. (3.109) can be written as

$$z_i = p_{1i}Y_1 + p_{2i}Y_2 + p_{3i}Y_3 + p_{4i}Y_4 + p_{5i}Y_5$$

a linear function of the transformed variables. Equations such as Eq. (3.111)

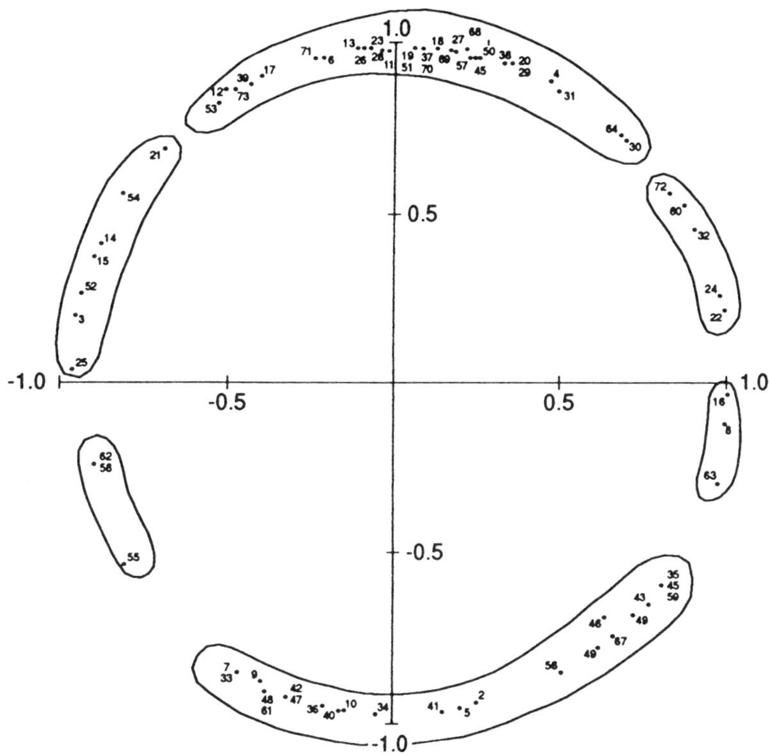

Figure 3.6 Principal components loadings for $p = 73$ linear and quadratic economic variables.

can be extended to any number of variables and any degree, although polynomials of degree higher than $k = 3$ are rare in practice because of interpretational difficulties. Figure 3.6 represents a two-dimensional scatter of loading coefficients for $p = 73$ linear and quadratic terms. Owing to the quadratic terms amongst the variables there is a clear tendency for the loadings to lie in a quadratic (circular) pattern. Also, since linear and quadratic terms tend to be highly intercorrelated, a PC transformation tends to concentrate variance in the first few latent roots. The PCs here find a natural interpretation as orthogonal polynomials.

Example 3.11. McDonald (1962) has considered orthogonal polynomials from the point of view of PCA. Consider

$$x = b_i z + c_i (z^2 - 1) + a_i e$$

$$= f(z) + a_i e \tag{3.112}$$

which is used to generate $p = 7$ variables by assigning particular values to the coefficients (Table 3.12). It is further assumed that $E(x_i) = E(z) = 0$ and

Table 3.12   Coefficients for Generating Eq. (3.110)

Variable	b_i	c_i	a_i
x_i	.170	1.115	.073
x_2	.026	1.126	.123
x_3	.995	.000	.099
x_4	−.076	−1.087	.276
x_5	.013	−1.128	.114
x_6	.894	.000	.447
x_7	.000	−1.108	.218

Source: McDonald, 1962; reproduced with permission.

$E(z^2) = 1$. We have $E(x) = f(z) = b_i z + c_i(z^2 - 1)$ with "correlation" matrix

$$
\mathbf{R} =
\begin{bmatrix}
.990 \\
.987 & .983 \\
.056 & .027 & .933 \\
-.958 & -.957 & .083 & .935 \\
-.990 & -.986 & .000 & .961 & .991 \\
-.032 & -.062 & .897 & .013 & .086 & .869 \\
-.971 & -.967 & .044 & .941 & .972 & .106 & .957
\end{bmatrix}
$$

which is identical to the usual correlation matrix except that unities on the main diagonal have been replaced by that proportion of the variance which is due to $f(z)$. The method is known as principal factor analysis and is discussed in more detail in Chapter 6. A linear decomposition into PCs results in the loadings of Table 3.13. Since observations are generated by a quadratic expression, it requires $r = 2$ dimensions to account for the variance due to $f(z)$. The loadings and the scores are plotted in Figures 3.7 and 3.8. Since x_i are simulated quadratic functions it is not possible to interpret the loadings in terms of specific physical attributes. The scores of Figure 3.8

Table 3.13   Factor Loadings for $p = 7$ Quadratic Functions

Variable	$\mathbf{z}_1$	$\mathbf{z}_2$
x_1	.994	.042
x_2	.991	.011
x_3	.014	.966
x_4	−.965	−.065
x_5	−.995	.015
x_6	−.071	.930
x_7	−.977	.050
sum-of-squares	4.853	1.807

Source: McDonald, 1962.

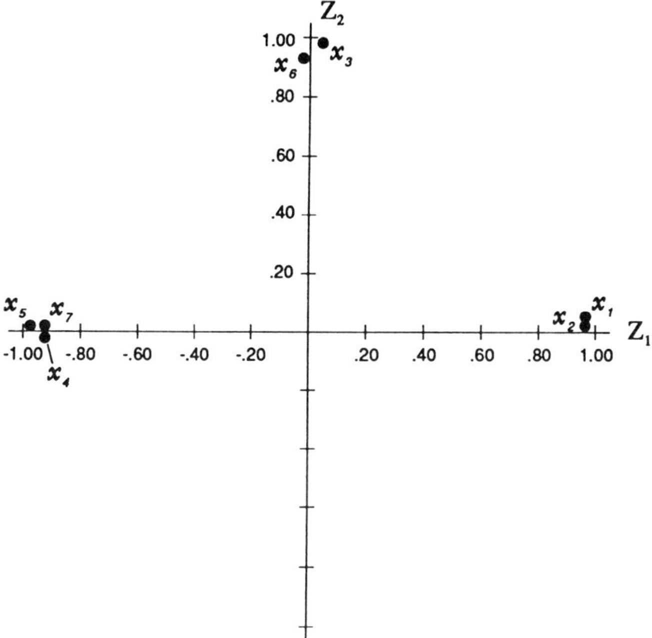

Figure 3.7 Principal factor loadings of Table 3.13.

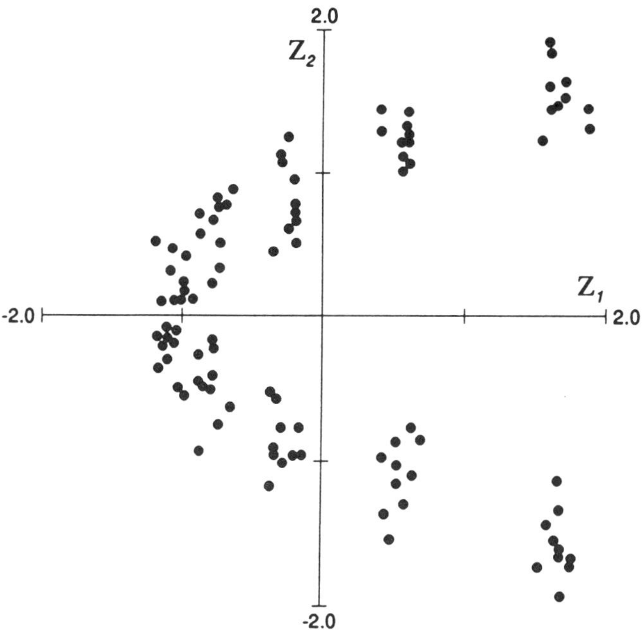

Figure 3.8 Principal factor scores for $n = 100$ simulated observations for Eq. (3.110).

however form a clear quadratic pattern which can be further smoothed by a parabolic least squares fit. This then provides an estimate of $f(z)$.

The second type of nonlinearity frequently encountered in practice is exponential, which occurs for example when dealing with rates of growth in time series or when measuring shape in biological allometry. When variables are exponential, linearity can be restored by a logarithmic transformation of the data. Consider p random variables $Y_1, Y_2, \ldots, Y_p$. When the variables vary exponentially and this is ignored, the loadings and scores are biased and it may require an unnecessarily larger number of components to account for the systematic variation. Assuming the variables are strictly positive, the jth principal component of (natural) logarithms is given by

$$z_j = p_{1j} \ln y_1 + p_{2j} \ln y_2 + \cdots + p_{pj} \ln y_p$$

$$= \ln(y_1^{p_{1j}} y_2^{p_{2j}} \cdots y_p^{p_{pj}}) \tag{3.113}$$

so that Eq. (3.111) can also be written as the exponential equation

$$\exp(z_j) = y_1^{p_{1j}} y_2^{p_{2j}} \cdots y_p^{p_{pj}} \tag{3.114}$$

A useful byproduct of a logarithmic transformation is that it tends to make results less dependent on the unit(s) of measure.

Equation (3.111) assumes the original variables are not expressed as differences about their means. In applied statistical work however random variables are almost always expressed as deviations from their means. Logarithms can also be adjusted in this fashion. Let $x_i = \ln y_i$ $(i = 1, 2, \ldots, n)$; then

$$x_i - \bar{x}_i = \ln y_i - (\overline{\ln y_i})$$

$$= \ln y_i - \frac{1}{n} (\ln y_{1i} + \ln y_{2i} + \cdots + \ln y_{ni})$$

$$= \ln y_i - \ln(y_{1i}, y_{2i} \cdots y_{ni})^{1/n}$$

$$= \ln y_i - \ln G_i$$

$$= \ln (y_i/G_i) \tag{3.115}$$

where G_i is the geometric mean of y_i. The jth PC can then be expressed as

$$\exp(2j) = \left(\frac{y_1}{G_1}\right)^{p_{1j}} \left(\frac{y_2}{G_2}\right)^{p_{2j}} \cdots \left(\frac{Y_p}{G_p}\right)^{p_{pj}} \tag{3.116}$$

$$z_j = \ln\left[\left(\frac{y_1}{G_1}\right)^{p_{1j}}\left(\frac{y_2}{G_2}\right)^{p_{2j}} \cdots \left(\frac{Y_p}{G_p}\right)^{p_{pj}}\right]$$

$$= p_{1j}\ln\left(\frac{y_1}{G_1}\right) + p_{2j}\ln\left(\frac{y_2}{G_2}\right) + \cdots + p_{pj}\ln\left(\frac{Y_p}{G_p}\right) \tag{3.117}$$

The mean-adjusted model (Eq. 3.115) is also dependent on whether the covariance or correlation matrix is used. In practice however when working with logarithms correlation loadings obtained from the correlation matrix often tend to be similar to the correlation loadings obtained from the covariance matrix, probably owing to the fact that the logarithmic transformation yields coefficients that are percentage rates of change. For an alternative approach to the use of logarithms in PCA see Amato (1980).

Example 3.12. Jolicoeur (1963) studied the relationship(s) between length and width of the humerus and femur bones of the North American Marten (*Martes americana*) to determine their "robustness." The following measurements (millimeters) were obtained for 92 males and 47 females.

Y_1 = Total length of the humerus, from the head to the medial condyle
Y_2 = Maximum epicondylar width of the distal end of humerus
Y_3 = Total length of femur, from head to the medial condyle
Y_4 = Maximum width of the distal end of the femur

Since growth tends to take place at a more or less exponential rate, the PCA is carried out using logarithms rather than the original measurements themselves. The means and covariance matrices of the logarithms (base 10) are given in Table (3.14).

A PCA is carried out on the two covariance matrices of Table 3.14. The first latent vector $\mathbf{P}_1 = (.4121, .5846, .3894, .5803)^T$ has positive elements, which are roughly equal to $1/\sqrt{4} = .50$, indicating the first PC is proportional to the geometric means of the four variates. It represents a measure of general size since the four variables tend to be equicorrelated (Section

Table 3.14 Means, Variances, and Covariances for $x_i = \log_{10}y_i$ of Length and Width of Humerus Bone of *Martes americana*

	92 Males	47 Females
$\bar{Y}$	$(1.8066, 1.1138, 1.8493, 1.1003)^T$	$(1.7458, 1.0365, 1.7894, 1.0244)^T$
S	$10^{-4}\begin{bmatrix} 1.1544 & .9109 & 1.0330 & .7993 \\ .9109 & 2.0381 & .7056 & 1.4083 \\ 1.0330 & .7056 & 1.2100 & .7958 \\ .7993 & 1.4083 & .7958 & 2.0277 \end{bmatrix}$	$10^{-4}\begin{bmatrix} .9617 & .2806 & .9841 & .6775 \\ .2806 & 1.8475 & .3129 & 1.2960 \\ .9841 & .3129 & 1.2804 & .7923 \\ .6775 & 1.2960 & .7923 & 1.7819 \end{bmatrix}$

Source: Jolicoeur, 1963; reproduced with permission.

Table 3.15   Latent Roots and Vectors of Covariance Matrices S of Table 3.14

	92 Males		47 Females	

$$
\mathbf{L} \quad 10^{-4}
\begin{bmatrix}
4.5482 & & & 0 \\
& 1.1164 & & \\
& & .6447 & \\
0 & & & .1210
\end{bmatrix}
\qquad
10^{-4}
\begin{bmatrix}
3.7749 & & & 0 \\
& 1.6047 & & \\
& & .3680 & \\
0 & & & .1240
\end{bmatrix}
$$

$$
\mathbf{P} \quad
\begin{bmatrix}
.4121 & .5846 & .3894 & .5803 \\
.5208 & -.4025 & .6411 & -.3947 \\
.1484 & .6804 & -.1276 & -.7057 \\
.7327 & -.1807 & -.6489 & .9071
\end{bmatrix}
\qquad
\begin{bmatrix}
.3520 & .5474 & .4104 & .6388 \\
.5025 & -.6091 & .5977 & -.1390 \\
.2168 & .5738 & .2265 & -.7567 \\
.7594 & -.0145 & -.6504 & .0119
\end{bmatrix}
$$

Source: Jolicouer, 1963; reproduced with permission.

3.3). We thus have

$$
Z_1 = \log_{10}\left[\left(\frac{y_1}{64.06}\right)^{.4121}\left(\frac{y_2}{13.00}\right)^{.5846}\left(\frac{y_3}{0.60}\right)^{.3894}\left(\frac{y_4}{12.60}\right)^{.5803}\right]
$$

The second latent vector contains both positive and negative elements. The second largest direction of variation therefore indicates that humerus and femur width increase when humerus length and femur length decrease, and vice versa. This represents variation in the relative width of the limb bones so that Z_2 can be interpreted as a shape factor; Z_3 can also be interpreted in a similar fashion. The fourth component accounts for just under 2% of the trace and is taken as a random residual factor. Correlation loadings are given in Table 3.16 where the PCA has partitioned bone length and width into two independent "directions"—size and shape. This helps to understand some of the more general aspects relating to physiological structure. The concepts of morphometry discussed above also find use in geology (Fordham and Bell, 1978). For greater detail concerning problems of biological morphology the reader is referred to Bookstein (1989), Sundberg (1989), and Somers (1986, 1989).

Table 3.16   Correlation Loadings for the Length and Width of Humerus Bone of *Martes americana*

Variables	Z_1	Z_2	Z_3
X_1 (humerus length)	.8181	.5122	.1105
X_2 (humerus width)	.8734	-.2979	.3816
X_3 (femur length)	.7549	.6157	-.0928
X_4 (femur width)	.8692	-.2929	-.3965
Percent variance	70.73	17.36	10.03

Source: Jolicouer, 1963; reproduced with permission.

Example 3.13. Another example of a logarithmic transformation, applied to a larger set of variables, concerns the external body measurements of the Humpback whale, taken between 1925 and 1931 at South Georgia and South Africa whaling stations (Matthews, 1938). The data are reproduced in Tables 3.17 and 3.18 for males and females respectively. To compare the results using a logarithmic transformation to those using raw observations, both sets of correlation loadings are computed (Table 3.19). Owing to the high intercorrelation between the variables the results do not differ substantially, although logarithms appear to give marginally better results. The untransformed data have already been analyzed by Machin and Kitchenham (1971; see also Machin, 1974) although their results differ from ours because of the treatment of missing data. Because of the difficulties of gathering data, some observations are missing, and the remaining may contain measurement error, for example, $Y_{18,13} = 7.12$ m, which appears to be out of line. This particular observation is deleted from the analysis and replaced by the mean value $\bar{Y}_{13}$. The missing values are likewise replaced by the sample means of the corresponding variables, which permits analysis of the entire sample (see Basilevsky et al., 1985). The variables are defined as follows:

Length	Y_1 = total length; tip of snout to notch of flukes
Head	Y_2 = tip of snout to blowhole
	Y_3 = tip of snout to angle of gape
	Y_4 = tip of snout to center of eye
	Y_5 = center of eye to center of ear
	Y_6 = length of head from condyle to tip
	Y_7 = greatest width of skull
Tail	Y_8 = notch of flukes to posterior emargination of dorsal fin
	Y_9 = width of flukes at insertion
	Y_{10} = notch of flukes to center of anus
	Y_{11} = notch of flukes to umbilicus
	Y_{12} = notch of flukes to end of system of ventral grooves
Flipper	Y_{13} = axilia to tip of flipper
	Y_{14} = anterior end of lower border to tip of flipper
	Y_{15} = greatest width of flipper
Dorsal Fin	Y_{16} = vertical height of dorsal fin
	Y_{17} = length of base of dorsal fin
Sex	Y_{18} = center of anus to center of reproductive aperture

Two sets of PCs are computed from joint male/female covariance matrices: the first is based on natural logarithms of the data and the second on the original unit of measure (m). Although the global structure of the correlation loadings of both analyses does not differ greatly, those based on logs discriminate to a better extent the first two size/shape components (Table 3.19). The correlation loadings are obtained from covariance

Table 3.17 External Body Measurements (m) of the Humpback Whale (Males) of South Georgia Bay and South Africa, 1925–1931

Y_1	Y_2	Y_3	Y_4	Y_5	Y_6	Y_7	Y_8	Y_9	Y_{10}	Y_{11}	Y_{12}	Y_{13}	Y_{14}	Y_{15}	Y_{16}	Y_{17}	Y_{18}
8.4	1.4	1.8	1.95	.38	2.85	.8	2.1	3.85	3.6	.25	.55	2.2	2.45	.65	2.15	1.45	1.0
12.8	2.7		3.5	.6	4.2	1.2	2.95		5.3	.32	1.4	3.6	4.1	1.0	4.1		.55
14.75	3.1	3.6	4.1	.67	4.76	1.25	3.45	5.8	5.5	.46	1.8	3.4	4.25	1.0	5.0	2.3	1.35
13.1	2.6	3.1	3.42	.61	4.21	1.1	2.9	5.6	5.1	.26	1.22	3.5	3.7	.95			1.3
13.0	2.65	3.1	3.48	.42	4.0	.92	3.05	5.25	5.1	.37	1.5	3.5	3.8	.9	4.4	1.85	1.2
12.55	2.7	3.07	3.42	.57	3.98	.96	2.86	5.35	5.2	.24	.8	3.0	3.62	.78			2.31
9.92	1.85	2.16	2.39	.51	3.35	1.0	2.55	4.2	4.48	.2	1.1	2.68	3.16	.72	2.85	1.4	.9
12.70	2.6		3.5	.6	3.95	1.02	2.95	5.35	5.35	.34	.8			.94			1.25
12.8	2.35		3.3			1.04	3.35	4.7								2.2	
11.5	2.05	2.7	2.9	.5	3.75	1.0	2.75	4.65	4.6	.25	1.1	2.8	3.0	.9		1.9	1.1
9.55	1.75	2.25	2.35	.47	3.1	.82		4.0	4.1	.15	.75	2.17	2.85	.72	3.15	1.6	.8
8.0	1.65	1.85	2.0	.4	2.35	.7	2.05	3.45	3.38	.13	.50			.58	2.26	1.27	.7
9.83	1.78	2.13	2.35	.44	3.2	.9	2.52	4.48	4.6		.85	2.65	3.22	.75	2.77	1.75	1.08
13.0	2.37	2.98	3.23	.58	4.51	1.08	3.19	5.77	5.7	.27	1.0	3.27	3.95	.96			1.45
14.07	2.86		3.67	.65	4.45		3.45	5.93	5.73	.32	1.13	3.54	4.07	.92	4.14		1.17
12.27	2.36	2.68	3.02	.58	4.1	.92	2.99	5.05	5.05	.29	1.0	3.84	3.93	.83	3.55	1.93	1.25
14.5	3.0	3.55	3.76		4.85	1.14	3.46	6.5	6.5	.28	1.0	4.4	4.63	.94	4.45		1.46
9.93	1.84	2.07	2.39		3.2	.71	3.11	4.3	4.44	.26	.71	7.12	3.2	.68		1.7	.95
10.95	1.95	2.44	2.53	.55	3.65	.95	2.55	4.6	4.65	.22	.7	3.15		.85			1.0
9.7	1.86	3.12	2.32		3.6	.76	2.3	4.2	4.05	.19	.97			.55	2.75	1.59	.8
11.9	2.4	2.88	3.07	.56	4.0	.95	3.0	5.2	5.0	.2	1.0			.78	3.63	1.92	1.3
13.15			3.53				2.98										
12.75			2.85				3.32										
11.35			2.95			.92	2.8	4.8									
14.13	2.85	3.22	3.56	.65	4.68	1.09	3.36	6.02	6.25	.42	1.25	3.75	4.06	1.08	4.22	2.18	1.42
13.41	2.66	3.42	2.95	.53	4.45	1.08	3.32	5.7	5.5	.3	.85	3.3	3.85	.93	4.38	2.41	1.27
11.37	2.32	2.78	2.96	.6	3.96	.98	2.83	4.98	5.03		1.0	3.75	3.50	.75	3.42	1.91	1.2
12.28	2.37	2.83	3.04	.58	3.97	.88	2.85	5.05	4.9	.3	1.35	3.8	3.95	.92	3.6	2.2	1.18
11.3	2.12	2.67	2.9	.5	3.75	.83	2.78	5.02	5.06	.25	.98	2.89	3.39	.9	3.51	2.17	1.2
13.16	2.82	3.34	3.56	.61		.98	3.01	5.36	5.2	.37	1.35	3.87	4.41	1.02	4.3	2.15	1.41
11.6	2.28	2.67	2.95	.48	4.14	.96	2.88	5.06	4.92	.24	.85	2.92	3.46	.91			1.2

Source: Matthews, 1938.

Table 3.18   External Body Measurements (m) of the Humpback Whale (Females) of South Georgia Bay and South Africa, 1925–1931

Y_1	Y_2	Y_3	Y_4	Y_5	Y_6	Y_7	Y_8	Y_9	Y_{10}	Y_{11}	Y_{12}	Y_{13}	Y_{14}	Y_{15}	Y_{16}	Y_{17}	Y_{18}
13.55	3.0	3.45	3.6		4.5	1.3	3.0	5.45	5.05	.28	.7		3.95	1.0	4.2	2.4	.45
14.1	3.1		3.9	.65	4.47	1.28	3.13	5.64	5.33	.32	1.5	3.93	4.45	1.0	4.55	2.4	.53
14.9	2.88	3.45	3.87	.7	4.9	1.15	3.45	5.85	5.6	.3	1.45	4.0	4.65	1.0			.5
12.6		3.1				1.3	2.85	5.9				3.64	4.03	.97			.32
13.15	2.5	3.0	3.35	.66	4.37	1.0	3.32	5.35	5.25	.35	.9	3.63	3.91	.88	3.85	2.01	.47
12.5	2.59	2.9	3.31	.53	4.1	1.15	3.13	5.26	5.05	.24	1.2	3.25	3.62	.87	3.8	2.11	.42
14.85	2.98	3.4	3.75	.75	4.88	1.14	3.7	6.45	6.3	.28	1.6	4.35	4.65	1.0	4.35	2.36	.55
13.55	2.85	3.18	3.6	.68	4.46	1.03	3.07	5.45	5.0	.29	1.5	3.38	4.05	.96	4.0	2.15	.38
14.9	3.15	3.5	3.84	.76		1.26	3.67	5.95				4.37	5.1	1.08	4.45	2.38	.35
10.5	1.94	2.45	2.6	.51	3.66	.82	2.68	4.65	4.5	.25	1.0			.77	3.06	1.67	.62
10.05	1.78	2.0	2.25	.47	3.35	.9	2.45		4.4	.21	1.0						.45
12.8	2.94	3.38	3.56		4.06	.98	2.72	5.2	5.09	.23	1.37						.41
11.65			2.75			1.1	2.7	4.6				3.1	3.75		3.5	1.9	.35
14.53	2.92	3.48	3.87	.63	4.35	1.16	3.35	6.0	5.8	.3	1.3	4.15		1.17	4.4	2.44	.58
11.27	2.5	2.95	3.1	.52	3.55	.84	2.8	4.9	5.0	.34	1.15	3.1	3.65	.84	3.5	1.64	.47
9.92	1.95	2.2	2.48	.47	3.4	.8	2.4	4.5	.27	.85	2.72	3.11	.73	2.95	1.64	.4	
11.8	2.33	2.60	2.82	.48	3.98	1.0	2.98	5.25	5.38	.25	.94	3.25	3.70	.85	3.35		.7
13.66	2.90	3.23	3.56	.63	4.13	1.11	3.02	4.58	4.73	.26	.67	3.75	4.20	.93	4.30	2.30	.47
13.05	2.65	3.25	3.42	.56	4.33	1.03	3.03	5.44		.3	1.55	3.93	3.86				.38
12.35	2.55	2.9	3.25	.61	4.12	.97	2.98	5.4		.22	1.05	3.57		.88			.69
13.9	2.95	3.5	3.8			1.06	3.32	5.96		.28	.95	4.25	4.41	.93	4.47	2.47	.5
14.0	2.8		3.6		4.5	1.08	5.33	5.9						.91			.65
9.5	1.75	2.2	2.36				2.32	3.92	3.9			2.65	3.2	.69			.39
9.9	2.0	2.45	2.7	.5	3.5	.87	2.65	4.65	4.6	.20	.10	2.73	3.2	.73	3.41	1.9	.45
14.2	2.23	3.38	4.09	.65	4.52	1.07	3.14	5.55	5.63	.3	1.2	4.08	4.36	1.1	4.77		.38
10.25	1.81	2.44	2.55	.47	3.55	.83	2.54	4.7	4.81	.27	.78	2.65	3.0	.76	3.0	1.9	.6
12.75	2.53	2.85	3.16		4.43	1.03	3.22	5.59	5.18	.25	1.0	3.67	4.07	.92	3.83	2.25	.55
13.16	2.5	3.04	3.31	.55	4.25	1.0	3.41	5.72	5.57	.3	1.1	3.64	3.78	.88	3.88	2.2	.47
13.5	3.7	3.16	3.54	.66	4.17	1.06	3.13	5.55	5.32			3.34	4.18	.87			.53
11.38	2.39	2.71	3.07	.49	3.8	.92	2.86	5.15	5.08	.2	.9	3.17	3.3	.9	3.42	2.19	.5
10.72	2.06	2.56	2.8	.44	2.56	.89	2.68	4.67	4.67	.19	.95	3.18	3.32	.8	3.27	2.04	.42

Source: Matthews, 1938.

Table 3.19   Correlation Loadings for the External Body Measurements of the Humpback Whale for Data in Log Meters and Meters (Male and Female Covariance Matrix)[a]

Variables		Log Meters Z_1	Z_2	Z_3	Z_4	Meters Z_1	Z_2	Z_3	Z_4
Length	Y_1	.962	.158	−.153	.014	.998	−.001	−.049	−.006
Head	Y_2	.933	.052	−.185	−.066	.924	−.149	−.068	−.103
	Y_3	.890	.093	−.146	−.110	.894	−.135	.048	−.183
	Y_4	.947	.106	−.113	−.030	.953	−.081	−.082	−.114
	Y_5	.797	.034	−.078	−.119	.774	−.189	.034	−.115
	Y_6	.880	.132	−.095	.073	.865	−.070	.020	−.270
	Y_7	.791	.024	−.215	.142	.767	−.055	−.075	−.124
Tail	Y_8	.868	.201	−.008	.069	.885	.183	−.007	−.081
	Y_9	.811	−.047	−.131	.098	.779	−.090	−.150	−.102
	Y_{10}	.751	.171	−.020	−.010	.745	.202	−327	.424
	Y_{11}	.886	.176	−.124	−.017	.916	.176	.115	.185
	Y_{12}	.840	.291	−.060	.037	.864	.332	.196	.034
Flipper	Y_{13}	.754	.032	−.200	−.276	.771	−.365	.402	.251
	Y_{14}	.805	.038	−.202	−.271	.819	−.374	.300	.110
	Y_{15}	.869	.090	−.084	.150	.864	−.002	−.005	−.103
Dorsal fin	Y_{16}	.678	.288	.315	.551	.612	.182	.048	−.268
	Y_{17}	.720	.025	.640	−.249	.594	−.063	.017	−.177
Sex	Y_{18}	−.179	.983	−.006	−.031	.037	.673	.466	−.219
Variance (%)		66.5	7.57	4.38	3.40	65.4	5.99	2.55	3.49

[1] Percentage variance equals sum of squares of correlation loadings for each component.

loadings by adjusting for the standard deviations of the variables (Table 3.8). A possible disadvantage of the logarithmic transformation however lies in the computation of the percentage of variance which is explained by each PC, when the covariance rather than the correlation matrix is used.* The reason for this is that the latent roots (Table 3.20) can provide misleading results in this respect. For example, the trace of the covariance matrix of logarithms is .66114, indicating that Z_2 accounts for $.22106/.66114 = .3343$ or 1/3 of the total variance, clearly a misleading result. The percentages of variance in Table 3.1 are therefore computed using the correlation loadings found in that table (see Theorem 3.13). Using Table 3.19 the PCs can be identified as follows. The first component is an index of total body size for both males and females, and is maximally correlated with vectors Y_1, Y_2, and Y_4. Since the first half of the sample represents males and the second half females, we can use the scores of Z_1 to compare the overall size of males and females. The mean male scores for males and females are −.315 and .284, respectively, indicating a larger overall body structure for females.

The second PC correlates with shape variables Y_8, Y_{12}, Y_{16}, and Y_{18}, and in view of the latter provides an effective discriminator for sex. Since values

---

*The disadvantage however is more a function of the standard computer statistical packages, which do not permit sufficient flexibility, but rather seem to assume the correlation matrix is of main interest.

Table 3.20 Latent Roots of Covariance Matrices in Data in Log Meters. Meters for Male and Female Whale Measurements Combined (see Table 3.19)

Log_e Meters	Meters
.30993	4.99623
.22106	.19893
.03736	.13748
.02271	.11778
.01446	.09468
.00987	.07683
.00927	.05890
.00874	.04342
.00655	.03685
.00550	.03191
.00393	.02983
.00307	.01980
.00250	.01322
.00190	.00988
.00181	.00519
.00122	.00235
.00096	.00176
.00032	.00130

of Y_{18} are larger for males, the correlation loading (.983) is positive although this is arbitrary since signs can be reversed for all loadings with a corresponding reversal in the interpretation of Z_2. The remaining components Z_3 and Z_4 account for shape peculiarities found primarily in the dorsal fin and which are independent of body size and those shape factors that are related to sex.

3.8 ALTERNATIVE SCALING CRITERIA

3.8.1 Introduction

It was seen in previous sections that for a covariance (correlation) matrix the latent vectors can be scaled in three distinct ways: (1) unit vectors so that $\mathbf{P}^T\mathbf{P} = \mathbf{P}\mathbf{P}^T = \mathbf{I}$, (2) elements as correlation loadings, and (3) elements as covariance loadings (Theorem 3.13). The first normalization is usually imposed for reasons of computational simplicity, whereas the latter two are used as interpretational aids since loading coefficients satisfy the relationship $\mathbf{A}^T\mathbf{A} = \mathbf{L}$ and $\mathbf{A}\mathbf{A}^T = \mathbf{X}^T\mathbf{X}$. Equivalent results also hold for the inner product and cosine matrices (Section 2.4). Owing to its scale and location invariance however the correlation matrix is the more popular Grammian form, and has retained a virtual monopoly in the more applied literature. It is possible however to scale PCs in alternative ways.

3.8.2 Standardized Regression Loadings

When using PCA to explain variation in a set of random variables, the latent vector elements are standardized to lie in the interval $[-1, 1]$. This yields the correlation loading matrix $\mathbf{A}^T$. When the objective is to express the PCs in terms of the observed variates, however, it is at times more convenient to use the matrix $\mathbf{A}$, as defined by Eq. (3.51). We can then write

$$\mathbf{Z}' = \mathbf{XA} \tag{3.118}$$

where $\mathbf{A}$ may be termed as the matrix of standardized regression loading coefficients. Note that $\mathbf{Z}' \neq \mathbf{Z}^* \neq \mathbf{Z}$ (see Section 3.4). The matrix $\mathbf{Z}'$ can be obtained as follows. Pre- and postmultiplying Eq. (3.44) by $\mathbf{L}^{1/2}$ we have

$$\mathbf{L}^{1/2}\mathbf{P}^T\mathbf{X}^T\mathbf{XPL}^{1/2} = \mathbf{L}^2 \tag{3.119}$$

so that $\mathbf{Z}' = \mathbf{ZL}^{1/2} = \mathbf{XPL}^{1/2} = \mathbf{XA}$. Since

$$\mathbf{A} = (\mathbf{X}^T\mathbf{X})^{-1}\mathbf{X}^T\mathbf{Z}' \tag{3.120}$$

the rationale for the standardization $\mathbf{Z}' = \mathbf{ZL}^{1/2}$ seems to be that it yields standardized regression coefficients when the PCs are regressed on the random variables (standardized or otherwise), that is, the regression coefficients satisfy Theorem 3.13. Regression standardization has been used by Hawkins (1974) for purposes of error detection (Section 5.8.2; see also Jackson, 1991).

Example 3.14. Consider the correlation loadings of Example 3.5. We have, using Eq. (3.116),

$$(\mathbf{Z}_1', \mathbf{Z}_2', \mathbf{Z}_3') = (\mathbf{X}_1, \mathbf{X}_2, \mathbf{X}_3)\begin{bmatrix} .9128 & -.5116 & .7842 \\ .0243 & .8355 & .5167 \\ .4076 & .2006 & -.3436 \end{bmatrix}$$

which yields the equations

$$\mathbf{Z}_1' = .9128\mathbf{X}_1 + .0243\mathbf{X}_2 + .4076\mathbf{X}_3$$

$$\mathbf{Z}_2' = -.5116\mathbf{X}_1 + .8355\mathbf{X}_2 + .2006\mathbf{X}_3$$

$$\mathbf{Z}_3' = .7842\mathbf{X}_1 + .5167\mathbf{X}_2 + -.3436\mathbf{X}_3$$

3.8.3 Ratio Index Loadings

An alternative method of scaling is to express latent vector elements in relation to the largest coefficient of a PC. This is achieved by dividing each element (loading) by the largest coefficient for that component, which then assumes the value of unity. The motivation for such a scaling seems to be to ignore the presence of residual variation and to examine each PC in isolation from the remaining ones (Geffers, 1967). The difficulty with such an approach is that it ignores the ranking of the PCs or the magnitudes of the individual correlation loadings.

The ratio loadings however can always be converted back into latent vectors or correlation (covariance) loadings. Let p_{ij} represent the latent vector element for the ith variable and the jth PC. Also, let

$$p = \max(p_{1j}, p_{2j}, \ldots, p_{kj}) \tag{3.121}$$

for k random variables. Then for some jth PC we have

$$\frac{p_{1j}^2 + p_{2j}^2 + \cdots + p_{kj}^2}{p^2} = \frac{1}{p} = c \tag{3.122}$$

the sum of squares of the ratio coefficients, where p^2 is also included in the numerator. It follows that the largest latent vector element (for the jth PC) is given by $p = 1/c$, and multiplying the ratio coefficients by p recovers the initial unit-length latent vectors, which can then be used to construct correlation (covariance) loadings. Although the original signs of the coefficients are lost, this is not of major importance since signs of the latent vectors are not uniquely determined. Of course the relative signs of the loadings, for the jth component, are preserved so that interpretation of the PCs is still possible.

Example 3.15. An example of the ratio index loading coefficients is Jeffer's (1967) study of timber used as pitprops in mines. The variables are defined as follows:

Y_1:	Topdiam	Top diameter of the prop (in)
Y_2:	Length	Length of the prop (in.)
Y_3:	Moist	Moisture content of the prop, expressed as a percentage of the dry weight
Y_4:	Testsg	Specific gravity of the timber at the time of the test
Y_5:	Ovensg	Oven-dry specific gravity of the timber
T_6:	Ringtop	Number of annual rings at the top of the prop
Y_7:	Ringbut	Number of annual rings at the base of the prop
Y_8:	Bowmas	Maximum bow (in.)
Y_9:	Bowdist	Distance of the point of maximum bow from the top of the prop (in.)

Y_{10}:	Whorls	Number of knot whorls
Y_{11}:	Clear	Length of clear prop from the top of the prop (in.)
Y_{12}:	Knots	Average number of knots per whorl
Y_{13}:	Diaknot	Average diameter of the knots (in.)

The ratio loadings for the first $T = 6$ Pcs are given in Table 3.21.

Note it is no longer possible to compare coefficients between PCs or to determine the percentage of variance explained by each PC. Otherwise the components are identified in the usual manner. Here $\mathbf{Z}_1$ is a size index for the props, $\mathbf{Z}_2$ picks up moisture content (presence or lack of) of the timber, which evidently determines the density of the wood, and so forth. The latent roots together with percentage of variance accounted for are given in Table 3.22.

A more informative analysis however is one which is based on the correlation loadings of Table 3.23, where Jeffer's ratio coefficients are converted using Eq. (3.122). Thus for Z_1, variable Y_2 has a unit ratio coefficient; also

$$\frac{p_{11}^2 + p_{21}^2 + \cdots + p_{k1}^2}{p_{21}^2} = \frac{1}{p_{21}^2} = 6.00$$

so that $p = p_{21} = .40824$, and using this value the first latent vector and correlation loadings can be reconstructed from Table 3.21. A similar technique can be used for the remaining PCs. Table 3.22 gives us a better idea as to the overall importance of the principal components.

Table 3.21 Scaled Latent Vectors for $r = 6$ Principal Components of $p = 13$ Physical Measurements of Pitprops

Variable	$\mathbf{Z}_1$	$\mathbf{Z}_2$	$\mathbf{Z}_3$	$\mathbf{Z}_4$	$\mathbf{Z}_5$	$\mathbf{Z}_6$
$\mathbf{X}_1$: Topdiam	0.96	0.40	−0.43	−0.11	0.14	0.19
$\mathbf{X}_2$: Length	1.00	0.34	−0.49	−0.13	0.19	0.26
$\mathbf{X}_3$: Moist	0.31	1.00	0.29	0.10	−0.58	−0.44
$\mathbf{X}_4$: Testsg	0.43	0.84	0.73	0.07	−0.59	−0.09
$\mathbf{X}_5$: Ovensg	0.14	−0.31	1.00	0.06	−0.29	1.00
$\mathbf{X}_6$: Ringtop	0.70	−0.26	0.99	−0.08	0.53	0.08
$\mathbf{X}_7$: Ringbut	0.99	−0.35	0.53	−0.81	0.36	0.00
$\mathbf{X}_8$: Bowmax	0.72	−0.35	−0.51	0.36	−0.31	−0.09
$\mathbf{X}_9$: Bowdist	0.88	0.32	−0.43	0.12	0.18	0.05
$\mathbf{X}_{10}$: Whorls	0.93	−0.46	−0.25	−0.26	−0.26	−0.28
$\mathbf{X}_{11}$: Clear	−0.03	0.38	−0.15	1.00	0.57	0.28
$\mathbf{X}_{12}$: Knots	−0.28	0.63	0.19	−0.37	1.00	−0.27
$\mathbf{X}_{13}$: Diaknots	−0.27	0.57	−0.68	−0.38	−0.13	1.00

Source: Jeffers, 1967; reproduced with permission.

Table 3.22   Latent Roots of the Correlation Matrix of $p = 13$ Physical Properties of Pitprops

Component	Eigenvalue	Percentage of Variability	
		Component	Cumulative
1	4.219	32.4	32.4
2	2.378	18.3	50.7
3	1.878	14.4	65.1
4	1.109	8.5	73.6
5	0.910	7.0	80.6
6	0.815	6.3	86.9
7	0.576	4.4	91.3
8	0.440	3.4	94.7
9	0.353	2.7	97.4
10	0.191	1.5	98.9

Source: Jeffers, 1967, reproduced with permission, last three omitted.

Table 3.23   Correlation Loadings for $r = 6$ Principal Components of the Physical Measurements of Pitprops

	Z_1	Z_2	Z_3	Z_4	Z_5	Z_6
X_1 : Topdiam	.805	.327	−.284	−.078	.080	.107
X_2 : Length	.839	.278	−.323	−.093	.109	.147
X_3 : Moisture	.260	.816	.191	.071	−.332	−.248
X_4 : Testsg	.361	.686	.481	.049	−.338	−.051
X_5 : Ovensg	.117	−.253	.659	.042	−.166	.565
X_6 : Ringtop	.587	−.212	.652	−.057	.304	.045
X_7 : Ringbut	.830	−.286	.349	−.575	.206	0
X_8 : Bowmax	.604	−.286	−.336	.255	−.177	−.051
X_9 : Bowdist	.738	−.261	−.283	.085	.103	.028
X_{10}: Whorls	.780	−.376	−.165	−.184	−.149	−.158
X_{11}: Clear	−.025	.310	−.099	.709	.327	.158
X_{12}: Knots	−.235	.514	.125	−.262	.573	−.153
X_{13}: Diaknots	−.226	.465	−.448	−.270	−.075	.565

3.8.4   Probability Index Loadings

A still different normalization rule is to scale the latent vectors so that their elements sum to unity. This is at times done when PCs are used to construct indices in socioeconomic research (Ram, 1982). Loading coefficients scaled in such a manner are then interpreted as mixture coefficients. Let $p_1, p_2, \ldots, p_k$ denote latent vector elements for some PC, such that $p_i \geq 0$ $(i = 1, 2, \ldots, k)$ and $\Sigma_{i=1}^{k} p_i^2 = 1$. Then the elements p_i can be rescaled to unit-sum probability index loadings by defining

$$q_i = \frac{p_i}{\left(\sum\limits_{i=1}^{k} p_i\right)} \qquad (i = 1, 2, \ldots, k) \tag{3.123}$$

so that $\sum_{i=1}^{k} q_i = 1$. Conversely given numbers q_i such that

$$\sum_{i=1}^{k} q_i^2 = \frac{\sum\limits_{i=1}^{k} p_i^2}{\left(\sum\limits_{i=1}^{k} p_i\right)^2}$$

$$= \frac{1}{\left(\sum\limits_{i=1}^{k} p_i\right)^2} \tag{3.124}$$

we have

$$\sum_{i=1}^{k} p_i = \frac{1}{\left(\sum\limits_{i=1}^{k} q_i\right)^{1/2}} \tag{3.125}$$

which permits us to recover the original latent vectors using eq. (3.123) since

$$p_i = \frac{q_i}{\left(\sum\limits_{i=1}^{k} q_i^2\right)^{1/2}} \tag{3.126}$$

EXERCISES

3.1 Let $\mathbf{P}_x = \mathbf{X}(\mathbf{X}^T\mathbf{X})^{-1}\mathbf{X}^T$ be a projection matrix. Verify that $\mathbf{P}_x$ is idempotent and symmetric, that is, $\mathbf{P}_x\mathbf{P}_x = \mathbf{P}_x$ and $\mathbf{P}_x^T = \mathbf{P}_x$.

3.2 Prove, by induction, that Eq. (3.66) represents an orthogonal set of variables. How is the Gram–Schmidt transformation related to least squares regression?

3.3 Let $\mathbf{\Pi}^T\mathbf{P}\mathbf{\Pi} = \mathbf{\Lambda}$ where $\mathbf{P}$ is a $(p \times p)$ population correlation matrix. Show that

(a) $\text{tr}(\mathbf{\Lambda}) = p$

(b) $0 \le |\mathbf{P}| \le 1$. What do the two bounds represent in terms of relationships between the random variables?

3.4 Prove that $\alpha^T\alpha = \Lambda$ and $\alpha\alpha^T = P$ where α is a matrix of correlation loadings. (Equation 3.17)

3.5 Using Eq. (3.14) verify that $\zeta^*\zeta^{*T} = I$.

3.6 Let A be a $(p \times p)$ matrix whose elements are given by Eq. (3.30). Show that the determinant of A is given by Eq. (3.31).

3.7 Consider the (4×4) covariance (correlation) matrix of Example 3.2. Verify the results of Theorem 3.8. Also verify that latent vectors of P are orthogonal unit vectors.

3.8 Prove that columns of a $(n \times p)$ matrix X generate the same vector space as the columns of X^TX

3.9 Let $S = \frac{1}{n-1} X^TX$ be the sample covariance matrix where L and P are latent roots and vectors of X^TX respectively. Prove that S has latent roots $\frac{1}{n-1} L$ and latent vectors P. Using this result show that correlation loadings of S are given by

$$a_{ij} = \frac{1}{n-1} p_{ij}(s_i)^{1/2}$$

where s_i is the ith diagonal element of S, and consequently correlation loadings are the same for both X^TX and S. Show also that X^TX and S possess the same standardized PC scores, but that unstandardized scores differ.

3.10 Using Eqs. (3.103a and b) prove the results of Theorem 3.17.

3.11 Prove that latent vectors which correspond to zero latent roots are not necessarily zero (see also Theorem (5.14).

3.12 Prove that for the covariance matrix

$$\Sigma = \begin{bmatrix} \sigma^2 & \sigma^2\rho & \cdots & \sigma^2\rho \\ \sigma^2\rho & \sigma^2 & \cdots & \sigma^2\rho \\ \vdots & & \ddots & \\ \sigma^2\rho & \sigma^2\rho & \cdots & \sigma^2 \end{bmatrix}$$

we can write $\Sigma = (\lambda_1 - \lambda)\pi_1\pi_1^T + \lambda I$ where λ_1 is the non-isotropic root and λ is the common isotropic root of multiplicity $p - 1$ (section 3.3).

3.13   Let $\zeta = \Pi^T X$ and let $Z = P^T X$ be the sample realization of ζ where X is a $(p \times l)$ vector. Show that $\zeta = WZ$ where W is a $(p \times p)$ orthogonal matrix.

3.14   Let $r_{ij.1} = \dfrac{r_{ij} - r_{i1}r_{j1}}{[(1 - r_{i1}^2)(1 - r_{j1}^2)]^{1/2}}$ be the partial correlation coefficient between the ith and jth variable when the effect of the first component has been removed. Show for the equal correlation model (Eq. 3.29), that

$$r_{ij,1} = \frac{1}{p+1}$$

where $\alpha = r_{il} = r_{jl}$, the loadings for the first component.

3.15   The following data represent averages of 10 independent measurements, from the Herbarius at the University of British Columbia (source: Orloci, 1967)

Characters	Specimens (0.1-mm units)																	Mean (x_1)	Standard Deviation (s_1)
	1	2	3	4	5	6	7	8	9	10	11	12	13	14	15	16	17		
Corolla (Y_1)	75	61	53	77	41	56	60	57	48	53	58	49	40	60	66	47	50	55.94	10.29
Calyx (Y_2)	27	17	20	19	16	18	19	28	29	35	37	29	30	36	45	36	41	28.35	9.05
Style (Y_3)	65	54	64	57	33	62	64	46	53	54	54	64	31	48	39	31	30	49.94	12.78
Corolla lobe (Y_4)	11	11	13	15	12	19	21	11	10	8	10	13	4	9	7	4	4	10.71	4.78
Pedicel (Y_5)	15	16	13	12	11	14	21	17	16	14	18	26	6	13	16	10	7	14.41	4.82

Compute correlation loading coefficients using (a) the correlation matrix, and (b) the covariance matrix. How do the two sets of coefficients compare to each other? Are both sets equally informative? Explain (see also Exercise 3.23).

3.16   Let $n(X) = (X^T X)^{1/2}$, the so-called nucleus of the $(n \times p)$ matrix X. Show that

(a) $\operatorname{tr}[n(X)] = \lambda_1^{1/2} + \lambda_2^{1/2} + \cdots + \lambda_p^{1/2}$
(b) $|n(X)| = \lambda_1^{1/2}\lambda_2^{1/2} + \cdots + \lambda_p^{1/2}$

3.17   Using Eqs. (3.103a and b) prove
(a) $X(X^T X)^{-1} = ZL^{-1/2}P^T = ZA^{-1}$
(b) $X(X^T X)^{-1/2} = ZP^T$
(c) $X(X^T X)^{-1}X^T = ZZ^T$

3.18   The use of the ratio $l_i/\operatorname{tr}(X^T X)$ as the proportion of total variance accounted for by the ith sample PC has the defect that it lies in the half-closed interval $1 \geq l_i \geq 0$. Show that the statistic

$$\nu = \frac{l_i - s_m^2}{\operatorname{tr}(\mathbf{X}^{\mathrm{T}}\mathbf{X}) - s_m^2}$$

lies in the interval $1 \geq \nu \geq 0$ where s_m^2 is the largest diagonal element of $\mathbf{X}^{\mathrm{T}}\mathbf{X}$ (Kloek and Bannink, 1962).

3.19 Using Theorem 3.15 show that PC correlation loadings extracted from a covariance matrix are identical to those from a correlation matrix if the variances of the variables are equal.

3.20 Principal components can be used as multivariate indices, for example, when evaluating toxicity in phase III clinical trials for cancer treatment. The following table of $n = 6$ observations for $p = 3$ random variables is given by Herson (1980).

Toxicity Index			Coefficients				
	Wbcs	Lymphocytes	Polymorphonuclear Cells	Blast Cells	Platelets	Hemoglobin	Variance (%)
Y_1	0.597	0.278	0.329	0.552	0.037	0.390	37.6
Y_2	−0.326	0.408	0.432	−0.418	0.460	0.395	26.8
Y_3	0.117	−0.457	−0.091	0.187	0.848	−0.122	14.1
Total							78.5

(a) Compute the PC correlation loadings, given that all three variables are standardized to unit length.

(b) Compute the percentage of variance which all three PCs explain of the observed variables.

(c) Compute the percentage of covariance which all three PCs explain of the observed variables.

CHAPTER 4

Statistical Testing of the Ordinary Principal Components Model

4.1 INTRODUCTION

The PC model finds diverse application in many disciplines since its relatively straightforward structure makes it easy to interpret in various settings and according to differing requirements. When using PCs with sampled data, however, many practitioners seem to be unaware of the statistical nature of the model. This in turn is reinforced by the lack of statistical significant testing capabilities of some computer packages, although statistical tables are now readily available for many relevant distributions (see Kres, 1983). This seems to be a glaring drawback, particularly when a comparison is made with other multivariate techniques such as least squares regression, discriminant analysis, and canonical correlation. To be sure, "rules of thumb" are frequently employed; for example, latent roots smaller than unity are often treated as "insignificant," and correlation loadings smaller than .20 are routinely omitted in computer software packages. Such practice is statistically arbitrary, and seems to be prompted more by intuitive concepts of practicality and "parsimony" than by probabilistic requirements of sample–population inference.

In this chapter we consider the topic of statistical significance testing for the ordinary PCA model of Chapter 3. Although the PCA model can itself be used in statistical inference (Dauxoi's et al., 1982) this aspect is not considered in the present chapter. Although traditionally the objective of factor analysis has been to identify the "true" model, a recent emphasis has shifted from this position to the more pragmatic one of attempting to find that model that gives the best approximation to the true model. In this latter view we ask the question "how much identifiable information does the sample at hand contain," rather than "how can we use a sample to estimate (corroborate) a true population model." Thus in the context of factor analysis an appropriate question might be not what the number of correct factors is, but how many factors can be reliably extracted, given a set of data. The two methodological orientations are not as opposed as they may appear at first sight, since the main difference lies in their respective starting points rather than any basic disagreement on the relevance or feasibility of statistical testing. Whereas the latter approach starts with the sample at hand, the former begins with a prespecified target population and then attempts to verify whether the sample data accords with the null hypothesis. The difference was already encountered in the Introduction under the guise of exploratory statistical data analysis versus a confirmatory establishment of formal scientific hypotheses. Clearly in the limit both approaches have the common task of reconciling theory with sample evidence, that is demonstrating consisting between theory and data a requirement that forms the basis of scientific statistical testing.

More specifically, two sets of methodologies are considered—those based on maximum-likelihood criteria and those that utilize Baysian criteria or other less formal data-analytic procedures. Since the theory of large sample significance testing is well developed for the multivariate normal, this distribution looms prominently in maximum-likelihood methodology although clearly normality is not always the best assumption. Even when the assumption of normality is tenuous, however, it may still provide a close approximation to reality. It is also useful to compute normality-based test criteria as supplements to the usual computer output, in order to provide comparisons with other criteria that may be used. In addition, even though optimality of the PC model is not conditional on normality, if normality is satisfied then the PC loadings and scores become ML estimators, with the additional desirable property of asymptotic efficiency.

THEOREM 4.1. Let $\mathbf{Y}$ be a $(n \times p)$ data matrix consisting of n independent samples observed for p multivariate normal variates Y_1, $Y_2, \ldots, Y_p$. Then the sample covariance matrix $\hat{\mathbf{\Sigma}} = (1/N)\mathbf{X}^T\mathbf{X}$ is a ML estimator of $\mathbf{\Sigma}$, where $N = n - 1$ and $\lambda_1 > \lambda_2 > \cdots > \lambda_p$. In addition, solutions of $|\hat{\mathbf{\Sigma}} - l_i\mathbf{I}| = 0$ and $(\hat{\mathbf{\Sigma}} - l_i\mathbf{I})\mathbf{P}_i = \mathbf{0}$ are ML estimators of the population latent roots and latent vectors λ_i and $\mathbf{\Pi}_i$, respectively.

The proof of the theorem consists of noting that latent vectors corresponding to the roots $\lambda_1 > \lambda_2 > \cdots > \lambda_p$ are unique (up to multiplication by -1) so that the $l_1 > l_2 > \cdots > l_p$ must be ML estimators, given $\hat{\mathbf{\Sigma}} = (1/N)\mathbf{X}^T\mathbf{X}$ is a ML estimator of $\mathbf{\Sigma}$ (Anderson, 1984). Alternatively, on the assumption that $\hat{\mathbf{\Sigma}} = (1/N)\mathbf{X}^T\mathbf{X}$ is a Wishart distribution, one may obtain the normal equations for $\mathbf{\Lambda}$ and $\mathbf{\Pi}$ (Girshick, 1936; Flury, 1988). The ML estimators of $\mathbf{\Lambda}$ and $\mathbf{\Pi}$ cannot be derived by maximizing the likelihood function with respect to these parameters, since the likelihood is constant under orthogonal transformations (Kendall, 1957)—such maximization simply permits us to derive estimators of $\mathbf{\Lambda}$ and $\mathbf{\Pi}$, without necessarily demonstrating that they are ML. Note also that ML estimation for a sample is not strictly parallel to a population decomposition into PCs (Section 3.2). Whereas a theoretical population PCA begins with a covariance matrix $\boldsymbol{\Sigma}$, the starting point for a sample analysis is a data matrix $\mathbf{Y}$.

The following theorem is of fundamental importance for statistical testing and for the derivation of ML estimators.

THEOREM 4.2.   Let $\mathbf{X}_1, \mathbf{X}_2, \ldots, \mathbf{X}_p$ be $(n \times 1)$ vectors sampled from a $N(\mathbf{0}, \boldsymbol{\Sigma})$ multivariate normal distribution. Then the $(p \times p)$ matrix $\mathbf{X}^{\mathrm{T}}\mathbf{X}$ is distributed as the central Wishart distribution

$$f(\mathbf{X}^{\mathrm{T}}\mathbf{X}) = \frac{1}{2^{Np/2} \pi^{1/4p(p-1)} |\boldsymbol{\Sigma}|^{N/2} \prod_{i=1}^{p} \Gamma\left(\frac{N-i+1}{2}\right)} |\mathbf{X}^{\mathrm{T}}\mathbf{X}|$$

$$\times \frac{N-p-1}{2} \exp(-\tfrac{1}{2} \operatorname{tr} \boldsymbol{\Sigma}^{-1}\mathbf{X}^{\mathrm{T}}\mathbf{X}) \tag{4.1}$$

A special case arises for p uncorrelated, standardized random variables. Given a sample of size n, the correlation matrix $\mathbf{R} = \mathbf{X}^{\mathrm{T}}\mathbf{X}$ is distributed as

$$f(\mathbf{R}) = \frac{[\Gamma(N/2)]^p}{\pi^{1/4p(p-1)} \prod_{i=1}^{p} \Gamma\left(\frac{N-i+1}{2}\right)} |\mathbf{R}|^{\frac{1}{2}(N-p-2)} \tag{4.2}$$

When $p = 1$, the Wishart distribution is equivalent to the chi-squared distribution Eq. (1.10).

In what follows we consider significance tests for covariance and correlation matrices of multivariate normal densities, together with their latent roots and latent vectors (loadings). Most of the tests are based on the likelihood ratio (LR) criterion, whose exact distributions are unfortunately not always known. Even for the case of known distributions (see Consul, 1969), percentage points do not seem to have been tabulated. Thus in practice one must resort to asymptotic chi-squared approximations, which only hold for large multivariate normal samples. Although this limits the usefulness of LR tests, nevertheless this represents an improvement over a total reliance on informal rules of thumb or intuition. More recent work on robustness and testing for multivariate normality is considered in the final section.

4.2   TESTING COVARIANCE AND CORRELATION MATRICES

The starting point for a sample PCA is a $(n \times p)$ data matrix $\mathbf{Y}$, which is used to compute a covariance (correlation) matrix between the p random variables. A PCA of such a matrix may then reveal a particular feature or informative dimensions of the data which were not suspected to be present. Before carrying out a full-fledged PCA, however, it is usually advisable to perform preliminary testing of the covariance (correlation) matrix in order to avoid unnecessary computation. For example, we may first wish to test for independence between the random variables or to determine whether more than one sample (for the same variables) possess identical population covariance matrices.

4.2.1 Testing For Complete Independence

When all random variables are uncorrelated it does not make sense to perform a PCA since the variables already possess an independent (uncorrelated) distribution. In practice, however, sample covariances (correlations) are not likely to be identically zero, and sample dispersion matrices will not be perfectly diagonal. A PCA of accidentally correlated data will in reality yield meaningless results, and it is often a good idea to first test for complete independence of the random variables.

Consider p linearly independent normal variates and their sample realizations $\mathbf{Y}_1, \mathbf{Y}_2, \ldots, \mathbf{Y}_p$ such that $\rho(\mathbf{Y}) = p$. Let $\mathbf{X} = \mathbf{Y} - \bar{\mathbf{Y}}$. We wish to test hypotheses of the form

$$H_0: \Sigma = \mathbf{D} \quad \text{or} \quad H_0: \mathbf{P} = \mathbf{I}$$

$$H_a: \Sigma \neq \mathbf{D} \quad \text{or} \quad H_a: \mathbf{P} \neq \mathbf{I} \tag{4.2a}$$

where $\mathbf{P}$ is a population correlation matrix and $\mathbf{D}$ is a diagonal covariance matrix. A test statistic can be derived which maximizes the LR criterion (Section 1.2.2). Let $\mathbf{x}_i$ denote the ith row of $\mathbf{X}$, so that $\sum_{i=1}^{n} \mathbf{x}_i \mathbf{x}_i^{\mathrm{T}} = \mathbf{X}^{\mathrm{T}}\mathbf{X}$. When H_0 is true, the maximum of the likelihood is given by

$$L(\hat{\Omega}_0) = \frac{2}{(2\pi)^{np/2}|\mathbf{D}|^{n/2}} \exp\left[-\frac{1}{2}\sum_{i=1}^{n} (\mathbf{y}_i - \bar{\mathbf{Y}})\hat{\mathbf{D}}^{-1}(\mathbf{y}_i - \bar{\mathbf{Y}}^{\mathrm{T}})\right]$$

$$= \frac{1}{(2\pi)^{np/2}\left(\prod_{i=1}^{p} \hat{\sigma}_i^2\right)^{n/2}} \exp\left[-\frac{1}{2}\sum_{i=1}^{n} \mathbf{x}_i\hat{\mathbf{D}}^{-1}\mathbf{x}_i^{\mathrm{T}}\right]$$

$$= \frac{1}{(2\pi)^{np/2}\left(\prod_{i=1}^{p} \hat{\sigma}_i^2\right)^{n/2}} \exp\left\{-\frac{1}{2}\sum_{i=1}^{n} \mathrm{tr}[(\mathbf{x}_i\mathbf{x}_i^{\mathrm{T}})\hat{\mathbf{D}}^{-1}]\right\}$$

$$= \frac{1}{(2\pi)^{np/2}\left(\prod_{i=1}^{p} \hat{\sigma}_i^2\right)^{n/2}} \exp\left[-\frac{n}{2}\sum_{i=1}^{n} \text{tr}\left[\left(\frac{1}{n}\mathbf{X}^{\mathrm{T}}\mathbf{X}\right)\hat{\mathbf{D}}^{-1}\right]\right]$$

$$= \frac{1}{(2\pi)^{np/2}\left(\prod_{i=1}^{p} \hat{\sigma}_i^2\right)^{n/2}} \exp\left[-\frac{n}{2}\sum_{i=1}^{n} \text{tr}(\hat{\mathbf{D}}\hat{\mathbf{D}}^{-1})\right]$$

$$= \frac{1}{(2\pi)^{np/2}\left(\prod_{i=1}^{p} \hat{\sigma}_i^2\right)^{n/2}} \exp[-\tfrac{1}{2}np] \tag{4.3}$$

where $\hat{\mathbf{D}} = \text{diag}\,(\hat{\sigma}_1^2,\,\hat{\sigma}_2^2,\ldots,\hat{\sigma}_p^2)$ is the ML estimator of $\mathbf{D}$. Also, taken over the entire set of values of $\boldsymbol{\mu}$ and $\boldsymbol{\Sigma}$, the maximum of the likelihood is

$$L(\hat{\Omega}) = \frac{1}{(2\pi)^{np/2}|\hat{\boldsymbol{\Sigma}}|^{n/2}} \exp\left[-\frac{1}{2}\sum_{i=1}^{n} (\mathbf{y}_i - \bar{\mathbf{Y}})\hat{\boldsymbol{\Sigma}}^{-1}(\mathbf{y}_i - \bar{\mathbf{Y}})^{\mathrm{T}}\right]$$

$$= \frac{1}{(2\pi)^{np/2}|\hat{\boldsymbol{\Sigma}}|^{n/2}} \exp\left[-\frac{1}{2}\text{tr}\sum_{i=1}^{n} (\mathbf{x}_i\mathbf{x}_i^{\mathrm{T}})\hat{\boldsymbol{\Sigma}}^{-1}\right]$$

$$= \frac{1}{(2\pi)^{np/2}|\hat{\boldsymbol{\Sigma}}|^{n/2}} \exp\left[-\frac{n}{2}\hat{\boldsymbol{\Sigma}}\hat{\boldsymbol{\Sigma}}^{-1}\right]$$

$$= \frac{1}{(2\pi)^{np/2}|\hat{\boldsymbol{\Sigma}}|^{n/2}} \exp[-\tfrac{1}{2}np] \tag{4.4}$$

so that the ratio of the two likelihoods is maximized by

$$\lambda = \frac{L(\hat{\Omega}_0)}{L(\hat{\Omega})}$$

$$= \frac{(2\pi)^{-np/2}\left(\prod_{i=1}^{p} \hat{\sigma}_i^2\right)^{-n/2} \exp(-\tfrac{1}{2}np)}{(2\pi)^{-np/2}|\hat{\boldsymbol{\Sigma}}|^{-n/2} \exp(-\tfrac{1}{2}np)}$$

$$= \frac{|\hat{\boldsymbol{\Sigma}}|^{n/2}}{\left(\prod_{i=1}^{p} \hat{\sigma}_i^2\right)^{n/2}} \tag{4.5}$$

Although the distribution of λ is unknown $-2\ln\lambda$ can be shown to be distributed asymptotically as the chi-squared distribution, that is, as $n\to\infty$

we have

$$\chi^2 = -2 \ln \lambda$$

$$= -n \left[\ln|\hat{\Sigma}| - \ln \left(\prod_{i=1}^{p} \hat{\sigma}_i^2 \right) \right] \tag{4.6}$$

approaches the chi-squared distribution with $\frac{p}{2}(p-1)$ degrees of freedom. The rate of convergence improves when n is replaced by the correction factor $[n - (1/6p)(2p^2 + p + 2)]$ (Bartlett, 1954). Also, since

$$\lambda = \frac{|\hat{\Sigma}|^{n/2}}{\left(\prod_{i=1}^{p} \hat{\sigma}_i^2 \right)^{n/2}}$$

$$= |\mathbf{R}|^{n/2} \tag{4.7}$$

(Box, 1949) Eq. (4.6) can be written as $\chi^2 = -n \ln|\mathbf{R}|$ where $\mathbf{R}$ is the sample correlation matrix. Convergence is more rapid if n is replaced by $[n - 1/6(2p + 5)]$. The distribution of

$$-[n - 1/6(2p + 5)] \ln|\mathbf{R}| \tag{4.8}$$

then approaches the chi-squared distribution with $\frac{p}{2}(p-1)$ degrees of freedom. When $\mathbf{R}$ approaches diagonal form, $|\mathbf{R}|$ approaches unity and we tend not to accept $\mathbf{H}_0$ for large values of Eq. (4.8) and (4.6). Given multivariate normality the chi-squared approximation is usually satisfactory for $n - p \geq 50$. Since the test is asymptotic, the biased ML estimator $\hat{\Sigma}$ can usually be used in place of $\mathbf{S}$. The exact distribution of the likelihood ratio statistic has recently been tabulated by Mathai and Katiyar (1979). Mudholkar and Subbaiah (1981) provide a Monte Carlo evaluation of the likelihood ratio test for complete independence against several alternative tests, and conclude that the likelihood ratio criterion generally performs well.

Example 4.1.   From Example 3.1 we have the covariance matrix

$$\hat{\Sigma} = \begin{bmatrix} 471.51 & 324.71 & 73.24 & 4.35 \\ 324.71 & 224.84 & 50.72 & 2.81 \\ 73.24 & 50.72 & 11.99 & 1.23 \\ 4.35 & 2.81 & 1.23 & .98 \end{bmatrix}$$

Assuming the variables are drawn from a four-dimensional normal distribution, we have $|\hat{\Sigma}| = 86.32$, $\Pi_{i=1}^{4} \hat{\sigma}_i^2 = 1,245,689.3$, $n = 11$, and $p = 4$.

Using Bartlett's correction factor we have, from Eq. (4.6),

$$\chi^2 = -\left[(n-1) - \frac{1}{6p}(2p^2 + p + 2)\right]\left[\ln|\hat{\Sigma}| - \ln\left(\prod_{i=1}^{n} \hat{\sigma}_i^2\right)\right]$$

$$= -[10 - \tfrac{1}{24}(32 + 4 + 2)][\ln(86.32) - \ln(1245689.3]$$

$$= -8.4167\,(4.4581 - 14.0352)$$

$$= 80.61$$

Since $\frac{p}{2}(p-1) = 6$, we have $\chi^2_{.05,6} = 12.6$ and conclude that Σ is probably not diagonal. To illustrate the test using Eq. (4.8) we convert Σ into the sample correlation matrix

$$\mathbf{R} = \begin{bmatrix} 1.00 & .9973 & .9740 & .2023 \\ .9973 & 1.00 & .9767 & .1893 \\ .9740 & .9767 & 1.00 & .3590 \\ .2023 & .1893 & .3590 & 1.00 \end{bmatrix}$$

where

$$\chi^2 = -[10 - 1/6\,(8 + 5)]\ln(.0000693)$$

$$= -7.8333\,(-9.5771)$$

$$= 75.02$$

which is again larger than $\chi^2_{.05,6} = 12.6$ and we tend to reject H_0. $\qquad\square$

Example 4.2.   The independence test is illustrated by Vierra and Carlson (1981) using simulated data. Given a random sample from an independent p-variate normal distribution, it is always possible to obtain nonzero loadings since $\mathbf{R}$ (or Σ) can indicate an apparent departure from diagonality. The magnitudes of the loadings can at times be surprisingly high, and this may create false impressions concerning a PC analysis. The data and results are shown in Tables 4.1–4.4 where the correlation loadings are generally low except for several large values, which are certainly large enough to be retained by most informal rules of thumb. When latent roots are known the LR test is particularly easy to compute. Using Eq. (4.6) we have

$$\chi^2 = -[(n-1) - 1/6\,(2p + 5)]\ln|\mathbf{R}|$$

$$= -[(n-1) - 1/6\,(2p + 5)]\ln(l_1, l_2, \ldots, l_p)$$

$$= -[47 - 1/6\,(22 + 5)]\ln[(1.8404)(1.5657)\cdots(.4496)]$$

Table 4.1   Input Random Data for $n = 48$ Samples and $p = 11$ Independent Normal Variates

Cases	Y_1	Y_2	Y_3	Y_4	Y_5	Y_6	Y_7	Y_8	Y_9	Y_{10}	Y_{11}
1	45	65	34	49	56	46	44	52	51	50	52
2	45	35	30	36	57	31	67	22	60	47	51
3	53	31	74	45	95	66	64	43	83	64	88
4	75	43	43	36	21	30	37	34	55	76	54
5	66	62	61	35	28	27	20	47	61	18	32
6	62	30	68	11	74	50	66	44	42	76	52
7	34	37	35	25	52	40	46	23	63	39	63
8	47	33	45	48	23	71	71	75	39	57	42
9	56	56	52	47	55	32	42	55	84	59	78
10	50	58	53	22	17	48	60	33	26	67	55
11	65	76	50	28	73	30	65	62	34	33	43
12	22	48	62	58	62	52	67	79	13	50	70
13	45	45	40	75	63	37	18	58	64	39	65
14	42	50	52	41	15	45	52	50	59	61	45
15	45	52	30	12	54	54	6	34	73	46	53
16	43	61	27	85	46	72	45	55	78	44	35
17	41	46	81	40	38	46	48	57	65	51	53
18	57	51	65	46	85	45	69	52	34	72	63
19	24	41	75	69	17	16	53	18	43	40	47
20	31	25	52	25	56	56	22	83	33	44	43
21	46	40	51	74	52	72	9	23	53	43	7
22	62	61	73	49	54	76	39	49	42	41	63
23	66	19	49	45	74	41	56	45	51	39	37
24	37	52	52	57	52	50	68	44	22	59	39
25	38	59	48	44	53	67	61	49	68	71	42
26	51	38	48	54	44	67	54	57	50	45	52
27	67	51	53	57	86	47	73	44	59	60	31
28	41	59	57	53	24	85	56	23	34	72	69
29	29	41	67	55	48	66	72	41	58	52	57
30	80	58	37	58	57	53	32	34	21	48	61
31	30	55	52	54	51	37	62	44	45	52	61
32	22	62	62	64	29	51	23	35	45	44	27
33	54	30	25	71	58	28	68	54	67	49	83
34	14	67	58	62	59	66	20	35	73	72	15
35	68	21	64	76	28	14	36	57	36	69	59
36	81	44	13	32	40	61	65	66	43	70	62
37	52	59	60	35	71	41	33	37	54	51	55
38	17	47	73	45	64	63	58	66	62	4	68
39	73	63	32	67	29	43	20	53	37	34	33
40	62	55	47	55	50	59	28	55	74	56	53
41	54	35	21	58	38	51	68	62	34	29	61
42	67	64	25	46	73	48	56	40	29	49	67
43	35	18	69	68	36	55	67	55	66	75	40
44	45	36	82	19	54	82	49	40	68	61	56
45	45	63	52	34	58	15	26	68	37	73	54
46	29	57	37	49	27	63	33	46	18	65	40
47	68	38	39	49	34	19	40	32	61	26	36
48	66	31	54	28	52	53	49	14	53	81	51

Source: Vierra and Carlson, 1981; reproduced with permission.

Table 4.2   Sample Correlation Matrix of the $p = 11$ Independent Normal Variates

X_1	1.00000										
X_2	−.04125	1.00000									
X_3	−.33646	−.12844	1.00000								
X_4	−.15420	.00858	−.07771	1.00000							
X_5	.09997	−.00102	.07737	−.14453	1.00000						
X_6	−.23473	.04781	.11924	.00208	.06843	1.00000					
X_7	−.01007	−.27203	.08298	−.05531	.18106	.07611	1.00000				
X_8	−.00491	−.02097	−.06750	.06446	.10542	.00716	.09059	1.00000			
X_9	−.07462	−.13710	.09665	.02159	.16349	.05971	−.14651	−.16515	1.00000		
X_{10}	.07407	−.10432	.12435	−.15301	−.01333	.14561	.18907	−.13320	−.05044	1.00000	
X_{11}	.12407	−.15870	.02143	−.16290	.24314	−.06673	.36298	.16018	.00976	.08284	1.00000

Source: Vierra and Carlson, 1981; reproduced with permission.

Table 4.3   Latent roots, Percentage of Variance, and Cumulative Percentage of Variance of $p = 11$ Normal Variates

Factor	Latent Roots	Variance (%)	Cumulative Percentage
1	1.84042	16.7	16.7
2	1.56568	14.2	31.0
3	1.26055	11.5	42.4
4	1.19852	10.9	53.3
5	1.10770	10.1	63.4
6	.9183	8.3	71.7
7	.76670	7.0	78.7
8	.69619	6.3	85.0
9	.66689	6.1	91.1
10	.52933	4.8	95.9
11	.44963	4.1	100.0

Source: Vierra and Carlson, 1981; reproduced with permission.

Table 4.4   Factor Loadings for the $p = 11$ Normal Variates

	Components				
Variables	Z_1	Z_2	Z_3	Z_4	Z_5
X_1	.16635	−.71306	−.39459	.00966	−.00644
X_2	−.45693	−.12573	.03502	−.04976	.70432
X_3	.22364	.68361	.00923	.02021	−.02931
X_4	−.37734	.09022	.47469	.05174	−.41020
X_5	.48425	−.02315	.01899	.54138	.41816
X_6	.07382	.54256	.13468	−.14290	.44059
X_7	.71061	.03610	.25245	−.24161	−.14362
X_8	.17458	−.25950	.71277	.07183	.14340
X_9	.00243	.33959	−.33086	.71133	−.17150
X_{10}	.37947	.20823	−.40280	−.53786	.05361
X_{11}	.70154	−.20912	.12741	.14336	−.00470

Source: Vierra and Carlson, 1981; reproduced with permission.

$$= -(47 - 4.5)\ln(.3752)$$

$$= -42.5\,(-.98029)$$

$$= 41.66$$

Since we have $1/2p(p-1) = 1/2(11)(10) = 55$ degrees of freedom, using $\alpha = .05$ the LR criterion is compared to $\chi^2_{.05,55} = 73.3$, which indicates a nonrejection of the null hypothesis, that is, there is no reason to suppose the existence of a multivariate (normal) distribution so that a PCA is not warranted in this instance. $\qquad\qquad\square$

4.2.2 Testing Sphericity

A special case of the test for independence is the so-called sphericity test, which attempts to determine whether in addition to being diagonal a covariance matrix also possesses equal (diagonal) elements, that is, whether a set of random variables is both independent and homoscedastic. The test derives its name from the fact that given multivariate normality the distribution of a set of p independent and homoscedastic random variables will have the form of a p-dimensional sphere. Note that the sphericity test only applies to covariance matrices since a correlation matrix is by definition homoscedastic, and a test for sphericity in this case is identical to the independence test of the previous section.

The test can be derived by using the LR principle. We test, for some unknown σ^2,

$$H_0 : \Sigma = \sigma^2 \mathbf{I}$$

$$H_a : \Sigma \neq \sigma^2 \mathbf{I} \qquad\qquad (4.9)$$

where we reject H_0 when at least two diagonal elements are not equal. Since Eq. (4.9) is a special case of Eq. (4.2a) the LR statistic can be derived as follows. From Eq. (4.5) we have

$$\lambda = \frac{|\hat{\Sigma}|^{n/2}}{\left[\prod_{i=1}^{p} \hat{\sigma}^2\right]^{n/2}}$$

$$= \frac{|\hat{\Sigma}|^{n/2}}{[(\hat{\sigma}^2)^p]^{n/2}}$$

$$= \frac{|\hat{\Sigma}|^{n/2}}{(\hat{\sigma}^2)^{np/2}}$$

and replacing $\hat{\sigma}^2$ by its ML estimate $(1/p)\,\mathrm{tr}(\hat{\Sigma}) = (1/pn)\,\mathrm{tr}(\mathbf{X}^T\mathbf{X})$ yields

$$\lambda = \frac{\left|n\mathbf{X}^T\mathbf{X}\right|^{n/2}}{\left[\dfrac{1}{np}\,\mathrm{tr}(\mathbf{X}^T\mathbf{X})\right]^{np/2}}$$

$$= \frac{\left|\mathbf{X}^T\mathbf{X}\right|^{n/2}}{\left[\dfrac{1}{p}\,\mathrm{tr}(\mathbf{X}^T\mathbf{X})\right]^{np/2}} \tag{4.10}$$

When the latent roots of $\mathbf{X}^T\mathbf{X}$ are known the statistic Eq. (4.10) is particularly easy to compute and can be expressed as

$$\lambda = \left[\frac{\displaystyle\prod_{i=1}^{p} l_i^{1/p}}{\displaystyle\sum_{i=1}^{p}\frac{l_i}{p}}\right]^{np/2} \tag{4.11}$$

where the expression in square brackets is the ratio of the geometric and arithmetic means of latent roots, and where we use n in place of degrees of freedom because of the asymptotic nature of the test. As $n \to \infty$, the distribution of $-2\ln\lambda$ under H_0 tends to the chi-squared with $1/2\,(p+2)(p-1)$ degrees of freedom. Note that Eq. (4.10) does not change when $\mathbf{X}^T\mathbf{X}$ is replaced by the sample covariance matrix. Since convergence is usually improved by Bartlett's correction factor, the criterion is normally expressed as

$$\chi^2 = -\left[n - \frac{1}{6p}(2p^2 + p + 2)\right]\left[\ln\left|\mathbf{X}^T\mathbf{X}\right| - p\ln\left(\frac{1}{p}\,\mathrm{tr}\,\mathbf{X}^T\mathbf{X}\right)\right] \tag{4.12}$$

Example 4.3.   The test for sphericity can be illustrated by referring to the random data of Table 4.1, which has covariance matrix as in Table 4.5 where $\ln|\mathbf{S}| = 61.05473$, $\mathrm{tr}(\mathbf{S}) = 3134.32$, $n = 48$, and $p = 11$. The chi-squared

Table 4.5   Covariance Matrix of $p = 11$ Normal Random Variates (see Table 4.1)

$$\mathbf{S} = \begin{bmatrix}
282.01 & & & & & & & & & & \\
-9.85 & 195.17 & & & & & & & & & \\
-92.47 & -28.53 & 266.97 & & & & & & & & \\
-44.12 & 2.38 & -22.02 & 291.10 & & & & & & & \\
32.03 & .27 & 24.93 & -47.84 & 363.86 & & & & & & \\
-68.84 & 11.89 & 34.06 & 0.00 & 23.19 & 301.76 & & & & & \\
-2.51 & -70.17 & 1.49 & -17.02 & 65.53 & 25.95 & 348.63 & & & & \\
-1.34 & -4.23 & -17.46 & 17.96 & 31.94 & .55 & 27.69 & 254.35 & & & \\
-20.05 & -33.61 & 26.39 & 5.39 & 54.93 & 18.30 & -47.87 & -46.20 & 308.29 & & \\
21.42 & -23.99 & 32.07 & -41.86 & -3.74 & 42.33 & 58.03 & -33.13 & -14.65 & 267.53 & \\
33.23 & -34.55 & 4.95 & 45.20 & 73.66 & -19.13 & 109.65 & 39.45 & 3.08 & 20.88 & 254.65
\end{bmatrix}$$

statistic Eq. (4.12) is then

$$\chi^2 = -[48 - \tfrac{1}{66}(242 + 11 + 2)][61.055 - 11\ln(\tfrac{3134.32}{11})]$$

$$= -44.1364\,(61.055 - 62.1750)$$

$$= 49.43$$

which is compared to $\chi^2_{\alpha,1/2(p+1)(p-1)} = \chi^2_{.05,65} = 84.8$. We tend to accept H_0 and conclude that the population covariance matrix is of the spherical form $\Sigma = \sigma^2 \mathbf{I}$. $\qquad\square$

Example 4.4. In their study of arthropod infestation in stored grain bulks of Canadian prairie wheat, Sinha and Lee (1970) measure three environmental variables and the extent of infestation by six types of arthropods. The nine variables are defined as follows:

Y_1 = Grade of cereal (wheat, oats, and barley, according to the Canadian grading system) on a scale of 1 (highest quality) to 6
Y_2 = Moisture content by percentage weight.
Y_3 = Dockage (presence of weed, broken kernels, and other foreign material)
Y_4 = Acarus
Y_5 = Cheyletus
Y_6 = Glycyphagus
Y_7 = Tarsonemus
Y_8 = Cryptolestes
Y_9 = Procoptera $\qquad\square$

Before proceeding to carry out a PCA of the nine variables, the correlational structure is first tested to see whether the correlation matrix of Table 4.6

Table 4.6 Correlation Matrix for Nine Arthropod Infestation Variables ($n = 165$)

	Y_1	Y_2	Y_3	Y_4	Y_5	Y_6	Y_7	Y_8	Y_9
	1.000								
	.441	1.000							
	.441	.342	1.000						
	.107	.250	.040	1.000					
$\mathbf{R} =$	.194	.323	.060	.180	1.000				
	.105	.400	.082	.123	.220	1.000			
	.204	.491	.071	.226	.480	.399	1.000		
	.197	.158	.051	.019	.138	−.114	.154	1.000	
	−.236	−.220	−.073	−.199	−.084	−.304	−.134	−.096	1.000

Source: Sinha and Lee, 1970; reproduced with permission.

differs significantly from the unit matrix. Using Eq. (4.8) we have

$$\chi^2 = [164 - 1/6\,(18 + 5)]\ln(.18792)$$

$$= -(160.17)(-1.67169)$$

$$= 267.7$$

which for $\alpha = .05$ and $1/2\,(9)(8) = 36$ degrees of freedom is highly significant, implying the correlation matrix of Table 4.6 differs from the identity
matrix.

4.2.3   Other Tests for Covariance Matrices

Several other tests of significance can be used to determine data homogeneity or to test for equal correlation.

Covariance Matrix Equality
The same variables may at times be observed in different samples, and a
question arises as to whether the samples can be considered to have
originated from the same population. Should this be the case, there is an
advantage to pooling the samples since this results in an increase of degrees
of freedom and yields more reliable estimates of PC loadings and scores.
Given k samples, each containing an identical set of p multivariate normal
variables with n_g $(g = 1, 2, \ldots, k)$ observations, we wish to test the null
hypothesis

$$H_0\colon \mathbf{\Sigma}_1 = \mathbf{\Sigma}_2 = \cdots = \mathbf{\Sigma}_k$$

against the alternative that at least two covariance matrices are not equal.
Let

$$\mathbf{X}^T\mathbf{X} = \sum_{g=1}^{k} \mathbf{X}_{(g)}^T \mathbf{X}_{(g)}, \qquad n = \sum_{g=1}^{k} n_g, \qquad \hat{\mathbf{\Sigma}} = \sum_{g=1}^{k} \hat{\mathbf{\Sigma}}_g$$

where $\mathbf{X}_{(1)}, \mathbf{X}_{(2)}, \ldots, \mathbf{X}_{(k)}$ are $(n_k \times p)$ data matrices.
   Then using the LR method it can be shown (Srivastava and Carter, 1983;
Anderson, 1984) that the corresponding criterion is given by

$$\lambda = \prod_{g=1}^{k} \frac{|\hat{\mathbf{\Sigma}}_g|^{\frac{n_g}{2}}}{|\hat{\mathbf{\Sigma}}|^{\frac{n}{2}}}$$

$$= \prod_{g=1}^{k} \frac{|\mathbf{X}_{(g)}^T \mathbf{X}_{(g)}|^{\frac{n_g}{2}}}{|\mathbf{X}^T\mathbf{X}|^{\frac{n}{2}}} \circ \frac{n^{1/2pn}}{\prod\limits_{g=1}^{k} n^{\frac{p n_g}{2} g}} \tag{4.13}$$

For large n, the criterion $-2 \ln \lambda$ has an approximate chi-squared distribution with $1/2(k-1)p(p+1)$ degrees of freedom. Convergence is more rapid if 2 is replaced by

$$m = \frac{2}{n}(n - 2a)$$

where

$$a = \frac{\left(\sum\limits_{g=1}^{k} \dfrac{n}{n_g} - 1\right)(2p^2 + 3p - 1)}{12(p+1)(k-1)}$$

Example 4.5. Consider the male and female turtle data of Example 3.3. We wish to decide whether a PCA can be carried out using pooled male and female data, that is, we wish to test the hypothesis

$$H_0: \mathbf{\Sigma}_1 = \mathbf{\Sigma}_2$$

$$H_a: \mathbf{\Sigma}_1 \neq \mathbf{\Sigma}_2$$

The sample means and covariance matrices for the two groups are summarized in Table 4.7. Using Eq. (4.13) we have $k = 2$, $n_1 = n_2 = 24$, and $n = 4$ so that

$$\lambda = \frac{\prod\limits_{g=1}^{k} |\hat{\mathbf{\Sigma}}_g|^{\frac{n_g}{2}}}{|\hat{\mathbf{\Sigma}}|^{\frac{n}{2}}}, \qquad a = \left(\frac{n^2}{n_1 n_2} - 1\right)\frac{(2p^2 + 3p - 1)}{12(p+1)(k-1)}$$

where we use sample sizes rather than degrees of freedom due to the asymptotic nature of the test.

We have the determinants

$$|\hat{\mathbf{\Sigma}}_1| = \begin{vmatrix} 138.77 & 79.15 & 37.38 \\ 79.15 & 50.04 & 21.65 \\ 37.38 & 21.65 & 11.26 \end{vmatrix}, \quad |\hat{\mathbf{\Sigma}}_2| = \begin{vmatrix} 451.39 & 271.17 & 168.70 \\ 271.17 & 171.73 & 103.29 \\ 168.70 & 103.29 & 66.65 \end{vmatrix}$$

$$= 792.64 \qquad\qquad\qquad = 12648.64$$

Table 4.7   Mean Vectors X and Covariance Matrices $\hat{\mathbf{\Sigma}}_1$ and $\hat{\mathbf{\Sigma}}_2$ for the Turtle Carapace Data of Table 3.1

	24 Males			24 Females		
Statistic $\bar{\mathbf{X}}$	Length $(\mathbf{X}_1)$ $(\bar{X}_1 = 113.38)$	Width $(\mathbf{X}_2)$ $(\bar{X}_2 = 88.29)$	Height $(\mathbf{X}_3)$ $(\bar{X}_3 = 40.71)$	Length $(\mathbf{X}_1)$ $(\bar{X}_1 = 136.00)$	Width $(\mathbf{X}_2)$ $(\bar{X}_2 = 102.58)$	Height $(\mathbf{X}_3)$ $\bar{X}_3 = 51.96)$
$\hat{\mathbf{\Sigma}}_g$	138.77 79.15 37.38	79.15 50.04 21.65	37.38 21.65 11.26	451.39 271.17 168.70	271.17 171.73 103.29	168.70 103.29 66.65

Source: Jolicoeur and Mosimann, 1960; reproduced with permission.

and the pooled determinant is

$$|\hat{\Sigma}| = |\hat{\Sigma}_1| + |\hat{\Sigma}_2|$$

$$= \begin{vmatrix} 590.16 & 350.32 & 206.08 \\ 350.32 & 221.77 & 124.94 \\ 206.08 & 124.94 & 77.91 \end{vmatrix}$$

$$= 44493.8$$

Using sample covariance matrices the natural log of Eq. (4.13) if given by

$$\ln \lambda = \sum_{g=1}^{2} \tfrac{1}{2} n_g \ln|\hat{\Sigma}_g| - \tfrac{1}{2} n \ln|\hat{\Sigma}|$$

$$= \tfrac{1}{2}[24(6.6754) + 24(9.4453)] - \tfrac{48}{2}(10.7031)$$

$$= 193.45 - 256.87$$

$$= -63.42$$

where

$$a = \frac{(2+2-1)(2 \cdot 3^2 + 3 \cdot 3 - 1)}{12(3+1)(2-1)}$$

$$= \frac{3(26)}{48}$$

$$= 1.625$$

and the correction factor becomes

$$m = \frac{2}{n}(n - 2a)$$

$$= \tfrac{2}{48}[48 - 2(1.625)]$$

$$= 1.8646$$

The chi-squared approximation is then given by

$$\chi^2 = -m \ln \lambda$$

$$= -1.86(-63.42)$$

$$= 117.96$$

which is distributed approximately as $\chi^2_{1/2(p-1)(p+1)} = \chi^2_4$. For a Type I error, $\alpha = .05$, we have $\chi^2_{.05,4} = 9.49$, and H_0 is rejected, and we conclude that male and female covariance matrices differ significantly.

***Example* 4.6.**   For the whale data of Example 3.13 both males and females are included in the PCA, so that both are assumed to possess an identical variance/covariance matrix. Since Y_{18} is a discriminant variable for males and females, it is omitted from the analysis. Using Eq. (4.13) to test for equality of the two covariance matrices we have

$$
\begin{aligned}
\ln \lambda &= \sum_{g=1}^{2} \tfrac{1}{2} n_g \ln|\hat{\Sigma}_g| - \tfrac{1}{2} n \ln|\hat{\Sigma}| \\
&= \tfrac{1}{2}[n_1 \ln|\hat{\Sigma}_1| + n_2 \ln|\hat{\Sigma}_2| - n \ln|\hat{\Sigma}|] \\
&= \tfrac{1}{2}\left[n_1 \sum_{i=1}^{p} \ln l_{1i} + n_2 \sum_{i=1}^{p} \ln l_{2i} - n \sum_{i=1}^{p} \ln l_i \right] \\
&= \tfrac{1}{2}[31(-92.60) + 31(-92.76) - 62(-76.25)] \\
&= -509.42
\end{aligned}
$$

where $m = 1.4364$. The asymptotic chi-squared statistic is then

$$
\begin{aligned}
\chi^2 &= -m \ln \lambda \\
&= -1.4364(-509.42) \\
&= 731.72
\end{aligned}
$$

which is significant at the $\alpha = .01$ level, indicating a difference in the two covariance matrices. The latent roots and PC correlation loadings for the male and female covariance matrices are given in Tables 4.8 and 4.9

Table 4.8 Latent Roots for Male and Female Covariance Matrices (Variables in Natural Logarithms)

Male	Female
.38545	.22667
.03474	.03991
.02825	.02055
.01919	.01585
.00914	.01308
.00837	.00925
.00743	.00743
.00616	.00509
.00443	.00472
.00340	.00297
.00194	.00252
.00143	.00218
.00103	.00118
.00088	.00109
.00062	.00069
.00047	.00051
.00018	.00013

Table 4.9   Correlation Loadings for Male and Female Covariance Matrices (Variables in Natural Logarithms)[a]

	Males			Females			
	Z_1	Z_2	Z_3	Z_1	Z_2	Z_3	Z_4
X_1	.970			.983			
X_2	.956			.915			
X_3	.892			.900			
X_4	.957			.951			
X_5	.751	.214		.848			
X_6	.919			.836			
X_7	.855			.644	.347	.364	
X_8	.944			.783		.307	
X_9	.810			.780	.199		.309
X_{10}	.827			.733			.555
X_{11}	.929			.830		.227	.261
X_{12}	.920			.775			.279
X_{13}	.649	.342	.340	.871			
X_{14}	.752	.361	.339	.864			
X_{15}	.871			.877			
X_{16}	.827			.526		.782	
X_{17}	.820	.410	.381	.505	.848		

[a]Loadings smaller than .10 are omitted.

respectively. The first component is a size dimension, whereas the second, third, and fourth components represent shape dimensions. Note also (Table 4.8) that the males' body structure is more closely related to their general body size than is the case for the females.   □

Testing for Proportionality

A more general test is to verify whether $k = 2$ covariance matrices are proportional, that is, whether hypothesis

$$H_0: \Sigma_1 = c\Sigma_2$$

holds, where $c > 0$ is a scalar constant. The pooled ML estimator is then

$$\hat{\Sigma} = \frac{1}{n}(S_1 + cS_2)$$

where the Wishart distribution Eq. (4.1) is given by

$$k|\Sigma|^{-n/2} c^{pn/2} \exp[-1/2 \operatorname{tr} \Sigma^{-1}(S_1 + cS_2)]$$

The test is based on the fact that when H_0 is true the latent roots of $\Sigma_2\Sigma_1^{-1}$ must all equal c^{-1}. It is possible to derive the asymptotic distribution of c, which can then be used to test H_0. For further detail see Rao (1983) and

Guttman et al. (1985). Flury (1986, 1988) has generalized the test for $k > 2$ groups (samples).

Testing for Equal Correlation

One of the simplest structures for a PCA occurs when all variables are correlated equally (Section 3.3). In this situation a single dominant PC explains all of the nonisotropic variance for the population. It is therefore at times desirable to test for such a structure before computing PCs.

Consider a population covariance matrix Σ with an equicorrelation structure. Here Σ can also be expressed as

$$\Sigma = \sigma^2[(1 - \rho)\mathbf{I} + \rho\mathbf{ee}^T] \tag{4.14}$$

where ρ is the population correlation coefficient and $\mathbf{e}$ is a column vector of unities. We wish to test the hypothesis

$$H_0: \Sigma = \sigma^2[(1 - \rho)\mathbf{I} + \rho\mathbf{ee}^T]$$

$$H_a: \Sigma \neq \sigma^2[(1 - \rho)\mathbf{I} + \rho\mathbf{ee}^T] \tag{4.15}$$

Let $\hat{\sigma}^2$ and r be the ML estimators of σ^2 and ρ respectively. Then under H_0 the ML estimator of $|\Sigma|$ is $|\hat{\Sigma}| = (\sigma^2)^p(1 - r)^{p-1}[1 + (p - 1)r]$, whereas under H_a it is of the general form $\Sigma = (1/n)\mathbf{X}^T\mathbf{X}$. The ratio of the two likelihoods is then (see Srivastava and Carter, 1983).

$$\lambda = \frac{L(\hat{\Omega}_0)}{L(\hat{\Omega})} = \frac{|\hat{\Sigma}|}{(\hat{\sigma}^2)^p(1 - r)^{p-1}[1 + (p - 1)r]} \tag{4.16}$$

The test is from Wilks (1946). Here $-2 \ln \lambda$ is approximately chi-squared with $\frac{p}{2}(p + 1) - 2$ degrees of freedom. Box (1949) has shown that the approximation is improved by using the factor

$$m = \left[\frac{n - p(p + 1)^2(2p - 3)}{6(p - 1)(p^2 + p - 4)}\right] \tag{4.17}$$

so that $-m \ln \lambda$ is approximately chi-squared with $\frac{p}{2}(p + 1) - 2$ degrees of freedom. Since the test is asymptotic we have retained the biased ML estimator $\hat{\Sigma}$.

The ML ratio test Eq. (4.16) is only applicable to covariance matrices so that its counterpart using the correlation matrix is not available. The hypothesis

$$H_0: \mathbf{R} = \rho\mathbf{I}$$

$$H_a: \mathbf{R} \neq \rho\mathbf{I} \tag{4.18}$$

is therefore usually tested using Lawley's (1963) heuristic test, which is based on the off-diagonal elements of the sample correlation matrix. Let

$$\bar{r}_j = \frac{1}{p-1} \sum_{\substack{i=1 \\ i \neq j}}^{p} r_{ij}, \qquad \bar{r} = \frac{2}{p(p-1)} \sum_{i<j}^{p}\sum^{p} r_{ij} \qquad (4.19)$$

that is, $\bar{r}_j$ is the mean of all off-diagonal elements for the jth column of $\mathbf{R}$, and $\bar{r}$ is the overall mean of all off-diagonal elements. Since the diagonal elements are fixed at unity, they do not enter into the computations. Also, let

$$h = \frac{(p-1)^2[1-(1-\bar{r})^2]}{p-(p-2)(1-\bar{r})^2} \qquad (4.20)$$

Then Lawley (1963) has suggested that in large samples the statistic

$$\chi^2 = \frac{n}{(1-\bar{r})^2} \left[\sum_{i<j}^{p}\sum^{p} (r_{ij} - \bar{r})^2 - h \sum_{j=1}^{p} (\bar{r}_j - \bar{r})^2 \right] \qquad (4.21)$$

is approximately chi-squared with $1/2(p+1)(p-2)$ degrees of freedom. When the null hypothesis is accepted, we conclude that there exists a single nonisotropic dimension (PC). It will be seen in the next section that existence of a single isotropic component can also be determined by a direct test of the latent roots.

Example 4.7.   To illustrate the use of Eq. (4.16) we consider the Painted Turtle carapace measurements (females) of Example 4.5. We wish to test whether $\mathbf{\Sigma}$ possess the equal correlation and homoscedastic structure $\rho\sigma^2$. Using the unbiased estimator $\mathbf{S}$ we have $|\mathbf{S}| = 12{,}648.636$, $p = 3$, and $n = 24$ where the unknowns σ^2, ρ are estimated from $\mathbf{S}$ as

$$s^2 = \frac{1}{p} \sum_{i=1}^{p} s_{ii}^2 = \tfrac{1}{3}(689.77) = 229.923$$

$$s^2 r = \frac{2}{p(p-1)} \sum_{i<j}\sum s_{ij} = \tfrac{1}{3}(543.16) = 181.053$$

so that

$$r = \frac{181.053}{229.923} = .7875$$

Taking natural logs, Eq. (4.16) becomes

$$\ln \lambda = \ln|\mathbf{S}| - p \ln(s^2) - (p-1)\ln(1-r) - \ln[1 + (p-1)r]$$
$$= 9.4453 - 16.3132 + 3.0976 - 3.5417$$
$$= -7.312$$

where

$$m = \frac{23 - 3(16)(3)}{6(2)(8)} = 21.5$$

so that

$$\chi^2 = -m \ln \lambda = -21.5(-7.312) = 157.21 \tag{4.22}$$

is approximately chi-squared with $\frac{p}{2}(p+1) - 2 = 4$ degrees of freedom. Since $\chi^2_{.05,4} = 9.49$ we conclude that $\mathbf{\Sigma}$ does not have the form $\sigma^2 \rho$.

Equation (4.16) does not indicate whether equality is violated for the diagonal elements of $\mathbf{\Sigma}$, the off-diagonal elements, or both. A situation may arise therefore whereby H_0 is rejected due to unequal diagonal elements of $\mathbf{\Sigma}$. If variance however is of no interest, $\mathbf{S}$ may be converted to the correlation matrix

$$\mathbf{R} = \begin{bmatrix} 1.00 & & \\ .974 & 1.00 & \\ .973 & .965 & 1.00 \end{bmatrix}$$

and equality of correlation coefficients tested using Eq. (4.21). We have $p = 3$, $n = 24$, $\bar{r}_1 = .9730$, $\bar{r}_2 = .9695$, $\bar{r}_3 = .9690$, $\bar{r} = 1/3(2.912) = .9707$, and $h = 1.336$ as defined by Eqs. (4.19) and (4.20). Then

$$\sum_{i<j}\sum (r_{ij} - \bar{r})^2 = (.974 - .9707)^2 + (.973 - .9707)^2 + (.965 - .9707)^2$$
$$= .0000487$$

$$\sum_{j=1}^{3} (\bar{r}_j - \bar{r})^2 = (.9730 - .9707)^2 + (.9695 - .9707)^2 + (.9690 - 9707)^2$$
$$= .0000096$$

and Eq. (4.21) becomes

$$\chi^2 = \frac{23}{(1 - .9707)^2} [(.0000487 - 1.336\,(.0000096)]$$
$$= .9618$$

Since $\chi^2_{.05,2} = 5.66$, we fail to reject H_0 at level $\alpha = .05$ and conclude that when the carapace measurements are standardized to unit variance, the resulting correlation matrix exhibits an equicorrelational structure.   □

4.3 TESTING PRINCIPAL COMPONENTS BY MAXIMUM LIKELIHOOD

Once preliminary testing has been carried out on the covariance (correlation) matrix the next stage is to determine how many PCs have arisen as a result of random sampling, measurement error, or residual (individual) variation. The tests used to determine significant PCs involve latent roots rather than loadings, since it turns out that the roots tell us in a more direct fashion the number of isotropic dimensions which are present, and thus the number of stable loading coefficients. Note that it does not make sense to test whether roots are significantly different from zero if the population multivariate distribution is p-dimensional (nondegenerate). Zero roots arise from exact linear dependencies amongst the variables, which are at time of interest in regression-type Problems (Chapter 10). For a PCA however they are of little practical interest since they correspond to PCs, which explain precisely zero percent of the population variance. The discarding of such components is thus tantamount to removing mathematically deterministic relationships amongst the variables. The analytical strategy proposed in this section consists of two stages. First, reject PCs that correspond to isotropic roots, since these cannot possibly be of any interest. Second, discard those PCs that possess no evident interpretation, those that possess both small correlation loadings and that also explain a small percentage of variance. The two stages are distinct since the second stage (if required) must by necessity remain somewhat arbitrary and conditional on the actual application and objective of a PCA. More recent criteria are discussed in Section 4.3.5, which may obviate the two-stage selection process by the use of penalty functions which penalize high-dimensional models. Thus before any attempts are made at behavioral interpretation, residual components should be discarded in order to reduce PCs to a smaller set of nontrivial dimensions. A statistical test of significance therefore should precede any discarding of PCs if we have reasonable guarantees that data are multivariate normal. For a general review of tests for latent roots and latent vectors see also Tyler (1981).

4.3.1 Testing Equality of All Latent Roots

Given a multivariate sample from $N(\mathbf{\mu}, \mathbf{\Sigma})$ the distribution of all the latent roots has been given by Roy (1957) and James (1964). The functional form for arbitrary $\mathbf{\Sigma}$ and n is involved and is difficult to use for testing purposes. When $\mathbf{\Sigma} = \lambda \mathbf{I}$ however the exact joint distribution of the sample latent roots

$l_1 > l_2 > \cdots > l_p$ of $\mathbf{X}^T\mathbf{X}$ can be written as

$$f(L) = c \exp\left[-\frac{n}{2\lambda}\sum_{i=1}^{p} l_i\right]\prod_{i=1}^{p} l_i^{(n-p-1)/2}\prod_{i<j}^{p}(l_i - l_j) \qquad (4.23)$$

where

$$c = \frac{\pi^{p^2/2}}{\left(\dfrac{n}{2\lambda}\right)^{\frac{-np}{2}}}\prod_{i=1}^{p}\Gamma[\tfrac{1}{2}(n+1-i)]\prod_{i=1}^{p}\Gamma[\tfrac{1}{2}(p+1-i)]$$

for n degrees of freedom. It can also be shown that the sample roots l_i are distributed independently of the sample latent vectors. When $p = 1$, Eq. (4.23) reduces to the univariate beta distribution. Percentage points for the largest root l_1 have been tabulated for $p = 2$–5. For further reference to distributional properties of the latent roots of $\mathbf{X}^T\mathbf{X}$ see Muirhead (1982) and Anderson (1984).

The test for equality of all the latent roots is based on the LR criterion from Mauchly (1940). In fact it is identical to the sphericity test Eq. (4.11), but is expressed in terms of the latent roots. It is handier to use after a PCA has already been carried out and the roots are known. From Eq. (4.11) we have

$$\lambda = \left[\frac{|\mathbf{X}^T\mathbf{X}|}{\left[\dfrac{\text{tr}(\mathbf{X}^T\mathbf{X})}{p}\right]^p}\right]^{n/2}$$

$$\left[\frac{\prod\limits_{i=1}^{p} l_i}{\left[\dfrac{1}{p}\sum\limits_{i=1}^{p} l_i\right]^p}\right]^{n/2} \qquad (4.24)$$

so that the sphericity test is equivalent to testing the hypothesis

$$H_0: \lambda_1 = \lambda_2 = \cdots = \lambda_p$$

$$H_a: \text{not all equal}$$

where $-2\ln\lambda$ is approximately chi-squared with $1/2(p+2)(p-1)$ degrees of freedom. The test criterion can be written as

$$\chi^2 = -\left[n - \frac{1}{6p}(2p^2 + p + 2)\right]\left[\ln\prod_{i=1}^{p} l_i + p\ln\left(\frac{1}{p}\sum_{i=1}^{p} l_i\right)\right] \qquad (4.25)$$

which is identical to Eq. (4.12) where again we use n rather than $n-1$. Since we tend to accept H_0 for small values of χ^2 the sphericity test is formally equivalent to testing for equality of the geometric and arithmetic means of the latent roots. Also when using the correlation matrix, a criterion equivalent to Eq. (4.8) is

$$\chi^2 = -[n - \tfrac{1}{6}(2p + 5)] \ln \prod_{i=1}^{p} l_i \qquad (4.26)$$

but which is more handy to use when the latent roots are known.

4.3.2 Testing Subsets of Principal Components

When the null hypothesis of sphericity (total independence) is rejected it is still possible for a smaller number of $p - r$ latent roots to be equal. As noted earlier (Section 3.3) this corresponds to an isotropic $p - r$ dimensional sphere embedded in the p-dimensional normal ellipsoid, and any orthogonal rotation of the principal axes cannot increase or decrease length (variance) of the axes. Consider a sample of size n from a multivariate normal $N(\mu, \Sigma)$. We wish to test

$$H_0: \lambda_{r+1} = \lambda_{r+2} = \cdots = \lambda_p$$

$$H_a: \text{not all } (p - r) \text{ roots are equal}$$

When $r = p - 1$, the test is equivalent to testing for existence of a single nonisotropic PC. An appropriate test is the extension of the LR test for complete independence,

$$\lambda = \left[\frac{\prod_{i=r+1}^{p} l_i}{\left(\dfrac{1}{q} \sum_{i=r+1}^{p} l_i \right)^q} \right]^{n/2} = \left[\frac{\prod_{i=r+1}^{p} l_i}{(\bar{l}_q)^q} \right]^{n/2} \qquad (4.27)$$

where $\bar{l}_q$ is the arithmetic mean of the last $q = p - r$ sample roots (Bartlett, 1950; Anderson, 1963a). Lawley (1956) has shown that the chi-squared approximation is improved slightly if Bartlett's multiplier is increased by the amount

$$\sum_{i=1}^{r} \frac{(\bar{l}_q)^2}{(l_i - \bar{l}_q)^2}$$

The statistic

$$\chi^2 = -\left[n - r - \frac{1}{6q}(2q^2 + q + 2) + \sum_{i=1}^{r} \frac{(\bar{l}_q)^2}{(l_i - \bar{l}_q)^2} \right]$$

$$\times \left[\sum_{i=r+1}^{p} \ln l_i - q \ln \left(\frac{1}{q} \sum_{i=r+1}^{p} l_i \right) \right] \tag{4.28}$$

is then approximately chi-squared with $\frac{q}{2}(q+1) - 1$ degrees of freedom where

$$\frac{1}{q} \sum_{i=r+1}^{p} l_i = \bar{l}_q$$

Lawley's correction may be conservative (James, 1969). The test can be used to check for equality of any $1 < r \le p$ adjacent latent roots of a sample covariance matrix. When the correlation matrix is used, Eq. (4.28) does not possess an asymptotic chi-squared distribution. However the test is at times employed for the last $p - r$ latent roots of a correlation matrix when these account for a small percentage of total variance, and when n is large. When complete sphericity is rejected, Eq. (4.28) can be applied to the last 2, 3, ..., $p - 1$ roots in a sequential manner. The testing procedure often results in more PCs than can be meaningfully interpreted, and a careful inspection of the retained loadings is normally required in addition to the test. Anderson (1963a) has shown that the LR criterion can also be used to test the equality of any r adjacent roots. At times Lawley's correction factor is omitted, in which case Eq. (4.28) can also be written as

$$\chi^2 = -[n - r - \tfrac{1}{6q}(2q^2 + q + 2)]\left[\sum_{i=1}^{r} \ln l_i - q \ln \left(\frac{1}{q} \sum_{i=1}^{r} l_i \right) \right] \tag{4.29}$$

Alternatively, it is possible to test whether the first r roots account for a sufficient percentage of the variance. This is achieved by testing the null hypothesis

$$H_0 : \frac{\sum\limits_{i=r+1}^{p} \lambda_i}{\sum\limits_{i=1}^{p} \lambda_i} = H \tag{4.30}$$

for some $0 < H < 1$. Let

$$T = \sum_{i=r+1}^{p} l_i - H \sum_{i=1}^{p} l_i$$

$$= (1 - H) \sum_{i=r+1}^{p} l_i - H \sum_{i=1}^{r} l_i \tag{4.31}$$

Assuming the latent roots of Σ are distinct, it can be shown that as $n \to \infty$ the limiting distribution of

$$n^{1/2} \left[T + H \sum_{i=1}^{r} \lambda_i - (1 - H) \sum_{i=r+1}^{p} \lambda_i \right] \tag{4.32}$$

is normal with mean zero and variance

$$\tau = 2(H)^2 \sum_{i=1}^{r} \lambda_i^2 + 2(1 - H)^2 \sum_{i=r+1}^{p} \lambda_i^2 \tag{4.33}$$

When population roots λ_i are replaced by l_i it is possible to test Eq. (4.30), that is, to test whether the first r PCs explain a significant proportion of variance, and to construct confidence intervals for expressions of the form (see Saxena, 1980)

$$\sum_{i=r+1}^{p} \lambda_i - H \sum_{i=1}^{p} \lambda_i \tag{4.34}$$

Example 4.8. We use the whale measurements of Example 3.11 to illustrate Eq. (4.28). Since the last 14 roots account for just over 10% of the total variance, we wish to determine whether they can be considered to represent isotropic dimensions. We have

$$H_0: \lambda_5 = \lambda_6 = \cdots = \lambda_{18}$$

$$H_a: \text{not all equal}$$

where $n = 62$, $p = 18$, $q = 14$, $\sum_{i=5}^{18} \ln l_i = -80.11085\, q$ and $\ln(l_q) = 14 \ln(.00506) = -74.15965$

$$\sum_{i=1}^{4} \frac{\bar{l}_q^2}{(l_i - \bar{l}_q)} = \frac{(.005006)^2}{(.309925 - .005006)^2} + \cdots + \frac{(.005006)^2}{(.022712 - .005006)^2}$$

$$= .10484$$

$$n - r - \frac{1}{6q}(2q^2 + q + 2) = 62 - 4 - \frac{1}{84}(392 + 14 + 2) = 53.1428$$

so that

$$\chi^2 = -(53.1428 + .10484)(-80.11085 + 74.15965)$$

$$= 316.9$$

For $\alpha = .05$ and $\frac{q}{2}(q+1) - 1 = 104$ degrees of freedom, the test statistic indicates a significant difference among at least two of the last 14 latent roots. The last 14 PCs however appear to be uninterpretable, and for most practical intents and purposes can be omitted from the analysis. The example also illustrates the known fact that LR chi-squared tests of dimensionality typically retain more components than can be interpreted in a sensible manner (see also Section 4.3.5). Note that the rejection of H_0 does not necessarily imply that the last subset of $n - r$ components represent systematic variation—it simply indicates that the residual variance structure is not isotropic at some significance level. This can occur, for example, when different sources of error exist such as sampling variation, measurement error, missing values, natural variation, and so forth. $\square$

4.3.3 Testing Residuals

The main reason for testing whether the smallest $p - r$ latent roots are isotropic is to be able to determine the effective dimensionality of a data matrix, that is, the number of components required to reproduce the explainable part of matrix $\mathbf{X}$. An alternative approach is to consider the residuals of a principal components model (Jackson and Mudholkar, 1979; see also Gnanadesikan and Kettenring, 1972) which is based on testing PC residuals. Let $\mathbf{P}_{z(r)} = \mathbf{Z}_{(r)}(\mathbf{Z}_{(r)}^T\mathbf{Z}_{(r)})^{-1}\mathbf{Z}_{(r)}^T = \mathbf{Z}_{(r)}\mathbf{Z}_{(r)}^T$ be the projection matrix that projects orthogonally columns of $\mathbf{X}$ onto the first r principal components (Section 3.5). Then the matrix that projects $\mathbf{X}$ onto the residual space, or the last $p - r$ components, is given by $(\mathbf{I} - \mathbf{P}_{z(r)})$. The matrix of residuals is then

$$\mathbf{Q}^2 = (\mathbf{X} - \hat{\mathbf{X}})$$

$$= (\mathbf{I} - \mathbf{P}_{z(r)})\mathbf{X}$$

$$= \mathbf{Z}_{(p-r)}\mathbf{Z}_{(p-r)}^T\mathbf{X}$$

$$= \mathbf{Z}_{(p-r)}\mathbf{A}_{(p-r)}^T$$

$$= \mathbf{Z}\mathbf{A}^T - \mathbf{Z}_{(r)}\mathbf{A}_{(r)}^T \tag{4.35}$$

and the residual variance matrix is given by

$$(\mathbf{X} - \hat{\mathbf{X}})^{\mathrm{T}}(\mathbf{X} - \hat{\mathbf{X}}) = \mathbf{A}_{(p-r)}\mathbf{Z}_{(p-r)}^{\mathrm{T}}\mathbf{Z}_{(p-r)}\mathbf{A}_{(p-r)}^{\mathrm{T}}$$

$$= \mathbf{A}_{(p-r)}\mathbf{A}_{(p-r)}^{\mathrm{T}}$$

$$= \mathbf{L}_{(p-r)} \tag{4.35a}$$

the diagonal matrix of the last (smallest) latent roots. The total residual variance is then

$$\mathrm{tr}(\mathbf{X} - \hat{\mathbf{X}})^{\mathrm{T}}(\mathbf{X} - \hat{\mathbf{X}}) = \mathrm{tr}(\mathbf{L} - \mathbf{L}_{(r)})$$

$$= \sum_{i=r+1}^{p} l_i \tag{4.36}$$

Let $\mathbf{x}_i$ and $\mathbf{z}_i$ represent the ith row (observation) vector of $\mathbf{X}$ and $\mathbf{Z}$ respectively. Once the first r PCs have been obtained, the adequacy of the model can be tested by using the predicted observation vector

$$\hat{\mathbf{x}}_i = \mathbf{z}_i \mathbf{A}_{(r)}^{\mathrm{T}} \tag{4.37}$$

which yields the residual sums of squares

$$\mathbf{Q} = (\mathbf{x}_i - \hat{\mathbf{x}}_i)^{\mathrm{T}}(\mathbf{x}_i - \hat{\mathbf{x}}_i) \tag{4.38}$$

Equation (4.38) can be used as a measure of overall fit of the p-dimensional observation vector. Let

$$\theta_1 = \sum_{i=r+1}^{p} l_i, \qquad \theta_2 = \sum_{i=r+1}^{p} l_i^2, \qquad \theta_3 = \sum_{i=r+1}^{p} l_i^3$$

where $h_0 = 1 - (2\theta_1\theta_2\theta_3/3\theta_2^2)$. Then Jackson and Mudholkar (1979) have shown that the statistic

$$c = \frac{\theta_1[(Q/\theta_1)h_0 - \theta_2 h_0(h_0 - 1)/\theta_1^2 - 1)]}{(2\theta_2 h_0^2)^{1/2}} \tag{4.39}$$

is distributed approximately as $N(0, 1)$, and can thus be used to test for the significance of departure of the ith observation vector from the predicted values (Eq. 4.37). Let p_i denote the one-tailed probability associated with c_i, the value of c for the ith observation. Then under the null hypothesis H_0: the last $(p - r)$ roots are isotropic, the expression

$$Q_0 = -2\sum_{i=1}^{n} \ln p_i$$

is distributed approximately as the chi-squared distribution with $2n$ degrees of freedom and can be used to test for significance of residuals for the n observations (Jackson, 1981).

4.3.4 Testing Individual Principal Components

The previous sections provide global asymptotic tests for the principal components model and can serve as useful overall indicators of the structure of a sample covariance (correlation) matrix. It is also possible however to derive asymptotic tests for the individual latent roots and vectors (loadings), which allows for a more detailed or local analysis of the individual roots and loading coefficients.

Testing Individual Latent Roots

THEOREM 4.3 (Anderson, 1963a). Consider n observations from $N(\boldsymbol{\mu}, \boldsymbol{\Sigma})$, and let $\lambda_1 \geq \lambda_2 \geq \cdots \geq \lambda_p$ and $\boldsymbol{\Pi}_1, \boldsymbol{\Pi}_2, \ldots, \boldsymbol{\Pi}_p$ be the latent roots and latent vectors of $\boldsymbol{\Sigma}$. Also, let $l_1 > l_2 > \cdots > l_p$ and $\mathbf{P}_1, \mathbf{P}_2, \ldots, \mathbf{P}_p$ be latent roots and vectors of $\hat{\boldsymbol{\Sigma}} = \frac{1}{n} \mathbf{X}^{\mathrm{T}} \mathbf{X}$. Then, as $n \to \infty$,

(i) every l_i is distributed independently of its corresponding latent vector $\mathbf{P}_i$.

(ii) For a distinct latent root λ_i the expression

$$\frac{(l_i - \lambda_i)}{\lambda_i \sqrt{\dfrac{2}{n}}} \tag{4.40}$$

is distributed approximately as $N(0, 1)$.

Theorem 4.3 was first considered by Girshick (1939) who used a Taylor series expansion of l_i (about λ_i) to show that the sample variance of l_i is approximately $2\lambda_i^2/n$ (see also Waternaux, 1976). In matrix form we have the approximations

$$l_i \simeq \lambda_i + \boldsymbol{\Pi}_i^{\mathrm{T}}(\hat{\boldsymbol{\Sigma}} - \boldsymbol{\Sigma})\boldsymbol{\Pi}_i \tag{4.40a}$$

$$\mathbf{P}_i \simeq \boldsymbol{\Pi}_i + \sum_{i \neq k} w_{ik} \boldsymbol{\Pi}_i \tag{4.40b}$$

where $w_{ik} = \boldsymbol{\Pi}_i^{\mathrm{T}}(\hat{\boldsymbol{\Sigma}} - \boldsymbol{\Sigma})\boldsymbol{\Pi}_i/(\lambda_k - \lambda_i)$. Equation (4.40a and b) can be used to study point estimation of the latent roots and vectors (see Skinner et al. 1986). Theorem 4.3 can also be used to establish asymptotic confidence intervals and tests of hypotheses concerning individual roots. A two-sided normal $100(1 - \alpha)$ percentage confidence interval is given by

$$\frac{l_i}{1 + \mathbf{Z}_{\alpha/2}\left(\dfrac{2}{n}\right)^{1/2}} \leq \lambda_i \leq \frac{l_i}{1 - \mathbf{Z}_{\alpha/2}\left(\dfrac{2}{n}\right)^{1/2}} \tag{4.41}$$

Bounds can also be placed on the last $p - r$ isotropic roots of $\hat{\mathbf{\Sigma}}$. Let λ be a population latent root of multiplicity $q = p - r$. Then a sample estimate of λ is

$$\bar{l}_q = \frac{1}{q} \sum_{i=r+1}^{p} l_i \tag{4.42}$$

where from Theorem 4.3 we know the distribution of

$$\frac{\bar{l}_q - \lambda}{\lambda\left(\dfrac{2}{nq}\right)^{1/2}} \tag{4.43}$$

approaches $N(0, 1)$. Using Eq. (4.41), a two-sided confidence interval for λ is then

$$\frac{\bar{l}_q}{1 + \mathbf{Z}_{a/2}\left(\dfrac{2}{nq}\right)^{1/2}} \leq \lambda \leq \frac{\bar{l}_q}{1 - \mathbf{Z}_{a/2}\left(\dfrac{2}{nq}\right)^{1/2}} \tag{4.44}$$

which can be used to estimate λ after the null hypothesis of q isotropic roots has been accepted. There also seems to be some evidence that Theorem 4.3 carries over (asymptotically) to nonnormal populations (Davis, 1977).

Example 4.9.   For Example 4.6, a 95% confidence interval for the first (male) latent root is

$$\frac{38545}{1 - 1.96\left(\dfrac{2}{31}\right)^{1/2}} \leq \lambda_1 \leq \frac{38545}{1 - 1.96\left(\dfrac{2}{31}\right)^{1/2}}$$

so that $.25733 \leq \lambda_1 \leq .76758$, which does not include the female latent root .22667. For an application to Palaeoecology see Reyment (1963).

Testing An Entire Latent Vector

Tests for individual PCs of a covariance matrix can also be carried out using latent vectors. The reason for considering tests of an entire latent vector (see also Example 4.10) is that in practice an adjusted latent vector $\mathbf{C}_i$ may be used in place of the observed vector $\mathbf{P}_i$; for example, values close to zero may be set identically equal to zero in the belief that these are the true population values. A test is then required to determine whether the adjusted latent vector $\mathbf{C}_i$ represents a population vector. We first state a theorem due to Girshick.

THEOREM 4.4   (Girshick, 1939). Consider a sample of size n drawn from a normal $N(\boldsymbol{\mu}, \boldsymbol{\Sigma})$ population with latent roots $\lambda_1 > \lambda_2 > \cdots > \lambda_p$ and latent vectors $\boldsymbol{\Pi}_1, \boldsymbol{\Pi}_2, \ldots, \boldsymbol{\Pi}_p$. If $l_1 > l_2 > \cdots > l_p$ and $\mathbf{P}_1, \mathbf{P}_2, \ldots, \mathbf{P}_p$ are latent roots and latent vectors of $\mathbf{S} = (1/n)\mathbf{X}^T\mathbf{X}$, then as $n \to \infty$ the distribution of

$$\sqrt{n}\,(\mathbf{P}_i - \boldsymbol{\Pi}_i)$$

approaches a multivariate normal with mean vector 0 and covariance matrix

$$\sum_{\substack{s=1 \\ s \neq 1}}^{p} \frac{\lambda_s \lambda_i}{(\lambda_s - \lambda_i)^2} \, \boldsymbol{\Pi}_s \boldsymbol{\Pi}_s^T \qquad (i = 1, 2, \ldots, p) \tag{4.46}$$

Equation (4.46) is asymptotically independent of l_i.

Anderson (1963) considers the null hypothesis

$$H_0 \colon \boldsymbol{\Pi}_i = \mathbf{C}_i$$

against the alternative that at least one element of $\boldsymbol{\Pi}_i$ is not equal to an element of some arbitrary constant vector $\mathbf{C}_i$. Using Theorem 4.4 it can be shown that the distribution of

$$\chi^2 = n \left(l_i \mathbf{C}_i^T \mathbf{S}^{-1} \mathbf{C}_i + \frac{1}{l_i} - \mathbf{C}_i^T \mathbf{S} \mathbf{C}_i - 2 \right) \tag{4.47}$$

approaches a chi-squared distribution with $p - 1$ degrees of freedom. Adjustments other than n (or $n - 1$) are also possible (Schott, 1987). For an extension to the correlation matrix see Schott (1991). Anderson's test is asymptotic and cannot be used for a small sample. A different approach is taken by Mallows (1960) (see also Srivastava and Khatri, 1979, p. 296) and for a more special case by Kshirsagar (1961). Mallows considers H_0: hypothetical linear combination $\mathbf{XC} = \mathbf{Z}_{(c)}$ is a PC against the alternative H_a that $\mathbf{XC}$ is not a PC. When H_0 is not rejected we accept $\mathbf{C}$ as a latent vector of $\mathbf{S}$. Using multiple regression theory an exact test of H_0 is shown to be based on the statistic

$$F = \frac{(n - p + 1)}{p - 1} \, [(\mathbf{C}^T \mathbf{S}^{-1} \mathbf{C})(\mathbf{C}^T \mathbf{S} \mathbf{C}) - 1] \tag{4.48}$$

which under H_0 is the F distribution with $p - 1$ and $n - p + 1$ degrees of freedom. The same test is also proposed by Jolicoeur (1984). Kshirsagar (1961, 1966) considers the null hypothesis: $\mathbf{XC} = \mathbf{Z}_{(c)}$ is a PC ($\mathbf{C}$ is a latent vector) given a covariance matrix has the isotropic structure of Eq. (3.29).

The test is also exact and is based on multiple correlation theory. Let λ_1 and λ be the nonisotropic and isotropic roots, respectively, and let $\mathbf{\Pi}_1$ be the latent vector corresponding to λ_1. Let $\mathbf{X}$ represent a sample of size n from a p-dimensional normal distribution where the PCs are normal independent $N(0, 1)$ variates but $\mathbf{Z}_1 \sim N(0, \lambda_1)$. Then the expression

$$\chi^2 = \sum_{j=2}^{p} \sum_{i=1}^{n} z_{ij}^2$$

$$= \sum_{j=1}^{p} \sum_{i=1}^{n} z_{ij}^2 - \sum_{j=1}^{n} z_{ij}^2$$

$$= \text{tr}(\mathbf{Z}^\mathsf{T}\mathbf{Z}) - \text{tr}(\mathbf{Z}_1^\mathsf{T}\mathbf{Z}_1)$$

$$= \text{tr}(\mathbf{X}^\mathsf{T}\mathbf{X}) - \mathbf{C}^\mathsf{T}\mathbf{X}^\mathsf{T}\mathbf{X}\mathbf{C} \tag{4.49}$$

under the null hypothesis $\mathbf{\Pi}_1 = \mathbf{C}$ is chi-squared with $n(p-1)$ degrees of freedom. Let $\delta = \mathbf{C}^\mathsf{T}\mathbf{X}^\mathsf{T}\mathbf{X}\mathbf{C}$. Then Eq. (4.49) can be expressed as

$$\chi^2 = \text{tr}(\mathbf{X}^\mathsf{T}\mathbf{X}) - \delta \tag{4.50}$$

which may be used to test overall departure from H_0. Such departure can be due to two main reasons: (1) there exists more than a single nonisotropic PC, or (2) the hypothetical vector $\mathbf{C}$ is not the true vector $\mathbf{\Pi}_1$. Following Bartlett (1951b) the overall chi-squared statistic can be partitioned as

$$\chi^2 = \chi_0^2 + \chi_d^2 \tag{4.51}$$

and the directional contribution χ_d^2 can be tested given χ_0^2, that is, given the existence of $p-1$ isotropic roots. The coefficient of regression of $\mathbf{X}\mathbf{\Pi}_i$ on $\mathbf{X}\mathbf{C}$ is $(\mathbf{C}^\mathsf{T}\mathbf{X}^\mathsf{T}\mathbf{X}\mathbf{C})^{-1}\mathbf{C}^\mathsf{T}\mathbf{X}^\mathsf{T}\mathbf{X}\mathbf{\Pi}_i$ where $\text{var}(\mathbf{X}\mathbf{C}) = (1/n)(\mathbf{C}^\mathsf{T}\mathbf{X}^\mathsf{T}\mathbf{X}\mathbf{C})$, and standardizing yields the scalar

$$\beta = \frac{\mathbf{C}^\mathsf{T}(\mathbf{X}^\mathsf{T}\mathbf{X})\mathbf{\Pi}_i}{(\mathbf{C}^\mathsf{T}\mathbf{X}^\mathsf{T}\mathbf{X}\mathbf{C})^{1/2}}$$

which is distributed as $N(0, 1)$. Thus

$$\sum_{i=2}^{p} [\mathbf{C}^\mathsf{T}(\mathbf{X}^\mathsf{T}\mathbf{X})\mathbf{\Pi}_i]^2 / \mathbf{C}^\mathsf{T}\mathbf{X}^\mathsf{T}\mathbf{X}\mathbf{C}$$

is chi-squared with $(p-1)$ degree of freedom or

$$\chi_\alpha^2 = \sum_{i=2}^{p} \frac{(\mathbf{C}^\mathsf{T}\mathbf{X}^\mathsf{T}\mathbf{X}\mathbf{\Pi}_i)(\mathbf{\Pi}_i^\mathsf{T}\mathbf{X}^\mathsf{T}\mathbf{X}\mathbf{C})}{\mathbf{C}^\mathsf{T}\mathbf{X}^\mathsf{T}\mathbf{X}\mathbf{C}}$$

is chi-squared with $(p - 1)$ degrees of freedom. Thus under

$$H_0: \mathbf{\Pi}_1 = \mathbf{C}$$

we have

$$\sum_{i=2}^{p} \frac{(\mathbf{C}^T\mathbf{X}^T\mathbf{\Pi}_i)(\mathbf{\Pi}_i^T\mathbf{X}^T\mathbf{X}\mathbf{C})}{\mathbf{C}^T\mathbf{X}^T\mathbf{X}\mathbf{C}} = \sum_{i=1}^{p} \frac{(\mathbf{C}^T\mathbf{X}^T\mathbf{\Pi}_i)(\mathbf{\Pi}_i^T\mathbf{X}^T\mathbf{X}\mathbf{C})}{\mathbf{C}^T\mathbf{X}^T\mathbf{X}\mathbf{C}}$$

$$- \frac{(\mathbf{C}^T\mathbf{X}^T\mathbf{X}\mathbf{\Pi}_1)(\mathbf{C}^T\mathbf{X}^T\mathbf{X}\mathbf{\Pi}_1)}{\mathbf{C}^T\mathbf{X}^T\mathbf{X}\mathbf{C}}$$

$$= \sum_{i=1}^{p} \frac{(\mathbf{C}^T\mathbf{X}^T\mathbf{X}\mathbf{\Pi}_i)(\mathbf{\Pi}_i^T\mathbf{X}^T\mathbf{X}\mathbf{C})}{\mathbf{C}^T\mathbf{X}^T\mathbf{X}\mathbf{C}} - \mathbf{C}^T\mathbf{X}^T\mathbf{X}\mathbf{C}$$

$$= \frac{\mathbf{C}^T(\mathbf{X}^T\mathbf{X})^2\mathbf{C}}{\mathbf{C}^T(\mathbf{X}^T\mathbf{X})\mathbf{C}} - \mathbf{C}^T\mathbf{X}^T\mathbf{X}\mathbf{C}$$

so that

$$\chi_d^2 = \{[\mathbf{C}^T(\mathbf{X}^T\mathbf{X})^2\mathbf{C}][\mathbf{C}^T(\mathbf{X}^T\mathbf{X})\mathbf{C}]^{-1} - \mathbf{C}^T\mathbf{X}^T\mathbf{X}\mathbf{C}\}$$

$$= (n - 1)[(\mathbf{C}^T\mathbf{S}^2\mathbf{C})(\mathbf{C}^T\mathbf{S}\mathbf{C})^{-1} - \mathbf{C}^T\mathbf{S}\mathbf{C}] \tag{4.52}$$

is chi-squared with $(p - 1)$ degrees of freedom. The test assumes that λ is known (so that the isotropic components can be standardized to unit variance), which is rarely the case in practice. When λ is not known Eqs. (4.49) and (4.52) can be used to define the F statistic, for unstandardized components, as

$$F = \hat{\lambda} \frac{\chi_\alpha^2/p - 1}{\chi^2/n(p - 1)}$$

$$= \frac{\hat{\lambda}(n - 1)[(\mathbf{C}^T\mathbf{S}^2\mathbf{C})(\mathbf{C}^T\mathbf{S}\mathbf{C})^{-1} - \mathbf{C}^T\mathbf{S}\mathbf{C}]}{n(\operatorname{tr}\mathbf{S} - \mathbf{C}^T\mathbf{S}\mathbf{C})} \tag{4.53}$$

which is distributed with $p - 1$ and $(n - 1)(p - 1)$ degrees of freedom (Srivastava and Khatri, 1979). The case of an isotropic distribution when the number of variables leads to ∞ is considered by Yin and Krishnaiah (1985). A numerical example is given by Kshirsager (1961).

Testing Elements of the Latent Vectors

Once $r \le p$ components of a covariance matrix have been accepted as accounting for a significant portion of the variance, the next step lies in determining which individual loadings are significantly different from zero. This can be of importance when the retained PCs are not expected to be significantly correlated with each and every variable. Since $\mathbf{A} = \mathbf{P}\Lambda^{1/2}$, testing for zero loadings is equivalent to testing

$$H_0: \pi_{ij} = 0$$

$$H_a: \pi_{ij} \neq 0$$

Using Eq. (4.46) it can be shown (Girshick, 1939) that variance/covariance matrices for elements within latent vectors are given by

$$\mathbf{\Gamma}_1 = E(\mathbf{P}_1 - \mathbf{\Pi}_1)(\mathbf{P}_1 - \mathbf{\Pi}_1)^{\mathrm{T}} = \frac{\lambda_1}{(n-1)} \sum_{s=2}^{p} \frac{\lambda_s}{(\lambda_s - \lambda_1)^2} \mathbf{\Pi}_1 \mathbf{\Pi}_1^{\mathrm{T}}$$

$$\mathbf{\Gamma}_2 = E(\mathbf{P}_2 - \mathbf{\Pi}_2)(\mathbf{P}_2 - \mathbf{\Pi}_2)^{\mathrm{T}} = \frac{\lambda_2}{(n-1)} \sum_{\substack{s=1 \\ s \neq 2}}^{p} \frac{\lambda_s}{(\lambda_s - \lambda_2)^2} \mathbf{\Pi}_2 \mathbf{\Pi}_2^{\mathrm{T}}$$

$$\mathbf{\Gamma}_p = E(\mathbf{P}_p - \mathbf{\Pi}_p)(\mathbf{P}_p - \mathbf{\Pi}_p)^{\mathrm{T}} = \frac{\lambda_p}{(n-1)} \sum_{s=1}^{p-1} \frac{\lambda_s}{(\lambda_s - \lambda_{p-1})^2} \mathbf{\Pi}_{p-1} \mathbf{\Pi}_{p-1}^{\mathrm{T}}$$

$$(4.54)$$

The equations can also be expressed in matrix form. Thus for $\mathbf{\Gamma}_1$ we have

$$\mathbf{\Gamma}_1 = [\mathbf{\Pi}_2 \vdots \mathbf{\Pi}_3 \vdots \cdots \vdots \mathbf{\Pi}_p] \begin{bmatrix} \dfrac{\lambda_1 \lambda_2}{(\lambda_2 - \lambda_1)^2} & & & \mathbf{0} \\ & \dfrac{\lambda_1 \lambda_3}{(\lambda_3 - \lambda_1)^2} & \ddots & \\ & & & \dfrac{\lambda_1 \lambda_p}{(\lambda_p - \lambda_1)^2} \\ \mathbf{0} & & & \end{bmatrix} \begin{bmatrix} \mathbf{\Pi}_2^{\mathrm{T}} \\ \cdots \\ \mathbf{\Pi}_3^{\mathrm{T}} \\ \cdots \\ \vdots \\ \cdots \\ \mathbf{\Pi}_p^{\mathrm{T}} \end{bmatrix}$$

$$= \mathbf{\Pi}_{(-1)} \mathbf{D}^{2(-1)} \mathbf{\Pi}_{(-1)}^{\mathrm{T}}$$

where subscript (-1) denotes that the first latent vector and the first diagonal term of $\mathbf{D}^2$ (corresponding to λ_1) are omitted. The remaining covariance matrices are given by

$$\mathbf{\Gamma}_i = \mathbf{\Pi}_{(-i)} \mathbf{D}_{(-i)}^2 \mathbf{\Pi}_{(-i)}^{\mathrm{T}} \qquad (i = 2, 3, \ldots, p) \qquad (4.55)$$

Thus the covariance between any gth and hth elements is

$$\operatorname{cov}(p_{gi}, p_{hi}) = \gamma_{gi,hi}^2 = \frac{\lambda_i}{(n-1)} \sum_{\substack{s=1 \\ s \neq 1}}^{p} \frac{\lambda_s}{(\lambda_s - \lambda_i)^2} \pi_{gs} \pi_{hs} \qquad (i = 1, 2 \ldots, p)$$

$$(4.56)$$

It can also be shown that cross-variances between the gth element of the ith latent vector and the hth element of the jth latent vector is given by

$$\text{cov}(p_{gi}, p_{hj}) = \gamma^2_{gi,hj} = \frac{-\lambda_i \lambda_j}{(n-1)(\lambda_i - \lambda_j)^2} \pi_{gi} \pi_{hj} \qquad (4.57)$$

for $i \neq j$. Equation (4.57) indicates that latent vector elements are correlated between different latent vectors, even though the latent vectors are orthogonal. Such correlation is not usually evident, and is frequently ignored in practice. Since $\gamma^2_{gi,hj}$ increases as $\lambda_i \to \lambda_j$, it follows that covariance is largely a function of the differences between latent roots. Since roots are usually ranked in decreasing order, most of the correlation will occur between elements of adjacent latent vectors. It follows that PC loadings which are associated with insignificantly different latent roots cannot be distinguished in practice and can safely be ignored (see also Section 4.3.2). This provides additional rationale for first testing latent root differences before proceeding with a PCA.

Not only is it possible to estimate the covariance structure between the loadings, but Eq. (4.56) can also be used to provide asymptotic estimates of their standard deviations. This in turn permits the testing of hypotheses for individual loading coefficients. When the parent distribution is multivariate normal, the asymptotic distribution of

$$z = \frac{p_{ij} - \pi_{ij}}{\gamma_{ij}} \qquad (4.58)$$

is also normal and Eq. (4.58) can be used to test hypotheses concerning population parameters π_{ij}. This provides an objective procedure for discarding individual loading coefficients. Of course the magnitudes of the retained loadings may not be very high, and further deletion may be required using auxiliary information. A statistical test of significance however should always precede any deletion of coefficients.

Example 4.10. Reyment (1969) uses the Anderson (1963a) chi-squared statistic (Eq. 4.47) to test whether two covariance matrices possess equal (collinear) dominant latent vectors. Here $p = 3$ body dimensions of freshwater African ostracods drawn from individuals cultured in two different environments are obtained using the variables

$$Y_1 = \log_{10} \text{ (carapace length)}$$
$$Y_2 = \log_{10} \text{ (carapace height)}$$
$$Y_3 = \log_{10} \text{ (carapace breadth)}$$

The sample covariance matrices and their latent roots and vectors for the two respective samples (groups) are:

(a) The sample from the first environment $(n_1 = 365)$:

$$\mathbf{S}_1 = \begin{bmatrix} .0003390 & .002137 & .0003069 \\ .0002137 & .0003393 & .0002552 \\ .0003069 & .0002552 & .0005396 \end{bmatrix}$$

$$\mathbf{P}_1 = \begin{bmatrix} .525445 & .045020 & -.849636 \\ .481346 & .807700 & .340480 \\ .701580 & -.587872 & .402731 \end{bmatrix}$$

$$l_{11} = .0009445 , \qquad l_{21} = .0001655 , \qquad l_{31} = .0001079$$

(b) The sample from the second environment $(n_2 = 908)$:

$$\mathbf{S}_2 = \begin{bmatrix} .0005074 & .0002332 & .0002084 \\ .0002332 & .0003311 & .0002448 \\ .0002084 & .0002448 & .0004515 \end{bmatrix}$$

$$\mathbf{P}_2 = \begin{bmatrix} .628000 & -.733153 & .260973 \\ .515343 & .140505 & -.845387 \\ .583130 & .665392 & .466062 \end{bmatrix}$$

$$l_{12} = .0010223 , \qquad l_{22} = .0001371 , \qquad l_{32} = .0000377$$

Here the objective is to test whether the two covariance matrices are homogeneous with respect to the (dominant) size dimension. Letting $\mathbf{C}_1 = (.0005074, .0002332, .0002084)^{\mathrm{T}}$ (the first latent vector of $\mathbf{S}_2$) we have $l_1 \equiv l_{11} = .0009445$, $\mathbf{S} \equiv \mathbf{S}_1$, and

$$l_{11}\mathbf{C}_1^{\mathrm{T}}\mathbf{S}_1^{-1}\mathbf{C}_1 = .0009445(.0005074, .0002332, .0002084)$$

$$\times \begin{bmatrix} .0003390 & .0002137 & .0003069 \\ .0002137 & .0003393 & .0002552 \\ .0003069 & .0002552 & .0005396 \end{bmatrix}^{-1} \begin{bmatrix} .0005074 \\ .0002332 \\ .0002084 \end{bmatrix}$$

$$= .0009445(1235.0164)$$

$$= 1.166477$$

$$\frac{1}{l_{11}}\mathbf{C}_1^{\mathrm{T}}\mathbf{S}_1\mathbf{C}_1 = 1058.7613(.0005074, .002332, .002084)$$

$$\times \begin{bmatrix} .0003390 & .0002137 & .0003069 \\ .0002137 & .0003393 & .0002552 \\ .0003069 & .0002552 & .0005396 \end{bmatrix} \begin{bmatrix} .0005074 \\ .0002332 \\ .0002084 \end{bmatrix}$$

$$= 1058.7613(.0009238)$$

$$= .97808$$

so that

$$\chi^2 = n\left(l_{11}\mathbf{C}_1^{\mathrm{T}}\mathbf{S}_1^{-1}\mathbf{C}_1 + \frac{1}{l_{11}}\mathbf{C}_1^{\mathrm{T}}\mathbf{S}_1\mathbf{C}_1 - 2\right)$$

$$= 365(1.16647 + .97808 - 2)$$

$$= 52.76$$

Comparing the theoretical chi-squared value with $p - 1 = 2$ degrees of freedom indicates the rejection of H_0, so that the two covariance matrices appear to be heterogenous with respect to the first (size) dimension. $\square$

Example 4.11. (Jackson and Hearne, 1973). One of the most important properties of rocket fuel performance is the impulse produced during firing. An estimate of this number is normally obtained by a method known as static testing, whereby the fired rocket is firmly fastened to the ground and the resultant impulse measured by gages attached to its head. To improve reliability, two identical gages are attached to the rocket's head. Each gage is connected with two separate systems of recording: (1) an electronic integrator which determines total impulse, and (2) an oscilloscope and camera which record the rocket's thrust as a function of time. The photographic record is then planimetered to obtain total impulse. This results in the following variables:

$Y_1 = $ Gage 1: integrator reading
$Y_2 = $ Gage 1: planimeter reading
$Y_3 = $ Gage 2: integrator reading
$Y_4 = $ Gage 2: planimeter reading

In the study $n = 40$ readings are recorded. Although the sample size is too small for use with asymptotic theory, and multivariate normality may not apply, the example should nevertheless serve as an illustration of the theory.

The covariance matrix for the four variables is

$$\hat{\mathbf{\Sigma}} = \begin{bmatrix} 102.74 & 88.67 & 67.04 & 54.06 \\ 88.67 & 142.74 & 86.56 & 80.03 \\ 67.04 & 86.56 & 84.57 & 69.42 \\ 54.06 & 80.03 & 69.42 & 99.06 \end{bmatrix}$$

with latent roots

$$\mathbf{L} = \begin{bmatrix} 335.35 & & & \mathbf{0} \\ & 48.04 & & \\ & & 29.33 & \\ \mathbf{0} & & & 16.42 \end{bmatrix}$$

and latent vector

$$\mathbf{P} = \begin{bmatrix} .49 & -.62 & -.57 & -.26 \\ .61 & -.18 & .76 & -.15 \\ .46 & .14 & -.17 & .86 \\ .45 & .75 & -.26 & -.41 \end{bmatrix}$$

we then have, from Eq. (4.56),

$$\operatorname{var}(p_{gi}) = \gamma^2_{gi} = \frac{\lambda_i}{(n-1)} \sum_{\substack{s=1 \\ s \neq i}}^{p} \frac{\lambda_s}{(\lambda_s - \lambda_i)^2} \pi^2_{gs} \tag{4.59}$$

and setting $g = i = 1$, we obtain for the first loading

$$\operatorname{var}(p_{11}) = \gamma^2_{11} = \frac{l_1}{(n-1)} \sum_{s=2}^{4} \frac{l_s}{(l_s - l_1)^2} p^2_{is}$$

$$= \frac{335.35}{39} \left[\frac{(48.04)(.62)^2}{(48.04 - 335.35)^2} + \frac{(29.33)(.57)^2}{(29.33 - 335.35)^2} \right.$$

$$\left. + \frac{(16.42)(.26)^2}{(16.42 - 335.35)^2} \right]$$

$$= 8.599[.000224 + .000102 + .000011]$$

$$= .002898 .$$

Likewise the covariance between, say, p_{11} and p_{21} is given by

$$\operatorname{cov}(p_{11}, p_{21}) = \gamma^2_{11,21} = \frac{l_1}{(n-1)} \sum_{s=2}^{4} \frac{l_s}{(l_s - l_1)^2} p_{1s} p_{2s}$$

$$= \frac{335.35}{39} \left[\frac{(48.04)(.62)(-.18)}{(48.04 - 335.35)^2} + \frac{(29.33)(-.57)(.76)}{(29.33 - 335.35)^2} \right.$$

$$\left. + \frac{(16.42)(-.26)(-.15)}{(16.42 - 335.35)^2} \right]$$

$$= 8.599[-.00006495 - .00013567 + .0000063]$$

$$= -.001670$$

Also, to illustrate the use of Eq. (4.57) the covariance between $p_{23} = .76$

and $p_{14} = -.26$ is given by

$$\operatorname{cov}(p_{23}, p_{14}) = \gamma^2_{23,14} = -\frac{l_3 l_4}{n(l_3 - l_4)^2} p_{23} p_{14}$$

$$= \frac{-(29.33)(16.42)}{39(29.33 - 16.42)^2} (.76)(-.26)$$

$$= .0146$$

Transforming the covariances into correlation coefficients then results in the $(p^2 \times p^2) = (16 \times 16)$ correlation matrix (Jackson and Hearne, 1973).

	p_{11}	p_{21}	p_{31}	p_{41}	p_{12}	p_{22}	p_{32}	p_{42}	p_{13}	p_{23}	p_{33}	p_{43}	p_{14}	p_{24}	p_{34}	p_{44}
p_{11}	1.0															
p_{21}	−.2	1.0														
p_{31}	−.3	−.4	1.0													
p_{41}	−.6	−.5	.1	1.0												
p_{12}	.1	.1	0	−.2	1.0											
p_{22}	.1	.1	0	−.2	−.9	1.0										
p_{32}	.2	.1	−.1	−.2	.2	−.4	1.0									
p_{42}	.2	.1	−.1	−.3	.9	−.7	−.1	1.0								
p_{13}	.1	−.1	0	0	−.9	.9	−.4	−.7	1.0							
p_{23}	.2	−.4	.1	.1	−.7	.7	−.3	−.6	.9	1.0						
p_{33}	.1	−.1	0	0	.2	−.2	.1	.1	−.5	−.6	1.0					
p_{43}	.0	−.1	0	0	.9	−.9	.4	.7	−.7	−.4	−.2	1.0				
p_{14}	0	0	−.1	0	−.1	0	.4	−.3	−.3	−.4	.8	−.4	1.0			
p_{24}	0	0	−.1	0	0	0	.1	−.1	.3	.5	−1.0	.4	−.8	1.0		
p_{34}	.1	0	−.3	.1	.1	0	−.3	.2	−.3	−.5	.8	−.4	.6	−.9	1.0	
p_{44}	0	0	−.1	0	.2	.1	−.7	.4	−.2	−.3	.5	−.2	.1	−.6	.8	1.0

The correlation loadings of the four variables are

	$\mathbf{Z}_1$	$\mathbf{Z}_2$	$\mathbf{Z}_3$	$\mathbf{Z}_4$
$\mathbf{X}_1$	.885	−.424	−.305	−.104
$\mathbf{X}_2$	.935	−.104	.345	−.051
$\mathbf{X}_3$	.916	.106	−.100	.379
$\mathbf{X}_4$	.828	.522	−1.41	−.167

where each latent vector is multiplied by the square root of its latent root and divided by the standard deviation of the corresponding variable. To test whether loadings are significantly different from zero, the standard deviations of the latent vector elements are computed using Eq. (4.56). The matrix of latent vectors can then be written as

$$
\mathbf{P} = \begin{bmatrix}
.49 & -.62 & -.57 & -.26 \\
(.05) & (.19) & (.21) & (.18) \\
.61 & -.18 & .76 & -.15 \\
(.04) & (.25) & (.08) & (.21) \\
.46 & .14 & -.17 & .86 \\
(.03) & (.14) & (.24) & (.05) \\
.45 & .75 & -.26 & -.41 \\
(.06) & (.11) & (.26) & (.13)
\end{bmatrix}
$$

where standard deviations appear in brackets. With the aid of Eq. (4.58) we see that a_{22}, a_{32}, a_{33}, a_{43}, a_{14}, and a_{24} are insignificantly different from zero. The revised table of loadings appears in Table 4.10. The first PC is an estimate of total trust, whereas the remaining three PCs measure contrasts between gages, indicating a possible lack of synchronization. For further illustration of the test(s) see Jackson (1981).

Table 4.10   Significant Loading Coefficients for the Four Rocket Propulsion Variables

	$\mathbf{Z}_1$	$\mathbf{Z}_2$	$\mathbf{Z}_3$	$\mathbf{Z}_4$
$\mathbf{X}_1$	.885	$-.424$	$-.305$	—
$\mathbf{X}_2$	.935	—	.345	—
$\mathbf{X}_3$	.916	—	—	.379
$\mathbf{X}_4$	.828	.522	—	.167

4.3.5   Information Criteria of Maximum Livelihood Estimation of the Number of Components

More recently attempts have been made to utilize entropy information statistics (Section 1.6) to estimate the number of explanatory factors that can be extracted from a covariance matrix. The main reason for this is that the classical ML principle often leads to choosing a higher number of dimensions of a model than can be interpreted, and thus may not be the appropriate procedure for implementing the intuitive notion of the "right" model. Information statistics are also based on the assumption of multivariate normality, and are similar to Mallows' (1973) C_p statistic, which is well-known in regression analysis. The general objective can be understood in terms of a model-selection criterion, which takes into account both the goodness of fit (likelihood) of a model as well as the number of parameters used to achieve the fit. Such criteria take the form of a penalized likelihood function, specifically the negative log likelihood plus a penalty term which increases with the number of parameters fitted. Some penalty functions also depend on the sample size in order to incorporate the concept of consistent estimation.

The first attempts in this direction, in the context of factor analysis, are from Akaike (1971a, 1987), who utilized his AIC criterion (Akaike, 1974a; 1974b) to estimate the number of factors or dimensions required to obtain an adequate (penalized) fit. Akaike (1971a) evaluates the quality of the estimate by the expectation of its log likelihood, which in turn is based on the Kullback–Leibler mean information statistic (Kullback and Leibler, 1951). Let $\hat{\theta}$ be an estimate of a parameter θ of a probability distribution with density function $f(y \mid \theta)$, where the random variable Y is distributed independently of $\hat{\theta}$. Then the quality of the estimate $\hat{\theta}$ is evaluated by

$$E[\ln f(y \mid \hat{\theta})] = E \int f(y \mid \theta) \ln f(y \mid \hat{\theta}) \, dy \qquad (4.60)$$

It can be shown (Akaike, 1971a) that the resultant criterion is given by

$$\text{AIC} = -2(\ln L - m)$$

$$= -2(\ln \text{ of maximum likelihood} - \text{number of free parameters in the}$$

$$\text{model}) \qquad (4.61)$$

(see also Bozdogan, 1987). Here m is also the expected value (degrees of freedom) of the asymptotic chi-squared distribution associated with the likelihood ratio criterion, and is equal to the dimension of the model (parameter vector θ). The procedure consists of varying the number of parameters and generating a set of alternative models. We then chose the model with minimum AIC. Alternatively, we can consider the equivalent problem of maximizing the expression

$$C_1 = \ln L - m \qquad (4.62)$$

which is more convenient to work with.

Let a set of p random variables have distribution $N(\mu, \Sigma)$. Thus omitting terms which are functions of the constant n, the likelihood function is given by

$$L = -\frac{n}{2}[\ln \Sigma + \text{tr}(S\Sigma^{-1})]$$

Also, let $\Sigma = \alpha\alpha^T + \Delta$ where α is a matrix of PC loadings and Δ is a diagonal matrix with equal nonnegative diagonal entries. The AIC criterion is then implemented by replacing L into Eq. (4.62) with $\Sigma = \alpha\alpha^T + \Delta$ and varying the number r of PCs until the expression is maximized, where $m = [2p(r + 1) - r(r - 1)]$. The testing procedure is a special case of the ML factor model and is pursued further in Chapter 6.

The attractive feature of the AIC criterion is that it reduces the number of significant factors when compared to the ML chi-squared procedure discussed in Section 4.3.2. A difficulty with Eq. (4.62) however lies in the penalty term m = number of free parameters, which does not depend on the sample size n. The AIC criterion therefore is an inconsistent estimator of the true dimensionality of a model. The criterion is usually seen as not being motivated by the objective of estimating "the correct" model (number of explanatory PCs), but rather by that of finding the number of PCs which can be estimated given a single multivariate normal sample. Since consistency is an asymptotic property, the argument goes, and since all samples are of finite size, it matters little whether the AIC criterion provides us with consistent estimates. The view clearly misses an essential point. Given that repeated samples can be taken, of varied size, it is clearly of some importance that values of an estimator should converge to a single "true" value of the population, since without such a criterion it is difficult to relate sample information to population structure.

A different approach to the problem of estimating dimensionality is provided by Schwartz (1978; see also Rissanen, 1978), who uses Bayes estimation under a special class of priors to derive the alternative criterion

$$C_2 = \ln(L) - \frac{m}{2}\ln(n) \qquad (4.63)$$

[the Bayesian nature of the factor model is also recognized by Akaike (1987)]. Schwartz's criterion differs from Eq. (4.62) in the penalty function since C_2 is now also a function of the sample size n. Clearly as n increases the two criteria produce different results, with C_2 yielding a smaller number of PCs.

Other modifications use penalty terms that are logarithmic functions of n. Hannan and Quinn (1979), for example, have proposed the criterion

$$C_3 = \ln(L) - c\ln(\ln n) \qquad (4.64)$$

for some constant $c > 2$. However, C_3 has a slow rate of increase with n, possibly not a desirable feature in factor analysis. Other penalty functions are discussed by Bozdogan (1987), who provides a comprehensive review of the background theory of AIC and proposes a number of alternative criteria, one of them being

$$C_4 = 2\ln(L) - m[\ln(n) + 2] - \ln(I) \qquad (4.65)$$

where I is Fisher's information which depends on the reduced number m of estimated parameters.

The criteria discussed in this section are all general criteria for estimating dimensionality of a linear model, and do not apply only to factor analysis. This is a further illustration of our claim, that estimating dimensionality is not a problem that only concerns factor analysis. As witnessed by the bibliography, only recently has the statistical generality of the problem been addressed in any systematic fashion. None of the above criteria however seem to have been evaluated for their relative ability to pick out the true dimension of a PC factor model, or for robustness against nonnormality and other sampling difficulties such as stability, presence of outliers, or missing data.

4.4 OTHER METHODS OF CHOOSING PRINCIPAL COMPONENTS

When large samples are available from multivariate normal populations the statistical tests considered in the preceding sections represent valid procedures for determining the number of significant components and/or individual loadings. For PCA, however, experience has shown that statistical significance is not necessarily synonomous with meaningful interpretation, particularly when the population is nonnormal and when outliers are present in the sample. A difficulty associated with ML tests lies in the assumption of multivariate normality and large samples. As the sample size increases, the Bartlett–Anderson chi-squared tests tend to retain an excessive number of PCs, although this can be circumvented by the use of information-based criteria (Section 4.3.5). Even the AIC criterion however may overestimate true dimensionality when n (or p) is large. Thus trivial PCs are frequently retained by ML tests together with those whose interpretation is more straightforward but by no means certain. This implies that ML tests should be used only as a first stage in what may become a multistage decision process as to which PCs are to be retained, that is, which components can be considered trivial from a substantive as well as a statistical point of view. Such a strategy seems to be less subjective than those traditionally employed (particularly in the social sciences), which seek to minimize the number of PCs using informal rules of thumb (Section 4.4.3) together with the questionable "principle of parsimony." Unfortunately, statistical testing is still not widely employed (or available) with many statistical packages, and experience seems to be insufficient to establish the utility of the various tests described in Section 4.3.

In an attempt to circumvent the at times tenuous assumption of multivariate normality (and large samples), workers have recently turned to non-ML estimation/tests based on resampling schemes. The advantage of such nonparametric procedures is they do not require the assumption of normality, at least in moderate to large samples.

4.4.1 Estimates Based on Resampling

The three best-known resampling procedures used in conjunction with PCA are (1) the jackknife, (2) the bootstrap, and (3) cross validation. Although these are by no means the only nonparametric resampling methods (see

Efron, 1981), they have been used in practice with the PC model and evidence exists for their usefulness. The only drawback for such methods seems to be the heavy reliance on electronic computing time, although with present technology this does not seem to be a major obstacle.

The Jackknife

Consider a random sample $x_1, x_2, \ldots, x_n$ taken from an unknown population with parameter θ. Jackknife estimation is a procedure for obtaining an estimate $\hat{\theta}$ of θ, together with its standard deviation. If $\hat{\theta}$ is a biased estimator, the bias of the jackknife can be shown to decrease as a polynomial function of n (see Kendall and Stuart, 1979). The estimator is obtained by the following device. Using the original sample, n subsamples, each of size $n - 1$, are formed by deleting systematically each observation in turn. Let $\hat{\theta}_n$ be a statistic calculated using the entire sample and let $\hat{\theta}_i$ ($i = 1$, $2, \ldots, n$) be the statistic calculated with the ith observation removed. A set of n different values (the so-called pseudovalues) of the statistic can be obtained as

$$\hat{\theta}_i^* = n\hat{\theta}_n - (n - 1)\hat{\theta}_i \tag{4.66}$$

and the jackknife is then the overall estimator

$$\hat{\theta}^* - \frac{1}{n} \sum_{i=1}^{n} \hat{\theta}_i^* \tag{4.67}$$

or the mean of the n partial estimates. When n is large, economy in computation may be achieved by deleting more than a single observation at a time. The variance of the pseudovalues (Eq. 4.66) is then

$$s^2 = \frac{1}{n-1} \sum_{i=1}^{n} (\hat{\theta}_i^* - \hat{\theta}^*)^2 \tag{4.68}$$

and the variance of $\hat{\theta}^*$ is obtained as $(s^2)^* = s^2/n$.

The simple device of generating n subsamples and then estimating a series of values of estimates can be applied to obtaining jackknifed PC loading coefficients, together with their standard deviations. This is done by computing a set of n loadings and averaging to obtain the final estimates. The same applies to scores, latent roots, and vectors. Although the process requires extensive computation, it has the advantage of being applicable to relatively small samples. However, for testing purposes we still require the assumption of (approximate) normality. Jackknife estimation has recently being applied to PCA by Reyment (1982), Gibson et al. (1984), and McGillivray (1985).

Example 4.12.   The jackknife is used by McGillivray (1985) to compute standard errors of latent vector elements for logarithmically transformed

Table 4.11 Jackknifed PC (± 2 SD) Derived for Males from the Covariance Matrix (Cov) of Logarithmically Transformed Data and the Correlation Matrix (Cor) of the Original Data

Variable	PC 1		PC 2	
	Cov	Cor	Cov	Cor
Mandible	0.24 ± 0.04	0.23 ± 0.03	-0.06 ± 0.10	0.24 ± 0.08
Skull width	0.21 ± 0.04	0.22 ± 0.03	-0.01 ± 0.10	0.32 ± 0.14
Skull length	0.31 ± 0.12	0.24 ± 0.03	-0.35 ± 0.16	-0.17 ± 0.09
Coracoid	0.30 ± 0.03	0.31 ± 0.02	-0.01 ± 0.13	0.03 ± 0.04
Sternum length	0.40 ± 0.09	0.23 ± 0.05	-0.11 ± 0.13	0.52 ± 0.08
Keel	0.45 ± 0.11	0.23 ± 0.04	-0.26 ± 0.16	0.48 ± 0.06
Sternum width	0.28 ± 0.25	0.03 ± 0.04	0.50 ± 0.16	0.36 ± 0.15
Humerus	0.29 ± 0.05	0.32 ± 0.02	-0.02 ± 0.13	0.20 ± 0.06
Ulna	0.28 ± 0.06	0.32 ± 0.02	-0.02 ± 0.12	0.26 ± 0.09
Carpometacarpus	0.30 ± 0.08	0.33 ± 0.02	-0.07 ± 0.10	0.22 ± 0.10
Femur	0.23 ± 0.05	0.32 ± 0.02	-0.02 ± 0.09	0.15 ± 0.04
Tibiotarsus	0.26 ± 0.06	0.29 ± 0.03	-0.15 ± 0.13	0.26 ± 0.08
Tarsometatarsus length	0.28 ± 0.06	0.31 ± 0.02	-0.03 ± 0.07	0.20 ± 0.07
Tarsometatarsus width	0.29 ± 0.05	0.23 ± 0.04	0.10 ± 0.12	0.36 ± 0.07
Synsacrum	0.08 ± 0.34	0.03 ± 0.08	0.98 ± 0.15	-0.23 ± 0.15
Scapula	0.16 ± 0.05	0.15 ± 0.09	0.10 ± 0.12	0.02 ± 0.15

Source: McGillivray, 1985; reproduced with permission.

body measurements of the Great Horned Owl (males). The results are given in Table 4.11.

Since the logarithmic transformation tends to equalize variance, the PCs based on the covariance matrix are not very different from those based on the correlation matrix (Section 3.7), especially when interval estimates are considered. Here assumptions of normality and a large sample are not required for calculation purposes, but approximate normality (which may hold by the Central Limit Theorem) is needed for testing purposes. Thus normality of the PCs should be verified (Section 4.6) before proceeding to test significance. A definite advantage of the jackknife is that it may be applied equally to both covariance as well as correlation matrices, which is more difficult to do with ML methods. $\qquad\square$

The Bootstrap

Another resampling procedure is the bootstrap estimator, which is also a "nonparametric" technique for estimating parameters and their standard errors within a single sample (Efron, 1979, 1981). The idea is to mimic the process of selecting many samples of size n by duplicating each sample value m times, mixing the resultant mn values, and randomly selecting a sequence of independent samples, each of size n. This yields a set of independent estimates for the parameter(s) in question, which allows an overall mean estimate and the calculation of the standard deviation(s). Preliminary results for the correlation coefficient of a bivariate normal sample indicates that the bootstrap performs better than other resampling schemes, including the jackknife (Efron, 1981), although it requires more computation. Indeed for PCA and other multivariate methods the computation required may be prohibitive. The bootstrap however has been used to obtain interval estimates for PCs by Diaconis and Efron (1983) and Stauffer et al. (1985), who compare ecological data to those generated "randomly" by computing interval estimates for the latent roots and vectors. As with the jackknife, the bootstrap interval estimates may be obtained for both the covariance and correlation matrices. The difficulty again is that at least approximate normality is required for significance testing, so that normality (or at least symmetry of distribution) of the PCs should be assessed before testing is carried out.

Cross Validation

Cross validation is another resampling scheme and is based on the idea that a set of data can be subdivided into groups, with the model estimated in one group(s) and then evaluated in the remaining group(s) to verify goodness of fit, forecasting properties, and so forth. If the omitted observations can be estimated from the remaining data and then compared to the actual omitted values, that model is chosen which provides the best predictor of these values (Mosteller and Wallace, 1963; Stone, 1974). Recently, cross validation has been applied to PCA as a possible alternative to normality-based ML tests to obviate assumptions of normality and large samples. Let x_{ij} be any element of a data matrix $\mathbf{X}$. Then predicted values are given by

$$\hat{x}_{ij} = \sum_{h=1}^{r} a_{jh} z_{ih} \tag{4.69}$$

(Section 3.4) where r is the number of components to be tested. The rows of $\mathbf{X}$ are divided into G groups, the first is omitted, and a PCA is carried on the remaining $G-1$ groups. The "missing" observations are estimated by Eq. (4.69), assuming the z_{ih} values exist for the missing group ($h = 1, 2, \ldots, r$), and the estimates are then compared to actual values by the predictive sum of squares

$$\text{PRESS}(r) = \frac{1}{np} \sum_{i=1}^{n} \sum_{j=1}^{p} (x_{ij} - \hat{x}_{ij})^2 \tag{4.70}$$

where n is the number of elements in the omitted group. The process is repeated for each of the G omitted groups, and the total predictive sum of squares is then computed as

$$\text{TPRESS} = \sum_{g=1}^{G} \text{PRESS}_g(r) \tag{4.71}$$

for r PCs. Theoretically the optimal number of groups G should equal the total sample size (Stone, 1974), but this is frequently computationally involved. Wold (1978) has therefore suggested that 4–7 groups be used, which appears to work well in practice and does not decrease statistical efficiency to any great extent.

A difficulty arises using Eq. (4.69) since PC values (scores) are normally not available for the omitted observations. Eastment and Krzanowski (1982) suggest that both rows and columns of X be omitted to allow the predicted values $\hat{x}_{ij}$ to be computed from all the data, except for the ith row, jth column. Let z_{ij}^* denote the ith observation for the jth PC when the jth column of X is omitted, and let a_{ij} denote the loading for the jth PC and the ith variable. Then predicted values can be obtained as

$$\hat{x}_{ij} = \sum_{h=1}^{r} a_{jh} z_{ih}^* \tag{4.72}$$

Note that the sign of $\hat{x}_{ij}$ cannot be determined without a decomposition of the entire matrix X. The choice of the optimum value of r then depends on a suitable function of Eq. (4.71). One such choice is the statistic

$$W(r) = \frac{\text{TPRESS}(r-1) - \text{TPRESS}(r)}{\text{TPRESS}(r)} \frac{D_r}{D_R} \tag{4.73}$$

where D_r is the number of degrees of freedom required to fit the rth component and D_R is the number of degrees of freedom remaining after the rth component has been fitted, that is, $D_r = n + p - 2r$ and $D_R = D - D_r$, where $D = p(n-1)$ is the total number of degrees of freedom; r is then equal to the number of PCs for which $W(r) > 1$.

Cross validation in the context of PCA is relatively recent and does not seem to have been widely used. The lack of a known distribution for $W(r)$ precludes statistical inference. Recent Monte Carlo simulations (Krzanowski 1983) seem to indicate that the procedure retains less PCs than the Bartlett chi-squared test, and in fact frequently yields a similar number of retained PCs as the "greater than or equal to the mean root" rule (Section 4.4.3).

Example 4.13. Using McReynolds' gas chromatography data (McReynolds, 1970), Wold (1978) and Eastment and Krzanowski (1982) perform a PCA of chemical retention indices for $p = 10$ compounds and $n = 226$ liquid phases. The values of $W(r)$ are given in Table 4.12.

An examination of the $W(r)$ values reveals that $r = 3$ components (Z_1, Z_2,

Table 4.12   A Cross-Validatory Analysis Using $W(r)$ to Select the Number of PCs

Component	Latent Root	D_r	D_R	$W(r)$
1	306,908	234	2,016	279.69
2	3,054	232	1,784	2.18
3	1,572	230	1,554	.20
4	1,220	228	1,326	1.99
5	487	226	1,100	.66
6	265	224	876	.22
7	140	222	654	.11
8	84	220	434	.05
9	55	218	216	.02
10	30	216	0	—

Source: Eastment and Krzanowski, 1982; reproduced with permission.

and $\mathbf{Z}_4$) correspond to $W(r) > 1$. Since $W(r)$ is not a monotonically decreasing function of r, the procedure does not necessarily select the first r components. Also, although components may account for an equal percentage of variance, they need not possess equal predictive (explanatory) power. $\square$

4.2.2   Residual Correlations Test

Velicer (1976) has proposed an alternative method for selecting nontrivial PCs based on partial (residual) correlation after $r < p$ components have been extracted. It also does not permit sample–population significance testing and does not require distributional assumptions. Let

$$\mathbf{X} = \mathbf{Z}_{(r)}\mathbf{A}_{(r)}^{\mathrm{T}} + \boldsymbol{\epsilon}_{(r)}$$

where $\boldsymbol{\epsilon}_{(r)}$ is a $(n \times p)$ residual matrix after $r < p$ PCs have been extracted. The residual sums of squares and product matrix is

$$\boldsymbol{\epsilon}_{(r)}^{(\mathrm{T})}\boldsymbol{\epsilon}_{(r)} = \mathbf{X}^{\mathrm{T}}\mathbf{X} - \mathbf{X}^{\mathrm{T}}\mathbf{P}_{Z_{(r)}}\mathbf{X}$$

$$= \mathbf{X}^{\mathrm{T}}\mathbf{X} - \mathbf{X}^{\mathrm{T}}\mathbf{Z}_{(r)}(\mathbf{Z}_{(r)}^{\mathrm{T}}\mathbf{Z}_{(r)})^{-1}\mathbf{Z}_{(r)}^{\mathrm{T}}\mathbf{X}$$

$$= \mathbf{X}^{\mathrm{T}}\mathbf{X} - \mathbf{X}^{\mathrm{T}}\mathbf{Z}_{(r)}\mathbf{Z}_{(r)}^{\mathrm{T}}\mathbf{X}$$

$$= \mathbf{X}^{\mathrm{T}}\mathbf{X} - \mathbf{A}_{(r)}\mathbf{A}_{(r)}^{\mathrm{T}} \tag{4.74}$$

which can be computed for both covariance and correlation matrices. Let

$$\mathbf{R}^* = \mathbf{D}^{-1/2}\boldsymbol{\epsilon}_{(r)}^{\mathrm{T}}\boldsymbol{\epsilon}_{(r)}\mathbf{D}^{-1/2}$$

be the matrix of partial correlations, where $\mathbf{D} = \mathrm{diag}(\boldsymbol{\epsilon}_{(r)}^{T}\boldsymbol{\epsilon}_{(r)})$. Let r_{ij}^{*} represent the off-diagonal elements of $\mathbf{R}^{*}$ and let

$$f_r = \sum_{i \neq j} \sum \frac{r_{ij}^{*}}{p(p-1)} \tag{4.75}$$

which lies in the interval 0–1. The stopping rule proposed by Velicer (1976) consists of accepting components up to and including those that correspond to the minimum value of f_r, since small values of Eq. (4.75) indicate that the r retained PCs are uniformly correlated with most of the variables, and in this sense capture nonresidual variation. Velicer (1976) gives several values of f_r for known data in psychology, and points out that for these data his procedure results in a smaller number of PCs than would be retained by the commonly used rules of thumb.

4.4.3 Informal Rules of Thumb

Finally, several other procedures have been proposed (mainly in the psychometric literature) as possible stopping rules for selecting components. These methods however are statistically (or mathematically) flawed (see McDonald, 1975), or else appear to be based on ad hoc, subjective reasoning and cannot be recommended as replacements for criteria considered in the above sections. Since they appear to be widely known however and have made their way into standard statistical computer packages, they are considered here for the sake of completeness.

The first method, and perhaps the best known, is to reject PCs that correspond to latent roots smaller than or equal to the mean of all latent roots; that is, we retain roots $l_1 > l_2 > \cdots l_r$ such that

$$l_1 > l_2 > \cdots > l_r \geq \frac{\sum_{i=1}^{p} l_i}{p} = \bar{l}$$

When the correlation matrix is used $\bar{l} = 1$, and this corresponds to the usual "rule of parsimony" frequently encountered in applications. For the correlation matrix the rule can also be rationalized by the notion that a PC which does not account for more trace than any single variable cannot possibly be of any interest. Two objections may be raised against such a practice. First, a PC with latent root smaller than the mean root may nevertheless possess a meaningful interpretation, since it can correlate highly with one, or perhaps two, observed variables. Such information may be of interest in certain applications. Second, it is precisely the low variance components which could be of primary interest, for example, when examining residuals for outliers (Section 4.7) or estimating an orthogonal regression (Section 10.2.3).

The second procedure is graphical in nature, and is usually based on a visual inspection of the latent roots or on logarithms of the latent roots. The procedure has been called the "scree test" by Cattell (1966), although it also cannot be considered as a statistical test in any true sense of the word, since at best it simply provides a graphical supplement to procedures considered in previous sections. The method consists of plotting latent roots against their rank numbers and observing whether, at some point (latent root), the slope becomes "markedly" less steep, that is, the latent roots tend to be isotropic. Thus a characteristic plot of the latent roots will frequently exhibit exponential decline, perhaps with lower-order roots decreasing in a linear fashion. Although artificial simulations based on "random" data do indicate that plots can be useful guides or indicators of correct dimensionality, real data appear to give less clear-cut results (Joliffe, 1986; Farmer, 1971). The real usefulness of the "test" however is probably as a graphical guide when carrying out statistical testing.

A third procedure based on Horn (1965) which has become more prominent in the psychometric literature is to use regression analysis on the latent roots of a correlation matrix of normal data in order to predict mean latent roots (for a specified range of n and p values) and to use the estimated mean root as a selection criterion for real data matrices. As such the method can be considered as a generalization of the two rules of thumb considered above. The procedure, known as parallel analysis, is based on Allen's and Hubbard's (1986) equation.

$$\ln(l_i) = a + b_i \ln(n-1) + c_i \ln[(p-i-1)(p-i+2)/2] + d_i \ln(l_{i-1})$$

where i is the ordinal position of the ith latent root.

Although the procedure is of a more advanced nature, it nevertheless suffers from a lack of a clear statistical rationale and represents a somewhat ad hoc approach to the problem of estimating dimensionality. It cannot be expected therefore to perform well in general empirical situations (see also Lautenschlager, 1989).

Example 4.14. The random (independent) normal data of Example 4.2 indicate complete sphericity of the latent roots when using the chi-squared test. The latent roots of the covariance matrix are given in Table 4.13 (Figs. 4.1 and 4.2). Plotting l_i against i exhibits the characteristic exponential decline common for random normal data. The plot can also be linearized by transforming to logarithms. Thus the linearity of $\ln l_i$ Fig. 4.2) would seem to confirm the chi-squared test of complete sphericity. Linearity can be confirmed further by least squares regression or else more simply by a plot of $\ln l_{i-1} - \ln l_i$ $(i = 2, 3, \ldots, r)$ against rank number, such as in Figure 4.3, where approximate linearity is indicated by a random scatter of residuals. For random, independent (normal) data, linearity is simply a byproduct of

**Table 4.13   Latent Roots (Natural Logarithms)
of Random, Independent Normal Data (Example 4.2)**

Number	Latent Roots l_i	$\ln l_i$
1	553.4971	6.3163
2	440.5440	6.0880
3	388.3756	5.9620
4	342.4686	5.8362
5	294.6565	5.6858
6	263.6820	5.5747
7	212.8715	5.3607
8	197.6678	5.2866
9	168.8241	5.1289
10	147.6623	4.9949
11	124.0577	4.8207

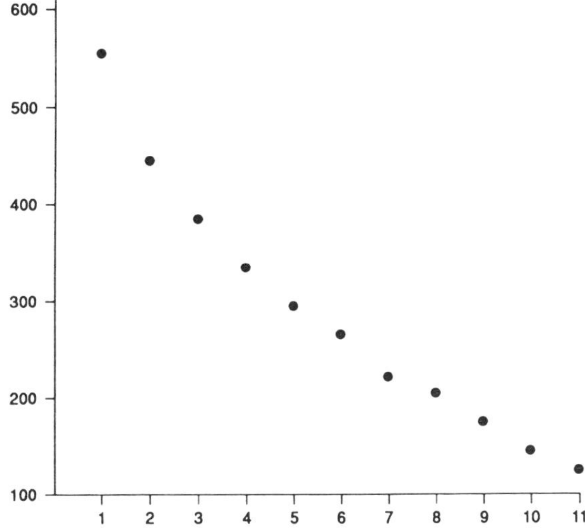

Figure 4.1 Latent roots of Table 4.12 derived from the covariance matrix of $p = 11$ independent, normal variates.

the arbitrary ordering of the latent roots, and can always be removed by a random permutation.

4.5   DISCARDING REDUNDANT VARIABLES

The intent behind a PCA is to reduce the dimensionality of a set of observed variables and at the same time to maximize the retained variance. In situations where a large set of variables is available, we may ask ourselves a

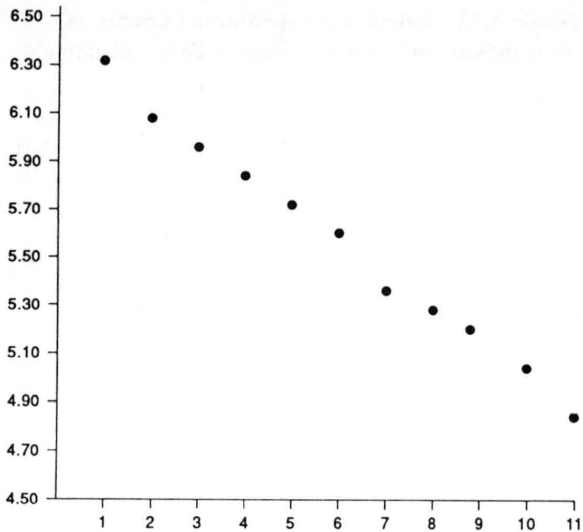

Figure 4.2 Logarithmic transformation of latent roots of Table 4.13 of independent normal variates.

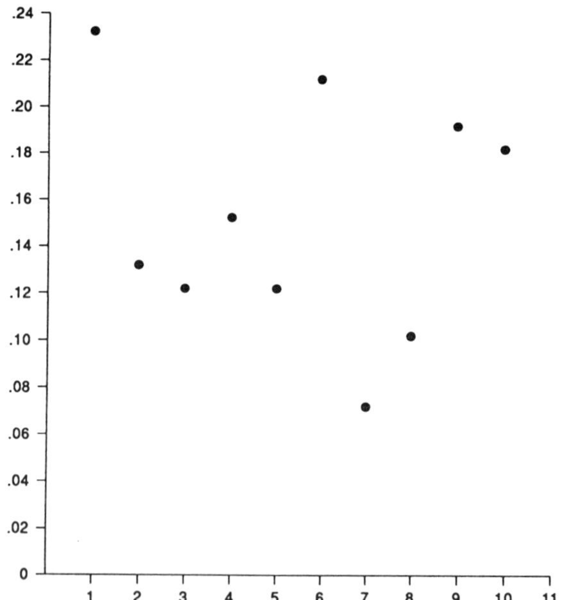

Figure 4.3 The $\ln l_{i-1} - \ln l_i$ plotted against rank number (Table 4.12).

somewhat different question: which variables can we omit from the set and still retain most of the essential information? For instance, the external body measurements of the humpback whale (Example 3.11) correlate mainly with a single size dimension so that most of the variables except body length may be considered as redundant. This can result in considerable economy of measurement by reducing the number of variables required to capture the concept of "size." Also, if chosen haphazardly redundant variables can result in a nonsensical interpretation of a PCA (see Ramsey, 1986).

At times the decision to retain certain variables rests on some specific need, or perhaps on purely theoretical grounds. Such variables will be retained automatically and will not be affected by the selection process. For the remaining variables however statistical criteria can be used in the selection process. Several criteria have been investigated by Joliffe (1972, 1973) and McCabe (1984).

Let

$$\mathbf{X}^T\mathbf{X} = \left[\begin{array}{c|c} \mathbf{X}_1^T\mathbf{X}_1 & \mathbf{X}_1^T\mathbf{X}_2 \\ \hline \mathbf{X}_2^T\mathbf{X}_1 & \mathbf{X}_2^T\mathbf{X}_2 \end{array}\right]$$

be the partitioned $(p \times p)$ matrix $\mathbf{X}^T\mathbf{X}$ where $\mathbf{X}_1$ is the $(n \times k)$ matrix of retained variables. We have $C(^p_k)$ choices for $\mathbf{X}_1^T\mathbf{X}_1$ given k, and $2^p - 1$ choices for k. Given k, the conditional dispersion of the rejected set $\mathbf{X}_2$ given $\mathbf{X}_1$ is

$$\begin{aligned}
(\mathbf{X}^T\mathbf{X})_{22.1} &= \mathbf{X}_2^T\mathbf{X}_2 - \mathbf{X}_2^T\mathbf{X}_1(\mathbf{X}_1^T\mathbf{X}_1)^{-1}\mathbf{X}_1^T\mathbf{X}_2 \\
&= \mathbf{X}_2^T\mathbf{X}_2 - \mathbf{X}_2^T\mathbf{P}_{x1}\mathbf{X}_2
\end{aligned} \tag{4.76}$$

where

$$|\mathbf{X}^T\mathbf{X}| = |\mathbf{X}_1^T\mathbf{X}_1|\,|(\mathbf{X}^T\mathbf{X})_{22.1}| \tag{4.77}$$

Let $c_1, c_2, \ldots, c_{p-k}$ be the latent roots of $(\mathbf{X}^T\mathbf{X})_{22.1}$. Then McCabe (1984) shows that

$$\min|(\mathbf{X}^T\mathbf{X})_{22.1}| = \min \prod_{i=1}^{p-k} c_i \tag{4.78}$$

$$\min \operatorname{tr}(\mathbf{X}^T\mathbf{X})_{22.1} = \min \sum_{i=1}^{p-k} c_i \tag{4.79}$$

$$\min|(\mathbf{X}^T\mathbf{X})_{22.1}|^2 = \min \sum_{i=1}^{p-k} c_1^2 \tag{4.80}$$

are optimal for discarding set $\mathbf{X}_2$, where $|\cdot|^2$ denotes the sum of squares of matrix elements. Note that it follows from Eq. (4.78) that minimizing $|(\mathbf{X}^T\mathbf{X})_{22.1}|$ (the "residual" variance) is equivalent to maximizing $|\mathbf{X}_1^T\mathbf{X}_1|$, the retained variance. In practice, the choice of which criterion is used will depend on the specific objective(s) at hand.

The criteria (Eqs. 4.78–4.80) make use of the conditional sums of squares and products matrix $(\mathbf{X}^{\mathrm{T}}\mathbf{X})_{22.1}$. An alternative method for rejecting variables is to use PCs. Joliffe (1972, 1973) has considered several methods. One that seems to work well is to use the first r PCs to reject $1 < r < p$ variables. We simply note which variables possess the highest loadings for the first component, the second component, and so on until r PCs are used up. Alternatively, we can use large loadings for the last $p - r$ components to reject variables, that is, the variable which possesses maximal correlation with one of the last $p - r$ principal components is rejected. Results for simulated data are given by Joliffe (1972) and for real data by Machin (1974), who considers the whale measurements of Example 3.11.

4.6 ASSESSING NORMALITY

It was seen in Section 4.3 that tests using maximum likelihood require at least approximate multivariate normality if they are to be used to estimate the number of significant parameters (dimensions). In this section we consider several tests which can be used to assess univariate and multivariate normality.

4.6.1 Assessing Univariate Normality

The first step when assessing a multivariate sample for normality is to test the marginal distributions, since nonnormality of the latter precludes normality in higher dimensions (Section 2.8). Also certain types of nonnormality can become more apparent from the shape(s) of the marginal distributions. A PCA of marginal normal variables may either exhibit less multivariate nonnormality, be more robust to departures from multivariate normality, or both. Many tests are available for univariate normality and a review is provided by Mardia (1980). The most promising of these seem to be based on quantiles, which also lend themselves well to graphical methods and which are available with several large statistical computer packages.

Consider a random independent sample $x_1, x_2, \ldots, x_n$ taken from a univariate distribution $f(x)$. Also let the n order statistics $y_{(1)} < y_{(2)} < \cdots < y_{(n)}$ represent the original sample values x_i ranked in increasing order. A percentile then divides a sample into two groups of a given percentage. Certain commonly used percentiles are given specific names, for example, the median and quartile, which are the fiftieth and twenty-fifth percentiles respectively. When percentages are converted to probabilities (proportions) they are known as quantiles, so that for any probability p we can define the corresponding quantile $Q(p)$. In a given sample quantiles are obtained from the ranked observations, that is $Q(p)$ is taken as $y_{(i)}$ whenever p is one of the fractions

$$p_i = \frac{(i - .50)}{n} \qquad (i = 1, 2, \ldots, n) \qquad (4.81)$$

Other intervals can also be employed (see Barnett, 1975)—for example, the BMDP statistical programs use the intervals

$$p_i = \frac{100(3i - 1)}{3n + 1} \qquad (i = 1, 2, \ldots, n) \qquad (4.82)$$

The quantiles $y_{(i)}$ can then be plotted against p_i to yield what is known as a quantile plot. Let $F(x)$ be the cumulative distribution of x. Then a quantile plot is simply the sample estimate of $F^{-1}(x)$. When sample quantiles are plotted against their theoretical values $F^{-1}(x)$, we obtain a theoretical quantile–quantile or a probability plot. It follows that when sample estimates approach their theoretical values, a quantile–quantile plot approaches a straight line with $45°$ at the origin. This provides a straightforward graphical technique for assessing univariate normality, since values of the cumulative normal can be readily obtained from tables. Marked departures of points from the $45°$ diagonal indicate outlier departure from univariate normality. Thus by studying the patterns of such departure it is possible to detect outlier observations and to determine direction of skew if any exists (Chambers et al., 1983).

4.6.2 Testing for Multivariate Normality

Testing for univariate normality can only reveal whether random variables are *not* multivariate normal (Exercise 2.10; Section 2.8.1). When all marginals are univariate normal, however, we still require a test for multivariate normality. Quantile–quantile plots can be used to assess multivariate normality (Healy, 1968, Easton and McCulloch, 1990) by computing sample values of the multivariate normal exponent and comparing the resulting quantiles with those of the chi-squared distribution.

Consider the $(n \times p)$ matrix

$$\mathbf{Y} = \begin{bmatrix} y_{11} & y_{12} & \cdots & y_{1p} \\ y_{21} & y_{22} & \cdots & y_{2p} \\ \cdot & \cdot & & \cdot \\ \cdot & \cdot & & \cdot \\ \cdot & \cdot & & \cdot \\ y_{n1} & y_{n2} & & y_{np} \end{bmatrix}$$

where rows are multivariate sample points $\mathbf{y}_1 = (y_{11}, y_{12}, \ldots, y_{1p})^{\mathrm{T}}$, $\mathbf{y}_2 = (y_{21}, y_{22}, \ldots, y_{2p})^{\mathrm{T}}$ and $\mathbf{y}_n = (y_{n1}, y_{n2}, \ldots, y_{np})^{\mathrm{T}}$. Let $\bar{\mathbf{Y}} = (\bar{y}_1, \bar{y}_2,$

$\ldots, \bar{y}_p)^{\mathrm{T}}$ and $\mathbf{S} = \frac{1}{n-1}(\mathbf{X}^{\mathrm{T}}\mathbf{X})$. Then

$$d^2 = (\mathbf{y}_i - \bar{\mathbf{Y}})^{\mathrm{T}}\mathbf{S}^{-1}(\mathbf{y}_i - \bar{\mathbf{Y}})$$

$$= \mathbf{z}_i^{\mathrm{T}}\mathbf{z}_i \qquad (i = 1, 2, \ldots, n) \tag{4.83}$$

are squared generalized interpoint distances* where

$$\mathbf{z}_i = \mathbf{S}^{-1/2}(\mathbf{y}_i - \bar{\mathbf{Y}})$$

The d_i^2 can be converted into quantiles and plotted against theoretical chi-squared quantiles, which must yield a straight line when samples are taken from a multivariate normal distribution. Recently Easton and McCulloch (1990) devised a general multivariate approach to quantile plots. In practice exact straight lines are rare owing to sampling variability, and judgment will be required as to whether the plot is linear. The subjective element can always be removed of course by the further use of regression or first differences, if so desired.

Theoretical quantile plots can be constructed in one of three ways:

1. Ordered d_i^2 values can be plotted against the expected order statistics of the chi-squared distribution with p degrees of freedom, which on the null hypothesis of multivariate normality results in a straight line through the origin at 45°.

2. For small samples, better results may be obtained by plotting ordered d_i^2 values against the expected order statistics of a corrected beta distribution (Small, 1978).

3. Ordered $(d_i^2)^{1/2}$ and alternatively $(d^2)^{1/3}$ values are plotted against expected order statistics of the standardized normal distribution. (Healy, 1968).

The last approach uses the normal approximation to the chi-squared, and should therefore not be used in small samples. Departures from normality need not occur in p-dimensional space, but may occur in lower dimensions and can thus be masked in a p-dimensional quantile–quantile plot. Subspaces should therefore be investigated as well, which for large p may become impractical. Andrews et al. (1973) have suggested that plots of d_i^2 be supplemented by plots of angular position. An alternative plotting technique based on the Shapiro–Wilks statistic has also been proposed by Royston (1983). Small (1980) has suggested an approximate method of assessing multivariate normality using univariate marginals. Still a different approach, from Mardia (1970) and Malkovich and Afifi (1973), is to use the multivariate extension of the skewness and kurtosis tests, which has been

---

* Also known as squared radii, centroids, or squared Mahalanobis distance.

used in biological morphometrics by Reyment (1971). A measure for multivariate skewness can be defined as

$$b_i = \frac{1}{n^2} \sum_{i=1}^{n} \sum_{j=1}^{n} [(\mathbf{y}_i - \bar{\mathbf{Y}})^{\mathrm{T}} \mathbf{S}^{-1} (\mathbf{y}_j - \bar{\mathbf{Y}})]^3 \tag{4.84}$$

and of multivariate kurtosis by

$$b_2 = \frac{1}{n} \sum_{i=1}^{n} [(\mathbf{y}_i - \bar{\mathbf{Y}})^{\mathrm{T}} \mathbf{S}^{-1} (\mathbf{y}_i - \bar{\mathbf{Y}})]^2 \tag{4.85}$$

Then the asymptotic distribution of

$$\chi^2 = \frac{nb_1}{6} \tag{4.86}$$

is chi-squared with $1/6\, p(p + 1)(p + 2)$ degrees of freedom. Also since b_2 is asymptotically normal with expected value $p(p + 2)$ and variance $1/n[8p(p + 2)]$,

$$z = \frac{b_2 - p(p + 2)}{\left[\dfrac{1}{n} 8p(p + 2)\right]^{1/2}} \tag{4.87}$$

is asymptotically $N(0, 1)$. Gnanadesikan (1977, p. 175), however, reports that investigations indicate large samples are required to obtain good approximations. Reyment (1971) also suggests large samples of at least $n = 100$ when using Eq. (4.86) and $n = 180$ for Eq. (4.87). A basic weakness of using Eqs. (4.86) and (4.87) is that insignificant values of b_1 and b_2 do not necessarily imply normality. An algorithm for skewness and kurtosis is given by Mardia and Zemroch (1975).

Example 4.15. Royston (1983) has considered the following hematological variables measured for $n = 103$ workers, using the square root normal approximation of squared distances (Eq. 4.83) (see Table 4.1):

Y_1 = Hemoglobin concentration
Y_2 = Packed cell volume
Y_3 = White blood cell count
Y_4 = Lymphocyte count
Y_5 = Neutrophil count
Y_6 = Serum lead concentration

Univariate, bivariate, trivariate … quantile–quantile plots reveal normality in all subspaces and for all sample points, except for individuals 21, 47, and 52 which are nonnormal outliers in three-dimensional space defined by $\mathbf{Y}_3$,

Table 4.14 Hematological Measurements for $n = 103$ Workers

Case No.	Y_1	Y_2	Y_3	Y_4	Y_5	Y_6
1	13.4	39	4100	14	25	17
2	14.6	46	5000	15	30	20
3	13.5	42	4500	19	21	18
4	15.0	46	4600	23	16	18
5	14.6	44	5100	17	31	19
6	14.0	44	4900	20	24	19
7	16.4	49	4300	21	17	18
8	14.8	44	4400	16	26	29
9	15.2	46	4100	27	13	27
10	15.5	48	8400	34	42	36
11	15.2	47	5600	26	27	22
12	16.9	50	5100	28	17	23
13	14.8	44	4700	24	20	23
14	16.2	45	5600	26	25	19
15	14.7	43	4000	23	13	17
16	14.7	42	3400	9	22	13
17	16.5	45	5400	18	32	17
18	15.4	45	6900	28	36	24
19	15.1	45	4600	17	29	17
20	14.2	46	4200	14	25	28
21	15.9	46	5200	8	34	16
22	16.0	47	4700	25	14	18
23	17.4	50	8600	37	39	17
24	14.3	43	5500	20	31	19
25	14.8	44	4200	15	24	19
26	14.9	43	4300	9	32	17
27	15.5	45	5200	16	30	20
28	14.5	43	3900	18	18	25
29	14.4	45	6000	17	37	23
30	14.6	44	4700	23	21	27
31	15.3	45	7900	43	23	23
32	14.9	45	3400	17	15	24
33	15.8	47	6000	23	32	21
34	14.4	44	7700	31	39	23
35	14.7	46	3700	11	23	23
36	14.8	43	5200	25	19	22
37	15.4	45	6000	30	25	18
38	16.2	50	8100	32	38	18
39	15.0	45	4900	17	26	24
40	15.1	47	6000	22	33	16
41	16.0	46	4600	20	22	22
42	15.3	48	5500	20	23	23
43	14.5	41	6200	20	36	21
44	14.2	41	4900	26	20	20
45	15.0	45	7200	40	25	25
46	14.2	46	5800	22	31	22

Table 4.14 (*Continued*)

Case No.	Y_1	Y_2	Y_3	Y_4	Y_5	Y_6
47	14.9	45	8400	61	17	17
48	16.2	48	3100	12	15	18
49	14.5	45	4000	20	18	20
50	16.4	49	6900	35	22	24
51	14.7	44	7800	38	34	16
52	17.0	52	6300	19	21	16
53	15.4	47	3400	12	19	18
54	13.8	40	4500	19	23	21
55	16.1	47	4600	17	28	20
56	14.6	45	4700	23	22	27
57	15.0	44	5800	14	39	21
58	16.2	47	4100	16	24	18
59	17.0	51	5700	26	29	20
60	14.0	44	4100	16	24	18
61	15.4	46	6200	32	25	16
62	15.6	46	4700	28	16	16
63	15.8	48	4500	24	20	23
64	13.2	38	5300	16	26	20
65	14.9	47	5000	22	25	15
66	14.9	47	3900	15	19	16
67	14.0	45	5200	23	25	17
68	16.1	47	4300	19	22	22
69	14.7	46	6800	35	25	18
70	14.8	45	8900	47	36	17
71	17.0	51	6300	42	19	15
72	15.2	45	4600	21	22	18
73	15.2	43	5600	25	28	17
74	13.8	41	6300	25	27	15
75	14.8	43	6400	36	24	18
76	16.1	47	5200	18	28	25
77	15.0	43	6300	22	34	17
78	16.2	46	6000	25	25	24
79	14.8	44	3900	9	25	14
80	17.2	44	4100	12	27	18
81	17.2	48	5000	25	19	25
82	14.6	43	5500	22	31	19
83	14.4	44	4300	20	20	15
84	15.4	48	5700	29	26	24
85	16.0	52	4100	21	15	22
86	15.0	45	5000	27	18	20
87	14.8	44	5700	29	23	23
88	15.4	43	3300	10	20	19
89	16.0	47	6100	32	23	26
90	14.8	43	5100	18	31	19
91	13.8	41	8100	52	24	17
92	14.7	43	5200	24	24	17

Table 4.14   (Continued)

Case No.	Y_1	Y_2	Y_3	Y_4	Y_5	Y_6
93	14.6	44	9899	69	28	18
94	13.6	42	6100	24	30	15
95	14.5	44	4800	14	29	15
96	14.3	39	5000	25	20	19
97	15.3	45	4000	19	19	16
98	16.4	49	6000	34	22	17
99	14.8	44	4500	22	18	25
100	16.6	48	4700	17	27	20
101	16.0	49	7000	36	28	18
102	15.5	46	6600	30	33	13
103	14.3	46	5700	26	20	21

Source: Royston, 1983; reproduced with permission.

Y_4 and Y_5 (Fig. 4.4*a*). Once the three sample points are removed and Y_3–Y_6 replaced by their logarithms, the remaining 100 appear to be sampled from the multivariate normal (Fig. 4.4*b*). □

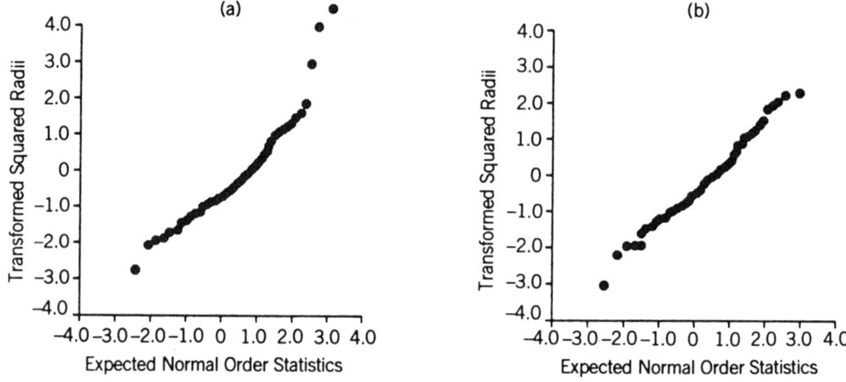

Figure 4.4   Normal plots of (square roots) of squared distances of the hematology data using logarithms with: (a) outliers present, and (b) outliers removed (Royston, 1983; reproduced with permission).

Example 4.16.   Reyment (1971) presents several examples of the use of multivariate skewness and kurtosis for determining possible normality. A pooled sample of $n = 302$ observations on Swedish grasshoppers (Omocstus haemorrhoidalis) for four variables is available:

$Y_1 = $ length of hind femur

$Y_2 = $ pronotal length

$Y_3 = $ elytron length

$Y_4 = $ least width between pronotus ridges

Univariate analysis indicates significant skewness for $\mathbf{Y}_2$ and $\mathbf{Y}_4$ as well as kurtosis for $\mathbf{Y}_4$. The null hypothesis of multivariate normality can therefore be rejected without recourse to a multivariate test. To illustrate the use of Eqs. (4.86) and (4.87) however we have (Reyment, 1971) $b_1 = 1.21$ and $b_2 = 27.25$ so that

$$\chi^2 = \tfrac{302}{6}(1.21) = 60.9$$

with 20 degrees of freedom, and

$$z = \frac{27.25 - 24}{[\tfrac{32}{302}(6)]^{1/2}} = 4.08$$

Thus the data exhibit significant multivariate skewness and kurtosis, again rejecting the null hypothesis of multivariate normality. $\square$

4.6.3 Retrospective Testing for Multivariate Normality

Generally speaking, a test for multivariate normality will precede a PCA if significance tests are to be carried out on the loadings and/or latent roots. However if significance testing of the PCs is optional, normality tests can also be carried out using the PCs themselves. Thus a PCA can be performed first, followed by an assessment of the latent roots and vectors and of multivariate normality. Here we may wish to test multivariate normality retrospectively in order to decide whether likelihood ratio tests can be used. Since the PCs are mutually orthogonal, multivariate normality can easily be assessed by testing the individual PCs themselves, for example, using normal theoretical quantile–quantile plots (Srivastava and Carter, 1983). It now becomes unnecessary to examine pairs, triplets, and so on since marginal normality of the principal components implies their joint normality, and that of the data. Also, even though the original p-dimensional distribution may not be normal, testing the principal components can reveal whether normality holds in certain subspaces. For example, normality may hold only for the last $p - r$ residual components, which suffices for purposes of testing for isotropic variation.

Finally, since the latent roots are conventionally defined as order statistics, they may be tested directly by means of theoretical chi-squared quantile plots. This approach seems particularly appropriate when the number of variables is large.

4.7 ROBUSTNESS, STABILITY AND MISSING DATA

Unbiased, asymptotically efficient estimates are obtained when (1) the data represent an uncontaminated random multivariate normal sample, and (2) no data are missing. Also, when multivariate normality is accepted, statistical testing of the PCs can proceed unimpeded. However, since sampling difficulties can frequently destroy an ideal modeling situation, modification(s) to a PCA may become necessary. A prior question of some interest therefore is the extent to which the PCs (as well as their test statistics) are affected by departure(s) from normality, homogeneity of sample, and missing data; that is, how robust or stable is a PCA to departure from standard assumptions and to variations in sampling conditions?

4.7.1 Robustness

A number of studies have been conducted on the effects of nonnormality on a principal components analysis. The robustness of the Bartlett–Anderson chi-squared test has been investigated by Waternaux (1976) and Davis (1977) (see also Ruymgaart, 1981).

Waternaux (1976) considers four trivariate population distributions; normal, short and long-tailed nonnormal, and exceptionally long-tailed nonnormal. Monte Carlo simulations indicate that the chi-squared tests are not very robust, particularly in small samples and when testing the largest (or smallest) latent root(s). Nonrobustness to small sample departure from multivariate normality is also confirmed by Layard (1974), who tested equality of covariance matrices (Section 4.2). On a more theoretical level Davis (1977) derived results that indicated conditions under which the Bartlett–Anderson test is conservative. An empirical investigation of the effect(s) of sample size on latent roots and vectors was also conducted by Dudzinski et al. (1975) using geological data, but the results are not conclusive. Indeed there seems to be a lack of evidence concerning robustness in large samples. Also the effect(s) of ancillary conditions such as approximate univariate normality or skewness on robustness has not been investigated to any great extent. The Mallows–Akaike type dimension estimation criteria considered in Section 4.3.5 also have not been evaluated for robustness to either nonnormality or small sample size. The question of assessing multivariate normality seems to be of some importance since it is not yet possible to transform nonnormal data to approximate normality when $p > 2$ (e.g., see Andrews et al., 1971).

The second important consideration is whether PCA is robust against (nonnormal) outliers. Nonnormal outlier sample points are introduced when observations from long-tailed distributions contaminate a normal sample, or when we commit nonnormal error(s) of measurement. Since a PCA utilizes Euclidian norms and inner products, it can be expected to be heavily influenced by outliers because the larger the outlier, the disproportionately greater the influence. Three broad approaches can be taken to increase robustness of PCAs to outliers. First, outlying observations can be eliminated from the sample; second, outliers can be modified by replacing them with more appropriate values; third, robust versions of covariance/correlation matrices can be used.

An alternative procedure is to replace the covariance matrix by a matrix of weighted sums of squares, and products (about weighted means) when carrying out a PCA, the so-called M-estimators (Campbell, 1980). Thus Matthews (1984), for example, has used M-estimators in conjunction with Royston's hematological data to correct for outlier individuals 21, 47, and 52 (Example 4.14). The M-estimators are applied directly to the PCs, which are then used to estimate the covariance matrix. The procedure is iterative and yields a weighted dispersion matrix together with its latent roots and latent vectors. Devlin et al. (1981) carried out Monte Carlo simulations of the effect(s) of various types of robust correlation matrices on a PCA, and recommended several robust dispersion matrices (see also Rousseeuw and Van Zomeren, 1990). Finally, a projection-pursuit approach to robust estimation of dispersion matrices and principal components can also be used (Li and Chen, 1985). Since no estimator appears to be globally best, each robust estimator should be used with the specific conditions under which it is optimal.

4.7.2 Sensitivity of Principal Components

A closely related but perhaps a more general question to ask is how sensitive is a PCA to changes in the variances of the components? That is, given a change in some latent root, how much change can be expected in the corresponding correlation loadings? Let $v = v(\mathbf{c})$ be a function of $\mathbf{c}$ to be maximized, and let $\bar{v} = v(\bar{\mathbf{c}})$ be the maximum of the function achieved at $\mathbf{c} = \bar{\mathbf{c}}$. Consider a small departure $\bar{v} - v = e$ from the maximum. Then $\{\mathbf{c} | \bar{v} - v \le e\}$ defines values of $\mathbf{c}$ in the arbitrarily small region about $\bar{v}$, the "indifference region with boundary e." Using a Taylor series expansion we obtain the second-order approximation

$$v \simeq \bar{v} + \mathbf{g}^{\mathrm{T}}\mathbf{r} + \tfrac{1}{2}\mathbf{r}^{\mathrm{T}}\mathbf{H}\mathbf{r} \tag{4.88}$$

where

$r = c - \bar{c}$

$\mathbf{g}$ = gradient vector of $v(\mathbf{c})$ evaluated at $\mathbf{c} = \bar{\mathbf{c}}$

$\mathbf{H}$ = Hessian matrix of $v(c)$ of second derivatives evaluated at $c = \bar{c}$

and where $\mathbf{H}$ is negative (semi) definite at $\mathbf{c} = \bar{\mathbf{c}}$. Since at the maximum

$\mathbf{g} = 0$, the region e about $\bar{v}$ can be approximated by

$$|\mathbf{r}^T\mathbf{H}\mathbf{r}| \le 2e \tag{4.89}$$

(de Sarbo et al. 1982; Krzanowski, 1984). Let $\mathbf{A} = -\mathbf{H}$ so that $\mathbf{A}$ is positive (semi) definite (Section 2.2). Then $\mathbf{r}^T\mathbf{A}\mathbf{r} = 2e$ is the equation of a p-dimensional ellipsoid, which defines a region of the coefficient space within which differences $\mathbf{r} = \mathbf{c} - \bar{\mathbf{c}}$ result in a reduction of at most e in the criterion function v. It follows that the maximum change (perturbation) that can be induced in the coefficients without decreasing $\bar{v}$ by more than e is the maximum of $\mathbf{r}^T\mathbf{r}$ subject to the constraint $\mathbf{r}^T\mathbf{A}\mathbf{r} = 2e$. Differentiating the Lagrange expression

$$\phi = \mathbf{r}^T\mathbf{r} - (\mathbf{r}^T\mathbf{A}\mathbf{r} - 2e) \tag{4.90}$$

and setting to zero yields

$$(\mathbf{A}^{-1} - \lambda\mathbf{I})\mathbf{r} = \mathbf{0} \tag{4.91}$$

(Section 2.7). The appropriate value of $\mathbf{r} = \mathbf{c} - \bar{\mathbf{c}}$ is thus the eigenvector corresponding to the largest latent root of $\mathbf{A}^{-1}$ (smallest latent root of $\mathbf{A}$), normalized such that $|\mathbf{r}^T\mathbf{A}\mathbf{r}| = 2e$. This is the same as finding the component $\mathbf{c}$ whose angle θ with $\bar{c}$ (in p-dimensional space) is maximum, but where variance is no more than e of that of $\bar{c}$. Using Eq. (4.88) Krzanowski (1984) developes a sensitivity analysis for PCA. Let $\mathbf{S}$ be the sample covariance (correlation) matrix. Then (Section 3.4) the function to be maximized (in present notation) is

$$v = \mathbf{c}^T\mathbf{S}\mathbf{c} - l(\mathbf{c}^T\mathbf{c} - 1) \tag{4.92}$$

so that the maximum is achieved at $\bar{\mathbf{c}} = \mathbf{c}_1$, the latent vector which corresponds to the largest latent root $l = l_1 = \mathbf{c}_1^T\mathbf{S}\mathbf{c}_1$. Now at the maximum the Hessian matrix of second derivatives of v is $2\mathbf{S} - 2l_1\mathbf{I}$, where $l_1 > l_2 > \cdots > l_p$ are latent roots of $\mathbf{S}$ with corresponding latent vectors $\mathbf{c}_1, \mathbf{c}_2, \ldots, \mathbf{c}_p$. The latent values of $\mathbf{H}$ are therefore $2(l_i - l_1)$ with corresponding latent vectors $\mathbf{c}_1, \mathbf{c}_2, \ldots, \mathbf{c}_p$, and the smallest latent root of $\mathbf{A} = -\mathbf{H}$ is therefore $2(l_1 - l_2)$ with latent vector $\mathbf{c}_2$. The maximum perturbation that can be applied to $\mathbf{c}_1$ while ensuring that the variance of the resulting component is within e of l_1 therefore depends on $\mathbf{r} = k\mathbf{c}_2$, where

$$k = \pm\frac{e}{(l_1 - l_2)^{1/2}} \tag{4.93}$$

The PC that is "maximally e different" from $\mathbf{c}_1$ is then given by

$$\mathbf{c} = \mathbf{c}_1 + \mathbf{r} = \mathbf{c}_1 \pm \mathbf{c}_2[e/(l_1 - l_2)]^{1/2} \tag{4.94}$$

and imposing the normalization of $c^T c = 1$ we have

$$\mathbf{c}_{(1)} = \{\mathbf{c}_1 \pm \mathbf{c}_2 [e/(l_1 - l_2)]^{1/2} / \{1 + e(l_1 - l_2)\}^{1/2} \tag{4.95}$$

the component that differs maximally from $\mathbf{c}_1$ but whose variance is at most e less than that of $\mathbf{c}_1$. Since $l_1 \neq l_2$ with unit probability (Theorem 3.12), the component (Eq. 4.95) is defined for all sample covariance matrices $\mathbf{S}$. The cosine of the angle θ between $\mathbf{c}_{(1)}$ and $\mathbf{c}_1$ is then

$$\cos \theta = [1 + e(l_1 - l_2)]^{-1/2} \tag{4.96}$$

Equation (4.96) can be generalized to any jth or $(j + 1)$th latent root. As was the case for Girshick's covariance matrix (Theorem 4.4), the stability of the loading coefficients are functions of the differences of the latent roots, and this reinforces once again the necessity for discarding PCs with similar variance. Recently Critchley (1985) has developed the so-called influence functions for the detection of influential observations for principal components.

Example 4.17. Using previously published data, Krzanowski (1984) considers the effect(s) of a 5% and 10% change in the latent roots on the PC correlation loadings. The loadings and their perturbed values are summarized in Table 4.15 where $e = l_j / 10$ and $e = l_j / 20$. Since l_1 accounts for most of the variance, the interpretation of $\mathbf{Z}_1$ is not altered in terms of the general size component, although the angle of perturbation is not small. The largest effect(s) of the perturbations is felt in low-order PCs because of the closeness of the corresponding latent roots. A statistical test of significance

Table 4.15 Sensitivity Analysis of White Leghorn Fowl Latent Vectors

Component	Variance			Coefficients				Angle θ
		$\mathbf{X}_1$	$\mathbf{X}_2$	$\mathbf{X}_3$	$\mathbf{X}_4$	$\mathbf{X}_5$	$\mathbf{X}_6$	
1	4.568	0.35	0.33	0.44	0.44	0.43	0.44	
Perturbed 5%	4.352	0.47	0.48	0.39	0.37	0.36	0.37	14
Perturbed 10%	4.159	0.50	0.54	0.36	0.33	0.32	0.34	19
2	0.714	0.53	0.70	−0.19	−0.25	−0.28	−0.22	
Perturbed 5%	0.682	0.26	0.87	−0.16	−0.24	−0.24	−0.20	19
Perturbed 10%	0.656	0.15	0.90	−0.15	−0.23	−0.22	−0.18	26
3	0.412	−0.76	0.64	0.05	−0.02	0.06	0.05	
Perturbed 5%	0.393	−0.72	0.61	−0.11	−0.15	0.21	0.18	16
Perturbed 10%	0.377	−0.69	0.59	−0.16	−0.20	0.25	0.22	23
4	0.173	0.05	0.00	−0.52	−0.49	0.51	0.47	
Perturbed 5%	0.165	0.04	0.00	−0.55	−0.42	0.30	0.65	17
Perturbed 10%	0.156	0.03	0.00	−0.55	−0.39	0.21	0.71	
5	0.076	−0.04	−0.00	−0.19	0.15	−0.67	0.70	
Perturbed 5%	0.073	−0.03	−0.03	0.13	−0.15	−0.67	0.72	24
Perturbed 10%	0.068	−0.02	−0.04	0.22	−0.25	−0.64	0.70	33

(Section 4.3) can also precede (accompany) a sensitivity analysis of this type, to obtain a better grasp of the variations that could represent residual error. $\square$

4.7.3 Missing Data

Another difficulty that can cause estimation problems and upset multivariate normality is when a portion of the data is missing. This was the case in Example 3.13, where some of the measurements were not recorded. The simplest solution to the problem is to delete sample points for which at least one variable is missing, if most of the data are intact. The "listwise" deletion of observations however can cause further difficulties. First, a large part of the data can be discarded even if many variables have but a single missing observation. Second, the retained part of the data may no longer represent a random sample if the missing values are missing systematically. Third, discarding data may result in a nonnormal sample, even though the parent population is multivariate normal. Of course, for some data sets deleting sample points is out of the question—for example, skeletal remains of old and rare species.

An alternative approach is to use the available data to estimate missing observations. For example variate means (medians) can be used to estimate the missing values. The problem with such an approach is its inefficiency, particularly in factor analysis where the major source of information is ignored—the high intercorrelations that typically exist in a data matrix which is to be factor analyzed. Two types of multivariate missing data estimators can be used, even in situations where a large portion of the data is missing: multivariate regression and iterative (weighted) **PCA**. For a review of missing data estimators see Anderson et al. (1983) and Basilevsky et al. (1985). A recent approach, the so-called EM algorithm, also appears to be promising particularly in the estimation of factor scores (Section 6.8).

Generally, for a given data matrix not all sample points will have data missing. Assume that m individuals have complete records that are arranged as the first m rows of $\mathbf{Y}$, and $(n - m)$ individuals have missing data points in the last $(n - m)$ rows. If an observation is missing, it can be estimated using a regression equation computed from the complete portion of the sample. Without loss of generality, assume that the ith individual has a missing observation on the jth variable. The dependent variable in this case is Y_j and we have the estimate

$$\hat{y}_{ij} = \hat{\beta}_0 + \sum_{k=1}^{j-1} \hat{\beta}_k y_{ik} + \sum_{k=j+1}^{p} \hat{\beta}_k y_{ik} \tag{4.97}$$

Since the method does not utilize all of the sample information when estimating regression equations, a more general approach is to use the entire data matrix when estimating the regression equation.

Another procedure which can be used is the PC model itself; that is, PCs and the missing data can be estimated simultaneously. Let $\mathbf{I}_y = (w_{ij})$ denote the $(n \times p)$ indicator matrix where

$$w_{ij} = \begin{cases} 0 & \text{if } x_{ij} \text{ is observed} \\ 1 & \text{if } x_{ij} \text{ is not observed} \end{cases}$$

Also let $\mathbf{J}$ be the $(n \times p)$ matrix whose elements are ones and let $\otimes$ denote the direct product of two matrices. Then $\mathbf{Y}$ can be expressed as

$$\mathbf{Y} = [(\mathbf{I} - \mathbf{I}_y) \otimes \mathbf{Y}] + [\mathbf{I}_y \otimes \mathbf{Y}] \tag{4.98}$$

where

$$\mathbf{Y}^{(k)} = (\mathbf{I} - \mathbf{I}_y) \otimes \mathbf{X}$$

$$\mathbf{Y}^{(u)} = \mathbf{I}_y \otimes \mathbf{Y} \tag{4.99}$$

are the known and unknown parts, respectively. The procedure is equivalent to replacing the unknown values by zeros. Let $\mathbf{Y}^{(k)} = \mathbf{Z}_{(r)} \mathbf{P}_{(r)}^{\mathrm{T}} + \mathbf{e}$ for some $1 \le r < p$. Then new estimates for the missing values (as well as those present) are given by $\hat{\mathbf{Y}}^{(k)} = \mathbf{Z}_{(r)} \mathbf{P}_{(r)}^{\mathrm{T}}$. The process is continued until satisfactory estimates are obtained. Iterative least squares algorithms have also been proposed by Wiberg (1976) and de Ligny et al. (1981). A better procedure is probably to replace the missing entries with the variable means and iterate until stable estimates are obtained for some suitable value of k (see Woodbury and Hickey, 1963). Also, the variables can be weighted to reflect differential accuracy due to an unequal number of missing observations (see Section 5.6). The advantage of the iterative method is that it allows the estimation of missing values in situ, that is, within the PC model itself. The regression and PC estimation procedures do not require the assumption of normality. Finally, correlations may be computed using available pairs of observations (see Yawo et al., 1981).

A closely related problem is as follows. Having calculated the latent roots and latent vectors of $\mathbf{X}^{\mathrm{T}}\mathbf{X} = \mathbf{A}$, how can we use these computations to obtain the roots and latent vectors of a modified version of $\mathbf{A}$ (or vice versa), where the modifications are defined as follows:

$\mathbf{A} = \begin{bmatrix} \mathbf{A} \\ \mathbf{a}^{\mathrm{T}} \end{bmatrix}$, that is, appending (deleting a row $\mathbf{a}^{\mathrm{T}}$

$\mathbf{A} = [\mathbf{A} : \mathbf{a}]$, that is, appending (deleting) a column $\mathbf{a}$

A discussion and solutions are provided by Bunch and Neilson (1978) and Bunch et al. (1978). The results can be used for prediction purposes using time series or other ordered data, or when performing a sensitivity analysis of the effect(s) of deleting (adding) rows or columns of $\mathbf{A}$.

EXERCISES

4.1 Using Girshick's test (Eq. 4.56) show that for the matrix

$$\Sigma = \sigma^2 \begin{bmatrix} 1 & \rho & \cdots & \rho \\ \rho & 1 & & \cdot \\ \cdot & \cdot & \cdot & \cdot \\ \cdot & \cdot & \cdot & \cdot \\ \cdot & \cdot & \cdot & \cdot \\ \rho & \rho & \cdots & 1 \end{bmatrix}$$

the variances and covariances of the elements of the first latent vector $\mathbf{P}_1$ of Σ are given by

$$\operatorname{var}(p_{i1}) = \frac{[1 + (k-1)\rho](k-1)(1-\rho)}{nk^3\rho^2}$$

$$\operatorname{cov}(p_{r1}, p_{s1}) = \frac{-[1 + (k-1)\rho][1-\rho]}{nk^3\rho^2}$$

so that correlation between p_{r1} and p_{s1} is given by

$$\operatorname{cor}(p_{r1}, p_{s1}) = \frac{1}{k-1}$$

4.2 The following data have been published by Goldstein (1982)

Ward	$\mathbf{Y}_1$	$\mathbf{Y}_2$	$\mathbf{Y}_3$	$\mathbf{Y}_4$	$\mathbf{Y}_5$	$\mathbf{Y}_6$	$\mathbf{Y}_7$
1	28	222	627	86	139	96	20
2	53	258	584	137	479	165	31
3	31	39	553	64	88	65	22
4	87	389	759	171	589	196	84
5	29	46	506	76	198	150	86
6	96	385	812	205	400	233	123
7	46	241	560	83	80	104	30
8	83	629	783	255	286	87	18
9	112	24	729	225	108	87	26
10	113	5	699	175	389	79	29
11	65	61	591	124	252	113	45
12	99	1	644	167	128	62	19
13	79	276	699	247	263	156	40
14	88	466	836	283	469	130	53
15	60	443	703	156	339	243	65
16	25	186	511	70	189	103	28
17	89	54	678	147	198	166	80
18	94	749	822	237	401	181	94
19	62	133	549	116	317	119	32
20	78	25	612	117	201	104	42
21	97	36	673	154	419	92	29

for the $n = 21$ wards of Hull (England) where

Y_1 = Overcrowding, per 1000 households
Y_2 = No inside toilet, per 1000 households
Y_3 = Do not possess a car, per 1000 households
Y_4 = Number of males per 1000 in unskilled or semiskilled work.
Y_5 = Number with infectious jaundice, per 100,000.
Y_6 = Number with measles, per 100,000.
Y_7 = Number with scabies, per 100,000.

Y_1, Y_2, Y_3, and Y_4 pertain to the 1971 census whereas Y_5, Y_6, and Y_7 are rates as reported between 1968 and 1973.

(a) Using an appropriate dispersion matrix compute latent roots, latent vectors, and correlation loadings for the seven variables.

(b) Ignoring the small sample size, test for equality of the latent roots using the Bartlett–Anderson likelihood ratio criterion (Sections 4.3.1 and 4.3.2).

(c) Ignoring the small sample size, use Girshick's criterion (Example 4.11) to test the significance of correlation loadings for the last three PCs.

(d) Using the F-test (Eq. 4.53), determine whether the covariance matrix possesses isotropic structure.

4.3 Consider the latent roots of Table 4.13 in Example 4.14

(a) Using the chi-squared criterion (Eq. 4.6), test whether the covariance matrix is of diagonal form.

(b) Using Eq. (4.11) test whether the covariance matrix exhibits spherical form.

CHAPTER 5

Extensions of the Ordinary Principal Components Model

5.1 INTRODUCTION

The PC model described in Chapter 3 represents the classic approach to the problem of decomposing a set of correlated random variables into a smaller orthogonal set. As such it is relatively straightforward to interpret in most situations, and is used in a wide range of disciplines. The PC model can nevertheless be generalized and extended in several directions. This gives it greater flexibility and makes it applicable to a wider range of situations.

5.2 PRINCIPAL COMPONENTS OF SINGULAR MATRICES

The ordinary principal components model is defined in terms of the decomposition of nonsingular Grammian matrices into real positive latent roots and real orthogonal latent vectors. At times singular matrices are also of interest, and this requires an extension of PCA to a more general case. This was already encountered in part in Chapter 3 (Theorem 3.17). Singular Grammian matrices can arise from three principal causes. First, for a $(n \times p)$ data matrix $\mathbf{Y}$ where $p < n$, it may be the case that one random variable (or more) is a perfect linear combination of one or more other variables. This may happen, for example, when a new variable is created as a linear function of several other variables and is then included as a column of $\mathbf{Y}$. Or we may be given compositional (percentage) data where each of the n samples is broken down into its constituent parts, which are then expressed as percentages (proportions) of the total sample (Section 5.9.1). When these percentages are included as columns of $\mathbf{Y}$, the matrix becomes singular, since each row of $\mathbf{Y}$ has the constant sum proportional to 100. In this case $\rho(\mathbf{X}^T\mathbf{X}) = r < p < n$ and $\rho(\mathbf{X}\mathbf{X}^T) = r < p < n$ (Theorem 3.11). A singularity of this type, should it occur, results in zero latent roots, but otherwise a

PCA is unaffected, that is, the model of Chapter 3 also applies to singular matrices.

The second situation in which a singular Grammian matrix arises is when we consider $\mathbf{X}^T\mathbf{X}$ such that $p > n$. This can occur, for example, when sample points are difficult (or costly) to obtain, and as many variables as possible are included to extract the maximal amount of information from the sample. The same situation arises when the sample points are of primary interest, and more variables are then included than sample points to reverse (formally) the roles of sample/random variable spaces.

The third situation which results in singularity is when interest again centers on the Grammian matrix $\mathbf{X}\mathbf{X}^T$, that is, on the row structure of $\mathbf{Y}$, but $r \le p < n$. We know from Theorem 3.17 that it is not necessary to decompose $\mathbf{X}\mathbf{X}^T$ when latent roots and latent vectors of $\mathbf{X}^T\mathbf{X}$ are available, given $\mathbf{X} = \mathbf{Y} - \bar{\mathbf{Y}}$. However, when the main interest lies in the sample rather than in the variable space, it no longer makes sense to center the columns at zero. When alternative centering is employed, a PCA can no longer be based on $\mathbf{X}^T\mathbf{X}$ (Section 5.3). Finally, we may require the spectrum of a singular matrix when computing generalized matrix inverses.

5.2.1 Singular Grammian Matrices

When only $r < p$ columns of $\mathbf{X}$ are linearly independent, the expansion of $\mathbf{X}^T\mathbf{X}$ assumes the more general form

$$
\mathbf{X}^T\mathbf{X} = [\mathbf{P}_1, \mathbf{P}_2, \ldots, \mathbf{P}_r, \mathbf{0} \ldots \mathbf{0}]
\begin{bmatrix} l_1 & & & & \vdots & \\ & l_2 & & & \vdots & \mathbf{0} \\ \mathbf{0} & & \ddots & & \vdots & \\ & & & l_r & \vdots & \\ \hline \mathbf{0} & & & & \vdots & \mathbf{0} \end{bmatrix}
\begin{bmatrix} \mathbf{P}_1^T \\ \mathbf{P}_2^T \\ \vdots \\ \mathbf{P}_r^T \\ \mathbf{0} \\ \vdots \\ \mathbf{0} \end{bmatrix}
$$

$$
= l_1\mathbf{P}_1\mathbf{P}_1^T + l_2\mathbf{P}_2\mathbf{P}_2^T + \cdots l_r\mathbf{P}_r\mathbf{P}_r^T
$$

$$
= \mathbf{A}_1\mathbf{A}_1^T + \mathbf{A}_2\mathbf{A}_2^T + \cdots + \mathbf{A}_r\mathbf{A}_r^T
$$

$$
= \mathbf{A}_{(r)}\mathbf{A}_{(r)}^T \tag{5.1}
$$

so that the (i, j)th element of $\mathbf{X}^T\mathbf{X}$ is given by

$$
\mathbf{X}_i^T\mathbf{X}_j = a_{1i}a_{1j} + a_{2i}a_{2j} + \cdots + a_{ri}a_{rj} \tag{5.2}
$$

$\mathbf{A}_{(r)}$ is the $(p \times r)$ matrix of loadings of the singular matrix $\mathbf{X}^T\mathbf{X}$. The last $(p - r)$ columns are identically zero since they correspond to $(p - r)$ linearly dependent columns of $\mathbf{X}$. The r PCs are given by

$$\mathbf{Z}_{(r)} = \mathbf{X}\mathbf{A}_{(r)} \tag{5.3}$$

where $\mathbf{A}_{(r)} = \mathbf{P}_{(r)}\mathbf{L}_{(r)}^{-1/2}$ so that

$$\mathbf{X} = \mathbf{Z}_{(r)}\mathbf{A}_{(r)}^{T} \tag{5.4}$$

Equation (5.4) does not imply that $\mathbf{X}$ has no residual variation due to sampling and/or error measurement, but simply expresses the fact that the last $(p - r)$ PCs are zero owing to $(p - r)$ redundant (linearly dependent) random variables. Theorem 3.17 can thus be generalized to the form

$$\mathbf{X}\mathbf{X}^{T} = [\mathbf{Q}_1, \mathbf{Q}_2, \ldots, \mathbf{Q}_r]\begin{bmatrix} l_1 & & & \\ & l_2 & & \mathbf{0} \\ \mathbf{0} & & \ddots & \\ & & & l_r \end{bmatrix}\begin{bmatrix} \mathbf{Q}_1^{T} \\ \mathbf{Q}_2^{T} \\ \vdots \\ \mathbf{Q}_r^{T} \end{bmatrix}$$

$$= l_1\mathbf{Q}_1\mathbf{Q}_1^{T} + l_2\mathbf{Q}_2\mathbf{Q}_2^{T} + \cdots + l_r\mathbf{Q}_r\mathbf{Q}_r^{T} \tag{5.5}$$

where $\rho(\mathbf{X}\mathbf{X}^{T}) = r < p < n$. It should be pointed out however that latent vectors which correspond to zero latent roots need not be necessarily zero (Section 5.9.1).

5.2.2 Rectangular Matrices and Generalized Inverses

When the latent roots and latent vectors of $\mathbf{X}^{T}\mathbf{X}$ and $\mathbf{X}\mathbf{X}^{T}$ are known, the rectangular matrix $\mathbf{X}$ can be decomposed in terms of these roots and vectors (Theorem 3.17). A similar expansion exists when $\mathbf{X}$ is not of full rank, the so-called singular value decomposition of a matrix.

THEOREM 5.1   (The Singular Value Decomposition Theorem). Let $\mathbf{X}$ be a $(n \times p)$ data matrix of reduced rank $r < p < n$. Then there exists a real $(n \times r)$ matrix $\mathbf{Q}$, a real $(p \times r)$ matrix $\mathbf{P}$, and a real $(r \times r)$ diagonal matrix $\mathbf{\Delta}_{(r)}$ such that

    (i) $\mathbf{Q}^{T}\mathbf{X}\mathbf{P} = \mathbf{\Delta}_{(r)}$ where $\mathbf{\Delta}_{(r)}^2 = L_{(r)}$ is the diagonal matrix of r nonzero latent roots of $\mathbf{X}^{T}\mathbf{X}$.

    (ii) $\mathbf{X}^{T}\mathbf{X} = \mathbf{P}\mathbf{\Delta}_{(r)}^2\mathbf{P}^{T}$.

    (iii) $\mathbf{X}\mathbf{X}^{T} = \mathbf{Q}\mathbf{\Delta}_{(r)}^2\mathbf{Q}^{T}$.

    (iv) $\mathbf{P}^{T}\mathbf{P} = \mathbf{Q}^{T}\mathbf{Q} = \mathbf{I}_{(r)}$ and $\mathbf{Q} = \mathbf{Z}$, the PCs of $\mathbf{X}^{T}\mathbf{X}$.

    (v) $\mathbf{X}^{+} = \mathbf{P}\mathbf{\Delta}_{(r)}^{+}\mathbf{Q}^{T}$ where $\mathbf{X}^{+}$ and $\mathbf{\Delta}^{+}$ are Moore–Penrose generalized inverses.

    (vi) $\mathbf{P}\mathbf{P}^{T} = \mathbf{X}^{+}\mathbf{X}$ and $\mathbf{Q}\mathbf{Q}^{T} = \mathbf{X}\mathbf{X}^{+}$ are projection matrices that project vectors orthogonally onto the r-dimensional column space of $\mathbf{X}$.

(vii) $(\mathbf{X}^T\mathbf{X}) = \mathbf{P}\mathbf{L}_{(r)}^+\mathbf{P}^T$ and $(\mathbf{X}\mathbf{X}^T) = \mathbf{Q}\mathbf{L}_{(r)}^+\mathbf{Q}^T$ where $(\mathbf{X}^T\mathbf{X})^+$ and $(\mathbf{X}\mathbf{X}^T)^+$ are Moore–Penrose generalized inverse of the corresponding singular Grammian matrices.

PROOF

(i) We have $\rho(\mathbf{X}) = \rho(\mathbf{X}^T\mathbf{X}) = \rho(\mathbf{X}\mathbf{X}^T)$ so that

$$\mathbf{X} = l_1^{1/2}\mathbf{Q}_1\mathbf{P}_1^T + l_2^{1/2}\mathbf{Q}_2\mathbf{P}_2^T + \cdots + l_r^{1/2}\mathbf{Q}_r\mathbf{P}_r^T \qquad (5.6)$$

where $\mathbf{Q}_i$ and $\mathbf{P}_i$ $(i = 1, 2, \ldots, r)$ are latent vectors of $\mathbf{X}\mathbf{X}^T$ and $\mathbf{X}^T\mathbf{X}$ respectively and

$$\mathbf{\Delta} = \mathbf{L}^{1/2} = \begin{bmatrix} l_1^{1/2} & & & & & \vdots & \\ & l_2^{1/2} & & & \mathbf{0} & \vdots & \\ & & \ddots & & & \vdots & \mathbf{0} \\ & \mathbf{0} & & & & \vdots & \\ & & & l_r^{1/2} & & \vdots & \\ \hdashline & & \mathbf{0} & & & \vdots & \mathbf{0} \end{bmatrix} = \begin{bmatrix} \mathbf{L}_{(r)}^{1/2} & \vdots & \mathbf{0} \\ \hdashline \mathbf{0} & \vdots & \mathbf{0} \end{bmatrix}$$

$$(5.7)$$

It follows

$$\mathbf{Q}^T\mathbf{X}\mathbf{P} = \mathbf{\Delta}_{(r)} \qquad (5.8)$$

where $\mathbf{\Delta}_{(r)} = \mathbf{L}_{(r)}^{1/2}$.

(ii)–(iv) Using Eq. (5.8) we have

$$\mathbf{X}^T\mathbf{X} = \mathbf{P}\mathbf{\Delta}_{(r)}\mathbf{Q}^T\mathbf{Q}\mathbf{\Delta}_{(r)}\mathbf{P}^T$$

$$= \mathbf{P}\mathbf{\Delta}_{(r)}^2\mathbf{P}^T$$

$$= \mathbf{P}\mathbf{L}_{(r)}\mathbf{P}^T \qquad (5.9)$$

$$\mathbf{X}\mathbf{X}^T = \mathbf{Q}\mathbf{\Delta}_{(r)}\mathbf{P}^T\mathbf{P}\mathbf{\Delta}_{(r)}\mathbf{Q}^T$$

$$= \mathbf{Q}\mathbf{\Delta}_{(r)}^2\mathbf{Q}^T$$

$$= \mathbf{Q}\mathbf{L}_{(r)}\mathbf{Q}^T \qquad (5.10)$$

where columns of $\mathbf{P}$ and $\mathbf{Q}$ are orthogonal latent vectors so that

$$\mathbf{P}^T\mathbf{P} = \mathbf{Q}^T\mathbf{Q} = \mathbf{I}_{(r)} \qquad (5.11)$$

(v and vi) The Moore–Penrose generalized inverse of the singular (diagonal) matrix $\mathbf{\Delta}$ is

$$\mathbf{\Delta}^{+} = \begin{bmatrix} \mathbf{\Delta}_{(r)}^{-1} & \vdots & \mathbf{0} \\ \hdashline \mathbf{0} & \vdots & \mathbf{0} \end{bmatrix} \tag{5.12}$$

where diagonal elements δ_i^{+} of $\mathbf{\Delta}_{(r)}^{+}$ are defined as

$$\delta_i^{+} = \begin{cases} \dfrac{1}{\delta_i} & \delta_i \neq 0 \\ 0 & \delta_i = 0 \end{cases} \qquad (i = 1, 2, \ldots, r)$$

The Moore–Penrose generalized inverse of $\mathbf{X}$ can then be defined as

$$\mathbf{X}^{+} = \mathbf{P}\mathbf{\Delta}_{(r)}^{+}\mathbf{Q}^{\mathrm{T}}$$

$$= \frac{1}{\delta_1}\mathbf{P}_1\mathbf{Q}_1^{\mathrm{T}} + \frac{1}{\delta_2}\mathbf{P}_2\mathbf{Q}_2^{\mathrm{T}} + \cdots + \frac{1}{\delta_r}\mathbf{P}_r\mathbf{Q}_r^{\mathrm{T}} \tag{5.13}$$

since

$$(\mathbf{P}\mathbf{P}^{\mathrm{T}})(\mathbf{P}\mathbf{P}^{\mathrm{T}}) = \mathbf{P}(\mathbf{P}^{\mathrm{T}}\mathbf{P})\mathbf{P}^{\mathrm{T}}$$

$$= \mathbf{P}\mathbf{P}^{\mathrm{T}}$$

$$(\mathbf{Q}\mathbf{Q}^{\mathrm{T}})(\mathbf{Q}\mathbf{Q}^{\mathrm{T}}) = \mathbf{Q}(\mathbf{Q}^{\mathrm{T}}\mathbf{Q})\mathbf{Q}^{\mathrm{T}}$$

$$= \mathbf{Q}\mathbf{Q}^{\mathrm{T}}$$

Since matrices $\mathbf{P}\mathbf{P}^{\mathrm{T}}$ and $\mathbf{Q}\mathbf{Q}^{\mathrm{T}}$ are symmetric and idempotent they are orthogonal projection matrices. Also using Eq. (5.13) we have

$$\mathbf{X}\mathbf{X}^{+} = (\mathbf{Q}\mathbf{\Delta}_{(r)}\mathbf{P}^{\mathrm{T}})(\mathbf{P}\mathbf{\Delta}_{(r)}^{+}\mathbf{Q}^{\mathrm{T}}) = \mathbf{Q}\mathbf{Q}^{\mathrm{T}} \tag{5.14}$$

$$\mathbf{X}^{+}\mathbf{X} = (\mathbf{P}\mathbf{\Delta}_{(r)}\mathbf{Q}^{\mathrm{T}})(\mathbf{Q}\mathbf{\Delta}_{(r)}\mathbf{P}^{\mathrm{T}}) = \mathbf{P}\mathbf{P}^{\mathrm{T}} \tag{5.15}$$

(vii) The Moore–Penrose inverse of $\mathbf{L}$ is

$$\mathbf{L}^{+} = \begin{bmatrix} 1/l_1 & & & & \vdots & \\ & 1/l_2 & & \mathbf{0} & \vdots & \mathbf{0} \\ & & \ddots & & \vdots & \\ & \mathbf{0} & & 1/l_r & \vdots & \\ \hdashline & & \mathbf{0} & & \vdots & \mathbf{0} \end{bmatrix} = \begin{bmatrix} \mathbf{L}_{(r)}^{-1} & \vdots & \mathbf{0} \\ \hdashline \mathbf{0} & \vdots & \mathbf{0} \end{bmatrix}$$

where elements of $\mathbf{L}^{+}$ are defined similarly to those of $\mathbf{\Delta}^{+}$.

It is then straightforward to show that $(\mathbf{X}^{\mathrm{T}}\mathbf{X})^{+}$ and $(\mathbf{X}\mathbf{X}^{\mathrm{T}})^{+}$ obey the four Penrose conditions since

$$(\mathbf{X}^{\mathrm{T}}\mathbf{X})^{+} = \mathbf{P}\mathbf{L}^{+}\mathbf{P}^{\mathrm{T}}$$

$$= \mathbf{P}\mathbf{L}_{(r)}^{-1}\mathbf{P}^{\mathrm{T}} \tag{5.16a}$$

$$(\mathbf{X}\mathbf{X}^{\mathrm{T}})^{+} = \mathbf{Q}\mathbf{L}^{+}\mathbf{Q}^{\mathrm{T}}$$

$$= \mathbf{Q}\mathbf{L}_{(r)}^{-1}\mathbf{Q} \tag{5.16b}$$

Example 5.1.   Klovan (1975) published a (20×10) data matrix constructed such that $\rho(\mathbf{Y}) = 3$. The matrix consists of 10 geological variables observed for 20 locations (Table 5.1). As a result only three columns of the matrix are linearly independent. The variables are defined as follows:

Y_1 $ $ = Magnesium in calcite
Y_2 $ $ = Iron in sphalerite
Y_3 $ $ = Sodium in muscovite
Y_4 $ $ = Sulphide
Y_5 $ $ = Crystal size of carbonates
Y_6 $ $ = Spacing of cleavage
Y_7 $ $ = Elongation of ooliths
Y_8 $ $ = Tightness of folds
Y_9 $ $ = Veins/meter^2
Y_{10} = Fractures/meter^2 $\qquad\qquad\qquad\qquad\qquad\qquad\qquad$ $\square$

Both matrices $\mathbf{X}^{\mathrm{T}}\mathbf{X}$ and $\mathbf{X}\mathbf{X}^{\mathrm{T}}$ possess three nonzero latent roots and latent vectors (Table 5.3). The columns of $\mathbf{X}$ are standardized to unit length so that $\mathbf{X}^{\mathrm{T}}\mathbf{X}$ represents a correlation matrix of the 10 geological variables (Table 5.2). Since $\mathrm{tr}(\mathbf{X}^{\mathrm{T}}\mathbf{X}) = \mathrm{tr}(\mathbf{L})$ and $|\mathbf{X}^{\mathrm{T}}\mathbf{X}| = |\mathbf{L}|$, the first three latent roots contain the entire variance, since they all lie in a three-dimensional subspace. The latent roots of $\mathbf{X}$ are then

$$\Delta = \begin{bmatrix} (5.46321)^{1/2} & & & \vdots & \\ & (3.191644)^{1/2} & & \vdots & \mathbf{0} \\ \mathbf{0} & & (1.345070)^{1/2} & \vdots & \\ \hdashline & & \mathbf{0} & \vdots & \mathbf{0} \end{bmatrix}$$

$$= \begin{bmatrix} 2.337 & & \mathbf{0} & \vdots & \\ & 1.787 & & \vdots & \mathbf{0} \\ \mathbf{0} & & 1.160 & \vdots & \\ \hdashline & \mathbf{0} & & \vdots & \mathbf{0} \end{bmatrix}$$

Table 5.1   Geological Data Matrix with Three Linearly Independent Columns

Locality	Y_1	Y_2	Y_3	Y_4	Y_5	Y_6	Y_7	Y_8	Y_9	Y_{10}
1	1175	999	975	625	158	262	437	324	431	433
2	936	820	813	575	267	379	478	413	411	428
3	765	711	716	599	457	548	579	558	491	513
4	624	598	600	542	471	515	531	520	490	500
5	417	422	422	432	444	441	437	439	437	437
6	401	403	375	401	405	270	317	290	515	465
7	520	504	488	469	427	370	410	386	507	482
8	661	626	618	553	462	466	506	480	529	523
9	877	787	773	594	354	401	493	434	500	498
10	1060	932	898	656	315	312	468	370	580	552
11	1090	960	935	681	334	375	518	427	567	555
12	896	811	790	629	403	411	511	448	570	555
13	748	688	672	560	401	399	472	426	525	512
14	617	573	553	477	360	315	385	342	487	462
15	436	424	389	393	361	207	277	236	514	455
16	664	587	560	419	212	182	287	221	397	369
17	750	665	651	484	259	299	387	331	399	396
18	903	787	791	573	291	396	486	427	421	437
19	998	888	887	657	366	499	583	527	480	506
20	1162	999	994	671	252	404	539	450	449	471

Source: Klovan, 1975; reproduced with permission.

Table 5.2   Correlation Matrix of 10 Geological Variables of Table 5.1

1.0000									
.9979	1.0000								
.9944	.9981	1.0000							
.9077	.9327	.9416	1.0000						
−.5760	−.5224	−.4976	−.1798	1.0000					
.1303	.1826	.2342	.4786	.6153	1.0000				
.5809	.6251	.6636	.8342	.2574	.8803	1.0000			
.2826	.3341	.3823	.6102	.5186	.9873	.9442	1.0000		
.0122	.0573	.0352	.2860	.5402	.1811	.2161	.2091	1.0000	
.2601	.3157	.3141	.5914	.5495	.5237	.6044	.5742	.9085	1.0000

Source: Klovan, 1975; reproduced with permission.

Example 5.2.   To find the (unique) Moore–Penrose inverse of a singular Grammian matrix consider

$$
\mathbf{X}\mathbf{X}^{\mathrm{T}} = \begin{bmatrix}
.6068 & -.3819 & .3649 & -.5898 \\
-.3819 & 1.0985 & -.6532 & -.0633 \\
.3649 & -.6532 & .4650 & -.1766 \\
-.5898 & -.0633 & -.1767 & .8297
\end{bmatrix}
$$

Table 5.3 Latent Roots and Vectors of the Correlation Matrix $R = X^T X$

		Latent Vectors		
		Z_1	Z_2	Z_3
X_1:	Mg	.34360	−.32982	.07613
X_2:	Fe	.35887	−.30012	.08135
X_3:	Na	.36705	−.28667	.03533
X_4:	Sulphide	.41762	−.10964	.08093
X_5:	Crystal	.00738	.55964	−.00733
X_6:	Spacing	.27950	.33595	−.39794
X_7:	Elongation	.39770	.13380	−.24201
X_8:	Tightness	.32715	.28106	−.34827
X_9:	Veins	.13996	.30297	.66790
X_{10}:	Fractures	.28471	.30364	.44208
Latent roots		5.46321	3.191644	1.345070

Source: Klovan, 1975; reproduced with permission.

The matrix has rank 3 and its generalized inverse is

$$(\mathbf{XX}^T)^+ = \mathbf{QL}_{(r)}^+ \mathbf{Q}^T$$

$$= \begin{bmatrix} .4970 & .3500 & .6162 \\ -.6438 & .5451 & -.1965 \\ .4782 & -.1478 & -.7073 \\ -.3314 & -.7473 & .2852 \end{bmatrix} \begin{bmatrix} 1/1.84595 & 0 & 0 \\ 0 & 1/1.11718 & 0 \\ 0 & 0 & 1/.03682 \end{bmatrix}$$

$$\times \begin{bmatrix} .4970 & -.6438 & .4782 & -.3314 \\ .3500 & .5451 & -.1478 & -.7473 \\ .6162 & -.1965 & -.7073 & .2852 \end{bmatrix}$$

$$= \begin{bmatrix} .7492 & -.2505 & -.2505 & -.2505 \\ -.2505 & .7502 & -.2494 & -.2500 \\ -.2505 & -.2494 & .7508 & -.2497 \\ -.2505 & -.2500 & -.2497 & .7496 \end{bmatrix} \qquad \Box$$

5.3 PRINCIPAL COMPONENTS AS CLUSTERS: LINEAR TRANSFORMATIONS IN EXPLORATORY RESEARCH

The examples used in Chapter 3 conform largely to a structured "a priori" approach to the PC model—for example, the measurement of size and shape of biological organisms or mineral/ore types based on core sample composition. A less structured exploratory approach to PCA is also possible, particularly when we wish to identify patterns and/or causes of intercorrelations amongst the variables. Here PCA can provide a natural extension of binary measures of association, in terms of nonhierarchical clusters that contain highly intercorrelated variables. The use of PCs as clusters is made possible by an indeterminacy in an otherwise unique solution, since the loadings (and scores) can be altered arbitrarily by a linear transformation of the coordinates. The indeterminacy is inherent in the nature of coordinate axes, which from a general mathematical viewpoint are arbitrary and whose choice is usually guided by simplicity of form, ease of manipulation of algebraic expressions, and other practical considerations. Thus Cartesian coordinates can be rotated to any position, and may intersect at any angle without affecting the configuration of a given set of points. In statistical data analysis the location of a set of coordinate axes is also arbitrary, in the sense that variance, angles, and distance are invariant with respect to their position. The $r < p$ dimensional subspace of the ordinary principal components model is, however, chosen so as to satisfy two general constraints—orthogonality of axes and the stepwise optimization of variance.

When searching for clusters, however, different constraints are usually required. Once the optimal "correct" r-dimensional subspace is found we may wish to introduce an additional condition, namely, that each PC be maximally correlated with a single subset or cluster of the random variables. This permits a straightforward identification of the PCs in terms of the clusters if such clusters exist. This generally implies the elimination, from a PC solution, of the initial conditions of variance maximization and component orthogonality since a second linear transformation is applied to the principal components, that is, to the orthogonally rotated original variables. Such secondary transformations or rotations can be orthogonal or more generally oblique.

5.3.1 Orthogonal Rotations

If the primary function of a secondary linear transformation is to locate clusters in a linear subspace by rotating loading coefficients to "simple structure" the first step in the analysis is to decide upon the number of components to retain (see Chapter 4). This is because rotated principal components are not invariant with respect to a change in their number; that is, adding or removing a principal component will change the outcome of the rotation by altering the magnitudes and/or signs of the loadings. The second point to keep in mind is that a rotation of the latent vectors $\mathbf{P}$ (and corresponding unstandardized principal components) will not yield the same results as rotating the correlation loadings $\mathbf{A}^T$ (and standardized components). Although it is more usual to rotate correlation loadings, it is instructive to first begin with the orthonormal latent vectors. Consider the sample principal components model (Section 3.4) where $1 < r < k$ components are retained as being significant, and $\mathbf{X} = \mathbf{Z}_{(r)}\mathbf{P}_{(r)}^T + \boldsymbol{\delta}$, where $\mathbf{P}_{(r)}$ is the $(p \times r)$ matrix of orthonormal latent vectors, $l_1 > l_2 > \cdots > l_r$, and $\boldsymbol{\delta}$ represents the last $(p - r)$ principal component. Rotating the r orthogonal principal components to a new orthogonal position is then equivalent to multiplying the latent vectors and the scores by an $(r \times r)$ orthogonal matrix

$\mathbf{T}$ and its inverse $\mathbf{T}^{-1}$, that is, the model can always be expressed as

$$\mathbf{X} = \mathbf{Z}_{(r)}\mathbf{T}\mathbf{T}^{-1}\mathbf{P}_{(r)}^{\mathrm{T}} + \boldsymbol{\delta}$$

$$= \mathbf{V}_{(r)}\mathbf{Q}_{(r)}^{\mathrm{T}} + \boldsymbol{\delta} \tag{5.17}$$

where $\mathbf{V}_{(r)} = \mathbf{Z}_{(r)}\mathbf{T}$ and $\mathbf{Q}_{(r)} = \mathbf{P}_{(r)}\mathbf{T}$ are the new rotated scores and latent vectors respectively, and matrix $\mathbf{Q}_{(r)}$ is not to be confused with that of the previous section. Then $\mathbf{T}$ is a matrix of direction cosines of the new axes with respect to the old set, that is,

$$\mathbf{P}_{(r)}^{\mathrm{T}}\mathbf{Q}_{(r)} = \mathbf{P}_{(r)}^{\mathrm{T}}\mathbf{P}_{(r)}\mathbf{T}$$

$$= \mathbf{T} \tag{5.18}$$

and similarly $\mathbf{Z}_{(r)}^{\mathrm{T}}\mathbf{V}_{(r)} = \mathbf{L}_{(r)}\mathbf{T}$. Clearly $\mathbf{T}$ is not unique since an infinite number of orthogonal rotations is possible unless an additional criterion is introduced to fix the location of the axes. We first consider two theorems which outline the general properties of orthogonal rotations.

THEOREM 5.2   Let $\mathbf{X} = \mathbf{Z}_{(r)}\mathbf{P}_{(r)}^{\mathrm{T}} + \boldsymbol{\delta}$ be a PC model and let $\mathbf{X} = \mathbf{V}_{(\mathrm{T})}\mathbf{Q}_{(r)}^{\mathrm{T}} + \boldsymbol{\delta}$ as in Eq. (5.17). Then

(i) Predicted values $\hat{\mathbf{X}}$ and their sum of squares remain unchanged by $\mathbf{T}$.

(ii) Rotated latent vectors remain orthogonal, that is, $\mathbf{Q}_{(r)}^{\mathrm{T}}\mathbf{Q}_{(r)} = \mathbf{I}_{(r)}$.

(iii) Rotated principal components are no longer orthogonal, that is, $\mathbf{V}_{(r)}^{\mathrm{T}}\mathbf{V}_{(r)}$ is not a diagonal matrix.

PROOF.   Clearly

(i) The transformation $\mathbf{T}$ cannot alter the predicted values since $\hat{\mathbf{X}} = \mathbf{V}_{(r)}\mathbf{Q}_{(r)}^{\mathrm{T}} = \mathbf{Z}_{(r)}\mathbf{P}_{(r)}^{\mathrm{T}}$. It follows that the predicted sums of squares also remain unchanged, which can be seen from

$$\hat{\mathbf{X}}^{\mathrm{T}}\hat{\mathbf{X}} = (\mathbf{V}_{(r)}\mathbf{Q}_{(r)}^{\mathrm{T}})^{\mathrm{T}}(\mathbf{V}_{(r)}\mathbf{Q}_{(r)}^{\mathrm{T}})$$

$$= \mathbf{P}_{(r)}\mathbf{T}\mathbf{T}^{-1}\mathbf{L}_{(r)}\mathbf{T}\mathbf{T}^{-1}\mathbf{P}_{(r)}^{\mathrm{T}}$$

$$= \mathbf{P}_{(r)}\mathbf{L}_{(r)}\mathbf{P}_{(r)}^{\mathrm{T}}$$

$$= \mathbf{Z}_{(r)}^{\mathrm{T}}\mathbf{Z}_{(r)}$$

the original predicted sum of squares.

(ii) The rotated latent vectors remain orthogonal since

$$\mathbf{Q}_{(r)}^{\mathrm{T}}\mathbf{Q}_{(r)} = (\mathbf{P}_{(r)}\mathbf{T})^{\mathrm{T}}(\mathbf{P}_{(r)}\mathbf{T})$$

$$= \mathbf{T}^{-1}\mathbf{P}_{(r)}^{\mathrm{T}}\mathbf{P}_{(r)}\mathbf{T}$$

$$= \mathbf{I}$$

since $\mathbf{P}_{(r)}$ and $\mathbf{T}$ are orthogonal.

(iii) Rotated PC scores are no longer uncorrelated since

$$\mathbf{V}_{(r)}^{\mathrm{T}}\mathbf{V}_{(r)} = (\mathbf{Z}_{(r)}\mathbf{T})^{\mathrm{T}}(\mathbf{Z}_{(r)}\mathbf{T})$$

$$= \mathbf{T}^{-1}\mathbf{Z}_{(r)}^{\mathrm{T}}\mathbf{Z}_{(r)}\mathbf{T}$$

$$= \mathbf{T}^{-1}\mathbf{L}_{(r)}\mathbf{T}$$

which is not a diagonal matrix when the PC scores are not standardized to unit length.

The third part of Theorem 5.2 is perhaps somewhat unexpected since the orthogonal rotation is applied to orthogonal principal components. The correlation occurs because the original principal components are not standardized to unit length. Also, the new components no longer successively account for maximum variance (Exercise 5.2), although from the first part of the theorem it is clear that total explained variance remains unchanged. Jackson (1991) provides a numerical example of correlated, orthogonally rotated principal components. The lack of orthogonality and variance maximization leads some authors to caution against a routine use of orthogonally rotated principal components (e.g., see Rencher, 1992), but the difficulties are more formal than real. First, component correlation can be removed by employing standardized component scores. Second, variance maximization is a mathematical constraint imposed on the model to obtain a convenient solution* and is not necessarily "data driven." As a result, it may not reveal interesting configurations of the variables in certain subspaces.

The usual practice, when carrying out a principal component analysis, is to use the correlation loading coefficients $\mathbf{A}^{\mathrm{T}}$ and standardized component scores. This is because interpretation is made easier and the rotated component scores maintain orthogonality, although this is not necessarily so for the loading coefficients. In terms of correlation loadings and stan-

---

* Variance maximization is an outcome of the arbitrary ranking of the latent roots, which may or may not be of substantive interest.

dardized principal components the model becomes

$$\mathbf{X} = \mathbf{Z}_{(r)}\mathbf{P}_{(r)}^{\mathrm{T}} + \boldsymbol{\delta}$$

$$= \mathbf{Z}_{(r)}\mathbf{L}_{(r)}^{-1/2}\mathbf{L}_{(r)}^{-1/2}\mathbf{P}_{(r)}^{\mathrm{T}} + \boldsymbol{\delta}$$

$$= \mathbf{Z}_{(r)}^{*}\mathbf{A}_{(r)}^{\mathrm{T}} + \boldsymbol{\delta} \qquad\qquad (5.18a)$$

and rotating the axes orthogonally through some angle yields

$$\mathbf{X} = \mathbf{Z}_{(r)}^{*}\mathbf{T}\mathbf{T}^{-1}\mathbf{A}_{(r)}^{\mathrm{T}} + \boldsymbol{\delta}$$

$$= \mathbf{Z}_{(r)}^{*}\mathbf{T}(\mathbf{A}_{(r)}\mathbf{T})^{\mathrm{T}} + \boldsymbol{\delta}$$

$$= \mathbf{G}_{(r)}\mathbf{B}_{(r)}^{\mathrm{T}} + \boldsymbol{\delta} \qquad\qquad (5.18b)$$

where $\mathbf{G}_{(r)} = \mathbf{Z}_{(r)}^{*}\mathbf{T}$ and $\mathbf{B}_{(r)} = (\mathbf{A}_{(r)}\mathbf{T})$ are the new scores and loadings, respectively. The transformed principal components model has the following properties.

THEOREM 5.3.   Let $\mathbf{X} = \mathbf{Z}_{(r)}^{*}\mathbf{A}_{(r)}^{\mathrm{T}} + \boldsymbol{\delta}$ be a principal components model and let $\mathbf{X} = \mathbf{G}_{(r)}\mathbf{B}_{(r)}^{\mathrm{T}} + \boldsymbol{\delta}$ as in Eq. (5.18b). Then
   (i) Predicted values $\hat{\mathbf{X}}$ and their sum of squares remain unchanged by $\mathbf{T}$.
   (ii) Rotated correlation loadings $\mathbf{A}_{(r)}^{\mathrm{T}}$ are no longer orthogonal, that is, $\mathbf{B}_{(r)}^{\mathrm{T}}\mathbf{B}_{(r)}$ is not a diagonal matrix.
   (iii) Rotated principal components remain orthonormal, that is, $\mathbf{G}_{(r)}^{\mathrm{T}}\mathbf{G}_{(r)} = \mathbf{I}_{(r)}$.

PROOF
   (i) The proof is left as an exercise (Exercise 5.5).
   (ii) We have

$$\mathbf{B}_{(r)}^{\mathrm{T}}\mathbf{B}_{(r)} = (\mathbf{A}_{(r)}\mathbf{T})^{\mathrm{T}}(\mathbf{A}_{(r)}\mathbf{T})$$

$$= \mathbf{T}^{-1}\mathbf{A}_{(r)}^{\mathrm{T}}\mathbf{A}_{(r)}\mathbf{T}$$

$$= \mathbf{T}^{-1}\mathbf{L}_{(r)}^{1/2}\mathbf{P}_{(r)}^{\mathrm{T}}\mathbf{P}_{(r)}\mathbf{L}^{1/2}\mathbf{T}$$

$$= \mathbf{T}^{-1}\mathbf{L}_{(r)}\mathbf{T}$$

which is not diagonal since $\mathbf{L}_{(r)}$ is not generally proportional to the unit matrix.

(iii) Rotated component scores are still orthonormal, that is,

$$\mathbf{G}_{(r)}^{T}\mathbf{G}_{(r)} = (\mathbf{Z}_{(r)}^{*}\mathbf{T})^{T}(\mathbf{Z}_{(r)}^{*}\mathbf{T})$$

$$= \mathbf{T}^{-1}\mathbf{Z}_{(r)}^{*T}\mathbf{Z}_{(r)}^{*}\mathbf{T}$$

$$= \mathbf{T}^{-1}\mathbf{T}$$

$$= \mathbf{I}$$

since the columns of $\mathbf{Z}_{(r)}^{*}$ are orthonormal.

The main properties of orthogonal rotations may be summarized as follows (see also Example 5.3):

1. An orthogonal transformation $\mathbf{T}$ of unstandardized principal components and standardized latent vectors only preserves orthogonality of the latent vectors but not of the principal components. The covariance matrix of the rotated component scores is given by $\mathbf{V}_{(r)}^{T}\mathbf{V}_{(r)} = \mathbf{T}^{-1}\mathbf{L}_{(r)}\mathbf{T} = \mathbf{T}^{-1}\mathbf{P}_{(r)}^{T}\mathbf{X}^{T}\mathbf{X}\mathbf{P}_{(r)}\mathbf{T}$ which is nondiagonal since $\mathbf{T}$ does not contain latent vectors of the matrix $\mathbf{P}_{(r)}^{T}\mathbf{X}^{T}\mathbf{X}\mathbf{P}_{(r)}$. An orthogonal transformation $\mathbf{T}$ of standardized principal components and loading coefficients only preserves orthogonality of the principal components but not of the loadings. The covariance matrix of the rotated loadings is given by $\mathbf{B}_{(r)}^{T}\mathbf{B}_{(r)} = \mathbf{T}^{-1}\mathbf{L}_{(r)}\mathbf{T} = \mathbf{T}^{-1}\mathbf{P}_{(r)}^{T}\mathbf{X}^{T}\mathbf{X}\mathbf{P}_{(r)}\mathbf{T}$ which is again nondiagonal.

2. Since matrix $\hat{\mathbf{X}}^{T}\hat{\mathbf{X}}$ is preserved by the orthogonal transformation $\mathbf{T}$, the variance/covariance structure of the original variables is left unaltered within the r-dimensional subspace. Also distance is preserved between the variables in the subspace.

3. Rotated principal components no longer preserve the constraint that they successively maximize variance of the observed variables, that is, we no longer have the ordering $l_1 > l_2 > \cdots > l_r$. Thus the largest variance may be associated with any one of the rotated principal components since the total variance explained by the r components is repartitioned between them. This point should be kept in mind when selecting the appropriate explanatory subspace. If the objective of an orthogonal rotation is to locate orthogonal variable clusters (if they exist), the choice of r itself may be aided by such rotations—successively increasing values of r are chosen, each being accompanied by a rotation to determine whether a "meaningful" subspace has been located.

4. Since the rotated (standardized) principal components are orthogonal, the new loading coefficients $\mathbf{B}_{(r)}$ still represent correlations between the variables and the principal components. Also, the new component scores are still orthogonal to the residual error term δ (Exercise 5.3).

In practice components may either be rotated simultaneously or two at a time (see Horst, 1965). For a clockwise rotation of two axes through an angle θ, the general transformation matrix is particularly simple to express since

$$\mathbf{T} = \begin{bmatrix} \cos\theta & -\sin\theta \\ \sin\theta & \cos\theta \end{bmatrix}$$

For r components the elements of $\mathbf{T}$ are embedded in an $(r \times r)$ orthogonal matrix with diagonal elements equal to unity and the off-diagonal elements equal to zero. For example, let $r = 4$ and let $\mathbf{T}_{lk}$ be a (4×4) orthogonal matrix which rotates the lth and kth axes in a four-dimensional space. Then the matrix

$$\mathbf{T}_{14} = \begin{bmatrix} \cos\theta & 0 & 0 & -\sin\theta \\ 0 & 1 & 0 & 0 \\ 0 & 0 & 1 & 0 \\ \sin\theta & 0 & 0 & \cos\theta \end{bmatrix}$$

rotates the first and fourth axes, and so forth. The signs of $\sin\theta$ are determined by the direction of the rotation, and θ is chosen to maximize a predefined criterion. Thus given r axes we have a total of

$$c\binom{r}{2} = \frac{r(r-1)}{2}$$

pairwise rotations for any given direction. The orthogonal transformation which rotates all r axes can then be obtained as

$$\mathbf{T} = \mathbf{T}_{12}, \mathbf{T}_{13}, \ldots, \mathbf{T}_{r-1,r}$$

To achieve optimal interpretability we require an additional criterion, which could indicate when an optimal position of the component axes has been reached. Several criteria are employed depending on the broad requirements of the problem at hand (see Harman, 1967, Horst, 1965)

The Varimax Criterion

Originally from Kaiser (1958), several versions of the varimax procedure are available (see Horst, 1965, Chapter 18; Lawley and Maxwell, 1971) depending on whether components are rotated simultaneously or in pairs. By far it is the most popular criterion for rotating PC axes. Let $\mathbf{B}^{\mathrm{T}}$ denote the $(r \times p)$ new (rotated) loading matrix with typical element b_{ij}, and consider the expression

$$V_j^* = \frac{1}{p} \sum_{i=1}^{p} (b_{ij}^2)^2 - \frac{1}{p^2} \left(\sum_{i=1}^{p} b_{ij}^2 \right)^2 \tag{5.19}$$

for $j = 1, 2, \ldots, r$. Equation (5.19) represents the variance of the (squared) loadings for the jth PC. Squared loadings are used to avoid negative signs, but they also represent contribution to variance of the variables. Also, b_{ij}^2 is the contribution to the total variance explained by the jth component for the ith variable. The purpose is to maximize the sum

$$V^* = \sum_{j=1}^{r} V_j^* \qquad (5.20)$$

which results in a pattern of elements of $\mathbf{B}^{\mathrm{T}}$ where some are made as small as possible, and others are made as large as possible (in absolute value). Actually the varimax criterion tries to obtain PCs with a high correlation for some variables or no correlation at all with others. For this reason it minimizes the number of PCs and is well suited for locating clusters that lie at right angles to each other.

Since rotations are carried out in an r-dimensional subspace the values of the rotated loadings will depend on the number of components which are retained. Since in practice r is not always known, rotations are often repeated, varying r until a satisfactory result is obtained. Indeed rotations can be used as aids in selecting the value of r, but should not replace the statistical testing criteria described in Chapter 4. Since Eq. (5.20) depends on the total percentage of variance accounted for by the r components, less reliable variables are given less weight than those whose variance is well explained by the r components, a desirable feature in statistical estimation. An adjusted criterion however is also at times used, the so-called normal varimax criterion given by

$$V_j = \frac{1}{p} \sum_{i=1}^{p} \left(\frac{b_{ij}^2}{h_i^2} \right)^2 - \frac{1}{p^2} \left(\sum_{i=1}^{p} \frac{b_{ij}}{h_i^2} \right)^2 \qquad (5.21)$$

and we maximize $V = \sum_{j=1}^{r} V_j$ where h_i^2 is the proportion of variance of the ith variable explained by the first r PCs. Since V is adjusted for the differential effects of the total variance h_i^2 it does not depend on the total percentage of variance explained by the components, and both reliable and unreliable variables are given equal weight in the criterion.

Example 5.3. Consider the geological data of Example 5.1. Since $\rho(\mathbf{X}) = 3$ there is no ambiguity concerning the selection of the number of components to be rotated. Tables 5.4–5.6 contain the correlation loadings, latent roots, and scores of the initial and rotated solution, respectively. Maximizing the varimax criterion (Eq. 5.20) leads to the orthogonal transformation matrix

$$\mathbf{T} = \begin{bmatrix} .75862 & .57799 & .30070 \\ -.63877 & .56888 & .51803 \\ .12835 & -.58507 & .80076 \end{bmatrix}$$

and the new loadings $\mathbf{B}_{(3)}^{\mathrm{T}} = (\mathbf{AT})^{\mathrm{T}}$ (Table 5.5) are portrayed in Figure (5.1). We can see that there is a general tendency for loadings to both increase and

Table 5.4 Unrotated Loadings of the Correlation Matrix of 10 Geological Variables for Three Nonzero Components

		$\mathbf{Z}_1$	$\mathbf{Z}_2$	$\mathbf{Z}_3$
$\mathbf{X}_1$:	Mg	.8031	−.5892	.0883
$\mathbf{X}_2$:	Fe	.8388	−.5362	.0944
$\mathbf{X}_3$:	Na	.8579	−.5121	.0410
$\mathbf{X}_4$:	Sulphide	.9761	−.1959	.0939
$\mathbf{X}_5$:	Crystal	.0172	.9998	−.0085
$\mathbf{X}_6$:	Space	.6533	.6002	−.4615
$\mathbf{X}_7$:	Elongation	.9296	.2390	−.2807
$\mathbf{X}_8$:	Tightness	.7647	.5021	−.4039
$\mathbf{X}_9$:	Veins	.3271	.5413	.7746
$\mathbf{X}_{10}$:	Fractures	.6655	.5425	.5127
Latent Roots $\mathbf{L}$		5.4632	3.1916	1.3451

Source: Klovan, 1975; reproduced with permission.

Table 5.5 Varimax Rotated Loadings ($\mathbf{B}^T$) and Rotated Latent Roots

		$\mathbf{G}_1$	$\mathbf{G}_2$	$\mathbf{G}_3$
$\mathbf{X}_1$:	Mg	.9970	.0773	.0070
$\mathbf{X}_2$:	Fe	.9909	.1246	.0500
$\mathbf{X}_3$:	Na	.9832	.1806	.0255
$\mathbf{X}_4$:	Sulphide	.8777	−.3979	.2672
$\mathbf{X}_5$:	Crystal	−.6267	.5837	.5163
$\mathbf{X}_6$:	Space	.0530	.9890	.1378
$\mathbf{X}_7$:	Elongation	.5165	.8375	.1786
$\mathbf{X}_8$:	Tightness	.2075	.9639	.1666
$\mathbf{X}_9$:	Veins	.0019	.0438	.9990
$\mathbf{X}_{10}$:	Fractures	.2241	.3933	.8917
Latent Roots $\mathbf{K}$		4.46852	3.31843	2.21298

Source: Klovan, 1975; reproduced with permission.

decrease in magnitude for each PC, which permits a clearer identification of the components. The loading coefficients lie in the positive octant (with the exception of one) and are thus positive, which precludes orthogonality. The new orthogonal axes also do not coincide, in this case, with any variable clusters which may exist but the rotation does represent the closest orthogonal fit to the configuration of points in three-dimensional vector space. From Table 5.5 we can see that $\mathbf{G}_1$ represents mineral variables $\mathbf{X}_1$–$\mathbf{X}_4$, which in turn are negatively related to the crystal size of the carbonates. Similarly $\mathbf{G}_2$ may be taken as an index of deformation $(\mathbf{X}_6, \mathbf{X}_7, \mathbf{X}_8)$, and $\mathbf{G}_3$ represents an index of permeability (Klovan, 1975). The corresponding rotated varimax scores $\mathbf{G}_{(3)} = \mathbf{Z}^*_{(3)}\mathbf{T}$, together with the original scores, are shown in Table 5.6.

The data in this Example are artificial and are meant only to serve as a

Table 5.6   Original and Varimax Rotated Scores for the Geological Data of Table 5.1

Location	Z_1	Z_2	Z_3	G_1	G_2	G_3
1	.3887	−2.2838	.1359	1.697	−1.101	−.904
2	.1989	−.9798	−1.1363	.614	.218	−1.322
3	.9083	1.2182	−1.0941	−.223	1.811	.027
4	.2926	1.3964	−.9847	−.780	1.503	.022
5	−.9595	1.0959	−1.4845	−1.580	.911	−.881
6	−1.6185	.6760	.9116	−1.503	−1.053	.573
7	−.7464	.9150	.1871	−1.092	−.021	.382
8	.2992	1.2964	.0020	−.585	.888	.742
9	.4987	.0377	.1130	.359	.238	.250
10	.9111	−.4040	2.1310	1.192	−.926	1.724
11	1.2932	−.1946	1.5206	1.265	−.248	1.479
12	.8987	.6101	1.1914	.434	.163	1.499
13	.1935	.5946	.4475	−.171	.182	.711
14	−.8158	.1323	.3074	−.648	−.568	.079
15	−1.8532	.1726	1.3474	−1.308	−1.715	.599
16	−1.7613	−1.5741	−.2310	−.353	−1.731	−1.497
17	−.8554	−1.0525	−.9248	−.092	−.542	−1.503
18	.2300	−.6998	−1.1155	.465	.378	−1.160
19	1.3189	.1578	−.7823	.779	1.277	−.141
20	1.1785	−1.1592	.5417	1.531	.335	−.672

Source: Klovan, 1975, reproduced with permission.

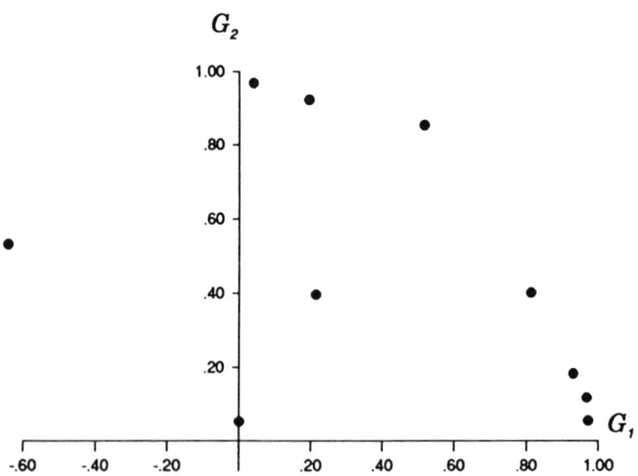

Figure 5.1   Varimax-rotated correlation loadings (Table 5.5).

numerical example. PCA is applied widely in geology, however, to help gain greater insight into the distribution of elements in mineral groups, analyze mixtures, and so forth. For an example of real data the reader is referred to Webb and Briggs (1966). □

Example 5.4.   The varimax criterion is frequently employed in exploratory research to uncover patterns of intercorrelations which are not visible from an inspection of the data, or of the correlation matrix. In addition, if theory is lacking, PCA can be a valuable tool for uncovering relationships and/or data structures. The following is an example from Andrews (1948), in the field of allergy research, who performed a rotated PCA of the correlation matrix of 36 antigens observed for 291 patients. The objective here is to supplement the pairwise correlation information obtained in the first phase of the analysis, that is, to determine whether or not patients' allergic sensitivities would group themselves into related clusters or "families." If key members of the cluster(s) are viewed as symptoms, the PC associated with the cluster has the interpretation of a particular type of allergy infection. The next step would be to see whether objects in the same cluster possess common chemical properties which do not appear in other clusters, but which could be tested for their ability to cause allergic reactions. Because of the nature of the problem, the objective is to carry out an orthogonal rotation of the axes.

Ten PCs are selected to account for the significant variance structure of the 36 original variables. The varimax-rotated correlation loadings, grouped into clusters, are shown in Table 5.7. The antigen reactions do not seem to group themselves according to any simple or apparent scheme of basic, common proteins. This is the typical situation in exploratory research, where prior knowledge is scant and a preliminary statistical analysis is typically the beginning of an investigation. Further examples of exploratory research may be found in Sparling and Williams (1978), who use the varimax rotation to isolate differences in avian vocalizations, and Drury and Daniels (1980), who employ rotated components to identify orthogonal component-clusters of bicycle riding skills. For an empirical assessment of PCA and hierarchical clustering see Fordham and Bell (1977).   □

Table 5.7   The $r = 10$ Clusters Obtained by Orthogonal Varimax Rotation of $p = 36$ Allergy Variables. Since Some Variables Appear in More than a Single Cluster the Groupings Overlap

G_1		G_2	
LePage glue	.68	Ragweed	.58
Horse epithelium	.52	Aspergillus	.55
Raw silk	.47	Timothy	.42
Alternaria	.36	Cat epithelium	.40
Pyrethrum	.34	Kapok	.40
Flaxseed	.30	Dog epithelium	.33
		Pyrethrum	.30

G_3		G_4	
Banana	.69	Feathers	.67
Peanut	.53	Shellfish	.63
Orange	.44	Dust	.48

Table 5.7   (*Continued*)

Lima bean	.39		Fish	.35
Kapok	.37		Rice	.34
Mustard	.34		Raw silk	.34
	G_5			**G_6**
Tomato	.60		Tomato	.68
Horse serum	.48		Green pea	.62
White potato	.46		Lima bean	.52
Cat epithelium	.45		Alternaria	.45
Milk (casein and whey)	.42		Orange	.45
Dog epithelium	.32		Ragweed	.42
Rice	,32		Chicken	.41
Flaxseed	.32		Rice	.41
Pork	.30		Mustard	.36
			Timothy	.35
			Rabbit epithelium	.35
			Wheat	.33
	G_7			**G_8**
Beef	.56		Cat epithelium	.50
Dog epithelium	.51		Fish	.46
Horse epithelium	.50		Rabbit epithelium	.37
Egg white	.47		Aspergillus	.35
Mustard	.47		Beef	.33
Cat epithelium	.44		Pork	.31
Alternaria	.36		Dog epithelium	.30
Fish	.32			
Banana	.32			
	G_9			**G_{10}**
Chocolate	.62		Rice	.43
Rice	.45		Horse epithelium	.39
Chicken	.40		Egg white	.38
Green pea	.40		Horse serum	.35
Peanut	.38		White potato	.35
Pork	.32		Alternaria	.34
Wheat	.30		Pyrethrum	.30

Source: Andrews, 1948.

The Quartimax Criterion

The varimax criterion seeks to maximize the variance of the loadings across the variables. An older criterion in previous use and which at times is still employed is the so-called quartimax criterion, which seeks to maximize variance across the PCs. Let

$$Q^* = \frac{1}{r}\sum_{j=1}^{r}(b_{ij}^2)^2 - \frac{1}{r^2}\left(\sum_{j=1}^{r}b_{ij}^2\right)^2 \qquad (i = 1, 2, \ldots, p) \qquad (5.22)$$

Table 5.8   Quartimax Rotated Loadings and Latent Roots

		G_1	G_2	G_3
X_1:	Mg	.9995	.0312	$-.0105$
X_2:	Fe	.9963	.0809	.0301
X_3:	Na	.9907	.1357	.0026
X_4:	Sulphide	.8991	.3704	.2335
X_5:	Crystal	$-.5911$	.6391	.4921
X_6:	Space	.1003	.9915	.0831
X_7:	Elongation	.5570	.8209	.1259
X_8:	Tightness	.2540	.9608	.1112
X_9:	Veins	.0196	.0973	.9951
X_{10}:	Fractures	.2559	.4297	.8653
Latent Roots **K**		4.5814	3.3458	2.0728

where b_{ij} are the new loadings. Equation (5.22) represents the "variance" of the squared loadings, which in turn represents contribution to variance of the variables. The quartimax maximizes

$$Q = \sum_{i=1}^{p} Q_i^*$$ (5.23)

the sum of variances of the rotated loadings. Since the quartimax criterion attempts to maximize variance across the components, it tends to produce a dominant component, which makes it appropriate for growth/size studies and related areas (Section 3.3), but undesirable for a cluster analysis. Although there is a tendency to concentrate variance in the first component, the results at times do not differ greatly from the variable clustering criteria such as the varimax. Table 5.8 gives the loading coefficients rotated by Eq. (5.23), which may be compared to those of Table 5.6.

Other Orthogonal Criteria

In recent years alternative criteria have been proposed, many of which consist of linear combinations of the varimax and quartimax criteria. They possess the general form

$$R = c_1 V + c_2 Q$$ (5.24)

where V and Q are the varimax and quartimax criteria respectively and c_1 and c_2 are constants. The use of criterion (Eq. 5.24) requires a priori knowledge of c_1 and c_2, however, which are essentially arbitrary.

A somewhat different orthogonal rotation criterion has been proposed by McCammon (1966). Let b_{ij} denote the rotated correlation loadings. Then we minimize the entropy expression

$$H = -\sum_{i=1}^{p} \sum_{j=1}^{r} b_{ij}^2 \ln(b_{ij}^2) \tag{5.25}$$

The advantage of Eq. (5.25) over the varimax criterion appears to be that the minimum entropy solution produces a higher proportion of coefficients whose absolute values are closer to zero. This gives a better resolution to the rotated loadings and enhances component-cluster identification.

5.3.2 Oblique Rotations

The previous section illustrates the use of orthogonal rotations in exploratory research as an attempt to locate, with as little ambiguity as possible, orthogonal variable clusters or groupings. Orthogonal rotations however may be viewed as a special case of a more general class of factor models, namely, those that remove the last constraint used in the initial PC solution–that of component orthogonality. For this reason oblique rotations are more effective in uncovering natural variable clusters, since variable groupings need not be orthogonal to each other. Even in the case of cluster orthogonality, an oblique rotation will still yield orthogonal (or approximately orthogonal) axes as a special case. Oblique rotations also make possible a particularly straightforward interpretation of the PC factors, in terms of the variables forming the clusters that represent the components.

THEOREM 5.4.   Let Eq. (5.18b) be a PC model with respect to an oblique basis $\mathbf{G}_1, \mathbf{G}_2, \ldots, \mathbf{G}_r$. Then the following properties hold in the $r < p$ space:
   (i) $\hat{\mathbf{X}}^{\mathrm{T}}\hat{\mathbf{X}} = \mathbf{B}\boldsymbol{\Phi}\mathbf{B}^{\mathrm{T}}$ where $\boldsymbol{\Phi} = \mathbf{G}^{\mathrm{T}}\mathbf{G} = \mathbf{T}^{\mathrm{T}}\mathbf{T}$ is the correlation matrix of the oblique components.
   (ii) $\hat{\mathbf{X}}\hat{\mathbf{X}}^{\mathrm{T}} = \mathbf{G}\mathbf{B}^{\mathrm{T}}\mathbf{B}\mathbf{G}^{\mathrm{T}}$.
   (iii) $\mathbf{B}^{\mathrm{T}} = (\mathbf{G}^{\mathrm{T}}\mathbf{G})^{-1}\mathbf{G}^{\mathrm{T}}\mathbf{X}$ and $\hat{\mathbf{X}} = \mathbf{G}\mathbf{B}^{\mathrm{T}} = \mathbf{P}_{\mathrm{G}}\mathbf{X}$ where $\mathbf{P}_{\mathrm{G}}$ is idempotent and symmetric.

PROOF
   (i) From Eq. (5.18b) we have

$$\mathbf{X}^{\mathrm{T}}\mathbf{X} = (\mathbf{G}\mathbf{B}^{\mathrm{T}} + \boldsymbol{\delta})^{\mathrm{T}}(\mathbf{G}\mathbf{B}^{\mathrm{T}} + \boldsymbol{\delta})$$

$$= \mathbf{B}\mathbf{G}^{\mathrm{T}}\mathbf{G}\mathbf{B}^{\mathrm{T}} + \mathbf{B}\mathbf{G}^{\mathrm{T}}\boldsymbol{\delta} + \boldsymbol{\delta}^{\mathrm{T}}\mathbf{G}\mathbf{B}^{\mathrm{T}} + \boldsymbol{\delta}^{\mathrm{T}}\boldsymbol{\delta}$$

$$= \mathbf{B}\boldsymbol{\Phi}\mathbf{B}^{\mathrm{T}} + \boldsymbol{\delta}^{\mathrm{T}}\boldsymbol{\delta}$$

where $\mathbf{G}^{\mathrm{T}}\boldsymbol{\delta} = \boldsymbol{\delta}^{\mathrm{T}}\mathbf{G} = 0$ owing to the initial orthogonality of the PCs. Since $\boldsymbol{\delta}^{\mathrm{T}}\boldsymbol{\delta}$ is the matrix of residual errors, $\hat{\mathbf{X}}^{\mathrm{T}}\hat{\mathbf{X}} = \mathbf{B}\boldsymbol{\Phi}\mathbf{B}^{\mathrm{T}}$

represents the variance/covariance structure accounted for by the first r components.

(ii)
$$\mathbf{XX}^\mathrm{T} = (\mathbf{GB}^\mathrm{T} + \boldsymbol{\delta})(\mathbf{GB}^\mathrm{T} + \boldsymbol{\delta})^\mathrm{T}$$

$$= \mathbf{GB}^\mathrm{T}\mathbf{BG}^\mathrm{T} + \boldsymbol{\delta}^\mathrm{T}\boldsymbol{\delta}$$

and the explained portion is then $\hat{\mathbf{X}}\hat{\mathbf{X}}^\mathrm{T} = \mathbf{GB}^\mathrm{T}\mathbf{BG}^\mathrm{T}$.

(iii) Premultiplying Eq. (5.18b) by $\mathbf{G}$ we have

$$\mathbf{G}^\mathrm{T}\mathbf{X} = \mathbf{G}^\mathrm{T}\mathbf{GB}^\mathrm{T} + \mathbf{G}^\mathrm{T}\boldsymbol{\delta}$$

$$= \mathbf{G}^\mathrm{T}\mathbf{GB}^\mathrm{T}$$

so that

$$\mathbf{B}^\mathrm{T} = (\mathbf{G}^\mathrm{T}\mathbf{G})^{-1}\mathbf{G}^\mathrm{T}\mathbf{X} \qquad (5.26)$$

is the $(r \times p)$ matrix of regression coefficients of the original random variables on r oblique PCs. Premultiplying Eq. (5.26) by $\mathbf{G}$ we then obtain

$$\mathbf{GB}^\mathrm{T} = \mathbf{G}(\mathbf{G}^\mathrm{T}\mathbf{G})^{-1}\mathbf{GX}$$

$$= \mathbf{P}_\mathrm{G}\mathbf{X}$$

$$= \hat{\mathbf{X}} \qquad (5.27)$$

the predicted values of $\mathbf{X}$.

Equation (5.26) yields an important relationship between the oblique correlation loading coefficients, component correlations, and correlations between the oblique components and observed random variables. Since components are no longer orthogonal the correlation loadings obtained (say) from the correlation matrix need not be equal to the correlation coefficients between the variates and the PCs. In the psychometric literature the former are known as the "pattern" and the latter as the "structure." In what follows we refer to the two sets as the regression loading coefficients and the correlation loading coefficients. From Eq. (5.26) we have

$$\boldsymbol{\Phi}\mathbf{B}^\mathrm{T} = \mathbf{G}^\mathrm{T}\mathbf{X} \qquad (5.27a)$$

where $\boldsymbol{\Phi}$ is the $(r \times r)$ component correlation matrix, $\mathbf{B}^\mathrm{T}$ is a $(p \times r)$ matrix of regression loading coefficients (coordinates of the variables $\mathbf{X}$ with respect to the oblique components $\mathbf{G}$), and $\mathbf{G}^\mathrm{T}\mathbf{X}$ is the correlation matrix of the variables and oblique components, that is, the matrix of correlation loading coefficients. For the sake of simplicity we assume both variables and components are standardized to unit length. In general, the oblique PC model can be interpreted in terms of least squares regression theory where both the coefficients and the independent variables (i.e., the components) are to be estimated from the data. However, since the elements of $\mathbf{B}$ do not necessarily lie in the interval $[-1, 1]$, component identification is usually made easier by consulting the correlation loading coefficients $\mathbf{G}^T\mathbf{X}$.

As was the case for orthogonal rotations, the oblique transformation matrix $\mathbf{T}$ is arbitrary from a general mathematical point of view. There exist an infinite number of matrices which can rotate orthogonal components to an oblique form. Since we wish to rotate the PCs so as to obtain the clearest identification possible (given the data), a further optimization criterion is necessary. Two criteria that are frequently used are the oblimin and the quartimax criteria.

The Oblimin Criterion

The oblimin criterion minimizes the expression

$$Q = \sum_{j=1}^{r} \sum_{h=1}^{r} \left[\sum_{i=1}^{p} b_{ij}^2 b_{ih}^2 - \frac{1}{p} \left(\sum_{i=1}^{p} b_{ij}^2 \right)\left(\sum_{i=1}^{p} b_{ih}^2 \right) \right] \qquad (5.28)$$

that is, it minimizes, across all pairs of axes, the sum of covariances of squared loadings while summing for each pair of axes across the p variables. Equation (5.28) is also known as the oblique varimax criterion. The computational algorithm is iterative and transforms the columns of $\mathbf{A}$ one at a time, until a matrix $\mathbf{T}$ is obtained such that $\mathbf{B}^T = (\mathbf{AT})^T$. Since the oblimin criterion maximizes the spread of loading magnitudes per component, and minimizes the correlations between the components, it is suitable for use as a technique of nonhierarchical cluster analysis. Once the loadings $\mathbf{B}$ are known, they can be transformed into correlation loadings (pattern) by the use of eq. (5.27a).

Many other oblique rotations have been developed, particularly in the psychological literature, for which the reader is referred to Harman (1967), Mulaik (1972), Gorsuch (1974), and Mcdonald (1985). Not a great deal of statistical/numerical work seems to have been done with respect to oblique criteria and many seem to continue to suffer from statistical defects and excessive subjectivity. The oblimin criterion however has been employed with real data and appears to perform satisfactorily. The following examples use the criterion to enhance interpretation of the factors and to carry out exploratory statistical analysis.

EXAMPLE 5.5.   We again consider the geological data used in Examples 5.3 and 5.4. It was seen the orthogonal rotations improved the interpretability of the PCs. To verify whether an orthogonal component structure is appropriate however we compute oblimin regression loadings and correlation loadings, which are given in Tables 5.9 and 5.10 respectively. The

Table 5.9   Oblimin Regression Loading Coefficients B (pattern) for 10 Geological Variables

		G_1	G_2	G_3
X_1:	Mg	1.002	$-.012$	.019
X_2:	Fe	.990	.029	.055
X_3:	Na	.975	.096	.017
X_4:	Sulphide	.840	.283	.228
X_5:	Crystal	$-.709$	.566	.416
X_6:	Space	$-.069$	1.033	$-.055$
X_7:	Elongation	.420	.815	.032
X_8:	Tightness	.090	.984	$-.015$
X_9:	Veins	$-.006$	$-.196$	1.062
X_{10}:	Fractures	.177	.188	.881

Table 5.10   Oblimin Correlation Loading (Structure) Between Oblique Components and 10 Geological Variables

		G_1	G_2	G_3
X_1:	Mg	1.000	.196	.053
X_2:	Fe	.997	.249	.105
X_3:	Na	.995	.299	.093
X_4:	Sulphide	.906	.543	.375
X_5:	Crystals	$-.579$	.590	.614
X_6:	Space	.136	.997	.355
X_7:	Elongation	.584	.912	.374
X_8:	Tightness	.287	.996	.382
X_9:	Veins	$-.003$	.228	.984
X_{10}:	Fractures	.249	.575	.963

oblique components (axes) are shown in Figure 5.2 where

$$\Phi = \begin{bmatrix} 1.000 & & \\ .201 & 1.000 & \\ .039 & .400 & 1.000 \end{bmatrix}$$

is the matrix of component correlations. Thus whereas the first and third components are, to all practical intents and purposes orthogonal, the second component is somewhat correlated with the other two. Comparing Table 5.10 with Table 5.5 we see that although the intrepetation of the PCs is preserved, the oblimin components yield a slightly clearer interpretation. The two sets of correlation loadings do not differ to a great extent because of the low degree of intercorrelations amongst the PCs. The oblique component scores are shown in Table 5.11. Oblique rotation has also been used by McElroy and Kaesler (1965) to study lithologic relations within rock strata. □

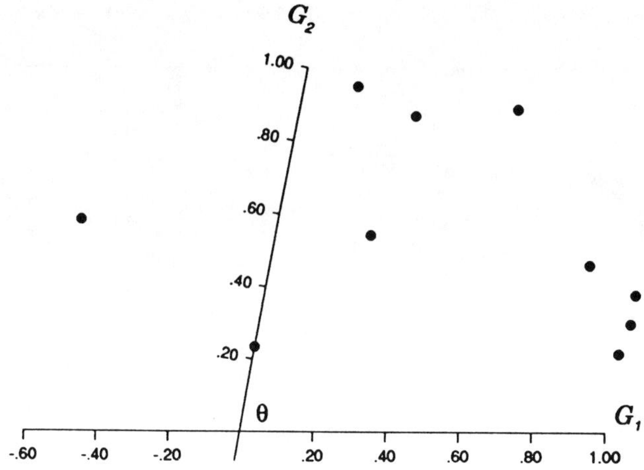

Figure 5.2 Oblimin-rotated correlation loadings of Table 5.9 where $\theta = 78.4°$.

Table 5.11   Oblimin Component Scores for 10 Geological Variables

Location	G_1	G_2	G_3
1	1.605	−1.038	−1.076
2	.642	.041	−1.222
3	−.068	1.745	.421
4	−.649	1.376	.331
5	−1.488	.536	−.703
6	−1.592	−1.101	.282
7	−1.093	−.080	.336
8	−.513	.934	.903
9	.375	.322	.307
10	1.093	−.442	1.510
11	1.226	.182	1.416
12	.433	.488	1.510
13	−.162	.289	.729
14	−.695	−.617	−.068
15	−1.454	−1.718	.166
16	−.486	−2.008	−1.853
17	−.124	−.817	−1.587
18	.505	.209	−1.033
19	.886	1.313	.168
20	1.560	.386	−.537

Example 5.6.   Example 5.5 illustrates an oblique rotation using artificial data. An oblique rotation has been used by Mukherjee (1974) to study discrepancies of response to demographic surveys. Since it is found in practice that retesting the same respondents after several months does not produce identical responses, a degree of nonsampling error would seem to

Table 5.12   Oblique PCA for Survey Reporting Discrepancies

| Discrepancy Measure | Regression Loadings (Pattern) | | | | Correlation Loadings (Structure) | | | |
	G_1 Age	G_2 Children	G_3 Pregnancy	G_4 Education	G_1 Age	G_2 Children	G_3 Pregnancy	G_4 Education
Respondent's education	−0.080	0.130	−0.114	0.176	−0.130	0.151	−0.151	0.193
Spouse's education	−0.069	−0.136	0.096	−0.477	−0.045	0.176	0.096	0.459
Number of pregnancies	−0.108	0.134	0.696	0.082	−0.141	−0.008	0.664	0.064
Female child	0.108	0.326	−0.051	−0.079	0.112	0.334	−0.115	−0.114
Male child	−0.008	0.673	0.178	−0.027	−0.036	0.638	0.040	−0.059
Age at consummation	0.809	0.072	−0.107	−0.033	0.815	0.055	−0.127	−0.232
Age at marriage	0.918	0.155	−0.108	0.149	0.874	0.126	−0.157	−0.081
Respondent's age	0.451	−0.221	0.265	−0.127	0.492	−0.295	0.314	−0.248

Source: Mukherjee, 1974.

Table 5.13   Intercorrelations Among Four Primary Oblique Factors

| Code | Primary Factor Pattern | Primary Factor Pattern | | | |
		G_1	G_2	G_3	G_4
G_1	Age-related discrepancy	1.00			
G_2	No. of children discrepancy	.05	1.00		
G_3	Discrepancy in reporting pregnancy	−.01	.21	1.00	
G_4	Discrepancy in reporting education	.25	−.03	.05	1.00

Source: Mukherjee (1974).

be present in the survey. Defining the test–retest discrepancies for $p = 8$ variables as the difference between the two sets of answers, Mukherjee (1974) obtains $r = 4$ oblique factors (Table 5.12 and 5.13) for Indian data. Since the first PC does not exhibit isotropy, it is concluded that the variates tend to cluster. The first oblique factor is identified as inconsistency in reporting age-related data, the second is a related to discrepancies with respect to number of children, and the last two define discrepancy scores in reporting pregnancies and educational qualification of spouse, respectively. The nature of each discrepancy is in turn related to cultural aspects of Indian society.   □

Positive Quadrant Rotations

The oblimin oblique rotation is useful when the objective is to associate, as closely as possible, oblique PCs with as many nonoverlapping subsets of variables as possible. A second purpose may be to extract a dominant size or growth component, followed by oblique secondary PCs which express more local groupings of the variables. Other objectives are also possible. For example, given chemical or geological mixtures, we may wish to rotate the axes obliquely in such a way that all correlation loadings become positive, and the axes pass through certain coordinate points in order to gain insight into compositions of mixtures (see Section 5.9).

5.3.3 Grouping Variables

Once clusters have been identified by a linear transformation of the axes, a further step may be to group or aggregate all the variables, within a given cluster, into a composite variable. For example, we may have an excess of variables, and rather than deleting the variables or replacing the original variables by a smaller set of PCs, we may wish to retain all the variables by aggregating them into a smaller number of "supervariables." Oblique rotations are particularly useful for such a purpose, and result in a reduction of the original data set while at the same time retaining much of the original information.

Consider the following PC solution:

$$\mathbf{X}_1 = b_{11}\mathbf{G}_1 + b_{12}\mathbf{G}_2 + b_{13}\mathbf{G}_3 + \boldsymbol{\delta}_1$$

$$\mathbf{X}_2 = b_{21}\mathbf{G}_1 + b_{22}\mathbf{G}_2 + b_{23}\mathbf{G}_3 + \boldsymbol{\delta}_2$$

$$\mathbf{X}_3 = b_{31}\mathbf{G}_1 + b_{32}\mathbf{G}_2 + b_{33}\mathbf{G}_3 + \boldsymbol{\delta}_3$$

$$\mathbf{X}_4 = b_{41}\mathbf{G}_1 + b_{42}\mathbf{G}_2 + b_{43}\mathbf{G}_3 + \boldsymbol{\delta}_4$$

$$\mathbf{X}_5 = b_{51}\mathbf{G}_1 + b_{52}\mathbf{G}_2 + b_{53}\mathbf{G}_3 + \boldsymbol{\delta}_5$$

$$\mathbf{X}_6 = b_{61}\mathbf{G}_1 + b_{62}\mathbf{G}_2 + b_{63}\mathbf{G}_3 + \boldsymbol{\delta}_6$$

$$\mathbf{X}_7 = b_{71}\mathbf{G}_1 + b_{72}\mathbf{G}_2 + b_{73}\mathbf{G}_3 + \boldsymbol{\delta}_7 \tag{5.29}$$

where $p = 7$ and $r = 3$ and we have $g = 2$ oblique clusters $\mathbf{X}_\mathrm{I}$ and $\mathbf{X}_\mathrm{II}$. Since variables within each cluster are highly correlated, they may be aggregated as

$$\hat{\mathbf{X}}_\mathrm{I} = \mathbf{X}_1 + \mathbf{X}_2 + \mathbf{X}_3$$

$$= (b_{11} + b_{21} + b_{31})\mathbf{G}_1 + (b_{12} + b_{22} + b_{32})\mathbf{G}_2 + (b_{13} + b_{23} + b_{33})\mathbf{G}_3$$

$$= c_1\mathbf{G}_1 + c_2\mathbf{G}_2 + c_3\mathbf{G}_3$$

$$\hat{\mathbf{X}}_\mathrm{II} = (b_{41} + b_{51} + b_{61} + b_{71})\mathbf{G}_1 + (b_{42} + b_{52} + b_{62} + b_{72})\mathbf{G}_2$$

$$+ (b_{43} + b_{53} + b_{63} + b_{73})\mathbf{G}_3$$

$$= d_1\mathbf{G}_1 + d_2\mathbf{G}_2 + d_3\mathbf{G}_3 \tag{5.30}$$

To convert the coefficients into correlation loadings we have

$$\frac{\hat{\mathbf{X}}_\mathrm{I}}{s_\mathrm{I}} = \frac{c_1}{s_\mathrm{I}}\mathbf{G}_1 + \frac{c_2}{s_\mathrm{I}}\mathbf{G}_2 + \frac{c_3}{s_\mathrm{I}}\mathbf{G}_3$$

$$\frac{\hat{\mathbf{X}}_\mathrm{II}}{s_\mathrm{II}} = \frac{d_1}{s_\mathrm{II}}\mathbf{G}_1 + \frac{d_2}{s_\mathrm{II}}\mathbf{G}_2 + \frac{d_3}{s_\mathrm{II}}\mathbf{G}_3 \tag{5.31}$$

where $s_I^2 = \mathbf{X}_I^T \mathbf{X}_I$ and $s_{II}^2 = \mathbf{X}_{II}^T \mathbf{X}_{II}$ are the variances of the two aggregates. Equation (5.30) may then be used in place of the original system (Eq. 5.29).

Example 5.7. Variable aggregation is frequently carried out using economic variables. Consider the following variables with represent prices of $p = 10$ commodities:

Y_1:   Agriculture
Y_2:   Mining and quarrying
Y_3:   Food, drink, and tobacco
Y_4:   Chemicals
Y_5:   Engineering
Y_6:   Textiles
Y_7:   Other manufacturers
Y_8:   Gas and electricity
Y_9:   Services
Y_{10}: Noncompetitive imports   □

Since prices tend to move jointly over time, the 10 variables are highly correlated, and we wish to group them into a smaller number of aggregate variables. The correlation loadings, together with the latent roots, are exhibited in Table 5.14 were the original data are first transformed into logarithms to linearize the relationships and reduce differences in the variances. All price variables (except X_5) are highly correlated, which is consistent with the existence of a single commodity group or cluster, with $\mathbf{X}_5$ forming a cluster of its own.

Table 5.14   Correlation Loadings from the Covariance Matrix of 10 Price Indices; United Kingdom, 1955–1968

		Z_1	Z_2	Z_3	R^2
$\mathbf{X}_1$:	Agriculture	.9798	−.1078		.9716
$\mathbf{X}_2$	Mining	.9947			.9894
$\mathbf{X}_3$	Food	.9756	−.1682		.9801
$\mathbf{X}_4$	Chemicals	.9924			.9849
$\mathbf{X}_5$	Engineering	−.3468	.4146	.8030	.9370
$\mathbf{X}_6$	Textiles	.9768	−.1770		.9854
$\mathbf{X}_7$	Manufacturing	.9861	−.1265		.9884
$\mathbf{X}_8$	Gas and electricity	.9176	.3817		.9877
$\mathbf{X}_9$:	Services	.9817		−.1188	.9778
$\mathbf{X}_{10}$:	Imports	.8815	−.1777	.3550	.9346
Latent roots		5.182	.137	.073	
Variance (%)		94.87	2.51	1.34	

The variables can therefore be aggregated into two groups, as

$$\hat{\mathbf{X}}_{\mathrm{I}} = (.9798 + .9947 + \cdots + .8815)\mathbf{Z}_1 + (-.1188 + .3550)\mathbf{Z}_3$$

$$= 8.6862\mathbf{Z}_1 + .2362\mathbf{Z}_3$$

$$\hat{\mathbf{X}}_{\mathrm{II}} = -.3468\mathbf{Z}_1 + .8030\mathbf{Z}_2$$

where $s_1 = 8.897$ from the covariance matrix. Standardizing to unit variance then yields the equations

$$\hat{\mathbf{X}}_{\mathrm{I}}^* = \frac{1}{8.8970}\,\hat{\mathbf{X}}_{\mathrm{I}} = \frac{8.6862}{8.8970}\,\mathbf{Z}_1 + \frac{.2362}{8.8970}\,\mathbf{Z}_3$$

$$= .9763\mathbf{Z}_1 + .0265\mathbf{Z}_3$$

$$\hat{\mathbf{X}}_{\mathrm{II}} = -.3468\mathbf{Z}_1 + .8030\mathbf{Z}_3$$

which represent a more compact system. The aggregated equations can be further rotated to oblique form if so desired.

5.4 ALTERNATIVE MODES FOR PRINCIPAL COMPONENTS

A $(n \times p)$ data matrix $\mathbf{Y}$ represents a two-way classification, consisting of n observations measured for p random variables. In classical PCA interest lies in an analysis of the structure of the random variables, with an analysis of the sample spaces as a byproduct if the observations are centered about the column (variable) means. At times however the main interest may lie in the sample points or the row space of $\mathbf{Y}$. Here it no longer makes sense to center the data by column means. Or else data may be given in the form of a three-way (or higher) classification, for example, variables, individuals (objects), and time. The interest would then be to observe how the sample point measurements vary over time. In both cases we have a different mode of data presentation, which does not conform to the classic PC set-up and thus requires further modification and adjustment if a PCA is to provide meaningful results.

5.4.1 Q-Mode Analysis

When sample points rather than variables are of interest it is usually more fitting to base a PC analysis on a $(n \times n)$ product-moment matrix of the observations, rather than the variables. This is particularly appropriate when values other than variable means are to be used for centering purposes, or when more variables than sample points are available. The latter situation can arise, for example, when the variables are of interest but sample points are expensive (difficult to obtain), in which case the available sample will be subjected to intense scrutiny by using as many variables as can be obtained under the circumstances. A PCA carried out on the sample points is known as a Q-mode factor analysis, in contrast to the usual analysis of random variables which is then termed an R-mode factor analysis. The main practical rationale however for using Q-mode analysis is to study the sample points, and perhaps to group them into more homogeneous subsets. This objective can be achieved in several ways depending on the type of association matrix defined for the observations.

The more straightforward method of analyzing sample points is to make use of Theorem 3.17, since we know $(\mathbf{XX}^T)\mathbf{Z} = \mathbf{ZL}$ when $\mathbf{X} = \mathbf{Y} - \bar{\mathbf{Y}}$ and $\mathbf{Z}$ is a matrix of scores obtained from $\mathbf{X}^T\mathbf{X}$. An examination of the columns of $\mathbf{Z}$ then indicates which sample points lie close to each other, that is, which sample points possess similar profiles or characteristics in terms of the random variables (e.g., see McCammon, 1966; Zhou et al., 1983). When $n > p$ it is easier to obtain $\mathbf{Z}$, $\mathbf{P}$, and $\mathbf{L}$ from $\mathbf{X}^T\mathbf{X}$ whereas when $n < p$ it is computationally more efficient to decompose $\mathbf{XX}^T$. When $n = p$, any one of the two association matrices may be used. Since a Q-mode analysis using Theorem 3.17 is a direct outcome of the usual PC decomposition, only the variables are adjusted to a common (zero) mean. A parallel situation exists when the variables are not centered, that is, when a PC decomposition is based on the matrix $\mathbf{Y}^T\mathbf{Y}$ or $\mathbf{YY}^T$.

The simplest departure from a usual PCA occurs when the rows of $\mathbf{Y}$ are centered by the means

$$\bar{Y}_i = \frac{1}{p}\sum_{j=1}^{p} y_{ij} \qquad (i = 1, 2, \ldots, n) \tag{5.32}$$

that is, the sample points are expressed as deviations from the mean vector $\bar{\mathbf{Y}} = (\bar{Y}_1, \bar{Y}_2, \ldots, \bar{Y}_n)^T$, thus setting the row means to zero. Other values can also be used—for example adjusting the row means to equal .50 (see Miesch, 1980). Owing to the different centering procedures, however, as well as to the nature of sample points (as opposed to random variables), it becomes more difficult to justify the general use of measures such as covariances or correlation coefficients which after all are designed for random variables. As a special case, however, Gower (1967) has pointed out that adjustments of the form of Eq. (5.32) can be rationalized when sample points lie on a sphere, and it becomes more natural to use a circle as a measurement of shortest distance (Fig. 5.4). Let $\mathbf{y} = (y_1, y_2, \ldots, y_p)^T$ and $\mathbf{x} = (x_1, x_2, \ldots, x_p)^T$ be any two row vectors (sample observations) of a $(n \times p)$ data matrix. Then the distance between the two points along a circle is the radius r, where θ is the angle separating the two points. When both $\mathbf{y}$ and $\mathbf{x}$ are adjusted by Eq. (5.32) we have

$$\sum_{i=1}^{p} y_i = \sum_{i=1}^{p} x_i = 0$$

and

$$\cos \theta = \sum_{i=1}^{p} \frac{x_i y_i}{\left(\sum_{i=1}^{p} x_i^2 \sum_{i=1}^{p} y_i^2 \right)^{1/2}}$$

$$= r_{xy} \tag{5.33}$$

so that the correlation coefficient r_{xy} can be interpreted as the cosine of the angle between $\mathbf{x}$ and $\mathbf{y}$. The "distance" between the two points can also be expressed in terms of θ, and we have

$$\theta = \cos^{-1}(r_{xy}) \tag{5.34}$$

Thus given n sample points either the cosine measure (Eq. 5.33) or angles (Eq. 5.34) between the points can be used in place of covariance or correlation coefficients, and a principal components analysis (perhaps accompanied by rotation) will reveal pattern(s) of interelationship(s) which may exist between the sample points. Since the objective however is to maximize the mean sum of squared distances between the sample points, this again leads us back to a standard R-mode PCA of the matrix $\mathbf{X}^T\mathbf{X}$, as is seen from the following Theorem.

THEOREM 5.5　(Rao, 1964). Let $\mathbf{Y}$ be a $(n \times p)$ data matrix with row vectors $\mathbf{y}_1 = (y_{11}, y_{12}, \ldots, y_{1p})^T$ and $\mathbf{y}_2 = (y_{21}, y_{22}, \ldots, y_{2p})^T, \ldots, \mathbf{y}_n = (y_{n1}, y_{n2}, \ldots, y_{np})^T$. Then the mean sum of squares of the $c(\tbinom{n}{2})$ interpoint distance is given by $\mathrm{tr}(\mathbf{X}^T\mathbf{X})$, that is

$$\frac{1}{n} \sum_{i,j=1}^{n} (\mathbf{y}_i - \mathbf{y}_j)^T (\mathbf{y}_i - \mathbf{y}_j) = \mathrm{tr}(\mathbf{X}^T\mathbf{X})$$

where $\mathbf{X} = \mathbf{Y} - \bar{\mathbf{Y}}$.

The theorem may be proved by induction. An optimal least squares fit in an r-dimensional subspace is thus still provided by the first $r < p$ principal components, which correspond to the r largest latent roots.

Theorem 5.5 assumes that the original n points are represented in a p-dimensional Euclidian space with orthogonal coordinate axes. There are situations however where oblique axes may be more appropriate (Section 5.3.2). In such a case the Euclidian distance between two sample points $\mathbf{y}_i$ and $\mathbf{y}_j$ is the quadratic form $(\mathbf{y}_i - \mathbf{y}_j)^T \mathbf{\Gamma}^{-1}(\mathbf{y}_i - \mathbf{y}_j)$, with $\mathbf{\Gamma}$ positive definite, and Theorem 5.5 can then be expressed in a more general form (Section 5.6).

Example 5.8.　An example of Q-analysis using the cosine measure is given in Miesch (1976) (Tables 5.15 and 5.16). Two and three-dimensional

Table 5.15   Hypothetical Data on Olivine Compositions (Weight %)

Sample	SiO_2	FeO	MgO
1	40.048	14.104	45.848
2	38.727	21.156	40.117
3	36.085	35.260	28.655
4	34.764	42.312	22.924
5	33.443	49.364	17.193
6	30.801	63.468	5.731

Source: Miesch, 1976b; reproduced with permission.

**Table 5.16   Correlation Loadings (Varimax) of
Q-Analysis Based on Cosines Between Samples**

Sample	G_1	G_2
1	0.2089	0.7911
2	0.2869	0.7131
3	0.4429	0.5571
4	0.5209	0.4791
5	0.5989	0.4011
6	0.7549	0.2451

Source: Miesch, 1976b; reproduced with permission.

spherical coordinates of a principal components analysis are depicted in
Figures 5.3 and 5.4. The objective of the graphical representation is to
locate "key" sample points in terms of their basic (percentage) composition,
and the samples rather than variables (minerals) are therefore of main

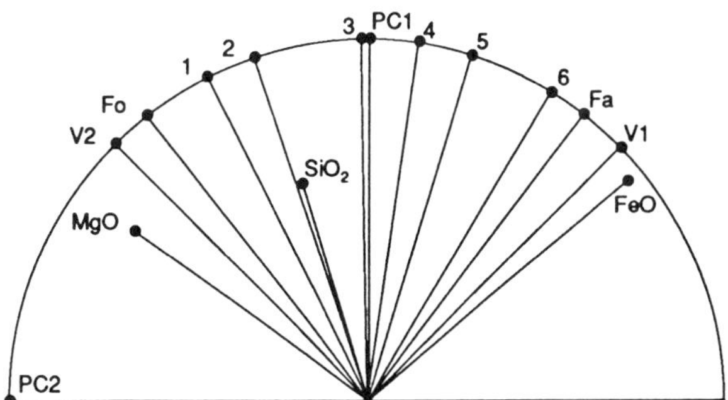

Figure 5.3 Olivine compositions represented as vectors in two and three dimensions. Vectors
V_1 and V_2 are varimax axes. Vectors PC_1 and PC_2 are principal components axes. Other vectors
represent compositions of fayalite (Fa) and forsterite (Fo) and oxide constituents of olivine
SiO_2, FeO, and MgO (Miesch, 1976; reproduced with permission).

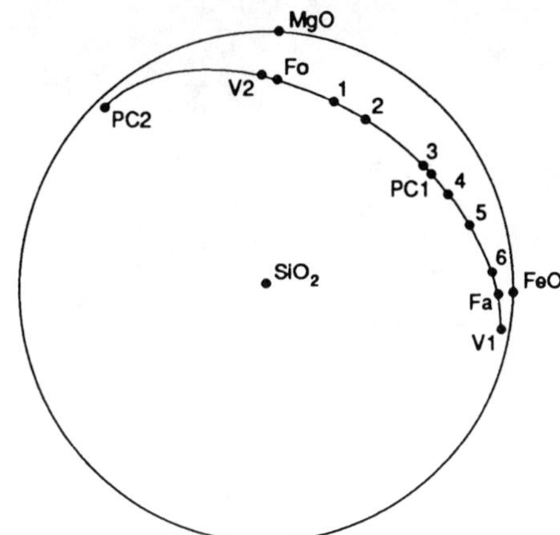

Figure 5.4 Three-dimensional vector system as a sterogram (Miesch, 1976; reproduced with permission).

interest. For a further comparison of R and Q-mode factor analysis in the context of Geology see Klovan (1975). □

5.4.2 Multidimensional Scaling and Principal Coordinates

The initial input into a standard factor analysis consists of a data matrix. At times however the original data are not available, and our initial observations consist of interpoint distances or measures of similarity between the variables (sample points). The problem now becomes to estimate, within a reduced r-dimensional space, the original vectors or coordinates which had given rise to the interpoint distances. This is known as multidimensional scaling (MDS). Since non-Euclidian and nonmetric distances can also be used, MDS at times possesses a degree of flexibility which may not be found in a standard PCA. Care however must be taken when interpreting the results, since although a distance matrix is symmetrix it cannot be positive (semi) definite, owing to zero diagonal entries. As is well known, the latent roots and vectors of a symmetric matrix are real, but the former need not be nonnegative. Also, it can be difficult to establish statistical properties of certain MDS procedures or solutions since these are algorithmic (numerical) in nature and cannot be expressed in algebraic or closed form. For further detail see Coombs and Kao (1960) and Gnanadesikan (1977).

The classic version of MDS uses Euclidian distance as similarity measures and can be related, in a straightforward manner, to PCA (e.g., see Dobson et al., 1974). Two cases appear in practice—when observed distances are identical to true distances (similarities) and when distances are subject to errors of observation. When data (the distances) are observed without measurement error the solution has been given by Young and Householder (1938; see also Eckart and Young, 1936). Consider n points, together with their matrix of squared distances

$$\mathbf{D}^2 = \begin{bmatrix} 0 & d_{12}^2 & \cdots & d_{1n}^2 \\ d_{21}^2 & 0 & \cdots & d_{2n}^2 \\ \vdots & \vdots & \ddots & \\ d_{n1}^2 & d_{n2}^2 & \cdots & 0 \end{bmatrix}$$

The points, generally speaking, may represent samples (individuals) or random variables. We assume that the origin of the coordinate system is placed at one of the unknown points, so that in effect we only have $n-1$ points (row vectors) to consider. Let $\mathbf{Y}$ be the unknown $(n \times p)$ data matrix and let $\mathbf{y}_i$ be the ith row vector of $\mathbf{Y}$. Then from the cosine law of the triangle any (i, j)th element of $\mathbf{YY}^T$ can be expressed as

$$\mathbf{y}_i \mathbf{y}_j^T = \frac{1}{2} (\mathbf{y}_i \mathbf{y}_i^T + \mathbf{y}_j \mathbf{y}_j^T - d_{ij}^2)$$

$$= \frac{1}{2} (d_{in}^2 + d_{jn}^2 - d_{ij}^2) \tag{5.35}$$

Since $\mathbf{Z}^T(\mathbf{YY}^T)\mathbf{Z} = \mathbf{L}$ from Theorem 3.17 the unknown points can be computed as

$$\mathbf{Y} = \mathbf{ZL}^{1/2} \tag{5.36}$$

where $\mathbf{Z}$ is the $(n-1) \times p$ matrix of adjusted latent vectors of $\mathbf{YY}^T$ and $\mathbf{L}$ is the diagonal matrix of latent roots. The solution to the MDS problem then consists in reconstructing the unknown data matrix $\mathbf{Y}$, which can be achieved in the following two steps:

1. Taking the nth point as arbitrary origin, construct the matrix of inner products using eq. (5.35).
2. Compute the latent roots and vectors of $\mathbf{YY}^T$ and take the coordinates of the $(n-1)$ unknown points as $\mathbf{ZL}^{1/2}$.

It can be shown that the choice of point selected as origin does not affect the result. Young and Householder (1938) also show that dimensionality of the n points with mutual distances d_{ij} is equal to $\rho(\mathbf{YY}^T)$. The procedure is therefore a variation of a PCA of $\mathbf{YY}^T$, where one of the points is taken as the origin. As in PCA, a smaller number of dimensions may be retained if these correspond to small (lower) order latent roots. Finally, a necessary and sufficient condition for a set of numbers $d_{ij} = d_{ji}$ to be mutual distances between a real set of points in Euclidian space is that the symmetric matrix $\mathbf{YY}^T$ be positive (semi) definite (Grammian), which clearly is the case here.

The Young and Householder model has a drawback in that it is affected by errors of measurement. Since the origin is translated to one of the n points, differential errors of observation render the model dependent on which point is selected as origin. Torgerson (1952) has proposed a more robust model where the origin is translated to the mean of the n observations. Following Torgerson (1952, 1958) and Rao (1964), let $\mathbf{y}_1, \mathbf{y}_2, \ldots, \mathbf{y}_n$ be n points (row vectors) and let $\bar{\mathbf{y}}$ be the mean point. Then given any two points $\mathbf{y}_i$ and $\mathbf{y}_j$ we have

$$(\mathbf{y}_i - \bar{\mathbf{y}})(\mathbf{y}_j - \bar{\mathbf{y}})^T = \left[\mathbf{y}_i - \frac{1}{n}(\mathbf{y}_1 + \mathbf{y}_2 + \cdots + \mathbf{y}_n)\right]\left[\mathbf{y}_i - \frac{1}{n}(\mathbf{y}_1 + \mathbf{y}_2 + \cdots + \mathbf{y}_n)\right]^T$$

$$= \frac{1}{n^2}[(\mathbf{y}_i - \mathbf{y}_1) + \cdots + (\mathbf{y}_i - \mathbf{y}_n)][(\mathbf{y}_j - \mathbf{y}_1) + \cdots + (\mathbf{y}_j - \mathbf{y}_n)]^T$$

$$= \frac{1}{n^2}\sum_{g=1}^{n}\sum_{h=1}^{n}[(\mathbf{y}_j - \mathbf{y}_g)(\mathbf{y}_i - \mathbf{y}_g)^T + (\mathbf{y}_i - \mathbf{y}_h)(\mathbf{y}_i - \mathbf{y}_h)^T$$

$$- (\mathbf{y}_i - \mathbf{y}_j)(\mathbf{y}_i - \mathbf{y}_j)^T - (\mathbf{y}_g - \mathbf{y}_h)(\mathbf{y}_g - \mathbf{y}_h)^T]$$

$$= \frac{1}{2n^2}\sum_{g=1}^{n}\sum_{h=1}^{n}(d_{jg}^2 + d_{ih}^2 - d_{ij}^2 - d_{hg}^2)$$

$$= \frac{1}{2n^2}\left[-d_{ij}^2 + \frac{1}{n}(d_{i.}^2 + d_{.j}^2) - \frac{1}{n^2}d_{..}^2\right] \tag{5.37}$$

where d_{ij}^2 is the squared distance between sample points $\mathbf{y}_i$ and $\mathbf{y}_j$ and

$$d_{i.}^2 = \sum_{h=1}^{n} d_{ih}^2, \qquad d_{.j}^2 = \sum_{g=1}^{n} d_{gj}, \qquad d_{..}^2 = \sum_{g=1}^{n}\sum_{h=1}^{n} d_{gh}^2 \tag{5.38}$$

Equation (5.37) is similar in form to the expression $y_{ij} - \bar{y}_{i.} - \bar{y}_{.j} + \bar{y}$ which is used to estimate row and column effects in a two-way (fixed effects) ANOVA. Using Eq. (5.37) is thus equivalent to adjusting the row and column means, and the overall mean, to zero. Gower (1966, 1967; see also Reyment et al., 1984) has provided a more recent but equivalent version of the model, known as the principal coordinate (principal axes) model. When the number of variables exceeds the number of sample points, and the matrix $\mathbf{Y}$ is of full rank, all roots are nonzero and retaining all the PCs will reproduce the original distances d_{ij}. More generally the number of nonzero roots equals $\rho(\mathbf{Y})$, and when lower-order nonzero roots are omitted, the principal coordinates model provides in a lower dimensional space, the closest fit possible to the original $(1/2)n(n-1)$ interpoint distances. The model can be operationalized as follows:

1. Given a $(n \times p)$ data matrix $\mathbf{Y}$ consisting of n sample points and p

random variables form the $(n \times n)$ matrix $\mathbf{E}$ with elements

$$
e_{ij} = \begin{cases} -\dfrac{1}{2} d_{ij}^2 & i \neq j \\ 0 & i = j \end{cases} \tag{5.39}
$$

2. Form the $(n \times n)$ matrix $\mathbf{F}$ with elements

$$
f_{ij} = \begin{cases} e_{ij} - \bar{e}_{i.} - \bar{e}_{.j} + \bar{e}_{..} & i \neq j \\ 0 & i = j \end{cases} \tag{5.40}
$$

where

$$
\bar{e}_{i.} = \frac{1}{n} \sum_{j=1}^{n} e_{ij}, \qquad \bar{e}_{.j} = \frac{1}{n} \sum_{i=1}^{n} e_{ij}
$$

and

$$
\bar{e}_{..} = \frac{1}{n} \sum_{i=1}^{n} \bar{e}_{i.} = \frac{1}{n} \sum_{j=1}^{n} \bar{e}_{.j} \tag{5.41}
$$

3. Carry out a PCA of the positive semidefinite matrix $\mathbf{F}$. Then the rth latent vector of $\mathbf{F}$, normalized such that its sum of squares equals the rth latent root, yields the desired coordinates along the rth principal axis.

It is also easy to show that

$$
\begin{aligned} d_{ij}^2 &= f_{ii} + f_{jj} - 2f_{ij} \\ &= e_{ii} + e_{jj} - 2e_{ij} \end{aligned} \tag{5.42}
$$

(Exercise 5.6). Equation (5.40) is identical to Eq. (5.37) since $\mathbf{F}$ represents a distance matrix adjusted such that row, column, and overall means are zero. Torgerson's classical MDS (principal coordinate) decomposition can be summarized by the following theorem.

THEOREM 5.6. Let $\mathbf{D} = (d_{ij})$ be an $(n \times n)$ distance matrix defined in Euclidian space, and let

$$
\mathbf{F} = \mathbf{HEH} \tag{5.43}
$$

where $\mathbf{H} = \mathbf{I}_n - (1/n)\,\mathbf{11}^{\mathrm{T}}$ and $\mathbf{F} = (f_{ij})$ as defined by Eq. (5.40). Suppose $\mathbf{F}$ is positive semidefinite of rank $p \leq n - 1$. Let $\mathbf{L}$ and $\mathbf{Q}$ be the nonzero latent roots and latent vectors of $\mathbf{F}$, respectively, such that $\mathbf{Q}^{\mathrm{T}}\mathbf{Q} = \mathbf{L}$. Then the configuration of n points in p space is given by the columns of $\mathbf{Q}$, that is, the latent vectors of $\mathbf{F}$ normalized such that $\mathbf{Q}^{\mathrm{T}}\mathbf{Q} = \mathbf{L}$ and with means at the origin.

For a proof of the theorem see Gower (1966). Theorem 5.6 yields a "full rank" solution to the MDS problem, that is, the points are fully reconstructed within the $p < n - 1$ subspace defined by the p variables if all of the nonzero latent roots and latent vectors are kept (Theorem 5.1). When $r < p$ latent vectors are retained, however, the distances $\hat{d}_{ij}$ within this subspace are still optimal estimates of d_{ij}, in the sense

$$\phi = \sum_{i=1}^{n} \sum_{j=1}^{n} (d_{ij} - \hat{d}_{ij})^2$$

is minimized (see Theorem 3.9). Theorem 5.6 can be extended to non-Euclidian distances.

THEOREM 5.7 (Mardia, 1978). Let $\mathbf{B}$ be an arbitrary (symmetric) distance matrix adjusted as in Eqs. (5.39)–(5.41). Then for a given $l \leq r < n$ there exists a (fitted) positive semidefinite matrix $\hat{\mathbf{B}}$, of rank at most r, such that

$$\sum_{i=1}^{n} \sum_{j=1}^{n} (b_{ij} - \hat{b}_{ij}) = \text{tr}(\mathbf{B} - \hat{\mathbf{B}})^2 \qquad (5.44)$$

PROOF.   Let $l_1 \geq l_2 \geq \cdots \geq l_n$ be the latent roots of $\mathbf{B}$. Then we minimize

$$\text{tr}(\mathbf{B} - \hat{\mathbf{B}})^2 = \sum_{i+1}^{n} (l_i - \hat{l}_i)^2 \qquad (5.45)$$

where $\hat{l}_i$ are latent roots of $\hat{\mathbf{B}}$ and the minimum is taken over nonnegative roots only. Thus when there are r positive latent roots in $\hat{\mathbf{B}}$ we have $l_i = \hat{l}_i$ $(i = 1, 2, \ldots, r)$, and only the first r roots are used when $\hat{\mathbf{B}}$ is positive semidefinite. The minimum of Eq. (5.45) defines a measure of distortion in $\mathbf{B}$, and is given by $\Sigma_{i=1}^{r} l_i^2$. Mardia (1978; see also Mardia et al., 1979) gives other results, including a treatment of missing values (missing distances). The MDS procedure can also be applied to more general Euclidian distances such as the Mahalanobis distance (see Krzanowski, 1976).

5.4.3   Three-Mode Analysis

An immediate extention of the R- and Q-mode models is the so-called three-mode principal components model, which attempts to combine the objectives of both the R- and the Q-mode types of analyses. For the R and Q modes the observed data are represented by a single matrix array with variables as columns (rows) and sample points as rows (columns). A generalization can be imagined however in the form of a three-way rectangular array with typical element y_{ijk}, where the third subscript represents an additional mode or method of classification and yields a total of $n \times p \times t$ observations ($i = 1, 2, \ldots, n$; $j = 1, 2, \ldots, p$; $k = 1, 2, \ldots, t$). Thus the third mode, which is denoted by the extra subscript k, may for example represent different occasions, time periods, geographical areas, or experimental conditions (e.g., see Mills et al., 1978; Sinha et al., 1973). Although three-way data are at times portrayed as a three-dimensional matrix array, this can be misleading since such entities are undefined within the usual matrix algebra which underlies the PC model. Rather we must consider such data as sets of t distinct ($n \times p$) matrix arrays, which are to be analyzed jointly.

Consider a set of t ($n \times p$) data matrices $\mathbf{Y}_{(1)}, \mathbf{Y}_{(2)}, \ldots, \mathbf{Y}_{(t)}$, each consisting of the same random variables and sample points, but at different occasions. Since occasions are not necessarily independent, the purpose is to combine the t matrices into a single PCA, which could exhibit the interrelationships between the three modes of classifying the data. A number of methods have been proposed to achieve this end, for a review (together with an extensive bibliography) see Kroonenberg (1983a and b).

Two-stage Principal Components

Consider t separate principal component analyses carried out on, say, the columns of $\mathbf{Y}_{(1)}, \mathbf{Y}_{(2)}, \ldots, \mathbf{Y}_{(t)}$. Since the variables will generally be correlated across the t occasions, the t sets of loadings should also exhibit correlation, although loading coefficients will still be uncorrelated within each set of variables. It is therefore possible to carry out a second-stage PCA, treating the first-stage components as second-stage input variables, and obtain joint axes within which the original loadings can be plotted for the t occasions. Bouroche and Dussaix (1975), for example, have considered several alternatives for three-way data using such an approach. We briefly describe one of their procedures for illustrative purposes. Assuming the correlational structure among the variables is of interest, the first-stage analysis consists of

$$\mathbf{P}_{(1)}^{T}(\mathbf{X}_{(1)}^{T}\mathbf{X}_{(1)})\mathbf{P}_{(1)} = \mathbf{L}_{(1)}$$

$$\mathbf{P}_{(2)}^{T}(\mathbf{X}_{(2)}^{T}\mathbf{X}_{(2)})\mathbf{P}_{(2)} = \mathbf{L}_{(2)}$$

$$\overline{\mathbf{P}_{(t)}^{T}(\mathbf{X}_{(t)}^{T}\mathbf{X}_{(t)})\mathbf{P}_{(t)} = \mathbf{L}_{(t)}} \tag{5.46}$$

where $\mathbf{X}_{(k)}^{T}\mathbf{X}_{(k)}$ represents a ($p \times p$) association matrix for the kth occasion and $\mathbf{P}_{(k)}$ and $\mathbf{L}_{(k)}$ are corresponding latent vectors and latent roots ($k = 1, 2, \ldots, t$). At this stage all p latent vectors are retained for each occasion.

The second stage consists of correlating the first (dominant) latent vectors $\mathbf{P}_{1(1)}, \mathbf{P}_{1(2)}, \ldots, \mathbf{P}_{1(t)}$ the second latent vectors $\mathbf{P}_{2(1)}, \mathbf{P}_{2(2)}, \ldots, \mathbf{P}_{2(t)}, \ldots,$ the pth latent vectors $\mathbf{P}_{p(1)}, \mathbf{P}_{p(1)}, \ldots, \mathbf{P}_{p(t)}$ across all t occasions. This yields p ($t \times t$) association matrices, each of which are diagonalized in turn and the following latent vectors are retained:

$\mathbf{V}_1 =$ the latent vector associated with the largest latent root of $\mathbf{P}_{1(1)}$, $\mathbf{P}_{1(2)}, \ldots, \mathbf{P}_{1(t)}$

$\mathbf{V}_2 =$ the latent vector associated with the largest latent root of $\mathbf{P}_{2(1)}$, $\mathbf{P}_{2(2)}, \ldots, \mathbf{P}_{2(t)}$

$\mathbf{V}_p =$ the latent vector associated with the largest latent root of $\mathbf{P}_{p(1)}$, $\mathbf{P}_{p(2)}, \ldots, \mathbf{P}_{p(t)}$

In practice the process is terminated when $q < p$ components have been extracted. The projections (coordinates) of the variables onto the second-stage PCs can then be computed in the form of correlation loading coefficients. This yields a summary representation of the p variables across t occasions, within a reduced number of q axes. Clearly the procedure is most appropriate when at the second stage all p correlation matrices possess isotropic latent roots. When this is not the case, the procedure results in information loss.

Example 5.9   (Bouroche and Dussaix, 1975). Annual sample surveys are carried out by SOFRES in France to determine characteristics that influence automobile demand. The questionnaire consists in part of the following five variables:

$Y_1 =$ Price of automobile

$Y_2 =$ Inside space

$Y_3 =$ Trunk size

$Y_4 =$ Index measuring comfort of car

$Y_5 =$ Sound proofing

The variables are measured for $n = 10$ brands of cars over a period of $t = 3$ consecutive years. The data matrices appear in Table 5.17, which represents an automobile popularity typology over a period of three years, and can be used to trace consumer automobile choice. Since the purpose of the analysis is to trace behavior of the five variables over time, the data are centered by variable means, and PCs are computed from correlation matrices. The three sets of loadings for three PCs are given in Table 5.18. Although components are orthogonal within years, they are clearly correlated between years. Carrying out a second-stage PCA of the three sets of components of Table 5.18 then yields a reduced set of latent vectors (Table 5.19). The latent vectors can be standardized in the usual way to yield correlation loadings.

$\square$

Mapping Covariance Matrices

An alternative method, with somewhat similar objectives, has been developed by Escoufier (1980a and b). Consider n individuals, each of which are observed to possess an association matrix $\mathbf{S}_i$ $(i = 1, 2, \ldots, n)$. The

Table 5.17 Three-way Market Data Cross Classified by Car Brand, Reason for Choice, and Year

Car Brand	Y_1	Y_2	Y_3	Y_4	Y_5
			Year 1		
A	338.0	118.0	195.0	173.0	16.0
B	210.0	185.0	210.0	342.0	1.0
C	180.0	404.0	310.0	684.0	38.0
D	142.0	212.0	203.0	708.0	48.0
E	102.0	110.0	52.0	330.0	41.0
F	33.0	313.0	152.0	273.0	39.0
G	207.0	188.0	76.0	456.0	32.0
H	45.0	245.0	61.0	520.0	60.0
I	325.0	35.0	60.0	64.0	0
J	63.0	85.0	14.0	437.0	0
			Year 2		
A	397.0	102.0	445.0	142.0	3.0
B	294.0	145.0	440.0	273.0	13.0
C	239.0	282.0	270.0	518.0	18.0
D	92.0	160.0	301.0	435.0	12.0
E	315.0	78.0	110.0	230.0	30.0
F	205.0	193.0	401.0	125.0	0
G	161.0	286.0	148.0	170.0	15.0
H	114.0	171.0	42.0	391.0	32.0
I	509.0	30.0	95.0	93.0	3.0
J	408.0	39.0	31.0	334.0	5.0
			Year 3		
A	313.0	136.0	447.0	98.0	9.0
B	202.0	261.0	489.0	206.0	16.0
C	172.0	359.0	337.0	313.0	33.0
D	118.0	373.0	404.0	231.0	33.0
E	164.0	215.0	28.0	168.0	50.0
F	65.0	282.0	414.0	130.0	23.0
G	165.0	239.0	53.0	139.0	41.0
H	60.0	346.0	118.0	231.0	27.0
I	375.0	80.0	85.0	147.0	32.0
J	120.0	253.0	49.0	196.0	11.0

Source: Bouroche and Dussaix, 1975; reproduced with permission.

objective consists of the following: (1) to find a graphical representation of the n matrices, (2) to define a new matrix which is a good "compromise" between the original matrices, and (3) to obtain a joint mapping of both the individuals and their covariance matrices S_i.

To achieve the first objective consider n covariance matrices S_i ($i =$

Table 5.18 Principal Component Loadings for Data of Table 5.17

		Year 1			Year 2			Year 3		
		$Z_{1(1)}$	$Z_{2(1)}$	$Z_{3(1)}$	$Z_{1(2)}$	$Z_{2(2)}$	$Z_{3(2)}$	$Z_{1(3)}$	$Z_{2(3)}$	$Z_{3(3)}$
X_1:	Price	.6132	.4766	.3621	−.9342	−.0474	.2439	−.3892	.8397	.2814
X_2:	Space	.8092	.2881	.3759	.7990	.3198	−.2421	.6012	−.7826	.0317
X_3:	Trunk size	.3679	.7408	.4737	.0883	.9612	.2066	.8989	.3774	.2111
X_4:	Comfort	.8840	.2859	−.3445	.7173	−.3017	.6243	.3366	−.5755	.1717
X_5:	Sound proof	.8524	−.4110	.2467	.5507	−.6227	−.1205	−.4892	−.5226	.6855
Latent roots		4.43	1.04	.77	2.48	1.59	.62	1.88	1.59	.47

Source: Bouroche and Dussaix, 1975; reproduced with permission.

Table 5.19 Second-Stage Latent Vectors of the Three Principal Components

		V_1	V_2	V_3
X_1:	Price	−.4498	−.1983	.2167
X_2:	Space	.4895	−.1757	.4377
X_3:	Trunk size	.1838	−.8539	.1889
X_4:	Comfort	.5854	−.0603	−.6552
X_5:	Sound proof	.4262	.4438	.5445

Source: Bouroche and Dussaix, 1975; reproduced with permission.

$1, 2, \ldots, n$), together with the matrix $\mathbf{C}$ with typical element $c_{ij} = \mathrm{tr}(\mathbf{S}_i\mathbf{S}_j)$ $(i = 1, 2, \ldots, n)$. This yields a symmetric, positive definite matrix containing inner products of elements of the original n matrices. The idea is then to compute latent roots and vectors of $\mathbf{C}$, and thus obtain a plot of the $\mathbf{S}_i$.

Second, the "compromise" matrix $\mathbf{S}$ is defined as

$$\mathbf{S} = \sum_{i=1}^{n} p_{i1}\mathbf{S}_i \tag{5.47}$$

a linear combination of the original covariance matrices, where p_{i1} are elements of the first latent vector $\mathbf{P}_1$ of $\mathbf{C}$. To determine how well $\mathbf{S}$ summarizes the n covariance matrices let $\mathbf{Q}$ denote the matrix of latent vectors of $\mathbf{S}$ such that $\mathbf{Q}\mathbf{Q}^T = \mathbf{S}$, and let $\mathbf{Q}_i$ be the matrix of latent vectors of $\mathbf{S}_i$ where $\mathbf{Q}_i\mathbf{Q}_i^T = \mathbf{S}_i$. The latent vectors are thus in effect correlation loadings. Projecting each $\mathbf{Q}_i$ onto the column space of $\mathbf{Q}$ yields the n matrices

$$\mathbf{Q}^{(i)} = \mathbf{Q}(\mathbf{Q}^T\mathbf{Q})^{-1}\mathbf{Q}^T\mathbf{Q}_i$$

$$= \mathbf{Q}_i\boldsymbol{\beta} \qquad (i = 1, 2, \ldots, n)$$

where $\boldsymbol{\beta}_i$ is the vector of regression coefficients of the latent vectors of $\mathbf{Q}$ on those of $\mathbf{Q}_i$.

Example 5.10 (Escoufier, 1980a). Consider $n = 4$ individuals with covariance matrices

$$\mathbf{S}_1 = \begin{bmatrix} 2 & 0 & -2 & 0 \\ 0 & 2 & 0 & -2 \\ -2 & 0 & 2 & 0 \\ 0 & -2 & 0 & 2 \end{bmatrix} \qquad \mathbf{S}_2 = \begin{bmatrix} 5 & 3 & -5 & -3 \\ 3 & 5 & -3 & -5 \\ -5 & -3 & 5 & 3 \\ -3 & -5 & 3 & 5 \end{bmatrix}$$

$$\mathbf{S}_3 = \begin{bmatrix} 5 & -3 & -5 & 3 \\ -3 & 5 & 3 & -5 \\ -5 & 3 & 5 & -3 \\ 3 & -5 & -3 & 5 \end{bmatrix} \qquad \mathbf{S}_4 = \begin{bmatrix} 8 & 0 & -8 & 0 \\ 0 & 8 & 0 & -8 \\ -8 & 0 & 8 & 0 \\ 0 & -8 & 0 & 8 \end{bmatrix}$$

Then

$$
\mathbf{C} = \begin{bmatrix}
32 & 80 & 80 & 128 \\
80 & 272 & 128 & 320 \\
80 & 128 & 272 & 320 \\
128 & 320 & 320 & 512
\end{bmatrix}
$$

Table 5.20   Latent Roots and Vectors of Matrix C with Typical Element $c_{ij} = \mathrm{tr}(\mathbf{S}_i\mathbf{S}_j)$

$\mathbf{P}_1$	$\mathbf{P}_2$
.1842	0
.4602	.7072
.4602	−.7072
.7365	0
$l_1 = 944.0$	$l_2 = 144.0$

with latent roots and vectors as in Table 5.20. Since for the present example $\rho(\mathbf{C}) = 2$, the last two latent roots are zero and we have an exact representation in a two-dimensional space (Fig. 5.5), from which we can see that $\mathbf{S}_1$ and $\mathbf{S}_2$ are proportional while $\mathbf{S}_2$ and $\mathbf{S}_3$ are equidistant from the origin. Also, using Eq. (5.47) we have

$$
\mathbf{S} = \begin{bmatrix}
10.86 & 0 & -10.86 & 0 \\
0 & 10.86 & 0 & -10.86 \\
-10.86 & 0 & 10.86 & 0 \\
0 & -10.86 & 0 & 10.86
\end{bmatrix}
$$

with the nonzero latent roots $l_1 = l_2 = 21.72$ and corresponding latent vectors

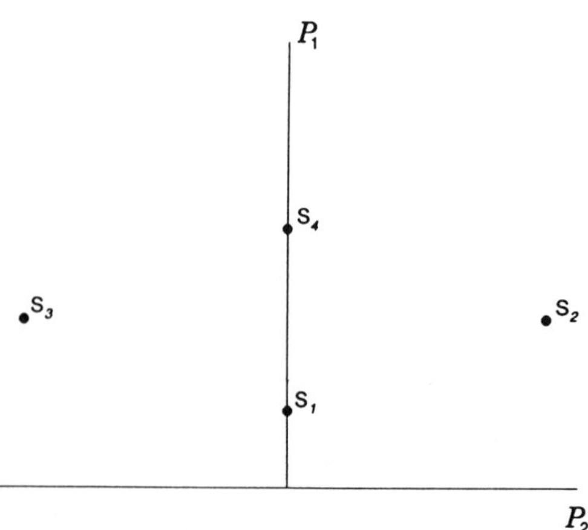

Figure 5.5   The $n = 4$ covariance matrices in a reduced space.

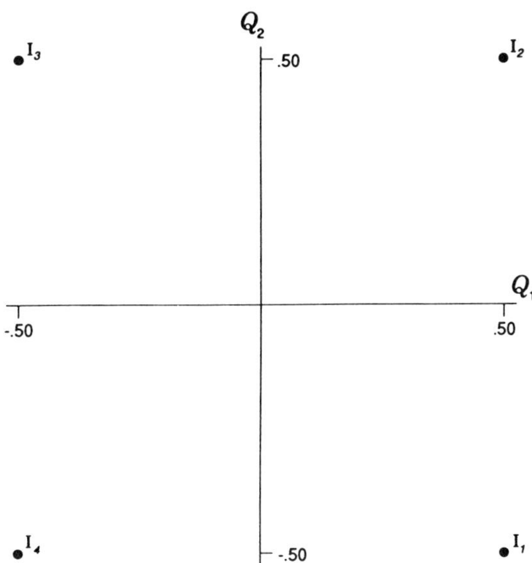

Figure 5.6 The $n = 4$ individuals mapped on the space defined by the latent vectors $\mathbf{Q}_1$ and $\mathbf{Q}_2$ of matrix $\mathbf{S}$.

$(.50, .50, -.50, -50)^{\mathrm{T}}$ and $(-.50, .50, .50, -50)^{\mathrm{T}}$, which form vertices of a square (Fig. 5.6). The discrepancy between the representation of the individuals given by the compromise matrix and the representation obtained from the initial matrices $\mathbf{S}_i$ can then be seen from Figure 5.7.

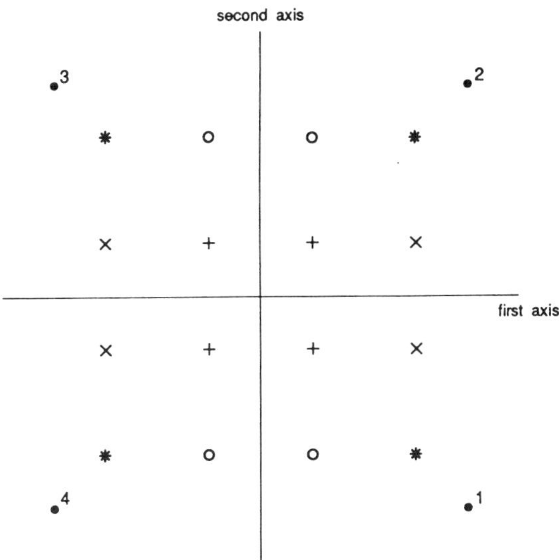

Figure 5.7 Graphical representation of the matrices $\hat{\mathbf{Q}}^{(i)}$ in the space of the compromise matrix ($\mathbf{S}_1 = +$; $\mathbf{S}_2 = \times$; $\mathbf{S}_3 = \bigcirc$; $\mathbf{S}_4 = *$).

Tucker's Method

A better known methodology for three-mode PC is from Tucker (1966, 1967). The method consists of three sets of latent vectors plus the so-called "core matrices" which relate loadings to each of the three data modes. Before PCs can be computed, however, the three-way data must be arranged in a matrix array. This may be done in one of several ways. For example, the t $(n \times p)$ matrices may be "stacked" on top of each other to yield an augmented (partitioned) data supermatrix, which is then used to compute a covariance (correlation) matrix. Alternatively the modes can be "nested" within each other. The method adopted by Tucker is as follows. Assuming the three-way data have been centered in an appropriate way, they can then be expressed in the form of three distinct matrices; the $(n \times pt)$ matrix $_n\mathbf{X}$, the $(p \times nt)$ matrix $_p\mathbf{X}$, and the $(t \times np)$ matrix $_t\mathbf{X}$. The matrices are decomposed into principal components and the three outputs related by means of the "core" matrices. We have

$$_n\mathbf{X} = \mathbf{AF}(\mathbf{B} \otimes \mathbf{C})$$

$$_p\mathbf{X} = \mathbf{B}^\mathrm{T}\mathbf{G}(\mathbf{A}^\mathrm{T} \otimes \mathbf{C})$$

$$_t\mathbf{X} = \mathbf{C}\,\mathbf{H}(\mathbf{A} \otimes \mathbf{B}) \tag{5.49}$$

where $\otimes$ denotes the Kronecker product (Section 2.9) and

$\mathbf{A}$ = the $(n \times n)$ matrix of latent vectors of $_n\mathbf{X}_n\mathbf{X}^\mathrm{T}$
$\mathbf{B}$ = the $(p \times p)$ matrix of latent vectors of $_p\mathbf{X}_p\mathbf{X}^\mathrm{T}$
$\mathbf{C}$ = the $(t \times t)$ matrix of latent vectors of $_t\mathbf{X}_t\mathbf{X}^\mathrm{T}$

The matrices $\mathbf{F}$, $\mathbf{G}$, and $\mathbf{H}$ are the core matrices whose orders are always the same as the left-hand side data matrices of Eq. (5.49). A three-mode core matrix relates the three types of factors, and as its name implies, is considered to contain the basic relations within the three mode data under study. The core matrices are computed from the data matrices and the latent vectors—for example, the matrix $\mathbf{F}$ is obtained as

$$\mathbf{F} = \mathbf{A}_n^\mathrm{T}\mathbf{X}(\mathbf{B}^\mathrm{T} \otimes \mathbf{C}^\mathrm{T}) \tag{5.50}$$

and similarly for $\mathbf{G}$ and $\mathbf{H}$. Although Eq. (5.49) utilize all the latent vectors of the three data matrices, the idea is to use a reduced number. Also as in the usual two-mode PCA, the loadings can be rotated to aid interpretation.

The three-mode PC model can be extended to the general n-mode case in a more-or-less routine fashion. The notation, particularly subscripting, becomes somewhat tedious and to overcome this Kapteyn et al. (1986) introduce an alternative development of the model.

Example 5.11. Hohn (1979) has presented an example of Tucker's three-mode model using Oudin's (1970) geological data, consisting of elemental (chloroform) analysis of heavy fractions of organic extracts of Jurassic shales of the Paris basin. The percentage composition of the extracts are given in Table 5.21. The data arrangement is equivalent to a three-way, completely cross-classified design with a single replication per cell. The purpose of the analysis is to compare composition of chemical elements. Thus prior to the elemental analysis the organic extract of each of the four samples is fractioned into three categories; resin, CCl_4 soluble asphaltenes, and CCl_4 insoluble asphaltenes. The three modes are therefore fraction, locality, and elemental composition. The three sets of PC latent vectors are shown in Tables 5.22–5.24. Since interpretation is more difficult using the orthonormal latent vectors, they are standardized to yield correlation loadings. The first component of Table 5.22 represents the joint

Table 5.21 Compositional Analyses of Elements of Heavy Fractions of Organic Extracts; Jurassic shales of the Paris Basin, France

Fraction	Locality	Elemental Percentage Composition					
		C	H	C/H	O	N	S
Resin	Echanay	72.44	9.24	7.83	8.16	.56	2.53
	Ancerville	76.14	9.39	8.11	6.18	1.23	3.68
	Essises	77.61	8.68	8.94	5.58	1.12	.29
	Bouchy	81.46	7.63	10.67	4.59	.88	.42
CCl_4	Echanay	65.03	7.14	9.10	10.90	1.09	2.38
soluble	Ancerville	71.42	7.68	9.29	11.61	1.39	1.69
asphaltenes	Essises	79.44	7.22	11.09	8.65	1.07	.18
	Bouchy	78.34	6.56	11.92	5.96	1.66	.75
CCl_4	Echany	66.38	6.61	10.04	18.32	.00	1.65
insoluble	Ancerville	73.20	7.61	9.62	10.82	2.67	.00
asphaltenes	Essises	75.61	6.46	11.70	11.76	1.54	.80
	Bouchy	79.90	6.65	12.01	7.46	1.92	.65

Source: Oudin (1970).

Table 5.22 The First Three Latent Vectors of Elemental Percentages for Three Fractions and Four Locations After Varimax Rotation

Element	1	2	3
C	−.2057	.6602	−.0728
H	.5881	.2899	.0365
C/H	−.6201	.0832	−.0817
O	−.1240	−.6721	−.0571
N	−.0201	−.0180	.9784
S	.4599	−.1454	−.1620

Source: Hohn, 1979; reproduced with permission.

Table 5.23   First Two Latent Vectors of Fractions After Varimax Rotation

Fraction	1	2
Resin	.0267	.9563
Soluble asphalt	.4914	.2410
Insoluble asphalt	.8705	−.1654

Source: Hohn, 1979; reproduced with permission.

Table 5.24   First Three Latent Vectors of Locations After Varimax Rotation

Location	1	2	3
Echanay	.9661	.0141	−.0036
Ancerville	−.0031	.0019	.9997
Essises	.1867	.6516	.0181
Bouchy	−.1783	.7584	−.0179

Source: Hohn 1979; reproduced with permission.

presence of H and S (which is also accompanied by a lack of C/H), the second reveals a contrast between C and O, while the third represents the single element S. The remaining two tables are interpreted in a similar manner. The fractions yield two components (insoluble asphalt and resin, respectively) while the locations (listed in order of increasing depth) reveal three PCs correlated to Echanay, Essises and Bouchy, and Ancerville.

The first element component (high H, S; low C/W) has a high weight (.8670) for the second fraction component (resins) and the first locality component (Echanay), but a high negative weight (−.9433) for the same elements for the first fractions component (insoluble asphaltenes) for the second locality component (Essises and Bouchy) as seen from the core matrix (Table 5.25). Thus we have high H, S (low C/H) in the resins for the

Table 5.25   Core Matrix with Components for Elements, Fractions, and Localities

	1	2	3
		Element	
	1		
Fraction 1	.0841	−.9433	.0233
2	.8670	−.0291	.8957
	2		
Fraction 1	−1.0819	.2639	−.2327
2	.0276	.7281	.1798
	3		
Fraction 1	−.5856	.2414	.6120
2	−.1716	−.1282	−.1512

Source: Hohn, 1979; reproduced with permission.

shallowest sample (Esheney) but low H, S (high C/H) in the insoluble asphaltenes for the deeper samples (Essises, Bouchy). Similarly the second element component (high C, low O) weights positively for the combination: resins (fraction component 2) and sample component 2 (Essises, Bouchy), but negatively for the combination of fraction component 1 (insoluble asphaltenes) and sample component 1 (Eschanay). Thus Hohn (1979) concludes that the shallowest sample is characterized by high H and S in the resins, and low C and high O in the insoluble asphaltenes. With depth the resins become richer in C and poorer in O, whereas the insoluble asphaltenes become poorer in both H and S. The composition of a given sample appears to be a function of both depth and fraction, but the fractions do not appear to behave in an identical manner. The three-mode PCA thus serves as a useful exploratory tool and may provide information concerning interaction effects and possible contrasts. The three-mode PC model can also be used to perform an MDS analysis (Tucker, 1972).					□

5.4.4 Joint Plotting of Loadings and Scores

The loading coefficients of $\mathbf{X}^T\mathbf{X}$ describe the variable space and the scores contain information concerning the sample space. At times it is of considerable interest to be able to compare the two spaces. For example, we may wish to know which sample points are particularly influenced by certain variable clusters as represented by the first sets of PC loadings, and vice versa. Since the spectral decompositions of $\mathbf{X}^T\mathbf{X}$ and $\mathbf{X}\mathbf{X}^T$ are duals, a PCA can provide useful information as to the interrelationship(s) between the sample space and the variable space.

The most straightforward method of comparing the two spaces is by a visual inspection of the loadings and the scores (Example 3.6). Here large loading coefficients are matched with relatively high PC scores, with matches of equal signs denoting presence and those of opposite signs denoting absence of some variable space attribute in a particular sample point. Since both loadings and scores form orthogonal sets, unidimensional and multidimensional comparisons are possible. A two-dimensional joint plot of the (unit length) latent vectors of $\mathbf{X}^T\mathbf{X}$ and $\mathbf{X}\mathbf{X}^T$ (loadings and scores of $\mathbf{X}^T\mathbf{X}$) is shown in Figure 5.8. Such plots have also been called "biplots" (Gabriel 1971) and are most appropriate for small data matrices for which most of the information is concentrated in low-dimensional subspaces. Also since rows and columns of $\mathbf{X}$ represent different vector spaces, care must be taken when interpreting such plots, that is, when comparing overall proximities of the loadings and the scores. Although both sets of vectors are standardized to common (unit) length, the common coordinate system represents a somewhat artificial device. A joint plot, on the other hand, is not without utility. Frequently relative position is of interest, and here biplots can be used to compare relative distances between two sample points and all the variables, as well as the relative distances between any two

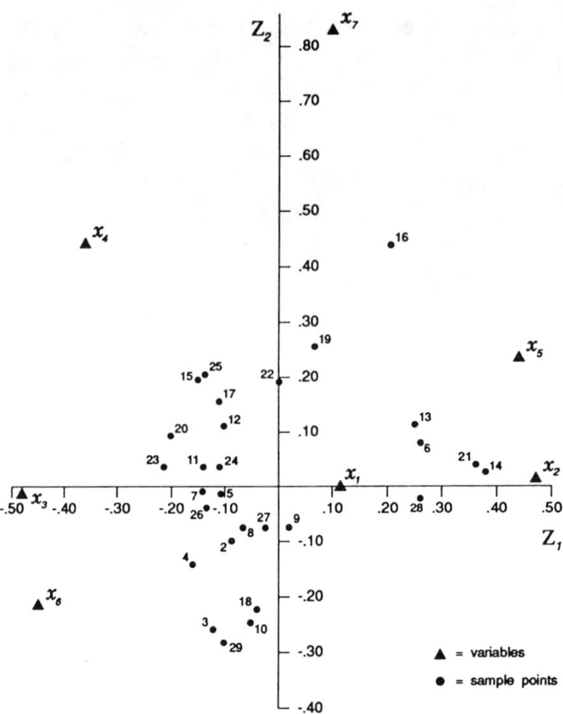

Figure 5.8   Joint plot of standardized principal component loadings and scores of London data of Example 3.6.

variables and all the sample points. Such plots are particularly useful in exploratory analysis since they can indicate broad, qualitative structure(s) of the data, model specification, and so forth (Bradu and Grine, 1979; Weber and Berger, 1978).

The plot of Figure 5.8 is asymmetric, in the sense that rows and columns of $\mathbf{X}$ are not treated equally with respect to scale and origin. To obtain a symmetric joint plot of the loadings and the scores a variation of principal axes (coordinates) adjustment is at times used (Section 5.4.2). Introduced by Benzecri (1970) the method has become known as "analyses factoriéle des correspondences" or "correspondence analysis" in English. It consists of adjusting the rows and columns of $\mathbf{Y}$ to the same origin and scale, which facilitates their joint plotting in a reduced space, usually two-dimensional. The distinctive feature of correspondence analysis however is that it considers a data matrix as a bidimensional probability distribution. First, each entry of $\mathbf{Y}$ is divided by the sum of all the entries, that is, $\mathbf{Y}$ is first converted to the matrix $\mathbf{F}$ with typical element

$$f_{ij} = \frac{y_{ij}}{\displaystyle\sum_{i=1}^{n}\sum_{j=1}^{p} y_{ij}}$$

Next, the relative frequencies f_{ij} are adjusted for scale and location, which yields a matrix with elements

$$\frac{f_{ij}}{\sqrt{f_{i.}f_{.j}}} - \sqrt{f_{i.}}\,\sqrt{f_{.j}}$$

where $f_{i.}$ and $f_{.j}$ are row and column totals respectively. A matrix $\mathbf{V}$ of sums of squares and products can then be defined with elements

$$v_{jk} = \sum_{i=1}^{n}\left(\frac{f_{ij}}{\sqrt{f_{i.}}\,\sqrt{f_{.j}}} - \sqrt{f_{i.}}\,\sqrt{f_{.j}}\right)\left(\frac{f_{ik}}{\sqrt{f_{i.}}\,\sqrt{f_{.k}}} - \sqrt{f_{i.}}\,\sqrt{f_{.k}}\right)$$

Owing to the symmetric adjustment of the variables and the sample points, the loadings and the scores of the matrix $\mathbf{V}$ can be compared mathematically on a more equal footing. Because of the constant-sum rows and columns the matrix $\mathbf{V}$ is of rank $r = \min(p-1, n-1)$, and the smallest latent root is therefore zero (Section 5.9.1). For numerical examples and computer programs see David et al. (1977), Lebart et al. (1984), Greenacre (1984). In some countries, particularly France, the method has become a popular procedure for "data analysis (see Deville and Malinvaud, 1983). The method is pursued further in Chapter 9 in the context of discrete data.

Several limitations of the method and its rationale emerge when it is applied to continuous random variables. First, because of the nature of the adjustments, the entries of the matrix $\mathbf{F}$ are considered in terms of "probabilities," an undefined concept when dealing with specific values of a continuous variable. To be theoretically meaningful the columns of $\mathbf{Y}$ must first be broken down into discrete intervals, thus effectively destroying the continuous nature of the data. Also, the data must be nonnegative in order to be discretized so that positive frequencies can be obtained. The biggest objection to correspondence analysis of data matrices however lies in its implicit mispecification of the matrix $\mathbf{Y}$, when columns constitute the variable space and rows represent a sample space. Owing to the symmetric nature of the adjustments, the rows and columns of $\mathbf{Y}$ are assumed to represent a bidimensional sample, much in the same way as a contingency table. Such a representation is clearly of questionable validity for a sample taken from a multidimensional distribution. A comment of a more historical nature also seems to be in order. Although for discrete data correspondence analysis represents a straightforward (graphic) extension of PC/canonical correlation-type models (Section 8.5), this seems to have been somewhat obscured in the past (e.g., see Theil, 1975), perhaps because of the literal, nonstandard translation of statistical terms from French into English.* The result has been an overinflated claim of originality for the procedure.

---

* For example, a scatter diagram is referred to as a "cloud," (weighted) variance becomes "inertia," and so forth.

5.5   OTHER METHODS FOR MULTIVARIABLE AND MULTIGROUP PRINCIPAL COMPONENTS

The previous section dealt with an extension of PCA to data sets that are cross-classified according to three criteria or modes—p random variables, n sample points, and t occasions or time periods. The PC loading coefficients can then be computed so that we can observe the interrelated and interactive structure of the data. Two other extensions of PCA are possible: when we have more than one set of variables, and when there are two or more samples. In this section we describe several methods that can be used ·to analyze simultaneously more than one data set.

5.5.1   The Canonical Correlation Model

Due to Hotelling (1936b), the canonical correlation model can be viewed as a generalization of PCA since it seeks to account for the multidimensional correlational structure between two sets of variates observed for the same sample. Consider two sets of variates $\mathbf{X}_{(1)}$ and $\mathbf{X}_{(2)}$, each containing p_1 and p_2 random variables respectively. We assume that the variables are measured about their means, $p_1 < p_2$ and $p_1 + p_2 = p$ and both sets are observed for the same n sample points. The $(n \times p)$ data matrix can then be written in the vertically partitioned form $\mathbf{X} = [\mathbf{X}_{(1)} : \mathbf{X}_{(2)}]$, where $\mathbf{X}_{(1)}$ is $(n \times p_1)$ and $\mathbf{X}_{(2)}$ is $(n \times p_2)$. The matrix $\mathbf{X}^T\mathbf{X}$ can be expressed in the partitioned form as

$$\mathbf{X}^T\mathbf{X} = \begin{bmatrix} \mathbf{X}^T_{(1)} \\ \cdots \\ \mathbf{X}^T_{(2)} \end{bmatrix} \begin{bmatrix} \mathbf{X}_{(1)} & : & \mathbf{X}_{(2)} \end{bmatrix}$$

$$\begin{bmatrix} \mathbf{X}^T_{(1)}\,\mathbf{X}_{(1)} & \vdots & \mathbf{X}^T_{(1)}\,\mathbf{X}_{(2)} \\ \cdots\cdots\cdots & \vdots & \cdots\cdots\cdots \\ \mathbf{X}^T_{(2)}\,\mathbf{X}_{(1)} & \vdots & \mathbf{X}^T_{(2)}\,\mathbf{X}_{(2)} \end{bmatrix} \tag{5.51}$$

where $\mathbf{X}^T_{(2)}\mathbf{X}_{(1)} = (\mathbf{X}^T_{(1)}\mathbf{X}_{(2)})^T$ is a $(p_2 \times p_1)$ matrix. The objective of the canonical correlation model is to provide an internal analysis of the correlational structure between the two sets of variables. To derive the model we consider the p-component partitioned vector of population random variables:

$$\mathbf{X} = [\mathbf{X}_{(1)} \cdot \mathbf{X}_{(2)}]^T = [X_{(1)1}X_{(1)2} \cdots X_{(1)p1} : X_{(2)1}X_{(2)2} \cdots X_{(2)p2}]^T$$

with covariance matrix

$$\boldsymbol{\Sigma} = \begin{bmatrix} \boldsymbol{\Sigma}_{11} & \vdots & \boldsymbol{\Sigma}_{12} \\ \cdots & \vdots & \cdots \\ \boldsymbol{\Sigma}_{21} & \vdots & \boldsymbol{\Sigma}_{22} \end{bmatrix} \tag{5.52}$$

Then given $\mathbf{X}$ and $\mathbf{\Sigma}$ the problem of canonical correlation can be stated more precisely as follows; how can we compute a linear combination

$$u = \alpha_1 X_{(1)1} + \alpha_2 X_{(1)2} + \cdots + \alpha_{p1} X_{(1)p1} = \boldsymbol{\alpha}^{\mathrm{T}} \mathbf{X}_{(1)}$$

in the first group, and a linear combination

$$v = \beta_1 X_{(2)1} + \beta_2 X_{(2)2} + \cdots + \beta_{p2} X_{(2)p2} = \boldsymbol{\beta}^{\mathrm{T}} \mathbf{X}_{(2)}$$

in the second, such that the correlation between the two linear combinations is maximized. Hotelling's (1936b) solution to the problem can be stated in the following theorem (see Anderson, 1958).

THEOREM 5.8. Let $\mathbf{\Sigma}$ be a $(p \times p)$ partitioned covariance matrix. The coefficients that maximize the correlation between the linear combinations $u = \boldsymbol{\alpha}^{\mathrm{T}} \mathbf{X}_{(1)}$ and $v = \boldsymbol{\beta}^{\mathrm{T}} \mathbf{X}_{(2)}$ are the latent vector solutions of the systems of equations.

$$(\mathbf{\Sigma}_{11}^{-1} \mathbf{\Sigma}_{12} \mathbf{\Sigma}_{22}^{-1} \mathbf{\Sigma}_{21} - \lambda^2) \boldsymbol{\alpha} = \mathbf{0}$$

$$(\mathbf{\Sigma}_{22}^{-1} \mathbf{\Sigma}_{21} \mathbf{\Sigma}_{11}^{-1} \mathbf{\Sigma}_{12} - \mu^2) \boldsymbol{\beta} = \mathbf{0}$$

where the matrices are as in Eq. (5.52) and $\lambda = \mu = \boldsymbol{\alpha}^{\mathrm{T}} \mathbf{\Sigma}_{12} \boldsymbol{\beta}$ is the maximum correlation.

PROOF. For simplicity assume both linear combinations are standardized to unit variance, that is,

$$\mathrm{var}(u) = E(u^2) = E(\boldsymbol{\alpha}^{\mathrm{T}} \mathbf{X}_{(1)} \mathbf{X}_{(1)}^{\mathrm{T}} \boldsymbol{\alpha} = \boldsymbol{\alpha}^{\mathrm{T}} \mathbf{\Sigma}_{11} \boldsymbol{\alpha} = 1$$

$$\mathrm{var}(v) = E(v^2) = E(\boldsymbol{\beta}^{\mathrm{T}} \mathbf{X}_{(2)} \mathbf{X}_{(2)}^{\mathrm{T}} \boldsymbol{\beta}) = \boldsymbol{\beta}^{\mathrm{T}} \mathbf{\Sigma}_{22} \boldsymbol{\beta} = 1$$

The correlation between u and v is then given by

$$E(uv) = E(\boldsymbol{\alpha}^{\mathrm{T}} \mathbf{X}_{(1)} \mathbf{X}_{(2)}^{\mathrm{T}} \boldsymbol{\beta}) = \boldsymbol{\alpha}^{\mathrm{T}} \mathbf{\Sigma}_{12} \boldsymbol{\beta}$$

which is to be maximized subject to the constraint that u and v are unit vectors. We have the Lagrangian expression

$$\phi = \boldsymbol{\alpha}^{\mathrm{T}} \mathbf{\Sigma}_{12} \boldsymbol{\beta} - \frac{1}{2} \lambda (\boldsymbol{\alpha}^{\mathrm{T}} \mathbf{\Sigma}_{11} \boldsymbol{\alpha} - 1) - \frac{1}{2} \mu (\boldsymbol{\beta}^{\mathrm{T}} \mathbf{\Sigma}_{22} \boldsymbol{\beta} - 1) \qquad (5.53)$$

and differentiating with respect to the coefficients and setting to zero yields the normal equations

$$\frac{\partial \phi}{\partial \boldsymbol{\alpha}} = \boldsymbol{\Sigma}_{12}\hat{\boldsymbol{\beta}} - \hat{\lambda}\boldsymbol{\Sigma}_{11}\hat{\boldsymbol{\alpha}} = \mathbf{0} \tag{5.54}$$

$$\frac{\partial \phi}{\partial \boldsymbol{\beta}} = \boldsymbol{\Sigma}_{12}^{T}\hat{\boldsymbol{\alpha}} - \hat{\mu}\boldsymbol{\Sigma}_{22}\hat{\boldsymbol{\beta}} = \mathbf{0} \tag{5.55}$$

Multiplying Eq. (5.54) by λ and Eq. (5.55) by $\boldsymbol{\Sigma}_{22}^{-1}$ and rearranging yields

$$\hat{\lambda}\boldsymbol{\Sigma}_{12}\hat{\boldsymbol{\beta}} = \hat{\lambda}^{2}\boldsymbol{\Sigma}_{11}\hat{\boldsymbol{\alpha}} \tag{5.56}$$

$$\boldsymbol{\Sigma}_{22}^{-1}\boldsymbol{\Sigma}_{21}\hat{\boldsymbol{\alpha}} = \hat{\mu}\hat{\boldsymbol{\beta}} \tag{5.57}$$

and letting $\hat{\mu} = \hat{\lambda}$ and substituting Eq. (5.57) into Eq. (5.56) we then have

$$(\boldsymbol{\Sigma}_{11}^{-1}\boldsymbol{\Sigma}_{12}\boldsymbol{\Sigma}_{22}^{-1}\boldsymbol{\Sigma}_{21} - \hat{\lambda}^{2})\hat{\boldsymbol{\alpha}} = \mathbf{0} \tag{5.58}$$

Alternatively we can write

$$(\boldsymbol{\Sigma}_{22}^{-1}\boldsymbol{\Sigma}_{21}\boldsymbol{\Sigma}_{11}^{-1}\boldsymbol{\Sigma}_{12} - \hat{\lambda}^{2})\hat{\boldsymbol{\beta}} = \mathbf{0} \tag{5.59}$$

where $\hat{\mu}^{2} = \hat{\lambda}^{2} = \hat{\rho}^{2}$, the generalized multiple correlation coefficient(s) between the two sets of variates.

Thus the largest (positive) correlation between the linear combinations u and v is the positive square root of the largest latent root $\hat{\lambda}_{1}^{2}$, the second largest is the positive square root of $\hat{\lambda}_{2}^{2}$, and so forth until all $\hat{\lambda}_{1}^{2} \geq \hat{\lambda}_{2}^{2} \geq \cdots \geq \hat{\lambda}_{p}^{2}$ roots are known. The number of nonzero latent roots is equal to $\rho(\boldsymbol{\Sigma}_{12})$. In practice both vectors $\hat{\boldsymbol{\alpha}}$ and $\hat{\boldsymbol{\beta}}$ can be computed from Eq. (5.58) since

$$\hat{\boldsymbol{\alpha}} = \frac{\boldsymbol{\Sigma}_{11}^{-1}\boldsymbol{\Sigma}_{12}\hat{\boldsymbol{\beta}}}{\hat{\lambda}}, \qquad \hat{\boldsymbol{\beta}} = \frac{\boldsymbol{\Sigma}_{22}^{-1}\boldsymbol{\Sigma}_{21}\hat{\boldsymbol{\alpha}}}{\hat{\lambda}} \tag{5.60}$$

(see Exercise 5.7) and the normal equations (Eq. 5.58) suffice to carry out a canonical correlation analysis. In particular, one of the sets may be viewed as dependent and the other as independent, in which case the canonical correlation model can be considered as an extention of multiple regression. In this situation the largest root $\hat{\lambda}_{1}^{2}$ is viewed as the (largest) coefficient of multiple determination and the canonical weights as regression coefficients of one set upon the other.

To observe better the geometric properties of the model, and to relate it to PCA, consider the partitioned sample matrix (Eq. 5.51). Let $\mathbf{R}^{2}$ denote the diagonal matrix of sample latent roots r_{i}^{2} $(i = 1, 2, \ldots, p_{1})$ where r_{i}^{2} are sample equivalents of $\hat{\lambda}_{i}^{2}$, that is r_{i}^{2} are solutions of the determinantal equation

$$|(\mathbf{X}_{(1)}^{\mathrm{T}}\mathbf{X}_{(1)})^{-1}(\mathbf{X}_{(1)}^{\mathrm{T}}\mathbf{X}_{(2)})(\mathbf{X}_{(2)}^{\mathrm{T}}\mathbf{X}_{(2)})^{-1}(\mathbf{X}_{(2)}^{\mathrm{T}}\mathbf{X}_{(1)}) - r^2\mathbf{I}| = 0 \qquad (5.61)$$

Let $\mathbf{A}$ and $\mathbf{B}$ represent sample equivalents of α and β. Then the normal equation for $\mathbf{A}$ is

$$\mathbf{X}_{(1)}^{\mathrm{T}}\mathbf{X}_{(2)}(\mathbf{X}_{(2)}^{\mathrm{T}}\mathbf{X}_{(2)})^{-1}\mathbf{X}_{(2)}^{\mathrm{T}}\mathbf{X}_{(1)}\mathbf{A} = \mathbf{X}_{(1)}^{\mathrm{T}}\mathbf{X}_{(1)}\mathbf{A}\mathbf{R}^2$$

or

$$\mathbf{A}^{\mathrm{T}}\mathbf{X}_{(1)}^{\mathrm{T}}\mathbf{X}_{(2)}(\mathbf{X}_{(2)}^{\mathrm{T}}\mathbf{X}_{(2)})^{-1}\mathbf{X}_{(2)}^{\mathrm{T}}\mathbf{X}_{(1)}\mathbf{A} = \mathbf{A}^{\mathrm{T}}\mathbf{X}_{(1)}^{\mathrm{T}}\mathbf{X}_{(1)}\mathbf{A}\mathbf{R}^2 = \mathbf{R}^2 \qquad (5.62)$$

since $\mathbf{A}^{\mathrm{T}}\mathbf{X}_{(1)}^{\mathrm{T}}\mathbf{X}_{(1)}\mathbf{A} = \mathbf{I}$ by the normalization rule. An equivalent result holds for $\mathbf{B}$. The two sets of coefficients are related by the regression relations

$$\mathbf{A} = \mathbf{R}^{-1}(\mathbf{X}_{(1)}^{\mathrm{T}}\mathbf{X}_{(1)})^{-1}\mathbf{X}_{(1)}^{\mathrm{T}}\mathbf{X}_{(2)}\mathbf{B}$$

$$\mathbf{B} = \mathbf{R}^{-1}(\mathbf{X}_{(2)}^{\mathrm{T}}\mathbf{X}_{(2)})^{-1}\mathbf{X}_{(2)}^{\mathrm{T}}\mathbf{X}_{(1)}\mathbf{A} \qquad (5.63)$$

The number of nonzero roots r_i^2 is equal to $\rho(\mathbf{X}_{(1)}^{\mathrm{T}}\mathbf{X}_{(2)})$, and corresponding to these values we obtain the two sets of standardized canonical variates

$$\mathbf{Z}_{(1)} = \mathbf{X}_{(1)}\mathbf{A} , \qquad \mathbf{Z}_{(2)} = \mathbf{X}_{(2)}\mathbf{B} \qquad (5.64)$$

which are analogous to PCs except that they maximize multiple correlation coefficients between the two sets rather than the variance within sets. The matrices $\mathbf{A}$, $\mathbf{B}$, and $\mathbf{R}^2$ provide a spectral decomposition of the portioned submatrices of $\mathbf{X}^{\mathrm{T}}\mathbf{X}$.

THEOREM 5.9. Let $\mathbf{X}^{\mathrm{T}}\mathbf{X}$ be a partitioned dispersion matrix as in Eq. (5.51). Then the submatrices of $\mathbf{X}^{\mathrm{T}}\mathbf{X}$ have the following spectral representation.

(i) $\mathbf{X}_{(1)}^{\mathrm{T}}\mathbf{X}_{(1)} = (\mathbf{A}^{\mathrm{T}})^{-1}\mathbf{A}^{-1}$

(ii) $\mathbf{X}_{(2)}^{\mathrm{T}}\mathbf{X}_{(2)} = (\mathbf{B}^{\mathrm{T}})^{-1}\mathbf{B}^{-1}$

(iii) $\mathbf{X}_{(1)}^{\mathrm{T}}\mathbf{X}_{(2)} = (\mathbf{A}^{\mathrm{T}})^{-1}\mathbf{R}\mathbf{B}^{-1}$

The first two parts of the theorem follow from the normalization rule while part iii is a direct consequence of the spectral decomposition Theorem 5.7.

We can thus write

$$
\mathbf{Z}^{\mathrm{T}}\mathbf{Z} = \begin{bmatrix} \mathbf{Z}_{(1)}^{\mathrm{T}} \\ ---- \\ \mathbf{Z}_{(2)}^{\mathrm{T}} \end{bmatrix} [\mathbf{Z}_{(1)} \vdots \mathbf{Z}_{(2)}] = \left[\begin{array}{c:c} \mathbf{Z}_{(1)}^{\mathrm{T}}\mathbf{Z}_{(1)} & \mathbf{Z}_{(1)}^{\mathrm{T}}\mathbf{Z}_{(2)} \\ \hdashline \mathbf{Z}_{(2)}^{\mathrm{T}}\mathbf{Z}_{(1)} & \mathbf{Z}_{(2)}^{\mathrm{T}}\mathbf{Z}_{(2)} \end{array}\right]
$$

$$
= \left[\begin{array}{c:c} \mathbf{A}^{\mathrm{T}}\mathbf{X}_{(1)}^{\mathrm{T}}\mathbf{X}_{(1)}\mathbf{A} & \mathbf{A}^{\mathrm{T}}\mathbf{X}_{(1)}^{\mathrm{T}}\mathbf{X}_{(2)}\mathbf{A} \\ \hdashline \mathbf{B}^{\mathrm{T}}\mathbf{X}_{(2)}^{\mathrm{T}}\mathbf{X}_{(1)}\mathbf{A} & \mathbf{B}^{\mathrm{T}}\mathbf{X}_{(2)}^{\mathrm{T}}\mathbf{X}_{(2)}\mathbf{B} \end{array}\right]
$$

$$
+ \left[\begin{array}{c:c} \mathbf{I}_{p1} & \mathbf{R} \\ \hdashline \mathbf{R} & \mathbf{I}_{p2} \end{array}\right] \tag{5.65}
$$

where $\mathbf{I}_{p1}$ and $\mathbf{I}_{p2}$ are $(p_1 \times p_1)$ and $(p_2 \times p_2)$ unit matrices respectively, and $\mathbf{R}$ is a $(p_1 \times p_2)$ matrix with p_1 diagonal elements $1 \geq r_1 > r_2 > \cdots > r_{p1} \geq 0$ and the remaining elements are zero. Roots corresponding to linear dependence assume values of unity and those that correspond to orthogonality are identically zero. The diagonal elements of $\mathbf{R}$ can therefore be considered as latent roots of the bilinear form $\mathbf{B}^{\mathrm{T}}\mathbf{X}_{(2)}^{\mathrm{T}}\mathbf{X}_{(1)}\mathbf{A}$.

An interpretation which is more in accord with regression analysis can also be placed on the diagonal matrix $\mathbf{R}^2$. Replacing Eq. (5.64) into Eq. (5.62) we have

$$
\mathbf{R}^2 = \mathbf{Z}_{(1)}^{\mathrm{T}}\mathbf{X}_{(2)}(\mathbf{X}_{(2)}^{\mathrm{T}}\mathbf{X}_{(2)})^{-1}\mathbf{X}_{(2)}^{\mathrm{T}}\mathbf{Z}_{(1)}
$$

$$
= \mathbf{Z}_{(1)}^{\mathrm{T}}\mathbf{P}_{(2)}\mathbf{Z}_{(1)} \tag{5.66}
$$

where $\mathbf{P}_{(2)}$ is a symmetric projection matrix. The diagonal elements of $\mathbf{R}^2$ are therefore percentages of variance of the set $\mathbf{X}_{(1)}$ accounted for by the set $\mathbf{X}_{(2)}$. Also, it can be shown that

$$
\mathbf{R}^2 = \mathbf{Z}_{(2)}^{\mathrm{T}}\mathbf{P}_{(1)}\mathbf{Z}_{(2)} \tag{5.67}
$$

where $\mathbf{P}_{(1)} = \mathbf{X}_{(1)}(\mathbf{X}_{(1)}^{\mathrm{T}}\mathbf{X}_{(1)})^{-1}\mathbf{X}_{(1)}^{\mathrm{T}}$. We also have the determinant

$$
|(\mathbf{X}_{(1)}^{\mathrm{T}}\mathbf{X}_{(1)})^{-1}(\mathbf{X}_{(1)}^{\mathrm{T}}\mathbf{X}_{(2)})^{-1}\mathbf{X}_{(2)}^{\mathrm{T}}\mathbf{X}_{(1)}|
$$

$$
= \frac{|\mathbf{X}_{(1)}^{\mathrm{T}}\mathbf{X}_{(2)}(\mathbf{X}_{(2)}^{\mathrm{T}}\mathbf{X}_{(2)}^{-1})\mathbf{X}_{(2)}^{\mathrm{T}}\mathbf{X}_{(1)}|}{|\mathbf{X}_{(1)}^{\mathrm{T}}\mathbf{X}_{(1)}|}
$$

$$
= \frac{|\mathbf{X}_{(1)}^{\mathrm{T}}\mathbf{P}_{(2)}\mathbf{X}_{(1)}|}{|\mathbf{X}_{(1)}^{\mathrm{T}}\mathbf{X}_{(1)}|} \tag{5.68}
$$

which is the generalized multiple correlation coefficient representing the variance of $\mathbf{X}_{(1)}$ accounted for by the set $\mathbf{X}_{(2)}$. A similar expression exists for the variance of $\mathbf{X}_{(2)}$ accounted for by $\mathbf{X}_{(1)}$. Thus either set of variables can be taken as the set of dependent (independent) variables.

The matrices $\mathbf{A}$ and $\mathbf{B}$ which contain the latent vectors are akin to loading coefficients, since they relate the observed variates in both sets to their respective cannonical variates. They may be standardized to yield correlations between the observed and the latent variables, much in the same way as the PCs. Using Eq. (5.64) we have

$$\mathbf{X}_{(1)}^T\mathbf{Z}_{(1)} = (\mathbf{X}_{(1)}^T\mathbf{X}_{(1)})\mathbf{A}$$

$$\mathbf{X}_{(2)}^T\mathbf{Z}_{(2)} = (\mathbf{X}_{(2)}^T\mathbf{X}_{(2)})\mathbf{B}$$

which can be used to obtain correlation loadings.

Finally, a likelihood ratio test can be developed to test the equality of the latent roots much in the same way as for PCA, and this leads to a test of independence between the two sets of variates. We have

$$H_0: \mathbf{\Sigma}_{21} = \mathbf{0}$$

$$H_a: \mathbf{\Sigma}_{21} \neq \mathbf{0}$$

or in terms of the latent roots (canonical correlations),

$$H_0: \rho_1^2 = \rho_2^2 = \cdots = \rho_{p1}^2 = 0$$

$$H_a: \text{at least one not zero}$$

The LR statistic is then given by (not to be confused with the term of Eq. (5.53)

$$\lambda = \frac{|(\mathbf{X}_{(1)}^T\mathbf{X}_{(1)})^{-1}\mathbf{X}_{(1)}^T\mathbf{X}_{(2)}(\mathbf{X}_{(2)}^T\mathbf{X}_{(2)})^{-1}\mathbf{X}_{(2)}^T\mathbf{X}_{(1)}}{|\mathbf{X}_{(1)}^T\mathbf{X}_{(1)}|}$$

$$= \prod_{i=1}^{p_i} (1 - r_i^2) \tag{5.69}$$

where $-2\ln\lambda$ approaches the chi-squared distribution as $n\to\infty$. Bartlett (1954) has developed a multiplying factor which increases the rate of convergence. The approximate chi-squared statistic, given multivariate normality, is then

$$\chi^2 = -\left[(n-1) - \frac{1}{2}(p_1 + p_2 + 1)\right]\ln\lambda. \tag{5.70}$$

with p_1p_2 degrees of freedom. Equation (5.70) can be used to test for complete independence (all roots equal zero). To test the last $p_1 - q$ roots let

$$\lambda_q = \prod_{i=q+1}^{p_1} (1 - r_i^2) \tag{5.71}$$

Then

$$\chi_q^2 = -\left[(n - 1) - \frac{1}{2}(p_1 + p_2 + 1)\right] \ln \lambda_r \tag{5.72}$$

is approximately chi-squared with $(p_1 - q)(p_2 - q)$ degrees of freedom. A further refinement in the approximation has been obtained by Fujikoshi (1977) and Glynn and Muirhead (1978; see also Muirhead, 1982) whereby the distribution of

$$\chi^2 = -\left[n - q - \frac{1}{2}(p_1 + p_2 + 1) + \sum_{i=1}^{q} r_i^{-2}\right] \ln \lambda_q \tag{5.73}$$

is approximately chi-squared with $(p_{1-}q)(p_{2-}q)$ degrees of freedom. Evidence suggests however that the canonical correlation model is not very robust against nonnormality (Muirhead and Waternaux, 1980).

Example 5.12. Sinha et al. (1986) have used canonical correlation analysis to determine the intercorrelation(s) between a set of dependent and independent variables (seed germination, presence of fungi, and environmental conditions) using $n = 8135$ measurements taken at grain bins in Winnipeg (Manitoba) during 1959–1967 (see also Sinha et al., 1969). The two sets of variables are defined as follows for $p = p_1 + p_2 = 6 + 6 = 12$.

Dependent: Set I (germination, field fungi)

Y_1 = Percentage germinability of seed
Y_2 = Amount of *alternaria*
Y_3 = Amount of *cochliobolus*
Y_4 = Amount of *cladosporium*
Y_5 = Amount of *nigrospora*
Y_6 = Amount of *gonatobotrys*

Independent: Set II (environment)

Y_7 = Month of collection
Y_8 = Bin number; one of two bins each containing 500 bushels of wheat
Y_9 = Depth of sample
Y_{10} = Grain temperature C° at each depth
Y_{11} = Location of sample (peripheral or inner area of bin)
Y_{12} = Percent moisture of grain at each depth and location

Table 5.26   Latent Roots and Correlations and Bartlett's chi-squared Test for Wheat Data

Pair of Canonical Variates	Latent Roots (r_i^2)	Canonical Correlations (r_i)	χ^2 Test	Degrees of Freedom
1	.836	.91	16049.3	36
2	.107	.32	1351.7	25
3	.034	.18	425.2	16
4	.013	.11	142.7	9
5	.003	.05	34.6	4
6	.000	.02	6.1	1

Source: Sinha et al., 1968; reproduced with permission.

The principal results appear in Tables 5.26 and 5.27. Since variables are measured in noncomparable units of measure, the analysis is based on the correlation matrix. All chi-squared values of Table 5.26 are significant at the $\alpha = .05$ level, no doubt due in part to the large sample size. To identify the canonical variates the authors correlate these variables with the observed set (Table 5.27). Since there are two groups of observed variates (the dependent and independent set) and two canonical variables per each multiple correlation, we have in all two sets of pairwise correlations per latent root. The authors conclude that location (with somewhat lower moisture) influences germination and reduces *alternaria* and *nigrospora* as well as *gonatobotrys* (first pair). Also the second canonical pair reveals some aspects of interdomain relationship not exposed by the first arrangement.

Table 5.27   Correlations Between the First Two Pairs of Cannonical Variates and Wheat Storage Variables[a]

Variables		Canonical Set A		Canonical Set B	
		Set I: Dependent	Set II: Independent	Set I: Dependent	Set II: Independent
Y_1:	Germination	.57	.10	−.52	−.10
Y_2:	Alternaria	−.49	−.09	−.15	
Y_3:	Cochliobolus	−.16		−.28	−.05
Y_4:	Cladosporium			.64	.12
Y_5:	Nigrospora	−.37	−.07	.29	.05
Y_6:	Gonatobotrys	−.25		.15	
Y_7:	Month	−.09	.06	.43	.13
Y_8:	Bin	.05		−.13	−.46
Y_9:	Depth	−.15			
Y_{10}:	Temperature		.24	−.07	−.81
Y_{11}:	Moisture	−.26	−.26	.12	.12
Y_{12}:	Location	.15	.92	−.07	.13
Multiple correlations		$r_1 = .91$		$r_2 = .32$	

Source: Sinha et al. (1986).

We may conclude that month, bin, temperature and location are involved with germination, *alternaria cladosporium*, and *nigrospora* in a bulk grain ecosystem. Canonical variables can also be rotated to enhance interpretation (Cliff and Krus, 1976). Unlike PCA, however, it does not make sense to use correlation "loadings" as an interpretive aid since in canonical correlation analysis the correlations between an individual variable and the canonical variate are redundant because they merely show how the variable by itself relates to the other set of variables. Thus all information about how the variables in one set contribute jointly to canonical correlation with the other set is lost—see Rancher (1988) for detail.   □

5.5.2 Modification of Canonical Correlation

The canonical correlation model of the previous section is normally employed to study direct correlations between two sets of variables, including the case when one set is viewed as dependent and the other as independent. Indeed the model can be obtained by an iterative regression procedure (see Lyttkens, 1972) and has been employed in simultaneous equation theory (Hooper, 1959) and discrimination (Glahn, 1968). Canonical variates can also be used in place of classification functions (Falkenhagen and Nash, 1978). At times a different objective arises, namely, to estimate those underlying and unobserved variates that have produced the intercorrelation between the two sets of observed random variables. Thus we may require an overall measure of relationship between two sets of variables. The procedure has been termed "redundancy analysis" (see Rencher, 1992). That is, analogous to PCA, we wish to maximize the correlation between v and $\mathbf{X}_{(1)} = (X_1, X_2, \ldots, X_{p1})^{\mathrm{T}}$ and u and $\mathbf{X}_{(2)} = (X_{p1+1}, X_{p1+2}, \ldots, X_p)^{\mathrm{T}}$ in order to maximize the explanatory power of the latent variables u and v. The classical canonical correlation model does not necessarily achieve this. Consider the linear combinations $u = \boldsymbol{\alpha}^{\mathrm{T}}\mathbf{X}_{(1)}$ and $v = \boldsymbol{\beta}^{\mathrm{T}}\mathbf{X}_{(2)}$ such that $\boldsymbol{\alpha}^{\mathrm{T}}\boldsymbol{\alpha} = \boldsymbol{\beta}^{\mathrm{T}}\boldsymbol{\beta} = 1$, but u and v are not necessarily unit vectors. The covariances between u and $\mathbf{X}_{(2)}$ and v and $\mathbf{X}_{(1)}$ are then given by $\mathrm{cov}(\boldsymbol{\alpha}^{\mathrm{T}}\mathbf{X}_{(1)}, \mathbf{X}_{(2)}) = \boldsymbol{\alpha}^{\mathrm{T}}\boldsymbol{\Sigma}_{12}$ and $\mathrm{cov}(\boldsymbol{\beta}^{\mathrm{T}}\mathbf{X}_{(2)}, \mathbf{X}_{(1)}) = \boldsymbol{\beta}^{\mathrm{T}}\boldsymbol{\Sigma}_{21}$ and to maximize explanatory power we can maximize the sum of squares of the covariances $\boldsymbol{\alpha}^{\mathrm{T}}\boldsymbol{\Sigma}_{12}\boldsymbol{\Sigma}_{21}\boldsymbol{\alpha}$ and $\boldsymbol{\beta}^{\mathrm{T}}\boldsymbol{\Sigma}_{21}\boldsymbol{\Sigma}_{12}\boldsymbol{\beta}$ subject to constraints $\boldsymbol{\alpha}^{\mathrm{T}}\boldsymbol{\alpha} = \boldsymbol{\beta}^{\mathrm{T}}\boldsymbol{\beta} = 1$. This is equivalent to maximizing the Lagrangian expressions

$$\phi = \boldsymbol{\alpha}^{\mathrm{T}}\boldsymbol{\Sigma}_{12}\boldsymbol{\Sigma}_{21}\boldsymbol{\alpha} - \mu(\boldsymbol{\alpha}^{\mathrm{T}}\boldsymbol{\alpha} - 1)$$

$$\psi = \boldsymbol{\beta}^{\mathrm{T}}\boldsymbol{\Sigma}_{21}\boldsymbol{\Sigma}_{12}\boldsymbol{\beta} - \lambda(\boldsymbol{\beta}^{\mathrm{T}}\boldsymbol{\beta} - 1) \tag{5.74}$$

(see Tucker, 1958). Differentiating and setting to zero yields the normal equations

$$\partial\phi/\partial\boldsymbol{\alpha} = 2\boldsymbol{\Sigma}_{12}\boldsymbol{\Sigma}_{21}\hat{\boldsymbol{\alpha}} - 2\hat{\mu}\hat{\boldsymbol{\alpha}} = \mathbf{0}$$

$$\partial\psi/\partial\boldsymbol{\beta} = 2\boldsymbol{\Sigma}_{21}\boldsymbol{\Sigma}_{12}\hat{\boldsymbol{\beta}} - 2\hat{\lambda}\hat{\boldsymbol{\beta}} = \mathbf{0}$$

or

$$(\mathbf{\Sigma}_{12}\mathbf{\Sigma}_{21} - \hat{\mu})\hat{\alpha} = \mathbf{0}$$

$$(\mathbf{\Sigma}_{21}\mathbf{\Sigma}_{12} - \hat{\lambda})\hat{\beta} = \mathbf{0} \tag{5.75}$$

that is, $\hat{\alpha}$ and $\hat{\beta}$ are latent vectors of $\mathbf{\Sigma}_{12}\mathbf{\Sigma}_{21}$ and $(\mathbf{\Sigma}_{12}\mathbf{\Sigma}_{21})^{\mathrm{T}}$. Let $\mathbf{A}$ and $\mathbf{B}$ represent sample latent vector matrices. Then the sample equivalent of Eq. (5.75) is

$$(\mathbf{X}_{(1)}^{\mathrm{T}}\mathbf{X}_{(2)})(\mathbf{X}_{(2)}^{\mathrm{T}}\mathbf{X}_{(1)}) = \mathbf{AL}^{2}\mathbf{A}^{\mathrm{T}}$$

$$(\mathbf{X}_{(2)}^{\mathrm{T}}\mathbf{X}_{(1)})(\mathbf{X}_{(1)}^{\mathrm{T}}\mathbf{X}_{(2)}) = \mathbf{BL}^{2}\mathbf{B}^{\mathrm{T}} \tag{5.76}$$

so that

$$\mathbf{B}^{\mathrm{T}}(\mathbf{X}_{(2)}^{\mathrm{T}}\mathbf{X}_{(1)})\mathbf{A} = \mathbf{L} \tag{5.77}$$

The matrices $\mathbf{A}$ and $\mathbf{B}$ contain left and right latent vectors, respectively, of the matrix $\mathbf{X}_{(2)}^{\mathrm{T}}\mathbf{X}_{(1)}$ and $\mathbf{L}$ is the matrix of p_1 nonzero latent roots. The sample linear combinations with maximum correlation are then given by

$$\mathbf{U} = \mathbf{X}_{(1)}\mathbf{A} \qquad \mathbf{V} = \mathbf{X}_{(2)}\mathbf{B} \tag{5.78}$$

where $\mathbf{V}^{\mathrm{T}}\mathbf{U} = \mathbf{L}$, that is, the linear combinations are biorthogonal. Using eq. (5.76) and (5.77) it can also be shown that

$$\mathbf{U}^{\mathrm{T}}\mathbf{X}_{(2)}\mathbf{X}_{(2)}^{\mathrm{T}}\mathbf{U} = \mathbf{L}^{2}$$

$$\mathbf{V}^{\mathrm{T}}\mathbf{X}_{(1)}\mathbf{X}_{(1)}^{\mathrm{T}}\mathbf{V} = \mathbf{L}^{2} \tag{5.79}$$

that is, $\mathbf{U}$ and $\mathbf{V}$ are latent vectors of $\mathbf{X}_{(2)}\mathbf{X}_{(2)}^{\mathrm{T}}$ and $\mathbf{X}_{(1)}\mathbf{X}_{(1)}^{\mathrm{T}}$ respectively (see Exercise 5.9). The decomposition (Eq. 5.79) can be viewed as the simultaneous PCA of the sets $\mathbf{X}_{(1)}$ and $\mathbf{X}_{(2)}$ since Eq. (5.78) implies that

$$\mathbf{X}_{(1)} = \mathbf{UA}^{\mathrm{T}} + \mathbf{\delta}_{(1)}$$

$$\mathbf{X}_{(2)} = \mathbf{VB}^{\mathrm{T}} + \mathbf{\delta}_{(2)} \tag{5.80}$$

where $\mathbf{\delta}_{(1)}^{\mathrm{T}}\mathbf{\delta}_{(2)} = \mathbf{0}$. By derivation it is clear that the vectors $\mathbf{U}$ and $\mathbf{V}$ are maximally correlated.

The linear combinations u and v are not necessarily unit vectors. Van den Wallenberg (1977) has proposed a model which is identical to that of Tucker (1958) with the exception that u and v are normalized to unit length. We have

$$\mathbf{E}(u^{2}) = \mathbf{E}(\mathbf{UU}^{\mathrm{T}}) = \mathbf{\alpha}^{\mathrm{T}}\mathbf{\Sigma}_{11}\mathbf{\alpha} = 1$$

$$\mathbf{E}(v^{2}) = \mathbf{E}(\mathbf{VV}^{\mathrm{T}}) = \mathbf{\beta}^{\mathrm{T}}\mathbf{\Sigma}_{22}\mathbf{\beta} = 1$$

which replace the constraints in Tucker's model. The change in constraints is equivalent to maximizing correlations between the observed variates and the linear combinations (PCs). Differentiating the Lagrangian expressions and setting to zero yields the normal equations (Exercise 5.10)

$$(\boldsymbol{\Sigma}_{12}\boldsymbol{\Sigma}_{21} + \mu\boldsymbol{\Sigma}_{11})\boldsymbol{\alpha} = \mathbf{0}$$

$$(\boldsymbol{\Sigma}_{21}\boldsymbol{\Sigma}_{12} - \lambda\boldsymbol{\Sigma}_{22})\boldsymbol{\beta} = \mathbf{0} \tag{5.81}$$

5.5.3　Canonical Correlation for More than Two Sets of Variables

The classical canonical correlation model described in Section 5.5.1 seeks linear combinations (canonical variates), one from each of the two groups, that are maximally correlated. As it stands the canonical correlation model cannot be generalized to more than two sets of variables since correlation is intrinsically a binary concept. Rather than define the model in terms of finding linear combinations that are maximally correlated, it is possible to define canonical correlation in terms of finding a single auxilliary linear combination (canonical variate), together with two different linear combinations (one from each set) that are maximally correlated with the auxilliary canonical variate. Couched in this form, canonical correlation can be readily extended to any finite number of variable groupings, much in the same way as scalar coorelation is generalized by PCA.

Consider p sets of random variables observed in a given sample. We then have the observation matrices $\mathbf{X}_{(1)}, \mathbf{X}_{(2)}, \ldots, \mathbf{X}_{(p)}$ where each matrix is $(n \times p_i)$. Let $\mathbf{Z}$ be a $(n \times 1)$ vector, which is not observed directly but is assumed to be related to the observed sets by the regression relations

$$\mathbf{Z} = \mathbf{X}_{(1)}\mathbf{A}_1 + \mathbf{e}_1, \qquad \mathbf{Z} = \mathbf{X}_{(2)}\mathbf{A}_2 + \mathbf{e}_2, \ldots, \qquad \mathbf{Z} = \mathbf{X}_{(p)}\mathbf{A}_p + \mathbf{e}_p$$

$$\tag{5.82}$$

Let $\mathbf{P}_{(i)} = \mathbf{X}_{(i)}(\mathbf{X}_{(i)}^{T}\mathbf{X}_{(i)})^{-1}\mathbf{X}_{(i)}^{T}$ be the orthogonal projection matrix for the ith equation. Then the variance explained by the set $\mathbf{X}_{(i)}$ is given by $\mathbf{Z}^{T}[\mathbf{X}_{(i)}(\mathbf{X}_{(i)}^{T}\mathbf{X}_{(i)})^{-1}\mathbf{X}_{(i)}^{T}]\mathbf{Z}$. Also, we know $\mathbf{X}_{(i)}(\mathbf{X}_{(1)}^{T}\mathbf{X}_{(i)})^{-1}\mathbf{X}_{(i)}^{T} = \mathbf{Z}_{(i)}\mathbf{Z}_{(i)}^{T}$ where $\mathbf{Z}_{(i)}$ is the $(r \times p_i)$ matrix of standardized PCs of $\mathbf{X}_{(i)}^{T}\mathbf{X}_{(i)}$ (Exercise 3.20). Now, let $\mathbf{Z}^*$ be a latent vector of one of the matrices $\mathbf{X}_{(i)}(\mathbf{X}_{(i)}^{T}\mathbf{X}_{(i)})^{-1}\mathbf{X}_{(i)}^{T}$, such that it maximizes explained variance. Then $\mathbf{Z}^{*T}\mathbf{Z}^* = 1$ and

$$R^2 = \frac{\mathbf{Z}^{*T}\mathbf{X}_{(i)}(\mathbf{X}_{(i)}^{T}\mathbf{X}_{(i)})^{-1}\mathbf{X}_{(i)}^{T}\mathbf{Z}^*}{\mathbf{Z}^{*T}\mathbf{Z}^*} = 1 \tag{5.83}$$

since the latent roots of an idempotent symmetric matrix take on the values of zero or unity (e.g., Basilevsky, 1983). Also, as is known from regression theory $0 \le R^2 \le 1$, so that the latent roots of the projection matrix can be viewed as the two extreme values of R^2. Clearly, in practice $\mathbf{Z}^*$ will rarely account for the entire variance of the p sets of variables. We then seek the best compromise possible, in the form of that vector $\mathbf{Z}$ which accounts for the maximal variance. Such a vector is given by the latent vector that corresponds to the largest latent root of the sum of projection matrices. That is, we select $\mathbf{Z}$ such that

$$\mathbf{Z}^{\mathrm{T}}\left\{\sum_{i=1}^{p}\mathbf{X}_{(i)}(\mathbf{X}_{(i)}^{\mathrm{T}}\mathbf{X}_{(i)})^{-1}\mathbf{X}_{(i)}^{\mathrm{T}}\right\}\mathbf{Z} = l_1 \tag{5.84}$$

More generally, the sum in Eq. (5.84) can also be replaced by a weighted sum. When $\mathbf{Z}$ accounts for the entire variance of the p sets of random variables we have $l_1 = p$ and $\mathbf{Z}$ in addition becomes the (common) latent vector of the p projection matrices. This approach to canonical correlation is from Carroll (1968). A review of this and other approaches is also given by Kettenring (1971). An advantage of this approach is that it links canonical correlation to the familiar concept of PCA when more than a single set of variables is present.

5.5.4 Multigroup Principal Components

Canonical correlation-type models consider a vertical partition (augmentation) of a data matrix. An alternative situation is presented when we have a horizontal partition, that is, when a common set of random variables is observed in two or more distinct groups or samples. The objective is then to compare PCs computed from the same variables but in different sample spaces.

Several methods can be used to handle the situation of multigroup PCA, depending on the situation and objective(s) of the analysis. For example, when using PCA for classification or taxonomic purposes with g distinct groups of individuals, each group can be replaced by the mean vector, resulting in a $(g \times p)$ matrix which can be analyzed using R or Q-mode analysis. The procedure however should only be used when a broad analysis is of interest since all of the within-group variation is destroyed in the process. Since g groups could also represent samples from different populations, a PCA would presumably be preceded by a suitable test, for example, a test for equality of the covariance matrices (Section 4.2).

A different objective presents itself when we wish to compare the similarity (disimilarity) of PCs for g distinct groups (populations)—for example, in the case of the painted turtle data (Example 3.3), when comparing morphological size and shape factors for males and females (see Pimentel, 1979). In this situation PC loadings (scores) can be computed for each separate group and then correlated across groups to reveal dimensions that are similar for the groups. For large g, and in the event of high intercorrelation between PCs from different groups, a secondary PCA may further reduce the complexity of the data and reveal common structure(s) between the populations. If the groups represent samples from a common population, the data may of course be pooled into a single sample and a single PCA carried out for the pooled group. This however can be carried out in two distinct ways. First, the original g data matrices may be pooled (averaged) and a PCA carried out on the resultant covariance (correlation) matrix (e.g., see Lindsay, 1986). Note that if the groups do not represent samples from the same population, the resultant PCs will be determined by both within and between group variation, resulting in a hybrid set of PCs which are difficult to interpret. Second, it is possible to combine the g sample covariance matrices into a single pooled covariance matrix. The advantage of this procedure is at times viewed as avoiding the mixing up of within and between group variation (see Thorpe, 1983). Since pooling is only legitimate when groups represent a common population, the advantage is probably more illusory than real. Finally, it should be pointed out that any form of pooling invariably results in a loss of information, and perhaps a better approach to the problem lies in stacking the g data matrices into a large r supermatrix.

More recently attention has shifted to multigroup PC structures for groups that do not necessarily represent an identical population. First consider the case for two groups. The problem, in the present context, was first considered by Cliff (1966) and more recently by Krzanowski (1979). Let $X_1, X_2, \ldots, X_p$ be a set of random vectors observed in two distinct samples of size n and m, respectively. We then have two $(n \times p)$ and $(m \times p)$ data matrices, which can be viewed as partitions of a larger data matrix $\mathbf{X}$. Let the two matrices be denoted as $\mathbf{X}_{(1)}$ and $\mathbf{X}_{(2)}$. Then if both represent samples from similar populations, they should possess similar PCs. Let

$$\mathbf{X}_{(1)} = \mathbf{Z}_{(1)}\mathbf{A}_{(1)}^{\mathrm{T}} + \boldsymbol{\delta}_{(1)}, \qquad \mathbf{X}_{(2)} = \mathbf{Z}_{(2)}\mathbf{A}_{(2)}^{\mathrm{T}} + \boldsymbol{\delta}_{(2)} \tag{5.85}$$

where

$$\mathbf{A}_{(1)}^{\mathrm{T}} = \mathbf{L}_{(1)}^{1/2}\mathbf{P}_{(1)}^{\mathrm{T}}, \qquad \mathbf{A}_{(2)}^{\mathrm{T}} = \mathbf{L}_{(2)}^{1/2}\mathbf{P}_{(2)} \quad \text{and} \quad \mathbf{P}_{(1)}^{\mathrm{T}}\mathbf{X}_{(1)}^{\mathrm{T}}\mathbf{X}_{(1)}\mathbf{P}_{(1)} = \mathbf{L}_{(1)},$$

$$\mathbf{P}_{(2)}^{\mathrm{T}}\mathbf{X}_{(2)}^{\mathrm{T}}\mathbf{X}_{(2)}\mathbf{P}_{(2)} = \mathbf{L}_{(2)} \tag{5.86}$$

If the common latent vectors $\mathbf{P}_{(1)}$ and $\mathbf{P}_{(2)}$ are drawn from the same population, their degree of relationship can be measured by inner products or by cosines of the angles between them. This leads to the following theorem.

THEOREM 5.10   (Krzanowski, 1979). Let

$$\mathbf{H}^{\mathrm{T}}(\mathbf{P}_{(1)}^{\mathrm{T}}\mathbf{P}_{(2)})(\mathbf{P}_{(1)}^{\mathrm{T}}\mathbf{P}_{(2)})^{\mathrm{T}}\mathbf{H} = \mathbf{M} \tag{5.87}$$

and $U = P_{(1)}H$ where U is a unit vector, M is a diagonal matrix of r nonzero latent roots, and H represents orthogonal latent vectors.

(i) The ith minimum angle between an arbitrary vector in the space of the first r principal components of the first sample, and the one most nearly parallel to it in the space of the first r components of the second sample is $\cos^{-1}(m_i^{1/2})$ where m_i is the ith latent root of $(P_{(1)}^T P_{(2)})(P_{(1)}^T P_{(2)})^T$.

(ii) The columns $U_1, U_2, \ldots, U_r$ of $U = P_{(1)}H$ can be chosen to be mutually orthogonal vectors which are embedded in the subspace of the first sample. Similarly $(P_{(2)}P_{(2)}^T)U_1$ and $(P_{(2)}P_{(2)}^T)U_2$, $\ldots, (P_{(2)}P_{(2)}^T)U_r$ form a corresponding set of mutually orthogonal vectors in the subspace of the second sample.

PROOF. From Eq. (5.87) we have vectors $U = P_{(1)}H$ which are generated by columns of $P_{(1)}$. Also $P_{(2)}P_{(2)}^T$ is the projection matrix which projects vectors orthogonally onto the subspace spanned by the orthogonal columns of $P_{(2)}$.

Let $\hat{U} = (P_{(2)}P_{(2)}^T)U$ be the projection of U onto columns of $P_{(2)}$, that is $\hat{U}$ contains the projections of U in the subspace generated by the first r PCs of the first sample onto the subspace generated by r PCs of the second sample. Then from Eq. (5.87) we have

$$U^T(P_{(2)}P_{(2)}^T)U = U^T\hat{U} = M \tag{5.88}$$

so that the cosines of the angles θ_i that lie between U and $\hat{U}$ are given by

$$\cos\theta_i = \frac{|\hat{U}_i|}{|U_i|} = (\hat{U}_i^T\hat{U}_i)^{1/2}$$

$$= [U_i^T(P_{(2)}P_{(2)}^T)(P_{(2)}P_{(2)}^T)U_i]^{1/2}$$

$$= [U_i^T(P_{(2)}P_{(2)}^T)U_i]^{1/2}$$

$$= [H_i^T P_{(1)}^T(P_{(2)}P_{(2)}^T)P_{(1)}H_i]^{1/2} \tag{5.89}$$

since U is a unit vector. It follows from Eqs. (5.87) and (5.89) that the ith diagonal element of M contains $\cos^2\theta_i$ and the minimal angle θ_i $(i = 1, 2, \ldots, r)$ between the two r-dimensional subsets is given by

$$\theta_i = \cos^{-1}(m_i)^{1/2}$$

where $0 \le \theta \le \pi$.

The proof of the second part is left as Exercise 5.11.

Thus the minimum angle between a vector in the space of the first r PCs of the first sample, and the one most nearly parallel to it (in the space of the first r PCs) of the second sample, is given by $\theta_1 = \cos^{-1}(m_1)^{1/2}$ where $0 \le m_1 \le 1$ is the largest latent root of $(\mathbf{P}_{(1)}^{\mathrm{T}}\mathbf{P}_{(2)})(\mathbf{P}_{(1)}^{\mathrm{T}}\mathbf{P}_{(2)})^{\mathrm{T}}$. The latent roots m_i can therefore be used as measures of similarity between corresponding pairs of PCs in the two samples. Also $\Sigma_{i=1}^{r}, m_i$ measures global similarity between the two sets of PCs since

$$\sum_{i=1}^{r} m_i = \mathrm{tr}(\mathbf{P}_{(1)}^{\mathrm{T}}\mathbf{P}_{(2)})(\mathbf{P}_{(1)}^{\mathrm{T}}\mathbf{P}_{(2)})^{\mathrm{T}}$$

$$= \sum_{i=1}^{r}\sum_{j=1}^{r} \cos^2\theta_{ij} \tag{5.90}$$

where θ_{ij} is the angle between the ith PC of the first sample and the jth PC of the second sample. The sum (Eq. 5.90) varies between r (coincident subspaces) and 0 (orthogonal subspaces). The similarities between two samples can also be exhibited through the pairs $\mathbf{U} = \mathbf{P}_{(1)}\mathbf{H}$ and $\hat{\mathbf{U}} = (\mathbf{P}_{(2)}\mathbf{P}_{(2)}^{\mathrm{T}}\mathbf{U}$, and $\mathbf{P}_{(1)}$ and $\mathbf{P}_{(2)}$ are interchangeable. The methodology can also be extended to more than two groups.

THEOREM 5.11 (Krzanowski, 1979). Let $\mathbf{h}$ be an arbitrary vector in a p-dimensional space and let θ_k be the angle between $\mathbf{h}$ and the vector most parallel to it in the space generated by r PCs of group k ($k = 1, 2, \ldots, g$). Then the value of h that minimizes

$$\nu = \sum_{k=1}^{g} \cos^2\theta_k$$

is given by the latent vector $\mathbf{h}_1$ which corresponds to the largest latent root of $\Sigma_{k=1}^{g} \mathbf{A}_k\mathbf{A}_k^{\mathrm{T}}$ where $\mathbf{A}_k$ is the ($p \times r$) matrix of loadings for the kth sample.

PROOF. We have

$$\cos^2\theta_k = \mathbf{h}\mathbf{A}_k\mathbf{A}_k^{\mathrm{T}}\mathbf{h}^{\mathrm{T}}$$

so that

$$\sum_{k=1}^{g} \cos^2\theta_k = \mathbf{h}\left(\sum_{k=1}^{g} \mathbf{A}_k\mathbf{A}_k^{\mathrm{T}}\right)\mathbf{h}^{\mathrm{T}}$$

where $\mathbf{h}$ can be taken as the latent vector that corresponds to the largest latent root of $\Sigma_{k=1}^{g} \mathbf{A}_k\mathbf{A}_k^{\mathrm{T}}$. For greater detail, together with a numerical example, see Krzanowski (1979).

This procedure for comparing PCs in several samples utilizes a two-stage procedure (Section 5.4.2). Recently Flury (1984, 1988) proposed an alternative single-stage procedure for generalizing the PCA model to more than a single sample. Let

$$H_0\colon \mathbf{\Pi}^T \mathbf{\Sigma}_k \mathbf{\Pi} = \mathbf{\Lambda}_k \qquad (k = 1, 2, \ldots, g) \tag{5.91}$$

represent the null hypothesis that all g populations possess p common latent vectors $\mathbf{\Pi}$ and common PCs

$$\mathbf{U}_i = \mathbf{\Pi}^T \mathbf{X}_k$$

where $\mathbf{X}_k$ is a vector of p random variables in the kth population. Given sample covariance matrices $\mathbf{S}_1, \mathbf{S}_2, \ldots, \mathbf{S}_g$, the common likelihood of $\mathbf{\Sigma}_k$, under appropriate assumptions, is

$$L(\mathbf{\Sigma}_1, \mathbf{\Sigma}_2, \ldots, \mathbf{\Sigma}_g) = c \prod_{k=1}^{g} \mathrm{etr}\left(-\frac{1}{2} n_k \mathbf{\Sigma}_k^{-1} \mathbf{S}_k\right) |\mathbf{\Sigma}_k^{\frac{-n_k}{2}}| \tag{5.92}$$

where c is a constant and etr denotes the exponential trace (exponent of the trace). The log-likelihood ratio statistic for H_0 is the asymptotic chi-squared statistic

$$\chi^2 = -2 \ln \frac{L(\hat{\mathbf{\Sigma}}_1 \hat{\mathbf{\Sigma}}_2 \ldots \hat{\mathbf{\Sigma}}_g)}{L(\mathbf{S}_1, \mathbf{S}_2, \ldots, \mathbf{S}_g)}$$

$$= \sum_{k=1}^{g} n_k \ln \left|\frac{\hat{\mathbf{\Sigma}}_k}{\mathbf{S}_k}\right| \tag{5.93}$$

with $(1/2)(g-1)p(p-1)$ degrees of freedom where $\hat{\mathbf{\Sigma}}_k$ is the ML estimator of $\mathbf{\Sigma}$ in the kth population. Let $\mathbf{P}_1, \mathbf{P}_2, \ldots, \mathbf{P}_g$ be latent vectors common to $\mathbf{S}_1, \mathbf{S}_2, \ldots, \mathbf{S}_g$. Then

$$\mathbf{P}_k^T \left[\sum_{i=1}^{g} n_i \frac{(l_{ik} - l_{ij})}{l_{ik} l_{ij}} \mathbf{S}_i\right] \mathbf{P}_j = \mathbf{0} \qquad \left(\begin{matrix} k, j = 1, 2, \ldots, p \\ k \neq j \end{matrix}\right) \tag{5.94}$$

where $\mathbf{P}^T \mathbf{P} = \mathbf{I}$ and the group PCs are given by $\mathbf{U}_k = \mathbf{P}_k^T \mathbf{X}$. The latent vectors are not necessarily orthogonal across the g groups; also, the sample covariance matrix for the common PCs is given by

$$\mathbf{C}_k = \mathbf{P}^\mathsf{T}\mathbf{S}_k\mathbf{P} \qquad (k = 1, 2, \ldots, g) \tag{5.95}$$

and the correlation matrix by

$$\mathbf{R}_k = \mathbf{D}_k^{-1/2}\mathbf{C}_k\mathbf{D}_k^{-1/2} \qquad (k = 1, 2, \ldots, g) \tag{5.96}$$

where $\mathbf{D}_k = \mathrm{diag}(\mathbf{C}_k)$. Correlation matrices close to $\mathbf{I}$ would then imply departure from H_0, that is, lack of a common PC structure. Since the analysis is carried out using sample covariance matrices, the eigenvectors have to be standardized to yield correlation loadings. The model has been termed by Flury (1984) as the Common Principal Components model since the underlying common dimensions represent a compromise solution for the g groups. In the event that covariance matrices are proportional (Section 4.23) a common PC subspace can be defined for the g groups. Flury (1986a) has also provided statistical criteria which can be used for testing purposes.

Example 5.13. Flury (1984) presents a numerical example of his common principal components approach using Anderson's (1935) iris data (see also Fisher, 1936). Sample covariance matrices, ML estimates $\hat{\mathbf{\Sigma}}_k$ (under the restriction of common latent vectors), and common latent vectors are given in parts a–c and variances of the common PCs, latent roots of the $\mathbf{S}_k$, and correlation matrices of common PCs appear in d and e. If the hypothesis of common PCs is to be maintained for all three populations, the l_{ij} values must lie close to the latent roots of the $\mathbf{S}_k$. This is the case for sample 1 (versicolor) but less so for samples 2 (virginica) and 3 (setosa). The significance of the chi-squared statistic $\chi^2 = 63.91$ with 12 degrees of freedom confirms this initial impression. Further information concerning the intercorrelations between the common PCs is provided by correlation matrices $\mathbf{R}_1$, $\mathbf{R}_2$, and $\mathbf{R}_3$.

Table 5.28   A Common Principal Component Analysis of Anderson's (1935) Iris Data[a]

(a) Sample Covariance Matrices

Versicolor ($n_1 = 50$)

$$
S_1 = \begin{bmatrix}
26.6433 & 8.5184 & 18.2898 & 5.5780 \\
8.5184 & 9.8469 & 8.2653 & 4.1204 \\
18.2898 & 8.2653 & 22.0816 & 7.3102 \\
5.5780 & 4.1204 & 7.3102 & 3.9106
\end{bmatrix}
$$

Table 5.28   (Continued)

Virginica ($n_2 = 50$)

$$S_2 = \begin{bmatrix} 40.4343 & 9.3763 & 30.3290 & 4.9094 \\ 9.3763 & 10.4004 & 7.1380 & 4.7629 \\ 30.3290 & 7.1380 & 30.4588 & 4.8824 \\ 4.9094 & 4.7629 & 4.8824 & 7.5433 \end{bmatrix}$$

Setosa ($n_3 = 50$)

$$S_3 = \begin{bmatrix} 12.4240 & 9.9216 & 1.6355 & 1.0331 \\ 9.9216 & 14.3690 & 1.1698 & .9298 \\ 1.6355 & 1.1698 & 3.0159 & .6069 \\ 1.0331 & .9298 & .6069 & 1.1106 \end{bmatrix}$$

(b) MLEs of Population Covariance Matrices

$$\hat{\Sigma}_1 = \begin{bmatrix} 29.5860 & 7.3004 & 18.6600 & 4.6667 \\ 7.3004 & 7.4546 & 6.6121 & 2.8309 \\ 18.6600 & 6.6121 & 21.2145 & 6.2692 \\ 4.6667 & 2.8309 & 6.2692 & 3.2273 \end{bmatrix}$$

$$\hat{\Sigma}_2 = \begin{bmatrix} 40.6417 & 11.5005 & 27.8263 & 7.9275 \\ 11.5005 & 11.0588 & 8.8976 & 2.8603 \\ 27.8263 & 8.8976 & 29.6478 & 7.0677 \\ 7.9275 & 2.8603 & 7.0677 & 7.4885 \end{bmatrix}$$

$$\hat{\Sigma}_3 = \begin{bmatrix} 9.4477 & 3.5268 & 4.5255 & 1.2613 \\ 3.5268 & 10.2264 & -2.5687 & .2601 \\ 4.5255 & -2.5687 & 9.5669 & 2.1149 \\ 1.2613 & .2601 & 2.1149 & 1.6793 \end{bmatrix}$$

(c) Coefficients of Common Principal Components

$$P = \begin{bmatrix} .7367 & -.6471 & -.1640 & .1084 \\ .2468 & .4655 & -.8346 & -.1607 \\ .6047 & .5003 & .5221 & -.3338 \\ .1753 & .3382 & .0628 & .9225 \end{bmatrix}$$

(d) Variances l_{ii} of CPCs and Eigenvalues of S_j

Versicolor: $l_{1j} = 48.4602 \quad 7.4689 \quad 5.5394 \quad 1.0139$
Eigenvalues $= 48.7874 \quad 7.2384 \quad 5.4776 \quad .9790$
Virginica: $l_{2j} = 69.2235 \quad 6.7124 \quad 7.5367 \quad 5.3642$
Eigenvalues $= 69.5255 \quad 5.2295 \quad 10.6552 \quad 3.4266$
Setosa: $l_{3j} = 14.6444 \quad 2.7526 \quad 12.5065 \quad 1.0169$
Eigenvalues $= 23.6456 \quad 2.6796 \quad 3.6969 \quad .9053$

Table 5.28 *(Continued)*

(e) Covariance and Correlation Matrices of CPCs

$$
\mathbf{C}_1 = \begin{bmatrix}
48.4602 & 3.4072 & -1.1931 & .7172 \\
3.4972 & 7.4689 & -.3776 & .2049 \\
-1.1931 & -.3776 & 5.5394 & -.3278 \\
.7172 & .2049 & -.3278 & 1.0139
\end{bmatrix}
$$

$$
\mathbf{R}_1 = \begin{bmatrix}
1.000 & .1791 & -.0728 & .1023 \\
.1791 & 1.0000 & -.0587 & .0745 \\
-.0728 & -.0587 & 1.0000 & -.1383 \\
.1023 & .0745 & -.1383 & 1.0000
\end{bmatrix}
$$

$$
\mathbf{C}_2 = \begin{bmatrix}
69.2235 & -1.6211 & 2.6003 & -2.9062 \\
-1.6211 & 6.7124 & -1.9278 & 2.3514 \\
2.6003 & -1.9278 & 7.5367 & -2.2054 \\
-2.9062 & 2.3514 & -2.2054 & 5.3642
\end{bmatrix}
$$

$$
\mathbf{R}_2 = \begin{bmatrix}
1.0000 & -.0752 & .1138 & -.1508 \\
-.0752 & 1.0000 & -.2710 & .3919 \\
.1138 & -.2710 & 1.0000 & -.3468 \\
-.1508 & .3919 & -.3468 & 1.0000
\end{bmatrix}
$$

$$
\mathbf{C}_3 = \begin{bmatrix}
14.6444 & -.5682 & -9.9950 & -.2106 \\
-.5682 & 2.7526 & .0487 & -.4236 \\
-9.9950 & .0487 & 12.5065 & .4235 \\
-.2160 & -.4236 & .4235 & 1.0169
\end{bmatrix}
$$

$$
\mathbf{R}_3 = \begin{bmatrix}
1.000 & -.0895 & -.7385 & -.0546 \\
-.0895 & 1.0000 & .0083 & -.2532 \\
-.7385 & .0083 & 1.0000 & .1188 \\
-.0546 & -.2532 & .1188 & 1.0000
\end{bmatrix}
$$

Source: Flury, 1984; reproduced with permission.

[a] The sample covariance matrices reported here were multiplied by 10^2.

5.6 WEIGHTED PRINCIPAL COMPONENTS

When estimating the model $\mathbf{X} = \mathbf{Z}_{(r)}\mathbf{A}_{(r)}^{\mathrm{T}} + \boldsymbol{\delta}$, the r common components $\mathbf{Z}_{(r)}$ do not always account for the same percentage of variance of each variable. Thus some variables may have most of their variation explained by r common components, whereas others may witness a substantial part of their variance relegated to the residual matrix $\boldsymbol{\delta}$. This is because not all variables will be equally affected by sampling variation, measurement error, or natural variation in the population. Such error heteroscedasticity however is not taken into account when computing the PCs, and the values of the first r PCs are independent of r. Thus both precise as well as error-prone variables are treated equally since they receive the same weight, and this results in a misspecification of the model.

Let

$$\mathbf{X} = \boldsymbol{\chi} + \boldsymbol{\Delta} \tag{5.97}$$

where $\boldsymbol{\chi}$ and $\boldsymbol{\Delta}$ represent the true and error parts, respectively. The covariance can thus be decomposed into two parts as

$$\boldsymbol{\Sigma} = \boldsymbol{\Sigma}^* + \boldsymbol{\Psi} \tag{5.98}$$

in conformity with Eq. (5.97), where the true part is a linear combination of $r < p$ PCs, that is,

$$\boldsymbol{\chi} = \boldsymbol{\zeta}\boldsymbol{\alpha} \tag{5.99}$$

However rather than decompose $\boldsymbol{\Sigma}$ it is more valid to decompose the product

$$\boldsymbol{\Psi}^{-1}\boldsymbol{\Sigma} = \boldsymbol{\Psi}^{-1}\boldsymbol{\Sigma}^* + \mathbf{I} \tag{5.100}$$

since variates with greater (smaller) error variance should be given less (greater) weight in the analysis. Also, premultiplying by $\boldsymbol{\Psi}^{-1}$ results in a unit residual "covariance" matrix, that is, homoscedastic and uncorrelated residual terms. As can be seen in Chapter 6 this bears a greater resemblance to a "proper" factor analysis model.

Let $\boldsymbol{\Pi}$ be a $(p \times 1)$ vector of coefficients. We wish to maximize the quadratic form

$$\lambda = \frac{\boldsymbol{\Pi}^{\mathrm{T}}\boldsymbol{\Sigma}\boldsymbol{\Pi}}{\boldsymbol{\Pi}^{\mathrm{T}}\boldsymbol{\Psi}\boldsymbol{\Pi}} \tag{5.101}$$

Cross-multiplying and differentiating with respect to $\boldsymbol{\Pi}$ yields

$$2\lambda\boldsymbol{\Psi}\boldsymbol{\Pi} + \frac{\partial\lambda}{\partial\boldsymbol{\Pi}}\boldsymbol{\Pi}^{\mathrm{T}}\boldsymbol{\Psi}\boldsymbol{\Pi} = 2\boldsymbol{\Sigma}\boldsymbol{\Pi} \tag{5.102}$$

so that the necessary condition for a maximum is

$$\frac{\partial\lambda}{\partial\boldsymbol{\Pi}} = \frac{2(\boldsymbol{\Sigma} - \lambda\boldsymbol{\Psi})\boldsymbol{\Pi}}{\boldsymbol{\Pi}^{\mathrm{T}}\boldsymbol{\Psi}\boldsymbol{\Pi}} = \mathbf{0} \tag{5.103}$$

where we assume $\boldsymbol{\Pi}^{\mathrm{T}}\boldsymbol{\Psi}\boldsymbol{\Pi} > 0$ for all $\boldsymbol{\Pi} \neq \mathbf{0}$. The normal equations can then be written as

$$(\boldsymbol{\Sigma} - \hat{\lambda}\boldsymbol{\Psi})\hat{\boldsymbol{\Pi}} = \mathbf{0} \tag{5.104}$$

so that $\hat{\boldsymbol{\Pi}}$ is the latent vector of $\boldsymbol{\Sigma}$ (in the metric of $\boldsymbol{\Psi}$) which is associated with the largest root $\hat{\lambda}$ (Section 2.10). Alternatively Eq. (5.104) can be expressed as

$$(\boldsymbol{\Psi}^{-1}\boldsymbol{\Sigma} - \hat{\lambda}\mathbf{I})\hat{\boldsymbol{\Pi}} = \mathbf{0} \tag{5.105}$$

where $\hat{\boldsymbol{\Pi}}$ is the latent vector associated with the largest latent root of the weighted matrix $\boldsymbol{\Psi}^{-1}\boldsymbol{\Sigma}$. Using Eq. (5.98), the normal equations can be expressed in yet a third form as

$$[\boldsymbol{\Psi}^{-1}(\boldsymbol{\Sigma}^* + \boldsymbol{\Psi}) - \hat{\lambda}\mathbf{I}]\hat{\boldsymbol{\Pi}} = [\boldsymbol{\Psi}^{-1}\boldsymbol{\Sigma}^* - (\hat{\lambda} - 1)]\hat{\boldsymbol{\Pi}} = \mathbf{0} \tag{5.106}$$

from which it is evident that only linear combinations that correspond to $\hat{\lambda} > 1$ need be considered. Also, the error covariance matrix $\boldsymbol{\Psi}$ need not be restricted in any fashion—for example, it need not be diagonal or correspond to homoscedastic error terms. When $\boldsymbol{\Psi}$ is constrained to be diagonal however we obtain an important class of factor analysis models discussed in Chapter 6. Note also that the quantity

$$\mathbf{I} = \frac{1}{2}\ln\lambda$$

$$= \frac{1}{2}\ln\frac{\boldsymbol{\Pi}^{\mathrm{T}}\boldsymbol{\Sigma}\boldsymbol{\Pi}}{\boldsymbol{\Pi}^{\mathrm{T}}\boldsymbol{\Psi}\boldsymbol{\Pi}} \tag{5.107}$$

can be interpreted as a measure of the degree of information between $\mathbf{X}$ and $\boldsymbol{\chi}$.

When using weighted PCs with sample data a difficulty arises with respect to the estimation of $\boldsymbol{\Psi}$. Generally speaking, prior information is required, which usually takes the form of specifying a value for r, although for multivariate normal data the latent roots can be tested for isotropic structure and the number of common factors can therefore be determined (Section 4.3). Once the number of common components is known (given), the matrix $\boldsymbol{\Psi}$ can be estimated jointly with the latent vectors, usually by iteration. Note that the loading coefficients are no longer independent of the number r of common PCs as is the case for the ordinary principal components model of Chapter 3. An advantage of the weighted principal component model is that it is invariant with respect to changes in the units of measure, so that both the covariance and correlation matrices, for example, yield the same set of correlation loadings. Once the loadings and the scores are known, nothing new arises as to the interpretation or the use of the weighted principal component—for an example see Datta and Ghosh (1978).

An alternative way of considering a weighted principal components analysis is in terms of an oblique Euclidian space, which generalizes the usual orthogonal Cartesian coordinate system (see also Section 5.4.1).

THEOREM 5.12. Let $\mathbf{y}_1 = (y_{11}, y_{12}, \ldots, y_{1p})^{\mathrm{T}}$, $\mathbf{y}_2 = (y_{21}, y_{22}, \ldots, y_{2p})^{\mathrm{T}}$, $\ldots$, and $\mathbf{y}_n = (y_{n1}, y_{n2}, \ldots, y_{np})^{\mathrm{T}}$ be n points in an oblique Euclidian space, with squared interpoint distances given by $(\mathbf{y}_i - \mathbf{y}_j)^T\boldsymbol{\Gamma}^{-1}(\mathbf{y}_i - \mathbf{y}_j)$ for $\boldsymbol{\Gamma}$ positive definite. Then the optimal-fitting subspace of the n points is given by the solution of $(\mathbf{X}^T\mathbf{X} - l_i\boldsymbol{\Gamma})\mathbf{Q}_i = \mathbf{0}$, where $i = 1, 2, \ldots, r$ and the goodness of the fit is measured by the ratio $\mathbf{R}^2 = \Sigma_{i=1}^{r} l_i / \Sigma_{i=1}^{p} l_i$.

PROOF. Let $\boldsymbol{v}_i = \mathbf{Q}\mathbf{y}_i$ represent a linear transformation of the n points, where $\mathbf{Q}$ is a matrix of latent vectors of $\mathbf{X}^T\mathbf{X}$ in the metric of $\boldsymbol{\Gamma}$. Then $\mathbf{y}_i = \mathbf{Q}^T\boldsymbol{v}_i$ and we have, for the ith and jth points,

$$(\mathbf{y}_i - \mathbf{y}_j)^T\boldsymbol{\Gamma}^{-1}(\mathbf{y}_i - \mathbf{y}_j) = (\mathbf{Q}^T\boldsymbol{v}_i - \mathbf{Q}^T\boldsymbol{v}_j)^T\boldsymbol{\Gamma}^{-1}(\mathbf{Q}^T\boldsymbol{v}_i - \mathbf{Q}^T\boldsymbol{v}_j)$$

$$= (\boldsymbol{v}_i - \boldsymbol{v}_j)^T\mathbf{Q}\boldsymbol{\Gamma}^{-1}\mathbf{Q}^T(\boldsymbol{v}_i - \boldsymbol{v}_j)$$

$$= \operatorname{tr}(\boldsymbol{\Gamma}^{-1}\mathbf{X}^T\mathbf{X})$$

where the latent vectors can be chosen such that $\mathbf{Q}_i^T\boldsymbol{\Gamma}\mathbf{Q}_i = 1$ and $\mathbf{Q}_i^T\boldsymbol{\Gamma}\mathbf{Q}_j = 0$ for $i \neq j$ (Section 2.10). The sum of squares of interpoint distances are thus given by

$$\sum_{i,j=1}^{n} (\boldsymbol{v}_i - \boldsymbol{v}_j)^T(\boldsymbol{v}_i - \boldsymbol{v}_j) = (l_1 + l_2 + \cdots + l_p)$$

and the optimal fit is then provided by the first r latent roots and latent vectors of $\mathbf{X}^T\mathbf{X}$, in the metric of $\boldsymbol{\Gamma}$.

5.7   PRINCIPAL COMPONENTS IN THE COMPLEX FIELD

It was shown in Section 2.11 that we can define a multivariate normal distribution using complex random variables. A further question arises as to whether it is possible to extend PCA to cover the case of complex variables together with its general interpretation, distribution theory, and testing. It turns out that the formal extension of PCA to the complex field is straightforward. Let $\mathbf{X}$ be a vector of random variables with covariance matrix $\mathbf{E}(\mathbf{X}\mathbf{X}^T) = \boldsymbol{\Sigma}$. If a random sample is available, the unbiased estimator of $\boldsymbol{\Sigma}$ is $\mathbf{S} = \frac{1}{n-1}\mathbf{A}$ where $\mathbf{A}$ is the matrix of sums of squares and products of the sampled random variables. The population latent vectors are then obtained from the Hermitian matrix $\boldsymbol{\Sigma}$ together with the corresponding latent roots (Section 2.11.1), that is,

$$\bar{\boldsymbol{\Pi}}^T\boldsymbol{\Sigma}\boldsymbol{\Pi} = \boldsymbol{\Lambda}$$

where

$$\bar{\Pi}_i^T \Pi_j = \begin{cases} 1 & \text{if } i = j \\ 0 & \text{if } i \neq j \end{cases}$$

and Λ is a (real) diagonal, nonnegative matrix. Corresponding sample estimates are then obtained from matrix $\mathbf{S}$, that is

$$\bar{\mathbf{P}}^T \mathbf{SP} = \mathbf{L}$$

where (unstandardized) PCs are obtained as the linear combinations $\mathbf{P}_i^T \mathbf{X}_i$. Nothing new emerges in practice when using complex PCs—for time series application see Section 7.7. Hardy and Walton (1978) use principal components to analyze wind speed and wind direction where exponential notation is used to relate speed and direction within a single complex number $z = xe^{i\theta}$, where x is observed wind speed and θ is the observed wind direction. Gupta (1965) has also extended the usual hypothesis testing to the complex case including Kshirsager's test for a single nonisotropic PC (Section 4.3.3).

5.8 MISCELLANEOUS STATISTICAL APPLICATIONS

Aside from being used in analyzing data, the PC model can also be employed in conjunction with other statistical techniques, such as missing data estimation (Section 4.7.3) or regression (Chapter 10). In this section we consider several applications of PCA which are intended to solve specific statistical difficulties in areas other than factor analysis. First we consider certain optimality criteria.

5.8.1 Further Optimality Properties

Let $\mathbf{Y}$ be any $(n \times p)$ matrix. Then we wish to find a matrix $\hat{\mathbf{Y}}$ such that the error matrix $\mathbf{E} = \mathbf{Y} - \hat{\mathbf{Y}}$ is as small as possible. We chose the Euclidian (Frobenius) norm

$$\|\mathbf{E}\| = \left[\sum_{i=1}^{n} \sum_{j=1}^{p} e_{ij}^2 \right]^{1/2} \tag{5.108}$$

which is to be made a minimum.

THEOREM 5.13   Let $\mathbf{Y}$ be any $(n \times p)$ real matrix such that $\mathbf{Y} = \mathbf{Z}_{(i)} \mathbf{P}_{(i)}^T$ and $i = 1, 2, \ldots, r \leq p \leq n$. Then $\|\mathbf{E}\| = \|\mathbf{Y} - \hat{\mathbf{Y}}\|$ is a minimum when $\hat{\mathbf{Y}} = \mathbf{Z}_{(r)} \mathbf{P}_{(r)}^T$ where the columns of $\mathbf{Z}_{(r)}$ and $\mathbf{P}_{(r)}$ are the first r PCs and latent vectors respectively.

PROOF. We have (Exercise 5.12)

$$\|\mathbf{Y}\|^2 = \|\mathbf{Y} - \hat{\mathbf{Y}}\|^2$$

$$= \|\mathbf{Y} - \mathbf{Z}_{(r)}\mathbf{P}_{(r)}^{\mathrm{T}}\|^2$$

$$= \mathrm{tr}(\mathbf{Y} - \mathbf{Z}_{(r)}\mathbf{P}_{(r)}^{\mathrm{T}})^{\mathrm{T}}(\mathbf{Y} - \mathbf{Z}_{(r)}\mathbf{P}_{(r)}^{\mathrm{T}})$$

$$= \mathrm{tr}(\mathbf{Y}^{\mathrm{T}}\mathbf{Y} - \mathbf{Y}^{\mathrm{T}}\mathbf{Z}_{(r)}\mathbf{P}_{(r)}^{\mathrm{T}} - \mathbf{P}_{(r)}\mathbf{Z}_{(r)}^{\mathrm{T}}\mathbf{Y} + \mathbf{P}_{(r)}\mathbf{Z}_{(r)}^{\mathrm{T}}\mathbf{Z}_{(r)}\mathbf{P}_{(r)}^{\mathrm{T}})$$

$$= \mathrm{tr}(\mathbf{Y}^{\mathrm{T}}\mathbf{Y} - \mathbf{P}_{(r)}\mathbf{Z}_{(r)}^{\mathrm{T}}\mathbf{Z}_{(r)}\mathbf{P}_{(r)}^{\mathrm{T}}$$

$$= \mathrm{tr}(\mathbf{Y}^{\mathrm{T}}\mathbf{Y}) - \mathrm{tr}(\mathbf{P}_{(r)}\mathbf{Z}_{(r)}^{\mathrm{T}}\mathbf{Z}_{(r)}\mathbf{P}_{(r)}^{\mathrm{T}})$$

$$= \sum_{i=1}^{p} l_i - \mathrm{tr}(\mathbf{Z}_{(r)}^{\mathrm{T}}\mathbf{Z}_{(r)})\,\mathrm{tr}(\mathbf{P}_{(r)}^{\mathrm{T}}\mathbf{P}_{(r)})$$

$$= \sum_{i=1}^{p} l_i - \sum_{i=r}^{p} l_i$$

$$= \sum_{i=r+1}^{p} l_i$$

which is minimum for a PCA (Theorem 3.9) see also Okamoto (1976) and Ozeki (1979).

For each latent root a PCA provides the best unit rank least squares approximation for any real matrix $\mathbf{Y}$. The result is also given in Rao (1964); Gabriel (1978) provides several alternative but parallel formulations of Theorem 5.12. For further results within the context of quality control see Hanson and Norris (1981). A more general problem is to minimize the weighted norm

$$\|\mathbf{W}^*(\mathbf{Y} - \mathbf{Z}_{(r)}\mathbf{P}_{(r)}^{\mathrm{T}}\| \tag{5.109}$$

where the asterisk denotes the Hadamard (element by element) product. Equation (5.109) is related to the missing value problem when all elements of $\mathbf{W}$ assume the values 0 and 1 (Section 4.7.3). An iterative algorithm for minimizing Eq. (5.109) is provided by Gabriel and Zamir (1979). Gabriel and Odoroff (1983) have also considered a robust version of Theorem 5.12 by considering robust averages such as trimmed means and medians, which can be used in conjunction with a data matrix to carry out a (robust) PCA (see Section 4.7.1).

Optimal latent roots and latent vectors of a symmetric matrix can also be obtained under restrictions.

THEOREM 5.14   (Rao, 1964). Let $\mathbf{S}$ be a symmetric $(p \times p)$ matrix and $\mathbf{U}$ a $(p \times k)$ matrix of rank k. Also, let $\mathbf{V}_1, \mathbf{V}_2, \ldots, \mathbf{V}_r$ be p-dimensional vectors satisfying the restrictions

$$\text{(i)} \qquad \mathbf{V}_i^T \mathbf{V}_j = \begin{cases} 1 & i = j \\ 0 & i \neq j \end{cases}$$

$$\text{(ii)} \qquad \mathbf{V}_i^T \mathbf{U} = \mathbf{0} \qquad (i = 1, 2, \ldots, r)$$

Then the maximum of

$$\mathbf{V}_1^T \mathbf{S} \mathbf{V}_1 + \mathbf{V}_2^T \mathbf{S} \mathbf{V}_2 + \cdots + \mathbf{V}_r^T \mathbf{S} \mathbf{V}_r$$

is attained when $\mathbf{V}_i = \mathbf{R}_i$ $(i = 1, 2, \ldots, r)$, the ith latent vector of the matrix $[\mathbf{I} - \mathbf{U}(\mathbf{U}^T\mathbf{U})^{-1}\mathbf{U}^T]\mathbf{S} = [\mathbf{I} - \mathbf{P}_u]\mathbf{S}$. The maxima are then the sums of the latent roots of $[\mathbf{I} - \mathbf{P}_u]\mathbf{S}$.

The proof consists of showing that the latent roots of $(\mathbf{I} - \mathbf{P}_u)\mathbf{S}$ are the same as those of the symmetric matrix $\mathbf{S}^{1/2}(\mathbf{I} - \mathbf{P}_u)\mathbf{S}^{1/2}$. The result can be generalized to oblique (weighted) coordinates by considering the weighted expressions $\mathbf{V}_i^T \mathbf{\Omega} \mathbf{V}_i = 1$ and $\mathbf{V}_i^T \mathbf{\Omega} \mathbf{V}_j = 0$.

5.8.2   Screening Data

Principal components can be used as diagnostic aids in uncovering multivariate outliers and measurement errors in multivariate data. Consider the matrix $\mathbf{X}^+ = (\mathbf{X}^T\mathbf{X})^{-1}\mathbf{X}^T$ for nonsingular $\mathbf{X}^T\mathbf{X}$. Then $\mathbf{X}^+$ is known as the (unique) generalized inverse of the $(n \times p)$ matrix $\mathbf{X}$ (Section 5.2). Aside from its use in multivariate regression $\mathbf{X}^+$ can be employed to spot multivariate outliers in a data matrix, since a high value of the ith and jth elements of $\mathbf{X}^+$ indicates the ith sample point's value for variable j is not consistent with its values for the remaining variables. The inverse $\mathbf{X}^+$ is therefore sensitive to deviations from a multivariate pattern.

Example 5.14.   Consider the matrix

$$\mathbf{X} = \begin{bmatrix} -3 & -2 \\ -2 & -1 \\ -1 & 1.5 \\ 0 & -0.5 \\ 1 & 0 \\ 2 & 1 \\ 3 & 1 \end{bmatrix}$$

for $p = 2$ variables and $n = 7$ sample points (Gabriel and Haber, 1973). Although no observation appears to be an outlier, the third sample point is distinct from the rest. Forming the generalized inverse and transposing yields

$$(\mathbf{X}^{+})^{\mathrm{T}} = \begin{bmatrix} -.041 & -.161 \\ -.056 & -.037 \\ -.200 & +.400 \\ +.043 & -.105 \\ +.071 & -.086 \\ +.056 & +.037 \\ +.127 & -.048 \end{bmatrix}$$

where it is apparent that the third row is a multivariate outlier.   □

In Example 5.14 the matrix $\mathbf{X}$ is assumed to be of full rank. When $\rho(\mathbf{X}) = r < p \le n$ and $\mathbf{X}^{+}$ can no longer be computed in the usual way, a PC decomposition can be used to obtain the more general expression $\mathbf{X}^{+} = (\mathbf{X}^{\mathrm{T}}\mathbf{X})^{+}\mathbf{X}^{\mathrm{T}}$ where $(\mathbf{X}^{\mathrm{T}}\mathbf{X})^{+}$ is the generalized inverse of $\mathbf{X}^{\mathrm{T}}\mathbf{X}$. Let $\mathbf{Q}^{\mathrm{T}}\mathbf{X}\mathbf{P} = \mathbf{\Delta}_{(r)}$ be the singular value decomposition of $\mathbf{X}$ (Theorem 5.1) where $\rho(\mathbf{X}) = r < p < n$. Then it is easy to verify that $(\mathbf{X}^{\mathrm{T}}\mathbf{X})^{+} = \mathbf{P}\mathbf{\Delta}_{(r)}^{-2}\mathbf{P}^{\mathrm{T}}$ is a unique generalized inverse of $(\mathbf{X}^{\mathrm{T}}\mathbf{X})$ so that

$$\begin{aligned} \mathbf{X}^{+} &= (\mathbf{X}^{\mathrm{T}}\mathbf{X})^{+}\mathbf{X}^{\mathrm{T}} \\ &= \mathbf{P}\mathbf{\Delta}_{(r)}^{-2}\mathbf{P}^{\mathrm{T}}\mathbf{P}\mathbf{\Delta}_{r}\mathbf{Q}^{\mathrm{T}} \\ &= \mathbf{P}\mathbf{\Delta}_{(r)}^{-1}\mathbf{Q}^{\mathrm{T}} \end{aligned}$$

where $\mathbf{\Delta}_{(r)}$ is defined as in Theorem 5.1. The generalized inverse $\mathbf{X}^{+}$ can be used to estimate a regression equation when the explanatory variables are not linearly independent (Section 10.3).

A principal components analysis can also be used to devise a testing procedure for detecting errors of measurement in the observed variables. Consider the PC decomposition (Eq. 3.1) where the variables X_j are given by

$$X_j = \chi_j + \Delta_j \tag{5.110}$$

$(j = 1, 2, \ldots, p)$ where χ_j is the true part and Δ_j is the error part. We then may wish to test

$$H_0: \Delta_j = 0$$

$$H_a: \Delta_j \neq 0 \tag{5.111}$$

for some j. Consider the standardized PCs

$$
\begin{aligned}
\zeta_1 &= (\pi_{11}/\sqrt{\lambda_1})X_1 + (\pi_{12}/\sqrt{\lambda_1})X_2 + \cdots + (\pi_{1p}/\sqrt{\lambda_1})X_p \\
\zeta_2 &= (\pi_{21}/\sqrt{\lambda_2})X_1 + (\pi_{22}/\sqrt{\lambda_2})X_2 + \cdots + (\pi_{2p}/\sqrt{\lambda_2})X_p \\
\hline
\zeta_p &= (\pi_{p1}/\sqrt{\lambda_p})X_1 + (\pi_{p2}/\sqrt{\lambda_p})X_2 + \cdots + (\pi_{pp}/\sqrt{\lambda_p})X_p
\end{aligned}
\tag{5.112}
$$

where for the sake of convenience the roots are ranked as $\lambda_1 \le \lambda_2 \le \cdots \le \lambda_p$. Also note that the "standardized" coefficients $\pi_{ij}/\sqrt{\lambda_i}$ represent regression-type loadings rather than the usual correlations between the variates and components (Section 3.8.2). Clearly the first set of PCs correspond to *small* roots that contain most of the error information. When the unobserved values χ_j represent standardized $N(0,1)$ variates, replacing Eq. (5.112) into Eq. (5.110) yields the hypothesis

$$
H_0: \zeta_i = 0
$$

$$
H_a: \zeta_i = \sum_{i=1}^{p} (\pi_{ij}/\sqrt{\lambda_i})\Delta_j
\tag{5.113}
$$

which is equivalent to Eq. (5.111). The coefficients of Eq. (5.112) that correspond to small roots can also be rotated using an orthogonal transformation such as the varimax criterion in order to locate large error coefficients. Although of theoretical interest, the testing procedure cannot always be translated into practice since the true parts are generally not observed. However, if a previous "calibration" sample is available for which the true parts are known (or have been estimated), then the system of equations (Eq. 5.112) may be used to estimate the influence of errors in further samples.

5.8.3 Principal Components in Discrimination and Classification

The classical problem of discriminant analysis is well known. Given p random variables observed for g groups of sample points, each with n_t observations ($t = 1, 2, \ldots, g$), is it possible to compute functions of the random variables which "discriminate," in an optimal manner, between the groups? If so, and given a further set of observations, the discriminant function(s) can be used to classify the new observations with a minimal probability of error. Various solutions of the problem may be found in standard texts of multivariate analysis (e.g. see Anderson 1984a).

Principal components analysis can also be adapted to perform a similar role. Thus from the viewpoint of PCA the situation is as in Section 5.4.3, where p identical random variables are observed for t groups or classes, each with n_t observations. The idea is to first estimate $r < p$ principal components for each group, and then use least squares regression to compute t equations for the new observation vector $\mathbf{y} = (y_1, y_2, \ldots, y_p)^T$, using the loadings as independent variables (Wold, 1978; Wold et al., 1982). Group membership is then decided upon using a goodness of fit statistic of the regression equations. The procedure is also at times referred to as disjoint principal components analysis and has been used in medical and biochemical research (Dunn and Wold, 1980; Duewer et al., 1978) where it appears to possess advantages over classical discrimination. A major application lies in drug research, where the objective is to be able to relate chemical structure and function of organic molecules so that unobserved medical (biological) properties of molecules can be predicted without having to synthesize their structure (Dunn and Wold, 1978). The result is often a saving of time and effort in pursuing unfruitful lines of research.

Other variations of classification using PCA are also possible, where the loadings (scores) are used to cluster the variables (observations) (Saxena and Walters, 1974; Cammarata and Menon, 1976; Mager, 1980a). Also, when performing a hierarchical cluster analysis of the sample points, PCA can be used to reduce the dimensionality of the variable space. When using PCA in a cluster or classification analysis it is also of interest to consider heterogeneous populations that consist of mixtures of more fundamental, homogeneous populations. For example, let $\mathbf{Y}$ be a k-dimensional random variable distributed as a mixture of two normal distributions, with mixing parameters p and $q = 1 - p$, means $\boldsymbol{\mu}_1$ and $\boldsymbol{\mu}_2$, and common covariance matrix $\boldsymbol{\Sigma}$. Let $\boldsymbol{\Delta}$ denote the Mahalanobis distance between the two normal components of the mixture. Then $\mathbf{Y}$ has covariance matrix

$$\mathbf{V} = pq\mathbf{dd}^{\mathrm{T}} + \boldsymbol{\Sigma} \tag{5.114}$$

and

$$\Delta^2 = (\boldsymbol{\mu}_1 - \boldsymbol{\mu}_2)\boldsymbol{\Sigma}^{-1}(\boldsymbol{\mu}_1 - \boldsymbol{\mu}_2)^{\mathrm{T}}$$
$$= \mathbf{d}\boldsymbol{\Sigma}^{-1}\mathbf{d}^{\mathrm{T}}$$

where $\mathbf{d} = (\boldsymbol{\mu}_1 - \boldsymbol{\mu}_2)$.

Consider the spectral decomposition $\mathbf{V} = \lambda_1\boldsymbol{\pi}_1\boldsymbol{\pi}_1^{\mathrm{T}} + \lambda_2\boldsymbol{\pi}_2\boldsymbol{\pi}_2^{\mathrm{T}} + \cdots + \lambda_k\boldsymbol{\pi}_k\boldsymbol{\pi}_k^{\mathrm{T}}$ where $1 \le r < k$ terms are retained. Then the (squared) Mahalanobis distance in r-dimensional subspace is

$$\Delta_r^2 = \sum_{i=1}^{r} \frac{(\boldsymbol{\pi}_i^{\mathrm{T}}\mathbf{d})^2}{\lambda_i} \left[1 - pq \sum_{i=1}^{r} \frac{(\boldsymbol{\pi}_i^{\mathrm{T}}\mathbf{d})^2}{\lambda_i} \right]^{-1} \tag{5.115}$$

(Chang, 1983). The PCs that best discriminate between the two populations are therefore those that correspond to large values of $(\boldsymbol{\pi}_i^{\mathrm{T}}\mathbf{d})^2/\lambda_i$, assuming $\boldsymbol{\Sigma}$ possesses distinct roots. The distance between the mixture components based on the ith PC can then be tested as

$$H_0: \Delta_i = 0$$

$$H_a: \Delta_i \neq 0$$

using the likelihood ratio test. It can also be shown (Wolf, 1970) that sampling from discrete mixtures is related to the so-called latent class model (Section 6.13).

5.8.4 Mahalanobis Distance and the Multivariate T-Test

It was seen in Section 5.6 that distance in oblique Euclidian space is given by $(\mathbf{y}_i - \mathbf{y}_j)^T \boldsymbol{\Gamma}^{-1}(\mathbf{y}_i - \mathbf{y}_j)$ where $\boldsymbol{\Gamma}$ contains lengths and the nonorthogonal orientations of the coordinate axes. Let $\boldsymbol{\Gamma} = \mathbf{S}$, the sample covariance matrix. Then the expressions

$$\mathbf{d}_{ij} = (\mathbf{y}_i - \mathbf{y}_j)^T \mathbf{S}^{-1}(\mathbf{y}_i - \mathbf{y}_j) \tag{5.116}$$

and

$$\cos \theta = \frac{\mathbf{y}_i^T \mathbf{S}^{-1} \mathbf{y}_i}{[(\mathbf{y}_i^T \mathbf{S}^{-1} \mathbf{y}_i)(\mathbf{y}_j^T \mathbf{S}^{-1} \mathbf{y}_j)]^{1/2}} \tag{5.117}$$

are the Mahalanobis distance and cosines between the points $\mathbf{y}_i$ and $\mathbf{y}_j$, respectively. Equations (5.116) and (5.117) take into account the unequal variance and correlation amongst the variables. A special case of Eq. (5.116) is the multivariate T^2 statistic

$$(n-1)T^2 = (\bar{\mathbf{y}} - \boldsymbol{\mu})^T \mathbf{S}^{-1}(\bar{\mathbf{y}} - \boldsymbol{\mu}) \tag{5.118}$$

When $\mathbf{S}$ is not diagonal it may be advantageous to decompose $\mathbf{S}^{-1}$ into principal components. We have

$$(n-1)T^2 = (\bar{\mathbf{y}} - \boldsymbol{\mu})^T \mathbf{S}^{-1}(\bar{\mathbf{y}} - \boldsymbol{\mu})$$

$$= (\bar{\mathbf{y}} - \boldsymbol{\mu})^T (\mathbf{P} \mathbf{L}^{-1} \mathbf{P}^T (\bar{\mathbf{y}} - \boldsymbol{\mu}))$$

$$= [\mathbf{L}^{-1/2} \mathbf{P}^T (\bar{\mathbf{y}} - \boldsymbol{\mu})]^T [\mathbf{L}^{-1/2} \mathbf{P}^T (\bar{\mathbf{y}} - \boldsymbol{\mu})]$$

$$= \mathbf{Z}^T \mathbf{Z}$$

$$= \sum_{i=1}^{p} t_i^2 \tag{5.119}$$

where $\mathbf{Z}$ is a $(p \times 1)$ observation score vector for the components. The individual t_i are uncorrelated and are analogous to the usual univariate t statistic. An advantage of using t_i rather than T^2 is that for even moderately correlated random variables T^2 may consist of insignificant dimensions, that is, much (or most) of the significance may be concentrated in a few components. It is then possible to accept H_0 using T^2, in spite of significant mean difference(s) among some dimensions. On the assumption of multivariate normality, t_i^2 are univariate chi-squared with unit degrees of freedom, and the T^2 statistic can be replaced by either $t_i^2, t_2^2, \ldots, t_r^2$ or t_1^2, $t_1^2 + t_2^2, \ldots$ and $t_1^2 + t_2^2 + \cdots + t_r^2$ where the sums are distributed as $\chi_{(1)}^2$, $\chi_{(2)}^2, \ldots, \chi_{(r)}^2$. The decomposition can also be used to establish multivariate confidence intervals. Since the decomposition is not independent of the scaling of coordinates, the sample covariance matrix can be replaced by the correlation matrix in the event of large differences in the variances. The PC transformation has been used by Jackson (1959) in the context of multivariate quality control to test for significance in sample means.

Example 5.15. The following example is considered by Takemura (1985). We have $n = 60$ male students with scores for $p = 6$ different tests ($\mathbf{Y}_1 = $ English, $\mathbf{Y}_2 = $ reading comprehension, $\mathbf{Y}_3 = $ creativity, $\mathbf{Y}_4 = $ mechanical reasoning, $\mathbf{Y}_5 = $ abstract reasoning, $\mathbf{Y}_6 = $ mathematics). Grouping the students into two groups, depending on whether they intend to go to college, we have the following means and pooled sample covariance matrix:

	$\bar{\mathbf{Y}}_1$	$\bar{\mathbf{Y}}_2$	$\bar{\mathbf{Y}}_3$	$\bar{\mathbf{Y}}_4$	$\bar{\mathbf{Y}}_5$	$\bar{\mathbf{Y}}_6$
Group A	87.15	38.78	11.68	15.06	10.83	33.26
Group B	81.38	31.04	9.70	12.58	9.62	20.96

$$\mathbf{S} = \begin{bmatrix} 111.46 & & & & & \\ 56.96 & 67.22 & & & & \\ 18.87 & 17.28 & 14.47 & & & \\ 11.24 & 9.93 & 6.30 & 9.65 & & \\ 10.15 & 7.76 & 1.76 & 2.67 & 4.74 & \\ 60.08 & 49.64 & 14.55 & 12.86 & 10.94 & 81.41 \end{bmatrix}$$

The latent vectors, latent roots, and t_i^2 values for the six variables are then

	$\mathbf{P}_1$	$\mathbf{P}_2$	$\mathbf{P}_3$	$\mathbf{P}_4$	$\mathbf{P}_5$	$\mathbf{P}_6$
	.66	−.72	−.19	−.02	.02	.03
	.48	.24	.79	−.27	.08	.04
	.15	.03	.22	.76	−.54	−.24
	.10	.11	.01	.59	.72	.33
	.08	.06	−.06	−.00	.40	−.90
	.53	.63	−.53	−.06	−.15	.06
l_i	208.9	36.07	24.34	11.61	5.30	2.72
t_i^2	15.5	14.5	.81	.005	.07	1.69

where $T^2 = 32.56$ is significant, but only the first two components have significantly different means, at $\alpha = .05$. The first dimension differentiates academic subjects whereas the second contrasts English and mathematics.

5.9 SPECIAL TYPES OF CONTINUOUS DATA

In certain disciplines, for example geology, ecology, archaeology, and chemistry, the random variables may be expressed as either proportions or angular directions. Such data possess specific features not usually found in other types of continuous random variables, and consequently modify the structure and interpretation of a PCA.

5.9.1 Proportions and Compositional Data

Much of scientific data consists of concentration measurements such as parts per million, milligrams per liter, proportions (percentages), and so forth. When the concentrations do not add up to fixed constants ("closed arrays"), no special difficulties arise as to a PCA of the data, apart from the unit of measure problem and the effect(s) of additional transformations which may be required for approximate univariate normality of the variables. As was seen in Section (5.4.1), data expressed as proportions may also necessitate the use of cosine measures rather than covariances or correlations, particularly when performing a Q-analysis of the sample points (Imbrie and Purdy, 1962; Erez and Gill, 1977). When the data are nonlinear, for example, exponential, the logarithmic transformation will also tend to reduce variance differences as well as improve normality (see Hitchon et al., 1971). The transformation is particularly appropriate when measurements are distributed as the log-normal probability function, for example, sediment particle size (Klovan, 1966) or money income in economics (Aitchison and Brown, 1957). Here a rotation of axes will also be typically used for identification purposes (e.g., see Hudson and Ehrlich, 1980). Also when dealing with rates and proportions (Fleiss, 1973) the inverse sine transformation may be used in place of logs, but the practice does not seem to be common in factor analysis (see, however, Goodall, 1954).

Frequently proportions (percentages) represent an exhaustive analysis of a set of physical samples, for example, chemical composition of rocks or minerals (Le Maitre, 1968; Miesch, 1976a). Such numbers are said to form a (positive) simplex (Fig. 5.9). Here the row sums of $\mathbf{Y}$ are constant, and this introduces additional difficulties into a PCA, over and above those encountered when interpreting a correlation (covariance) between open arrays. Consider two variables $x_i > 0$ and $y_i > 0$ such that $x_i + y_i = 1$ $(i =$

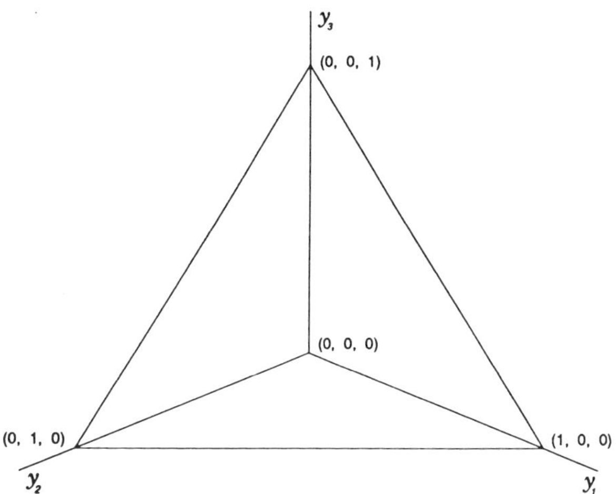

Figure 5.9 The equilateral triangle of possible points in a mixture $y_1 + y_2 + y_3 = 1$.

$1, 2, \ldots, n$). Then we have

$$\sum_{i=1}^{n} x_i y_i = \sum_{i=1}^{n} x_i - \sum_{i=1}^{n} x_i^2 \geq 0 \tag{5.120}$$

since $0 \leq x_i^2 \leq x_i \leq 1$, and for uncentered measures of association, constant-sums introduce positive association even when x_i and y_i are independent. Likewise for centered data the sums of products are

$$\sum_{i=1}^{n} x_i y_i = \sum_{i=1}^{n} x_i - \sum_{i=1}^{n} x_i^2$$

$$= -\sum_{i=1}^{n} x_i^2 \tag{5.121}$$

a negative quantity. It follows from Eq. (5.121) that the covariance function is always negative since

$$\sum_{i \neq q}^{k} \mathrm{cov}(x_q, x_i) = -\mathrm{var}(x_q) < 0 \qquad (q = 1, 2, \ldots, k) \tag{5.122}$$

The "bias" implied by Eq. (5.122) or Eq. (5.120) may be removed by deleting a column from **Y**, but this has the disadvantage of rendering a PCA dependent on the variable which has been deleted. Closed arrays are consequently analyzed either by considering the total data or by transforming the proportions to induce "independence."

THEOREM 5.15 Let $x_{i1} + x_{i2} + \cdots + x_{ik} = T$ $(i = 1, 2, \ldots, n)$ where $\mathbf{X}^T\mathbf{X} = (n-1)\mathbf{S}$. Then

(1) $\mathbf{X}^T\mathbf{X}$ possesses a zero latent root with a corresponding latent vector whose elements can all be chosen to equal $1/\sqrt{k}$.

(2) The latent vector elements that correspond to nonzero latent roots sum to zero.

PROOF

(1) Let $\mathbf{X}_1, \mathbf{X}_2, \ldots, \mathbf{X}_{k-1}$ be linearly independent. Then the set $\mathbf{X}_1$, $\mathbf{X}_2, \ldots, \mathbf{X}_{k-1}, \mathbf{X}_k$ is linearly dependent and $\rho(\mathbf{X}^T\mathbf{X}) = k - 1$, implying that $\mathbf{X}^T\mathbf{X}$ possesses a zero latent root. The statement also follows from the fact that the row (column) sums of $\mathbf{X}^T\mathbf{X}$ are identically zero. Let $\mathbf{V} = (v_1, v_2, \ldots, v_k)^T$ be the corresponding latent vector. Then $(\mathbf{X}^T\mathbf{X})\mathbf{V} = \mathbf{0}$ or

$$v_1(\mathbf{X}_1^T\mathbf{X}_1) + v_2(\mathbf{X}_1^T\mathbf{X}_2) + \cdots + v_k(\mathbf{X}_1^T\mathbf{X}_k) = 0$$
$$\underline{v_1(\mathbf{X}_2^T\mathbf{X}_1) + v_2(\mathbf{X}_2^T\mathbf{X}_2) + \cdots + v_k(\mathbf{X}_2^T\mathbf{X}_k) = 0}$$
$$v_1(\mathbf{X}_k^T\mathbf{X}_1) + v_2(\mathbf{X}_k^T\mathbf{X}_2) + \cdots + v_k(\mathbf{X}_k^T\mathbf{X}_k) = 0$$

where

$$(\mathbf{X}_1^T\mathbf{X}_1) = -\sum_{i \neq 1} (\mathbf{X}_1^T\mathbf{X}_i),$$

$$(\mathbf{X}_2^T\mathbf{X}_2) = \sum_{i \neq 2} (\mathbf{X}_2^T\mathbf{X}_i), \ldots, \qquad (\mathbf{X}_k^T\mathbf{X}_k) = -\sum_{i \neq k} \mathbf{X}_k^T\mathbf{X}_i$$

using the closure constraint (Eq. 5.122) where $\bar{\mathbf{X}}_1 = \bar{\mathbf{X}}_2 = \cdots = 0$. We than have

$$(v_1 - v_2)\mathbf{X}_1^T\mathbf{X}_2 + (v_1 - v_3)\mathbf{X}_1^T\mathbf{X}_3 + \cdots + (v_1 - v_k)\mathbf{X}_1^T\mathbf{X}_k = 0$$
$$\underline{(v_2 - v_1)\mathbf{X}_2^T\mathbf{X}_1 + (v_2 - v_3)\mathbf{X}_2^T\mathbf{X}_3 + \cdots + (v_2 - v_k)\mathbf{X}_2^T\mathbf{X}_k = 0}$$
$$(v_k - v_1)\mathbf{X}_k^T\mathbf{X}_1 + (v_k - v_2)\mathbf{X}_k^T\mathbf{X}_2 + \cdots + (v_k - v_{k-1})\mathbf{X}_k^T\mathbf{X}_{k-1} = 0$$

so that a solution can always be chosen such that $v_1 = v_2 = \cdots = v_k$. Standardizing we conclude that $\mathbf{V}^* = (1/\sqrt{k}, 1/\sqrt{k}, \ldots, 1/\sqrt{k})^T$ is a latent vector that corresponds to the zero latent root.

(2) Let $\mathbf{P}_1, \mathbf{P}, \ldots, \mathbf{P}_{k-1}$ be the latent vectors that correspond to the $k - 1$ nonzero latent roots of $\mathbf{X}^T\mathbf{X}$. Then $(\mathbf{X}^T\mathbf{X} - l_i\mathbf{I})\mathbf{P}_i = \mathbf{0}$, and for $l_i \neq 0$ we have

$$(\mathbf{X}_1^T\mathbf{X}_1)p_{1j} + (\mathbf{X}_1^T\mathbf{X}_2)p_{2j} + \cdots + (\mathbf{X}_1^T\mathbf{X}_k)p_{kj} = l_j p_{1j}$$
$$\underline{(\mathbf{X}_1^T\mathbf{X}_2)p_{1j} + (\mathbf{X}_2^T\mathbf{X}_2)p_{2j} + \cdots + (\mathbf{X}_2^T\mathbf{X}_k)p_{kj} = l_j p_{2j}}$$
$$(\mathbf{X}_1^T\mathbf{X}_k)p_{1j} + (\mathbf{X}_2^T\mathbf{X}_k)p_{2j} + \cdots + (\mathbf{X}_k^T\mathbf{X}_k)p_{kj} = l_j p_{kj}$$

Adding, we obtain

$$p_{1j} \sum_{h=1}^{k} \mathbf{X}_1^{\mathrm{T}}\mathbf{X}_h + p_{2j} \sum_{h=1}^{k} \mathbf{X}_2^{\mathrm{T}}\mathbf{X}_h + \cdots + p_{kj} \sum_{h=1}^{k} \mathbf{X}_k^{\mathrm{T}}\mathbf{X}_h = l_j \sum_{i=1}^{k} p_{ij}$$

or

$$p_{1j}(0) + p_{2j}(0) + \cdots + p_{kj}(0) = 0 = l_j \sum_{i=1}^{k} p_{ij}$$

so that $\Sigma_{j=1}^{k} p_{ij} = 0$ $(j = 1, 2, \ldots, k)$ when $l_j \neq 0$.

It follows from Theorem 5.14 that the columns of $\mathbf{X}$ lie in a $k-1$ dimensional hyperplane, with perpendicular vector $\mathbf{V}^* = (1/\sqrt{k}, 1/\sqrt{k}, \ldots, 1/\sqrt{k})^{\mathrm{T}}$. The situation is not unsimilar to that found in allometry when measuring overall size, except for compositional data, the constant-sum constraint does not account for any variance. The results of Theorem 5.14 hold for the covariance matrix $(\mathbf{X}^{\mathrm{T}}\mathbf{X})$ and do not necessarily extend to other association matrices.

It may also be difficult to interpret covariance/correlation coefficients for proportions (particularly when dealing with compositional data). To overcome the difficulty Aitchison (1983, 1984) employs a logarithmic transformation which at times improves linearity and normality. The transformed data are then

$$\ln x_j - \frac{1}{k} \sum_{i=1}^{k} \ln x_i \qquad (j = 1, 2, \ldots, k) \tag{5.123}$$

so that the transformed data matrix is given by

$$\mathbf{X} = \ln\left[\frac{1}{\tilde{y}}(\mathbf{Y})\right] \tag{5.124}$$

where $\tilde{y} = (y_1 y_2 \cdots y_p)^{1/p}$ is the geometric mean of an observation vector (Section 3.7). Equation (5.123) preserves the zero-sum property of the second part of Theorem 5.14. Although its application may be of theoretical interest, it does not necessarily result in very different PC loadings than would be obtained for the nontransformed proportions (see Joliffe, 1986). Finally, when analyzing compositional data important considerations may be (1) measurement or rounding error in the proportions, and (2) the type of association matrix to use. Small errors in the constant-sum constraint may have a disproportionate effect on the properties of Theorem 5.14 and may render the computations unstable. Also, even though proportions are unit-free numbers, using covariances may yield loadings and scores that are heavily influenced by the diagonal elements of **S**, that is, by the relative abundance or scarcity of the individual constituents that make up the compositions.

Example 5.16.   The following data represent samples of a rhyolite–basalt complex from the Gardiner River, Yellowstone National Park, Wyoming (Table 5.29). Using both the covariance and correlation matrices we have the latent roots and latent vector structure as in Table 5.30.   □

Converting the covariance latent vector elements of P_1 to correlation loading coefficients indicates the dependence of the vector on the variances of the oxides, that is, on their relative abundance (scarcity). Also, the departure of the elements of P_8 (Table 5.30) from the constant value .35355 also indicates the sensitivity of the constant sum vector to small errors which are present in sample points 6, 11, 15, and 16 of Table 5.29. For a related application see also Flores and Shideler, 1978).

Table 5.29   Compositional Percentage Data Consisting of $p = 8$ Measurements and $n = 15$ Samples

SiO_2	Al_2O_3	FeO	MgO	CaO	Na_2O	K_2O	H_2O	Totals
51.64	16.25	10.41	7.44	10.53	2.77	0.52	0.44	100.00
54.33	16.06	9.49	6.70	8.98	2.87	1.04	.53	100.00
54.49	15.74	9.49	6.75	9.30	2.76	.98	.49	100.00
55.07	15.72	9.40	6.27	9.25	2.77	1.13	.40	100.01
55.33	15.74	9.40	6.34	8.94	2.61	1.13	.52	100.01
58.66	15.31	7.96	5.35	7.28	3.13	1.58	.72	99.99
59.81	14.97	7.76	5.09	7.02	2.94	1.97	.45	100.01
62.24	14.82	6.79	4.27	6.09	3.27	2.02	.51	100.01
64.94	14.11	5.78	3.45	5.15	3.36	2.66	.56	100.01
65.92	14.00	5.38	3.19	4.78	3.13	2.98	.61	99.99
67.30	13.94	4.99	2.55	4.22	3.22	3.26	.53	100.01
68.06	14.20	4.30	1.95	4.16	3.58	3.22	.53	100.00
72.23	13.13	3.26	1.02	2.22	3.37	4.16	.61	100.00
75.48	12.71	1.85	.37	1.10	3.58	4.59	.31	99.99
75.75	12.70	1.72	.40	.83	3.44	4.80	.37	99.98

Variances

63.2939	1.4232	8.4027	5.899	9.7130	.1010	1.9360	.0106

Source: Miesch, 1976a.

Table 5.30 Latent Roots and Latent Vectors of the Compositional Data of Table 5.29 Using the Covariance and Correlation Matrices

P_1	P_2	P_3	P_4	P_5	P_6	P_7	P_8
			(a) Covariance Matrix				
−.835644	−.080680	.137053	−.028769	.192950	.292841	−.173304	−.350086
.124379	.432898	−.291339	−.120720	−.583357	.424091	−.225306	.356028
.304187	−.295150	.067320	.639876	.134390	.471175	.235729	.332789
.254413	−.626143	−.266213	−.545307	.127153	.009716	−.161878	.364038
.326904	.328751	.709615	−.216437	.264265	.042307	−.192517	.354922
−.029921	.406778	−.386978	−.149214	.403938	−.128041	.602835	.344456
−.145634	−.222519	.309422	−.003706	−.591419	−.392795	.449636	.356768
.001784	.062137	−.268122	.456809	.073883	−.582940	−.484546	.368115
$l_1 = 90.6367$	$l_2 = .0606$	$l_3 = .0353$	$l_4 = .0213$	$l_5 = 0.148$	$l_6 = .0067$	$l_7 = .0031$	$l_8 = .0000$
			(b) Correlation Matrix				
.382509	.004028	−.113068	.044658	.038187	−.338457	−.161409	.834762
.378941	.007632	.275387	.812091	.318431	.956994	.012881	.127293
.382671	−.000871	.042321	−.183269	−.136789	−.168152	.829130	.289088
.381818	−.028532	.049378	−.471083	.646937	.313657	−.204755	.264974
.381762	−.028570	.128552	.013153	−.668310	.421741	−.319930	.331487
−.350472	.153517	.900203	−.140696	.004855	.113456	.097338	.032723
−.380521	−.022029	−.245161	.250961	.115065	.751099	.363512	.148670
.067074	.987037	−.141965	.007624	−.001569	.022065	−.020715	.011348
$l_1 = 6.8176$	$l_2 = .9914$	$l_3 = .1715$	$l_4 = .0113$	$l_5 = .0051$	$l_6 = .0020$	$l_7 = .0012$	$l_8 = .0000$

5.9.2 Estimating Components of a Mixture

A physical interpretation of a linear combination of compositional data is that of a mixture, with vectors representing a specific composition or distribution of the mixture components. Thus when a PCA indicates the existence of clusters of mixture constituents, they can be interpreted as complexes or compounds formed from the basic elements of the composition. For example, a particular mineral or solid solution may be indicated by highly intercorrelated chemical elements or geochemical metallic oxides (Middleton, 1964; Saxena, 1969; Saxena and Walter, 1974). Here interest centers on the variable space, that is, on the elements or constituents of a given sample where either a covariance or correlation matrix may be used depending on the desired relative weighting of the variables. High PC scores, when rotated such that all are positive, can then be interpreted in terms of "pure form" samples which had given rise to the remaining samples as mixtures.

At times an analysis of mixtures is cast in the form of a Q-mode PCA. This may be due to computational reasons when more variables than observations are available. More frequently, however, a Q-mode analysis is easier to work with (assuming n is not large) if interest centers on the component scores. This is because compositional data (either with closed arrays or otherwise) can also be considered in terms of the rows of $\mathbf{Y}$ where the sample space is viewed as having arisen from a set of $n - r$ mixtures of a smaller number r of basic samples which contain the true proportion of the constituent compositions or "end members." That is, we can reproduce, in specified proportions, the entire set of samples in terms of physical mixtures of end members. This permits a description of the ingredients in terms of the end-member samples rather than in terms of the individual ingredients themselves (Fig. 5.10). The role of a PCA is then to provide an "unmixing" of the samples, in terms of the basic (end member) samples. It is assumed that these basic samples have been observed as part of the total sample, an assumption which is very nearly satisfied for large n.

A difficulty still remains in the identification of the true source of the mixtures because of the rotational arbitrariness of the factors. The arbitrariness can be removed, however, by the a priori condition that mixture propositions be nonnegative. This implies a rotation of the factors since nonnegativity for the loadings and the scores cannot be met for compositional data, owing to Theorem 5.14. The problem can be resolved by rotating the factor axes into the positive quadrant in such a way that each axis coincides with a plotted sample point, which is then taken as an end-member of the sample point mixtures. As indicated in Section 5.3.2, such a rotation is generally oblique (Fig. 5.11). The method appears to have originated in the analysis of mixtures encountered in geological samples (Imbrie and Purdy, 1962), and is today widely employed in geology and other earth sciences (e.g., Klovan, 1966; Hitchon et al., 1971; Miesch, 1976a; Butler, 1976; Hudson and Ehrlich, 1980) as well as chemistry and ecology (Kowalski et al., 1982; Gold et al., 1976). Expositions of the method may also be found in Joreskog et al. (1976) and Full et al. (1981), with variations in Klovan (1981) and Clarke (1978). Fortran programs

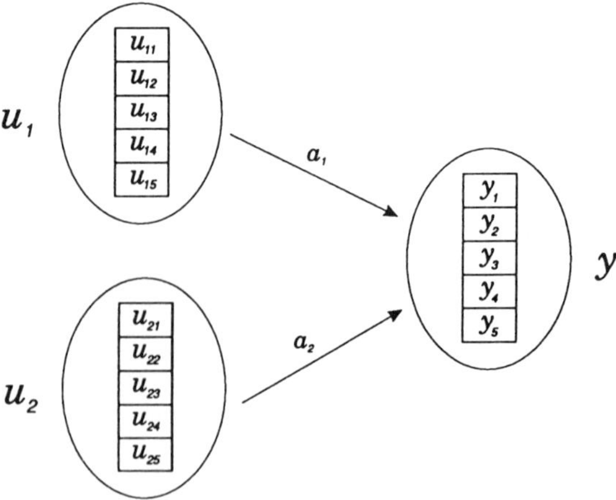

Figure 5.10 A compositional vector $\mathbf{Y}$ consisting of $p = 5$ ingredients from a mixture of $r = 2$ unknown sources such that $\mathbf{y} = a_1\mathbf{u}_1 + a_2\mathbf{u}_2$ and $0 \leq a_i \leq 1$.

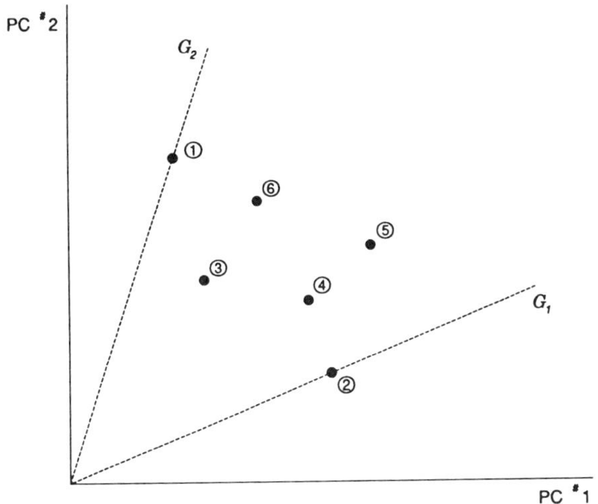

Figure 5.11 The $n - r = 4$ sample points described as linear combinations (mixtures) of $r = 2$ end members.

adapted to geological applications are given by Imbrie (1963) and Klovan and Miesch (1976).

A second major application of PCA in the analysis of mixtures is when dealing with continuous curves, for example, chemical spectroscopic distributions of absorbance at varying wavelength. Let $y_j(\lambda)$ be a continuous curve (distribution) which represents an additive mixture of r unknown and nonnegative linearly independent functions, that is,

$$y_j(\lambda) = \alpha_{1j} f_{1j}(\lambda) + \alpha_{2j} f_{2j}(\lambda) + \cdots + \alpha_{rj} f_{rj}(\lambda) \tag{5.125}$$

where the $f_{ij}(\lambda)$ may be normalized to unit area as

$$\int_{a_1}^{a_2} f_{ij}(\lambda)\, d\lambda = 1 \tag{5.126}$$

An example is Beer's law, which states that the spectrum of a mixture of r constituents (at wave length λ) is a linear function of the constituents' spectra in terms of molar absorptivity (percentage) at wave length λ. The coefficients α_i $(i = 1, \ldots, r)$ then represent concentrations of the r constituents (components) of the mixture (Howery, 1976; Weiner, 1977; Spjotvoll et al., 1982). In practice the curves will be sampled at n discrete points, resulting in

$$\mathbf{Y}_j = a_1 \mathbf{f}_1 + a_2 \mathbf{f}_2 + \cdots + a_r \mathbf{f}_r \tag{5.127}$$

where $\mathbf{Y}_j$ is a $(n \times 1)$ vector of observations on the jth mixture spectra, $\mathbf{f}_i$ is the $(n \times 1)$ vector of spectra absorbance of the ith constituent, and a_i is the concentration of constituent i. Neither the number, the spectra, nor concentrations of the constituents are known. Equation (5.127) can therefore be estimated by a PCA of $\mathbf{Y}_j$, where a_i are the loadings and f_j are the scores such that

$$\sum_{i=1}^{r} a_i = 1, \qquad a_i \geq 0, \qquad f_i > 0$$

The situation is not unlike considering a finite mixture of probability distributions. Examples from spectrophotometry and chromatography may be found in Ritter et al. (1976), Weiner and Howery (1972), and Cartwright (1986). Again, if a Q-mode is considered, the analysis is also at times carried out on the cosine matrix (Rozett and Petersen, 1976) or else simply on the matrix $\mathbf{Y}\mathbf{Y}^T$ (Burgard et al., 1977) instead of the covariance or correlation matrix.

Example 5.17   (Lawton and Sylvestre, 1971). We have $n = 5$ spectrophotometric curves (samples) (Fig. 5.12) measured at $p = 30$ wavelengths, resulting in the data matrix of Table 5.31. A PCA of the (uncentered) matrix $\mathbf{Y}\mathbf{Y}^T$ is carried out, which yields two dominant latent roots (the mixtures consist of two components) with latent vectors as given in Table 5.32. After suitable (generally oblique) rotation to remove negative values from $\mathbf{V}_2$ (see also Spjotvoll et al., 1982), the estimated spectra of the two pure constituents (here two dyes) is obtained (Table 5.33; Fig. 5.13).

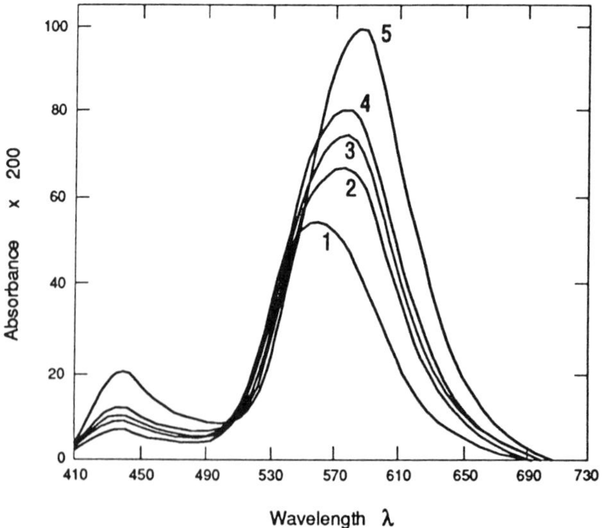

Figure 5.12   Five spectroscopic curves at wavelengths $410 \leq \lambda \leq 730$ (Lawton and Sylvestre, 1971; reproduced with permission).

Table 5.31   The $p = 5$ Spectrophotometric Curves Sampled at $n = 30$ wavelengths λ

Curve $\mathbf{Y}_1$	Curve $\mathbf{Y}_2$	Curve $\mathbf{Y}_3$	Curve $\mathbf{Y}_4$	Curve $\mathbf{Y}_5$
0.924	2.478	1.239	0.413	2.774
4.406	8.006	6.845	5.075	11.920
5.488	9.009	10.110	6.393	18.392
6.530	11.900	11.586	9.009	20.969
4.977	9.422	10.307	7.475	16.681
4.898	8.419	8.242	6.452	13.907
3.875	6.432	6.845	4.839	10.878
3.600	6.157	6.255	4.485	10.032
3.501	4.780	5.272	3.796	8.439
4.878	5.429	5.724	4.917	7.534
9.992	9.953	9.678	8.950	9.068
16.379	16.601	17.762	15.815	12.845
27.341	27.715	28.521	36.653	20.083
40.146	44.041	42.566	42.015	33.321
52.735	62.570	58.085	55.155	48.752
54.801	72.995	66.858	62.944	70.262
51.260	80.155	72.405	66.701	88.633
46.775	81.512	73.448	66.937	98.350
39.832	74.962	68.353	60.623	97.800
30.272	64.950	57.613	52.676	88.240
22.463	51.496	45.969	39.969	72.799
15.795	34.875	34.442	30.154	55.352
11.350	25.728	25.079	21.303	41.189
7.947	17.900	17.703	15.087	30.135
4.760	11.271	11.684	9.796	20.103
2.813	7.317	7.140	5.842	13.632
2.065	4.485	4.544	3.698	8.065
1.593	2.813	2.655	2.419	5.134
0.964	1.436	1.318	1.259	2.833
0.669	0.472	0.079	0.138	0.551

Source: Lawton and Sylvestre, 1971; reproduced with permission.

5.9.3   Directional Data

A general objective of PCA is to determine the magnitude and direction of the principal axes of variation. At times the data themselves are directions, for example, when attempting to determine the angular direction(s) of magnetized rock (Fisher, 1953; Creer, 1957) or the spatial origin of comets (Tyror, 1957). Since the directional component of the data is of interest, the distances are usually set arbitrarily to unity, which gives rise to a system of polar coordinates (Fig. 5.14). Converting the angles to direction cosines then yields a set of unit vectors $\mathbf{Y}_1, \mathbf{Y}_2, \ldots, \mathbf{Y}_p$ where $\mathbf{Y}_j^{\mathrm{T}}\mathbf{Y}_j = 1$ $(j =$

Table 5.32   The Orthonormal Latent Vectors that Correspond to the Two Largest Latent Vectors of YY^T

V_1	V_2
0.009	0.015
0.041	0.048
0.056	0.103
0.068	0.103
0.055	0.080
0.047	0.058
0.037	0.045
0.034	0.041
0.029	0.030
0.031	−0.004
0.050	−0.078
0.082	−0.169
0.134	−0.291
0.210	−0.400
0.290	−0.477
0.349	−0.302
0.389	−0.074
0.402	0.094
0.377	0.218
0.327	0.284
0.261	0.285
0.191	0.243
0.140	0.192
0.100	0.150
0.065	0.109
0.042	0.085
0.026	0.044
0.016	0.025
0.009	0.014
0.002	−0.002

Source: Lawton and Sylvestre, 1971; reproduced with permission.

$1, 2, \ldots, p$). For $p = 2$ we have

$$y_{i1} = \cos \theta , \qquad y_{i2} = \sin \theta \qquad (0 \le \theta < 2\pi)$$

and for $p = 3$

$$y_{i1} = \cos \theta , \qquad y_{i2} = \sin \theta \cos \varphi , \qquad y_{i3} = \sin \theta \sin \varphi$$

$$(0 \le \theta \le \pi; \ 0 \le \varphi \le 2\pi)$$

Table 5.33   Estimated Spectra of Two "Pure" Constituents of the Mixtures Using Rotated (Oblique) Components

F_1^*	F_{II}^*
0.0000	0.0004
0.0005	0.0015
0.0001	0.0025
0.0005	0.0028
0.0004	0.0023
0.0005	0.0018
0.0004	0.0014
0.0004	0.0013
0.0004	0.0010
0.0009	0.0007
0.0024	0.0003
0.0046	0.0001
0.0077	0.0000
0.0112	0.0006
0.0145	0.0017
0.0137	0.0051
0.0117	0.0087
0.0098	0.0109
0.0074	0.0117
0.0051	0.0112
0.0033	0.0096
0.0020	0.0074
0.0012	0.0056
0.0007	0.0042
0.0003	0.0028
0.0000	0.0020
0.0001	0.0011
0.0001	0.0007
0.0001	0.0004
0.0001	0.0000

Source: Lawton and Sylvestre, 1971; reproduced with permission.

In general, direction cosines can be expressed as

$$y_{ij} = \cos \theta \prod_{h=0}^{j-1} \sin \theta_h \qquad (5.128)$$

where $\sin \theta_0 = \cos \theta_p = 1$ (see Mardia, 1975).

In practice directional data are mainly encountered in physical space, that is, when $p = 2$ or 3. Here the data are distributed on the circumference (surface) of a circle (sphere) with unit radius, and interest lies in being able

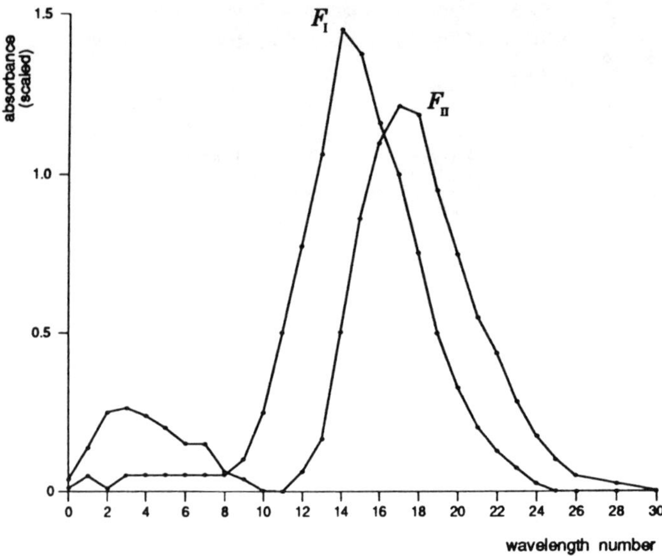

Figure 5.13 Estimated spectra of the two constituents of mixtures of Table 5.30.

to determine whether the points, that is, the direction cosines, form certain patterns or configurations on the circle (sphere). Thus one-dimensional formations consist of either a single cluster or two antipodal clusters lying on a common axis; a two-dimensional pattern consists of a band ("girdle") around the sphere; and a three-dimensional pattern is indicated by a uniform distribution of the points on the sphere's surface. A uniform distribution however does not imply a random configuration, since general spatial randomness is (usually) generated by the Poisson distribution. The convention here is to consider an alternative, but equivalent, system of three-dimensional polar coordinates (Fig. 5.15) where θ is the angle of declination (azimuth) and ϕ is the angle of inclination. The principal geographic or geologic directions are then described by the coordinate system (Table 5.34), both in angular and directional cosine form. Actual observations then assume values between -1 and $+1$, and the direction cosines are given by

$$y_{i1} = \cos\varphi\cos\theta , \qquad y_{i2} = \cos\varphi\sin\theta , \qquad y_{i3} = \sin\varphi \qquad (5.129)$$

The objective of locating configurations (clusters) on a shpere with the center at the origin is easily seen to be equivalent to carrying out a PCA of the matrix of sums of squares and products of the three direction cosine vectors with the number of clusters being indicated by the number of large latent roots (see also Watson, 1966). Note that the direction cosines are not corrected for the means. Furthermore, Eq. (5.129) can be used to estimate

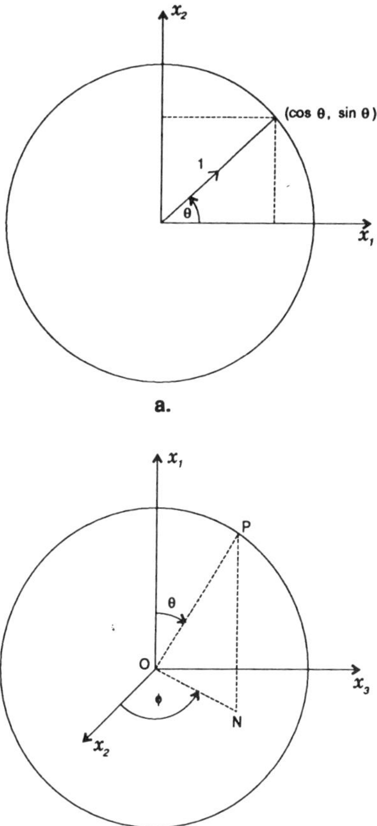

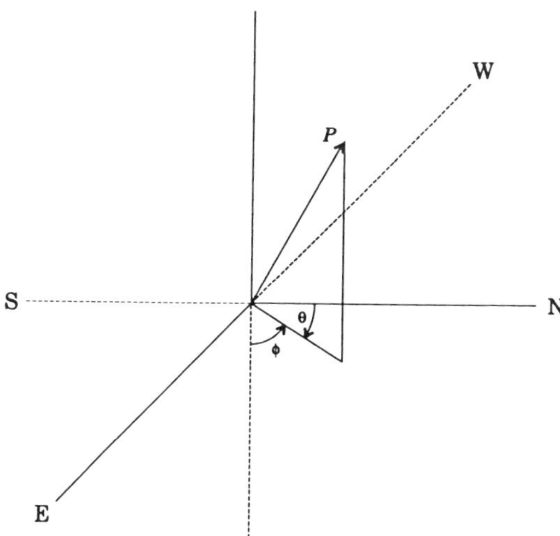

Figure 5.14 (*a*) Representation for a circular (two-dimensional) variable (*b*) Representation for a spherical (three-dimensional) variable.

Figure 5.15 Geographic orientation of a unit vector **P** in three-dimensional space.

the mean declination (azumith) and mean inclination using the latent vector elements. Solving Eq. (5.129) for θ and φ we have

$$\theta = \tan^{-1}\left(\frac{y_{i2}}{y_{i1}}\right), \qquad \varphi = \sin^{-1}y_{i3} \tag{5.130}$$

Table 5.34 The Principal Geographic Directions in Terms of the Polar Coordinate System of Figure 5.15 and Eq. (5.129)

Direction	Declination (θ)	Inclination (ϕ)	Direction y_1	y_2	Cosines y_3
North	0	0	1	0	0
South	180	0	-1	0	0
East	90	0	0	1	0
West	270	0	0	-1	0
Up	0	-90	0	0	-1
Down	0	90	0	0	1

Example 5.18   The following data are quoted in Fisher (1953) and refer to nine measurements of remanent magnetism of Icelandic lava flows during 1947 and 1948 (Table 5.35). The matrix of sums of squares and products is then

$$\mathbf{Y}^{\mathrm{T}}\mathbf{Y} = \begin{bmatrix} .90646 & .32502 & 2.35223 \\ .32502 & .42559 & 1.04524 \\ 2.35223 & 1.04524 & 4.66808 \end{bmatrix}$$

with latent roots and vectors as shown in Table 5.36. The existence of a single dominant latent root indicates the existence of a single cluster (or two

Table 5.35   Angles and Direction Cosines (Eq. 5.129) of Directions of Magnetization of Icelandic Lava, 1947–48

Sample	θ	ϕ	Y_1	Y_2	Y_3
1	343.2	66.1	.3878	$-.1171$	.9143
2	62.0	68.7	.1705	.3207	.9317
3	36.9	70.1	.2722	.2044	.9403
4	27.0	82.1	.1225	.0624	.9905
5	359.0	79.5	.1822	$-.0032$	.9833
6	5.7	73.0	.2909	.0290	.9563
7	50.4	69.3	.2253	.2724	.9354
8	357.6	58.8	.5176	$-.0217$	.8554
9	44.0	51.4	.4488	.4334	.7815

Source: Fisher, 1953; reproduced with permission.

Table 5.36 Latent Roots and Latent Vectors of Magnetized Lava (Direction Cosines)

	P_1	P_2	P_3
Y_1	.2966	−.0631	.9529
Y_2	.1335	.9908	.0240
Y_3	.9456	−.1201	−.3022
	$l_1 = 8.5533$	$l_2 = .2782$	$l_3 = .1680$

Source: Fisher, 1953; reproduced with permission.

antipodal clusters) whose direction can be estimated by the elements of P_1. We thus have, using Eq. (5.130),

$$\theta = \tan^{-1} \frac{.1335}{.2966} = 24.2°$$

Example 5.19. Creer (1957) analyzed $n = 35$ sandstone samples of Keuper Marls (England), taken from a 100-ft stratigraphical section, in order to measure the direction and intensity of the natural remanent magnetization. The angles, direction cosines, and latent roots and vectors of Y^TY are given in Tables 5.37 and 5.38.

$$Y^TY = \begin{bmatrix} 6.70657 & 4.12005 & 1.84181 \\ 4.12005 & 3.42493 & -1.30868 \\ 1.84181 & -1.30868 & 24.86850 \end{bmatrix}$$

The general orientation of the band can be measured by computing the angular direction of the perpendicular to the great circle passing through it, by using the latent vector associated with the smallest latent root. The

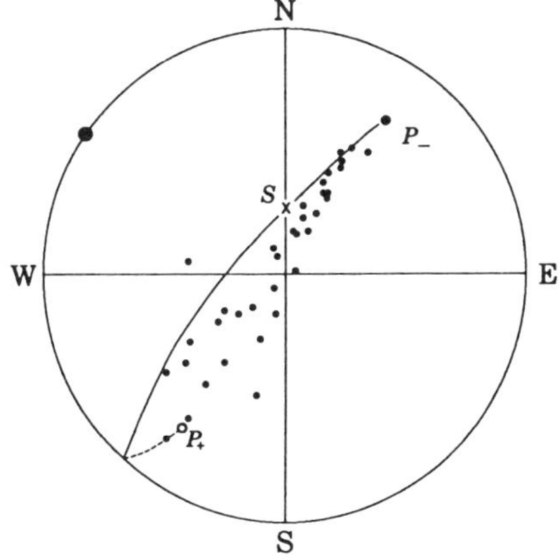

Figure 5.16 Polar equal-area projection of the directions of samples of Keuper Marls (Creer, 1957) where S is the direction of an axial dipole along the present geographical axis; P_- and P_+ represent the primary axis of magnetization; ● is the perpendicular pole to the primary axis of magnetization.

Table 5.37   Angles and Direction Cosines of the Data of Directions of Intensity of Natural Remanent Magnetization of Sandstone (Keuper Marls), Devonshire, England

Sample Number	θ	ϕ	Y_1	Y_2	Y_3
1	28	41	0.6664	0.3543	0.6561
2	19	70	0.3234	0.1114	0.9397
3	23	52	0.5667	0.2406	0.7880
4	35	40	0.6275	0.4394	0.6428
5	28	47	0.6022	0.3202	0.7134
6	24	60	0.4568	0.2034	0.8660
7	28	49	0.5793	0.3080	0.7547
8	29	67	0.3417	0.1894	0.9205
9	12	76	0.2366	0.0503	0.9703
10	17	66	0.3890	0.1189	0.9135
11	22	56	0.5185	0.2095	0.8290
12	28	60	0.4415	0.2347	0.8660
13	24	44	0.6571	0.2926	0.6947
14	27	60	0.4455	0.2270	0.8660
15	341	81	0.1479	−0.0509	0.9877
16	213	85	−0.0731	−0.0475	0.9962
17	235	64	−0.2514	−0.3591	0.8988
18	240	67	−0.1954	−0.3384	0.9205
19	192	77	−0.2200	−0.468	0.9744
20	193	49	−0.6392	−0.1476	0.7547
21	215	31	−0.7021	−0.4917	0.5150
22	216	55	−0.4640	−0.3371	0.8192
23	219	45	−0.5495	−0.4450	0.7071
24	234	51	−0.3699	−0.5091	0.7771
25	198	67	−0.3716	−0.1207	0.9205
26	229	45	−0.4639	−0.5337	0.7071
27	230	70	−0.2198	−0.2620	0.9397
28	231	37	−0.5026	−0.6207	0.6018
29	224	75	−0.1862	−0.1798	0.9659
30	217	19	−0.7551	−0.5690	0.3256
31	237	84	−0.0569	−0.0877	0.9945
32	276	58	0.0554	−0.5270	0.8480
33	30	73	0.2532	0.1462	0.9563
34	78	86	0.0145	0.0682	0.9976
35	13	76	0.2357	0.0544	0.9703

Source: Greer, 1957; reproduced with permission.

direction of the pole to the great circle is then given by the angles (Fig. 5.16)

$$\theta = \tan^{-1}(.82026/-.56538) = -55.4°$$

$$\phi = \sin^{-1}(.08660) = 4.97°$$

Table 5.38   Latent Roots, Latent Vectors of Direction Cosines of angulr Direction of Natural Remant Magnetization of Keuper Marls of Table 5.37

Directions	$\mathbf{P}_1$	$\mathbf{P}_2$	$\mathbf{P}_3$
$\mathbf{Y}_1$	.09006	.81989	$-.56540$
$\mathbf{Y}_2$	$-.04297$	.57037	.82016
$\mathbf{Y}_3$	.99501	$-.04959$	.08660
Latent roots	$l_1 = 25.0915$	$l_2 = 9.4612$	$l_3 = .4469$

Source: Creer, 1957; reproduced with permission.

Note that the perpendicular is also defined by the angles $\theta = 124.6°$ and $\phi = 4.97°$. Other material on directional data may be found in Upton and Fingleton 1989. $\square$

EXERCISES

5.1 Let $\mathbf{X}^+ = \mathbf{P}\Delta^+_{(r)}\mathbf{Q}^\mathrm{T}$ be the generalized inverse of a $(n \times p)$ matrix $\mathbf{X}$ (Theorem 5.1). Show that the following relations hold:

  (i) $\mathbf{XX}^+\mathbf{X} = \mathbf{X}$
  (ii) $\mathbf{X}^+\mathbf{XX}^+ = \mathbf{X}^+$
  (iii) $(\mathbf{XX}^+)^\mathrm{T} = \mathbf{XX}^+$
  (iv) $(\mathbf{X}^+\mathbf{X})^\mathrm{T} = \mathbf{X}^+\mathbf{X}$

5.2 Prove that the rotated principal components $\mathbf{V}_{(r)}$ of Theorem 5.2 no longer successively account for maximum variance.

5.3 Consider Eq. (5.18b) where $\mathbf{G}$ and $\mathbf{B}^\mathrm{T}$ are orthogonally rotated components and loadings respectively (Theorem 5.2a). Show that the new scores $\mathbf{G}$ are orthogonal to the residuals $\boldsymbol{\delta}$.

5.4 Let $\mathbf{B} = \mathbf{TA}$ be a linear transformation of loading coefficients $\mathbf{A}$. Show that $\mathbf{T}^{-1} = \mathbf{B}^\mathrm{T}\mathbf{AL}^{-1}$ where $\mathbf{L}$ is the diagonal matrix of latent roots.

5.5 Prove the first part of Theorem 5.3.

5.6 Prove Eq. (5.42).

5.7 Using Theorem 5.8 prove Eq. (5.60).

5.8 Show that relationships of Eq. (5.63) holds.

5.9 Derive the relationships of Eq. (5.79).

5.10 Derive the normal equations (Eq. 5.81) and compare the parameters with those of Tucker's model.

5.11 Prove the second part of Theorem 5.10.

5.12 Let $\mathbf{A}$ be a $(n \times m)$ matrix with Euclidian norm $\|\mathbf{A}\|$ as defined by Eq. (5.108). Show that $\|\mathbf{A}\|^2 = \text{tr}(\mathbf{A}^T\mathbf{A}) = \text{tr}(\mathbf{A}\mathbf{A}^T)$.

5.13 Prove Theorem 5.6.

5.14 Show that the canonical correlation model of Theorem 5.8 is equivalent to the model

$$\begin{bmatrix} \mathbf{0} & \Sigma_{12} \\ \Sigma_{21} & \mathbf{0} \end{bmatrix}\begin{bmatrix} \alpha \\ \beta \end{bmatrix} = \lambda \begin{bmatrix} \Sigma_{11} & \mathbf{0} \\ \mathbf{0} & \Sigma_{22} \end{bmatrix}\begin{bmatrix} \alpha \\ \beta \end{bmatrix}$$

5.15 The following data represent $p = 22$ physical and chemical measurements for $n = 8$ samples of crude oil (McCammon, 1966).

Properties	Samples							
1	0.9516	0.8936	0.8774	0.8805	0.8680	0.8231	0.8370	0.8161
2	1.5400	1.5010	1.4920	1.4935	1.4860	1.4615	1.4675	1.4570
3	2.39	1.46	1.37	0.52	0.34	0.49	0.11	0.34
4	89.6	83.8	77.0	81.8	82.7	67.1	73.3	60.4
5	27.8	15.0	19.0	17.8	17.2	6.4	12.1	8.3
6	11.5	9.6	9.1	8.9	8.6	4.6	6.0	4.6
7	14.5	5.3	6.2	3.3	1.3	0.6	1.2	0.2
8	44.7	66.8	61.4	65.4	65.7	83.4	73.4	79.7
9	1.5	3.3	4.3	4.6	7.2	5.0	7.3	7.2
10	6.1	15.9	17.2	17.8	20.2	25.3	29.2	27.9
11	30.2	28.2	29.1	30.6	30.6	31.9	30.3	31.4
12	19.9	17.4	15.8	16.6	16.6	16.3	15.0	15.1
13	11.7	10.9	10.7	10.2	9.0	8.1	8.4	8.6
14	5.8	4.6	4.9	4.6	4.1	3.2	3.1	2.7
15	3.2	2.5	2.4	2.1	2.1	1.6	1.4	1.5
16	5.8	5.8	5.6	5.1	5.0	3.9	4.0	4.0
17	2.6	2.2	2.2	2.0	1.8	1.5	1.3	1.3
18	3.5	2.9	2.7	2.5	2.3	1.6	1.5	1.5
19	1.1	0.8	0.9	0.7	0.6	0.5	0.5	0.5
20	5.5	4.7	4.6	4.3	4.1	3.5	2.9	3.2
21	4.6	3.9	3.8	3.5	3.4	2.5	2.3	2.3
22	6.8	6.0	6.7	5.7	5.0	3.7	3.7	3.4

(a) Using a Q-mode approach perform a PCA of the $n = 8$ samples, using the cosine measure.

(b) Carry out a PCA of the (8×8) correlation matrix of the samples. How do the scores differ from those in part a?

5.16 A chemical analysis of Hellenistic Black-Glazed Wares using attic and nonattic samples yields the following data (Hatcher et al., 1980):

Oxides of:	Al	Mg	Fe	Ti	Mn	Cr	Ca	Na	Ni
Attic									
	15.12	4.00	8.10	1.13	0.136	0.088	6.16	0.73	0.050
	15.59	4.15	8.10	1.05	0.097	0.080	4.06	1.11	0.046
	17.01	4.21	7.74	1.05	0.112	0.086	6.72	0.86	0.048
	17.48	4.87	8.66	1.10	0.094	0.076	5.60	0.89	0.050
	17.01	3.88	8.10	1.02	0.094	0.086	5.18	0.89	0.050
	18.43	4.21	8.10	1.07	0.087	0.076	5.74	0.94	0.044
	17.96	4.15	8.24	0.98	0.099	0.091	4.63	0.89	0.051
	17.01	3.81	7.74	1.07	0.124	0.076	8.95	0.78	0.048
$n = 9$	15.59	4.15	7.74	1.06	0.097	0.082	7.00	0.86	0.043
Nonattic									
	13.23	2.79	6.83	0.97	0.138	0.016	4.34	1.16	0.023
	18.43	1.69	7.39	1.07	0.133	+0.00	7.83	0.84	0.014
	17.48	2.52	6.48	1.02	0.120	<0.015	7.00	1.13	0.015
	16.07	2.92	7.39	1.02	0.151	0.016	7.00	1.16	0.025
	16.07	3.25	7.53	1.10	0.129	0.017	7.55	1.11	0.029
	20.32	3.55	8.10	0.98	0.097	0.018	3.22	0.67	0.024
	16.07	3.08	7.53	1.02	0.151	0.017	7.83	1.21	0.027
$n = 8$	17.96	2.09	7.04	0.97	0.094	<0.015	11.47	0.62	0.017

Using principal components determine whether it is possible to discriminate between the attic and nonattic measurements using a PCA of the matallic oxides.

5.17 The following correlation matrix is given by Allison and Cicchetti (1976) in a study of sleep in mammals:

	Y_1	Y_2	Y_3	Y_4	Y_5	Y_6	Y_7	Y_8	Y_9
Y_1	1.000								
Y_2	.582	1.000							
Y_3	−.377	−.342	1.000						
Y_4	−.712	−.370	.685	1.000					
Y_5	−.679	−.432	.777	.945	1.000				
Y_6	−.589	−.651	.682	.692	.781	1.000			
Y_7	−.369	−.536	.018	.253	.192	.158	1.000		
Y_8	−.580	−.591	.518	.662	.640	.588	.680	1.000	
Y_9	−.542	−.686	.226	.432	.377	.363	.930	.819	1.000

where the random variables are defined as

Y_1 = Slow-wave sleep
Y_2 = Paradoxical sleep
Y_3 = Life span
Y_4 = Body weight
Y_5 = Brain weight

Y_6 = Gestation time
Y_7 = Predation index
Y_8 = Sleep exposure
Y_9 = Overall danger

Carry out a cluster analysis of the variables by means of (a) the varimax rotation Eq. (5.20) and (b) the oblimin in criterion (Eq. 5.28) by using the first $r = 2$ principal components. Which rotation results in a clearer identification of the component axes?

CHAPTER 6

Factor Analysis

6.1 INTRODUCTION

Hitherto there has been very little discussion about the difference between the terms "principal components" and "factor analysis." This is in keeping with the established practice of applied statistical literature, where the term "factor analysis" is generally understood to refer to a set of closely related models intended for exploring or establishing correlational structure among the observed random variables. A difference in statistical and mathematical specification between the two models nevertheless exists, and in this chapter it becomes important to differentiate between principal components and factor analysis "proper." The difference is that principal components models represent singular value decompositions of random association matrices, whereas a factor analysis incorporates an a priori structure of the error terms. This is done in two ways: (1) the errors in the variables are assumed to be uncorrelated, and (2) $1 \le r < p$ common factors account for the entire covariance (correlation) between the variables, but not the variances. In this sense factor analysis resembles more a system of least squares regression equations than does the principal component model, although neither factors (the explanatory variables) nor loadings (the coefficients) are observed. Factor analysis is therefore motivated by the fact that measured variables can be correlated in such a way that their correlation may be reconstructed by a smaller set of parameters, which could represent the underlying structure in a concise and interpretable form. As was the case for the principal components decomposition, an important point is to decide which variables to include in a factor model since the number of factors and coefficients generally depend on the choice. If the aim of the analysis is purely exploratory, then virtually all variables under consideration can be "thrown into the pot." Here the objectives are normally limited to either a correlational analysis of the variables (in the presence of uncorrelated error terms), or the intention may be precisely to uncover which variables to retain as effective measures of the dimensions or phenomena of interest. A

351

haphazard choice of intercorrelated variables however has little utility for a causal or structural analysis, and here careful theoretical judgment employing both substantive as well as mathematical reasoning should accompany the selection process. Several sequential sets of estimates are usually required before a satisfactory outcome is reached.

Before any estimation procedure can begin we must first ask ourselves two statistical questions: (1) do unique common factors exist, from the mathematical point of view, and if so, (2) what is their precise status in the physical world? As the following sections show, the answer to the first question is generally "no." Indeed, it is the process of defining uniqueness restrictions that determines most factor models used in practice. The second question may be answered in two ways, depending on how we view or consider the factors in a given physical or nonmathematical setting. First, factors may be viewed as estimates of real or physical random variables which have given rise indirectly to the observed correlations. Once the influence of the factors is removed, the observed variables should become independent. Second, we can view factors as simply artificial mathematical constructs, which need not be real but which represent latent measurement scales or variables, which are by nature multidimensional and perhaps dynamic. Such composite multidimensional variables are generally not observable by themselves in toto, and therefore require a whole set or spectrum of more simple but intercorrelated variables in order to represent or capture them, each variable contributing an essential ingredient. Factor analysis can then be used to estimate or measure such composite variables. This is particularly common in the social or human sciences, when attempting to measure concepts such as "mental aptitude," "consumer tastes," "social class," "culture," "political orientation," and so forth. Clearly the question of the existential status of the factors is essentially empirical in nature, and cannot nor should not be a part of the a priori specification, but must rather form an integral part of the posterior analysis or investigation. Here an analysis of the loadings and the scores, or perhaps the use of further statistical models such as regression, may throw light on the factors, but such good fortune can by no means be guaranteed. This is because in the final analysis the estimated factors are primarily mathematical constructs, and as such do not rely on any process of identification or interpretation for their validity. Indeed, for certain applications, factors do not require interpretation, since they may also be used simply as instrumental variables, for example, to reduce multicollinearity and errors in variables in least-squares regression (Chapter 10) or to provide estimates of "true" dimensionality. The principal motivation for using factor analysis however does lie in the possibility of a "meaningful" interpretation of the data, particularly in exploratory analysis when using the so-called "unrestricted" factor models which, contrary to their name, embody restrictions of a fundamental nature.

6.2 THE UNRESTRICTED RANDOM FACTOR MODEL IN THE POPULATION

The general factor model can be written as

$$\mathbf{Y} = \boldsymbol{\mu} + \boldsymbol{\alpha}\boldsymbol{\Phi} + \boldsymbol{\epsilon} \tag{6.1}$$

or

$$\mathbf{X} = \boldsymbol{\alpha}\boldsymbol{\Phi} + \boldsymbol{\epsilon} \tag{6.2}$$

where $\mathbf{Y} = (y_1, y_2, \ldots, y_p)^{\mathrm{T}}$ is a vector of observable random variables, $\boldsymbol{\Phi} = (\phi_1, \phi_2, \ldots, \phi_r)^{\mathrm{T}}$ is a vector of $r < p$ unobserved or latent random variables called factors, $\boldsymbol{\alpha}$ is a $(p \times r)$ matrix of fixed coefficients (loadings), and $\boldsymbol{\epsilon} = (\epsilon_1, \epsilon_2, \ldots, \epsilon_p)^{\mathrm{T}}$ is a vector of random error terms. Usually for convenience the means are set to zero so that $E(\mathbf{X}) = E(\boldsymbol{\Phi}) = E(\boldsymbol{\epsilon}) = 0$. The random error term consists of errors of measurement, together with unique individual effects associated with each population variable y_j. When a random sample is taken, the residual errors also contain sampling variation. For the present model we assume that $\boldsymbol{\alpha}$ is a matrix of constant parameters and $\boldsymbol{\Phi}$ is a vector of random variables. Equation (6.1) resembles a regression model except the factors $\boldsymbol{\Phi}$ are not observed directly and must be estimated from the data together with the parameters. This introduces difficulties of identification which do not exist in other statistical models such as, for example, the classic single-equation regression model.

The following assumptions are usually made for the factor model (Eq. 6.2):

  (i) $\rho(\boldsymbol{\alpha}) = r < p$
  (ii) $E(\mathbf{X} \mid \boldsymbol{\Phi}) = \boldsymbol{\alpha}\boldsymbol{\Phi}$
  (iii) $E(\mathbf{X}\mathbf{X}^{\mathrm{T}}) = \boldsymbol{\Sigma}$, $E(\boldsymbol{\Phi}\boldsymbol{\Phi}^{\mathrm{T}}) = \boldsymbol{\Omega}$, and

$$\boldsymbol{\Psi} = E(\boldsymbol{\epsilon}\boldsymbol{\epsilon}^{\mathrm{T}}) = \begin{bmatrix} \sigma_1^2 & & & & 0 \\ & \sigma_2^2 & & & \\ & & \ddots & & \\ 0 & & & & \sigma_p^2 \end{bmatrix} \tag{6.3}$$

so that the errors are assumed to be uncorrelated. The factors however are generally correlated, and $\boldsymbol{\Omega}$ is therefore not necessarily diagonal. For the sake of convenience and computational efficacy the factors are usually assumed to be uncorrelated and of unit variance, so that $\mathbf{\Omega} = \mathbf{I}$ and

$$E(\phi_i \phi_j) = \begin{cases} 1 & i = j \\ 0 & i \neq j \end{cases} \tag{6.4}$$

When a nonorthogonal system is required the factors may be rotated to oblique form.

(iv) $E(\mathbf{\Phi}\boldsymbol{\epsilon}^T) = 0$ so that the errors and factors are uncorrelated.

Using assumptions i–iv we have, from Eq. (6.2),

$$
\begin{aligned}
E(\mathbf{XX}^T) = \mathbf{\Sigma} &= E(\boldsymbol{\alpha}\mathbf{\Phi} + \boldsymbol{\epsilon})(\boldsymbol{\alpha}\mathbf{\Phi} + \boldsymbol{\epsilon})^T \\
&= E(\boldsymbol{\alpha}\mathbf{\Phi}\mathbf{\Phi}^T\boldsymbol{\alpha}^T + \boldsymbol{\alpha}\mathbf{\Phi}\boldsymbol{\epsilon}^T + \boldsymbol{\epsilon}\mathbf{\Phi}^T\boldsymbol{\alpha}^T + \boldsymbol{\epsilon}\boldsymbol{\epsilon}^T) \\
&= \boldsymbol{\alpha}E(\mathbf{\Phi}\mathbf{\Phi}^T)\boldsymbol{\alpha}^T + \boldsymbol{\alpha}E(\mathbf{\Phi}\boldsymbol{\epsilon}^T) + E(\boldsymbol{\epsilon}\mathbf{\Phi}^T)\boldsymbol{\alpha}^T + E(\boldsymbol{\epsilon}\boldsymbol{\epsilon}^T) \\
&= \boldsymbol{\alpha}\mathbf{\Omega}\boldsymbol{\alpha}^T + E(\boldsymbol{\epsilon}\boldsymbol{\epsilon}^T) \\
&= \mathbf{\Gamma} + \mathbf{\Psi}
\end{aligned} \tag{6.5}
$$

where $\mathbf{\Gamma} = \boldsymbol{\alpha}\mathbf{\Omega}\boldsymbol{\alpha}^T$ and $\mathbf{\Psi} = E(\boldsymbol{\epsilon}\boldsymbol{\epsilon}^T)$ are the true and error covariance matrices, respectively. Also, postmultiplying Eq. (6.2) by $\mathbf{\Phi}^T$ we have

$$
\begin{aligned}
E(\mathbf{X}\mathbf{\Phi}^T) &= E(\boldsymbol{\alpha}\mathbf{\Phi}\mathbf{\Phi}^T + \boldsymbol{\epsilon}\mathbf{\Phi}^T) \\
&= \boldsymbol{\alpha}E(\mathbf{\Phi}\mathbf{\Phi}^T) + E(\boldsymbol{\epsilon}\mathbf{\Phi}^T) \\
&= \boldsymbol{\alpha}\mathbf{\Omega}
\end{aligned} \tag{6.6}
$$

using conditions iii and iv. For the special case of Eq. (6.4) we have $\mathbf{\Omega} = \mathbf{I}$, and the covariance between the manifest and the latent variables simplifies to $E(\mathbf{X}\mathbf{\Phi}^T) = \boldsymbol{\alpha}$.

An important special case occurs when $\mathbf{X}$ is multivariate normal, since in this case the second moments of Eq. (6.5) contain all the information concerning the factor model. It also follows that the factor model (Eq. 6.2) is linear, and the variables are conditionally independent given the factors $\mathbf{\Phi}$. Let $\mathbf{\Phi} \sim N(\mathbf{0}, \mathbf{I})$. Then using Theorem 2.14 the conditional distribution of $\mathbf{X}$ is

$$
\begin{aligned}
\mathbf{X} \mid \mathbf{\Phi} &\sim N[(\mathbf{\Sigma}_{12}\mathbf{\Sigma}_{22}^{-1}\mathbf{\Phi}), (\mathbf{\Sigma}_{11} - \mathbf{\Sigma}_{12}\mathbf{\Sigma}_{22}^{-1}\mathbf{\Sigma}_{12}^T)] \\
&\sim N[\boldsymbol{\alpha}\mathbf{\Phi}, (\mathbf{\Sigma} - \boldsymbol{\alpha}\boldsymbol{\alpha}^T)] \\
&\sim N[\boldsymbol{\alpha}\mathbf{\Phi}, \mathbf{\Psi}]
\end{aligned} \tag{6.6a}
$$

with conditional independence following from the diagonality of $\mathbf{\Psi}$. The common factors $\mathbf{\Phi}$ therefore reproduce all covariances (correlations) between the variables, but account for only a portion of the variance. It is also possible to derive ML estimators of α, Φ, and Ψ which have the advantage of asymptotic efficiency, and to conduct hypothesis testing and the construction of confidence intervals for the ML estimates of α. It turns out that it is always possible to select the coefficients α in such a way that they represent correlations between X and Φ (assuming the factors exist) whether Σ represents the covariance or correlation matrix. Finally, apart from the distribution of X, the factor model (Eq. 6.2) is assumed to be linear in the coefficients α, but not necessarily in the factors Φ. All these are properties not shared by the principal components model, but are found more frequently in regression-type specifications. Unlike the regression model the "independent variables" Φ are not known, thus the initial estimation of the model is based on Eq. (6.5) rather than on Eqs. (6.1) or (6.2), assuming that Σ (or its estimate) is known. This is normally achieved by setting $\Omega = I$ and estimating α and Ψ jointly, usually by iterative methods. When a finite sample is available, the factor scores Φ can also be estimated by one of several procedures (Section 6.8).

Since the factors are not observed directly, a question arises as to whether they are "real." Clearly, the question of the interpretability of factors cannot be established in complete generality for all possible cases and applications, and the matter of factor identification must remain, essentially, one of substantive appropriateness, depending on the nature of the application and type of variables and data used. Thus, although an arbitrary collection of ill-defined variables will hardly produce factors that can be interpreted as representing something real, a carefully selected set, guided by theoretical principles, may very well reveal an unexpected and empirically or theoretically meaningful structure of the data. Prior to the physical identification of the common factors however lies a more fundamental mathematical question—given a factor model, under what conditions (if any) can Σ be factored into two independent parts as implied by Eq. (6.5)? Furthermore, assuming a mathematically identifiable factorization exists, under what conditions is it possible to find a unique set of loadings α? Given $1 < r < p$ common factors it can be shown that it is not generally possible to determine α and Φ uniquely. Even in the case of a normal distribution this cannot be guaranteed, since although every factor model specified by Eq. (6.6a) leads to a multivariate normal, the converse is not necessarily true when $1 < r < p$. The difficulty is known as the factor identification or factor rotation problem, and arises from a basic property of the model not encountered in, for example, the common PC model of Chapter 3. Rather than effect a singular decomposition of a Grammian matrix we now have to contend with a system of quadratic equations in the unknown elements of $\Gamma = \alpha \Omega \alpha^T$ and Ψ, which must be solved in such a way that Γ is Grammian and Ψ is diagonal, with diagonal elements $\sigma_i^2 > 0$ ($i = 1, 2, \ldots, p$).

Assume there exist $1 < r < p$ common factors such that $\Gamma = \alpha \Omega \alpha^T$ and Ψ is Grammian and diagonal. The covariance matrix Σ has $C\binom{p}{2} + p = 1/2$

$p(p+1)$ distinct elements, which therefore equals the total number of normal equations to be solved. The number of solutions however is infinite, which can be seen as follows. Since Ω is positive definite, there must exist a non-singular $(r \times r)$ matrix $\mathbf{B}$ such that $\Omega = \mathbf{B}^\mathrm{T}\mathbf{B}$ and

$$\Sigma = \alpha\Omega\alpha^\mathrm{T} + \Psi$$

$$= \alpha(\mathbf{B}^\mathrm{T}\mathbf{B})\alpha^\mathrm{T} + \Psi$$

$$= (\alpha\mathbf{B}^\mathrm{T})(\alpha\mathbf{B}^\mathrm{T})^\mathrm{T} + \Psi$$

$$= \alpha^*\alpha^{*\mathrm{T}} + \Psi \tag{6.7}$$

Evidently both factorizations (Eqs. 6.5 and 6.7) of Σ leave the same residual errors Ψ and therefore must represent equally valid factor solutions. Also, we can effect the substitution $\alpha^* = \alpha\mathbf{C}$ and $\Omega^* = \mathbf{C}^{-1}\Omega(\mathbf{C}^\mathrm{T})^{-1}$, which again yields a factor model which is indistinguishable from Eq. (6.5). No sample estimator can therefore distinguish between such an infinite number of transformations, each of which is of potential interest. The coefficients α and α^* are thus statistically equivalent and cannot be distinguished from each other or identified uniquely. That is, both the transformed and untransformed coefficients, together with Ψ, generate Σ in exactly the same way and cannot be differentiated by any estimation procedure without the introduction of additional restrictions.

In view of the rotational indeterminacy of the factor model we require restrictions on Ω, the covariance matrix of the factors. The most straightforward and common restriction is to set $\Omega = \mathbf{I}$, that is, to define the factors as orthogonal unit vectors (Eq. 6.4) in much the same way as was done for the principal components model of Chapter 3. The number m of free parameters implied by the equations

$$\Sigma = \alpha\alpha^\mathrm{T} + \Psi \tag{6.8a}$$

is then equal to the total number $pr + p$ of unknown parameters in α and Ψ, minus the number of (zero) restrictions placed on the off-diagonal elements of Ω, which is equal to $1/2(r^2 - r)$ since Ω is symmetric (assuming r common factors). We then have

$$m = (pr + p) - 1/2(r^2 - r)$$

$$= p(r + 1) - 1/2r(r - 1) \tag{6.8b}$$

where the columns of α are assumed to be orthogonal. The number of degrees of freedom d is then given by the number of equations implied by Eq. (6.8a), that is, the number of distinct elements in Σ minus the number of free parameters m. We have

$$d = 1/2p(p + 1) - [pr + p - 1/2(r^2 - r)]$$
$$= 1/2[(p - r)^2 - (p - r)] \tag{6.9}$$

(Exercise 6.1) which for a meaningful (i.e., nontrivial) empirical application must be strictly positive. This places an upper bound on the number of common factors r which may be obtained in practice, a number which is generally somewhat smaller than the number of variables p (Exercise 6.2). Note also that Eq. (6.9) assumes that the normal equations are linearly independent, that is, Σ must be nonsingular and $\rho(\alpha) = r$. For a nontrivial set-up, that is, when $d > 0$, there are more equations than free parameters and the hypothesis of r common factors holds only when certain constraints are placed on the elements of Σ. Thus there could be problems of existence, but if parameters exist we may obtain unique estimates. These and other points are discussed further by Lawley and Maxwell (1971). Note that even when $\Omega = I$, the factor model (Eq. 6.8a) is still indeterminate, that is, the factor coefficients α can be rotated or transformed as in Eq. (6.7). This however is the only mathematical indeterminacy in the model (assuming that unique error variances exist) although others of a more applied nature also exist (see Elffers et al., 1978). The indeterminacy is usually resolved by initially fixing the coordinate system, that is, by "rotating" the factor loadings α such that they satisfy an arbitrary constraint. The nature of such constraints in turn defines the type of factor model (Section 6.4) assuming $r > 1$, since for a single common factor the loadings are always determinate.

Example 6.1. Consider the case where $p = 5$ and $r = 2$. The normal equations (Eq. 6.8a) may be expressed in terms of the elements as

$$\sigma_j = \alpha_{j1}^2 + \alpha_{j2}^2 + \cdots + \alpha_{jr}^2 + \psi_j \qquad (j = 1, 2, \ldots, p)$$

for the diagonal variance terms of Σ, and

$$\sigma_{ij} = \alpha_{i1}\alpha_{j1} + \alpha_{i2}\alpha_{j2} + \cdots + \alpha_{ir}\alpha_{jr} \qquad (i \neq j)$$

for the off-diagonal covariance terms. The number of free parameters is then given by

$$m = 5(2 + 1) - 1/2(2^2 - 2)$$
$$= 14$$

and the degrees of freedom are

$$d = 1/2[(5 - 2)^2 - (5 + 2)]$$
$$= 1$$

For $p = 5$ random variables we cannot therefore have more than $r = 2$ common factors. $\square$

Identification can also be considered through the perspective of sample-population inference, that is, inference from a particular realization of the model to the model itself (Reiersol, 1950; Anderson and Rubin, 1956). When Σ, Ω, and Φ are given in numerical form, we have what is known as a structure. A structure includes completely specified distributions of the factors, together with a set of equations and numerical coefficients that relate observed variables to the common factors. A structure is thus a particular realization of Eq. (6.5), which in turn represents the set of all structures compatible with the given specification. Given a structure Σ, Ω, and Ψ we can generate one, and only one, probability distribution of the observed random variables. However, there may exist several structures that generate the same distribution. Since these structures are all equivalent to each other, the theoretical factor model can not be identified, since the possibility that α and Ψ can possess different values in equivalent structures leads to a lack of uniqueness. Generally speaking, however, identification is often possible, as is demonstrated by the following theorems.

THEOREM 6.1 (Reiersol, 1950).   Let Σ be a $(p \times p)$ covariance matrix. A necessary and sufficient condition for the existence of r common factors is that there exist a diagonal matrix Ψ, with nonnegative elements, such that $\Sigma - \Psi = \alpha\alpha^T$ is positive definite and of rank r.

The proof is based on the well-known result that a necessary and sufficient condition for a $(p \times p)$ matrix to be expressed in the form $\mathbf{BB}^T$, where $\mathbf{B}$ is $(p \times r)$ and $r < p$, is that the matrix be positive definite and of rank r. Note that Theorem 6.1 does not assert the existence of nonnegative error variances—it simply relates the existence of such variance to the existence of r common factors. Note also that ψ_j need not be strictly positive. In practice negative residual variance can and does occur, at times with surprising frequency. Such cases are known as Heywood cases in the psychometric literature. When a Heywood case occurs, the factor model is not appropriate, and an alternative model such as principal components should be used.

THEOREM 6.2.   Let $\mathbf{X} = \alpha\Phi + \epsilon$ possess a structure such that $r = r_0 < p$, where r_0 is the minimum rank of α and Ψ is nonsingular. Then the factor model contains an infinite number of equivalent structures when $r = r_0 + 1$.

A proof may be found in Takeuchi et al. (1982). Thus for any single value of r we see again that there exist an infinite number of non-singular diagonal matrices Ψ with nonnegative elements which satisfy the factor model (Eq. 6.8a). Conversely, to determine a unique Ψ we require the value of r, and the minimum rank value $r = r_0$, is achieved only when $\boldsymbol{\Psi}$ is unique. Necessary and sufficient conditions for uniqueness of $\boldsymbol{\Psi}$ do not seem to be known. For other necessary or sufficient conditions see Anderson and Rubin (1956).

To achieve complete identification we thus require prior information in the form of knowledge of either $\boldsymbol{\Psi}$ or r, since it does not seem possible to obtain unique values of both simultaneously. Also, to achieve identification we require ancillary (but arbitrary) restrictions in order to fix the basis of a common factor space. The specific nature of these restrictions vary from model to model, and these are discussed in the following sections. It is important to note however that mathematical restrictions do not necessarily achieve what is, after all, a fundamental objective of factor analysis—not only to estimate the dimensionality of the explanatory subspace but also to identify the factors or dimensions that span it, in terms of substantive phenomena. On the other hand no solution is possible without these restrictions. In practice, therefore, since a priori information is not generally given—indeed such information is often a part of the objective of the analysis—a trial and error procedure is usually employed. At each stage we let $r_0 = 1, 2, \ldots, r < p$, compute the factor loadings, and then rotate these loadings (either orthogonally or obliquely) until a sensible result is obtained. In the event $\mathbf{X} \sim N(\mathbf{0}, \boldsymbol{\Sigma})$ statistical tests can also be employed to test for r. However, even though a unique (restricted) solution basis is found, it is unlikely to be retained once the initial loadings are computed. Owing to the inherent mathematical arbitrariness involving the transformation of axes, such a procedure can be open to abuse, and must be employed with some care. Note that the situation encountered here is somewhat different than that for the principal components model. Although axis rotations are also employed, the situation for principal components is less arbitrary since the a priori value of r will not alter the (unrotated) loadings. As was seen in Chapter 3, we simply compute p components and then select (or test for) r. However, the situation concerning factor identification only applies to the so-called exploratory factor model. When several samples are available, or a single large sample is divided into several parts, sample information may be used to impose prior restrictions on the loading, usually in the form of zeroes. Alternatively, zero restrictions may also be imposed on theoretical grounds. In this situation it is possible to obtain identification of the factors without further rotation. This is known as confirmatory factor analysis, and is discussed briefly in Section 6.11.

Factor analysis can be developed in greater generality by using theoretical probability distributions (Anderson, 1959; Martin and McDonald, 1975; Bartholomew, 1981, 1984). The starting point is the relationship

$$f(\mathbf{X}) = \int_R h(\boldsymbol{\Phi}) g(\mathbf{X} \mid \boldsymbol{\Phi}) \, d\boldsymbol{\Phi} \qquad (6.9a)$$

where $f(\mathbf{X})$ and $h(\boldsymbol{\Phi})$ are densities of $\mathbf{X}$ and $\boldsymbol{\Phi}$, $g(\mathbf{X} \mid \boldsymbol{\Phi})$ is the conditional density of $\mathbf{X}$ given values $\boldsymbol{\Phi}$, and R is the range space of $\boldsymbol{\Phi}$. Here it is understood that the vectors $\mathbf{X}$ and $\boldsymbol{\Phi}$ denote particular values of the corresponding vectors of random variables. Using Bayes' formula the conditional density of $\boldsymbol{\Phi}$, given $\mathbf{X}$, is

$$h(\boldsymbol{\Phi}\,|\,\mathbf{X}) = h(\boldsymbol{\Phi})g(\mathbf{X}\,|\,\boldsymbol{\Phi})/f(\mathbf{X}) \tag{6.9b}$$

Clearly $f(\mathbf{X})$ does not determine $g(\mathbf{X}\,|\,\boldsymbol{\Phi})$ and $h(\boldsymbol{\Phi})$ uniquely, and further assumptions must be made. The basic assumption is the conditional independence of $\mathbf{X}$, that is, the conditional distribution of $\mathbf{X}$ can be expressed as

$$g(\mathbf{X}\,|\,\boldsymbol{\Phi}) = \prod_{i=1}^{p} g(\mathbf{X}_i\,|\,\boldsymbol{\Phi}) \tag{6.9c}$$

so that

$$h(\boldsymbol{\Phi}\,|\,\mathbf{X}) = \frac{h(\boldsymbol{\Phi}) \prod_{i=1}^{p} g(\mathbf{X}_i\,|\,\boldsymbol{\Phi})}{f(\mathbf{X})} \tag{6.9d}$$

Further assumptions about the form of $g(\mathbf{X}_i\,|\,\boldsymbol{\Phi})$ and $h(\boldsymbol{\Phi})$ must be made in order to obtain $g(\mathbf{X}\,|\,\boldsymbol{\Phi})$, but this does not provide a unique solution to the factor problem since the prior distribution $h(\boldsymbol{\Phi})$ is still arbitrary. Thus a one-to-one transformation can be defined from $\boldsymbol{\Phi}$ to new factors $\boldsymbol{\eta}$, say, which does not affect $f(\mathbf{X})$, and no amount of empirical information can distinguish between the various transformations and $h(\boldsymbol{\Phi}\,|\,\mathbf{X})$. Prior assumptions must therefore be made about $h(\boldsymbol{\Phi})$, which could emerge from our assumptions about the nature of the common factors or be simply based on practical expedience if the sole objective lies in data reduction. Here a further common assumption is the linearity of the factor model (Eq. 6.2) although nonlinear functions of $\boldsymbol{\Phi}$ can also be considered. Also, if we assume that $\boldsymbol{\Phi} \sim N(\mathbf{0}, \mathbf{I})$, the conditional expectation of $\mathbf{X}$ (given $\boldsymbol{\Phi}$) is also normal with expectation $\boldsymbol{\alpha}\boldsymbol{\Phi}$ and covariance matrix $\boldsymbol{\Psi}$, that is, $\mathbf{X}\,|\,\boldsymbol{\Phi} \sim N(\boldsymbol{\alpha}\boldsymbol{\Phi}, \boldsymbol{\Psi})$. Thus since $\boldsymbol{\Psi}$ is diagonal we can characterize factor analysis by the general property of conditional uncorrelatedness of the variables, given the common factors. Using Lemma 6.5 it can also be shown using Baysian arguments that $\boldsymbol{\Phi}\,|\,\mathbf{X} \sim N[(\boldsymbol{\alpha}^{\mathrm{T}}\boldsymbol{\Sigma}^{-1}\boldsymbol{\alpha}\mathbf{X}), (\boldsymbol{\alpha}^{\mathrm{T}}\boldsymbol{\Psi}^{-1}\boldsymbol{\alpha} + \mathbf{I})]$. Thus Baysian sufficiency can be used to provide a statistically rigorous treatment of the factor identification problem. The theoretical merits of the approach are: (1) it indicates essential assumptions in order to resolve the identification problem, (2) it points the way to more general structures, and (3) it reveals possible inconsistencies between assumption and estimation procedure.

When a sample of size n is taken, the factor model (Eq. 6.2) can be expressed as

$$\mathbf{X} = \mathbf{FA} + \mathbf{e} \tag{6.10}$$

where $\mathbf{X}$ is now a $(n \times p)$ matrix consisting of n multivariate observations on the random variables $\mathbf{X}_1, \mathbf{X}_2, \ldots, \mathbf{X}_p$, each identically distributed. The unobserved explanatory factors are usually assumed to be random (in the population) but can also be assumed to be fixed (Section 6.7). Other specifications are also possible (see McDonald and Burr, 1967). If the factors or latent variables are further assumed to be orthonormal, and uncorrelated with the residual error term, we have from Eq. (6.10)

$$\mathbf{F}^{\mathrm{T}}\mathbf{X} = \mathbf{F}^{\mathrm{T}}\mathbf{FA} + \mathbf{F}^{\mathrm{T}}\mathbf{e}$$
$$= \mathbf{A} \tag{6.11}$$

and

$$\mathbf{X}^{\mathrm{T}}\mathbf{X} = (\mathbf{FA} + \mathbf{e})^{\mathrm{T}}(\mathbf{FA} + \mathbf{e})$$
$$= \mathbf{A}^{\mathrm{T}}\mathbf{F}^{\mathrm{T}}\mathbf{FA} + \mathbf{A}^{\mathrm{T}}\mathbf{F}^{\mathrm{T}}\mathbf{e} + \mathbf{e}^{\mathrm{T}}\mathbf{FA} + \mathbf{e}^{\mathrm{T}}\mathbf{e}$$
$$= \mathbf{A}^{\mathrm{T}}\mathbf{A} + \mathbf{e}^{\mathrm{T}}\mathbf{e} \tag{6.12}$$

In the following sections we consider the main factor models commonly used in practice, together with several extensions such as the latent class and latent profile models.

6.3 FACTORING BY PRINCIPAL COMPONENTS

Factor analysis differs from principal components in that $r < p$ unobservable or latent common factors are fitted to the observations, under the prior specification that the error terms are (1) mutually uncorrelated, (2) heteroscedastic, and (3) uncorrelated with the common factors. Since the error covariance matrix is diagonal, the common factors are assumed to reproduce the covariances (correlations) between the observed variables, but not the variances. Factor analysis can therefore be viewed as a special case of the weighted principal components model (Section 5.6), and historically principal components have been used to estimate factors. It turns out, however, that the unweighted principal components model can also be considered as a special case of the general factor model.

6.3.1 The Homoscedastic Residuals Model

The homoscedastic residuals factor model is a special case of Eq. (6.2) since it assumes that the error terms possess equal variance. It is also known as the principal factor model, but the name is confusing since the term is also applied to a somewhat different specification (Section 6.3.3). The homoscedastic residuals model has been considered by Whittle (1953) and Lawley (1953), and derives its advantage from the following property (Anderson and Rubin, 1956).

THEOREM 6.3.   Let $\boldsymbol{\Sigma} = \boldsymbol{\alpha}\boldsymbol{\alpha}^T + \boldsymbol{\Psi}$ be a factor model such that $\boldsymbol{\Psi} = \sigma^2\mathbf{I}$ for some scalar $\sigma^2 > 0$. Then the model $\boldsymbol{\Sigma} = \boldsymbol{\alpha}\boldsymbol{\alpha}^T + \sigma^2\mathbf{I}$ is identifiable.

PROOF.   The latent roots and latent vectors λ_i and $\boldsymbol{\Pi}_i$ of $\boldsymbol{\Sigma}$ are given by

$$
\begin{aligned}
(\boldsymbol{\Sigma} - \lambda_i\mathbf{I})\boldsymbol{\Pi}_i &= [(\boldsymbol{\alpha}\boldsymbol{\alpha}^T + \sigma^2\mathbf{I}) - \lambda_i\mathbf{I}]\boldsymbol{\Pi}_i \\
&= [\boldsymbol{\Gamma} - (\lambda_i - \sigma^2)\mathbf{I}]\boldsymbol{\Pi}_i \\
&= [\boldsymbol{\Gamma} - \lambda_i^*\mathbf{I}]\boldsymbol{\Pi}_i \\
&= \mathbf{0} \qquad\qquad\qquad (i = 1, 2, \ldots, r)
\end{aligned}
\qquad (6.13)
$$

where $\lambda_i^* = \lambda_i - \sigma^2$ are the latent roots of the "true" part $\boldsymbol{\Gamma} = \boldsymbol{\alpha}\boldsymbol{\alpha}^T$. We thus have a principal components decomposition of $\boldsymbol{\alpha}\boldsymbol{\alpha}^T$ and since $\lambda_i \geq \sigma^2$, the scalar σ^2 can be chosen as the minimal latent root of $\boldsymbol{\Sigma}$ with multiplicity $p - r$, and the factor model (Eq. 6.13) is identifiable. Since $\rho(\boldsymbol{\Gamma}) = r$, the principal components of $\boldsymbol{\Gamma}$ yield r common factors, which are unique except for sign changes, that is, multiplication by -1 (Section 3.2). Once the common factors are known, they may be rotated orthogonally or obliquely (Section 5.3) to enhance interpretability. None that to obtain correlation loadings the r latent vectors of $\boldsymbol{\Gamma}$ are standardized by λ_i^* $(i = 1, 2, \ldots, r)$.

Once the latent roots and vectors are known we have, in matrix form,

$$
\boldsymbol{\Pi}^T\boldsymbol{\Gamma}\boldsymbol{\Pi} = \boldsymbol{\Pi}^T\boldsymbol{\alpha}\boldsymbol{\alpha}^T\boldsymbol{\Pi} = \boldsymbol{\Lambda}^* \qquad (6.14)
$$

where $\boldsymbol{\alpha}\boldsymbol{\alpha}^T$ is $(p \times p)$ of rank r, $\boldsymbol{\Pi}$ is $(p \times r)$, and

$$
\boldsymbol{\Lambda}^* =
\begin{bmatrix}
\lambda_i^* & & & 0 \\
& \lambda_2^* & & \\
& & \ddots & \\
0 & & & \lambda_r^*
\end{bmatrix}
$$

is the diagonal matrix of nonzero latent roots of $\boldsymbol{\Gamma}$. The factor loadings are then given by

$$
\boldsymbol{\alpha} = \boldsymbol{\Pi}\boldsymbol{\Lambda}^{*1/2} \qquad (6.15)
$$

or $\boldsymbol{\alpha}_i = \boldsymbol{\Pi}_i\sqrt{\lambda_i^*}$ $(i = 1, 2, \ldots, r)$.

Given a sample of size n we have the decomposition

$$\mathbf{X} = \mathbf{X}^* + \mathbf{e}$$

$$= \mathbf{F}\mathbf{A}^{\mathrm{T}} + \mathbf{e}$$

where $\mathbf{X}^{\mathrm{T}}\mathbf{X} = \mathbf{A}^{\mathrm{T}}\mathbf{A} + \mathbf{e}^{\mathrm{T}}\mathbf{e}$ and $\mathbf{e}^{\mathrm{T}}\mathbf{e} = s^2\mathbf{I}$ is the homoscedastic sample error variance matrix. The sample analog of Eq. (6.13) is then

$$(\mathbf{X}^{\mathrm{T}}\mathbf{X} - l_i\mathbf{I})\mathbf{P}_i = (\mathbf{A}\mathbf{A}^{\mathrm{T}} - l_i^*\mathbf{I})\mathbf{P}_i = \mathbf{0} \tag{6.16}$$

where $l_i^* = l_i - s^2$ ($i = 1, 2, \ldots, r$). Since Eq. (6.16) provides the correct specification only when the last $(p - r)$ latent roots are equal, s^2 can be estimated as

$$s^2 = \frac{\displaystyle\sum_{i=r+1}^{p} l_i}{p - r} \tag{6.17}$$

Equation (6.16) can also be expressed as

$$\mathbf{P}^{\mathrm{T}}\mathbf{A}\mathbf{A}^{\mathrm{T}}\mathbf{P} = \mathbf{L}^* \tag{6.18}$$

where $\mathbf{L}^* = \mathbf{L} - s^2\mathbf{I}$ is diagonal. Equation (6.18) is thus a principal components model with $(p - r)$ isotropic roots and loadings $\mathbf{A}^{\mathrm{T}} = \mathbf{L}^{*1/2}\mathbf{P}^{\mathrm{T}}$. The difference between the ordinary principal components model of Chapter 3 and the more general factor analysis model is now clear. The latter contains heteroscedastic residual terms, whereas the former assumes the special case of equality of the residual variance terms. Note that since ordinary principal components do not adjust for error variance, the component loadings are inflated by a factor of σ^2, that is, principal component loading estimates are biased upward in the presence of uncorrelated residual errors. However, when the error terms are correlated, both principal components and factor analysis provide biased estimates.

6.3.2 Unweighted Least Squares Models

The main limitation of Eq. (6.13) is the assumption of equal error variance. A more general model is $\boldsymbol{\Sigma} = \boldsymbol{\alpha}\boldsymbol{\alpha}^{\mathrm{T}} + \sigma_i^2\mathbf{I}$, which implies the decomposition

$$(\boldsymbol{\Sigma} - \lambda_i\mathbf{I})\boldsymbol{\Pi}_i = [\boldsymbol{\Gamma} - (\lambda_i - \sigma_i^2)\mathbf{I}]\boldsymbol{\Pi}_i \tag{6.19}$$

where $\boldsymbol{\Psi} = \sigma_i^2\mathbf{I}$ is a diagonal heteroscedastic error variance matrix. Equation (6.19) provides a generalization of the homoscedastic model (Eq. 6.13). Note that $\lambda_i \geq \sigma_i^2$ since $\boldsymbol{\Gamma}$ is specified to be Grammian. Thus the practice of minimizing the residual sums of squares can be extended to a wider class of factor models, which may be solved using a principal components decomposition. Least squares models are also known as minimum residuals or minimum distance models, and include as special cases well-known factor models such as the principal factor and the "minres" models (Harman, 1967; Comrey, 1973). They all possess a common feature in that they minimize the criterion

$$U = \mathrm{tr}(\mathbf{S} - \boldsymbol{\Sigma})^2$$

$$= \sum_{i=1}^{p} (s_i^2 - \sigma_i^2)^2 \tag{6.20}$$

Following Joreskog (1977) the total derivative of $\mathbf{U}$ is

$$dU = d[\mathrm{tr}(\mathbf{S} - \boldsymbol{\Sigma})^2]$$

$$= \mathrm{tr}[d(\mathbf{S} - \boldsymbol{\Sigma})^2]$$

$$= -2\,\mathrm{tr}[(\mathbf{S} - \boldsymbol{\Sigma})\,d\boldsymbol{\Sigma}]$$

$$= -2\,\mathrm{tr}[(\mathbf{S} - \boldsymbol{\Sigma})(\boldsymbol{\alpha}\,d\boldsymbol{\alpha}^{\mathrm{T}} + d\boldsymbol{\alpha}\boldsymbol{\alpha}^{\mathrm{T}})]$$

$$= -4\,\mathrm{tr}[(\mathbf{S} - \boldsymbol{\Sigma})\boldsymbol{\alpha}\,d\boldsymbol{\alpha}^{\mathrm{T}}]$$

so that

$$\frac{\partial U}{\partial \boldsymbol{\alpha}} = -4(\boldsymbol{\Sigma} - \mathbf{S})\boldsymbol{\alpha} \tag{6.21}$$

using Lemmas 6.2 and 6.3 of Section 6.4.2. Setting to zero yields the normal equations

$$(\hat{\boldsymbol{\Sigma}} - \mathbf{S})\hat{\boldsymbol{\alpha}} = [(\hat{\boldsymbol{\alpha}}\hat{\boldsymbol{\alpha}}^{\mathrm{T}} + \hat{\boldsymbol{\Psi}}) - \mathbf{S}]\hat{\boldsymbol{\alpha}}$$

$$= \hat{\boldsymbol{\alpha}}(\hat{\boldsymbol{\alpha}}^{\mathrm{T}}\hat{\boldsymbol{\alpha}}) - (\mathbf{S}\hat{\boldsymbol{\alpha}} - \hat{\boldsymbol{\Psi}}\hat{\boldsymbol{\alpha}})$$

$$= \mathbf{0} \tag{6.22}$$

where $\hat{\boldsymbol{\alpha}}^{\mathrm{T}}\hat{\boldsymbol{\alpha}} = \mathbf{L}$ is diagonal. Equation (6.22) can be rewritten as

$$(\mathbf{S} - \hat{\boldsymbol{\Psi}})\hat{\boldsymbol{\alpha}} = \hat{\boldsymbol{\alpha}}\mathbf{L} \tag{6.23}$$

where $\mathbf{L}$ and $\hat{\boldsymbol{\alpha}}$ are latent roots and latent vectors of $(\mathbf{S} - \hat{\boldsymbol{\Psi}})$ respectively. Equation (6.23) therefore represents a principal component analysis of the corrected covariance (correlation) matrix $(\mathbf{S} - \hat{\boldsymbol{\Psi}})$ whose observed diagonal variance terms have been replaced by corrected values (and which can therefore possess only r nonzero latent roots). Note also that the model is not independent of scale. In addition, since the error terms are not known beforehand, both the loadings and error variances must be estimated simultaneously. This can be done in two ways. The first method is to use principal components to obtain the initial estimate of $\boldsymbol{\Psi}$, say $\hat{\boldsymbol{\Psi}}_{(1)}$, using the first r components. We then have

$$\hat{\boldsymbol{\Psi}}_{(1)} = \operatorname{diag}(\mathbf{S} - \hat{\boldsymbol{\alpha}}\hat{\boldsymbol{\alpha}}^T) \tag{6.23a}$$

where since residual errors are to be uncorrelated, only the diagonal elements of $\mathbf{S}$ are replaced by estimated values. Once the initial estimate $\hat{\boldsymbol{\Psi}}_{(1)}$ is known, it can be replaced in the normal equations (Eq. 6.23) to obtain an improved second round estimate $\hat{\boldsymbol{\Psi}}_{(2)}$, and so forth until the sequence of error variances and loadings converges to stable values. The second method leads to a model known at times as the "image factor model."

6.3.3 The Image Factor Model

Another common practice is to estimate residual error variances by regression analysis, where each variable (in turn) becomes the dependent variable, and to use the predicted variances (R^2 values) as diagonal elements of the matrix $\mathbf{S} - \hat{\boldsymbol{\Psi}}$. Such practice however is inadvisable and should be resisted since it leads to inconsistency of the statistical estimators. This can be seen from the following arguement. Let $\mathbf{Y}$ be a $(n \times 1)$ vector of observations on a dependent variable, and let $\mathbf{X}$ denote a $(n \times p)$ matrix of n observations on p independent variables such that $\mathbf{Y} = \boldsymbol{\chi}\boldsymbol{\beta} + \boldsymbol{\epsilon}$ and $\mathbf{X} = \boldsymbol{\chi} + \boldsymbol{\delta}$ where $\boldsymbol{\delta}$ is a matrix of measurement errors for the independent variables and $\boldsymbol{\chi}$ is the matrix of true (but unobserved) values of $\mathbf{X}$. The term $\boldsymbol{\epsilon}$ then represents error in $\mathbf{Y}$. For a random sample there will also exist additional terms in $\boldsymbol{\delta}$ and $\boldsymbol{\epsilon}$ such as sampling error, but (for the population) all variables must at least contain errors of measurement, together with natural random variation by the very nature of the factor hypothesis itself. Since in the regression set-up $\mathbf{Y}$ corresponds, in turn, to one of the variables of the set $\mathbf{X}_1, \mathbf{X}_2, \ldots, \mathbf{X}_p$, we have

$$\mathbf{Y} = \boldsymbol{\chi}\boldsymbol{\beta} + \boldsymbol{\epsilon}$$
$$= (\mathbf{X} - \boldsymbol{\delta})\boldsymbol{\beta} + \boldsymbol{\epsilon}$$
$$= \mathbf{X}\boldsymbol{\beta} + \boldsymbol{\eta} \tag{6.24}$$

where $\boldsymbol{\eta} = \boldsymbol{\epsilon} - \boldsymbol{\delta}\boldsymbol{\beta}$ is the residual error in both the dependent and independent variables and $\mathbf{X}$ represents the data matrix for the remaining $p - 1$ variables. The ordinary least squares estimator of $\boldsymbol{\beta}$ is then

$$\hat{\boldsymbol{\beta}} = (\mathbf{X}^T\mathbf{X})^{-1}\mathbf{X}^T\mathbf{Y}$$
$$= (\mathbf{X}^T\mathbf{X})\mathbf{X}^T(\mathbf{X}\boldsymbol{\beta} + \boldsymbol{\eta})$$

$$= (\mathbf{X}^{\mathrm{T}}\mathbf{X})^{-1}\mathbf{X}^{\mathrm{T}}\mathbf{X}\boldsymbol{\beta} + (\mathbf{X}^{\mathrm{T}}\mathbf{X})^{-1}\mathbf{X}^{\mathrm{T}}\boldsymbol{\eta}$$

$$= \boldsymbol{\beta} + (\mathbf{X}^{\mathrm{T}}\mathbf{X})^{-1}\mathbf{X}^{\mathrm{T}}\boldsymbol{\eta} \tag{6.25}$$

and taking expected values yields

$$E(\hat{\boldsymbol{\beta}}) = \boldsymbol{\beta} + E[(\mathbf{X}^{\mathrm{T}}\mathbf{X})^{-1}\mathbf{X}^{\mathrm{T}}\boldsymbol{\eta}]$$

$$\neq \boldsymbol{\beta} \tag{6.26}$$

since

$$E[(\mathbf{X}^{\mathrm{T}}\mathbf{X})^{-1}\mathbf{X}^{\mathrm{T}}\boldsymbol{\eta}] = E[(\mathbf{X}^{\mathrm{T}}\mathbf{X})^{-1}(\boldsymbol{\chi} + \boldsymbol{\delta})^{\mathrm{T}}(\boldsymbol{\epsilon} - \boldsymbol{\delta}\boldsymbol{\beta})]$$

$$= E[(\mathbf{X}^{\mathrm{T}}\mathbf{X})^{-1}(\boldsymbol{\chi}^{\mathrm{T}}\boldsymbol{\epsilon} - \boldsymbol{\chi}^{\mathrm{T}}\boldsymbol{\delta}\boldsymbol{\beta} + \boldsymbol{\delta}^{\mathrm{T}}\boldsymbol{\epsilon} - \boldsymbol{\delta}^{\mathrm{T}}\boldsymbol{\delta}\boldsymbol{\beta}] \tag{6.27}$$

Thus even when $\boldsymbol{\chi}^{\mathrm{T}}\boldsymbol{\epsilon} = \mathbf{X}^{\mathrm{T}}\boldsymbol{\delta} = \boldsymbol{\delta}^{\mathrm{T}}\boldsymbol{\epsilon} = 0$ we have $\boldsymbol{\delta}^{\mathrm{T}}\boldsymbol{\delta} \neq 0$, so that $E[(\mathbf{X}^{\mathrm{T}}\mathbf{X})^{-1}\mathbf{X}^{\mathrm{T}}\boldsymbol{\eta}] \neq 0$, and $\hat{\boldsymbol{\beta}}$ is inconsistent for $\boldsymbol{\beta}$. It follows that the predicted values (and thus the R^2 coefficients) are also inconsistent, and a principal components decomposition of the reduced principal factor correlation matrix will yield inconsistent estimators. Indeed, the bias can be worse than if an unreduced covariance (or correlation) matrix had been used. Moreover, using R^2 values can introduce secondary bias since R^2 is not corrected for degrees of freedom and is therefore a monotonically non-decreasing function of p, the number of explanatory variables. The model is also known as the principle factor model.

Another variant of image factor analysis is to use the weighted principal components model whereby a component analysis is performed on $\mathbf{X}^{\mathrm{T}}\mathbf{X}$ as

$$(\mathbf{X}^{\mathrm{T}}\mathbf{X} - l_i\mathbf{I})\mathbf{P}_i = [(\mathbf{X} - \mathbf{e})^{\mathrm{T}}(\mathbf{X} - \mathbf{e}) - l_i\mathbf{I}]\mathbf{P}_i$$

$$= (\mathbf{X}^{\mathrm{T}}\mathbf{X} - l_i\mathbf{e}^{\mathrm{T}}\mathbf{e})\mathbf{P}_i$$

$$= \mathbf{0} \tag{6.28}$$

with $\mathbf{e}^{\mathrm{T}}\mathbf{e}$ diagonal. The $(n \times p)$ matrix of residuals $\mathbf{e}$ is usually computed ahead of time by regression, but since all variables are still affected by errors in the regression equations, the weighted image factor model is also subject to estimation inconsistancy.

A number of other variants of the principal factor or image factor models have been proposed, but they differ only in the methods used to estimate the error terms. For example, iteration may be used with Eq. (6.28) in an attempt to improve the initial solution. As a general rule however models that combine principal components with ordinary least squares regression yield inconsistent estimators.

6.3.4 The Whittle Model

Principal component factors are usually regarded as unobserved variables, which vary randomly in the population. Whittle (1953) has shown that components (factors) can also be considered as fixed variates which are solutions of the weighted principal component model

$$(\Sigma - \lambda_i \Psi)\Pi_i = 0 \qquad (i = 1, 2, \ldots, r) \qquad (6.29)$$

where Ψ is a diagonal matrix of residual variances. The derivation is given in Section 3.7. When Ψ is not known, an iterative procedure suggested by Whittle (1953) can be used. Initially, $\Psi = I$ and $r < p$ components are computed and used to obtain the first-stage residual variances. The variance terms are then substituted into Eq. (6.29), second-stage common factors are obtained, and so forth until $\hat{\Psi}$ together with the loadings converges to stable values. A simulated numerical example may be found in Wold (1953).

6.4 UNRESTRICTED MAXIMUM LIKELIHOOD FACTOR MODELS

Neither the homoscedastic residuals model nor the principal component factor specifications resolve the difficulty of estimating residual variation since the former assumes an unrealistically simplified error structure and the latter represents least squares estimators that are statistically flawed. Also, it is not clear how to test the principal factor loadings and latent roots for statistical significance. As an alternative to least squares, when the population is multivariate normal, the principle of maximum likelihood can be used to derive the normal equations. The advantage of ML estimators is that they are efficient and consistent and permit statistical testing of parameters. The first significant contributions in this area are due to Lawley (1940, 1941) and Joreskog (1963, 1966), and in spite of initial difficulties of numerical convergence, ML estimation is today widely employed. Actually, several specifications of ML factor analysis have been proposed which lead to slightly different estimation procedures. Maximum likelihood estimation however can only be used for random factors since when factors are fixed, strictly speaking ML estimators do not exist (see Solari, 1969). The class of models is usually referred to as unrestricted maximum likelihood factor analysis, to differentiate it from the restricted or confirmatory factor models (see Lawley and Maxwell, 1971). It should be kept in mind however that the unrestricted factor models are also subject to restrictions or constraints, albeit of a more mathematical nature (Section 6.2).

6.4.1 The Reciprocal Proportionality Model

Consider a factor model like Eq. (6.2) where Ψ is a diagonal heteroscedastic error variance matrix. As noted at the outset, unique solutions do not generally exist, and in order to achieve identification further constraints are required. Since the primary role of constraints are to achieve identification, their choice is largely arbitrary. Assuming that initially orthogonal common factors are required, Lawley (1953) and Joreskog (1962, 1963) have proposed that the model be estimated in such a way that

$$\mathbf{\Psi} = \sigma^2 (\text{diag } \mathbf{\Sigma}^{-1})^{-1}$$

$$= \sigma^2 \mathbf{\Delta}^{-1} \tag{6.30}$$

where $\sigma^2 > 0$ is an arbitrary scalar and $\mathbf{\Delta} = \text{diag } \mathbf{\Sigma}^{-1}$ consists of diagonal elements of $\mathbf{\Sigma}^{-1}$. The constraint (Eq. 6.30) implies the residual variances are proportional to the reciprocal values of the diagonal terms of $\mathbf{\Sigma}^{-1}$, from which the model derives its name. The factor model can thus be expressed as

$$\mathbf{\Sigma} = \boldsymbol{\alpha}\boldsymbol{\alpha}^{\mathrm{T}} + \mathbf{\Psi}$$

$$= \mathbf{\Gamma} + \sigma^2 \mathbf{\Delta}^{-1} \tag{6.31}$$

and pre- and postmultiplying by $\mathbf{\Delta}^{1/2}$ we have

$$\mathbf{\Delta}^{1/2}\mathbf{\Sigma}\mathbf{\Delta}^{1/2} = \mathbf{\Delta}^{1/2}\mathbf{\Gamma}\mathbf{\Delta}^{1/2} + \sigma^2 \mathbf{I}$$

$$= \mathbf{\Sigma}^* \tag{6.32}$$

say, a weighted covariance matrix. The choice for the restriction (Eq. 6.30) is now clear—it converts the factor model to the equal residual variances model of Section 6.3.1.

THEOREM 6.4.   Let $\mathbf{\Sigma}$ be the population covariance matrix such that $\mathbf{\Sigma} = \boldsymbol{\alpha}\boldsymbol{\alpha}^{\mathrm{T}} + \sigma^2 \mathbf{\Delta}^{-1}$ and $\mathbf{\Delta} = \text{diag}(\mathbf{\Sigma}^{-1})$. Then the first $r < p$ roots $\lambda_1 > \lambda_2 > \cdots \lambda_r$ of $\mathbf{\Delta}^{1/2}\mathbf{\Sigma}\mathbf{\Delta}^{1/2}$ are distinct, and the remaining $(p - r)$ are equal to σ^2.

PROOF.   Let λ_i be the ith roots of $\mathbf{\Sigma}^* = \mathbf{\Delta}^{1/2}\mathbf{\Sigma}\mathbf{\Delta}^{1/2}$ and γ_i the ith root of $\mathbf{\Delta}^{1/2}\boldsymbol{\alpha}\boldsymbol{\alpha}^{\mathrm{T}}\mathbf{\Delta}^{1/2} = \mathbf{\Delta}^{1/2}\mathbf{\Gamma}\mathbf{\Delta}^{1/2}$. Then

$$|\mathbf{\Sigma}^* - \lambda_i \mathbf{I}| = |\mathbf{\Delta}^{1/2}\mathbf{\Sigma}\mathbf{\Delta}^{1/2} - \lambda_i \mathbf{I}|$$

$$= |(\mathbf{\Delta}^{1/2}\boldsymbol{\alpha}\boldsymbol{\alpha}^{\mathrm{T}}\mathbf{\Delta}^{1/2} + \sigma^2 \mathbf{I}) - \lambda_i \mathbf{I}|$$

$$= |\mathbf{\Delta}^{1/2}\boldsymbol{\alpha}\boldsymbol{\alpha}^{\mathrm{T}}\mathbf{\Delta}^{1/2} - (\lambda_i - \sigma^2)\mathbf{I}|$$

$$= \mathbf{\Delta}^{1/2}\mathbf{\Gamma}\mathbf{\Delta}^{1/2} - \gamma_i \mathbf{I}| \tag{6.33}$$

where $\gamma_i = \lambda_i - \sigma^2$ for $i = 1, 2, \ldots, p$. Since $\mathbf{\Gamma}$ is $(p \times p)$ and positive semidefinite of rank $r < p$, it must possess r nonzero and $(p - r)$ zero roots, that is, $\lambda_1 > \lambda_2 > \cdots > \lambda_r > \lambda_{r+1} = \lambda_{r+2} = \cdots = \lambda_p = \sigma^2$. Since scaling a covariance matrix by $\mathbf{\Delta}^{1/2}$ converts the model to that of equal residual variances, it follows that the reciprocal proportionality model (Eq. 6.31) is identifiable, if we are given the value of r. The model can therefore be viewed as a principal component analysis of the scaled matrix $\mathbf{\Sigma}^*$. Alternatively, the loadings may be obtained from the equation

$$\mathbf{\Sigma}\mathbf{\Delta}\boldsymbol{\alpha} = \boldsymbol{\alpha}\mathbf{\Lambda} \tag{6.33a}$$

The constraint that $\boldsymbol{\alpha}^T\mathbf{\Delta}\boldsymbol{\alpha}$ be diagonal therefore constitutes an arbitrary condition to ensure identifiability of the coefficients, that is, the constraint fixes the initial position of the coordinate axes. In practice when r is not known its value is usually decided on the basis of trial and error, accompanied by tests of significance. It can also be shown that Eq. (6.31) has another desirable feature—the elements of $\boldsymbol{\alpha}$ are always correlation loadings no matter whether the correlation, covariance, or sum-of-squares and products matrix $\mathbf{X}^T\mathbf{X}$ is used. The proof proceeds along the lines of Theorem 6.6 (Exercise 6.4).

The ML estimators for $\boldsymbol{\alpha}$ and σ^2 are obtained when the observations represent a sample from the $N(\boldsymbol{\mu}, \mathbf{\Sigma})$ distribution. Alternatively the factors $\mathbf{\Phi}_1, \mathbf{\Phi}_2, \ldots, \mathbf{\Phi}_r$ and the errors $\boldsymbol{\epsilon}_1, \boldsymbol{\epsilon}_2, \ldots, \boldsymbol{\epsilon}_p$ may be assumed to follow independent normal distributions with zero means and diagonal covariance matrices $\mathbf{I}$ and $\mathbf{\Psi}$, respectively. When the observed variates are multivariate normal, it can be shown that the sample variances and covariances follow the Wishart distribution

$$f(\mathbf{S}) = c|\mathbf{\Sigma}|^{-n/2}|\mathbf{S}|^{1/2(n-p-1)}\exp[\tfrac{1}{2}\operatorname{tr}(\mathbf{\Sigma}^{-1}\mathbf{S})]$$

where $\mathbf{S}$ can be replaced by $\mathbf{X}^T\mathbf{X}$ with a suitable adjustments in the constant of proportionality c. Replacing $\mathbf{\Sigma}$ by $\mathbf{\Sigma}^*$ and $\mathbf{S}$ by

$$\mathbf{S}^* = (\operatorname{diag}\mathbf{S}^{-1})^{1/2}\mathbf{S}(\operatorname{diag}\mathbf{S}^{-1})^{1/2}$$

$$= \mathbf{D}^{1/2}\mathbf{S}\mathbf{D}^{1/2}$$

$$= \frac{1}{n-1}\mathbf{D}^{1/2}(\mathbf{X}^T\mathbf{X})\mathbf{D}^{1/2} \tag{6.34}$$

however does not necessarily preserve the Wishart distribution, where $\mathbf{D} = \operatorname{diag}\mathbf{S}^{-1}$. Furthermore, the exact distribution of $\mathbf{S}^*$ is not known, although asymptotic results can be obtained. Assuming $\mathbf{D} \to \mathbf{\Delta}$ as $n \to \infty$, the approximate log-likelihood function is given by Joreskog (1963) as

$$L(\mathbf{\Sigma}^*) = k[-\ln|\mathbf{\Sigma}^*| + \operatorname{tr}(\mathbf{S}^*\mathbf{\Sigma}^{*-1})]$$

$$= k[-\ln|\boldsymbol{\alpha}\boldsymbol{\alpha}^T + \mathbf{\Psi}| + \operatorname{tr}[\mathbf{S}^*(\boldsymbol{\alpha}\boldsymbol{\alpha}^T + \mathbf{\Psi})^{-1}]] \tag{6.35}$$

where k is a constant of proportionality and ignoring expressions that depend on n. Differentiating Eq. (6.35) with respect to $\boldsymbol{\alpha}$ and $\boldsymbol{\Psi}$ and setting to zero yields the normal equations

$$(\mathbf{S}^* - \hat{\lambda}_i \mathbf{I})\hat{\boldsymbol{\alpha}}_i = \mathbf{0} \tag{6.36}$$

(see Section 6.4.2) where

$$\hat{\boldsymbol{\alpha}}_i^{\mathrm{T}} \hat{\boldsymbol{\alpha}}_i = \hat{\lambda}_i - \hat{\sigma}_i^2 \qquad (i = 1, 2, \ldots, r) \tag{6.37}$$

$$\hat{\sigma}^2 = \frac{1}{p-r}\left[\operatorname{tr}\mathbf{S}^* - \sum_{i=1}^{r}\hat{\lambda}_i\right]$$

$$= \frac{1}{p-r}\sum_{i=r+1}^{p}\hat{\lambda}_i \tag{6.38}$$

that is, $\hat{\sigma}^2$, $\hat{\boldsymbol{\alpha}}_i$, and $\hat{\lambda}_i$ are ML estimators. Alternatively, correlation loadings may also be obtained by solving the sample version of Eq. (6.33a), that is,

$$\mathbf{SD}\hat{\boldsymbol{\alpha}} = \hat{\boldsymbol{\alpha}}\hat{\boldsymbol{\Lambda}} \tag{6.39}$$

When the variables are not multivariate normal the estimators (Eqs. 6.36–6.38) are still optimal, but only in the least squares sense. A further advantage of the model is that it is not iterative in nature and is thus relatively easy to compute. It is not as optimal however as the full or iterative ML model, which generally provides more efficient estimators. Two well-known iterative models—due to Lawley and Rao—are frequently employed in practice.

6.4.2 The Lawley Model

The original ML factor model was introduced by Lawley (1940, 1941) and differs from the reciprocal proportionality model in the weighting scheme used for the dispersion matrix. It also is invariant with respect to scale change so that the correlation, covariance, and $\mathbf{X}^{\mathrm{T}}\mathbf{X}$ matrix yield identical correlation loading coefficients. Let $\mathbf{Y} = (y_1, y_2, \ldots, y_p)^{\mathrm{T}}$ be $N(\boldsymbol{\mu}, \boldsymbol{\Sigma})$. Then using the Wishart distribution (Eq. 6.35) the log likelihood can be expressed as

$$L(\boldsymbol{\Sigma}) = \ln c - \frac{n}{2}\ln|\boldsymbol{\Sigma}| + \frac{1}{2}(n-p-1)\ln|\mathbf{S}| + \frac{n}{2}\operatorname{tr}(\boldsymbol{\Sigma}^{-1}\mathbf{S}) \tag{6.40}$$

the maximum of which must be the same as that of

$$L = -\frac{n}{2}[\ln|\Sigma| + \text{tr}(S\Sigma^{-1})]$$

$$= -\frac{n}{2}[\ln|\alpha\alpha^T + \Psi| + \text{tr}(\alpha\alpha^T + \Psi)^{-1}S] \tag{6.41}$$

omitting constant functions of the observations, which do not alter the maximum.

In the random model the likelihood of (Eq. 6.40) is based on the ratio of the likelihood under the factor hypothesis (Eq. 6.5), to the likelihood under the hypothesis that Σ is any positive definite matrix. We have

$$H_0: \Sigma = \alpha\alpha^T + \Psi$$

$$H_a: \Sigma \text{ any positive definite matrix}$$

In practice it is numerically more convenient to optimize a different but equivalent function, as stated in the following lemma.

Lemma 6.1. Maximizing Eq. (6.40) is equivalent to minimizing

$$F(\Sigma) = \ln|\Sigma| + \text{tr}(S\Sigma^{-1}) - \ln|S| - p$$

$$= \text{tr}(\Sigma^{-1}S) - \ln|\Sigma^{-1}S| - p \tag{6.42}$$

$\square$

The proof consists of noting that Eqs. (6.41) and (6.42) differ only in terms of constant functions of n and p. It can also be shown that Eq. (6.42) is nonnegative and attains the value of zero if and only if $\Sigma = S$ (see Exercise 6.5). For this reason Eq. (6.42) is frequently used as a loss function, that is, as a measure of goodness-of-fit of Σ to S (Section 6.6). Although the minimization of the likelihood in the context of factor analysis was first considered by Lawley (1940; see also Lawley and Maxwell, 1971), the numerical technique employed did not always lead to convergence, until the introduction of the Fletcher–Powell (1963) algorithm by Joreskog (1966).

Although the maximization of Eq. (6.42) cannot be obtained in closed form, the ML normal equations can be derived by multivariate calculus. Ignoring functions of the observations and assuming Φ random and α fixed we maximize the expression

$$L = \ln|\Sigma| + \text{tr}(S\Sigma^{-1}) \tag{6.43}$$

which depends only on $\boldsymbol{\alpha}$ and $\boldsymbol{\Psi}$. We have

$$\frac{\partial L}{\partial \boldsymbol{\Psi}} = \frac{\partial \ln|\boldsymbol{\Sigma}|}{\partial \boldsymbol{\Psi}} + \frac{\partial(\operatorname{tr} \mathbf{S}\boldsymbol{\Sigma}^{-1})}{\partial \boldsymbol{\Psi}}$$

$$= \frac{\partial \ln|\boldsymbol{\alpha}\boldsymbol{\alpha}^{T} + \boldsymbol{\Psi}|}{\partial \boldsymbol{\Psi}} + \frac{\partial[\operatorname{tr} \mathbf{S}(\boldsymbol{\alpha}\boldsymbol{\alpha}^{T} + \boldsymbol{\Psi})^{-1}]}{\partial \boldsymbol{\Psi}}$$

$$= \operatorname{diag}(\boldsymbol{\alpha}\boldsymbol{\alpha}^{T} + \boldsymbol{\Psi})^{-1} - (\boldsymbol{\alpha}\boldsymbol{\alpha}^{T} + \boldsymbol{\Psi})^{-1}\mathbf{S}(\boldsymbol{\alpha}\boldsymbol{\alpha}^{T} + \boldsymbol{\Psi})^{-1}\frac{\partial \boldsymbol{\Psi}}{\partial \boldsymbol{\Psi}} \qquad (6.44)$$

where $\partial\boldsymbol{\Psi}/\partial\boldsymbol{\Psi} = \mathbf{I}$. Setting to zero we have

$$\frac{\partial L}{\partial \boldsymbol{\Psi}} = \operatorname{diag}(\hat{\boldsymbol{\alpha}}\hat{\boldsymbol{\alpha}}^{T} + \hat{\boldsymbol{\Psi}})^{-1} - \operatorname{diag}(\hat{\boldsymbol{\alpha}}\hat{\boldsymbol{\alpha}}^{T} + \hat{\boldsymbol{\Psi}})^{-1}\mathbf{S}(\hat{\boldsymbol{\alpha}}\hat{\boldsymbol{\alpha}}^{T} + \hat{\boldsymbol{\Psi}})^{-1} = \mathbf{0}$$

or

$$\operatorname{diag}[\hat{\boldsymbol{\Sigma}}^{-1}(\hat{\boldsymbol{\Sigma}} - \mathbf{S})\hat{\boldsymbol{\Sigma}}^{-1}] = \mathbf{0} \qquad (6.45)$$

where $\hat{\boldsymbol{\Sigma}} = \hat{\boldsymbol{\alpha}}\hat{\boldsymbol{\alpha}}^{T} + \hat{\boldsymbol{\Psi}}$. The normal equation (Eq. 6.45) is equivalent to

$$\operatorname{diag}(\hat{\boldsymbol{\Sigma}}) = \operatorname{diag}(\mathbf{S}) \qquad (6.46)$$

or the condition that the sample variances reproduced by the ML estimators $\hat{\boldsymbol{\alpha}}$ and $\hat{\boldsymbol{\Psi}}$ equal the sample variances observed in the sample.

Next, differentiating Eq. (6.43) with respect to $\boldsymbol{\alpha}$ we have

$$\frac{\partial L}{\partial \boldsymbol{\alpha}} = \frac{\partial \ln|\boldsymbol{\alpha}\boldsymbol{\alpha}^{T} + \boldsymbol{\Psi}|}{\partial \boldsymbol{\alpha}} + \frac{\partial \operatorname{tr}[\mathbf{S}(\boldsymbol{\alpha}\boldsymbol{\alpha}^{T} + \boldsymbol{\Psi})^{-1}]}{\partial \boldsymbol{\alpha}}$$

$$= \frac{1}{|\boldsymbol{\alpha}\boldsymbol{\alpha}^{T} + \boldsymbol{\Psi}|} \frac{\partial|\boldsymbol{\Sigma}|}{\partial \boldsymbol{\alpha}} + \operatorname{tr}\left[\mathbf{S}(\boldsymbol{\alpha}\boldsymbol{\alpha}^{T} + \boldsymbol{\Psi})^{-1} \frac{\partial}{\partial \boldsymbol{\alpha}} \mathbf{S}(\boldsymbol{\alpha}\boldsymbol{\alpha}^{T} + \boldsymbol{\Psi})^{-1} \right]$$

In terms of the elements α_{ij} the first term of the derivative becomes

$$\frac{\partial \ln|\boldsymbol{\alpha}\boldsymbol{\alpha}^{T} + \boldsymbol{\Psi}|}{\partial \alpha_{ij}} = -\frac{1}{|\boldsymbol{\alpha}\boldsymbol{\alpha}^{T} + \boldsymbol{\Psi}|} \sum_{g=1}^{p} \sum_{h=1}^{p} |\boldsymbol{\alpha}\boldsymbol{\alpha}^{T} + \boldsymbol{\Psi}|_{gh} \frac{\partial \sigma_{gh}}{\partial \alpha_{ij}}$$

$$= -\operatorname{tr}\left(\boldsymbol{\Sigma}^{-1} \frac{\partial \boldsymbol{\Sigma}}{\partial \alpha_{ij}} \right)$$

where the subscripts g, h denote the corresponding element of the determinant (matrix). For the (i, j)th loading the matrix $\partial\boldsymbol{\Sigma}/\partial\alpha_{ij}$ is symmetric, with zeroes everywhere except the ith row and jth column, which has elements $2\alpha_{ij}$ in the (i, j)th position. Direct multiplication and simplication then leads to

$$\frac{\partial \ln|\mathbf{\Sigma}|}{\partial \boldsymbol{\alpha}} = \mathbf{\Sigma}^{-1}\boldsymbol{\alpha} \tag{6.47}$$

The second term can also be simplified as

$$\mathrm{tr}\left[\mathbf{S}(\boldsymbol{\alpha}\boldsymbol{\alpha}^{\mathrm{T}} + \mathbf{\Psi})^{-1}\frac{\partial}{\partial \alpha_{ij}}\mathbf{S}(\boldsymbol{\alpha}\boldsymbol{\alpha}^{\mathrm{T}} + \mathbf{\Psi})^{-1}\right] = \mathrm{tr}\left[\mathbf{\Sigma}^{-1}\mathbf{S}\mathbf{\Sigma}^{-1}\frac{\partial \mathbf{\Sigma}}{\partial \alpha_{ij}}\right]$$

so that

$$\frac{\partial}{\partial \boldsymbol{\alpha}}\mathrm{tr}[\mathbf{S}(\boldsymbol{\alpha}\boldsymbol{\alpha}^{\mathrm{T}} + \mathbf{\Psi})^{-1}] = \mathbf{\Sigma}^{-1}\mathbf{S}\mathbf{\Sigma}^{-1}\boldsymbol{\alpha}$$

The *pr* derivatives of the likelihood, with respect to the loadings, can then be expressed as

$$\frac{\partial L}{\partial \boldsymbol{\alpha}} = -\mathbf{\Sigma}^{-1}\boldsymbol{\alpha} + \mathbf{\Sigma}^{-1}\mathbf{S}\mathbf{\Sigma}^{-1}\boldsymbol{\alpha} \tag{6.48}$$

and setting to zero we have the normal equations

$$(\hat{\mathbf{\Sigma}} - \mathbf{S})\hat{\mathbf{\Sigma}}^{-1}\hat{\boldsymbol{\alpha}} = \mathbf{0} \tag{6.49}$$

An alternative and a somewhat simpler derivation using Lemmas 6.2 and 6.3 has also been given by Joreskog (1977).

Lemma 6.2. Let f be a linear scalar function of a vector $\mathbf{X}$ and let $d\mathbf{X}$ denote the matrix of total differentials. If $df = \mathrm{tr}(\mathbf{c}\,d\mathbf{X}^{\mathrm{T}})$, where $\mathbf{c}$ depends on $\mathbf{X}$ but not on $d\mathbf{X}$, then $\partial f / \partial \mathbf{X} = \mathbf{c}$. $\qquad\square$

Lemma 6.3. Let $\mathbf{X}$ and $\mathbf{Y}$ be matrices. Then
    (i) $d(\mathbf{Y}\mathbf{X}) = d\mathbf{Y}\mathbf{X} + \mathbf{Y}\,d\mathbf{X}$
    (ii) $d\mathbf{X}^{-1} = -\mathbf{X}^{-1}\,d\mathbf{X}\mathbf{X}^{-1}$
    (iii) $d(\mathrm{tr}\,\mathbf{X}) = \mathrm{tr}(d\mathbf{X})$ $\qquad\square$

Using Eq. (6.42) we have the total derivative

$$dF = d\,\mathrm{tr}(\mathbf{\Sigma}^{-1}\mathbf{S}) - d\ln|\mathbf{\Sigma}^{-1}\mathbf{S}|$$

$$= \mathrm{tr}(d\mathbf{\Sigma}^{-1}\mathbf{S}) - \mathrm{tr}(\mathbf{S}^{-1}\mathbf{\Sigma}\,d\mathbf{\Sigma}^{-1}\mathbf{S})$$

$$= \mathrm{tr}[(\mathbf{S} - \mathbf{\Sigma})\,d\mathbf{\Sigma}^{-1}]$$

$$= \mathrm{tr}[(\mathbf{\Sigma} - \mathbf{S})\mathbf{\Sigma}^{-1}\,d\mathbf{\Sigma}\mathbf{\Sigma}^{-1}]$$

$$= \text{tr}[\boldsymbol{\Sigma}^{-1}(\boldsymbol{\Sigma} - \mathbf{S})\boldsymbol{\Sigma}^{-1}(d\boldsymbol{\alpha}\boldsymbol{\alpha}^{\mathrm{T}} + \boldsymbol{\alpha}\, d\boldsymbol{\alpha}^{\mathrm{T}})]$$

$$= 2\text{tr}[\boldsymbol{\Sigma}^{-1}(\boldsymbol{\Sigma} - \mathbf{S})\boldsymbol{\Sigma}^{-1}\boldsymbol{\alpha}\, d\boldsymbol{\alpha}^{\mathrm{T}}]$$

so that

$$\frac{\partial F}{\partial \boldsymbol{\alpha}} = 2\boldsymbol{\Sigma}^{-1}(\boldsymbol{\Sigma} - \mathbf{S})\boldsymbol{\Sigma}^{-1}\boldsymbol{\alpha}$$

which, equating to zero, leads to the normal equations (Eq. 6.49).

Equations (6.45) and (6.49) constitute the normal equations for the Lawley ML factor model. Since Eq. (6.49) does not have a unique solution, a further arbitrary constraint is required to render the model identifiable. The constraint which Lawley uses specifies that $\boldsymbol{\alpha}^{\mathrm{T}}\boldsymbol{\Psi}^{-1}\boldsymbol{\alpha} = \boldsymbol{\eta}$ be diagonal, and this fixes the basis of the solution space, thus yielding unique correlation loadings $\boldsymbol{\alpha}$. Since the constraint is arbitrary, it can always be removed once a unique initial solution is found, using any one of the orthogonal or oblique rotations (Section 5.3).

Lemma 6.4. Let $\boldsymbol{\Sigma} = \boldsymbol{\alpha}\boldsymbol{\alpha}^{\mathrm{T}} + \boldsymbol{\Psi}$ be a factor decomposition of $\boldsymbol{\Sigma}$. Then

$$\boldsymbol{\Sigma}^{-1} = \boldsymbol{\Psi}^{-1} - \boldsymbol{\Psi}^{-1}\boldsymbol{\alpha}(\mathbf{I} + \boldsymbol{\eta})^{-1}\boldsymbol{\alpha}^{\mathrm{T}}\boldsymbol{\Psi}^{-1} \qquad (6.50)$$

$\square$

Proof. Postmultiplying Eq. (6.50) by $\boldsymbol{\Sigma}$ we have

$$\boldsymbol{\Sigma}^{-1}\boldsymbol{\Sigma} = [\boldsymbol{\Psi}^{-1} - \boldsymbol{\Psi}^{1}\boldsymbol{\alpha}(\mathbf{I} + \boldsymbol{\eta})^{-1}\boldsymbol{\alpha}^{\mathrm{T}}\boldsymbol{\Psi}^{-1}](\boldsymbol{\alpha}\boldsymbol{\alpha}^{\mathrm{T}} + \boldsymbol{\Psi})$$

$$= \boldsymbol{\Psi}^{-1}(\boldsymbol{\alpha}\boldsymbol{\alpha}^{\mathrm{T}} + \boldsymbol{\Psi}) - \boldsymbol{\Psi}^{-1}\boldsymbol{\alpha}(\mathbf{I} + \boldsymbol{\eta})^{-1}\boldsymbol{\alpha}^{\mathrm{T}}\boldsymbol{\Psi}^{-1}(\boldsymbol{\alpha}\boldsymbol{\alpha}^{\mathrm{T}} + \boldsymbol{\Psi})$$

$$= \boldsymbol{\Psi}^{-1}\boldsymbol{\alpha}\boldsymbol{\alpha}^{\mathrm{T}} + \mathbf{I} - \boldsymbol{\Psi}^{-1}\boldsymbol{\alpha}(\mathbf{I} + \boldsymbol{\eta})^{-1}\boldsymbol{\alpha}^{\mathrm{T}}\boldsymbol{\Psi}^{-1}\boldsymbol{\alpha}\boldsymbol{\alpha}^{\mathrm{T}}$$

$$\quad - \boldsymbol{\Psi}^{-1}\boldsymbol{\alpha}(\mathbf{I} + \boldsymbol{\eta})^{-1}\boldsymbol{\alpha}^{\mathrm{T}}\boldsymbol{\Psi}^{-1}\boldsymbol{\Psi}$$

$$= \boldsymbol{\Psi}\boldsymbol{\alpha}\boldsymbol{\alpha}^{\mathrm{T}} + \mathbf{I} - \boldsymbol{\Psi}^{-1}\boldsymbol{\alpha}(\mathbf{I} + \boldsymbol{\eta})^{-1}\boldsymbol{\eta}\boldsymbol{\alpha}^{\mathrm{T}} - \boldsymbol{\Psi}^{-1}\boldsymbol{\alpha}(\mathbf{I} + \boldsymbol{\eta})^{-1}\boldsymbol{\alpha}^{\mathrm{T}}$$

$$= \boldsymbol{\Psi}^{-1}\boldsymbol{\alpha}\boldsymbol{\alpha}^{\mathrm{T}} + \mathbf{I} - \boldsymbol{\Psi}^{-1}\boldsymbol{\alpha}\boldsymbol{\alpha}^{\mathrm{T}}$$

$$= \mathbf{I}$$

where $\boldsymbol{\eta} = \boldsymbol{\alpha}^{\mathrm{T}}\boldsymbol{\Psi}^{-1}\boldsymbol{\alpha}$ is diagonal. $\square$

Postmultiplying Eq. (6.50) by $\boldsymbol{\alpha}$ leads to the further identity of Lemma 6.5.

Lemma 6.5. Let the conditions of Lemma 6.4 hold. Then

$$\mathbf{\Sigma}^{-1}\boldsymbol{\alpha} = \mathbf{\Psi}^{-1}\boldsymbol{\alpha}(\mathbf{I} + \boldsymbol{\eta})^{-1} \tag{6.51}$$

$\square$

Proof. Postmultiplying Eq. (6.50) by $\boldsymbol{\alpha}$ we have

$$\mathbf{\Sigma}^{-1}\boldsymbol{\alpha} = \mathbf{\Psi}^{-1}\boldsymbol{\alpha} - \mathbf{\Psi}^{-1}\boldsymbol{\alpha}(\mathbf{I} + \boldsymbol{\eta})^{-1}\boldsymbol{\alpha}^{\mathsf{T}}\mathbf{\Psi}^{-1}\boldsymbol{\alpha}$$

$$= \mathbf{\Psi}^{-1}\boldsymbol{\alpha} - \mathbf{\Psi}^{-1}\boldsymbol{\alpha}(\mathbf{I} + \boldsymbol{\eta})^{-1}\boldsymbol{\eta}$$

Since $\boldsymbol{\eta}$ is diagonal with nonzero diagonal elements, and post-multiplying by $\boldsymbol{\eta}^{-1}(\mathbf{I} + \boldsymbol{\eta})$ it follows that

$$\mathbf{\Sigma}^{-1}\boldsymbol{\alpha}\boldsymbol{\eta}^{-1}(\mathbf{I} + \boldsymbol{\eta}) = \mathbf{\Psi}^{-1}\boldsymbol{\alpha}\boldsymbol{\eta}^{-1}(\mathbf{I} + \boldsymbol{\eta}) - \mathbf{\Psi}^{-1}\boldsymbol{\alpha}$$

or

$$\mathbf{\Sigma}^{-1}\boldsymbol{\alpha}\boldsymbol{\eta}^{-1} + \mathbf{\Sigma}^{-1}\boldsymbol{\alpha} = \mathbf{\Psi}^{-1}\boldsymbol{\alpha}\boldsymbol{\eta}^{-1} + \mathbf{\Psi}^{-1}\boldsymbol{\alpha} - \mathbf{\Psi}^{-1}\boldsymbol{\alpha}$$

$$= \mathbf{\Psi}^{-1}\boldsymbol{\alpha}\boldsymbol{\eta}^{-1}$$

Postmultiplying by $\boldsymbol{\eta}$ yields

$$\mathbf{\Sigma}^{-1}\boldsymbol{\alpha} + \mathbf{\Sigma}^{-1}\boldsymbol{\alpha}\boldsymbol{\eta} = \mathbf{\Psi}^{-1}\boldsymbol{\alpha} \quad \text{or} \quad \mathbf{\Sigma}^{-1}\boldsymbol{\alpha} = \mathbf{\Psi}^{-1}\boldsymbol{\alpha}(\mathbf{I} + \boldsymbol{\eta})^{-1}$$

Substituting Eq. (6.51) into Eq. (6.49) we have

$$(\hat{\mathbf{\Sigma}} - \mathbf{S})\hat{\mathbf{\Psi}}^{-1}\hat{\boldsymbol{\alpha}}(\mathbf{I} + \hat{\boldsymbol{\eta}})^{-1} = \mathbf{0}$$

and postmultiplying by $(\mathbf{I} + \boldsymbol{\eta})$, substituting for $\mathbf{\Sigma}$, we obtain

$$\hat{\boldsymbol{\alpha}}\hat{\boldsymbol{\alpha}}^{\mathsf{T}}\hat{\mathbf{\Psi}}^{-1}\hat{\boldsymbol{\alpha}} + \hat{\boldsymbol{\alpha}} - \mathbf{S}\hat{\mathbf{\Psi}}^{-1}\hat{\boldsymbol{\alpha}} = \mathbf{0}$$

or

$$\mathbf{S}\hat{\mathbf{\Psi}}^{-1}\hat{\boldsymbol{\alpha}} = \hat{\boldsymbol{\alpha}}(\hat{\boldsymbol{\eta}} + 1) = \mathbf{0}$$

Premultiplying by $\hat{\mathbf{\Psi}}^{-1/2}$ and collecting terms we can write

$$[\hat{\mathbf{\Psi}}^{-1/2}\mathbf{S}\hat{\mathbf{\Psi}}^{-1/2} - (\hat{\boldsymbol{\eta}} + \mathbf{I})]\hat{\mathbf{\Psi}}^{-1/2}\hat{\boldsymbol{\alpha}} = \mathbf{0} \tag{6.52}$$

Let $\mathbf{S}^* = \hat{\mathbf{\Psi}}^{-1/2}\mathbf{S}\hat{\mathbf{\Psi}}^{-1/2}$ be a weighted sample covariance matrix. Then the final form normal equations can be expressed as

$$[\mathbf{S}^* - (\hat{\eta}_i + 1)\mathbf{I}]\hat{\mathbf{\Psi}}_i^{-1/2}\hat{\boldsymbol{\alpha}}_i = \mathbf{0} \qquad (i = 1, 2, \ldots, r) \tag{6.53}$$

where it is clear that $\hat{\eta}_i + 1$ is the ith latent root of $\mathbf{S}^*$. The loadings $\hat{\boldsymbol{\alpha}}_i$ are obtained from the latent vectors $\hat{\boldsymbol{\Psi}}_i^{1/2}\hat{\boldsymbol{\alpha}}_i$ such that $\hat{\boldsymbol{\alpha}}^T\hat{\boldsymbol{\Psi}}^{-1}\hat{\boldsymbol{\alpha}} = \hat{\boldsymbol{\eta}}$ is a diagonal matrix, a mathematical constraint introduced to render the loading co-efficients identifiable. The reciprocal proportionality model of the previous section and the ML model (Eq. 6.53) therefore differ mainly in the initial identifiability constraints used. For ML factor analysis when the parent distribution is normal, the constraint that $\boldsymbol{\alpha}^T\boldsymbol{\Psi}^{-1}\boldsymbol{\alpha}$ be diagonal is also equivalent to the a priori arbitrary constraint that the distribution of $\boldsymbol{\Phi}|\mathbf{X}$ have a diagonal covariance matrix (Section 6.2), that is, that the factors $\boldsymbol{\Phi}$ given the observed variables $\mathbf{X}$ be independent (Section 6.8). This may be seen from the fact that since $\boldsymbol{\alpha}^T\boldsymbol{\Psi}^{-1}\boldsymbol{\alpha}$ is diagonal, the off-diagonal terms are of the form

$$\sum_{i=1}^{p} \alpha_{ik}\alpha_{il}/\psi_i \tag{6.53a}$$

and must be zero, that is, when the variates are rescaled so that the residual variance of each is unity, the covariance between (k, l)th factor is zero. Also, the diagonal terms of the form α_{ik}^2/ψ_i represent the part of the variance of variate $\mathbf{X}_i$ which is explained by the kth factor φ_k. As for the principal components model, the diagonal elements of $\boldsymbol{\alpha}^T\boldsymbol{\Psi}^{-1}\boldsymbol{\alpha}$ are ranked in decreasing order so that φ_1 accounts for the maximal total variance, φ_2 accounts for the second maximal variance, and so forth until we reach φ_r, which explains the smallest portion of the total variance of the variates. $\square$

THEOREM 6.5.   Let $\mathbf{X} = (X_1, X_2, \ldots, X_p)^T$ represent a vector of p random variables such that $\mathbf{X} \sim N(\boldsymbol{\Phi}, \boldsymbol{\Sigma})$ and $\boldsymbol{\Sigma} = \boldsymbol{\alpha}\boldsymbol{\alpha}^T + \boldsymbol{\Psi}$. Then

(i) If there exists a unique diagonal matrix $\boldsymbol{\Psi}$ with positive diagonal elements such that r largest latent roots of $\boldsymbol{\Sigma}^* = \boldsymbol{\Psi}^{-1/2}\boldsymbol{\Sigma}\boldsymbol{\Psi}^{-1/2}$ are distinct and greater than unity, and the remaining $p - r$ roots are each unity, then $\boldsymbol{\alpha}$ can be uniquely defined such that

$$[\boldsymbol{\Sigma}^* - (\eta_i + 1)\mathbf{I}]\boldsymbol{\Psi}_i^{-1/2}\boldsymbol{\alpha}_i = \mathbf{0} \tag{6.54}$$

(ii) Let

$$\boldsymbol{\Sigma}^* = \boldsymbol{\Psi}^{-1/2}\boldsymbol{\Gamma}\boldsymbol{\Psi}^{-1/2} + \mathbf{I} \tag{6.55}$$

be the scaled model (Eq. 6.51). Then $\eta_i = \boldsymbol{\alpha}_i^T\boldsymbol{\Psi}_i^{-1}\boldsymbol{\alpha}_i$ $(i = 1, 2, \ldots, r)$ is the ith latent root of $\boldsymbol{\Psi}^{-1/2}\boldsymbol{\Gamma}\boldsymbol{\Psi}^{-1/2}$ such that $\eta_1 > \eta_2 > \cdots > \eta_r$ and $\eta_{r+1} = \eta_{r+2} = \cdots = \eta_p = 0$.

PROOF

(i) Since $\boldsymbol{\Sigma}^*$ is Grammian and nonsingular it can be expressed as

$$(\boldsymbol{\Sigma}^* - \lambda_i\mathbf{I})\boldsymbol{\Pi}_i = \mathbf{0} \tag{6.56}$$

for $\lambda_1 > \lambda_2 > \cdots \lambda_p$. When $\mathbf{\Psi}$ is unique, so are $\mathbf{\alpha}$ and $\mathbf{\Pi}$, and letting $\lambda_i = \eta_i + 1$ and $\mathbf{\Pi}_i = \mathbf{\Psi}_i^{-1/2}\mathbf{\alpha}_i$, we obtain Eq. (6.54).

(ii) Since $\mathbf{\Sigma}^* = \mathbf{\Psi}^{-1/2}\mathbf{\Gamma}\mathbf{\Psi}^{-1/2} + \mathbf{I}$ we have, from Eq. (6.52)

$$[(\mathbf{\Psi}^{-1/2}\mathbf{\Gamma}\mathbf{\Psi}^{-1/2} + \mathbf{I}) - (\eta_i + 1)]\mathbf{\Psi}_i^{-1/2}\mathbf{\alpha}_i = \mathbf{0} \quad \text{or}$$

$$(\mathbf{\Psi}^{-1/2}\mathbf{\Gamma}\mathbf{\Psi}^{-1/2} - \eta_i\mathbf{I})\mathbf{\Psi}\mathbf{X}_i^{-1/2}\mathbf{\alpha}_i = \mathbf{0} \tag{6.57}$$

Since $\mathbf{\Gamma} = \mathbf{\alpha}\mathbf{\alpha}^T$ is $(p \times p)$ and $\rho(\mathbf{\Gamma}) = r$, the last $p - r$ roots of $\mathbf{\Gamma}$ must equal zero, that is, corresponding to the zero roots of $\mathbf{\Psi}^{-1/2}\mathbf{\Gamma}\mathbf{\Psi}^{-1/2}$, the last $p - r$ roots of $\mathbf{\Sigma}^*$ are equal to unity.

Theorem 6.5 resembles Theorem 6.4 and indicates that the principle of maximum likelihood and normality are not essential to the derivation of Lawley's model. When the data are not sampled from $N(\mathbf{\mu}, \mathbf{\Sigma})$, the normal equations (Eq. 6.53) still yield optimal estimators in the least squares sense, although statistical testing is usually precluded. However, the latent roots of $\mathbf{\Sigma}^*$ must exceed unity if $\mathbf{\Gamma}$ is to be positive definite. Lawley's model can also be compared to a principal component decomposition of the weighted covariance matrix $\mathbf{\Sigma}^*$ where both $\mathbf{\Sigma}$ and the weights $\mathbf{\Psi}^{-1/2}$ are estimated by an iterative sequence such as the following:

1. Compute $\mathbf{S}$ (or $\mathbf{X}^T\mathbf{X}$), extract the first r principal components, and using Eq. (6.12) compute an estimate of $\mathbf{\Psi}$, say $\hat{\mathbf{\Psi}}_{(1)}$.

2. Construct the weighted covariance matrix $\mathbf{S}_{(1)} = \hat{\mathbf{\Psi}}_{(1)}^{-1/2}\mathbf{S}\hat{\mathbf{\Psi}}_{(1)}^{-1/2}$, compute principal components and obtain the second-order estimate of $\mathbf{\Psi}$, say $\hat{\mathbf{\Psi}}_{(2)}$.

3. Repeat the procedure until residual errors approach stable values $\hat{\mathbf{\Psi}}$. Now let

$$\mathbf{L}_{(r)} = \begin{bmatrix} l_1 & & & & \mathbf{0} \\ & l_2 & & & \\ & & \cdot & & \\ & & & \cdot & \\ \mathbf{0} & & & & l_r \end{bmatrix}$$

and let $\mathbf{P}_1, \mathbf{P}_2, \ldots, \mathbf{P}_r$ be the first r latent roots and latent vectors of $\mathbf{S}^* = \hat{\mathbf{\Psi}}^{-1/2}\mathbf{S}\hat{\mathbf{\Psi}}^{-1/2}$. Then

$$\mathbf{S}^*\mathbf{P}_{(r)} = \mathbf{P}_{(r)}\mathbf{L}_{(r)} \tag{6.58}$$

where $\mathbf{P}_{(r)}$ is the $(p \times r)$ matrix of latent vectors of $\mathbf{S}^*$. The ML estimator $\hat{\mathbf{\eta}}$ is obtained as

$$\hat{\mathbf{\eta}} = \mathbf{L}_{(r)} - \mathbf{I} \tag{6.59}$$

where

$$\hat{\boldsymbol{\alpha}}^{\mathrm{T}}\hat{\boldsymbol{\Psi}}^{-1}\hat{\boldsymbol{\alpha}} = \hat{\boldsymbol{\eta}}$$

and

$$\hat{\boldsymbol{\Psi}}^{-1/2}\hat{\boldsymbol{\alpha}} = \mathbf{P}_{(r)}\hat{\boldsymbol{\eta}}^{1/2}$$

$$= \mathbf{P}_{(r)}(\mathbf{L}_{(r)} - \mathbf{I})$$

so that

$$\hat{\boldsymbol{\alpha}} = \hat{\boldsymbol{\Psi}}^{1/2}\mathbf{P}_{(r)}(\mathbf{L}_{(r)} - \mathbf{I}) \tag{6.60}$$

is the $(p \times r)$ matrix of correlation loadings.

The algebraic derivation given above is mainly useful for reasons of exposition, since the numerical computations can be implemented in a number of different ways—see, for example, Johnson and Wichern (1982). A difficulty with the ML factor model is that the likelihood function (Eq. 6.42) may not possess a true maximum, which can be due to violations of conditions such as the strict positiveness of the error terms,* multivariate normality, or more generally unimodality. Even when the likelihood does possess a unique global maximum, a particular numerical algorithm may fail to locate it, since the iterative sequence may converge to a local maximum instead. This underlines the importance of having to verify assumptions such as multivariate normality of the population (see Section 4.6). The covariance matrix must also be nonsingular so that, for example, the number of variables cannot exceed the number of observations. An advantage of the model however lies in the improved properties of the estimator $\hat{\boldsymbol{\alpha}}$, such as greater efficiency and reduced bias, as well as invariance to the units of measurement.

THEOREM 6.6. Let $\mathbf{X}$ be a random vector with nonsingular covariance matrix $\boldsymbol{\Sigma}$. Then the weighted covariance matrix $\boldsymbol{\Sigma}^* = \boldsymbol{\Psi}^{-1/2}\boldsymbol{\Sigma}\boldsymbol{\Psi}^{-1/2}$ is invariant under a scale transformation of $\mathbf{X}$.

PROOF. Let $\mathbf{H}$ be a diagonal matrix such that $\mathbf{Z} = \mathbf{H}\mathbf{X}$. The matrix $\mathbf{Z}$ represents a change of scale of the random variables. The factor decomposi-

---

* For the case when some residual error variances are identically zero see Mardia et al., 1979, p. 277.

tion is then

$$\mathbf{H\Sigma H}^{\mathrm{T}} = \mathbf{H\Gamma H}^{\mathrm{T}} + \mathbf{H\Psi H}^{\mathrm{T}}$$
$$= \mathbf{H\Gamma H}^{\mathrm{T}} + (\mathbf{H\Psi}^{1/2})(\mathbf{H\Psi}^{1/2})^{\mathrm{T}}$$

and weighting by (rescaled) error variances yields

$$(\mathbf{H\Psi}^{1/2})^{-1}\mathbf{H\Sigma H}^{\mathrm{T}}(\mathbf{\Psi}^{1/2}\mathbf{H}^{\mathrm{T}})^{-1} = (\mathbf{H\Psi}^{1/2})^{-1}\mathbf{H\Gamma H}(\mathbf{\Psi}^{1/2}\mathbf{H}^{\mathrm{T}})^{-1} + \mathbf{I}$$

or

$$\mathbf{\Psi}^{-1/2}\mathbf{\Sigma}\mathbf{\Psi}^{-1/2} = \mathbf{\Psi}^{-1/2}\mathbf{\Gamma}\mathbf{\Psi}^{-1/2} + \mathbf{I}$$

the original factor model of the unscaled variables.

6.4.3 The Rao Canonical Correlation Factor Model

An alternative least squares derivations of the ML factor model has been provided by Rao (1955). The derivation is of interest in that it represents an alternative view of the factor model, and demonstrates the least squares optimality of Lawley's model without having to assume multivariate normality. Let $\mathbf{X}$ be a $(p \times 1)$ vector of random variables and $\mathbf{X}^* = E(\mathbf{X}) = \alpha\mathbf{\Phi}$, that is, that part which is predicted by the set of r common factors. Then $\mathbf{X} = \alpha\mathbf{\Phi} + \boldsymbol{\epsilon} = \mathbf{X}^* + \boldsymbol{\epsilon}$. We wish to compute coefficients that maximize the correlation(s) between $\mathbf{X} = (X_1, X_2, \ldots, X_p)^{\mathrm{T}}$ and $\mathbf{X}^* = (X_1^*, X_2^*, \ldots, X_p^*)^{\mathrm{T}}$. From Section 5.51 we know that the maximum correlation between $\mathbf{X}$ and $\mathbf{X}^*$ is the highest canonical correlation between the linear combinations $\mathbf{U} = \boldsymbol{\beta}^{\mathrm{T}}\mathbf{X}$ and $\mathbf{V} = \boldsymbol{\gamma}^{\mathrm{T}}\mathbf{X}^*$ where $\boldsymbol{\beta}$ and $\boldsymbol{\gamma}$ are unknown coefficients to be determined. We have

$$\mathrm{var}(U) = E(U^2) = E(\boldsymbol{\beta}^{\mathrm{T}}\mathbf{X}\mathbf{X}^{\mathrm{T}}\boldsymbol{\beta}) = \boldsymbol{\beta}^{\mathrm{T}}E(\mathbf{X}\mathbf{X}^{\mathrm{T}})\boldsymbol{\beta} = \boldsymbol{\beta}^{\mathrm{T}}\mathbf{\Sigma}\boldsymbol{\beta}$$

$$\mathrm{var}(V) = E(V^2) = E(\boldsymbol{\gamma}^{\mathrm{T}}\mathbf{X}^*\mathbf{X}^*\boldsymbol{\gamma}^{\mathrm{T}}) = \boldsymbol{\gamma}^{\mathrm{T}}E(\mathbf{X}^*\mathbf{X}^*{}^{\mathrm{T}})\boldsymbol{\gamma} = \boldsymbol{\gamma}^{\mathrm{T}}\mathbf{\Gamma}\boldsymbol{\gamma}$$

$$\mathrm{cov}(U, V) = E(UV) = E(\boldsymbol{\beta}^{\mathrm{T}}\mathbf{X}\mathbf{X}^*{}^{\mathrm{T}}) = \boldsymbol{\beta}^{\mathrm{T}}E(\mathbf{X}\mathbf{X}^*{}^{\mathrm{T}})\boldsymbol{\gamma} = \boldsymbol{\beta}^{\mathrm{T}}\mathbf{\Gamma}\boldsymbol{\gamma}$$

since

$$\mathbf{X}\mathbf{X}^*{}^{\mathrm{T}} = (\alpha\mathbf{\Phi} + \boldsymbol{\epsilon})(\alpha\mathbf{\Phi})^{\mathrm{T}}$$
$$= \alpha\mathbf{\Phi}\mathbf{\Phi}^{\mathrm{T}}\alpha^{\mathrm{T}} + \boldsymbol{\epsilon}\mathbf{\Phi}^{\mathrm{T}}\alpha$$
$$= \alpha\alpha^{\mathrm{T}}$$
$$= \mathbf{X}^*\mathbf{X}^*{}^{\mathrm{T}} \tag{6.61}$$

where $\mathbf{\Phi}\mathbf{\Phi}^{\mathrm{T}} = \mathbf{I}$ and $\boldsymbol{\epsilon}\mathbf{\Phi}^{\mathrm{T}} = \mathbf{0}$ by assumption. We thus maximize $\mathrm{cov}(U, V) = \boldsymbol{\beta}^{\mathrm{T}}\mathbf{\Gamma}\boldsymbol{\gamma}$ subject to the constraint of unit variance of U and V. From the canonical correlation model we have $\boldsymbol{\Sigma} = \boldsymbol{\Sigma}_{11}$, $\boldsymbol{\Sigma}_{12} = \boldsymbol{\Sigma}_{22} = \boldsymbol{\Sigma}_{21} = \boldsymbol{\Gamma}$, and

$$(\boldsymbol{\Sigma}_{11}^{-1}\boldsymbol{\Sigma}_{12}\boldsymbol{\Sigma}_{22}^{-1}\boldsymbol{\Sigma}_{21} - \lambda_i^2\mathbf{I})\boldsymbol{\Pi}_i = (\boldsymbol{\Sigma}^{-1}\boldsymbol{\Gamma}\boldsymbol{\Gamma}^{-1}\boldsymbol{\Gamma} - \lambda^2\mathbf{I})\boldsymbol{\Pi}_i$$

$$= (\boldsymbol{\Sigma}^{-1}\boldsymbol{\Gamma} - \lambda^2\mathbf{I})\boldsymbol{\Pi}_i$$

or

$$(\boldsymbol{\Gamma} - \boldsymbol{\Sigma}\lambda_i^2)\boldsymbol{\Pi}_i = \mathbf{0} \tag{6.62}$$

for $i = 1, 2, \ldots, r$ maximal correlations. The λ^2 are solutions of the determinental equation

$$|\boldsymbol{\Gamma} - \boldsymbol{\Sigma}\lambda_i^2| = |(\boldsymbol{\Sigma} - \boldsymbol{\Psi}) - \lambda_i^2\boldsymbol{\Sigma}|$$

$$= |(1 - \lambda_i^2)\boldsymbol{\Sigma} - \boldsymbol{\Psi}|$$

$$= \left|\boldsymbol{\Sigma} - \left(\frac{1}{1 - \lambda_i^2}\right)\boldsymbol{\Psi}\right|$$

$$= |\boldsymbol{\Sigma} - \xi_i\boldsymbol{\Psi}|$$

$$= 0$$

where we can let $\boldsymbol{\Pi}_1 = \boldsymbol{\beta}$ for maximal correlation and $\xi_i = 1/(1 - \lambda_i^2)$. Equation (6.62) can thus be expressed as

$$(\boldsymbol{\Sigma} - \xi_i\boldsymbol{\Psi})\boldsymbol{\beta}_i = \mathbf{0} \tag{6.64}$$

for $i = 1, 2, \ldots, r$ common factors. The normal equations (Eq. 6.64) can be reduced to those of Eq. (6.54) except for the normalization of the latent vectors, and the canonical correlation factor model (Eq. 6.64) is thus also a solution to Lawley's normal equations. The latent roots and latent vectors ξ_i and $\boldsymbol{\beta}_i$ of $\boldsymbol{\Sigma}$ (in the metric of $\boldsymbol{\Psi}$) thus maximize the correlation loading coefficients $\boldsymbol{\alpha}$ between the variates and the common factors, an optimality property not possessed by principal components when the residual errors are heteroscedastic. For a multivariate normal sample the solutions of

$$(\mathbf{S} - \hat{\xi}_i\hat{\boldsymbol{\Psi}})\mathbf{P}_i = \mathbf{0} \tag{6.65}$$

are also ML estimators. For the special case when $\boldsymbol{\Psi} = \sigma^2\mathbf{I}$ we obtain the PC model, and if the $(p \times p)$ matrix $\boldsymbol{\Sigma}$ is singular of rank $r < p$ we can write

$$\left|\boldsymbol{\Sigma} - \frac{1}{(1 - \lambda_i^2)}\boldsymbol{\Psi}\right| = \begin{cases} \left|\boldsymbol{\Sigma} - \dfrac{\sigma^2}{(1 - \lambda_i^2)}\mathbf{I}\right| & (1 \le i \le r) \\[2ex] |\boldsymbol{\Sigma} - \sigma^2\mathbf{I}| & (r < i \le p) \end{cases}$$

6.4.4 The Generalized Least Squares Model

The least squares or the so-called "minres" model (Eq. 6.23) can be generalized to minimize the weighted least squares criterion

$$
\begin{aligned}
\mathbf{G} &= \mathrm{tr}[\mathbf{S}^{-1}(\mathbf{S} - \boldsymbol{\Sigma})\mathbf{S}^{-1}(\mathbf{S} - \boldsymbol{\Sigma})] \\
&= \mathrm{tr}[\mathbf{S}^{-1}(\mathbf{S} - \boldsymbol{\Sigma})]^2 \\
&= \mathrm{tr}(\mathbf{I} - \mathbf{S}^{-1}\boldsymbol{\Sigma})^2
\end{aligned}
\tag{6.66}
$$

where $\mathbf{S}^{-1}$ is the weight matrix (Joreskog and Goldberger, 1972; Joreskog, 1977). The rationale for the criterion parallels that of generalized least squares regression for heteroscedastic error terms. We have the total differentials

$$
\begin{aligned}
d\mathbf{G} &= d[\mathrm{tr}(\mathbf{S}^{-1}\boldsymbol{\Sigma} - \mathbf{I})^2] \\
&= \mathrm{tr}[d(\mathbf{S}^{-1}\boldsymbol{\Sigma} - \mathbf{I})^2] \\
&= 2\,\mathrm{tr}[(\mathbf{S}^{-1}\boldsymbol{\Sigma} - \mathbf{I})\,d(\mathbf{S}^{-1}\boldsymbol{\Sigma} - \mathbf{I})] \\
&= 2\,\mathrm{tr}[(\mathbf{S}^{-1}\boldsymbol{\Sigma} - \mathbf{I})\mathbf{S}^{-1}\,d\boldsymbol{\Sigma}]
\end{aligned}
$$

and keeping $\boldsymbol{\Psi}$ fixed,

$$
\begin{aligned}
d\boldsymbol{\Sigma} &= d(\boldsymbol{\alpha}\boldsymbol{\alpha}^{\mathrm{T}} + \boldsymbol{\Psi}) \\
&= \boldsymbol{\alpha}(d\boldsymbol{\alpha}^{\mathrm{T}}) + (d\boldsymbol{\alpha})\boldsymbol{\alpha}^{\mathrm{T}} \\
&= 2\boldsymbol{\alpha}(d\boldsymbol{\alpha}^{\mathrm{T}})
\end{aligned}
$$

so that

$$
d\mathbf{G} = 4\,\mathrm{tr}[(\mathbf{S}^{-1}\boldsymbol{\Sigma} - \mathbf{I})\mathbf{S}^{-1}\boldsymbol{\alpha}(d\boldsymbol{\alpha}^{\mathrm{T}})]
$$

Using Lemma 6.2 we arrive at

$$
\frac{\partial \mathbf{G}}{\partial \boldsymbol{\alpha}} = 4\mathbf{S}^{-1}(\boldsymbol{\Sigma} - \mathbf{S})\mathbf{S}^{-1}\boldsymbol{\alpha}
$$

and setting to zero yields the normal equations

$$
\mathbf{S}^{-1}(\hat{\boldsymbol{\Sigma}} - \mathbf{S})\mathbf{S}^{-1}\hat{\boldsymbol{\alpha}} = \mathbf{0}
$$

$$
\hat{\boldsymbol{\Sigma}}\mathbf{S}^{-1}\hat{\boldsymbol{\alpha}} = \hat{\boldsymbol{\alpha}}
\tag{6.67}
$$

Premultiplying by $\hat{\boldsymbol{\Sigma}}^{-1}$ and rearranging terms leads to an expression which is equivalent to Lawley's normal equations (Eq. 6.49; see also Exercise 6.6). The conditional minimum of Eq. (6.67), given $\boldsymbol{\Psi}$, is however different from the conditional minimum of Lawley's ML model (see Joreskog, 1977). Nevertheless, the loadings obtained using Eq. (6.67) are also consistent estimators of $\boldsymbol{\alpha}$ and are asymptotically equivalent to the Lawley/Joreskog estimator (Browne, 1974)—in fact using an alternative weighting scheme at each iteration it is possible to force the solutions of Eq. (6.67) to converge to the Lawley/Joreskog solution (Lee and Jennrich, 1979). The generalized least squares model (Eq. 6.67) can also be compared to the ML factor model in terms of the diagonality constraint which is imposed to obtain identifiability of the model. Thus using Lemma 6.4 it can be shown (Exercise 6.7) that Eq. (6.67) is equivalent to

$$(\boldsymbol{\Psi}^{1/2}\mathbf{S}^{-1}\boldsymbol{\Psi}^{1/2})\boldsymbol{\Psi}^{-1/2}\boldsymbol{\alpha} = \boldsymbol{\Psi}^{-1/2}\boldsymbol{\alpha}(\mathbf{I} + \boldsymbol{\alpha}^{\mathrm{T}}\boldsymbol{\Psi}^{-1}\boldsymbol{\alpha})^{-1} \tag{6.68}$$

where, as for the ML model, we can take $\boldsymbol{\alpha}^{\mathrm{T}}\boldsymbol{\Psi}^{-1}\boldsymbol{\alpha}$ to be diagonal. The columns of $\boldsymbol{\Psi}^{-1/2}\boldsymbol{\alpha}$ are thus latent vectors of $\boldsymbol{\Psi}^{1/2}\mathbf{S}^{-1}\boldsymbol{\Psi}^{1/2}$, and nonzero diagonals $(\mathbf{I} + \boldsymbol{\alpha}^{\mathrm{T}}\boldsymbol{\Psi}^{-1}\boldsymbol{\alpha})^{-1}$ are the latent roots. Since the conditional minimum of $\mathbf{G}$ for given $\boldsymbol{\Psi}^{1/2}$ is obtained by choosing the smallest roots of $\boldsymbol{\Psi}^{1/2}\mathbf{S}^{-1}\boldsymbol{\Psi}^{1/2}$ the generalized least squares criterion can also be viewed as one which choses $\boldsymbol{\alpha}^{\mathrm{T}}\mathbf{S}^{-1}\boldsymbol{\alpha}$ to be diagonal.

6.5 OTHER WEIGHTED FACTOR MODELS

The ML and least squares estimators considered in Section 6.4 are not the only examples of weighted factor models used in practice. More specialized weighted models are also at times used to handle particular difficulties found in certain disciplines.

6.5.1 The Double Heteroscedastic Model

In the weighting schemes employed in Section 6.4 the residual error matrix $\boldsymbol{\Psi}$ is generally taken to be heteroscedastic, in the sense that two or more variables are assumed to have different error variances. The observations for each variable however are assumed to have equal error variance. A more general model is also possible, where the error variances of both the variables and observations are not constant, that is, for the (i, j)th element of $\boldsymbol{\epsilon}$ we have

$$\mathrm{var}(\epsilon_{ij}) = \sigma_{ij}^{2} \tag{6.69}$$

where $\mathrm{cov}(\epsilon_{ij}, \epsilon_{kl}) = 0$ for $i \neq k$ and $j \neq l$ so that the error terms are uncorrelated between the variables and the observations. Such an error structure, which can frequently exist for both experimental and nonexperimental data, has been incorporated into the factor model by Cochran and Horne (1977), who report superior results to those obtained using the principal components model of Chapter 3. As an example, consider Beer's Law, where $\mathbf{Y}$ is a $(n \times p)$ matrix consisting of r responses (molar absorbance, fluorescence, or chemiluminence) at n wavelength channels. Assuming the response at each wavelength channel to be a linear function of the concentration of each detectable species (plus error), we have the sample factor model $\mathbf{X} = \mathbf{FA} + \mathbf{E}$ where $\mathbf{F}$ and $\mathbf{A}$ are the $(n \times r)$ concentration and $(r \times p)$ static spectrum matrices, respectively. The objective is then to find the rank of the true data matrix $\mathbf{Y}$ where the variance of the error term varies across both spectra and wavelength channel. The doubly heteroscedastic error specification (Eq. 6.69) therefore implies a differential impact on the error term for the rows and columns of $\mathbf{Y}$. A solution is to weight the zero-mean data matrix $\mathbf{X}$ by the two diagonal matrices $\mathbf{C}^{-1/2}$ and $\mathbf{D}^{-1/2}$, so that errors assume the homoscedastic form $\sigma^2 \mathbf{I}$ where

$$\mathbf{C} = \mathrm{diag}(c_1, c_2, \ldots, c_n)$$

$$\mathbf{D} = \mathrm{diag}(d_1, d_2, \ldots, d_p)$$

We have

$$\mathbf{X}^* = \mathbf{C}^{-1/2}\mathbf{X}\mathbf{D}^{-1/2} = \mathbf{C}^{-1/2}\mathbf{FAD}^{-1/2} + \mathbf{C}^{-1/2}\mathbf{ED}^{-1/2}$$

or

$$\mathbf{X}^{*\mathrm{T}}\mathbf{X}^* = (\mathbf{D}^{-1/2}\mathbf{A}^{\mathrm{T}}\mathbf{F}^{\mathrm{T}}\mathbf{C}^{-1/2})(\mathbf{C}^{-1/2}\mathbf{FAD}^{-1/2}) + (\mathbf{D}^{-1/2}\mathbf{E}^{\mathrm{T}}\mathbf{C}^{-1/2}\mathbf{C}^{-1/2}\mathbf{ED}^{-1/2})$$

$$= \mathbf{D}^{-1/2}\mathbf{A}^{\mathrm{T}}\mathbf{F}^{\mathrm{T}}\mathbf{C}^{-1}\mathbf{FAD}^{-1/2} + \sigma^2\mathbf{I} \tag{6.70}$$

As for the reciprocal proportionality model (Section 6.4.1) the latent roots of the weighted covariance matrix (ignoring degrees of freedom) of Eq. (6.70) are given by the determinantal equation

$$|\mathbf{X}^{*\mathrm{T}}\mathbf{X}^* - \lambda_i\mathbf{I}| = |(\mathbf{D}^{-1/2}\mathbf{A}^{\mathrm{T}}\mathbf{F}^{\mathrm{T}}\mathbf{C}^{-1}\mathbf{FAD}^{-1/2} + \sigma^2\mathbf{I}) - \lambda_i\mathbf{I}|$$

$$= |(\mathbf{D}^{-1/2}\mathbf{A}^{\mathrm{T}}\mathbf{F}^{\mathrm{T}}\mathbf{C}^{-1}\mathbf{FAD}^{-1/2}) + (\sigma^2 - \lambda_i)\mathbf{I}|$$

$$= 0 \tag{6.71}$$

The covariance matrix of the true portion $\mathbf{D}^{-1/2}\mathbf{A}^{\mathrm{T}}\mathbf{F}^{\mathrm{T}}\mathbf{C}^{-1}\mathbf{FAD}^{-1/2}$ of the data is thus assumed to be of rank r, where $r < p < n$ so that $\mathbf{X}^{*\mathrm{T}}\mathbf{X}^*$ possesses r roots $l_1 > l_2 > \cdots > l_r$ and $(p - r)$ isotropic roots equal to σ^2. If the weights $\mathbf{C}$ and $\mathbf{D}$ are not known, the solution proceeds iteratively for successive choices of r, observing for each choice whether the last $(p - r)$ roots tend to the isotropic structure. Alternatively if σ^2 is known, for example $\sigma^2 = 1$, this fixes the number r of nonisotropic roots. The residuals e_{ij} can also be plotted on a graph (e.g., against time or wavelength in the chemical application) to examine any patterns that may occur, implying an incorrect choice of r since this yields a nonisotropic structure of the last $p - r$

roots. Also, let Q_r denote the sum of squares of the residuals when r common factors are used. Then when the true rank of $\mathbf{X}$ is r, the quantity

$$\hat{\sigma}^2 = \frac{Q_r}{(n-r)(p-r)} \tag{6.72}$$

should lie in the interval $[\sigma^2 + \delta, \sigma^2]$ for some small $\delta > 0$.

6.5.2 Psychometric Models

Since psychology and other social sciences have traditionally been frequent users of factor models, a number of variants of weighted factor analysis have appeared in these disciplines, of which the so-called "scaled image" and "alpha" models are well known. The scaled image model is defined by the normal equations

$$[\hat{\mathbf{\Psi}}^{-1/2}\hat{\mathbf{\Gamma}}\hat{\mathbf{\Psi}}^{-1/2} - \hat{\eta}_i\mathbf{I}]\hat{\mathbf{\Pi}}_i = 0 \qquad (i = 1, 2, \ldots, r) \tag{6.73}$$

where $\hat{\mathbf{\Psi}}$ and $\hat{\mathbf{\Gamma}}$ are first-stage estimates of Eq. (6.5). Although the model represents an attempt to control for bias arising out of error terms, it is not in fact optimal, either in the maximum likelihood or the least squares sense, unless further iteration is carried out, in which case it approaches the ML model (Eq. 6.54).

Another weighting scheme is to use "communality" (explained variance) as weights for the variables. The so-called alpha factor model is then based on the latent roots and latent vectors of the matrix

$$[\mathrm{diag}(\hat{\mathbf{\Gamma}})]^{-1/2}\hat{\mathbf{\Gamma}}[\mathrm{diag}(\hat{\mathbf{\Gamma}})]^{-1/2} \tag{6.74}$$

that is, the explained part $\hat{\mathbf{\Gamma}}$ of the correlation matrix is weighted inversely by the communalities. Statistical or sampling properties of this model however are not known, and the rationale for the weighting scheme appears to be based on a particular line of psychometric reasoning. For Monte Carlo comparisons between the alpha and other factor models see Velicer (1977), Acito and Anderson (1980), and Acito et al. (1980). Computer programs for the alpha model (as well as the minres and ML models) may be found in Derflinger (1979).

6.6 TESTS OF SIGNIFICANCE

When $\mathbf{X} \sim N(0, \mathbf{\Sigma})$, the hypothesis of r common factors can be tested using large sample theory, that is, we can test the null hypothesis

$$H_0: \mathbf{\Sigma} = \boldsymbol{\alpha}\boldsymbol{\alpha}^{\mathrm{T}} + \mathbf{\Psi}$$

against the alternative

$$H_a: \Sigma \neq \alpha\alpha^{\mathrm{T}} + \Psi$$

to determine whether Σ contains $r > 0$ common factors (plus a diagonal error variance matrix) versus the alternative that Σ is an arbitrary covariance matrix. The methodology and types of tests available for factor models are similar in scope to those used for principle components (Chapter 4).

6.6.1 The Chi-Squared Test

The classic test for r common ML factors is the large sample chi-squared test. First, consider the test for complete independence of the variates (Lawley and Maxwell, 1971). Under H_0 the likelihood function (Section 6.4) is

$$L(\omega) = c|\hat{\Sigma}|^{-n/2} \exp\left[-\frac{n}{2} \operatorname{tr}(\hat{\Sigma}^{-1}S)\right]$$

$$= c|\hat{\alpha}\hat{\alpha}^{\mathrm{T}} + \hat{\Psi}|^{-n/2} \exp[\operatorname{tr}(\hat{\alpha}\hat{\alpha}^{\mathrm{T}} + \hat{\Psi})S] \qquad (6.75)$$

and under H_a we have

$$L(\Omega) = c|S|^{-n/2} \exp\left[-\frac{n}{2} \operatorname{tr}(S^{-1}S)\right]$$

$$= c|S|^{-n/2} \exp\left(-\frac{np}{2}\right) \qquad (6.76)$$

since under $H_a \hat{\Sigma} = S$, the sample estimate. The likelihood ratio statistic for testing H_0 is $\lambda = L(\omega)/L(\Omega)$ where $-2\ln\lambda$ is asymptotically chi-squared.

For large n we have

$$\chi^2 \simeq -2\ln\lambda = -2\ln L(\omega) + 2\ln L(\Omega)$$

$$= n[\ln|\hat{\Sigma}| + \operatorname{tr}(S\hat{\Sigma}^{-1}) - \ln|S| - p]$$

Since from Eq. (6.45) diag $\hat{\Sigma}$ = diag S, the criterion reduced to

$$\chi^2 \simeq n[\ln|\hat{\Sigma}| + p - \ln|S| - p]$$

$$= n \ln\left(\frac{|\hat{\Sigma}|}{|S|}\right) \qquad (6.77)$$

Since the test for complete independence is equivalent to testing whether

$r = 0$ we have, from Eq. (6.77) for large samples

$$\chi^2 = -n \ln \left(\frac{|\mathbf{S}|}{|\hat{\boldsymbol{\Sigma}}|} \right) = n \ln |\mathbf{R}| \tag{6.78}$$

since when $r = 0$ we have $\hat{\boldsymbol{\Sigma}} = \text{diag} (\mathbf{S})$ and the determinantal ratio equals $|\mathbf{R}|$, the determinant of the correlation matrix. The statistic (Eq. 6.78) is valid for a large sample size n. For not-so-large samples the chi-squared approximation is improved if n is replaced by $(n - 1) - 1/6(2p + 5)$ (Box, 1949) for $d = (1/2)[(p - r)^2 - (p + r)]$ degrees of freedom, the difference between the number of parameters in $\boldsymbol{\Sigma}$ and the number of linear constraints imposed by the null hypothesis. In practice the likelihood ratio test will often indicate a larger number of common factors than what may be interpreted in a meaningful setting, and it is at times best to carry out a rotation of the loadings before deciding on the value of r. In any case, the number of common factors may not exceed that number for which d is nonpositive (Section 6.1).

When H_0 is accepted, that is, when the asymptotic chi-squared statistic indicates the presence of at least a single common factor, the test is repeated for larger values of r since the objective now becomes to estimate the "correct" number of common factors $\boldsymbol{\phi}_1, \boldsymbol{\phi}_2, \ldots, \boldsymbol{\phi}_r$ $(1 \le r < p)$. As was seen in Theorem 6.5 the latent roots of $\boldsymbol{\Sigma}^* = \boldsymbol{\Psi}^{-1/2} \boldsymbol{\Sigma} \boldsymbol{\Psi}^{-1/2}$ equal unity when those of the weighted or "reduced" correlation matrix $\boldsymbol{\Psi}^{-1/2} \boldsymbol{\Gamma} \boldsymbol{\Psi}^{-1/2}$ equal zero. The appropriate test for the existence of $0 < r < p$ common factors is therefore equivalent to testing whether the last $p - r$ roots $(\hat{\eta}_i + 1)$ of $\hat{\boldsymbol{\Psi}}^{-1/2} \mathbf{S} \boldsymbol{\Psi}^{-1/2}$ differ from unity. The rationale for the test is similar to that encountered in Section 4.3 when testing for the existence of r principle components. The criterion for testing for the existence of r common factors is given by (see Lawley and Maxwell, 1971)

$$\chi^2 = -\left[n - 1 - \frac{1}{6}(2p + 4r + 5) \right] \sum_{i=r+1}^{p} \ln(\eta_i + 1) \tag{6.78a}$$

since it is the low-order latent roots that provide a measure of the goodness of fit of the factor model to the data.

Several comments are in order when using the likelihood ratio statistic (Eq. 6.78a). First, the statistic is asymptotically chi-squared only when $\mathbf{X} \sim N(\mathbf{0}, \boldsymbol{\Sigma})$, and requires at least moderately large samples—Bartlett (1950), for example, suggests $n - r \ge 50$. Second, as Geweke and Singleton (1980) have pointed out, Wilks' (1946) theorem is applicable only under certain regularity conditions. Let $\boldsymbol{\theta}$ denote the vector of free parameters in $\boldsymbol{\alpha} \boldsymbol{\alpha}^T + \boldsymbol{\Psi}$ and let $\hat{\boldsymbol{\theta}}_n$ be the vector of ML estimators in a sample of size n. Then Wilks' (1946) theorem requires that the distribution of $\sqrt{n}(\hat{\boldsymbol{\theta}}_n - \boldsymbol{\theta})$ approaches a normal with zero mean and positive definite covariance matrix as $n \to \infty$, that is, $\hat{\boldsymbol{\theta}}_n$ must be a consistent estimator of $\boldsymbol{\theta}$. This is so only when we do not have a Heywood case, that is, when $\hat{\boldsymbol{\Psi}} > 0$ and when the number of common factors has been correctly estimated. When these regularity conditions do not hold the asymptotic ML theory can be misleading, particularly in a small sample. Finally, the chi-squared test is applicable only for covariance matrices and, strictly speaking, does not apply to correlation matrices.

6.2.2 Information Criteria

Information criteria described in Section 4.3.5 can also be used to test for the "correct" or "optimum" number of factors. Indeed, the AIC statistic was initially developed for time series and the ML factor model (Akaike 1974; 1987). The main idea is to use functions which are similar in spirit to Mallows' $\mathbf{C}_p$ statistic, which penalizes an excessive number of fitted parameters something which the chi-squared test does not do.

In what follows we base our discussion on Bozdogan (1987). Consider the absolutely continuous random vector $\mathbf{X}$, which has probability density $f(\mathbf{X}|\boldsymbol{\theta})$ and where $\boldsymbol{\theta}$ is some vector of parameters. If there exists a true vector of parameter $\boldsymbol{\theta}^*$, then the measure of goodness of fit used is the generalized Boltzmann information quantity

$$\mathbf{B} = E[\ln f(\mathbf{X}|\boldsymbol{\theta}) + \ln f(\mathbf{X}|\boldsymbol{\theta}^*)] = -\mathbf{I} \qquad (6.79)$$

also known in the statistical literature as the Kullback–Leibler information criterion. Under the principle of entropy maximization we wish to estimate $f(\mathbf{X}|\boldsymbol{\theta}^*)$ by means of $f(\mathbf{X}|\boldsymbol{\theta})$ such that the expected entropy

$$E_x(\mathbf{B}) = E_x\{E[\ln f(\mathbf{X}|\boldsymbol{\theta})] - E[\ln f(\mathbf{X}|\boldsymbol{\theta}^*)]$$

$$= E_x\{E[\ln f(\mathbf{X}|\boldsymbol{\theta})]\} \qquad (6.80)$$

is maximized (that is, the Kuelback–Leibler information $-\mathbf{I}$ is minimized since large values of this quantity imply that the model $f(\mathbf{X}|\boldsymbol{\theta})$ provides a good fit to $f(\mathbf{X}|\boldsymbol{\theta}^*)$. The expected value of Eq. (6.79) can therefore be used as a risk function to measure the average estimation error of the fitted model. The AIC criterion is then a sample estimator of $E[\ln f(\mathbf{X}|\boldsymbol{\theta})]$, the expected log likelihood or negentropy, and is given by the general expression

$$\mathrm{AIC}(r) = -2 \ln L(r) + 2m \qquad (6.81)$$

where m is the number of free parameters after a model has been fitted, and $L(r)$ is the likelihood. Equation (6.81) was originally developed by Akaike (1974a) in the context of time series, but it can be expressed in a form handy for maximum likelihood factor analysis (Akaike, 1987). Since for r common factors we have $L(r) = (n/2) \sum_{i=r+1}^{p} \ln \hat{\theta}_i$ and the number of free parameters is $m = p(r + 1) - (1/2)r(r - 1)$, Eq. (6.81) can be expressed as

$$\text{AIC}(r) = (-2) \left(\frac{n}{2} \sum_{i=r+1}^{p} \ln \hat{\theta}_i \right) + [2p(r + 1) - r(r - 1)] \qquad (6.81a)$$

where $\hat{\theta}_{r+1}$, $\hat{\theta}_{r+2} \cdots \hat{\theta}_p$ represent the last (smallest) latent roots. The idea of using Eq. (6.81a) is to vary the number of common factors, beginning with $r = 1$, and to chose that value of r which corresponds to the minimum of AIC(r). Although no statistical testing is involved as such, both Eqs. (6.78a) and (6.81a) depend on the value of $L(r)$, that is, on the last $(p - r)$ latent roots. Note however, that in Eq. (6.81a), m is not the number of degrees of freedom as for the chi-squared statistic, but corresponds to the number of free parameters in the system (Section 6.1).

As pointed out by Schwarz (1978), however, the penalty term $2m$ does not depend on the sample size n. This implies that the same number of common factors would be selected by Eq. (6.81a) for small as well as large samples, given a common factor structure. The AIC(r) criterion therefore is not a consistent estimator of the "correct" number of factors r. Schwarz's approach is to assume a prior distribution of a general form, where the observations are generated from a Koopman–Darmois family of densities, and to obtain a Baysian criterion which selects that model which is as probable as possible in the a posteriori sense. Schwarz's criterion may be expressed as

$$\text{SIC}(r) = -\frac{n}{2} \sum_{i=r+1}^{p} \ln \hat{\theta}_i + \frac{m}{2} \ln n \qquad (6.81b)$$

where the terms are as in Eq. (6.81a), but the penalty term $(m/2) \ln n$ also depends on the sample size n. The value of r is then selected such that SIC(r) is minimized. Actually Schwarz maximizes $-\text{SIC}(r)$, but Eq. (6.81b) is more handy to use when used in conjunction with Eq. (6.81a). It is also used in statistical computer packages such as SAS. The SIC(r) criterion selects a smaller number of common factors than does AIC(r) when $n > 8$ (Exercise 6.16). SIC(r) however is not the only alternative to Eq. (6.81)— for other formulations see Sclove (1987) and Bozdogan (1987).

Example 6.2. As an example of the testing procedures described in this section consider the following data for $p = 5$ characteristics of $n = 32$ brands of automobiles. Since the data are intended to serve only as a numerical illustration, the sample size and the number of variables is somewhat smaller than would normally be encountered in practice, and the testing procedure does not employ the full range of values of r. Also, the correlation rather than the covariance matrix is used since the ML factor model is invariant with respect to scale (unit) of measurement. The variables are defined as follows, and their values are given in Table 6.1.

Y_1 = Engine size (volume)
Y_2 = Horsepower
Y_3 = Carburator size (number of barrels)
Y_4 = Automobile weight (lbs.)
Y_5 = Time to achieve 60 miles per hour (sec)

Table 6.1   The $p = 5$ Characteristics of $n = 32$ Brands of Automobiles

Car No.	Y_1	Y_2	Y_3	Y_4	Y_5
1	160.0	110.0	4	2620	16.46
2	160.0	110.0	4	2875	17.02
3	108.0	93.0	1	2320	18.61
4	258.0	110.0	1	3215	19.44
5	360.0	175.0	2	3440	17.02
6	225.0	105.0	1	3460	20.22
7	360.0	245.0	4	3570	15.84
8	146.7	62.0	2	3190	20.00
9	140.8	95.0	2	3150	22.90
10	167.6	123.0	4	3440	18.30
11	167.6	123.0	4	3440	18.90
12	275.8	180.0	3	4070	17.40
13	275.8	180.0	3	3730	17.80
14	275.8	180.0	3	3780	18.00
15	472.0	205.0	4	5250	17.98
16	460.0	215.0	4	5424	17.82
17	440.0	230.0	4	5345	17.42
18	78.7	66.0	1	2200	19.47
19	75.7	52.0	2	1615	18.52
20	71.1	65.0	1	1835	19.90
21	120.1	97.0	1	2465	20.01
22	318.0	150.0	2	3520	16.87
23	304.0	150.0	2	3435	17.30
24	350.0	245.0	4	3840	15.41
25	400.0	275.0	2	3845	17.05
26	79.0	66.0	1	1935	18.90
27	120.3	91.0	2	2140	16.70
28	95.1	113.0	2	1513	16.92
29	351.0	264.0	4	3170	14.50
30	145.0	175.0	6	2770	15.50
31	301.0	335.0	8	3570	14.60
32	121.0	109.0	2	2780	18.80

Table 6.2 Principal Components Loading Coefficients (Unrotated) and Latent Roots of the Correlation Matrix of $p = 5$ Automobile Characteristics

Variables	Principal Components				
	Z_1	Z_2	Z_3	Z_4	Z_5
X_1	.875	.415	$-.209$	$-.029$	.131
X_2	.949	$-.057$	$-.106$	.286	$-.057$
X_3	.762	$-.432$	.479	$-.010$	.056
X_4	.776	.581	.176	$-.145$	$-.097$
X_5	$-.771$	.606	.306	.178	.038
Latent roots	3.3545	1.0671	.4090	.1352	.0342
Variance (%)	67.09	21.34	8.18	2.70	.68

First, a principal components analysis is carried out for the five variables to gain an initial idea of the number of principal dimensions required to represent the data adequately. Referring to Table 6.2 it appears that between two and three components are needed for the task. First, we test for complete independence, that is, whether one or more common factors are present. Using Eq. (6.78) (see also sec. 4.2.1) we have the approximate chi-squared criterion

$$\chi^2 = -(n - 1)\ln|\mathbf{R}|$$

$$= -31\ln[(3.3545)(1.0671)(.4090)(.1352)(.0342)]$$

$$= -31\ln(.0067695)$$

$$= -31(-4.9953225)$$

$$= 154.86$$

without using Box's correction factor. Replacing $(n - 1)$ by $(n - 1) - (1/6)(2p+5) = 31 - (1/6)(10+ 5) = 28.5$ we obtain $\chi^2 = 142.37$, a somewhat smaller figure. Using

$$d = \tfrac{1}{2}[(p - r)^2 - (p + r)]$$

$$= \tfrac{1}{2}[(5 - 0)^2 - (5)]$$

$$= 10$$

degrees of freedom we see that both chi-squared values are significant at the .01 level.

Next we commence the estimation process of the ML factor model, beginning with the $r = 1$ common factor and employing SAS statistical software. Surprisingly, an unexpected difficulty develops—we obtain a

Heywood case, that is, a single common factor attempts to explain more than 100% of the total observed variance. As was seen in the preceding sections this is possible in exploratory factor analysis. Although the irregularity or Heywood condition is known to occur, the frequency of its occurrence can at times be surprising.

Since the data cannot be represented by a single common factor (plus a regular error term) we continue for $r = 2$ factors. Here a regular solution is possible, and the loading coefficients are given in Table 6.3. Comparing these values with those of Table 6.2 we can see that both models yield loadings that are similar in value. The R^2 values (communalities) are also comparable, with the exception of X_3, which is a discrete random variable, and in this case does not have errors of measurement and consequently cannot be represented well by the factor model.

The latent roots of the weighted "reduced" correlation matrix $\hat{\Psi}^{-1/2}\hat{\Gamma}\hat{\Psi}^{-1/2}$ are $\hat{\eta}_1 = 33.108682$, $\hat{\eta}_2 = 12.343351$, $\hat{\eta}_3 = .686216$, $\hat{\eta}_4 = .121614$, and $\hat{\eta}_5 = -.807835$ and the latent roots of $\hat{\Psi}^{-1/2}\mathbf{S}\hat{\Psi}^{-1/2}$ are thus given by $\hat{\theta}_i = (\hat{\eta}_i + 1)$. The likelihood criterion (Eq. 6.78a) yields

$$\chi^2 = -[31 - \tfrac{1}{6}(10 + 8 + 5)] \sum_{i=3}^{5} \ln \hat{\theta}_i$$

$$= -27.167(\ln 1.686216 + \ln 1.121614 + \ln .192165)$$

$$= -27.167(-1.01216)$$

$$= 27.497$$

which is approximately chi-squared with $d = (1/2)(3^2 - 7) = 1$ degree of freedom. The value is still significant at the .01 level, and we may feel that $r = 3$ common factors are appropriate. This however is precluded because of negative degrees of freedom, and a three-factor model does not exist for $p = 5$.

Next, we illustrate the use of the Akaike AIC(r) and the Schwarz SIC(r) criteria. Once the latent roots $\hat{\theta}_i$ are known, the AIC criterion can be easily

Table 6.3 Maximum Likelihood Factor Loading Coefficients (Unrotated) of the Correlation Matrix of $p = 5$ Automobile Characteristics[a]

Variables	Maximum Likelihood Common Factors		
	$\hat{\varphi}_1$	$\hat{\varphi}_2$	R^2
X_1	.936	.206	91.84
X_2	.898	−.196	84.53
X_3	.622	−.353	51.14
X_4	.845	.469	93.35
X_5	−.620	.740	93.18

[a] The R^2 values are the communalities for the variables.

computed using Eq. (6.81a) as

$$\text{AIC}(r) = (-2)\left(\frac{n}{2}\sum_{i=3}^{5}\ln\hat{\theta}_i\right) + [2p(r+1) - r(r-1)]$$

$$= 32(1.01216) + [10(3) - 2(1)]$$

$$= 32.389 + 28$$

$$= 60.389$$

and Schwarz's criterion as

$$\text{SIC}(r) = -\frac{n}{2}\sum_{i=3}^{5}\ln\hat{\theta}_i + \frac{m}{2}\ln n$$

$$= -16(-1.01216) + 7(3.46574)$$

$$= 40.455$$

using Eq. (6.81b). Note that a comparison between the three criteria is not possible for our example since a constant number of $r = 2$ common factor is estimated. In a more realistic application, however, when using larger p the value of r for both information criteria is varied and that value is selected which coincides with the smallest value of the criteria. In general the chi-squared statistic yields the largest number of significant factors, followed by the AIC(r) with SIC(r), resulting in the most parsimonious model.

6.6.3 Testing Loading Coefficients

On the assumption of normality it is possible to derive exact asymptotic second moments of the estimated ML loadings of Section 6.4, which enables us to test hypotheses of the form

$$H_0: \alpha_{ij} = 0$$

$$H_a: \alpha_{ij} \neq 0$$

as well as to compute confidence intervals and estimate correlations amongst the α_{ij}. For the reciprocal proportionality model it can be shown (Joreskog, 1963) that if $\mathbf{D} \to \boldsymbol{\Delta}$ as $n \to \infty$, the sample variances and covariances are given by

$$nE[\hat{\boldsymbol{\alpha}}_s - \boldsymbol{\alpha}_s)(\hat{\boldsymbol{\alpha}}_s - \boldsymbol{\alpha}_s)^{\mathrm{T}}] = \frac{\lambda_s}{(\lambda_s - \sigma^2)}\left\{\boldsymbol{\Sigma} - \frac{\lambda_s}{2(\lambda_s - \sigma^2)}\boldsymbol{\alpha}_s\boldsymbol{\alpha}_s^{\mathrm{T}} + \sum_{j \neq s}\frac{\lambda_s}{(\lambda_j - \sigma^2)}\right.$$

$$\left. \times \left[\frac{(\lambda_j - \sigma^2)}{(\lambda_s - \lambda_j)} - 1\right]\boldsymbol{\alpha}_j\boldsymbol{\alpha}_j^{\mathrm{T}}\right\}$$

$$nE[(\hat{\boldsymbol{\alpha}}_s - \boldsymbol{\alpha}_s)(\hat{\boldsymbol{\alpha}}_t - \boldsymbol{\alpha}_t)^{\mathrm{T}}] = -\frac{\lambda_s\lambda_t}{(\lambda_s - \lambda_t)^2}\boldsymbol{\alpha}_s\boldsymbol{\alpha}_t^{\mathrm{T}}(s \neq t) \tag{6.82}$$

respectively. The results (Eq. 6.82) however are derived from an approximate rather than an exact distribution, although the $\hat{\boldsymbol{\alpha}}_j$ are asymptotically ML estimators. For the Lawley–Rao ML estimators consider the likelihood (Eq. 6.40) with expected matrix of second derivatives $E[\partial f/\partial \boldsymbol{\Psi}_i \partial \boldsymbol{\Psi}_j] = \mathbf{G}$. Also, let $\boldsymbol{\Lambda}$ be the diagonal matrix of the first r latent roots of $\boldsymbol{\Psi}^{-1/2}\boldsymbol{\Sigma}\boldsymbol{\Psi}^{-1/2}$ (Theorem 6.5). Then for a normal sample the distribution of $\hat{\boldsymbol{\Psi}}$ approaches $N[\boldsymbol{\Psi}, (2/n)\mathbf{G}^{-1/2}]$. Let $\mathbf{b}_{iq}$ be the column vector whose elements $b_{1,iq}$, $b_{2,iq}, \ldots, b_{p,iq}$ are the regression coefficients of the $\hat{\boldsymbol{\alpha}}_{iq}$ on $\hat{\boldsymbol{\Psi}}_1, \hat{\boldsymbol{\Psi}}_2, \ldots, \hat{\boldsymbol{\Psi}}_p$. The regression terms are corrected for the fact that the true $\boldsymbol{\Psi}_j$ are not known. Then Lawley has shown (Lawley and Maxwell, 1971; Chapter 5) that the covariance between any two loadings is

$$nE[(\hat{\boldsymbol{\alpha}}_{is} - \boldsymbol{\alpha}_{is})(\hat{\boldsymbol{\alpha}}_{jt} - \boldsymbol{\alpha}_{jt})^{\mathrm{T}}] = -\frac{\lambda_s \lambda_t}{(\lambda_s - \lambda_t)^2}\, \alpha_{is}\alpha_{jt} + 2\mathbf{b}_{is}^{\mathrm{T}}(\mathbf{G}^{-1/2})\mathbf{b}_{jt} \quad (6.83)$$

with variances given by setting $i = j$ and $r = s$ where it is understood that the denominator of the first term does not vanish. With the correction term of Jennrich and Thayer (1973), the regression coefficients $\hat{\mathbf{b}}_{j,iq}$ can be computed as

$$\hat{\mathbf{b}}_{j,iq} = -\hat{\boldsymbol{\alpha}}_{jq}(\hat{\lambda}_q - 1)^{-1}\hat{\boldsymbol{\Psi}}_j^{-2}$$

$$\times \left[\delta_{ij}\hat{\boldsymbol{\Psi}}_j - 1/2\hat{\alpha}_{iq}\hat{\alpha}_{jq}/(\hat{\lambda}_q - 1) + \hat{\lambda}_q \sum_{h \neq q}^{r} \hat{\alpha}_{ih}\hat{\alpha}_{jh}/(\hat{\lambda}_q - \hat{\lambda}_h) \right] \quad (6.84)$$

for r common factors, where δ_{ij} is the Kronecker delta. The derivation of the moments of the coefficients assumes that $\hat{\boldsymbol{\Sigma}}$ is the covariance matrix, but Lawley and Maxwell (1971) also give approximate results for the correlation matrix. Since ML factor models are invariant to changes in scale, however, there is no loss in generality by assuming the covariance matrix is used. The test then makes use of normal tables where $\sqrt{n}(\hat{\boldsymbol{\alpha}} - \boldsymbol{\alpha})$ approaches a multivariate normal with mean zero, assuming $\hat{\boldsymbol{\alpha}}$ is a consistent estimator of $\boldsymbol{\alpha}$. It is also possible to derive moments for rotated loadings (Archer and Jennrich, 1973; Jennrich, 1973), but this is presumably not required if coefficients are pretested before a rotation is used, with insignificant values being replaced by zeroes.

The preceding tests of significance require the assumption of multivariate normality. As for the PC model, however, it is also possible to use resampling schemes such as the jackknife, the bootstrap, or cross validation (Section 4.4.1) to test parameter variation and goodness of fit. The jackknife and bootstrap estimation however have potential difficulties in that the former may yield inconsistent coefficient estimates (e.g., correlation loadings greater than unity) and both require greater computing time than the parametric tests based on normality. Resampling methods however have not been widely used with factor models, and initial Monte Carlo investigations do not seem very informative (Pennel, 1972; Chatterjee, 1984). It is also not clear what effect, if any, improper solutions have on the variances. Again, variances for both the rotated and unrotated loadings can be estimated, but rotated loadings present a greater computational burden (see Clarkson, 1979). Cross validation on the other hand seems an attractive option, but it has not been used to any great extent with real data.

A few additional comments are now in order concerning Theorem 6.1, which lays down the necessary and sufficient condition for the existence of $1 \le r < p$ common factors, and which can thus be considered as the fundamental theorem of factor analysis. As was seen in previous sections, the condition is that there exists a diagonal matrix $\boldsymbol{\Psi}$ with entries $\phi_j > 0$ $(j = 1, 2, \ldots, p)$ such that $\boldsymbol{\alpha}\boldsymbol{\alpha}^{\mathrm{T}} = \boldsymbol{\Gamma}$ is positive semidefinite of rank r. Thus given a diagonal positive definite matrix $\boldsymbol{\Psi}$, the condition guarantees the mathematical existence of a decomposition $\boldsymbol{\Sigma} = \boldsymbol{\Gamma} + \boldsymbol{\Psi}$, subject to identification constraints, for $1 < r < p$. In practice however $\boldsymbol{\Psi}$ is almost never known and must therefore be estimated from the data, together with the coefficients $\boldsymbol{\alpha}$. As was seen in Example 6.2, the simultaneous existence of $1 \le r < p$ common factors and Grammian $\boldsymbol{\Gamma}$ and $\boldsymbol{\Psi}$ cannot generally be guaranteed. This implies zero and/or imaginary error terms, which clearly does not fit the factor specification (Section 6.1). The immediate consequences of such improper solutions are (1) the upper bound for the number of common factors implied by Eq. (6.9) is only a necessary condition, and (2) sample loadings become inconsistent estimators of the population parameters. The way out of the difficulty which is at times taken is to either force the factor model onto the data, perhaps using Baysian-type arguments (Martin and McDonald, 1975; Koopman, 1978) or else to relax the condition of positive (semi)definiteness of $\boldsymbol{\Gamma} = \boldsymbol{\alpha}\boldsymbol{\alpha}^{\mathrm{T}}$ and $\boldsymbol{\Psi}$ (vanDriel, 1978). Since improper solutions can occur with both populations as well as sample covariance matrices however, a Heywood case may be a signal that the factor model is not appropriate and perhaps should be replaced by a principal components model.

6.7   THE FIXED FACTOR MODEL

The models considered in the previous sections assume that the factors $\boldsymbol{\Phi}$ are random variables rather than fixed parameters. At times, however, the factors may be viewed as fixed in repeated sampling, so that in Eq. (6.1) $\mathbf{Y}$ and $\boldsymbol{\epsilon}$ are random but both $\boldsymbol{\alpha}$ and $\boldsymbol{\Phi}$ are fixed. This was the case, for example, in Sections 3.6 and 6.3.4 when dealing with Whittle's model, where the factors $\boldsymbol{\Phi}$ are considered as fixed parameters. Fixed factors could also be appropriate if the error terms are considered to consist only of measurement error rather than measurement error plus individual differences. For the fixed model we then have the specification $E(\boldsymbol{\epsilon}) = 0$, $E(\mathbf{Y}) = \boldsymbol{\mu} + \boldsymbol{\alpha}\boldsymbol{\Phi}$, $E(\boldsymbol{\epsilon}\boldsymbol{\epsilon}^{\mathrm{T}}) = \boldsymbol{\Psi}$, and $E\{[\mathbf{Y} - E(\mathbf{Y})][\mathbf{Y} - E(\mathbf{Y})]^{\mathrm{T}}\} = \boldsymbol{\Psi}$. The fixed model is at times also viewed in terms of matrix sampling rather than sample point sampling since the multivariate sample points must remain constant for repeated sampling. The fixed model appears to have been first considered by Lawley (1942) in the context of the normal log likelihood

$$L = \frac{n}{2}\left[c + \ln|\mathbf{\Phi}| + \mathrm{tr}(\mathbf{\epsilon}^{\mathrm{T}}\mathbf{\epsilon}\mathbf{\Phi}^{-1})\right] \qquad (6.85)$$

when attempting to obtain ML estimators of $\mathbf{\alpha}$ and $\mathbf{\Phi}$. It was subsequently proved by Solari (1969) however that the fixed case does not possess ML estimators.

Attempts to find estimators based on the likelihood or a function resembling the likelihood have persisted, based partly on the conjecture that parallel to the situation in regression analysis, factor loadings for the fixed case would resemble those of the random model. Recently McDonald (1979) has considered minimizing the loss function

$$L = \tfrac{1}{2}\left[\ln|\mathbf{\Delta}| + \mathrm{tr}\,\mathbf{\Delta}^{-1}\mathbf{\Phi}\right] \qquad (6.86)$$

which is essentially the log likelihood ratio defined on the basis of an alternative hypothesis

$$H_0\colon E(\mathbf{Y} - \mathbf{\mu} - \mathbf{\alpha}\mathbf{\Phi})(\mathbf{Y} - \mathbf{\mu} - \mathbf{\alpha}\mathbf{\Phi})^{\mathrm{T}} = \mathbf{\Psi}$$

$$H_a\colon E[\mathbf{Y} - E(\mathbf{Y})][\mathbf{Y} - E(\mathbf{Y})]^{\mathrm{T}} = \mathbf{\Delta} \qquad (6.87)$$

H_0, as in the random case, is the error covariance matrix under the factor hypothesis whereas H_a states that the error covariance matrix is any diagonal matrix $\mathbf{\Delta}$ with positive diagonal elements. Minimizing Eq. (6.86) then leads to the same estimators of $\mathbf{\alpha}$ and $\mathbf{\epsilon}$ as those obtained under Lawley's model (Section 6.4.2) except the fixed model yields least squares rather than ML estimators. Anderson and Rubin (1956) have shown that while ML estimators do not exist for fixed factors, estimates based on maximizing the noncentral Wishart likelihood function are asymptotically equivalent to the ML estimates for random factors. Thus for large samples it is possible to use asymptotic ML estimates for random factors to estimate nonrandom factors.

6.8 ESTIMATING FACTOR SCORES

The factor models of the previous sections provide estimates of the correlation loadings $\mathbf{\alpha}$, which for the most part can be obtained using either the correlation or the covariance matrix. Indeed, this frequently represents the primary objective of factor analysis since it is the loading coefficients which to a large extent determine the reduction of observed variables into a smaller set of common factors, and which enable the identification and interpretation of factors in terms of meaningful phenomenon. Although the loadings generally vary for each variable, they are nonetheless constant for each sample point, and in this sense do not provide information concerning the relative position of a given sample point vis a vis the factors. The situation is similar to that encountered with the principal component model (Chapter 3). Unlike principal component analysis, however, the factor scores are not unique, due to the a priori reduced dimensionality of the factor space. This in turn implies a different estimation strategy from that which was employed for principal components. Whereas in PCA the scores are obtained routinely as linear combinations $\mathbf{Z} = \mathbf{XPL}^{-1/2} = \mathbf{XA}$, where all p principal components are used, this is not possible for factor analysis "proper" due to the a priori specification $r < p$. Thus given the orthogonal factor model (Eq. 6.2) with r, $\boldsymbol{\alpha}$, and $\boldsymbol{\Psi}$ known (estimated), the factor model contains a further indeterminacy in the form of nonunique factor scores $\boldsymbol{\Phi}$ due to singularity of $\boldsymbol{\alpha}$. The final stage in a factor analysis of a data matrix is therefore that of finding optimal estimates of the scores or the "observations" for $\boldsymbol{\Phi}$. Two general avenues of approach are possible depending on whether $\boldsymbol{\Phi}$ is considered to be fixed or random.

6.8.1 Random Factors: The Regression Estimator

An estimator of the factor scores $\boldsymbol{\Phi}$ can be derived within the context of maximum likelihood theory by assuming the joint normality of $\boldsymbol{\Phi}$ and $\mathbf{X}$. Consider the augmented (partitioned) vector $\mathbf{Z} = [\boldsymbol{\Phi} : \mathbf{X}]^{\mathrm{T}}$, which by assumption is distributed as a $(r + p)$ dimensional normal distribution with mean vector $\boldsymbol{\mu} = \mathbf{0}$ and covariance matrix

$$
\begin{bmatrix} \boldsymbol{\Sigma}_{11} & \boldsymbol{\Sigma}_{12} \\ \boldsymbol{\Sigma}_{21} & \boldsymbol{\Sigma}_{22} \end{bmatrix} = E(\mathbf{ZZ})^{\mathrm{T}} = E \begin{bmatrix} \boldsymbol{\Phi} \\ \cdots \\ \mathbf{X} \end{bmatrix} [\boldsymbol{\Phi}^{\mathrm{T}} : \mathbf{X}^{\mathrm{T}}]
$$

$$
= \begin{bmatrix} E(\boldsymbol{\Phi}\boldsymbol{\Phi}^{\mathrm{T}}) & E(\boldsymbol{\Phi}\mathbf{X}^{\mathrm{T}}) \\ E(\mathbf{X}\boldsymbol{\Phi}^{\mathrm{T}}) & E(\mathbf{X}\mathbf{X}^{\mathrm{T}}) \end{bmatrix}
$$

It follows that the conditional distribution of $\boldsymbol{\Phi}$ given $\mathbf{X}$ is also normal, that is,

$$
\boldsymbol{\Phi} \mid \mathbf{X} \sim N[(\boldsymbol{\Sigma}_{12}\boldsymbol{\Sigma}_{22}^{-1}\mathbf{X}), \qquad (\boldsymbol{\Sigma}_{11} - \boldsymbol{\Sigma}_{12}\boldsymbol{\Sigma}_{22}^{-1}\boldsymbol{\Sigma}_{12}^{\mathrm{T}})]
$$

$$
\sim N[\boldsymbol{\alpha}^{\mathrm{T}}\boldsymbol{\Sigma}^{-1}\mathbf{X}, \qquad (\mathbf{I} - \boldsymbol{\alpha}^{\mathrm{T}}\boldsymbol{\Sigma}^{-1}\boldsymbol{\alpha})] \tag{6.88}
$$

where $E(\boldsymbol{\Phi}\mathbf{X}^{\mathrm{T}}) = \boldsymbol{\alpha}^{\mathrm{T}}$ and using Theorem 2.14. Using Lemma 6.5, the covariance matrix of $\boldsymbol{\Phi} \mid \mathbf{X}$ can also be put in an alternative form.

Lemma 6.6. Let the conditions of Theorem 6.5 hold. Then

$$\mathbf{I} - \boldsymbol{\alpha}^{\mathrm{T}}\boldsymbol{\Sigma}^{-1}\boldsymbol{\alpha} = (\mathbf{I} + \boldsymbol{\alpha}^{\mathrm{T}}\boldsymbol{\Psi}^{-1}\boldsymbol{\alpha}) \tag{6.89}$$

$$\square$$

Proof. It follows from Lemma 6.5 that

$$\boldsymbol{\alpha}^{\mathrm{T}}\boldsymbol{\Sigma}^{-1}\boldsymbol{\alpha} = \boldsymbol{\alpha}^{\mathrm{T}}[\boldsymbol{\Psi}^{-1}\boldsymbol{\alpha}(\mathbf{I} + \boldsymbol{\eta})^{-1}]$$

$$= \boldsymbol{\eta}(\mathbf{I} + \boldsymbol{\eta})^{-1}$$

where $\boldsymbol{\eta} = \boldsymbol{\alpha}^{\mathrm{T}}\boldsymbol{\Psi}^{-1}\boldsymbol{\alpha}$. We then can write

$$\mathbf{I} - \boldsymbol{\alpha}^{\mathrm{T}}\boldsymbol{\Sigma}^{-1}\boldsymbol{\alpha} = \mathbf{I} - \boldsymbol{\eta}(\mathbf{I} + \boldsymbol{\eta})^{-1}$$

$$= (\mathbf{I} + \boldsymbol{\eta})(\mathbf{I} + \boldsymbol{\eta})^{-1} - \boldsymbol{\eta}(\mathbf{I} + \boldsymbol{\eta})^{-1}$$

$$= (\mathbf{I} + \boldsymbol{\eta})^{-1}$$

$$= (\mathbf{I} + \boldsymbol{\alpha}^{\mathrm{T}}\boldsymbol{\Psi}^{-1}\boldsymbol{\alpha})^{-1}$$

The conditional distribution of $\boldsymbol{\Phi}$ given $\mathbf{X}$ is therefore

$$\boldsymbol{\Phi} \mid \mathbf{X} \sim N[\boldsymbol{\alpha}^{\mathrm{T}}\boldsymbol{\Sigma}^{-1}\mathbf{X}, (\mathbf{I} + \boldsymbol{\alpha}^{\mathrm{T}}\boldsymbol{\Psi}^{-1}\boldsymbol{\alpha}] \tag{6.90}$$

Equation (6.90) can also be viewed in Baysian terms as the posterior distribution of $\boldsymbol{\Phi}$ given the p variables $\mathbf{X}$ (Anderson, 1959; Bartholomew, 1981). It also follows that the expectation of the posterior distribution of the factor scores is given by

$$E(\boldsymbol{\Phi} \mid \mathbf{X}) = \boldsymbol{\alpha}^{\mathrm{T}}\boldsymbol{\Sigma}^{-1}\mathbf{X}$$

$$= (\mathbf{I} + \boldsymbol{\alpha}^{\mathrm{T}}\boldsymbol{\Psi}^{-1}\boldsymbol{\alpha})^{-1}\boldsymbol{\alpha}^{\mathrm{T}}\boldsymbol{\Psi}^{-1}\mathbf{X} \tag{6.91}$$

using Eq. (6.90). To obtain the maximum likelihood estimator consider a $(n \times p)$ data matrix $\mathbf{X}$ with known (estimated) loadings $\boldsymbol{\alpha}$, such that $\mathbf{X} = \boldsymbol{\Phi}\boldsymbol{\alpha} + \boldsymbol{\epsilon}$. The sample counterpart of Eq. (6.91) is then

$$\hat{\boldsymbol{\Phi}} = \mathbf{X}\hat{\boldsymbol{\Psi}}^{-1}\hat{\boldsymbol{\alpha}}^{\mathrm{T}}(\mathbf{I} + \hat{\boldsymbol{\alpha}}\hat{\boldsymbol{\Psi}}^{-1}\hat{\boldsymbol{\alpha}}^{\mathrm{T}})^{-1} \tag{6.92}$$

where $\hat{\boldsymbol{\Psi}} = \hat{\boldsymbol{\epsilon}}^{\mathrm{T}}\hat{\boldsymbol{\epsilon}}$ is the diagonal matrix of residual variances.

The maximum likelihood estimator (Eq. 6.92) is asymptotically efficient, although it is not unbiased. When normality of $\mathbf{X}$ is not assumed, the estimator (Eq. 6.92) possesses optimality in the least squares sense since it minimizes the mean squared error criterion. Let

$$\boldsymbol{\Phi} = \mathbf{X}\mathbf{B} + \boldsymbol{\delta} \tag{6.93}$$

where $\mathbf{B}$ is a $(p \times r)$ matrix of coefficients. Were $\boldsymbol{\Phi}$ known, $\mathbf{B}$ could be estimated by the ordinary least squares regression estimator

$$\hat{\mathbf{B}} = (\mathbf{X}^T\mathbf{X})^{-1}\mathbf{X}^T\boldsymbol{\Phi} \tag{6.94}$$

which can be derived by minimizing $\mathrm{tr}(\boldsymbol{\delta}^T\boldsymbol{\delta})$. Since $\boldsymbol{\Phi}$ is not known, however, it can be estimated as

$$\hat{\boldsymbol{\Phi}} = \mathbf{X}\hat{\mathbf{B}}$$

$$= \mathbf{X}(\mathbf{X}^T\mathbf{X})^{-1}\mathbf{X}^T\boldsymbol{\Phi}$$

$$= \mathbf{X}(\mathbf{X}^T\mathbf{X})^{-1}\hat{\boldsymbol{\alpha}}^T \tag{6.95}$$

Equation (6.95) is known as the regression estimator and is due to Thurstone (1935) and Thompson (1951). Actually it can be shown to be equivalent to the maximum likelihood estimator (Eq. 6.92) since using the identity (Eq. 6.50) of Lemma 6.4 we have

$$\hat{\boldsymbol{\Phi}} = \mathbf{X}\mathbf{S}^{-1}\hat{\boldsymbol{\alpha}}^T$$

$$= \mathbf{X}\hat{\boldsymbol{\Psi}}^{-1}\hat{\boldsymbol{\alpha}}^T(\mathbf{I} + \hat{\boldsymbol{\alpha}}\hat{\boldsymbol{\Psi}}^{-1}\hat{\boldsymbol{\alpha}}^T)^{-1} \tag{6.96}$$

keeping in mind we are dealing with sample estimates, and restricting $\hat{\boldsymbol{\alpha}}\hat{\boldsymbol{\Psi}}^{-1}\hat{\boldsymbol{\alpha}}^T$ to diagonal form (Exercise 6.17). The classic least squares estimator is therefore maximum likelihood when both the observed variables and the factors are assumed to be jointly normal. The estimator however is biased, since although Eq. (6.96) assumes the error to be in $\boldsymbol{\Phi}$, in reality error measurement belongs to $\mathbf{X}$—the regression estimator reverses the role of the dependent and independent variables. As $p \to \infty$, however, the sampling variance of $\hat{\boldsymbol{\Phi}}$ decreases. This is also true for factors that are associated with large terms in the diagonal matrix $\hat{\boldsymbol{\alpha}}\hat{\boldsymbol{\Psi}}\hat{\boldsymbol{\alpha}}^T$, that is, factors that account for large variance can be estimated more precisely than those that do not. Bartholomew (1981) however has pointed out that the criterion of unbiasedness may be inappropriate owing to the randomness of $\boldsymbol{\Phi}$, that is, the estimator (Eq. 6.96) represents an estimate of $E(\boldsymbol{\Phi}\,|\,\mathbf{X})$ rather than $\boldsymbol{\Phi}$, which is an appropriate criterion for random variables, see also Kano (1986).

6.8.2 Fixed Factors: The Minimum Distance Estimator

The estimator (Eq. 6.92) considers $\boldsymbol{\Phi}$ as random, to be estimated once $\mathbf{X}$ and $\boldsymbol{\alpha}$ are known. A different set of factor scores was derived by Bartlett (1937) using the alternative assumption that each sample point is characterized by a fixed vector of r parameters $\boldsymbol{\Phi}_i$. Then for the ith individual, $\mathbf{x}_i \sim N(\boldsymbol{\alpha}\boldsymbol{\Phi}_i, \boldsymbol{\Psi})$ where $\boldsymbol{\alpha}\boldsymbol{\Phi}_i + \boldsymbol{\epsilon}_i$ is now a $(p \times 1)$ vector of observations on p variables, maintaining the convention that vectors represent column arrays.

Since the $\mathbf{\Phi}_i$ are fixed population parameters, it is now more appropriate to seek linear estimators which posses optimal properties such as unbiasedness and least variance.

Let $\mathbf{x}_i$ be a $(p \times 1)$ vector of observations for the ith sample point such that $\mathbf{x}_i = \mathbf{\alpha}\mathbf{\Phi}_i + \mathbf{\epsilon}_i$, where we assume that $\mathbf{\alpha}$ is known. Then a reasonable optimality criterion might be to minimize the weighted (Mahalanobis) distance between $\mathbf{x}_i$ and its predicted value $\hat{\mathbf{x}}_i = \mathbf{\alpha}\mathbf{\Phi}_i$, that is, to minimize

$$
\begin{aligned}
d_i &= (\mathbf{x}_i - \mathbf{\alpha}\mathbf{\Phi}_i)^{\mathrm{T}}\mathbf{\Psi}^{-1}(\mathbf{x}_i - \mathbf{\alpha}\mathbf{\Phi}_i) \\
&= \mathbf{x}_i^{\mathrm{T}}\mathbf{\Psi}^{-1}\mathbf{x}_i - 2\mathbf{x}_i^{\mathrm{T}}\mathbf{\Psi}^{-1}\mathbf{\alpha}\mathbf{\Phi}_i + \mathbf{\Phi}_i^{\mathrm{T}}\mathbf{\alpha}^{\mathrm{T}}\mathbf{\Psi}^{-1}\mathbf{\alpha}\mathbf{\Phi}_i
\end{aligned}
\tag{6.97}
$$

Note that d_i also represents the weighted sum of squares of residual errors. Differentiating with respect to the unknown parameters $\mathbf{\Phi}_i$ and setting to zero yields the normal equations

$$
\frac{\partial d_i}{\partial \mathbf{\Phi}_i} = -2\mathbf{\alpha}^{\mathrm{T}}\mathbf{\Psi}^{-1}\mathbf{x}_i + 2\mathbf{\alpha}^{\mathrm{T}}\mathbf{\Psi}^{-1}\mathbf{\alpha}\mathbf{\Phi}_i = \mathbf{0}
$$

or

$$
\tilde{\mathbf{\Phi}}_i = (\mathbf{\alpha}^{\mathrm{T}}\mathbf{\Psi}^{-1}\mathbf{\alpha})^{-1}\mathbf{\alpha}^{\mathrm{T}}\mathbf{\Psi}^{-1}\mathbf{x}_i
\tag{6.98}
$$

The Bartlett estimator $\hat{\mathbf{\Phi}}_i$ is therefore the usual generalized least squares estimator of $\mathbf{x}_i$ on $\mathbf{\alpha}$. Minimizing Eq. (6.97) is also equivalent to minimizing

$$
\mathrm{tr}[(\mathbf{x}_i - \mathbf{\alpha}\mathbf{\Phi}_i)^{\mathrm{T}}\mathbf{\Psi}^{-1}(\mathbf{x}_i - \mathbf{\alpha}\mathbf{\Phi}_i)]
\tag{6.99}
$$

(Exercise 6.10). Note also that the estimator

$$
\hat{\mathbf{\Phi}}_i = (\mathbf{\alpha}^{\mathrm{T}}\mathbf{\alpha})^{-1}\mathbf{\alpha}^{\mathrm{T}}\mathbf{x}_i
$$

known as the least squares estimator, is misspecified when $\mathbf{\Psi} \neq \sigma^2\mathbf{I}$, and that $\mathrm{var}(\hat{\mathbf{\Phi}}_i) > \mathrm{var}(\tilde{\mathbf{\Phi}}_i)$. Solutions for $\tilde{\mathbf{\Phi}}_i$ are thus obtained by using the generalized inverse $(\mathbf{\alpha}^{\mathrm{T}}\mathbf{\Psi}^{-1}\mathbf{\alpha})^{-1}\mathbf{\alpha}^{\mathrm{T}}\mathbf{\Psi}^{-1}$. Although the estimator (Eq. 6.98) has the desirable properties of being unbiased and efficient, it does assume that scores are parameters rather than random variables. As pointed out in Section 6.7 this can introduce theoretical difficulties when using maximum likelihood estimation. Also note that the $\mathbf{\Phi}_i$ are correlated unless $(\mathbf{\alpha}^{\mathrm{T}}\mathbf{\Psi}^{-1}\mathbf{\alpha})$ is diagonal, as is the case for the maximum likelihood model of Section 6.4.2.

The third set of factor scores was proposed by Anderson and Rubin (1956, p. 139) who derived Bartlett's estimator subject to the constraint that scores be orthogonal. Minimizing

$$
\mathrm{tr}[(\mathbf{x}_i - \mathbf{\alpha}\mathbf{\Phi}_i)^{\mathrm{T}}\mathbf{\Psi}^{-1}(\mathbf{x}_i - \mathbf{\alpha}\mathbf{\Phi}_i)] - \lambda(\mathbf{\Phi}_i^{\mathrm{T}}\mathbf{\Phi}_i - 1)
\tag{6.100}
$$

leads to the estimator

$$\boldsymbol{\Phi}_i^* = [(\boldsymbol{\alpha}^{\mathrm{T}}\boldsymbol{\Psi}^{-1}\boldsymbol{\alpha})(\mathbf{I} + \boldsymbol{\alpha}^{\mathrm{T}}\boldsymbol{\Psi}^{-1}\boldsymbol{\alpha})]^{-1/2}\boldsymbol{\alpha}^{\mathrm{T}}\boldsymbol{\Psi}^{-1}\mathbf{x}_i \qquad (6.101)$$

for ML factors (Exercise 6.14). The estimator Eq. (6.101) possesses higher mean squared error than does the estimator Eq. (6.91).

When $\boldsymbol{\alpha}$ and $\boldsymbol{\Psi}$ are known, the estimators of fixed population factor scores are also ML estimators. When $\boldsymbol{\alpha}$ and $\boldsymbol{\Psi}$ are to be estimated, however, then ML estimators of the factor scores do not exist since the normal equations have no minimum (Anderson and Rubin, 1956). For large samples it is thus at times suggested that $\boldsymbol{\Phi}$ be treated as if it were random and ML estimators be used as approximations.

6.8.3 Interpoint Distance in the Factor Space

Once the factor scores are computed it may be of interest to examine the interpoint distances between the individual sample points in the reduced common factor space (Gower, 1966). For Thompson's estimator the distance between the (i, j)th sample point is

$$d_{\mathrm{T}} = (\boldsymbol{\varphi}_i - \boldsymbol{\varphi}_j)^{\mathrm{T}}(\boldsymbol{\varphi}_i - \boldsymbol{\varphi}_j)$$

$$= [(\mathbf{I} + \boldsymbol{\alpha}^{\mathrm{T}}\boldsymbol{\Psi}^{-1}\boldsymbol{\alpha})^{-1}\boldsymbol{\alpha}^{\mathrm{T}}\boldsymbol{\Psi}^{-1}(\mathbf{x}_i - \mathbf{x}_j)]^{\mathrm{T}}[(\mathbf{I} + \boldsymbol{\alpha}^{\mathrm{T}}\boldsymbol{\Psi}^{-1}\boldsymbol{\alpha})^{-1}\boldsymbol{\alpha}^{\mathrm{T}}\boldsymbol{\Psi}^{-1}(\mathbf{x}_i - \mathbf{x}_j)]$$

$$= (\mathbf{x}_i - \mathbf{x}_j)^{\mathrm{T}}\boldsymbol{\Psi}^{-1}\boldsymbol{\alpha}(\mathbf{I} + \boldsymbol{\alpha}^{\mathrm{T}}\boldsymbol{\Psi}^{-1}\boldsymbol{\alpha})^{-1}\boldsymbol{\alpha}^{\mathrm{T}}\boldsymbol{\Psi}^{-1}(\mathbf{x}_i - \mathbf{x}_j) \qquad (6.101a)$$

and using Bartlett's estimator we have

$$d_{\mathrm{B}} = (\boldsymbol{\varphi}_i - \boldsymbol{\varphi}_j)^{\mathrm{T}}(\boldsymbol{\varphi}_i - \boldsymbol{\varphi}_j)$$

$$= [(\boldsymbol{\alpha}^{\mathrm{T}}\boldsymbol{\Psi}^{-1}\boldsymbol{\alpha})^{-1}\boldsymbol{\alpha}^{\mathrm{T}}\boldsymbol{\Psi}^{-1}(\mathbf{x}_i - \mathbf{x}_j)]^{\mathrm{T}}[(\boldsymbol{\alpha}^{\mathrm{T}}\boldsymbol{\Psi}^{-1}\boldsymbol{\alpha})^{-1}\boldsymbol{\alpha}^{\mathrm{T}}\boldsymbol{\Psi}^{-1}(\mathbf{x}_i - \mathbf{x}_j)]$$

$$= (\mathbf{x}_i - \mathbf{x}_j)^{\mathrm{T}}\boldsymbol{\Psi}^{-1}\boldsymbol{\alpha}(\boldsymbol{\alpha}^{\mathrm{T}}\boldsymbol{\Psi}^{-1}\boldsymbol{\alpha})^{-2}\boldsymbol{\alpha}^{\mathrm{T}}\boldsymbol{\Psi}^{-1}(\mathbf{x}_i - \mathbf{x}_j) \qquad (6.102)$$

where both d_{T} and d_{B} represent Mahalanobis distance in r-dimensional space. It can be shown (Exercise 6.8) that $d_{\mathrm{T}} \le d_{\mathrm{B}}$ with equality holding if and only if $d_{\mathrm{T}} = d_{\mathrm{B}} = 0$.

6.9 FACTORS REPRESENTING "MISSING DATA:" THE EM ALGORITHM

A more recent innovation in maximum likelihood estimation is the so-called expectation maximization (EM) algorithm, which uses normal theory to estimate both the parameters and scores of the factor model simultaneously. The algorithm is iterative and does not necessarily yield the same estimates as the Lawley–Joreskog model, which uses the Fletcher–Powell numerical maximization procedure. The EM algorithm was originally developed to estimate missing data from the multivariate normal distribution (Demster et al., 1977), and in the context of factor analysis it simply treats the scores as data which are 100% missing.

Consider the ML normal factor model, where a multivariate normal sample is available in the form of a $(n \times p)$ matrix $\mathbf{X}$ where $\mathbf{X} = \mathbf{Y} - \bar{\mathbf{Y}}$. We wish to obtain an augmented matrix $(\mathbf{F}, \mathbf{X})$, but in the sampling situation the scores $\mathbf{F}$ are all missing. Consider the ith row of $(\mathbf{F}, \mathbf{X})$ where the row vectors are independently and normally distributed and the marginal distribution of every row of $\mathbf{F}$ has zero mean and unit variance. Then given a sample of n observations $x_1, x_2, \ldots, x_n$, we are to estimate the parameters $\boldsymbol{\alpha}$ and $\boldsymbol{\Psi}$, together with the missing variables $\mathbf{F}$. It is also convenient to assume an orthogonal factor structure, although this is not essential. Following Rubin and Thayer (1982) the likelihood of observing $(\mathbf{f}_1, \mathbf{x}_1)$, $(\mathbf{f}_2, \mathbf{x}_2), \ldots, (\mathbf{f}_n, \mathbf{x}_n)$ can be expressed as

$$
\begin{aligned}
L &= \prod_{i=1}^{n} g(x_i \mid \boldsymbol{\alpha}, \boldsymbol{\Psi}) g(\mathbf{f}_i \mid \boldsymbol{\alpha}, \boldsymbol{\Psi}) \\
&= \prod_{i=1}^{n} (2\pi)^{-p/2} |\boldsymbol{\Psi}|^{-1} \exp[-1/2(\mathbf{x}_i - \mathbf{f}_i\boldsymbol{\alpha})\boldsymbol{\Psi}^{-1}(\mathbf{x}_i - \mathbf{f}_i\boldsymbol{\alpha})^{\mathrm{T}}]\mathbf{X}(2\pi)^{-p/2} \\
&\quad \times \exp[-1/2\mathbf{f}_i\mathbf{f}_i^{\mathrm{T}}]
\end{aligned}
\tag{6.103}
$$

where $\mathbf{x}_i$ and $\mathbf{f}_i$ are rows of $\mathbf{X}$ and $\mathbf{F}$ respectively, $g(\mathbf{x}_i \mid \boldsymbol{\alpha}, \boldsymbol{\Psi})$ is the conditional distribution of $\mathbf{x}_i$ given $\mathbf{f}_i$ (viewed as a function of the parameters $\boldsymbol{\alpha}$ and $\boldsymbol{\Psi}$), and $g(\mathbf{f}_i \mid \boldsymbol{\alpha}, \boldsymbol{\Psi})$ is the joint distribution of the unobserved factor scores and is likewise considered to be a function of $\boldsymbol{\alpha}$ and $\boldsymbol{\Psi}$. The likelihood, based on the original data $\mathbf{X}$, is given by Eq. (6.40). The EM algorithm consists of two steps. First, in the E step we find the expectation of the likelihood (Eq. 6.103) over the distribution of the factor scores $\mathbf{F}$, given the data $\mathbf{X}$ and initial estimates of the parameters $\boldsymbol{\alpha}$ and $\boldsymbol{\Psi}$. This is done by calculating the expected values of the sufficient statistics $\mathbf{S}_{FF}$ and $\mathbf{S}_{XF}$ the variance and covariance matrices of $\mathbf{F}, \mathbf{X}$. The computations involve a least squares regression for estimating the factor scores given the initial values of $\boldsymbol{\alpha}$ and $\boldsymbol{\Psi}$. The second or maximization (M) step consists of maximizing the expected (log) likelihood just as if it were based on complete data. The computations involve least squares regression of the variables on the factor scores, which were estimated in the expectation step. The iterative process continues until stable values of the loadings and the scores are obtained. It is proved in Dempster et al. (1977) that each step increases the likelihood. An error in the proof is subsequently corrected by Wu (1983), who summarizes the theoretical properties of the convergence process of the algorithm.

The EM algorithm does not compute second derivatives, so that second-order sufficiency conditions for a maximum are not evaluated. Several other criticisms of the algorithm have also been raised (Demster et al., 1977; discussion), particularly concerning its claim of superiority over the Joreskog implementation of the Fletcher–Powell algorithm (Bentler and Tanaka, 1983; Rubin and Thayer, 1983) with respect to convergence. Certainly both algorithms can run into difficulties if certain regularity conditions are not met, and if the assumption of normality is violated. In this case the likelihood may not be unimodal, and if care is not taken convergence can be to a local rater than to a global maximum. Evidence to date seems to suggest however that both the EM as well as the Fletcher–Powell algorithms should not be used with multimodal likelihoods in the case of nonnormal data. Of course the best way to ensure a unimodal distribution is in the presence of normality, and this underlines the importance of first verifying multivariate normality (Section 4.6).

6.10 FACTOR ROTATION AND IDENTIFICATION

Unlike principal components, factor models require a priori restrictions in order to yield unique (or identifiable) solutions (Section 6.1). These restrictions are of a purely mathematical nature and thus do not guarantee that coordinate axes will assume positions that can enable the researcher to identity the "real" nature of the factors. In this sense therefore both the factors and principal components share the same identification problem. Thus since factors cannot be identified within nonsingular transformations, an arbitrary (orthogonal) position of the axes is initially chosen. Further rotation of the loadings and the scores may therefore be required to interpret the factors in a meaningful way, as was the case in Section 5.3. Evidently, rotations by themselves cannot enhance interpretability if the data do not contain relevant relationships amongst the variables. If the variables tend to cluster, however, a transformation of axes should reveal such a structure. An orthogonal, or more generally oblique, factor rotation will therefore normally accompany a factor analysis of data, particularly if the objective of the analysis is exploratory in nature. The main objective here is then identical to that of principal components, namely to locate or isolate clusters of variables.

Rotation of the common factors proceeds in the same fashion as encountered in Section 5.3. Once the loadings and the scores have been estimated, the factor model can be expressed as

$$\mathbf{X} = \hat{\boldsymbol{\phi}}\hat{\boldsymbol{\alpha}} + \hat{\boldsymbol{\epsilon}}$$

$$= \hat{\boldsymbol{\phi}}\mathbf{T}^{-1}\hat{\boldsymbol{\alpha}} + \hat{\boldsymbol{\epsilon}}$$

$$= \mathbf{G}\hat{\boldsymbol{\beta}} + \hat{\boldsymbol{\epsilon}} \tag{6.104}$$

where $\mathbf{T}$ is a $(r \times r)$ transformation matrix chosen to optimize some criterion, for example, the variance or the quartimax criterion. We have

$$\hat{\boldsymbol{\Sigma}} = \mathbf{X}^T\mathbf{X} = (\mathbf{G}\hat{\boldsymbol{\beta}} + \hat{\boldsymbol{\epsilon}})^T(\mathbf{G}\hat{\boldsymbol{\beta}} + \hat{\boldsymbol{\epsilon}})$$

$$= \hat{\boldsymbol{\beta}}^T\mathbf{G}^T\mathbf{G}\hat{\boldsymbol{\beta}} + \hat{\boldsymbol{\beta}}^T\mathbf{G}^T\hat{\boldsymbol{\epsilon}} + \hat{\boldsymbol{\epsilon}}^T\mathbf{G}\hat{\boldsymbol{\beta}} + \hat{\boldsymbol{\epsilon}}^T\hat{\boldsymbol{\epsilon}}$$

$$= \hat{\boldsymbol{\beta}}^T\mathbf{G}^T\mathbf{G}\hat{\boldsymbol{\beta}} + \hat{\boldsymbol{\epsilon}}^T\hat{\boldsymbol{\epsilon}}$$

$$= \hat{\boldsymbol{\beta}}^T\hat{\boldsymbol{\Omega}}\hat{\boldsymbol{\beta}} + \hat{\boldsymbol{\Psi}} \tag{6.105}$$

where $\mathbf{G}^T\mathbf{G}$ is the common factor covariance matrix. When the random variables are standardized, both $\hat{\boldsymbol{\Omega}}$ and $\mathbf{G}^T\mathbf{G}$ represent correlation matrices. Oblique rotations can also be used with the Bartlett estimator (Eq. 6.98) since oblique factor estimates are simply the original factors which have been rotated, that is, for the ith observation

$$\mathbf{G}_i^T = (\hat{\boldsymbol{\beta}}\hat{\boldsymbol{\Psi}}\hat{\boldsymbol{\beta}}^T)^{-1}\hat{\boldsymbol{\beta}}\hat{\boldsymbol{\Psi}}^{-1}\mathbf{x}_i^T$$

$$= [(\mathbf{T}^{-1})\hat{\boldsymbol{\alpha}}\hat{\boldsymbol{\Psi}}^{-1}\hat{\boldsymbol{\alpha}}^T(\mathbf{T}^{-1})^T]^{-1}\mathbf{T}^{-1}\hat{\boldsymbol{\alpha}}\hat{\boldsymbol{\Psi}}^{-1}\mathbf{x}_i^T$$

$$= \mathbf{T}^{-1}(\hat{\boldsymbol{\alpha}}\hat{\boldsymbol{\Psi}}^{-1}\hat{\boldsymbol{\alpha}}^T)^{-1}\mathbf{T}\mathbf{T}^{-1}\hat{\boldsymbol{\alpha}}\hat{\boldsymbol{\Psi}}^{-1}\mathbf{x}_i^T$$

$$= \mathbf{T}^{-1}\hat{\boldsymbol{\Phi}}_i^T \tag{6.106}$$

so that factor scores are automatically rotated by the same transformation matrix $\mathbf{T}$ as are the loadings. Since oblique positions of the axes represent the general case, they are frequently employed in exploratory cluster analysis (Example 6.3).

When $\mathbf{T}$ is orthogonal the new correlation loadings $\hat{\boldsymbol{\beta}}$ still correspond to correlation coefficients between the variables and rotated factors. For an oblique rotation, however, the so-called pattern coefficients $\hat{\boldsymbol{\beta}}$ become regression coefficients and must be further transformed to yield correlation loadings. From Eq. (6.104) we have

$$\hat{\boldsymbol{\beta}} = (\mathbf{G}^T\mathbf{G})^{-1}\mathbf{G}^T\mathbf{X} \tag{6.107}$$

where $\mathbf{G}^T\mathbf{X}$ is the matrix of correlations between the common factors and the variables given by

$$\mathbf{G}^T\mathbf{X} = (\mathbf{G}^T\mathbf{G})\hat{\boldsymbol{\beta}} \tag{6.108}$$

and $\mathbf{G}^T\mathbf{G}$ is the factor correlation matrix. The correlations (Eq. 6.108) are also known as structure coefficients and are generally different in magnitude to the regression coefficients $\hat{\boldsymbol{\beta}}$. Since the regression coefficients (Eq. 6.107) however are not bounded in the closed interval $[-1, 1]$, the identification of oblique factors is best accomplished by using the structure values (Eq. 6.108).

Once the values of the loadings and the scores are known, the identification of factors proceeds much in the way as for principal components. It is rarely that a single set of estimates will suffice to yield a thorough analysis of the data, and several passes are usually required before a satisfactory specification is obtained. In this respect factor analysis shares a common feature with most other multivariate techniques. The particular strategy and identification procedure will of course depend on the type of data and the general objective of the analysis. The following stepwise procedure, for example, may be employed when attempting to locate an underlying factor structure.

1. The first set of analysis are normally exploratory and tentative in nature. If the application is new and data are uncertain, it is difficult to make a priori assumptions with any degree of confidence, for example, if it has not been determined whether data are multivariate normal or if the exact number of common factors is not known. A combination of statistical testing and empirical evaluation is therefore almost always required even at the outset. Thus initial attempts should be made to ensure that data are not too highly skewed, and a preliminary principal component analysis may be conducted to help determine the number of factors.

2. Based on the results of part 1 an initial value of r is established and a factor analysis can be carried out followed by oblique (orthogonal) rotation. Again, it may require several experimental "passes" before the final value of r is decided upon, especially if the $AIC(r)$ or $SIC(r)$ criteria of Section 6.6.2 are used. Unlike the principal components model a change in the value of r will alter both the loading coefficients and the scores. In turn, the number of common factors retained will depend on a combination of statistical testing, empirical interpretation, and rotation of the axes.

Example 6.3. As a numerical illustration of Lawley's maximum likelihood model we consider an example from agricultural ecology (Sinha and Lee, 1970). Composite samples of wheat, oats, barley, and rye from various locations in the Canadian Prairie are collected from commercial and government terminal elevators at Thunder Bay (Ontario) during unloading of railway boxcars. The purpose is to study the interrelationships between arthropod infestation and grain environment. The following $p = 9$ variates are observed for $n = 165$ samples:

$Y_1 = $ Grade of sample indicating grain quality (1 highest, 6 lowest)

$Y_2 = $ Moisture content of grain (percentage)

$Y_3 = $ Dockage or presence of weed seed, broken kernels, and other foreign matter

Y_4 = Number of grain arthropods *Acarus*

Y_5 = Number of grain arthropods *Cheyletus*

Y_6 = Number of grain arthropods *Glycyphagus*

Y_7 = Number of grain arthropods *Larsonemus*

Y_8 = Number of grain arthropods *Cryptolestes*

Y_9 = Number of grain arthropods *Psocoptera*　　　　　□

Owing to univariate skewness of the sample histograms, the grade and dockage measurements were transformed by the square root transformation, and the discrete arthropod counts were subjected to the transformation $\log_{10}(Y_j + 1)$. The product-moment correlation matrix of the nine variables is shown in Table 6.4.

Initially a principal components analysis is carried out using the correlation matrix (Table 6.4) and the chi-squared statistic (Section 4.3) is computed as

$$\chi^2 = -N \sum_{i=1}^{r} \ln(l_i) + Nr \ln \left(\sum_{i=1}^{r} \frac{l_i}{r} \right)$$

$$= -164(-2.9072) + 656(-7089)$$

$$= 478.75 - 465.06$$

$$= 13.69$$

for $1/2 \, (5^2 - 13) = 6$ degrees of freedom where $N = n - 1$ and $r = 4$. The latent roots and correlation loadings are shown in Table 6.5. For $\alpha = .01$ the null hypothesis that the last five roots are isotropic cannot be rejected. Next a maximum likelihood factor analysis is carried out using $r = 3$ common factors (Table 6.6 and the loadings are given in Table 6.7. A comparison of Tables 6.5 and 6.7 indicates that principal components and ML factor loadings differ. The PCA shows that low grade, moisture, and to a lesser extent dockage content tend to encourage the presence of *Tarsonemus*, *Cheyletus*, and *Glycyphagus*, and to a lesser degree *Acarus*, but to discour-

Table 6.4　Correlation matrix for the $p = 9$ Variates of Arthropod Infestation Data

Y_1	1.000								
Y_2	.441	1.000							
Y_3	.441	.342	1.000						
Y_4	.107	.250	.040	1.000					
Y_5	.194	.323	.060	.180	1.000				
Y_6	.105	.400	.082	.123	.220	1.000			
Y_7	.204	.491	.071	.226	.480	.399	1.000		
Y_8	.197	.158	.051	.019	.138	−.114	.154	1.000	
Y_9	−.236	−.220	−.073	−.199	−.084	−.304	−.134	−.096	1.000

Source: Sinha and Lee, 1970; reproduced with permission.

Table 6.5   Unrotated Principal Component Correlation Loadings of Orthropod Infestation Data

	Z_1	Z_2	Z_3	Z_4
Y_1: Grade	.601	.573	−.070	−.010
Y_2: Moisture	.803	.080	−.046	.138
Y_3: Dockage	.418	.653	−.273	.280
Y_4: *Acarus*	.404	−.243	−.046	−.431
Y_5: *Cheyletus*	.577	−.274	−.430	.214
Y_6: *Glycyphagus*	.562	−.431	.407	.142
Y_7: *Tarsonemus*	.707	−.348	.270	.197
Y_8: *Cryptolestes*	.252	.337	.675	−.409
Y_9: *Procoptera*	−.439	.051	.384	.654
Latent roots	2.7483	1.3180	1.1107	.9833

Source: Sinha and Lee, 1970; reproduced with permission.

Table 6.6   Chi-Squared Test for Maximum Likelihood Factor Analysis of Arthropod Infestation Data

Number of Factors	χ^2	Degrees of Freedom	Probability
0	202.80	36	0.00
1	68.25	27	0.00
2	33.89	20	0.03
3	16.54	13	0.22

Source: Sinha and Lee, 1970; reproduced with permission.

Table 6.7   Maximum Likelihood Factor Loadings for Arthropod Infestation Data

	Unrotated Solution			Varimax Rotated Solution		
Variate	1	2	3	1	2	3
Grade	0.24	0.61	−0.40	0.22	0.74	0.06
Moisture	0.53	0.53	0.17	0.37	0.37	0.57
Dockage	0.02	0.57	−0.16	−0.05	0.57	0.17
Acarus	0.29	0.14	0.18	0.21	0.03	0.31
Cheyletus	0.79	0.03	0.07	0.73	0.00	0.32
Glycyphagus	0.28	0.37	0.56	0.06	0.02	0.73
Tarsonemus	1.00	0.00	0.00	0.95	0.02	0.31
Cryptolestes	0.35	0.08	−0.32	0.39	0.25	−0.11
Psocoptera	−0.15	−0.39	−0.16	−0.03	−0.25	−0.38

Source: Sinha and Lee, 1970; reproduced with permission.

age *Procoptera*. The second component indicates that low-grade cereals tend to contain more dockage, whereas the third and fourth represent contrasts between *Cheyletus* and *Glycyphagus*, *Cryptolestes*, and *Procoptera* on the one hand and *Cryptolestes* and *Procoptera* on the other. The (unrotated) ML factor loadings indicate that the presence of *Tarsonemus* and *Cheyletus* is primarily related to moisture (factor 1) and that low grade, moisture, and dockage is positively related to *Glycyphagus* and negatively to *Procoptera*. Rotated ML loadings indicate that the three environmental variables are positively related to *Cryptolestes* and negatively to *Procoptera*. The third factor (unrotated) relates negatively *Glycyphagus* to grade and to *Cryptolestes*, but rotated orthogonally the factor correlates positively with moisture and somewhat less to *Acarus*, *Cheyletus*, and *Tarsonemus*. Given the somewhat divergent interpretation which is possible, perhaps an oblique rotation should also be performed to attempt to reconcile the results. On the other hand, given the different mathematical specifications of the two models there are no reasons why principal component and the factor loadings should agree (given the present data) and in the final analysis the choice must depend on substantive considerations.

Example 6.4. A comparison between maximum likelihood factor analysis and principal components, using both orthogonal and oblique correlation loadings, is afforded by an analysis of Olympic decathlon data using eight Olympic decathlon championships since World War II and representing $n = 160$ complete starts (Linden, 1977). The objective is to explore motor performance functions (physical fitness) of athletes. The variables to be analyzed are the 10 events comprising the decathlon:

$$Y_1 = \text{100-m run}$$
$$Y_2 = \text{Long jump}$$
$$Y_3 = \text{Shotput}$$
$$Y_4 = \text{High jump}$$
$$Y_5 = \text{400-m run}$$
$$Y_6 = \text{100-m hurdles}$$
$$Y_7 = \text{Discus}$$
$$Y_8 = \text{Pole vault}$$
$$Y_9 = \text{Javelin}$$
$$Y_{10} = \text{1500-m run}$$

whose correlation matrix is given in Table 6.8. Both the rotated and unrotated correlation loadings are exhibited in Tables 6.9 and 6.10, respectively.

A comparison between PC and ML factor loadings indicates a different pattern of loadings for both initial solutions, but sets of rotated loadings tend to be similar. The principal component loadings appear to be more interpretable, perhaps because of a lack of significant measurement error, and account for a higher percentage of variance than do the factor loadings. Linden (1977) identifies four major clusters of variables that can be identified with the common factors. Factor 1 represents short-range running

Table 6.8 Correlation Matrix of Decathlon Variables

1.00									
.59	1.00								
.35	.42	1.00							
.34	.51	.38	1.00						
.63	.49	.19	.29	1.00					
.40	.52	.36	.46	.34	1.00				
.28	.31	.73	.27	.17	.32	1.00			
.20	.36	.24	.39	.23	.33	.24	1.00		
.11	.21	.44	.17	.13	.18	.34	.24	1.00	
−.07	.09	−.08	.18	.39	.00	−.02	.17	.00	1.00

Source: Linden, 1977; reproduced with permission.

Table 6.9 Principal Components Correlation Loadings for the Initial, Orthogonal (Varimax), and Oblique (Quartimin) Solutions for Decathlon Variables

	Initial				Orthogonal				Oblique			
	1	2	3	4	1	2	3	4	1	2	3	4
Y_1 : 100-m run	.69	.22	−.52	−.21	.88	.13	−.11	.16	.87	.06	−.15	.04
Y_2 : Long jump	.79	.18	−.19	.09	.63	.20	.00	.51	.53	.08	−.05	.45
Y_3 : Shotput	.70	−.54	.05	−.18	.25	.82	−.15	.22	.13	.82	−.16	.06
Y_4 : High jump	.67	.13	.14	.40	.24	.14	.04	.75	.08	.00	−.01	.77
Y_5 : 400-m run	.62	.55	−.08	−.42	.80	.07	.47	.11	.83	.03	.44	.02
Y_6 : 110-m hurdle	.69	.04	−.16	.35	.40	.16	−.15	.63	.26	.02	−.20	.62
Y_7 : Discus	.62	−.52	.11	−.23	.19	.81	−.08	.15	.09	.82	−.08	.01
Y_8 : Pole vault	.54	.09	.41	.44	−.04	.18	.23	.76	−.21	.06	.19	.81
Y_9 : Javelin	.43	−.44	.37	−.24	−.05	.74	.14	.11	−.13	.77	.15	−.01
Y_{10}: 1500-m run	.15	.60	.66	−.28	.05	−.04	.93	.09	.10	−.03	.93	.07
Total variance	37.9	15.2	11.1	9.1	21.3	20.2	12.3	19.4	18.9	19.5	12.1	18.4

Source: Linden, 1977; reproduced with permission.

Table 6.10 Maximum Likelihood (Rao Canonical) Correlation Loadings for the Initial, Orthogonal (Varimax), and Oblique Solutions for Decathlon Variables

	Initial				Orthogonal				Oblique			
	1	2	3	4	1	2	3	4	1	2	3	4
Y_1 : 100-m run	−.07	.35	.83	−.17	.81	.16	−.15	.28	.85	−.05	−.28	.06
Y_2 : Long jump	.09	.43	.60	.28	.47	.21	.00	.61	.29	−.03	−.08	.60
Y_3 : Shotput	−.08	1.00	.00	.00	.15	.83	−.10	.27	.03	.82	−.10	.06
Y_4 : Highjump	.18	.40	.34	.45	.14	.20	.11	.69	−.14	−.01	.07	.82
Y_5 : 400-m run	.39	.22	.67	−.14	.76	.07	.41	.20	.87	.00	.29	−.09
Y_6 : 110-m hurdle	.00	.36	.43	.39	.28	.22	−.05	.55	.07	.01	−.10	.60
Y_7 : Discus	−.02	.73	.03	.02	.12	.81	−.02	.15	.05	.86	−.01	−.10
Y_8 : Pole vault	.17	.25	.23	.39	.07	.17	.16	.48	−.12	.05	.14	.56
Y_9 : Javelin	.00	.44	−.01	.10	.03	.44	.04	.16	−.04	.46	.04	.06
Y_{10}: 1500-m run	1.00	.00	.00	.00	.04	−.05	.81	.12	.06	.07	.82	.03
Total variance	12.3	24.3	18.4	6.4	16.0	17.3	9.0	16.3	16.0	16.4	8.8	17.3

Source: Linden, 1977; reproduced with permission.

Table 6.11   Oblique Factor Regressions Loading Coefficients ("Patterns") for $n = 181$ Rheumatoid Arthritis Patients

Item	G_1 (Fatigue)	G_2 (Hostility)	G_3 (Friendliness)	G_4 (Vigor)	G_5 (Depression)	G_6 (Tension)	G_7 (Confusion)	Communalities
Fatigued	.907							.7364
Exhausted	.891							.8131
Bushed	.980							.7586
Worn out	.688							.6312
Weary	.684							.5948
Sluggish	.607			.275				.6278
Listless	.537							.4837
Miserable	.328				.260			.5707
Helpless	.259			.252				.4486
Furious		.739						.6012
Rebellious		.712						.5875
Bad tempered		.614						.5953
Ready to fight		.607						.3486
Spiteful		.591						.5242
Resentful		.545			.323			.6180
Peeved		.507						.5685
Annoyed		.481						.5981
Angry		.475					.251	.5500
Guilty		.304					.260	.3379
Good-natured			.737					.5850
Considerate			.709					.5123
Trusting			.561					.4058
Sympathetic			.557					.3353
Friendly			.549					.3973
Alert			.533					.4905
Cheerful			.467	.321				.6010
Helpful			.467	.334				.4672
Clear-headed			.415				.232	.2806
Carefree			.376					.2996
Energetic				.790				.6517
Active				.751				.6011
Vigorous				.695				.5169
Full of pep				.645				.6471
Lively				.614				.5147
Efficient			.305	.436				.4552
Sad					.615			.6205
Unhappy					.577			.6584
Blue					.571			.6314
Discouraged					.476			.5795
Bitter		.437			.456			.5760
Desperate		.397				.433		.5334
Hopeless				.429				.3480
Gloomy				.366				.6455
Unworthy					.315			.3274
Lonely				.302			.254	.3008
Worthless					.278			.3517
Terrified		.276						.3627
Nervous						.615		.5538
Tense						.580		.5063
On edge						.554		.6355
Shaky						.464	.279	.4880
Uneasy						.411		.5881
Restless						.409		.4357
Grouchy	.253	.309				.344		.5163

Table 6.11   *(Continued)*

Relaxed	.322	.344		.4420
Forgetful			.539	.3550
Muddled			.538	.6335
Bewildered			.510	.5655
Unable to concentrate			.461	.3094
Uncertain about things			.363	.4032
Panicky			.305	.4085
Confused		.267	.283	.3852
Anxious				.2193
Deceived				.3110
Sorry for things done				.2097

Source: Stitt et al., 1977; reproduced with permission.

speed (Y_1, Y_5), factor 2 is "explosive arm strength" or object projection (Y_3, Y_7, Y_9), factor 3 is running endurance (Y_{10}), and factor 4 correlates with "explosive leg strength" or body projection. Note that the long jump (Y_2) correlates with factor 1 and factor 4 since this event requires both running speed and body projection. The same can be said of the 100-m hurdle (Y_6), which also requires a combination of speed in running and jumping. Similar studies dealing with sports fitness and motor movements may be found in Cureton and Sterling (1964) and Maxwell (1976).

Example 6.5.   An exploratory oblique-cluster factor analysis is carried out by Stitt et al. (1977) to examine whether a "Profile of Mood States" questionnaire administered to $n = 181$ rheumatoid arthritis patients could reveal distinct mood categories. The rotated oblique factor coefficients, together with the factor correlation matrix, are presented in Tables 6.11 and 6.12, respectively. Here the oblique coefficients represent regression ("pattern") coefficients rather than correlations and their magnitudes should be interpreted with care. Nevertheless, a broad qualitative pattern seems to be present in the data (Fig. 6.1), which appears to reveal a "tree-like" structure of moods of the patients. Exploratory analyses of this type can then serve as a guide to further research. A general theoretical outline for the use of factor analysis in medical diagnostics may be found in Overall and Williams (1961; see also Shea (1978). Also a similar study dealing with smoking motives and nicotine addiction is given by Russell et al. (1974). For an analysis of aircraft accidents data see Gregg and Pearson (1961).   □

Table 6.12   Correlation Matrix for Factors of Table 6.8a

	Fatigue	Hostility	Friendliness	Vigor	Depression	Tension
Hostility	0.34	Hostility				
Friendliness	−0.17	−0.20	Friendliness			
Vigor	−0.50	−0.11	0.44	Vigor		
Depression	0.38	0.51	−0.18	−0.23	Depression	
Tension	0.53	0.38	−0.16	−0.21	0.41	Tension
Confusion	0.27	0.31	−0.09	−0.13	0.37	0.25

Source: Stitt et al., 1977; reproduced with permission.

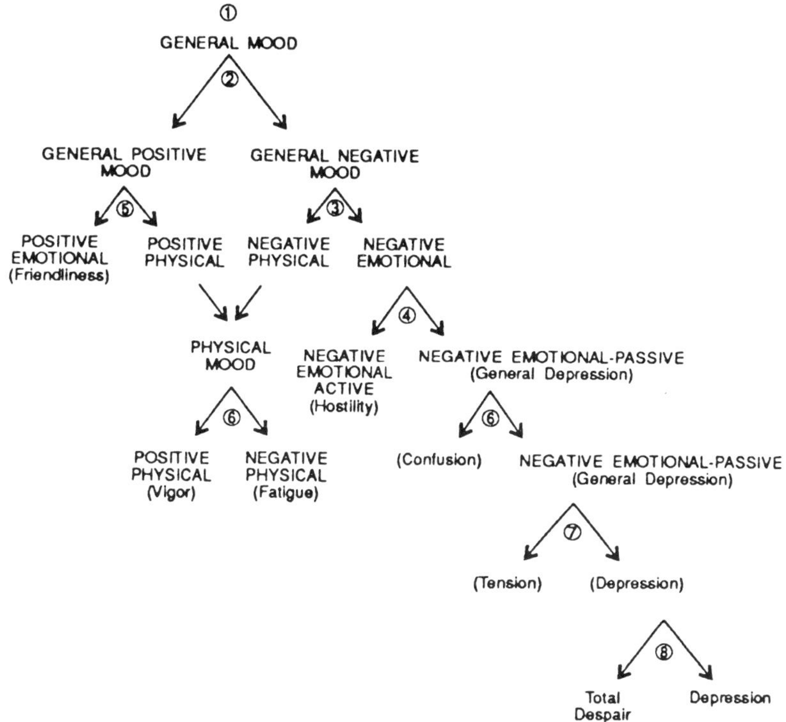

Figure 6.1 Tree diagram showing results for 1–8 factors (Stitt et al., 1977; reproduced with permission).

Example 6.6. A classic application of factor analysis to economic data is provided by Adelman and Morris (1965; 1967 – see also Adelman and Morris, 1970; Brookins, 1970; Rayner, 1970) in their analysis of economic development as it relates to social and demographic factors. An application of the methodology using more recent data is given by Abizadeh et al. (1990) (see also Abizadeh and Basilevsky, 1986). The data consist of $k = 20$ variables observed for $n = 52$ countries, and the objective is to measure the potential for economic development. Because of its heterogenous nature and method of collection the variables contain errors of measurement that vary from variable to variable and country to country, and the data thus provide a good vehicle for the implementation of ML factor analysis. The following variables are selected as good candidates for capturing the principal dimensions of the potential for economic development of 52 "free market" countries. Here, however, the main interest lies in utilizing the factor scores to provide a ranking for the economies viz a viz their relative states of "development."

Y_1 = Males enrolled in elementary schools (percentage of school-age group, 1981)

Y_2 = Females enrolled in elementary schools (percentage of school-age group, 1981)

Y_3 = Status of women (ratio of female to male enrollment in elementary schools, 1981)

Y_4 = Percent of government expenditure on education (1981)

Y_5 = Capital equipment (machinery) imports as percentage of total imports (1981)

Y_6 = Percent of population enrolled in higher education (1981)

Y_7 = Average rate of growth of services (1970–1982)

Y_8 = Average rate of growth of domestic investment (1970–1982)

Y_9 = Average rate of growth of primary exports (1970–1982)

Y_{10} = Average index of food production (1960–1970 = 100; 1980–1982)

Y_{11} = Average annual population growth (1970–1982)

Y_{12} = Percentage of physicians to population (1980)

Y_{13} = Ratio of US aid to total GDP (1981)

Y_{14} = Ratio of US military aid to total GDP (1981)

Y_{15} = Number of cities with population over 500,000 (1981)

Y_{16} = Net direct private investment

Y_{17} = Percentage of labor force to population working age (15–65 years of age; 1981)

Y_{18} = Resource balance

Y_{19} = Percent value of machinery as export (1981)

Y_{20} = Percent value of machinery as import (1981)   □

Five ML factors are computed and rotated to oblique form since on a priori grounds it is expected that the socioeconomic factors are correlated. The factor loadings are given in Table 6.13 and the factor scores in Tables 6.14 and 6.15 where only the first two sets of factor scores are considered.

First, the most important dimension is the economic base or infrastructure of development. The first factor ($\mathbf{F}_1$) correlates positively with the percentage of machinery exports ($\mathbf{Y}_{19}$), number of physicians per populations ($\mathbf{Y}_{12}$), percentage of population enrolled in higher education ($\mathbf{Y}_6$), number of cities with population over half a million ($\mathbf{Y}_{15}$), and average index of food production ($\mathbf{Y}_{10}$). It correlates negatively however with the percentage of government expenditure on education ($\mathbf{Y}_4$), average annual population growth ($\mathbf{Y}_{11}$), and average growth rate of primary exports ($\mathbf{Y}_9$). It therefore appears to represent the general development potential based on the existing infrastructural means of development of the countries. The scores on factor $\mathbf{F}_1$ for the 52 countries are given in Table 6.14 and are ranked in descending order. The results conform to our understanding and

Table 6.13 Oblique Factor Correlation Loadings for $p = 20$ Socioeconomic Variables

Variable	$\mathbf{F}_1$	$\mathbf{F}_2$	$\mathbf{F}_3$	$\mathbf{F}_4$	$\mathbf{F}_5$
$\mathbf{Y}_1$	0.18	0.46	−0.17	0.11	0.99
$\mathbf{Y}_2$	0.30	0.81	−0.21	—	0.85
$\mathbf{Y}_3$	0.34	0.97	−0.13	−0.26	0.49
$\mathbf{Y}_4$	−0.42	—	0.22	—	—
$\mathbf{Y}_5$	−0.32	—	−0.12	0.21	0.35
$\mathbf{Y}_6$	0.74	0.51	−0.26	−0.13	0.31
$\mathbf{Y}_7$	−0.19	−0.55	0.30	0.79	0.19
$\mathbf{Y}_8$	−0.12	−0.50	0.40	0.97	0.18
$\mathbf{Y}_9$	−0.82	−0.11	0.17	0.12	0.10
$\mathbf{Y}_{10}$	0.41	0.23	−0.26	—	0.13
$\mathbf{Y}_{11}$	−0.81	−0.39	0.29	0.32	−0.17
$\mathbf{Y}_{12}$	0.85	0.53	−0.32	−0.31	0.27
$\mathbf{Y}_{13}$	−0.35	−0.18	0.86	—	−0.20
$\mathbf{Y}_{14}$	−0.29	—	0.92	0.11	—
$\mathbf{Y}_{15}$	0.40	0.16	−0.14	—	—
$\mathbf{Y}_{16}$	—	—	—	0.34	0.10
$\mathbf{Y}_{17}$	—	—	−0.15	0.45	0.10
$\mathbf{Y}_{18}$	−0.21	−0.37	—	0.10	−0.16
$\mathbf{Y}_{19}$	0.88	0.22	−0.22	−0.14	—
$\mathbf{Y}_{20}$	—	—	−0.15	0.45	0.10

Source: Abizadeh et al., 1990; reproduced with permission.

experience of economic development. Setting aside for the moment the group of developed countries that have and will continue to have a good development potential (countries 1–14 in Table 6.14), South Korea ranks the highest in the remaining group followed by Singapore and others. Notably, these countries are the ones most expected to experience a relatively high rate of economic growth, *ceterus paribus*. Alternatively, beginning from the other end we find the poor African countries, which indicates a very low level of development potential.

The second factor $\mathbf{F}_2$ (Table 6.15) represents a social factor that emphasizes the status of women, an important social indicator of development potential. It is evident that women can contribute to economic development of a country by actively participating in its labor force, particularly if they are skilled and educated.

Again, the factor scores provide a ranking for this dimension of potential for economic developments, where it can be seen that countries like Panama, Thailand, the Philippines, and Singapore fall in the middle of the $\mathbf{F}_2$ scale as they did on the basis of $\mathbf{F}_1$ (Table 6.14). Also countries such as Syria, Ethiopia, and Sudan, which were low on $\mathbf{F}_1$, continue to rank low on $\mathbf{F}_2$ as well. The methodology can also be used to measure government size of various countries (Abizadeh and Basilevsky, 1990).

Table 6.14 Oblique Factor Scores for F_1, the Factor of General Potential for Economic Development

Country	F_1	Country	F_1
1. USA	1.861	27. Panama	−0.429
2. Germany	1.846	28. Thailand	−0.434
3. Japan	1.755	29. Philippines	−0.445
4. Italy	1.639	30. Peru	−0.462
5. Sweden	1.569	31. Turkey	−0.479
6. Austria	1.400	32. Ivory Coast	−0.516
7. Belgium	1.253	33. Mauritius	−0.561
8. France	1.250	34. Pakistan	−0.571
9. UK	1.214	35. Indonesia	−0.712
10. Spain	1.093	36. Senegal	−0.720
11. Finland	1.080	37. Morocco	−0.730
12. Denmark	1.063	38. El Salvador	−0.731
13. Canada	0.831	39. Venezuela	−0.746
14. Netherlands	0.804	40. Syria	−0.795
15. Korea	0.738	41. Jamaica	−0.838
16. Singapore	0.646	42. Papua NG	−0.873
17. New Zealand	0.274	43. Mexico	−0.936
18. Australia	0.215	44. Sudan	−0.946
19. Uruguay	0.103	45. Bolivia	−0.950
20. Yemen	0.005	46. Ethiopia	−0.985
21. India	−0.056	47. Zimbabwe	−1.084
22. Sri Lanka	−0.285	48. Tanzania	−1.148
23. Tunisia	−0.314	49. Nicaragua	−1.170
24. Chile	−0.327	50. Honduras	−1.203
25. Jordan	−0.406	51. Zambia	−1.228
26. Malaysia	−0.419	52. Kenya	−1.368

Source: Abizadeh et al., 1990; reproduced with permission.

6.11 CONFIRMATORY FACTOR ANALYSIS

The factor models described in preceding sections are usually referred to as exploratory factor models, since aside from a specification of r (or Ψ) no other a priori information is required. The objectives of estimating common factors in such models are thus not very different from those of a principal components analysis. By using a more general error structure, exploratory factor analysis seeks to gain further insight into the structure of multivariate data. At times however the investigator may already have carried out a factor analysis of a set of variables in a different sample, in which case prior information is available for use in further samples. Thus the number of common factors may already be known, or loading coefficients may have been tested and certain ones found to be insignificantly different from zero. If subsequent samples are taken it would be wasteful to ignore such prior

Table 6.15 Oblique Factor Scores for F_2, the Status of Women Factor for Potential Economic Development

Country	F_2	Country	F_2
1. Jamaica	0.951	27. Peru	0.289
2. Netherlands	0.797	28. Zambia	0.285
3. Chile	0.788	29. Thailand	0.284
4. Australia	0.752	30. Singapore	0.270
5. Denmark	0.731	31. Korea	0.203
6. Belgium	0.692	32. Tanzania	0.134
7. France	0.683	33. Malaysia	0.062
8. Spain	0.679	34. El Salvador	0.037
9. UK	0.668	35. Kenya	−0.009
10. Italy	0.635	36. Bolivia	−0.237
11. Sweden	0.635	37. Turkey	−0.319
12. New Zealand	0.632	38. Jordan	−0.350
13. Finland	0.593	39. Indonesia	−0.399
14. Germany	0.591	40. Papua NG	−0.430
15. Mexico	0.580	41. Ivory Coast	−0.622
16. USA	0.572	42. Syria	−0.787
17. Venezuela	0.531	43. Senegal	−1.082
18. Japan	0.518	44. Tunisia	−1.086
19. Austria	0.502	45. Nicaragua	−1.130
20. Panama	0.502	46. Sudan	−1.203
21. Canada	0.500	47. India	−1.320
22. Zimbabwe	0.394	48. Mauritius	−1.417
23. Uruguay	0.397	49. Ethiopia	−1.421
24. Honduras	0.363	50. Morocco	−1.851
25. Philippines	0.333	51. Pakistan	−2.497
26. Sri Lanka	0.317	52. Yemen	−4.444

Source: Abizadeh et al., 1990; reproduced with permission.

information, and we may then wish to impose zero restrictions on the loadings. Values other than zeroes can also be used, but zeroes are most common in practice. Other advantages are that restricted factor models can improve factor interpretation and render the factor model identifiable.

Let Ω be the factor correlation matrix. The total number of parameters to be estimated in α, Ω, and Ψ is

$$pr + \tfrac{1}{2}r(r+1) + p = \tfrac{1}{2}(2p+r)(r+1) \qquad (6.109)$$

Let n_α, n_Ω, and n_ψ be the number of specified (restricted) parameters in α, Ω, and Ψ respectively, where $m = n_\alpha + n_\Omega + n_\psi$. Then the number of unrestricted parameters to be estimated is

$$\tfrac{1}{2}(2p+r)(r+1) - m \qquad (6.110)$$

which should be smaller than $\frac{1}{2}p(p+1)$—the number of distinct elements of Σ—to yield a nontrivial solution. Thus

$$r^2 + m > \tfrac{1}{2}(p+r)(p+r+1) \tag{6.111}$$

and a necessary condition for factor parameters to be uniquely determined is that

$$n_\alpha + n_\Omega \geq r^2 \tag{6.112}$$

Sufficient conditions however are difficult to define, since not only the number but also the positioning of the zero-restricted parameters have to be taken into account (Joreskog and Lawley, 1968). Note that restrictions can include zero constraints on the off-diagonal elements of Ω, which results in mixed orthogonal/oblique factors. Clearly, if a sufficient number of restrictions are used, it becomes unnecessary to rotate the factors. The diagonal elements of Ψ can also be prespecified, but this is less usual in practice. For derivation and further detail concerning the confirmatory model see Lawley and Maxwell (1971). Just as for the exploratory model, confirmatory maximum likelihood factor analysis loadings are computed iteratively, but recently Bentler (1982) proposed a noniterative computational scheme for such a model. One difficulty with confirmatory factor analysis however, as with any other model incorporating a priori restrictions, is the introduction of bias into the estimates when incorrect restrictions are used. At times zero restrictions arise naturally in specific circumstances. Thus Anderson (1976), for example, has developed a factor model useful for confirming clinical trials, which uses a block triangular matrix of factor loadings to achieve a unique rotation of the axes.

Example 6.7. The data of Example 6.3 are divided into two groups using a random process, and a confirmatory ML factor analysis is carried out on half of the data. Since the exploratory factor analysis has indicated that *Tarsonemus*, grade, and *Glycyphagus* dominate the first, second, and third factors, respectively, these loadings are kept fixed, together with several other loadings. Coefficients that are set to zero are replaced by $-$. For biological interpretation and implications of the analysis see Sinha and Lee (1970) (Table 6.16). Here an oblique factor solution is obtained directly without any further rotation. $\qquad\qquad\Box$

6.12 MULTIGROUP FACTOR ANALYSIS

As for the principal components model, it is possible to extend factor analysis to the multimode and multigroup situations discussed in Sections

Table 6.16 Oblique Confirmatory ML solution with (Approximate) 95% Confidence Interval for $p = 9$ variates[a]

Variate	F_1	F_2	F_3
Grade	—	—	0.96 ± 0.16
Moisture	-0.14 ± 0.29	0.63 ± 0.31	0.43 ± 0.14
Dockage	-0.48 ± 0.32	0.51 ± 0.32	0.43 ± 0.14
Acarus	0.05 ± 0.24	0.05 ± 0.26	0.18 ± 0.12
Cheyletus	0.38 ± 0.30	-0.38 ± 0.32	0.27 ± 0.13
Glycyphagus	0.88 ± 0.18	—	—
Tarsonemus	—	0.74 ± 0.14	—
Cryptolestes	-0.17 ± 0.25	-0.00 ± 0.26	0.15 ± 0.12
Psocoptera	-0.58 ± 0.36	0.35 ± 0.36	-0.21 ± 0.13

Factor Correlations

1.00			
0.78 ± 0.15	1.00		$\chi^2 = 13.40$ with 13 degrees of
0.15 ± 0.13	0.22 ± 0.15	1.00	freedom, probability $= 0.34$.

[a]Fixed coefficients are replaced by dashes.

Source: Sinha and Lee, 1970; reproduced with permission.

5.4 and 5.5. An annotated bibliography for three-mode (and higher) models is given in Kroonenberg (1983b). In this section we describe briefly a generalization of the Lawley ML factor analysis model to the case of several groups of variables (from McDonald, 1970) (see also Joreskog, 1970, 1971; Browne, 1980) which is similar to a canonical correlation analysis of several groups of variables.

Consider t sets of variables $\mathbf{Y}_{(1)}, \mathbf{Y}_{(2)}, \ldots, \mathbf{Y}_{(t)}$ such that

$$\mathbf{Y}_{(1)} = \boldsymbol{\mu}_{(1)} + \boldsymbol{\alpha}_{(1)} \boldsymbol{\Phi}_{(1)} + \boldsymbol{\epsilon}_{(1)}$$

$$\mathbf{Y}_{(2)} = \boldsymbol{\mu}_{(2)} + \boldsymbol{\alpha}_{(2)} \boldsymbol{\Phi}_{(2)} + \boldsymbol{\epsilon}_{(2)}$$

$$\mathbf{Y}_{(t)} = \boldsymbol{\mu}_{(t)} + \boldsymbol{\alpha}_{(t)} \boldsymbol{\Phi}_{(t)} + \boldsymbol{\epsilon}_{(t)} \tag{6.113}$$

represent t factor models where $\mathbf{Y}_{(k)}$ is $(p_k \times 1)$, $\boldsymbol{\Phi}_{(k)}$ is $(r \times 1)$ and $\boldsymbol{\alpha}_{(k)}$ is a $(p_k \times r)$ matrix of loadings $(k = 1, 2, \ldots, t)$. Equation (6.113) may be expressed more compactly as

$$\mathbf{Y} = \boldsymbol{\mu} + \boldsymbol{\alpha}\boldsymbol{\Phi} + \boldsymbol{\epsilon} \tag{6.114}$$

where $p = \Sigma_{k=1}^{t} p_k$ and $\mathbf{Y}$ is $(p \times 1)$, $\boldsymbol{\alpha}$ $(p \times r)$, and $\boldsymbol{\Phi}$ $(r \times 1)$. The covariance matrix of the entire set is then

$$\boldsymbol{\Sigma} = \boldsymbol{\alpha}\boldsymbol{\alpha}^{\mathrm{T}} + \boldsymbol{\Psi} \tag{6.115}$$

where the $(p \times p)$ residual covariance matrix has the block diagonal form

$$\Psi = \begin{bmatrix} \Psi_{11} & & & 0 \\ & \Psi_{22} & & \\ & & \ddots & \\ 0 & & & \Psi_{tt} \end{bmatrix} \tag{6.116}$$

and $\mathrm{cov}(\Phi, \epsilon^T) = 0$, $\Phi \sim N(0, I)$, and $\epsilon \sim N(0, \Psi)$. Note that the error covariance matrix Ψ is no longer constrained to be strictly diagonal.

Given a total sample of size n, ML estimators can be obtained in a standard way. The sample estimator of the covariance matrix Σ is

$$S = \begin{bmatrix} S_{11} & S_{12} & \cdots & S_{1t} \\ S_{21} & S_{22} & \cdots & S_{2t} \\ \vdots & \vdots & & \vdots \\ S_{t1} & S_{t2} & \cdots & S_{tt} \end{bmatrix} \tag{6.117}$$

and minimizing the expression

$$F = \ln|\Sigma| + \mathrm{tr}(S\Sigma^{-1}) - \ln|S| - p$$

leads to the maximum likelihood estimators $\hat{\alpha}$ and $\hat{\Psi}$ such that

$$\hat{\Sigma} = \hat{\alpha}\hat{\alpha}^T + \hat{\Psi} \tag{6.118}$$

The actual solutions and numerical techniques are discussed by Browne (1980) who shows that multigroup factor analysis can be considered as a generalization of canonical correlation (see also Browne, 1979). Also Bentler and Lee (1979) have developed a multigroup factor model which is a more general version of Tucker's three-mode principal components model (Section 5.4).

6.13 LATENT STRUCTURE ANALYSIS

Both principal components as well as generalized least squares-type factor models standardize the observed variables to zero mean and employ second-order moment matrices, such as the covariance or correlation matrix, to obtain an initial set of coefficients. Although normality is generally not required (except for testing purposes), the usual factor models nevertheless assume that at most second moment statistics are sufficient to estimate the model. Factor models however are generally unidentified (no unique solution) and even in the presence of prior identification constraints there is no guarantee of avoiding a Heywood case. Two reasons for such a state of affairs are (1) skewed distributions and (2) bimodality of the data. This suggests a more general class of models, first suggested by Lazarsfeld (1950; see also Lazarsfeld and Henry, 1968; Anderson, 1959; Fielding, 1978; Bartholomew, 1987) and known generally as latent structure analysis. For a continuous (and continuously observed) multivariate population, latent structure models fall into two categories depending on whether the latent variables are considered to be discrete or continuous.

Consider the situation where a multivariate sample consists of r homogeneous groups such that the distribution is not necessarily symmetric. We then have a mixed sample consisting of subsamples drawn from r distinct populations, and the objective is to "unmix" the sample space into a small number of latent subsamples or classes. Since the distribution need not be symmetric, estimates of higher moments or "interactions" such as

$$\sum_{i=1}^{n} x_{ij} x_{ik} x_{ih} \tag{6.119}$$

must be taken into account. When the variables X_1, X_2, ..., X_p are standardized to unit length we have the corresponding higher order "correlation" coefficients ρ_{jk}, ρ_{jkh}, ..., and so forth to any desired order. In practice the existence of higher-order interaction terms is usually not known and their relevance must be determined together with the parameters of the model, defined as follows. Let a population consist of r latent classes A_1, A_2, ..., A_r each of relative size π_1, π_2, ..., π_r such that $\Sigma_{l=1}^{r} \pi_l = 1$. Let μ_{jl} denote the mean of variable Y_j within the latent class l. Then the normal equations for the so-called "latent profile" model is

$$1 = \sum_{l=1}^{r} \pi_l$$

$$\mu_j = 0 = \sum_{l=1}^{r} \pi_l \mu_{jl}$$

$$\rho_{jk} = \sum_{l=1}^{r} \pi_l \mu_{jl} \mu_{kl}$$

$$\rho_{jkh} = \sum_{l=1}^{r} \pi_l \mu_{jl} \mu_{kl} \mu_{hl} \tag{6.120}$$

and so on for higher order interactions. Although the basic model only considers means of the form μ_{jl}, within the latent classes higher moments such as variances can also be introduced. It is also interesting to note that a unique (identifiable) solution may be obtained by varying the number of interaction terms ρ_{jkh}, ..., given p observed random variables (see Anderson, 1959). Analogously to the factor model (Eq. 6.2), latent profile models use the concept of conditional independence, that is, latent classes are obtained in such a way that the variables within each class are independent, not merely orthogonal. The operative assumption is that second, third and higher moments have arisen because of the nonhomogeneous composition of the original sample.

Other models are also possible, for example, when the observed variables are discrete (Section 9.7) or when both the observed and latent spaces are continuous. Based on the assumption of local independence we can write

$$E(X_j X_k) = E(X_j \mid Z = z)E(X_k \mid Z = z)$$

for some latent variable Z. The specific type of model obtained is then dependent on the form of the functions $f_j(z)$ and $f_k(z)$. For example, in factor analysis we have the linear functions

$$f_j(z) = a_j + b_j z$$

$$f_k(z) = a_k + b_k z$$

but more general functions are also possible, such as polynomials or exponentials. Nonlinear continuous latent spaces thus provide a natural generalization of the factor model. For greater detail the reader is referred to Lazarsfeld and Henry (1968).

EXERCISES

6.1 Show that the number of degrees of freedom of a common factor model can be expressed as in Eq. (6.9).

6.2 Using Eq. (6.9) show that for $r = 1$ common factor we must have $p > 3$.

6.3 Prove that the matrix $\mathbf{\Gamma} = \boldsymbol{\alpha}\boldsymbol{\alpha}^{\mathrm{T}}$ (Theorem 6.3) can only have r nonzero latent roots.

6.4 Following Theorem 6.6 prove that the reciprocal proportionality model (Section 6.4.1) is invariant with respect to scaling of the observed random variables.

6.5 Prove that in Lemma 6.1

$$F(\mathbf{\Sigma}) = p(a - \ln g - 1)$$

where a and g are the arithmetic and geometric means of the latent roots of $\mathbf{\Sigma}^{-1}\mathbf{S}$.

6.6 Prove that Eq. (6.67) is equivalent to Lawley's normal equations (Eq. 6.49).

6.7 Prove Eq. (6.68).

6.8 Using Eqs. (6.101a) and (6.102), prove that $d_T \leq d_B$ (Mardia, 1977).

6.9 For Section 6.10, show that for an orthogonal matrix $\mathbf{T}$ the rotated loadings are correlation coefficients between the observed variables and the rotated factors.

6.10 Show that minimizing Eq. (6.99) leads to the same estimator as Eq. (6.98).

6.11 Show that the alpha factor model (Eq. 6.74) is equivalent to solving the equations

$$[\mathbf{\Gamma} - \lambda_i \, \mathrm{diag}(\mathbf{\Gamma})]\mathbf{\Pi}_i = \mathbf{0} \qquad (i = 1, 2, \ldots, r)$$

6.12 Prove Lemma 6.1.

6.13 Prove that the diagonal matrix of Eq. (6.3) is positive definite if and only if $\sigma_i^2 > 0$ for $i = 1, 2, \ldots, p$.

6.14 Show that minimizing Eq. (6.100) leads to the estimator Eq. (6.101).

6.15 Consider the expression

$$f(\mathbf{\Sigma}) = k[\ln|\mathbf{\Sigma}| + \mathrm{tr}(\mathbf{\Sigma}^{-1}\mathbf{A})]$$

where k is a constant of proportionality. Prove that for $\mathbf{\Sigma}$ and $\mathbf{A}$ positive definite $f(\mathbf{\Sigma})$ is minimized uniquely at $\mathbf{\Sigma} = \mathbf{A}$.

6.16 Show that for $n \geq 8$ the $\mathrm{SIC}(r)$ criterion (Eq. 6.81b) results in fewer common factors that the $\mathrm{AIC}(r)$ criterion (Eq. 6.81a).

6.17 Prove Eq. (6.96).

6.18 Stroud (1953) obtains the following correlation matrix for $p = 14$ body measurements of soldier termites (Table 6.17).

(a) Carry out a maximum likelihood factor analysis using the correlation matrix of Table 6.17. How many common factors are retained using (i) likelihood ratio chi-squared test, (ii) Akaike's criterion, and (iii) Schwartz's criterion. What do you conclude?

Table 6.17 Correlation Matrix of Body Measurements of Soldier Termites[a]

1	2	3	4	5	6	7	8	9	10	11	12	13	14
—	.685	.663	.857	.679	.741	.867	.802	.877	.865	.862	.682	.889	.657
	—	.761	.571	.350	.488	.560	.490	.535	.586	.537	.446	.548	.413
		—	.528	.389	.598	.494	.446	.587	.539	.653	.231	.473	.572
			—	.788	.850	.815	.813	.823	.927	.827	.801	.892	.597
				—	.862	.760	.756	.790	.807	.800	.668	.811	.500
					—	.728	.768	.790	.866	.812	.668	.800	.501
						—	.865	.939	.819	.910	.714	.888	.606
							—	.863	.840	.845	.844	.814	.528
								—	.864	.979	.647	.884	.694
									—	.895	.765	.886	.612
										—	.636	.874	.666
											—	.780	.357
												—	.579

[a] Y_1 = Length of right mandible; Y_2 = length of second antennal segment; Y_3 = length of third antennal segment; Y_4 = length of third tibia; Y_5 = width of third tibia; Y_6 = width of third femur; Y_7 = height of head; Y_8 = length of head; Y_9 = width of head; Y_{10} = length of pronotum; Y_{11} = width of pronotum; Y_{12} = length of postmentum; Y_{13} = width of postmentum; Y_{14} = maximum diameter of eye.

(b) Rotate the common factor loadings using (i) orthogonal rotation and (ii) oblique rotation. Do they aid analysis of the data? (See also Hopkins, 1966 and Sprent, 1972.)

CHAPTER 7

Factor Analysis of Correlated Observations

7.1 INTRODUCTION

In the previous chapters it is explicitly assumed that the data represent random, independent samples from a multivariate distribution. Here which observation appears first, second, . . . nth is strictly arbitrary since the rows of $\mathbf{Y}$ can be permuted without altering the outcome. Such a situation however is not characteristic of all forms of multivariate data. Thus, for certain types of data the observations appear in a specific order (sequence), and it is no longer permissible to interchange observations without a fundamental change in the outcome. This is generally true for observations which are ordered over time, spatial location, or both. The ordering induces correlation amongst neighboring data points, thus altering the perspective of the analysis—for example, by inducing lead-lag relationships between points of a time series. Now, rather than constituting a set of n observations on a random variable, sequential correlated data are viewed either as a realization of a stochastic process or else as a function that depends on physical–spatial variables such as time, spatial location, and so forth. Note that in the former view a time series is considered as consisting of n random variables observed jointly for one measurement or sample point, whereas for the latter the time points are considered as comprising the sample observations. A factor or a principal components analysis of correlated observations will feel the impact of such correlation, over and above the correlation that exits among the variables. Generally speaking, the analysis will depend on whether the serial correlation is taken into account or not, and whether the analysis is performed in the time or in the frequency domain.

7.2 TIME SERIES AS RANDOM FUNCTIONS

The most straight-forward case of correlated observations is that of a time series, where the observations are ordered over time at regular and discrete time points such as days, weeks, or months. A time series can also be observed at irregular time points or even continuously if time gaps between points can be made arbitrarily small. A unifying feature of such series is that they can be viewed as random time functions whose analytic form(s) are either not known or not specified. If p series $\mathbf{Y}_1(t)$, $\mathbf{Y}_2(t), \ldots, \mathbf{Y}_p(t)$ are available they can be represented in terms of the usual $(n \times p)$ data matrix

$$
\mathbf{Y}(t) = \begin{bmatrix} y_{11} & y_{12} & \cdots & y_{1p} \\ y_{21} & y_{22} & \cdots & y_{2p} \\ \vdots & \vdots & & \vdots \\ y_{n1} & y_{n2} & \cdots & y_{np} \end{bmatrix}
$$

where n is the number of time points and p is the number of characteristics or time series. If the series have arisen on an equal footing, that is, if none is considered to be functionally dependent on other series, then interest frequently lies in the correlational structure between such series, which may have arisen because of the presence of a smaller number of common time functions, for example, trends or cycles. Such regular behavior may not always be observable directly within the series, but may nevertheless exist, masked by random error or "noise." A factor analysis of multiple time series can then provide convenient estimates of the "hidden" time functions, in the form of factor or component scores (Sections 3.4 and 6.8). It is interesting to note that the existence of such functions usually removes much of the arbitrariness which at times is ascribed to factor identification since the existence of common factors in time series can be verified by further statistical analysis, for example, using graphical methods or by regression analysis. When correlation between the time series is of exclusive or primary interest, serial correlation can frequently be ignored. Here no change in the model is required since the series are simply viewed as a set of intercorrelated random variables, which happen to be observed over time. The principal areas of application of the methodology include the estimation of underlying time functions or data smoothing, construction of index numbers or "indicators," and data compression. Since the underlying time functions often possess a simpler structure than the original series, they can also be used for forecasting purposes (Otter and Schur, 1982). A unidimensional ordering of sample points also occurs in physical space represented, say, by perpendicular rock core drillings or horizontal lengths of geological rock formations (e.g., see, Harbaugh and Demirmen, 1964; Imbrie and Kipp, 1971).

In certain disciplines, for example the social sciences, time series are frequently considered as "variables," that is, observations on random variables which happen to be observed over time. Here the n time points are viewed as comprising the sample space whose dimensions are not necessarily orthogonal, owing to the serial correlation which usually exists among such observations. Multivariate analysis is then performed on the variables by ignoring the correlations among the time-ordered data points. Again, various Grammian association matrices may be used depending on the objective(s) of the analysis (Section 2.4) by correcting for the means, standard deviations or both. Time series variables however may at times require further correction owing to their time dependence since dominant movements such as trend may have to be removed if we are to perceive variance that is unique to each series, that is, variance that is independent of major components common to each time series. A simple translation of axes is therefore not always sufficient to eliminate the effects of "size" or level of the variables if size is an increasing (decreasing) function of time. Thus a preliminary regression analysis must at times be performed to eliminate dominant movement(s). Note however that it is not possible to eliminate time altogether from such variables as is at times advocated (e.g., see Walker 1967) since time series are intrinsically dependent on time. The following example illustrates some of the difficulties which may be involved when treating correlated time series as if they were ordinary variables.

Example 7.1. Using published data for $p = 18$ offence categories Ahamad (1967) has carried out a principal component analysis of the number of offences committed in each category, per year, during the period 1950–1963. The variables are defined as follows and the data matrix is given in Table 7.1. Since $n < p$

Y_1 = Homicide
Y_2 = Woundings
Y_3 = Homosexual offences
Y_4 = Heterosexual offences
Y_5 = Breaking and entering
Y_6 = Robbery
Y_7 = Larceny
Y_8 = Fraud and false pretence
Y_9 = Receiving
Y_{10} = Malicious injury to property
Y_{11} = Forgery
Y_{12} = Blackmail
Y_{13} = Assault
Y_{14} = Malicious damage
Y_{15} = Revenue laws
Y_{16} = Intoxication laws
Y_{17} = Indecent exposure
Y_{18} = Taking motor vehicle without consent

Table 7.1 Total Number of Offences for England and Wales, 1950–1963, for 18 Categories

Variate	1950	1951	1952	1953	1954	1955	1956	1957	1958	1959	1960	1961	1962	1963
Y_1	529	455	555	456	487	448	477	491	453	434	492	459	504	510
Y_2	5,258	5,619	5,980	6,187	6,586	7,076	8,433	9,774	10,945	12,707	14,391	16,197	16,430	18,655
Y_3	4,416	4,876	5,443	5,680	6,357	6,644	6,196	6,327	5,471	5,732	5,240	5,605	4,866	5,435
Y_4	8,178	9,223	9,026	10,107	9,279	9,953	10,505	11,900	11,823	13,864	14,304	14,376	14,788	14,722
Y_5	92,839	95,946	97,941	88,607	75,888	74,907	85,768	105,042	131,132	133,962	151,378	164,806	192,302	219,138
Y_6	1,021	800	1,002	980	812	823	965	1,194	1,692	1,900	2,014	2,349	2,517	2,483
Y_7	301,078	355,407	341,512	308,578	285,199	295,035	323,561	360,985	409,388	445,888	489,258	531,430	588,566	635,627
Y_8	25,333	27,216	27,051	27,763	26,267	22,966	23,029	26,235	29,415	34,061	36,049	39,651	44,138	45,923
Y_9	7,586	9,716	9,188	7,786	6,468	7,016	7,215	8,619	10,002	10,254	11,696	13,777	15,783	17,777
Y_{10}	4,518	4,993	5,003	5,309	5,251	2,184	2,559	2,965	3,607	4,083	4,802	5,606	6,256	6,935
Y_{11}	3,790	3,378	4,173	4,649	4,903	4,086	4,040	4,689	5,376	5,598	6,590	6,924	7,816	8,634
Y_{12}	118	74	120	108	104	92	119	121	164	160	241	205	250	257
Y_{13}	20,844	19,963	19,056	17,772	17,379	17,329	16,677	17,539	17,344	18,047	18,801	18,525	16,449	15,918
Y_{14}	9,477	10,359	9,108	9,278	9,176	9,460	10,997	12,817	14,289	14,118	15,866	16,399	16,852	17,003
Y_{15}	24,616	21,122	23,339	19,919	20,585	19,197	19,064	19,432	24,543	26,853	31,266	29,922	34,915	40,434
Y_{16}	49,007	55,229	55,635	55,688	57,011	57,118	63,289	71,014	69,864	69,751	74,336	81,753	89,709	89,149
T_{17}	2,786	2,739	2,598	2,639	2,587	2,607	2,311	2,310	2,371	2,544	2,719	2,820	2,614	2,777
Y_{18}	3,126	4,595	4,145	4,551	4,343	4,836	5,932	7,148	9,772	11,211	12,519	13,050	14,141	22,896

Source: Ahamad, 1967; reproduced with permission

Table 7.2 Correlation of $p = 18$ Offences for England and Wales, 1950–1963

1.00																	
−0.041	1.00																
−0.371	−0.133	1.00															
−0.174	0.969	−0.070	1.00														
0.128	0.947	−0.379	0.881	1.00													
0.013	0.970	−0.315	0.940	0.963	1.00												
0.074	0.961	−0.339	0.909	0.992	0.966	1.00											
0.099	0.923	−0.377	0.870	0.974	0.950	0.975	1.00										
0.157	0.900	−0.406	0.822	0.981	0.915	0.982	0.967	1.00									
0.347	0.469	−0.522	0.371	0.648	0.550	0.623	0.751	0.701	1.00								
0.089	0.954	−0.173	0.896	0.948	0.942	0.942	0.957	0.909	0.645	1.00							
0.183	0.941	−0.322	0.895	0.943	0.956	0.936	0.922	0.887	0.565	0.950	1.00						
0.175	−0.502	−0.534	−0.488	−0.385	−0.367	−0.389	−0.354	−0.346	−0.027	−0.535	−0.389	1.00					
−0.082	0.972	−0.234	0.962	0.930	0.964	0.944	0.885	0.875	0.412	0.893	0.922	−0.397	1.00				
0.264	0.876	−0.471	0.782	0.961	0.910	0.943	0.954	0.945	0.723	0.920	0.932	−0.272	0.834	1.00			
−0.022	0.976	−0.075	0.949	0.926	0.932	0.942	0.892	0.891	0.439	0.931	0.900	−0.590	0.958	0.816	1.00		
0.173	0.210	−0.545	0.109	0.339	0.282	0.335	0.448	0.414	0.677	0.304	0.286	0.428	0.144	0.498	0.074	1.00	
0.015	0.957	−0.167	0.893	0.959	0.916	0.954	0.919	0.926	0.542	0.944	0.905	−0.523	0.909	0.915	0.927	0.256	1.00

Source: Ahamad, 1967; reproduced with permission.

the correlation matrix of the 18 categories (Table 7.2) is singular. The data matrix has two further peculiarities which can modify a principal components analysis; first, the entries are counts (Chapter 8); second, the "sample space" consists of ordered time points. A principal components analysis of the correlation matrix (Table 7.3) reveals $r = 3$ components with at least a single loading coefficient whose magnitude is sufficiently large to be judged "significant," although the usual tests of significance are not applicable. Further possible difficulties also arise from using the correlation matrix with offence counts. First, since the rows of Table 7.1 are adjusted to zero means, the analysis treats infrequent offences such as homicides $(\mathbf{Y}_1)$ in the same way as the more common ones such as larceny $(\mathbf{Y}_7)$. Second, standardizing the categories to unit length tends to ignore differences among the variates due to time variability, an important consideration when attempting to forecast. Since the data of Table 7.1 represent a contingency table (Chapter 8), it is frequently useful to consider other measures of association, for example Euclidian measures such as cosines or the unadjusted inner product (Section 2.4) or non-Euclidian measures (Exercise 7.12).

Since the principal components loadings of Table 7.3 are computed from an association matrix of the variables they are independent of time, in the sense that they depend on the offence categories, rather than the years.

Table 7.3   Principal Component Loading Coefficients of $p = 18$ Offence Categories Using the Correlation Matrix of Table 7.2

	$\mathbf{Z}_1$	$\mathbf{Z}_2$	$\mathbf{Z}_3$	$\mathbf{R}^2$
$\mathbf{Y}_1$	0.085	0.540	0.797	93.4
$\mathbf{Y}_2$	0.971	−0.199	−0.054	98.6
$\mathbf{Y}_3$	−0.311	−0.809	0.118	76.6
$\mathbf{Y}_4$	0.917	−0.294	−0.162	95.3
$\mathbf{Y}_5$	0.992	0.051	0.021	98.6
$\mathbf{Y}_6$	0.976	−0.041	−0.091	96.2
$\mathbf{Y}_7$	0.992	0.013	−0.030	98.4
$\mathbf{Y}_8$	0.982	0.115	−0.030	97.9
$\mathbf{Y}_9$	0.966	0.126	0.024	94.9
$\mathbf{Y}_{10}$	0.642	0.566	0.112	74.6
$\mathbf{Y}_{11}$	0.974	−0.078	0.085	96.1
$\mathbf{Y}_{12}$	0.961	0.007	0.083	93.1
$\mathbf{Y}_{13}$	−0.422	0.734	−0.350	84.0
$\mathbf{Y}_{14}$	0.943	−0.176	−0.139	93.9
$\mathbf{Y}_{15}$	0.953	0.233	0.075	96.9
$\mathbf{Y}_{16}$	0.945	−0.280	0.040	97.3
$\mathbf{Y}_{17}$	0.337	0.751	−0.302	77.0
$\mathbf{Y}_{18}$	0.962	−0.129	0.016	94.2
l_j	12.898	2.715	.957	

Source: Ahamad, 1967; reproduced with permission.

Here the component scores are functions of time and estimate time behavior of the variables. The scores for the first two components are shown in Figure 7.1. The first component, which accounts for just over 71% of the total variance, seems to be highly correlated with the total number of individuals in the 13- to 19-year-old age group where $\hat{\mathbf{Z}}_1$ (Fig. 7.1a) represents that part of $\mathbf{Z}_1$ which is due to the age group, as determined by linear regression. The finding is clearly of some interest, but is perhaps not too surprising since the data of Table 7.1 are not corrected for population levels and the dominant principal component represents the general "size" effect, that is, an

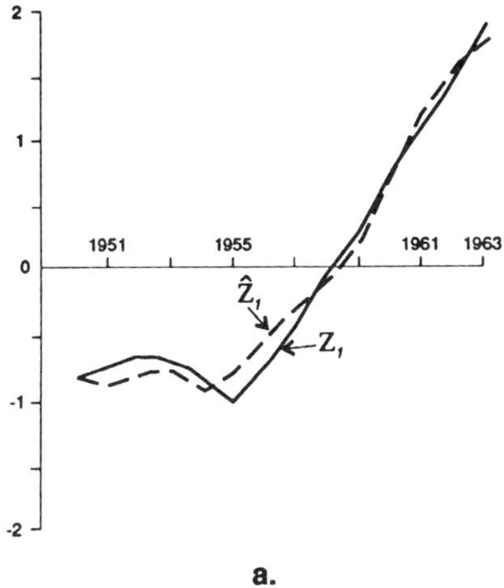

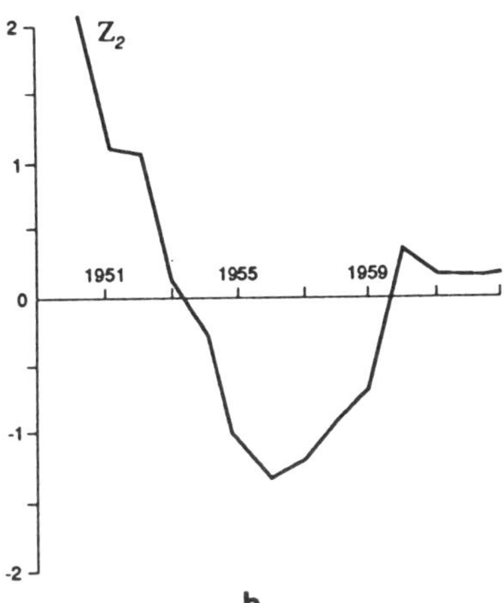

Figure 7.1 Principal component scores for the first two principal components of the offence categories of Table 7.1 (Ahamad, 1967; reproduced with permission).

increasing time trend perhaps reflecting population growth. The expectancy however is to a large extent retrospective in nature since it may not be clear on a priori grounds why Z_1 should reflect the 13–19 age group as opposed, for example, to some other group (see Ahamad, 1967; Walker, 1967). The high correlation should also be interpreted with care since it does not necessarily reflect direct causation, particularly when dealing with age cohorts over time (Section 7.3).

The second component Z_2, which accounts for 15% of the variance (Fig. 7.1b), reflects quadratic behavior over time and is correlated positively with Y_1, Y_{10}, Y_{13}, and Y_{17} and negatively with Y_3, variables that can be seen to possess quadratic movement over time (Table 7.1). Its exact identification however is not evident without a more intense scrutiny of the data. Finally, although Z_3 is correlated, to some extend, with the number of homicides (Y_1) (Table 7.3) it appears to reflect random, independent movement over time. This in itself may be of some interest since it represents evidence that homicides contain a large degree of purely random variation from year to year. The analysis should be viewed as tentative and exploratory, however, since more data and alternative association matrices which are more appropriate to integer count data (Chapters 8 and 9) should also be analyzed.

7.2.1 Constructing Indices and Indicators

One of the first applications of factor analysis to time series is from Rhodes (1937), who attempted to use factor scores to construct an index of general business activity of the British economy. Rhodes selected a set of leading indicators of business activity, and then used the series to compute a general index by taking the first (dominant) factor, which explained the major part of the variance. The factor model used by Rhodes (see Thurstone, 1935), however, is somewhat outdated and is no longer in use and has been replaced by the ordinary principal components model (Peters and Butler, 1970; Jaumotte et al., 1971; Bartels, 1977), although weighted models (Chapter 6) can also be used, depending on the specification of the error terms. Since the input variables are themselves dimensionless index numbers, either a covariance or correlation matrix can be used, keeping in mind that principal components are influenced by large differences in the variances of the variables. At times logarithmic transformations are also employed in an attempt to either linearize the series, reduce variance differences, or both. Since time series are functions of time, the correlation between them is often due to time variation resulting from the presence of (common) time functions, which may be estimated by the factor scores. The advantage of using factor analysis to estimate such functions is that the analytical form(s) of the functions need not be known a priori, since they can be viewed as empirical functions which maximize variance (correlation) amongst the observed series. The factor scores can thus be expressed as

$$\mathbf{F}_1 = f_1(t), \qquad \mathbf{F}_2 = f_2(t), \ldots, \qquad \mathbf{F}_r = f_r(t)$$

and may be plotted graphically to aid identification. Depending on the nature of the observed series $f_i(t)$ can represent time trends, seasonal (monthly) fluctuations, or longer cyclical movements and the factors are not restricted to linear functions such as (linear) trend. Another approach which may be used to identify the factors is to include exact analytical functions as "markers," keeping in mind that this can only be done by using principal components since exact (error-free) functions are not compatible with (heteroscedastic) factor models (Chapter 6). The resultant time factor scores can then be used as standardized indices which estimate the process(es) that gave rise to the observed time series. In addition to providing index estimates the factors utilize information inherent in the high correlation usually observed between time series, which regression cannot do due to multicollinearity.

Index numbers can also be constructed by using factors (principal components), in a more restricted sense. Thus a procedure which yields "best" linear index numbers of commodity prices *and* quantities has been suggested by Theil (1960). Consider K commodities which are exchanged over a period of T time intervals. Let p_{tj} and q_{tj} denote price and quantity of commodity j at time t. Then we can write

$$\mathbf{P} = \begin{bmatrix} p_{11} & p_{12} & \cdots & p_{1K} \\ p_{21} & p_{22} & \cdots & p_{2K} \\ \vdots & \vdots & \cdots & \vdots \\ p_{T1} & p_{T2} & \cdots & p_{TK} \end{bmatrix} \quad \mathbf{Q} = \begin{bmatrix} q_{11} & q_{12} & \cdots & q_{1K} \\ q_{21} & q_{22} & \cdots & q_{2K} \\ \vdots & \vdots & \cdots & \vdots \\ q_{T1} & q_{T2} & \cdots & q_{TK} \end{bmatrix}$$

where

$$\mathbf{C} = \mathbf{P}\mathbf{Q}^{\mathrm{T}} \tag{7.1}$$

is the $(T \times T)$ "value" matrix of total (aggregate) quantities for time periods $t = 1, 2, \ldots, T$. We wish to find those $(T \times 1)$ vectors $\mathbf{p}$ and $\mathbf{q}$ that provide the best least square fit to the matrix of bilinear forms $\mathbf{C}$. That is, we wish to minimize $\mathrm{tr}(\mathbf{E}\mathbf{E}^{\mathrm{T}})$ where

$$\mathbf{E} = \mathbf{C} - \mathbf{p}\mathbf{q}^{\mathrm{T}} \tag{7.2}$$

for some two unknown column vectors $\mathbf{p}$ and $\mathbf{q}$.

We have

$$\mathrm{tr}(\mathbf{E}\mathbf{E}^{\mathrm{T}}) = \mathrm{tr}(\mathbf{C} - \mathbf{p}\mathbf{q}^{\mathrm{T}})(\mathbf{C} - \mathbf{p}\mathbf{q}^{\mathrm{T}})^{\mathrm{T}}$$

$$= \mathrm{tr}(\mathbf{C}\mathbf{C}^{\mathrm{T}}) - 2\mathrm{tr}(\mathbf{C}\mathbf{p}\mathbf{q}^{\mathrm{T}}) + \mathrm{tr}(\mathbf{p}\mathbf{q}^{\mathrm{T}}\mathbf{q}\mathbf{p}^{\mathrm{T}})$$

$$= \mathrm{tr}(\mathbf{CC}^{T}) - 2\mathrm{tr}(\mathbf{p}^{T}\mathbf{Cq}) + \mathrm{tr}(\mathbf{p}^{T}\mathbf{pq}^{T}\mathbf{q})$$

$$= \mathrm{tr}(\mathbf{CC}^{T}) - 2\mathbf{p}^{T}\mathbf{Cq} + (\mathbf{p}^{T}\mathbf{p})(\mathbf{q}^{T}\mathbf{q}) \tag{7.3}$$

Differentiating Eq. (7.3) with respect to $\mathbf{p}$ and $\mathbf{q}$ and setting to zero yields

$$\mathbf{Cq} - (\mathbf{q}^{T}\mathbf{q})\mathbf{p} = \mathbf{0}$$

$$\mathbf{C}^{T}\mathbf{p} - (\mathbf{p}^{T}\mathbf{p})\mathbf{q} = \mathbf{0}$$

or

$$[(\mathbf{C}^{T}\mathbf{C}) - (\mathbf{p}^{T}\mathbf{p})(\mathbf{q}^{T}\mathbf{q})]\mathbf{q} = \mathbf{0} \tag{7.4a}$$

$$[(\mathbf{CC}^{T}) - (\mathbf{p}^{T}\mathbf{p})(\mathbf{q}^{T}\mathbf{q})]\mathbf{p} = \mathbf{0} \tag{7.4b}$$

using Eq. (7.1). Thus $\mathbf{q}$ and $\mathbf{p}$ are latent vectors of the Grammian matrices $\mathbf{C}^{T}\mathbf{C}$ and $\mathbf{CC}^{T}$ respectively and correspond to the dominant latent root

$$\lambda^{2} = (\mathbf{p}^{T}\mathbf{p})(\mathbf{q}^{T}\mathbf{q}) \tag{7.5}$$

Thus optimal price and quantity index numbers correspond to the principal component loadings and scores of $\mathbf{C}^{T}\mathbf{C}$. The procedure bears a close resemblance to estimating functional forms and growth curves (Section 10.6). The matrix $\mathbf{C}$ however is not adjusted for sample means so that the dominant root (Eq. 7.5) is mainly a function of these means, but the procedure can easily be modified should this be an undesired feature. Theil's (1960) Best Linear Index possesses several desirable properties which are not shared by the usual economic index numbers. For example, when $T = 2$ the best linear index number lies halfway between the Laspeyres and Paasche indices when quantities are equal in the two periods; closer to the Paasche index when current quantities are larger; and closer to the Laspeyres index when current quantities are smaller. This makes the Best Linear Index a useful compromise between the two commonly used indices. A difficulty with Theil's Best Linear Index is that it may be biased since elements of $\mathbf{pq}^{T}$ may systematically exceed corresponding elements of $\mathbf{C}$. In an attempt to correct for such potential bias Kloek and deWit (1961) have obtained a modified version of $\mathbf{p}$ and $\mathbf{q}$, such that $\mathrm{tr}(\mathbf{E}) = 0$. This is achieved by introducing a correction factor μ such that $\mathbf{C} - \mu\mathbf{I} = \mathbf{0}$. In practice μ may be computed iteratively, and the modified price and quantity indices are then known as Best Linear Unbiased Index numbers. The following two examples illustrate the use of factor analysis (principal components) in index number construction.

Example 7.2. Quarterly data are available for prices of consumer commodities in the United Kingdom, during the period 1955–1968 (first quarter of 1958 = 100). For the sake of simplicity the time series represent

$p = 10$ grouped consumer commodities, which are based on the industrial sector that produced them. Owing to the quarterly nature of the data we wish to determine the extent of seasonality of the series, together with any other systematic time behavior which may be present.

A principal components analysis is carried out on the logarithms of the $p = 10$ price series using the correlation matrix. Loadings for the first two components, which virtually explain the entire systematic variance of the series, are given in Table 7.4 and scores are plotted in Figures 7.2 and 7.3. Evidently, with the exception of Y_5 the price variables can be explained by the first component Z_1 whereas Z_2 reflects the price of the commodity group produced by the engineering sector. Hence the scores for Z_1 can be taken as a measure of the general consumer price index, excluding the engineering commodity which exhibits a different time pattern picked up by Z_2 (Fig. 7.3). Here Z_1 can also be understood as a random time function which represents the general inflationary increase in consumer prices. Note

Table 7.4 Principal Component Loadings for Natural Logarithms of $p = 10$ Quarterly Series, 1955–1968[a]

	Prices	Z_1	Z_2
Y_1:	Agriculture	.9872	
Y_2:	Mining and quarrying	.9855	
Y_3:	Food, drink and tobacco	.9867	
Y_4:	Chemicals	.9934	
Y_5:	Engineering	−.3686	.9268
Y_6:	Textiles	.9863	
Y_7:	Other manufacture	.9894	
Y_8:	Gas and electricity	.8937	.1517
Y_9:	Services	.9775	
Y_{10}:	Noncompetitive imports	.9022	.2317

[a] Loadings smaller than .10 are omitted.

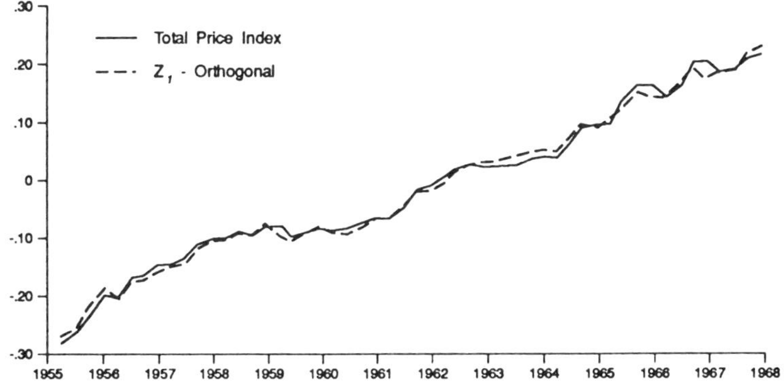

Figure 7.2 Component scores Z_1 for $p = 10$ economic price indices of Table 7.4.

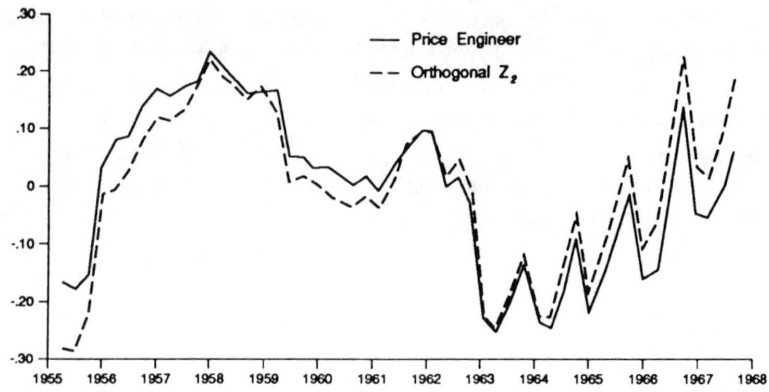

Figure 7.3 Component scores $\mathbf{Z}_2$ for $p = 10$ economic price indices of Table 7.4.

however that, as it stands, $\mathbf{Z}_2$ is a biased estimator of $\mathbf{Y}_5$ since the two components have been constrained to be orthogonal to each other, in spite of the correlation between them—the price of engineering goods is clearly influenced by the general price trend of all other commodities. The bias is readily removed by an oblique rotation of the loadings and scores of $\mathbf{Z}_1$ and $\mathbf{Z}_2$.

Example 7.3.   Kloek and deWit (1961) have applied Theil's (1960) Best Index numbers to published Dutch import and export data for the period 1921–1936, which consists of $p = 15$ commodity categories. Both the Best Linear (BL) and the Adjusted Best Linear and Unbiased (BLAU) price and quantity indices are computed and given in Table 7.5. Since the variables are not corrected for mean values, a dominant latent root exists which reflects the general mean levels of imports and exports. The fitting coefficient I^2 of Table 7.5 is a function of the dominant latent root and represents the total proportion of variance which is accounted for by the computed price and quantity indices.

7.2.2 Computing Empirical Time Functions

When given a set of correlated time series we may at times wish to investigate the presence of common time movements (functions) which can be present in the series. Since the functional form(s) of the unobserved time functions is generally not known, regression analysis cannot be used for the purpose. A common practice at times is to compute factor or component scores of the common factors, and then attempt to identify them by plotting the score values against time. Time functions of this type are also known as empirical time functions. Although rotation of the loadings and the scores is generally not required, in order to identify the time functions an orthogonal or oblique rotation may enhance interpretability (see Richman, 1981). Regression analysis, ARIMA, or other time series methods can also be

Table 7.5   BL and BLAU Price and Quantity Index Numbers p and q with Corresponding Roots λ^2 and Fitting Coefficients I^2.

	BL				BLAU			
	Imports		Exports		Imports		Exports	
λ^2	11180×10^{17}		5589×10^{17}		10994×10^{17}		5451×10^{17}	
I^2	0.99994		0.99982		0.99986		0.99963	
Years	p	q	p	q	p	q	p	q
1921	6900×10	3331×10	6048×10	2426×10	6900×10	3283×10	6054×10	2359×10
1922	5502	3701	4771	2691	5492	3667	4765	2643
1923	5408	3752	4460	2975	5398	3719	4448	2933
1924	5476	4313	4649	3615	5461	4282	4630	3575
1925	5302	4661	4423	4097	5283	4634	4395	4064
1926	4309	5021	4074	4322	4879	4999	4041	4295
1927	4945	5170	4096	4653	4920	5147	4059	4628
1928	5091	5302	4108	4868	5065	5279	4069	4843
1929	5044	5469	4116	4863	5017	5447	4076	4838
1930	4395	5480	3759	4598	4365	5464	3721	4576
1931	3639	5304	3191	4244	3607	5294	3154	4229
1932	2982	4515	2486	3506	2954	4508	2454	3495
1933	2768	4474	2176	3408	2740	4469	2144	3401
1934	2554	4116	2141	3357	2528	4111	2109	3349
1935	2515	3823	2104	3263	2491	3817	2073	3255
1936	2704	3883	2143	3455	2681	3876	2109	3448

Source: Kloek and deWit, 1961; reproduced with permission.

[a] Both p and q are expressed in square roots of guilders per year.

applied to each factor independently, either as an aid to identification of the basic functions (the scores) or in forecasting future values of the multivariate series—see, for example, Otter and Shur (1982).

Several types of data matrices can incorporate time. Such data occur frequently in the historical, social, biomedical, environmental, and atmospheric sciences when observing time records for a set of characteristics or variables ("historic data"), or when a single characteristic is observed simultaneously over a set of geographic regions or locations at regular time intervals (cross-sectional/time series data). Although irregularly spaced data can also be used, this may introduce distortion to the loadings and the scores (Karl et al., 1982). Not much evidence seems to exist on the point, however, although distortion would be expected on a priori grounds since irregularly spaced data can be considered as an instance of missing data. When columns of the data matrix $\mathbf{Y}$ correspond to the cross sections, it must be kept in mind that the loadings (which characterize the locations) cannot be plotted on an axis in the same way as the scores unless they also correspond to a natural ordering such as geographic direction. Thus ordering regions in alphabetical order, for example, does not necessarily induce a natural order amongst the regions, and a functional plot of the loadings is here meaningless (for an example of the latter see Jaumotte et al., 1971). Also, when using principal components in conjunction with time series it is important to distinguish between the various Grammian association matrices, since principal components are not unit invariant (Chapter 3). Thus failure to distinguish between the correlation and covariance matrix, for example, can lead to interpretational errors (e.g., Dyer, 1981). Time-distributed geographic data are particularly common in atmospheric and meteorological research, where they are intended to capture space/time variability (Craddock and Flood, 1969; Henry, 1977; Schickendanz, 1977; Barnett, 1978). When only monthly effects are of interest meterological data can also be presented in the form of a $(n \times 12)$ matrix where n is the number of years of available records (Craddock, 1965); Brier and Meltsen, 1976). It must be kept in mind however that with time-dependent data the outcome of a factor analysis will also depend on the degree of serial correlation within the series, as well as on correlation between the series—see Farmer (1971) for a simulated example. Also, the statistical testing procedures described in Chapter 4 and Section 6.6 are not, strictly speaking, applicable for dependent observations. The following example provides a general illustration of using factor scores to estimate empirical time functions for geographic regions.

Example 7.4.   Table 7.6 contains adjusted motor vehicle fatalities for the 10 Canadian Provinces, between 1961 and 1982, together with the total Canadian average. A ML factor analysis reveals the presence of a single dominant component, which explains 85.8% of the variance. The factor

Table 7.6   Number of Motor Vehicle Fatalities per 100 million km for 10 Canadian Provinces

Year	Canada	Nfld.	P.E.I.	N.S.	N.B.	Que.	Ont.	Man.	Sask.	Alta.	B.C.
1961	4.7	6.9	4.8	6.9	8.1	5.0	4.4	3.6	4.2	4.4	4.9
1962	5.1	6.0	6.1	7.2	7.2	5.9	4.7	3.7	4.4	4.2	5.6
1963	5.2	8.1	7.5	6.8	9.3	6.4	4.5	4.4	4.3	4.4	4.9
1964	5.4	6.9	7.3	7.2	9.6	7.3	4.2	4.5	4.6	4.8	4.8
1965	5.2	5.4	5.1	7.5	9.7	6.2	4.5	4.0	4.7	4.2	5.3
1966	5.2	6.1	6.0	7.6	10.0	6.5	4.2	4.4	5.6	4.2	5.3
1967	5.2	5.6	5.2	8.1	9.3	5.9	4.2	4.3	5.5	4.7	5.4
1968	4.7	4.3	7.4	6.9	7.8	5.7	3.7	4.1	4.9	4.5	5.1
1969	4.6	5.9	7.2	6.5	7.4	5.6	3.7	3.3	4.1	4.6	4.5
1970	4.0	3.9	5.4	5.4	6.6	4.9	3.2	3.0	3.8	3.7	4.4
1971	4.2	4.2	4.7	5.3	5.9	4.9	3.5	3.3	3.9	4.1	4.6
1972	4.2	4.9	6.5	5.6	5.7	5.3	3.4	2.8	4.5	3.6	4.8
1973	4.2	4.0	5.6	5.5	5.9	5.2	3.2	3.3	3.9	3.7	5.0
1974	3.7	4.2	5.2	5.0	6.1	4.3	2.7	2.9	4.5	3.8	4.7
1975	3.5	3.4	5.1	4.3	4.5	4.2	2.8	2.8	4.0	3.3	4.0
1976	3.0	3.3	3.4	3.5	4.5	3.4	2.3	3.1	3.4	2.9	3.4
1977	2.8	2.7	5.7	3.0	3.9	3.2	2.1	2.4	3.3	3.0	3.7
1978	2.8	2.2	3.1	3.5	3.9	3.6	2.1	2.5	3.6	2.3	3.0
1979	2.9	2.9	3.3	2.8	4.2	3.6	2.1	2.3	3.4	3.1	3.3
1980	2.7	2.4	3.6	3.0	3.2	3.0	2.1	2.2	3.0	3.0	3.4
1981	2.6	1.7	2.3	2.6	3.0	3.1	2.0	2.6	3.1	3.1	3.6
1982	2.3	1.8	1.9	2.9	3.6	2.7	1.7	2.1	2.5	2.3	2.8

Source: Statistics Canada; Transport Canada.

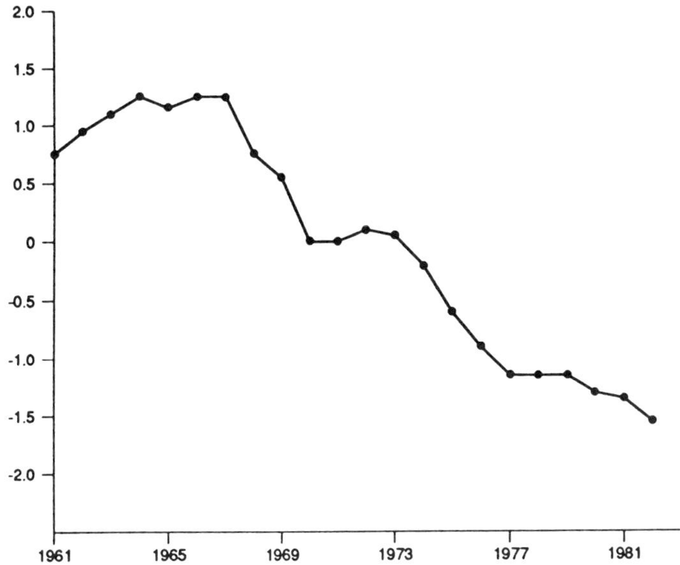

Figure 7.4 First (dominant) maximum likelihood factor score for Canadian motor vehicle fatalities, 1961–1982.

scores appear in Figure 7.4, where it can be readily verified that the dominant factor estimates the average fatality rate for Canada as a whole. A principal components analysis also reveals the presence of a dominant component, which accounts for a slightly higher percentage of variance. The component scores however are influenced by residual error variation and give a poorer fit to the overall fatality rate than do the maximum likelihood estimates. If temporal rather than regional correlation is of interest, a Q-mode factor analysis can also be performed, for example, by transposing the data of Table 7.6 (Exercise 7.12). An interesting historical analysis of climate in the western United States using ring data of drought-sensitive trees is also given by La Marche and Fritts (1971), where individual trees rather than provinces represent geographic location.

7.2.3 Pattern Recognition and Data Compression: Electrocardiograph Data

The first two sections deal with the extraction of time functions from observed data, which could then be used to characterize or compare the observed series. At times one may be given a large number of time series, where we are faced with the purely utilitarian task of reducing the size of the data to manageable proportions, at the same time retaining essential informational content. The problem of data reduction and smoothing occurs, for example, in electrocardiographic (ECG) research where interest lies in describing, from body surface readings, the electrical current sources within the heart. Since one of the main purposes of recording a large set of

ECG data is to distinguish normal from abnormal hearts, we would only be interested in retaining sufficient information to achieve this aim, for example, for therapeutic purposes, for automatic computerized screening of a population, or to obtain insight into the underlying biological processes at work (Van Bemmel, 1982).

The heart is a cyclic pump, with the cycle consisting of the heart pumping blood from the ventricles to the body, the lungs, and back again. The ECG is thus, most of the time, a continuous periodic function. The sinoatrial node causes the atrial muscle to depolarize and pump blood into the right ventricle. At the same time the left atrium is receiving blood from the lungs and pumping it into the left ventricle. The depolarization of the atria and ventricles is characterized in the ECG by the so-called P-wave and QRS complex (Fig. 7.5) where the letters P, Q, R, S, and T represent standard notation for the prominent features of the wave form. The electrical signal is relatively large (50 mV) and of moderate band width (0–100 Hz), and can be measured with low error (noise) which may emanate from either electrical (equipment) or physiological sources. It is also consistent from heartbeat to heartbeat but can vary somewhat from person to person. The heart, on the other hand, is not a simple electrical source but consists of a time-varying charge distribution that can be described by a time-varying current dipole moment, distributed throughout the volume of the heart.

Let E_p denote instantaneous voltages recorded at a body surface point p and consider k current generators $G_1, G_2, \ldots, G_k$. Then

$$E_p = c_1 G_1 + c_2 G_2 + \cdots + c_k G_k \tag{7.6}$$

where the constants $c_1, c_2, \ldots, c_k$ are determined by the geometry and resistivity of the body and are assumed to be constant throughout QRS. The basic aim of an ECG is to describe, from body surface recordings, the electrical current sources within the heart (see Young and Calvert, 1974; Ahmed and Rao, 1975). The number of recordings required to describe the equivalent generator is equal to the number of sources contributing to this generator. Thus for a k-function system we require k nonredundant recordings. If we have fewer than k ECG leads we miss information and if there are more we have redundancy. We thus need to determine the number of basic functions (voltage generators) needed to account for the voltages recorded on the body surface. Sher et al. (1960) have used principal

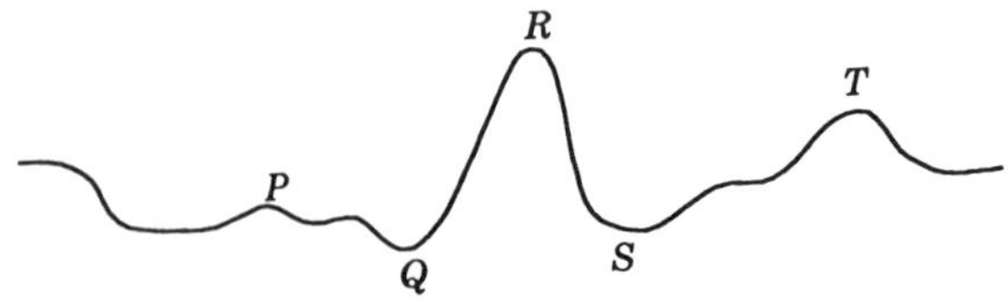

Figure 7.5 The *P* wave and the *QRS* complex of a normal ECG.

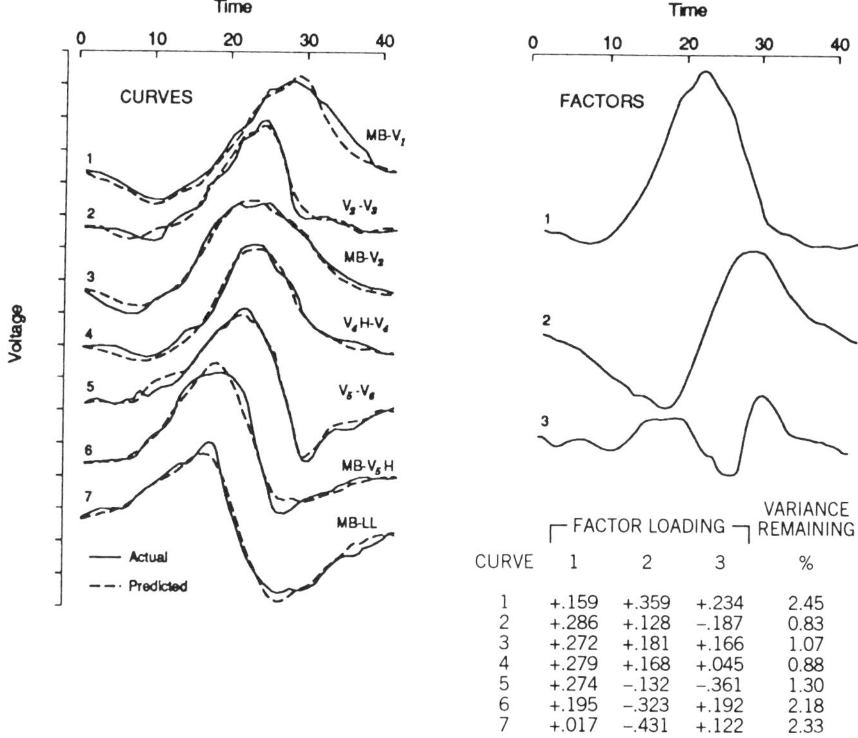

CURVE	FACTOR LOADING			VARIANCE REMAINING
	1	2	3	%
1	+.159	+.359	+.234	2.45
2	+.286	+.128	−.187	0.83
3	+.272	+.181	+.166	1.07
4	+.279	+.168	+.045	0.88
5	+.274	−.132	−.361	1.30
6	+.195	−.323	+.192	2.18
7	+.017	−.431	+.122	2.33

Figure 7.6 Orthogonal principal component functions of $p = 7$ electrocardiograph leads estimating the systematic portion of the data (Sher et al., 1960; reproduced with permission).

components to determine the number of basic time functions required to reproduce the recordings from seven leads (Figure 7.6), and found that $r = 3$ components are sufficient to reproduce the essential features of the ECG. Feature recognition, together with pattern classification, can also be applied to an ECG as well as other signals of biological origin such as the electroencephalogram, the spirogram, or hemodynamic signals, which can then become useful diagnostic aids. For a review of this area of study see Van Bemmel (1982) and Wold (1976).

7.3 DEMOGRAPHIC COHORT DATA

Owing to the scarcity of accurate data, many life tables are constructed artificially using statistical projection techniques. Thus a practice in common usage at the United Nations, for example, is to take life expectancy at birth and to project, in 5-year age intervals, the sex and age-specific mortality rates of the various countries. Such projections are possible because of the high serial correlation that exists between adjacent age groups. Principal component analysis is a particularly powerful and straightforward method of constructing model life tables, since it does not require a priori specification of the analytic functional forms of the component life tables, and can utilize all available data for all countries.

Let q denote the mortality rate for individuals who are between ages x_1 and x_2 and who were deceased during the interval $|x_1 - x_2|$. The ratio q is computed as a proportion of mortality with respect to those who had age x_1. The intervals $|x_1 - x_2|$ are typically taken as 5-year durations, and the base is customarily set at $n = 1000$ individuals. Thus we can write $_5q_{30}$, for example, to denote the mortality rate for individuals from 30 up to (but not including) 35 years of age, expressed as a proportion of 1000. The various values of q, when plotted against 5-year age intervals, then constitute the so-called life table, that is, mortality rates plotted as a function of age. Owing to differential mortalities between males and females, life tables are usually constructed separately for the two sexes. When sex-specific age groups are taken as the variables and countries as sample points a data matrix $\mathbf{Y}$ is obtained, which can be decomposed using factor analysis. Alternatively both sexes can be included in a three-mode analysis (Sections 5.4.3 and 6.12), but this does not appear to be current practice. Note that a factor model with a diagonal error covariance matrix may be inappropriate because of possibly correlated residual errors induced by serially correlated observations.

Example 7.5. Ledermann and Breas (1959) (see also United Nations Organization, 1962) use principle components to analyze a data matrix consisting of $p = 18$ age groups for $n = 157$ countries. The correlation matrix for males, together with correlation loadings for the first three components, are given in Tables 7.7 and 7.8. The first component $\mathbf{Z}_1$ represents general life expectancy, that is, that portion of life expectancy which can be predicted from l_0, the expectation of life at birth, generally a function of the widespread health conditions which are prevalent in the countries at that time. An examination of the loadings reveals it accounts for less variance of the older group (75 years of age and higher), indicating that older age groups are subject to greater age-specific risk. These influences appear to be associated with $\mathbf{Z}_2$ and $\mathbf{Z}_3$ and explain variation about $\mathbf{Z}_1$, that is, these components explain independent deviation of mortality from general health conditions.

The loading coefficients (Table 7.8) indicate age-specific effects while scores rank the countries in order of their importance vis-a-vis the components. A main objective for life table analysis is to reconstruct (estimate) the main features of the life tables for specific life expectancies. This can be done by computing the row vectors of the matrix of predicted values $\hat{\mathbf{X}} = \mathbf{Z}\mathbf{A}^{\mathsf{T}}$ where $\mathbf{Z}$ is the (157×3) matrix of scores. Thus for $l_0 = 70$ we obtain the estimated life table of Figure 7.8 using only the first component. Note that the mortality rates must first be linearized by taking logarithms.

Table 7.7 Correlation Matrix of Male Age-Specific Mortality Rates for $n = 157$ Countries (Logarithms)

l_0	0–1	1–4	5–9	10–14	15–19	20–24	25–29	30–34	35–39	40–44	45–49	50–54	55–59	60–64	65–69	70–74	75–79	80–84
1.000																		
.937	1.000																	
.948	.905	1.000																
.915	.844	.917	1.000															
.933	.842	.906	.929	1.000														
.911	.801	.872	.879	.949	1.000													
.898	.782	.856	.861	.916	.985	1.000												
.917	.793	.865	.869	.921	.967	.985	1.000											
.932	.810	.874	.875	.920	.940	.949	.982	1.000										
.904	.790	.841	.839	.888	.891	.890	.925	.939	1.000									
.933	.811	.841	.844	.890	.889	.888	.936	.974	.933	1.000								
.915	.790	.800	.805	.853	.846	.840	.891	.940	.920	.988	1.000							
.886	.758	.744	.757	.805	.800	.792	.844	.889	.887	.956	.982	1.000						
.850	.722	.697	.703	.753	.749	.740	.790	.832	.849	.910	.949	.983	1.000					
.828	.696	.676	.671	.736	.726	.715	.765	.809	.835	.888	.931	.970	.991	1.000				
.815	.702	.672	.647	.691	.704	.691	.735	.782	.796	.862	.907	.940	.951	.972	1.000			
.818	.713	.684	.653	.692	.705	.687	.726	.763	.780	.836	.875	.909	.939	.927	.963	1.000		
.715	.635	.597	.550	.602	.614	.591	.618	.647	.653	.714	.746	.777	.820	.844	.871	.946	1.000	
.520	.479	.456	.409	.419	.419	.384	.412	.441	.462	.498	.524	.542	.576	.610	.665	.749	.878	1.000

Source: Lederman and Breas, 1959.

Table 7.8 Principal Components Loadings for the First Three Components of Male Age-Specific Mortality Rates[a]

	Z_1	Z_2	Z_3
l_0	.9736	−.0887	.0734
0–1	.8839	−.0992	.2319
1–4	.9061	−.2296	.2421
5–9	.8978	−.2481	.1645
10–14	.9332	−.2399	.0758
15–19	.9245	−.2595	.0305
20–24	.9073	−.2888	−.0034
25–29	.9362	−.2425	−.0769
30–34	.9509	−.1832	−.0975
35–39	.9326	−.1096	−.1248
40–44	.9594	−.0385	−.1721
45–49	.9475	.0594	−.2307
50–54	.9261	.1607	−.2894
55–59	.9059	.3122	−.2363
60–64	.8817	.3098	−.2953
65–69	.8629	.3645	−.2230
70–74	.8659	.4199	−.1016
75–79	.7717	.5463	.0885
80–84	.5824	.6344	.3760
Variance (%)	81.5%	7.4%	3.7%

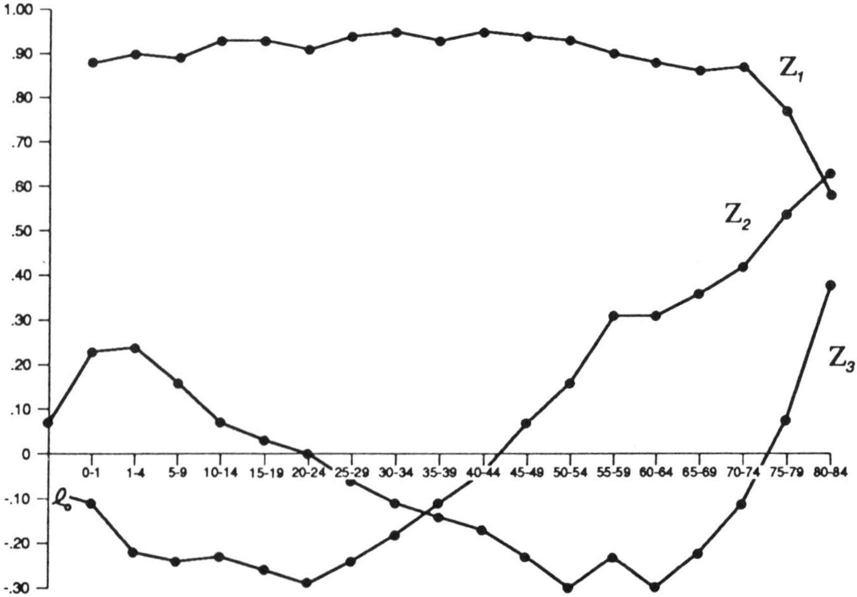

Figure 7.7 Principal component loadings of Table 7.8 for the first $r = 3$ components of male-specific mortality rates (Lederman and Breas, 1959).

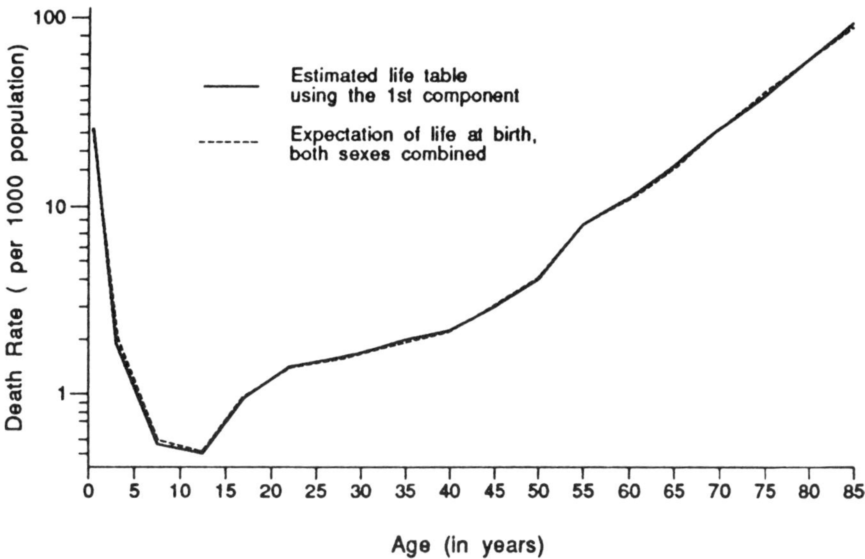

Figure 7.8 Estimated life table using the first principal component which is plotted against expectation of life at birth (both sexes combined; United Nations Bulletin 6, 1962).

7.4 SPATIAL CORRELATION: GEOGRAPHIC MAPS

The previous sections deal with one-dimensional time series. At times however, data are gathered over two- or three-dimensional physical space, for example, geographic variables distributed over a given land surface, or geological mineral exploration measurements obtained from 3-dimensional underground rock deposits. Since the sampling scheme for such data normally consists of an evenly spaced grid pattern, the observations for the variables will generally be correlated. The correlation can be induced, for example, by common land features shared by a number of adjacent observational sites. Generally speaking any feature(s) that overlaps several adjacent observation sites will induce correlation amongst the sample points. An example (Section 7.8) is satellite digital image processing, where digital energy information (light intensity, color, etc.) is stored in pixels which stand in a definite relationship to each other. As for time series the sample points are ordered, and the correlation cannot be removed by a random shuffling of the data points. When a factor analysis is carried out for a set of correlated variables distributed over adjacent geographical areas (space), the loadings are still interpreted in the usual way. Since the factor scores characterize an ordered sample space, however, their values are also ordered, and can therefore be represented physically on a sample space such as a geographic map, a picture, and so forth.

Example 7.6. Gordon and Whittaker (1972) use principal components in an attempt to identify fundamental sociological dimensions of private wealth and property in the south west region of England. Since average incomes may mask causes of poverty or deprivation, other variables that measure unemployment, housing, services, and so forth are introduced as follows:

Y_1 = Income

Y_2 = Unemployment rate

Y_3 = Unemployment seasonality

Y_4 = Female activity rate

Y_5 = Social class I and II (professional and intermediate occupations)

Y_6 = Social class III (skilled occupations)

Y_7 = Social class IV and V (partly skilled and unskilled occupations)

Y_8 = Migration of young males

Y_9 = Average domestic rateable value

Y_{10} = Houses with very low rateable value

Y_{11} = Households with all exclusive amenities

Y_{12} = Cars per household

Y_{13} = Telephone ownership

Y_{14} = Houses built postwar

Y_{15} = Owner occupation of building

Y_{16} = Employment growth (change in total employment, 1961–1966)

Y_{17} = Growth industries

Y_{18} = Industrial building

Y_{19} = Terminal age of education

Y_{20} = Movement of school leavers

Y_{21} = Doctors per head

Y_{22} = Accessibility to services

Y_{23} = Death rate

A principal component analysis is carried out, followed by the orthogonal varimax rotation. The resulting loadings (correlation matrix) are given in Table 7.9, where low-order magnitudes are omitted. The authors identify the following dimensions of "prosperity" in the southwest of England. Factor 1 denotes high incomes, skilled occupations, and houses built after World War II and factor 2 indicates property values in the areas.* The remaining components can be interpreted in the usual fashion using the loadings of Table 7.9. Since the sample points are spatially ordered the score coefficients have a similar ordering, and may be plotted on a map of the southwest region of England (Fig. 7.9). Again, we have reversed the signs

---

* We have altered the signs of the loadings in Table 7.9 to conform to this interpretation.

Table 7.9   Socioeconomic Indicators of Regional Prosperity in the Southwest of England

	Z_1	Z_2	Z_3	Z_4	Z_5	Z_6
Y_1	.7729					
Y_2						.3964
Y_3					−.3550	−.5729
Y_4	.8144					
Y_5	−.6922		−.5246			
Y_6	.6383		.5123			
Y_7		−.5480		−.5250		
Y_8				.7528		
Y_9		.8683				
Y_{10}		−.8050				
Y_{11}		.7385				
Y_{12}			−.8341			
Y_{13}	−.3670	.6393				
Y_{14}	.5822	.3774		.4330		
Y_{15}	−.4187			.5747		
Y_{16}				.5365		
Y_{17}					.4208	.4890
Y_{18}						−.6100
Y_{19}					.7199	
Y_{20}					−.5841	
Y_{21}						
Y_{22}			.3582		.5719	
Y_{23}			.6734			

Source: Gordon and Whittaker, 1972; reproduced with permission.

of the scores to maintain interpretational consistency with the loadings of Table 7.9. Comparing the loading and score coefficients it can be seen that the eastern portion of the region enjoys higher incomes and a more skilled labor force than does the western portion. A similar analysis can also be found in Hunter and Latif (1973) and Kuz et al., (1979). Geographic mappings are also common in meteorology and ecology (Knudson et al., 1977). Yanai et al. (1978) have also used factor analysis to study cancer mortality rates for 46 Japanese prefectures, using three time periods for males and females. Pirkle et al. (1982) use principal components to construct prediction indices for mineral exploration by identifying lithologic units which are favorable for the ore deposits.

7.5.   THE KARHUNEN–LOÈVE SPECTRAL DECOMPOSITION IN THE TIME DOMAIN

The usual multivariate analysis does not make use of the correlation between observations within time series or spatially distributed data. An

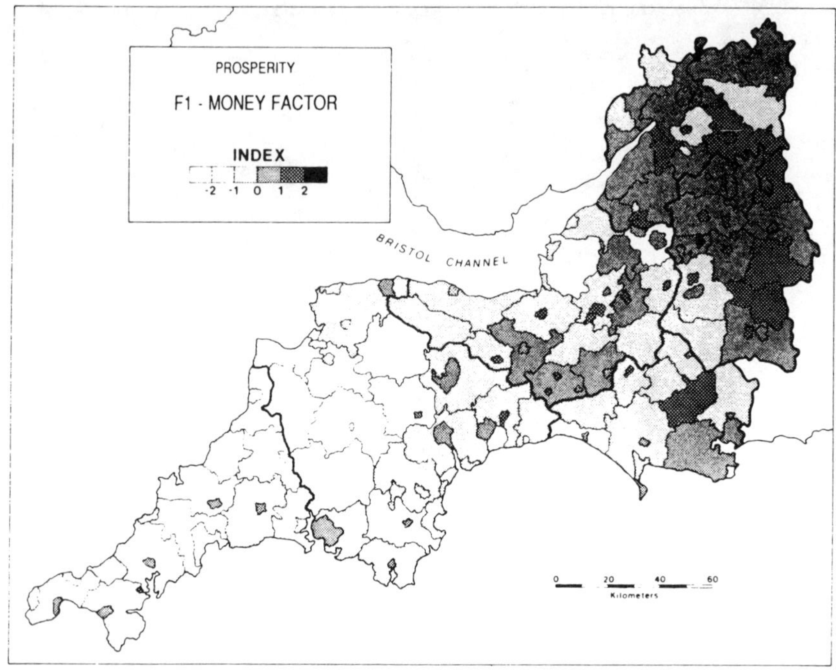

Figure 7.9 Spatial distribution of the first principal component of prosperity in the southwest region of England.

alternative approach is to use such correlations to effect a decomposition of a stochastic process into its more basic constituents. This can be done in both the time and the frequency domain. In this section we consider time-domain spectral analysis for both continuous and discrete processes. Stochastic processes defined over two-dimensional physical space are considered in Section 7.8.

7.5.1   Analysis of the Population: Continuous Space

We first consider analysis of a continuous random time process $X(t)$ defined within an interval in infinite dimensional space. The general mathematical theory is that of functional analysis in Hilbert space where we attempt to expand a continuous function $f(x)$, within a finite interval, into an infinite series of orthogonal functions (e.g., see Mercer, 1909). The statistical application of the theory comes independently from Karhunen (1947) and Loève (1945), although it seems to have been first considered, within the multivariate normal setting, by Kosambi (1943). The latter author also describes interesting computational machinery (hardware) which he suggests could be used to carry out the computations, although this was never done because of the author's difficult political circumstances. The principal objective of a spectral analysis is to expand $X(t)$ into an infinite linear series of functions, which in some sense reveal the structure of the process. Let $X(t)$ be a process such that $E[X(t)] = 0$. We wish to decompose $X(t)$ into a set of orthogonal functions $\phi_i(t)$, each with zero mean and unit variance, such that $X(t) = \sum_{i=1}^{\infty} z_i \phi_i(t)$. In particular, if $X(t)$ contains measurement error the objective is often to decompose $X(t)$ into two orthogonal parts as

$$X(t) = X^*(t) + N(t) \tag{7.7}$$

where $X^*(t)$ represents the true or the "signal" part of the series.

Consider a function $f(x)$ defined over a finite closed interval $[a, b]$, together with the linear expansion

$$f(x) = \sum_{i=1}^{\infty} z_i \phi_i(x) \tag{7.8}$$

Then if $f(x)$ is square integrable, that is,

$$\int_a^b [f(x)]^2 \, dx < \infty \tag{7.8a}$$

the magnitude of the squared error of the expansion can be defined as

$$\left\{ \int_a^b [f(x)] - \sum_{i=1}^{r} [z_i \phi_i(x)]^2 \, dx \right\}^{1/2} \tag{7.9}$$

where $\phi_i(x)$ are also square-integrable functions. Since the integral replaces the summation sign encountered in finite vector spaces, Eq. (7.9) can be considered as the infinite dimensional extension of the usual sum-of-squares criterion.

Definition 7.1.   Consider a sequence of functions $\phi_1(x), \phi_2(x), \ldots$ in the closed interval $[a, b]$. Then the system is said to be orthogonal if and only if

$$\int_a^b \phi_i(x)\phi_j(x) \, dx = \begin{cases} \lambda_i & i = j \\ 0 & i \neq j \end{cases} \tag{7.10}$$

Definition 7.2.   An orthogonal system of functions $\phi_1(x), \phi_2(x), \ldots$ in a closed interval $[a, b]$ is complete if, for every piecewise continuous function $f(x)$, the squared error criterion (Eq. 7.9) converges to zero as $r \to \infty$.

It is well known that given a complete orthogonal sequence $\phi_1(x), \phi_2(x), \ldots$ the function $f(x)$ can be expanded into a convergent infinite series in $[a, b]$, that is, we can write

$$f(x) = \sum_{i=1}^{\infty} z_i \phi_i(x) \tag{7.11}$$

In the mathematical literature (Eq. (7.11) is also known as a generalized Fourier expansion of $f(x)$ (see Kaplan, 1952). When Eq. (7.11) holds, the orthogonal functions $\phi_i(x)$ are said to form a basis of the space in the finite interval $[a, b]$.

Since a time series can be viewed as a function of time, the theory of generalized Fourier series can be applied to stochastic processes. Let $Y(t)$ be a continuous real-valued stochastic process (time-series) of second order, defined in the time interval $[0, T]$, with finite mean-value function $E[Y(t)] = m(t)$ and continuous autocovariance function (kernel) $C(t, s) = E[Y(t) - m(t)][Y(s) - m(s)]$ where s and t represent two points in time. The process $Y(t)$ is continuous in the mean in $[0, T]$ if

$$E[Y(t + h) - Y(t)]^2 \to 0 \qquad \text{for } h \to 0 \tag{7.12}$$

[note that although $C(t, s)$ must be continuous, $Y(t)$ need not be stationary]. Alternatively $Y(t)$ must be continuous in the mean. Loève (1945) and Karhunen (1947) have applied the theory of generalized Fourier expansions to show that if $Y(t)$ satisfies the conditions given above then

$$Y(t) = m(t) + \sum_{i=1}^{\infty} z_i \phi_i(t) \tag{7.13}$$

with convergence in the mean, that is, the partial sums approach $Y(t)$ in the sense of the squared criterion (Eq. 7.9). It is usual to define $X(t) = Y(t) - m(t)$ so that $E[X(t)] = 0$. The terms $\phi_i(t)$ are the eigenvectors of the homogeneous integral equation

$$\int_0^t C(t, s)\phi_i(t)\, dt = \lambda_i \phi_i(s) \tag{7.14}$$

for $i = 1, 2, \ldots.$ If $m(t) = E[X(t)] = 0$ then z_i are random variables such that $E(z_i) = 0$ and $E(z_i z_j) = 0$ for $i \neq j$ and $\phi_i(t)$ are fixed functions of time such that $E(z_i^2) = \lambda_i$ (Exercise 7.3).

THEOREM 7.1.   Let $\{\phi_i(t)\}$ be a complete orthogonal system of continuous functions in the interval $[0, T]$. Then the series $\sum_{i=1}^{\infty} z_i \phi_i(t)$ converges uniformly to $X(t)$, $t \in [0, T]$, and

$$z_r = \frac{1}{\lambda_r} \int_0^T X(t)\phi_r(t)\, dt \tag{7.15}$$

PROOF. Using Eq. (7.13) for $m(t) = 0$

$$X(t) = z_1\phi_1(t) + z_2\phi_2(t) + \cdots + z_r\phi_r(t) + \cdots$$

and multiplying by $\phi_r(t)$ and integrating in $[0, T]$ we have

$$\int_0^T X(t)\phi_r(t)\,dt = z_1 \int_0^T \phi_1(t)\phi_r(t)\,dt + z_2 \int_0^T \phi_2(t)\phi_r(t)\,dt$$

$$+ \cdots + z_r \int_0^T \phi_r^2(t)\,dt + \cdots$$

$$= z_r\lambda_r$$

using Eq. (7.10). Dividing by λ_r yields Eq. (7.15).

The λ_i, $\phi_i(t)$ are the eigenroots and eigenvectors of the integral equation (Eq. 7.14) and represent continuous analogues of the latent roots and latent vectors of the principal components model (Eq. 3.7). Likewise, the random functions $z_1, z_2, \ldots$ are the continuous principal components of the stochastic process $X(t)$ and Eq. (7.15) is the continuous analogue of Eq. (3.1). The orthogonality of z_i, in relation to the integral equation (Eq. 7.14), is given in the following theorem.

THEOREM 7.2. Consider Eq. (7.13) where $m(t) = 0$. Then
  (i) If the random variables z_i are orthogonal, then the functions $\phi_i(t)$ satisfy the integral equation (Eq. 7.14).
  (ii) If the orthogonal functions $\phi_i(t)$ satisfy Eq. (7.14), the z_i are orthogonal.

PROOF.
  (i) Let

$$X(t) = \sum_{i=1}^\infty z_i\phi_i(t) \qquad (0 \le t \le T)$$

$$X(s) = \sum_{i=1}^\infty z_i\phi_i(s) \qquad (0 \le s \le T)$$

Multiplying, and assuming orthogonality, we have

$$E[X(t)X(s)] = \sum_{i=1}^\infty E(z_i^2)\phi_i(t)\phi_i(s)$$

where $E(z_i^2) = \lambda_i$ and $E(z_iz_j) = 0$ for $i \ne j$. Since $C(t, s) =$

$E[X(t)X(s)]$ we have

$$C(t, s) = \sum_{i=1}^{\infty} \lambda_i \phi_i(t) \phi_i(s)$$

and multiplying by $\phi_i(s)$ and integrating yields

$$\int_0^T C(t, s)\phi_i(s)\, ds = \sum_{i=1}^{\infty} \int_0^T \lambda_i \phi_i(t)\phi_i(s)\, \phi_i(s)\, ds$$

$$= \sum_{i=1}^{\infty} \lambda_i \phi_i(t) \int_0^T \phi_i(s)\phi_i(s)\, ds$$

$$= \lambda_i \phi_i(t) \tag{7.16}$$

so that the eigenfunctions satisfy the integral equation (Eq. 7.14).

(ii) Conversely, assume the random variables z_i satisfy the integral equation (Eq. 7.14). Then when $\phi_i(t)$ are orthonormal we have, from Eq. (7.15),

$$X(s)z_i = \int_0^T X(s)X(t)\phi_i(t)\, dt$$

and

$$E[X(s)z_i] = \int_0^T E[X(s)X(t)]\phi_i(t)\, dt$$

$$= \int_0^T C(t, s)\phi_i(t)\, dt$$

$$= \lambda_i \phi_i(t)$$

Thus

$$E(z_i z_j) = \int_0^T E[X(t)z_i]\phi_i(t)\, dt$$

$$= \int_0^T \lambda_i \phi_i(t)\phi_j(t)\, dt$$

$$= \begin{cases} \lambda_i & i = j \\ 0 & i \neq j \end{cases} \tag{7.17}$$

and z_i form an orthogonal set of random variables with variance λ_i.

THEOREM 7.3.   Let $z_i = \int_0^T X(s)\phi_i(s)\, ds$ be the coefficients of the infinite expansion (Eq. 7.13) where $m(t) = 0$. Then $E(z_i^2) = \lambda_i$ are the latent roots of the integral equation (Eq. 7.14).

PROOF: We have

$$z_i = \int_0^T X(s)\phi_i(s)\, ds$$

so that

$$E(z_i^2) = \int_0^T \int_0^T E[X(t)X(s)]\phi_i(t)\phi_i(s)\, ds\, dt$$

$$= \int_0^T \left[\int_0^T C(t, s)\phi_i(t)\, dt\right]\phi_i(s)\, ds$$

$$= \int_0^T [\lambda_i\phi_i(s)]\phi_i(s)\, ds$$

$$= \lambda_i$$

since $\phi_i(s)$ are orthonormal.

In summary, given an orthogonal set $\{\phi_i(t)\}$ if the random variables z_i are also orthogonal then the problem of representing a stochastic process by a series of the form of Eq. (7.13) is equivalent to the condition that the integral equation (Eq. 7.14) possesses nonzero eigenfunctions $\phi_i(t)$, corresponding to eigenvalues $\lambda = \lambda_i$. In addition, when the process is stationary the autocovariance function only depends on the differences between the time points, that is, $C(t, s) = C(t - s)$. The integral equation (Eq. 7.14) can then be expressed as

$$\int_0^T C(t - s)\phi_i(t)\, ds = \lambda_i\phi_i(s) \tag{7.18}$$

When $C(t, s)$ can be expressed explicitly in terms of an analytic function of time, the eigenfunctions can also be written as analytical functions of time. In general, however, both the fixed eigenfunctions and the random principal components will be empirical functions. For use of the Karhunen-Loève expansion in nonstationary processes see Davis (1952). Fukunaga (1990) describes application in pattern recognition. In addition, the expansion can be shown to have the following properties.

1. Parseval's equation. If there exists a complete system of orthogonal

eigenfunctions $\{\phi_i(t)\}$ then

$$\int_0^T X(t)\,dt = \lambda_1 z_1^2 + \lambda_2 z_2^2 + \cdots \tag{7.19}$$

Conversely, if Parseval's equation (Eq. 7.19) holds then the system $\{\phi_i(t)\}$ is complete (Exercise 7.4).

2. The autocovariance function can be expanded into an infinite series. We have

$$C(t, s) = E[X(t)X(s)]$$

$$X(t) = \sum_{i=1}^{\infty} z_i \phi_i(t)$$

so that

$$C(t, s) = \sum_{i=1}^{\infty} \sum_{i=1}^{\infty} \phi_i(t)\phi_i(s)E(z_i z_j)$$

$$= \sum_{i=1}^{\infty} \lambda_i \phi_i(t)\phi_i(s) \tag{7.20}$$

since the random variables z_i are orthogonal (see Exercise 7.5).

3. The Karhunen–Loève coefficients z_i minimize the mean-squared error resulting from only using a finite number r of terms in Eq. (7.13), that is, they minimize

$$\int_0^T \left[X(t) - \sum_{i=1}^{r} z_i \phi_i(t) \right]^2 dt = \sum_{i=r+1}^{\infty} z_i^2 = 1 - \sum_{i=1}^{r} \lambda_i \tag{7.21}$$

using Eq. (7.19) where $X(t)$ and $\{\phi_i(t)\}$ are of unit length.

PROOF. The minimization can be established by showing that for any other set of coefficients $d_i \neq z_i$, $i = 1, 2, \ldots$ we have the inequality

$$\int_0^T \left[X(t) - \sum_{i=1}^{r} z_i \phi_i(t) \right]^2 \leq \int_0^T \left[X(t) - \sum_{i=1}^{r} d_i \phi_i(t) \right]^2 dt \tag{7.21a}$$

The right-hand side of Eq. (7.21a) is given by

$$\int_0^T \left[X(t) - \sum_{i=1}^r z_i \phi_i(t) + \sum_{i=1}^r z_i \phi_i(t) - \sum_{i=1}^r d_i \phi_i(t) \right]^2 dt$$

$$= \int_0^T \left[X(t) - \sum_{i=1}^r z_i \phi_i(t) + \sum_{i=1}^r (z_i - d_i)\phi_i(t) \right]^2 dt$$

$$= \int_0^T \left[X(t) - \sum_{i=1}^r z_i \phi_i(t) \right]^2 dt + \int_0^T \left[\sum_{i=1}^r (z_i - d_i)\phi_i(t) \right]^2 dt$$

$$+ 2 \int_0^T \left[(X(t) - \sum_{i=1}^r z_i \phi_i(t)) \right]\left[\sum_{i=1}^r (z_i - d_i)\phi_i(t) \right] dt \qquad (7.21b)$$

$$= \int_0^T \left[X(t) - \sum_{i=1}^r z_i \phi_i(t) \right]^2 dt + \sum_{i=1}^r (z_i - d_i)^2 \qquad (7.21c)$$

where for the last term of Eq. (2.71b) we have

$$\int_0^T X(t) \sum_{i=1}^r (z_i - d_i)\phi_i(t)\, dt - \int_0^T \sum_{i=1}^r z_i(z_i - d_i)\phi_i^2(t)\, dt$$

$$= \sum_{i=1}^r z_i(z_i - d_i) - \sum_{i=1}^r z_i(z_i - d_i)$$

$$= 0$$

The inequality (Eq. 7.21a) follows from Eq. (7.21c).

4. Let $\mathbf{Y} = (Y_1, Y_2, \ldots, Y_p)^T$ be a vector of random variables. Then the entropy function of a distribution $f(\mathbf{Y})$ is defined as (Section 1.6)

$$\mathbf{I} = -E\{\ln f(\mathbf{Y})\} \qquad (7.22a)$$

Entropy can be used in place of variance as a measure of variability of a distribution. When the components of $\mathbf{Y}$ are independently distributed Eq. (7.22a) becomes

$$\mathbf{I} = -\sum_{i=1}^p E[\ln f(Y_i)] \qquad (7.22b)$$

It can be shown that the Karhunen–Loève expansion maximizes the entropy measure

$$\mathbf{I}_\lambda = \sum_{i=1}^r \lambda_i \ln \lambda_i \qquad (7.22c)$$

for any finite number of the first largest eigenroots (Exercise 7.9).

Owing to the optimality properties the Karhunen–Loève expansion is a useful tool when considering expansions of a stochastic process, and from the development described above is easily seen to represent the continuous version of principal components analysis. Thus when a continuous process is sampled periodically at a finite number of occasions the Karhunen–Loève decomposition becomes identical to a principal components analysis of an autocovariance matrix (see Cohen and Jones, 1969). Although a distinction is at times made between the two models, particularly in the engineering literature (e.g., see Gerbrands, 1981), this is not of essence. The two models can also be obtained using more general methods such as co-ordinate-free theory (Ozeki, 1979). Karhunen–Loève analysis can be used with electrocardiograph data (Section 7.2.3) to obtain a description of the heart's function in terms of the so-called "intrinsic components" (Young and Calvert, 1974). Consider $v(r, \theta, \psi)$, the potential at a point on the body's surface at time t where (r, θ, ψ) denote a point on the body's surface in terms of spherical coordinates (r, θ, ψ). We then have $v(r, \theta, \psi; t) = \sum_{i=1}^{r} z_i(r, \theta, \psi)\varphi_i(t)$ where $\varphi_i(t)$ denotes the ith latent vector or intrinsic time component of the electrocardiograph and $z_i(r, \theta, \psi)$ depends only on location on the body's surface.

7.5.2 Analysis of a Sample: Discrete Space

Usually a stochastic process is observed in the form of one or more time series, each sampled at regular time intervals. An example would be daily, weekly, or monthly temperatures observed at n geographic locations, yielding matrices similar to those of Section 7.2.2 (Table 7.3). The interest in time series analysis however lies in the time points that is, locations, commodities, or other objects are simply viewed as convenient sample frames within which to estimate the variance/covariance or correlational structure of a stochastic process. Since interest here lies in the behavior of time a single time series consisting of n time points is normally viewed as a set of n random variables within a single observation, that is, a sample size of 1. When k "observations" or time series are available, each of length n, a $(n \times n)$ correlation (covariance) matrix can be computed which when decomposed into principal components yields the discrete analogue of the Karhunen–Loève decomposition (Eq. 7.14). The latent vectors (loading) then correspond to the time functions $\phi_i(t)$ ($i = 1, 2, \ldots, n$) and the scores correspond to the random variables z_i. The loadings can thus be plotted against time, where using the language of communication engineering the first r retained components are used as estimates of the true part or the "signal" within the series $\mathbf{X}(t)$. This maximizes the informational content of the sampled process and minimizes the variance of the "noise" component $\mathbf{N}(t)$. Little new arises, apart from what was seen to hold for the principal components model (Section 3.4). Note however that a principal components (Karhunen–Loève) analysis of time points implies homoscedastic, uncorrelated error terms for all time points. Furthermore, if a stochastic process is stationary (see Brillinger, 1981), the columns of the data matrix have approximately equal variance as well, and either a covariance or a correlation matrix can be used. Alternatively, a single time phenomenon can be measured simultaneously k times, at discrete points (or continuously), for example, a single earthquake or explosion may be measured by k different seismographs. Thus in terms of the electrocardiograph example of Section 7.5.1, if k simultaneous measurements of one heart beat of duration T are made on the surface of the body, we have k series $x_1(t), x_2(t), \ldots, x_k(t)$ where the $(k \times k)$ covariance matrix $\mathbf{C}$ has typical (population) elements given by

$$c_{ij} = \int_0^T x_i(t)x_j(t)\, dt \tag{7.22d}$$

The Karhunen–Loève analysis for the sample of k continuous series can be carried out by solving $\mathbf{P}^T\mathbf{S}\mathbf{P} = \mathbf{L}$ in the usual way.

A different situation arises when we only have a single time series at our disposal, that is, replications are not available to estimate the variance–covariance terms of $C(t, s)$. Much of social and historic data, for example, are available in the form of single time series observed at highly discrete points of time (annual, quarterly, monthly, etc.). In this situation all first and second moments must be estimated from a single series. The following procedure is found to work well in practice when the objective is to estimate empirical components of a series (Basilevsky and Hum, 1979). Consider a time series $\mathbf{Y}(t)$ of length $N = n + m + 1$ which is partitioned into $m + 1$ overlapping segments of length n, that is, we have the series $\mathbf{Y}(t) = (y_{-m+1}, y_{-m+2}, \ldots, y_1, \ldots, y_n)$ where negative subscripts are used to simplify subsequent notation. To compute a $[(m + 1) \times (m + 1)]$ sample covariance matrix $\mathbf{S}_m$ the series $\mathbf{Y}(t)$ is segmented into $m + 1$ lagged vectors $\mathbf{Y}(t - j) = (y_{1-j}, y_{2-j}, \ldots, y_{n-j}), (j = 0, 1, 2, \ldots, m)$, each of length $n > m$. Each vector therefore differs from its immediate neighbor by its first and last element.

A discrete analogue of Eq. (7.13) can then be written for each of the segments $\mathbf{Y}(t), \mathbf{Y}(t - 1), \ldots, \mathbf{Y}(t - m)$ as

$$\mathbf{Y}(t) = m(t) + \sum_{i=1}^{m+1} p_i(t)\mathbf{Z}_i$$

$$\mathbf{Y}(t - 1) = m(t - 1) + \sum_{i=1}^{m+1} p_i(t - 1)\mathbf{Z}_i \tag{7.23}$$

$$\mathbf{Y}(t - m) = m(t - m) + \sum_{i=1}^{m+1} p_i(t - m)\mathbf{Z}_i$$

where the sample means are computed as

$$m(t-j) = \bar{\mathbf{Y}}(t-j) = \frac{1}{n} \sum_{i=1-j}^{n-j} \mathbf{Y}_j \qquad (j = 0, 1, \ldots, m) \qquad (7.24)$$

Note that the original series does not have to be detrended. The coefficients $p_i(t-j)$ are elements of the orthonormal latent vectors $\mathbf{P}(t)$ and $\mathbf{P}(t-1), \ldots, \mathbf{P}(t-m)$ which correspond to the continuous eigenfunctions $\phi_i(t)$ of Eq. (7.13) and vector $\mathbf{Z}_i$ corresponds to the random variables z_i. In many applications $\mathbf{Z}_i$ values are of main interest, since they represent systematic components of the time series such as trend, cycle, and seasonal movements. The $p_i(t-j)$ and $\mathbf{Z}_i$ are computed from the covariance matrix $\mathbf{S}_m$ formed from the lagged vectors $\mathbf{Y}(t-j)$, that is,

$$s^2(w) = \frac{1}{n-1} \sum_{i=1}^{n} y_{i-w}^2 - n\bar{y}(t-w) \qquad (7.25)$$

$$= \frac{1}{n-1} \sum_{i=1}^{n} x_{i-w}^2 \qquad (w = 0, 1, \ldots, m) \qquad (7.26)$$

$$s(u,v) = \frac{1}{n-1} \sum_{i=1}^{n} y_{i-u}y_{i-v} - n\bar{y}(t-u)\bar{y}(t-v)$$

$$= \frac{1}{n-1} \sum_{i=1}^{n} x_{i-u}x_{i-v} \qquad (u, v = 0, 1, \ldots, m) \qquad (7.27)$$

where x denotes deviations from the sample means (Eq. 7.24). Thus elements of a common segment $\mathbf{Y}(t-j)$ are assumed to have constant mean and variance, but the moments can vary from segment to segment, that is, the time series is only "piecewise" or segment stationary. The computations are again essentially those of the principal (Karhunen–Loève) components model and can be implemented as follows:

1. Express the elements of the $(n \times 1)$ vectors $\mathbf{Y}(t)$ and $\mathbf{Y}(t-1)$, . . . , $\mathbf{Y}(t-m)$ as deviations about means (Eq. 7.24) and let the deviations form the columns of the $n \times (m+1)$ matrix $\mathbf{X}$.

2. Compute the eigenvectors $\mathbf{P}(t)$, $\mathbf{P}(t-1), \ldots, \mathbf{P}(t-m)$ and corresponding eigenvalues $\mathbf{L} = \text{diag} (l_1, l_2, \ldots, l_{m+1})$ of the matrix $\mathbf{X}^T\mathbf{X}$.

3. Let $\mathbf{A} = \mathbf{L}^{1/2}\mathbf{P}$ with column vectors $\mathbf{A}(t)$ and $\mathbf{A}(t-1), \ldots, \mathbf{A}(t-m)$ where $\mathbf{P}$ is the matrix of eigenvectors. Then

$$\mathbf{Y}(t-j) = m(t-j) + \sum_{i=1}^{m+1} a_i(t-j)\mathbf{Z}_i^* \qquad (7.28)$$

where $a(t-j)$ is the ith element of $\mathbf{A}(t-j)$ $(j = 0, 1, \ldots, m)$, that is, $a_i(t-j)$ are correlation loadings when $\mathbf{X}^T\mathbf{X} = \mathbf{R}$, the correlation matrix.

Since the variances of the different time segments are not assumed to be equal, either the covariance or the correlation matrix may be used with the usual provisos (Section 3.4.3).

Clearly, only the first r components are of interest since these contain the "signal" portion of the time series. For example, given a socioeconomic quarterly time series we can write

$$\mathbf{X}(t) = \mathbf{T}(t) + \mathbf{C}(t) + \mathbf{S}(t) + \mathbf{N}(t) \tag{7.29}$$

where $\mathbf{T}(t)$ is a polynomial trend, $\mathbf{C}(t)$ is a cycle with generally variable period and amplitude, $\mathbf{S}(t)$ represent seasonal (quarterly) within-year periodic variation (with possibly variable amplitude), and $\mathbf{N}(t)$ is the residual "noise" or error term. If $\mathbf{X}(t)$ possesses the terms $\mathbf{T}(t)$, $\mathbf{C}(t)$, and $\mathbf{S}(t)$ then the first r component scores should reflect these movements. However, it does not necessarily follow that each term of Eq. (7.29) corresponds to a single principal component. Thus if $\mathbf{X}(t)$ contains seasonal variation, then $\mathbf{S}(t)$ is generally represented by three orthogonal components. Likewise, the "cycle" $\mathbf{C}(t)$ can be picked up by several orthogonal components. For purposes of interpretation however it is usually preferable to represent similar time behavior by a single term. Initially, the first r principal component scores can be plotted against time, where a visual inspection will normally reveal their identity, so that rotation is generally not required to identify the components. In addition, assuming a normal process the likelihood ratio criterion can be used to test equality of all, or some, of the latent roots (Chapter 4) where components with insignificantly different roots will generally represent similar time behavior and thus may be aggregated into a single term.

Example 7.7 Given an annual series we may wish to investigate whether the time series contains nonstationary components such as trend and/or cycle(s). Figure 7.10 represents annual volume flows of the Red River, measured at Emerson (Manitoba) during the period 1913–1976. Here $N = 66$ where $m = 15$ lags prove adequate to capture the main movements of the series. We have the vectors $\mathbf{X}(t-15)$ $\mathbf{X}(t-14), \ldots, \mathbf{X}(1)$, $\mathbf{X}(0)$, and a

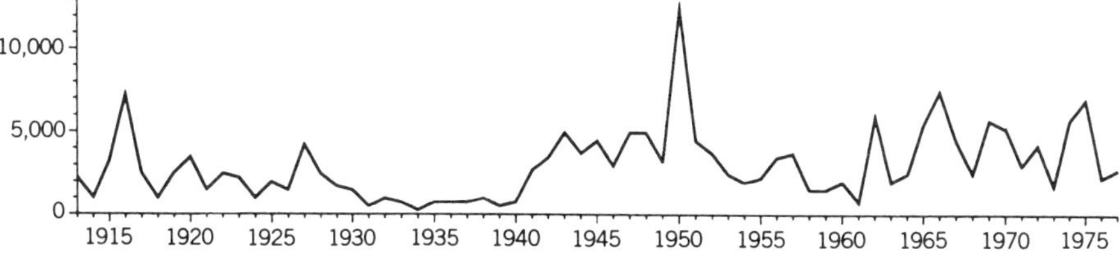

Figure 7.10 Red River mean annual discharge cubic feet per second, 1913–1976, at Emerson, Manitoba.

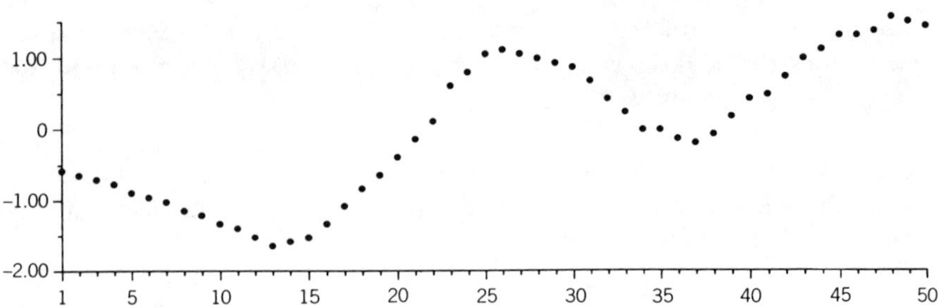

Figure 7.11 The cycle-trend component $\mathbf{Z}(1)$ of Red River discharge at Emerson, Manitoba.

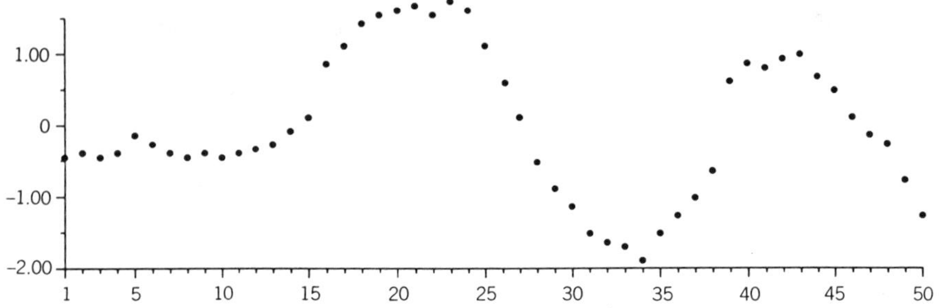

Figure 7.12 The cyclic component $\mathbf{Z}(2)$ of Red River discharge at Emerson, Manitoba.

principal components analysis of the (16×16) covariance matrix reveals the presence of two recognizable "signal" components in the series (Figs. 7.11 and 7.12) which account for 26.2 and 17.6% of the trace respectively. The first component $\mathbf{Z}(1)$ represents cyclic-like behavior superimposed on a linear trend, and further investigation reveals it corresponds to a moving average of the series. The second component $\mathbf{Z}(2)$ exhibits an irregular cyclic movement, which is uncorrelated with the moving average term and which represents deviations from the moving average. Consulting the correlation loadings (Table 7.10) indicates that the cycle-trend component is most highly and positively correlated with the segment 1921–1971, where we can perceive both trend and cyclical movements of the water flows. The second component, on the other hand, correlates negatively with the segment 1915–1965, its peaks indicating that the most dry period occurred during 1932–1938, and to a lesser extent during 1954–1961. Care, however, must be taken when interpreting correlation involving cyclical behavior since the nonlinear nature of periodic-like functions precludes associating in-dependence with lack of correlation. Also, since the signal components represent empirical functions, any extrapolation into the future carries some risk. Nevertheless the component functions reveal simpler behavior than the original series and are generally easier to understand and forecast. For comments on more general applications see Matalas and Reiher (1967).

Example 7.8. Given quarterly birth data we at times may wish to

Table 7.10   Correlation Loading Coefficients for $p = 16$ Lagged Variables

	Z(1)	Z(2)
X(0)	.059	−.511
X(1)	.104	−.585
X(2)	.172	−.597
X(3)	.297	−.593
X(4)	.452	−.548
X(5)	.560	−.418
X(6)	.598	−.302
X(7)	.656	−.216
X(8)	.698	−.072
X(9)	.691	.081
X(10)	.684	.213
X(11)	.608	.370
X(12)	.577	.390
X(13)	.538	.389
X(14)	.449	.449
X(15)	.366	.441

determine whether the series contains seasonal (and other) regular movements. Consider the series $Y(t)$ consisting of $N = 323$ quarterly time records of the number of registered births for two administrative Jamaica parishes (1880–1938), the mode of production of which is dominated by plantation organization with a highly seasonal workload (Basilevsky and Hum, 1979). The data for the births, as well as their estimated spectral components, are given in Figures 7.13–7.15. Since birthrates as such are not available, the series $Y(t)$ also reflects variation (trend) in the general population level of the parishes due perhaps to short-, medium-, or long-term migratory movements within the labor force, natural increase, and so forth.

Figures 7.14 and 7.15 reveal the presence of fairly regular movements in the total number of births, consisting of seasonal and cyclic behavior as well as a nonlinear trend, which represents a moving average of the series. To

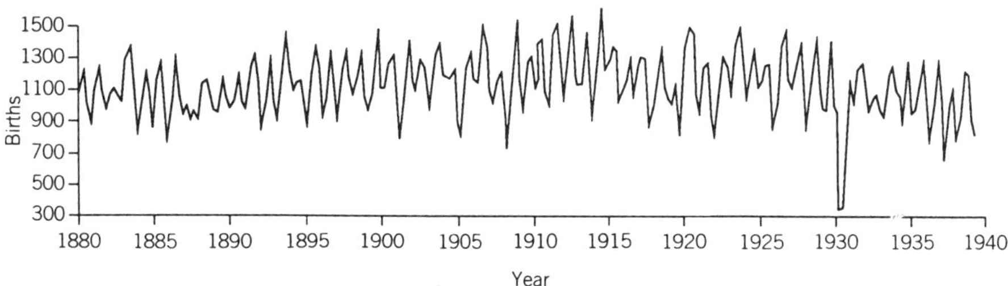

Figure 7.13   Jamaican plantation births, 1880–1938 (Basilevsky and Hum, 1979; reproduced with permission).

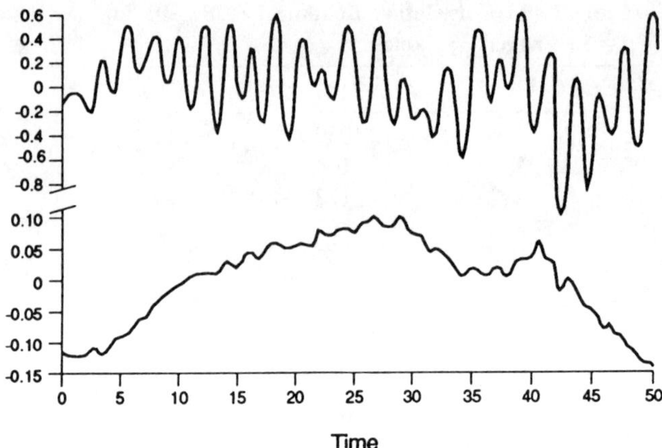

Figure 7.14 Estimated cycle $\mathbf{C}(t)$ and trend $\mathbf{T}(t)$ of Jamaican plantation births (Basilevsky and Hum, 1979; reproduced with permission).

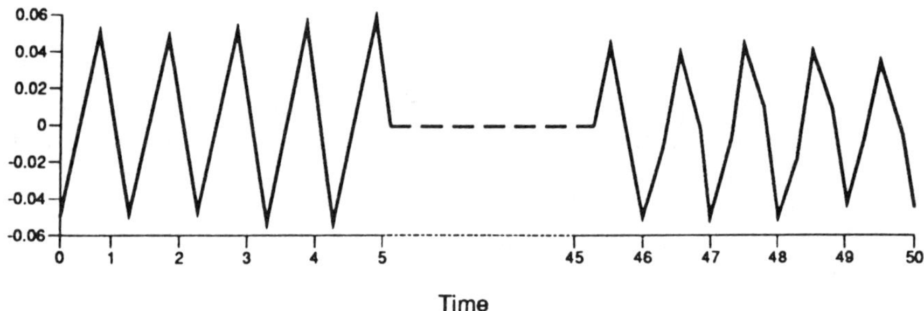

Figure 7.15 Estimated seasonality $\mathbf{S}(t)$ of Jamaican plantation births (Basilevsky and Hum, 1979; reproduced with permission).

illustrate the aggregation of component scores consider the first $r = 6$ components of $\mathbf{X}^T\mathbf{X}$, which yield the decomposition

$$\mathbf{X}(t) = (.322\mathbf{Z}_1 - .557\mathbf{Z}_2) + .263\mathbf{Z}_3 + (-.108\mathbf{Z}_4 + .335\mathbf{Z}_5 + .247\mathbf{Z}_6) + \mathbf{N}(t)$$

$$= \mathbf{S}(t) + \mathbf{T}(t) + \mathbf{C}(t) + \mathbf{N}(t) \tag{7.30}$$

Other columns of $\mathbf{X}$ yield similar decompositions. Since $m = 32$, $\mathbf{X}$ represents a (200×32) matrix. Only the first $r = 6$ components are retained since the remaining components fail to reveal globally meaningful variation. Also, an examination of the scores reveals the seasonal and the between-years cyclical terms broken down into two and three orthogonal components, respectively. This is further corroborated by Anderson's chi-squared criter-

Table 7.11 Latent Roots, Principal Components and Time Components of the Matrix $X^T X$

Latent Roots	Corresponding Principal Components	χ^2	Time Components	% Variance Explained
$l_1 = 286872.0$	Z_1^*	.10	$S(t)$: Seasonality	41.6
$l_2 = 284684.1$	Z_2^*			
$l_3 = 127786.6$	Z_3^*		$T(t)$: Trend	6.8
$l_4 = 58758.6$	Z_4^*			
$l_5 = 55943.6$	Z_5^*	4.0	$C(t)$: Cycle	18.5
$l_6 = 45054.2$	Z_6^*			

ion (Section 4.3)

$$\chi^2 = -n \sum_{j=1}^{r} \ln(l_j) + nr \ln\left[\sum_{i=1}^{r} \frac{l_j}{r} \right] \tag{7.31}$$

which reveals insignificant differences between the bracketed terms of the first line of Eq. (7.30). The results are summarized in Table 7.11.

The procedures of the present section can also be extended to more than a single time series. For example, a canonical correlation analysis can be carried out on two sets of lagged variables. For three variables or more Quenouille (1968) has also suggested the following procedure. Consider, for example, variables x, y, and z with correlation matrices $R(0)$, $R(1), \ldots, R(T)$ where $R(t)$ is the (3×3) matrix of correlations between present values and those at lag t $(t = 0, 1, \ldots, T)$. A principal components analysis is carried out on the t "quotient" matrices $R_t R_{t-1}^{-1}$ and the resulting latent roots are arranged in a matrix with t rows and 3 columns. A column of stable (large) values then indicates a strong trend component; a stable root that changes sign with increasing lag indicates a cyclical component, and an erratically variable root suggests a component that is independent of trend and cycle. Using simulated data Farmer (1971) has carried out experiments dealing with the estimation of sinusoidal components of serially correlated time series, using Quenouille's method.

7.5.3 Order Statistics: Testing Goodness of Fit

Let $y_1, y_2, \ldots, y_n$ be a random independent sample observed for a continuous random variable Y. The order statistics are those values arranged in the increasing order

$$y_{(1)} < y_{(2)}, \ldots < y_{(n)} \tag{7.32}$$

where $y_{(1)} = \min(y_1, y_2, \ldots, y_n)$ and $y_{(n)} = \max(y_1, y_2, \ldots, y_n)$ with intermediate values in between. Since Y is continuous, the probability of any two observations being equal is zero. A major use of order statistics lies in testing the hypothesis that n independent and identically distributed random variables have a specified continuous distribution function $F(y)$. The sample distribution function is defined by

$$F_n(y) = \begin{cases} 0 & y \le y_{(1)} \\ k/n & y_{(k)} \le y \le y_{(k+1)} \\ 1 & y_{(n)} \le y \end{cases} \qquad (7.33)$$

the proportion of observations which do not exceed y. A goodness of fit test can be defined by the use of the Cramer–von Mises statistic

$$W_n^2 = \int_{-\infty}^{\infty} [F_n(y) - F(y)]^2 \, dF(y) \qquad (7.34)$$

which is based on the Euclidian distance between $F_n(y)$ and $F(y)$ (see also Smirnov, 1936).

As for most goodness of fit tests, Eq. (7.34) is global in nature in that it does not reveal the source nor the nature of discrepancy between $F_n(y)$ and $F(y)$. It is at times desirable therefore to partition W_u^2 into a suitable set of components, each of which measures some distinctive aspect or feature of the data. One would then be able to test the significance of the individual components, in addition to assessing overall significance of W_n^2. First we observe that testing the null hypothesis of $y_1, y_2, \ldots, y_n$ having emanated from the continuous distribution function $F(y)$ is equivalent to testing the null hypothesis that $F(y_{(1)}) < F(y_{(2)}) < \cdots < F(y_{(n)})$ is an ordered sample of independent uniform $(0, 1)$ random variables. Let $u_1 = F(y_{(1)})$, $u_2 = F(y_{(2)}), \ldots$, and $u_n = F(y_{(n)})$ and let $G_n(u)$ be the empirical distribution derived from $u_1, u_2, \ldots, u_n$. Then the test can be expressed as

$$nW_n^2 = \int_0^1 [G_n(u) - u]^2 \, du$$

$$= \frac{1}{12n} + \sum_{k=1}^{n} \left[F(y_{(r)}) - \frac{k - \frac{1}{2}}{n} \right]^2$$

for $0 \le u \le 1$. Also, $X_n(u) = \sqrt{n}[G_n(u) - u]$ is a random variable the set of which may be considered as a stochastic process with parameter u. For fixed $u_1, u_2, \ldots, u_k$ the joint distribution of $X_n(u_1), X_n(u_2), \ldots, X_n(u_k)$ approaches a k-variate normal process specified by its mean and covariance function.

For finite n we have $\qquad E[X_n(u)] = 0$

$$E[X_n(u_i)X_n(u_j)] = \min(u_i, u_j) - u_i u_j \qquad (7.36)$$

We wish to decompose the process $X_n(u)$ into a set of orthogonal components $z_{n1}, z_{n2}, \ldots$, each with zero mean such that

$$X_n(u) = \sum_{j=1}^{\infty} z_{nj} \phi_j(u)$$

By Theorem 7.3 the continuous principal components of $X_n(u)$ are given by

$$z_{nj} = \int_0^1 \phi_j(u) X_n(u)\, du \qquad (j = 1, 2, \ldots) \qquad (7.37)$$

where $\phi_1(u)$, $\phi_2(u)$, and so on are normalized eigenfunctions obtained as solutions of Eq. (7.14). Putting

$$C(u_i, u_j) = \min(u_i, u_j) - u_i u_j$$

we have the integral equation

$$\int_0^1 [\min(u_i, u_j) - u_i u_j]\phi_j(u_i) = \lambda_j \phi_j(u_j) \qquad (7.38)$$

where λ_1, λ_2, and so on are the eigenvalues. It can be shown (Kac, 1951) that the solution to Eq. (7.38) is given by the eigenfunctions

$$\phi_1(u) = \sqrt{2}\sin(\pi u), \ \phi_2(u) = \sqrt{2}\sin(2\pi u), \ldots, \ \phi_j(u) = \sqrt{2}\sin(j\pi u)$$

and eigenvalues $\lambda_j = 1/j^2\pi^2$ $(j = 1, 2, \ldots,)$. Continuous principal components with zero mean and unit variance can therefore be defined as

$$z_{nj} = \sqrt{2}\, j\pi \int_0^1 \sin(j\pi u) X_n(u)\, du \qquad (j = 1, 2, \ldots) \qquad (7.39)$$

The inverse form

$$X_n(u) = \sqrt{2} \sum_{j=1}^{\infty} \frac{1}{j\pi} \sin(j\pi u) z_{nj} \qquad (0 \le u \le 1)$$

is therefore simply a Fourier sine series expansion of the function $X_n(u)$. The Cramer–von Mises statistic can then be expressed as

$$W_n^2 = \sum_{j=1}^{\infty} \frac{1}{j^2\pi^2} z_{nj}^2 \qquad (7.40)$$

where an examination of the individual z_{nj}^2 can yield local information concerning departure of the observations from H_0. An empirical process can also be expanded using the Karhunen–Loève expansion of Section 7.5.1 (see Wells, 1990). Percentage points for the components of Eq. (7.40) may be found in Durbin and Knott (1972). Rosenblatt (1952) has extended the results to the multivariate case.

7.6   ESTIMATING DIMENSIONALITY OF STOCHASTIC PROCESSES

Factor models can be used to estimate dimensionality of a stationary autoregressive (autoregressive-moving average) stochastic process. A stochastic process $\{x_t\}$ is said to follow a stationary autoregressive process of order p if

$$\sum_{j=0}^{p} a_j x_{t-j} = a_0 x_t + a_1 x_{t-1} + \cdots + a_p x_{t-p} = \epsilon_t \qquad (7.41)$$

where ϵ_t is a random error term and $x_t, x_{t-1}, \ldots, x_{t-p}$ possess a common probability distribution, and where without loss of generality we may impose the normalizing condition $a_0 \equiv 1$. Since the estimation of Eq. (7.41) is generally carried out on the assumption of normality, the weaker condition of second-order stationarity is usually imposed, that is, only the first two moments of $x_t, x_{t-1}, \ldots, x_{t-p}$ are required to be stationary. This implies that the random variables possess equal (say zero) means and equal variance and autocovariances which are independent of absolute time, that is, which are functions only of the lags or the relative proximities of the time points. A process is said to be autoregressive moving average (ARMA) if we have an autoregressive (AR) process (Eq. 7.41) where in addition the error term follows a moving average process of order q, that is,

$$\sum_{j=0}^{p} a_j x_{t-j} = \sum_{j=0}^{q} b_j \epsilon_{t-j}$$

or

$$a_0 x_t + a_1 x_{t-1} + \cdots + a_p x_{t-p} = b_0 \epsilon_t + b_1 \epsilon_{t-1} + \cdots + b_q \epsilon_{t-q} \qquad (7.42)$$

where the moving average process of $\{\epsilon_t\}$ follows the same restrictions as the AR process (Eq. 7.41).

When the orders p and q of Eq. (7.42) are known, the coefficients a_j and b_j can be estimated in one of several methods (e.g., see Wei 1990). Usually however the random variables $\{x_t\}$ of a stochastic process are not known, and instead we observe a single realization or a time series $x_1, x_2, \ldots, x_n$ at n discrete and equal intervals, that is, we obtain a single sample for the n random variables. In this situation both the orders as well as the coefficients must be estimated simultaneously, which gives rise to a situation not unlike that of factor analysis. In the following sections we describe several related estimation procedures using factor analysis (principal components) to estimate unknown stationary AR and ARMA processes as well as the so-called Kalman filter commonly used in control engineering and time series regression models.

7.6.1 Estimating a Stationary ARMA Model

Given an unknown ARMA process the most commonly used estimation procedures attempt to estimate the autoregressive parameters first. The methods can therefore also be applied to autoregressive processes as a special case. If the process is weakly (second-order) stationary, the (auto) correlations between the points $x_t, x_{t-1}, \ldots, x_{t-p}$ capture all of the required information concerning a_j, and the relevant association matrix is thus the correlation matrix. Given a particular realization of a stationary process, the sample autocorrelations of the time series are given by

$$r(k) = \sum_{t=k+1}^{n} x_t x_{t-k} \qquad (k = 0, 1, 2, \ldots) \qquad (7.43)$$

where $r(k)$ denotes the correlations between contemporaneous points and those lagged k periods. If the process is second-order stationary, its main characteristics are embodied in the autocorrelation sequence (Eq. 7.43), and the AR coefficients of the ARMA process may be estimated from the so-called extended Yule–Walker equations

$$\sum_{j=0}^{p} a_j r(t - j) = 0 \qquad (t \geq q + 1) \qquad (7.44)$$

(see Cadzow, 1982, 1984) where m distinct values of t satisfying $t \geq q + 1$ are considered. Such an overdetermined approach leads to the following normal equations:

$$\begin{bmatrix} r(q+1) & r(q) & \cdots & r(q-p+1) \\ r(q+2) & r(q+1) & \cdots & r(q-p+2) \\ \vdots & & \ddots & \\ r(q+m) & r(q+m-1) & \cdots & r(1-p+m) \end{bmatrix} \begin{bmatrix} 1 \\ a_1 \\ a_2 \\ \vdots \\ a_p \end{bmatrix} = \begin{bmatrix} 0 \\ 0 \\ \vdots \\ 0 \end{bmatrix} \qquad (7.45)$$

or, in matrix form, $\mathbf{R}_1 \mathbf{A} = \mathbf{0}$ where the "correlation" matrix $\mathbf{R}_1$ possesses a Toeplitz-like structure (Section 7.9.3) with the (i, j)th element given by

$$r(i, j) = r(q + 1 + i - j)$$

for $1 \leq i \leq m$ and $1 \leq j \leq p + 1$. Note that $\mathbf{R}_1$ is not a true correlation matrix since the number of rows do not necessarily equal the number of columns for $m > p$. In practice $r(i, j)$ is estimated by the unbiased estimator

$$r(i, j) = \frac{1}{n - |i - j|} \sum_{l=0}^{n} x(l + i - j)x(l) \qquad (7.46)$$

for $1 \le i \le p + 1$ and $1 \le j \le p + 1$. We have the following theorem (Cadzow, 1982).

THEOREM 7.4. If the autocorrelation lag entries of matrix $\mathbf{R}_1$ of Eq. (7.45) correspond to those of an ARMA process of order (p_1, q_1), the rank of $\mathbf{R}_1$ is p_1, provided that $p \ge p_1$ and $q \ge p_1$.

The proof of the theorem is left as an exercise (Exercise 7.3). When $\mathbf{R}_1$ has less than full rank, a nontrivial autoregressive solution $\mathbf{A} = (1, a_1, a_2, \ldots, a_p)^{\mathrm{T}}$ will always exist. In that case we can write, using Eq. (7.45),

$$\mathbf{R}_1^{\mathrm{T}} \mathbf{R}_1 \mathbf{A} = \mathbf{0}$$

and the required AR parameter vector $\mathbf{A}$ is then a latent vector (whose first component is unity) associated with a zero latent root. When $\mathbf{R}_1$ is of full rank, a trivial solution will not exist, and here we may wish to determine an ARMA model that provides the best fit to Eq. (7.42). This can be accomplished by choosing the latent vector $\mathbf{A}$ such that $\mathbf{A}^{\mathrm{T}} \mathbf{R}_1^{\mathrm{T}} \mathbf{R}_1 \mathbf{A}$ is minimized subject to the usual condition $\mathbf{A}^{\mathrm{T}} \mathbf{A} = 1$. Once the $p + 1$ parameters are known, those of the moving average part can also be computed. Alternatively, the principal component decomposition can be based on the association matrix of the form $\mathbf{X}_n^{\mathrm{T}} \mathbf{X}_n$ where elements of $\mathbf{X}_n$ are suitably lagged column vectors (Cadzow, 1982), the idea being similar to that of Section 7.5.2 except that the objective here is to identify the lag order p.

This discussion assumes that the orders p and q of the ARMA process are known. This generally is not the case and the integers p and q must be estimated, together with the coefficients of the process. A number of procedures are available to estimate p and q. One that appears to give good results is based on the principal components decomposition of $\mathbf{R}_1^{\mathrm{T}} \mathbf{R}_1$ where the initial orders p_e and q_e are set sufficiently high so that $p_e > p$, $q_e > q$ and $q_e - p_e \ge q - p$. Following Cadzow (1982), the initial arbitrary high value p_e is substituted into Eq. (7.44), which yields a new higher order matrix $\mathbf{R}_e$ of order $[m \times (p_e + 1)]$, whose true rank will be p, but whose actual rank is observed to be $\min[m, (p_e + 1)]$. The estimation of p is then identical to the principal components analysis of $\mathbf{R}_e^{\mathrm{T}} \mathbf{R}_e$, that is, we chose $p = r$, the number of common principal components considered to be present. The ARMA model is then completed by determining the model's associated moving average parameters.

7.6.2 Time Invariant State Space Models

Stationary time series can generally be represented by ARMA models whose dimensionality may be obtained as outlined in the previous section. An alternative procedure is to use the so-called state space representation which is fundamental to control engineering and statistical time series regression models, and which in many ways is a natural choice when modeling systems consisting of inputs and outputs. Consider an exact, linear, and time invariant transformation

$$X_t^* = \beta X_{t-1}^* \tag{7.47}$$

where the (one time period) true state of the system X_t^* is solely determined by its past state $\mathbf{X}_{t-1}^*$. Although in practice X_t^* and X_{t-1}^* are vectors (and β is a time invariant transition matrix of coefficients) for the moment we treat these terms as scalars. Here, beginning at time $t = 1$ all state values $X_2^*, X_3^*, \ldots, X_t^*$ are determined by the initial value X_1^*, since using Eq. (7.47) we have $X_t^* = \beta^t X_1^*$, and in this sense nothing new occurs within the system. Now, assume that a present value X_t^* is not only a function of its past value X_{t-1}^*, but that new occurrences or inputs u_t also occur, which influence the presence value X_t^* (but not the past value X_{t-1}^*). The u_t frequently represent random shock or noise at each time period, and are also known as "innovations" in the control literature owing to their property of creating novelty in the otherwise deterministic system.

Equation (7.47) can now be written as the so-called system equation

$$X_t^* = \beta X_{t-1}^* + \theta u_t \tag{7.48}$$

where the coefficient θ is time invariant. Now, assume that the present true value X_t^* undergoes a transformation due to measurement error, that is, we also have the observation equation

$$X_t = \alpha X_t^* + \epsilon_t \tag{7.49}$$

where ϵ_t is measurement error and X_t is the observed output of the system. The state-space system (Eq. 7.48) and (7.49) can be generalized to nonlinear equations with time-dependent coefficients (see Mehra, 1979; Young, 1985). The general mathematical–statistical theory underlying state-space equations and their estimation may be found in Jaswinski (1970) and their interpretation in terms of multivariate regression in Harrison and Stevens (1976), Boekee and Buss (1981), Meinhold and Singpurwalla (1983), and Brown et al. (1975). For a recent application see Anderson-Sprechter and Lederter, 1991.

An ARMA process (Eq. 7.42) represents a special case of a state-space model. As is well known, an arbitrary ARMA process of order p, q can be expressed in the moving average form

$$X_t = \epsilon_t + c_1\epsilon_{t-1} + c_2\epsilon_{t-2} + \cdots \tag{7.50}$$

where $\{\epsilon_t\}$ represents white noise and c_j are fixed coefficients. Any future value l periods ahead is then given by $X_{t+l} = \epsilon_{t+l} + c_1\epsilon_{t+l-1} + c_2\epsilon_{t+l-2} + \cdots$. Let $\hat{X}_t(l) = E[X_{t+l}|X_t, X_{t-1}, \ldots]$, the predicted or forecast value of X_{t+l} given present and past values. Subsequent forecasts can then be expressed in recursive form as

$$\hat{X}_{t+1}(l+1) = c_{l-1}\epsilon_{t+1} + c_l\epsilon_t + c_{l+1}\epsilon_{t-1} + \cdots$$

$$= c_{l-1}\epsilon_{t-1} + \hat{X}_t(l)$$

$$= a_t\hat{X}_t(0) + a_{t-1}\hat{X}_t(1) + \cdots + a_1\hat{X}_t(l-1) + c_{l-1}\epsilon_{t+1}$$

$$\tag{7.51}$$

where $\hat{X}_{t+1}(l+1)$ represents the next period forecast. The point here is that the forecast $\hat{X}_{t+1}(l+1)$ is a function of previous forecasts $\hat{X}_t(0)$, $\hat{X}_t(1), \ldots, \hat{X}_t(l-1)$. It can be shown that given a value $l = p - 1$ the following system of equations hold:

$$
\begin{bmatrix} \hat{X}_{t+1}(1) \\ \hat{X}_{t+1}(2) \\ \vdots \\ \hat{X}_{t+1}(p) \end{bmatrix}
=
\begin{bmatrix} 0 & 1 & 0 & \cdots & 0 \\ 0 & 0 & 1 & \cdots & 0 \\ \vdots & \vdots & \vdots & \ddots & 1 \\ a_p & a_{p-1} & a_{p-2} & \cdots & a_1 \end{bmatrix}
\begin{bmatrix} \hat{X}_t(0) \\ \hat{X}_t(1) \\ \vdots \\ \hat{X}_t(p-1) \end{bmatrix}
+
\begin{bmatrix} 1 \\ c_1 \\ \vdots \\ c_{p-1} \end{bmatrix} \epsilon_{t+1}
$$

or

$$\hat{\mathbf{X}}_{t+1} = \mathbf{A}\hat{\mathbf{X}}_t + \mathbf{C}\epsilon_{t+1} \tag{7.52}$$

which is of the same form as Eq. (7.49) and where the coefficients are defined as in Eqs. (7.42) and (7.50)—for an example see Pagan (1975). The state-space representation (Eq. 7.52) of an ARMA process was first considered by Akaike (1974), although it is based on the well-known fact that a difference (differential) equation of order p can always be reduced to a system of p first-order difference (differential) equations (Nagle and Saff, 1986). The matrix $\mathbf{A}$ has a special (structured) form and is known as the Frobenius or companion matrix (e.g., Basilevsky, 1983). The important point here is the generality of the seemingly narrowly defined first-order Markovian representation (Eq. 7.52), of which the stationary ARMA model is simply a special case. The state-space representation can also be extended to nonstationary and non-time-invariant processes.

As was seen in Section 7.6.1, the ARMA process is not fully specified unless the number of parameters p and q are known. The ARMA model is then said to be unidentified. Even when the model is put in the state-space form, it is still not identifiable since the basis of the predictor space is not unique. In general the state-space model (Eq. 7.52) is not identifiable since the space of linear least squares predictors of $\mathbf{X}_{t+1}$, $\mathbf{X}_{t+2}, \ldots$, given the present and past values $\mathbf{X}_t$, $\mathbf{X}_{t-1}$, $\mathbf{X}_{t-2}, \ldots, \mathbf{X}_{t-m}$ (for sufficiently large m), is also not fixed. The situation is similar to that of estimating least squares regression coefficients when the predictor variables are highly correlated (Section 10.3) or when deciding on the basis of a common factor space. A solution that yields the minimal or canonical realization has been proposed by Akaike (1974, 1976), which makes use of the canonical correlation model (Section 5.5). The idea is to compute canonical correlations between the data vectors $[\mathbf{X}_n, \mathbf{X}_{n-1}, \mathbf{X}_{n-2}, \ldots, \mathbf{X}_{n-p}]$, and the least squares predicted values $[\hat{\mathbf{X}}_n(0),\ \hat{\mathbf{X}}_n(1),\ \hat{\mathbf{X}}_n(2), \ldots, \hat{\mathbf{X}}_n(n+p)]$, where $\hat{\mathbf{X}}_n(0) \equiv \mathbf{X}_n$ is the common element between the two sets. The procedure assumes that p is known, which in practice is usually not the case. It may be estimated however as the order of an AR process using, for example, any one of the information criteria described in Section 6.6.2. Once p is known the analysis proceeds on a trial-and-error basis, beginning with the correlation analysis of $[\mathbf{X}_n, \mathbf{X}_{n-1}, \ldots, \mathbf{X}_{n-p}]$ and $[\mathbf{X}_n, \hat{\mathbf{X}}_n(1)]$. If the smallest canonical correlation is judged to be zero, then a linear combination of $[\mathbf{X}_n, \hat{\mathbf{X}}_n(1)]$ is uncorrelated with the data $[\mathbf{X}_n, \mathbf{X}_{n-1}, \ldots, \mathbf{X}_{n-p}]$ and $\hat{\mathbf{X}}_n(1)$ is not included in the state vector. The process continues by successively adding new components $\hat{\mathbf{X}}_n(2)$, $\hat{\mathbf{X}}_n(3)$, and so on until we fail to obtain significant canonical correlation. The estimation procedure is thus based on selecting the canonical form of state-space representation by choosing the state-space vector as the first maximum set of linearly independent elements within the sequence of predictors $\hat{\mathbf{X}}_t(0)$, $\hat{\mathbf{X}}_t(1)$, $\hat{\mathbf{X}}_t(2), \ldots$. As discussed above, the method is univariate in the sense that the letter $\mathbf{X}$ denotes a scalar quantity for each time period. The state-space model however, together with its canonical estimation, can be generalized in a straightforward manner to include two or more time series. Once the dimension and the elements of the state vector are known, the remaining parts of the state-space model can also be estimated—see Akaike (1974; 1976), Priestley (1981), and Wei (1990). The latter reference also provides numerical examples for the estimation of canonical state-space models.

7.6.3 Autoregression and Principal Components

The canonical correlation method described in the previous section for estimating stationary state-space form ARMA models is one possible approach to time series modeling. An alternative motivation for considering canonical correlation analysis of time series has been provided by Rao (1976a), within the context of an AR process. The method is based on the principal components analysis of random vectors obtained from the innovations of an observed process, and differs somewhat from that considered in Section 7.6.1. The basic idea here is to link the latent root theory of difference equations to that of the latent root decomposition of a Grammian correlation matrix, which demonstrates the close link between the concept of a "minimal realization" used in time series and that of a smallest common factor space. Consider two independent zero-mean, stationary, discrete stochastic processes $\mathbf{X}_t^*$ and $\mathbf{N}_t$, such that $\mathbf{X}_t = \mathbf{X}_t^* + \mathbf{N}_t$ as in Eq. (7.7), but with slightly altered notation. The observed process $\mathbf{X}_t$ is the result of a true process $\mathbf{X}_t^*$ having been contaminated by an independent error process $\mathbf{N}_t$, usually (but not always) specified to be normal. In the language of communications and control engineering $\mathbf{X}_t^*$ is an unobserved signal and $\mathbf{N}_t$ is unobserved white noise, such that

$$E(X_t^*) = E(N_t) = 0$$

$$E(X_t^* N_t) = E(N_t N_s) = 0$$

$$E(N_t^2) = \sigma^2 \tag{7.53}$$

although more general specifications are also possible. Given a semiinfinite realization of X_t, $-\infty < t \le T$, the objective is to obtain an optimal estimator or predictor (the "filter") of X_{t+m}^*. Since the observations of a time series are generally correlated, in order to reduce a normal time series to a set of independent observations it is usually first necessary to "prewhiten" it by regression analysis, that is, the series is replaced by a set of independent residuals. Let

$$\hat{X}_t(l) = E(X_{t+l} \mid X_s) \tag{7.54a}$$

$$\hat{X}_t^*(l) = E(X_{t+l}^* \mid X_s) \tag{7.54b}$$

$(-\infty < s \le t)$, the l-step prediction of X_{t+l} and X_{t+l}^*, respectively, using past and present values X_s. Then the prediction errors can be expressed as

$$Y_t(l) = X_{t+l} - \hat{X}_t(l)$$

$$= (X_{t+l}^* + N_{t+l}) - \hat{X}_t(l)$$

$$= (X_{t+l}^* - \hat{X}_t(l)) + N_{t+l} \tag{7.55}$$

since $X_{t+l} = X_{t+l}^* + N_{t+l}$ and $E(X_t \mid X_s) = E(X_t^* \mid X_s)$.

THEOREM 7.5.   Let $Y_t(l)$ be as defined by Eq. (7.55). Then
    (i) $E\{[X_{t+l}^* - \hat{X}_t^*(l)]N_{t+l}\} = 0 \qquad (l, k > 0)$
    (ii) $E\{[X_{t+l}^* - \hat{X}_t^*(l)][X_{t+l}^* - \hat{X}_t^*(m)]\} = E[Y_t(l)Y_t(m)]$
        $E\{[X_{t+l}^* - \hat{X}_t^*(l)]^2\} + \sigma^2 = E\{[Y_t(l)]^2\}$

For a proof the reader is referred to Rao (1976a). The first part of the theorem states that the prediction error of the true signal is uncorrelated with contamination noise N_{t+k}, whereas the second part shows that the covariance between any two prediction errors of process X_t is only due to the covariance between the corresponding prediction errors of the signal X_t^*. Consider the first p values of the index l. Then from Eq. (7.55) we have the p forecast errors for the observed values

$$Y_t(1) + [X_{t+1}^* - \hat{X}_t^*(1)] + N_{t+1}$$
$$Y_t(2) = [X_{t+2}^* - \hat{X}_t^*(2)] + N_{t+2}$$
$$\overline{Y_t(p) = [X_{t+p}^* - \hat{X}_t^*(p)] + N_{t+p}}$$

or, in vector form,

$$\mathbf{Y}_t = \mathbf{S}_t + \mathbf{M}_t \tag{7.56}$$

so that

$$\mathbf{\Sigma} = \mathbf{\Gamma} + \mathbf{\Psi} \tag{7.57}$$

where

$$\mathbf{\Sigma} = E(\mathbf{Y}_t\mathbf{Y}_t^{\mathrm{T}})$$
$$\mathbf{\Gamma} = E(\mathbf{S}_t\mathbf{S}_t^{\mathrm{T}})$$
$$\mathbf{\Psi} = E(\mathbf{M}_t\mathbf{M}_t^{\mathrm{T}}) = \sigma^2\mathbf{I} \tag{7.58}$$

When the signal process X_t^* follows a pth order autoregressive process,

$$X_t^* = \beta_1 X_{t-1}^* + \beta_2 X_{t-2}^* + \cdots + \beta_p X_{t-p}^* + \epsilon_t \tag{7.59}$$

where ϵ_t represents a sequence of $N(0, 1)$ independent random variables, we have from Eq. (7.59)

$$X_{t+p}^* = \sum_{i=1}^{p} \beta_i X_{t+(p-i)}^* + \epsilon_{t+p} \tag{7.60}$$

and taking conditional expectations with respect to X_s and using Eq. (7.54a) we have

$$\hat{X}^*(p) = \sum_{i=1}^{p} \beta_i \hat{X}_t^*(p - i) \tag{7.61}$$

Subtracting Eq. (7.60) from Eq. (7.61) yields

$$X_{t+p}^* - \hat{X}_t^*(p) = \sum_{i=1}^{p} \beta_i [X_{t+p-i}^* - \hat{X}_t^*(p-i)] + \epsilon_{t+p} \qquad (7.62)$$

Let the forecast errors for the true values be given by the vector

$$\mathbf{F}_t = \{[X_t^* - \hat{X}_t^*(0)], [X_{t+1}^* - \hat{X}_t^*(1)], \ldots, [X_{t+p-1}^* - \hat{X}_t^*(p-1)]\}^T$$

where

$$\mathbf{E}_t = (0, 0, \ldots, \epsilon_{t+p})^T \qquad (7.63)$$

and define the so-called companion (Frobenius) coefficient matrix

$$\mathbf{B} = \begin{bmatrix} 0 & 1 & 0 & \cdots & 0 \\ 0 & 0 & 1 & \cdots & 0 \\ 0 & 0 & 0 & \cdots & 0 \\ \vdots & \vdots & \vdots & & \vdots \\ & & & & 1 \\ \beta_p & \beta_{p-1} & \beta_{p-2} & \cdots & \beta_1 \end{bmatrix}$$

Then from Eqs. (7.56) and (7.62) we have a system of equations, similar to the state-space equations of the previous section, given by

$$\mathbf{S}_t = \mathbf{BF}_t + \mathbf{E}_t$$

$$\mathbf{Y}_t = \mathbf{S}_t + \mathbf{M}_t \qquad (7.64)$$

which is stable only if the latent roots of $\mathbf{B}$ possess moduli less than unity. Combining the two equations yields

$$\mathbf{Y}_t = \mathbf{BF}_t + (\mathbf{E}_t + \mathbf{M}_t) \qquad (7.65)$$

that is,

$$\begin{bmatrix} X_{t+1} - \hat{X}_t(1) \\ X_{t+2} - \hat{X}_t(2) \\ \hline X_{t+p} - \hat{X}_t(p) \end{bmatrix} = \begin{bmatrix} 0 & 1 & 0 & \cdots & 0 \\ 0 & 0 & 1 & \cdots & 0 \\ \vdots & \vdots & & \vdots & \vdots \\ & & & & 1 \\ \beta_p & \beta_{p-1} & & \cdots & \beta_1 \end{bmatrix}$$

$$\begin{bmatrix} X_t^* - \hat{X}_t(0) \\ X_{t+1}^* - \hat{X}_t(1) \\ \hline X_{t+p-1}^* - \hat{X}_t^*(p-1) \end{bmatrix} + \begin{bmatrix} 0 \\ 0 \\ \vdots \\ 0 \\ \beta_{t+p} \end{bmatrix} + \begin{bmatrix} N_{t+1} \\ N_{t+2} \\ \vdots \\ N_{t+p} \end{bmatrix}$$

Equation (7.65) is formally equivalent to the factor analysis model considered in Chapter 6 since the coefficients $\mathbf{B}$ can be taken as the factor loadings and $\mathbf{F}$ can be defined as the matrix of factor scores. Since the initial purpose is to estimate the order p of an AR process (assuming stability), interest centers on the intrinsic dimensionality of Eq. (7.65).

THEOREM 7.6. If an AR process (eq. 7.59) is of order p, then vector $\mathbf{F}_t$ is of dimension p.

PROOF. In the terminology of control engineering literature the system (Eq. 7.65) is observable if and only if for the ranks of successively higher powers of $\mathbf{B}$ we have

$$\rho(\mathbf{B}, \mathbf{B}^2, \ldots, \mathbf{B}^p) = p$$

and using the Caley–Hamilton theorem we know that $\mathbf{F}_t$ cannot be of dimension less than p. To show that $\mathbf{F}_t$ cannot be of dimension greater than p, consider the augmented version of $\mathbf{S}_t$,

$$\tilde{\mathbf{S}}_t = \{[X^*_{t+1} - \hat{X}^*_t(1)], \ldots, [X^*_{t+p} - \hat{X}^*_t(p)], \ldots, [X^*_{t+m} - \hat{X}^*_t(m)]\}$$

such that $\tilde{\mathbf{S}}_t = \tilde{\mathbf{B}}\tilde{\mathbf{F}}_t + \hat{\mathbf{E}}_t$, where $\tilde{\mathbf{B}}$ is the $(m \times m)$ augmented matrix

$$\tilde{\mathbf{B}} = \left[\begin{array}{c|c} \mathbf{B} & \mathbf{0} \\ \hline \mathbf{0} & \mathbf{I} \end{array}\right]$$

and $\tilde{\mathbf{F}}_t$ and $\tilde{\mathbf{F}}_t$ are correspondingly augmented vectors. Since the last $(m - p)$ latent roots of $\tilde{\mathbf{B}}$ must lie on the unit circle, the augmented system becomes unstable, which is contrary to assumption. Thus $\mathbf{F}_t$ cannot be of dimension greater than p, so that the dimension of the AR process must be precisely p, that is, the dimension of $\mathbf{F}_t$ (see Kalman and Bucy, 1961; Kalman et al., 1969).

Theorem 7.6 permits us to identify the order of an AR process by means of factor analysis. From the system (Eq. 7.64) we have

$$
\begin{aligned}
E(\mathbf{Y}_t\mathbf{S}_t^\mathrm{T}) &= E(\mathbf{S}_t\mathbf{S}_t^\mathrm{T}) + E(\mathbf{M}_t\mathbf{S}_t^\mathrm{T}) \\
&= E(\mathbf{S}_t\mathbf{S}_t^\mathrm{T}) \\
&= E[(\mathbf{BF}_t + \mathbf{E}_t)(\mathbf{BF}_t + \mathbf{E}_t)^\mathrm{T}] \\
&= \mathbf{B}[E(\mathbf{F}_t\mathbf{F}_t^\mathrm{T})]\mathbf{B}^\mathrm{T} + E(\mathbf{E}_t\mathbf{E}_t^\mathrm{T}) \\
&= \mathbf{B}\boldsymbol{\Phi}\mathbf{B}^\mathrm{T} + E(\mathbf{EE}_t^\mathrm{T})
\end{aligned}
\tag{7.66}
$$

where $\boldsymbol{\Gamma} = E(\mathbf{Y}_t\mathbf{S}_t^\mathrm{T}) = E(\mathbf{S}_t\mathbf{S}_t^\mathrm{T})$ so that the number of common factors equals the rank of $\boldsymbol{\Gamma}$, which is the same as the dimension of $\mathbf{F}_t$ (see Section 6.3). Thus the order of an AR process equals the number of nonzero roots of the determinantal equation

$$|\mathbf{\Gamma} - \lambda_i^2 \mathbf{\Sigma}| = \left| \mathbf{\Sigma} - \frac{1}{(1 - \lambda_i^2)} \mathbf{\Psi} \right|$$

$$= 0 \tag{7.67}$$

as given by Eq. (6.63). Since $\mathbf{\Psi} = \lambda^2 \mathbf{I}$ we have from Eq. (6.65) that if $\rho(\mathbf{\Sigma}) = p < m$, then identifying the order p of an AR process is the same as testing for the equality of the last $(m - p)$ roots of $\mathbf{\Sigma}$.

In practice the implementation of the method of estimating p is as follows. First, define the augmented column vector $\mathbf{Y}_t^{\mathrm{T}} = [(X_{t+1} - \hat{X}_t(1)), (X_{t+2} - \hat{X}_t(2)), \ldots, (X_{t-m} - \hat{X}_t(m))]$ where $m \gg p$, and then compute the usual sample covariance (correlation) matrix

$$\hat{\mathbf{\Sigma}} = \frac{1}{n} \sum_{t=1}^{n} (\mathbf{Y}_t - \bar{\mathbf{Y}})(\mathbf{Y}_t - \bar{\mathbf{Y}})^{\mathrm{T}} \tag{7.68}$$

In practice the observed series X_t is prewhitened to eliminate the serial correlation between the elements, for example, by fitting an autoregressive model of adequate order. Also, since the elements of the series Y_t may also be correlated, Rao (1976a) suggests the use of a subsequence of Y_t, consisting of elements that are sufficiently far apart. Next, the latent roots $\hat{\lambda}_1, \hat{\lambda}_2, \ldots, \hat{\lambda}_m$ of $\hat{\mathbf{\Sigma}}$ are obtained and the last $m - r$ tested for equality (Section 4.3.2). If equality of the last $m - r$ roots is accepted, the procedure of computing latent roots is repeated, but this time using the vector $\mathbf{Y}_t^{\mathrm{T}} = \{[X_{t+1} - \hat{X}_t(1)], [X_{t+2} - \hat{X}_t(2)], \ldots, (X_{t+q} - \hat{X}_t(q)]\}$ where $q = m - r$. The process is continued until the number of distinct latent roots becomes stable. This number is then the estimated order (dimension) of the AR model, and the residual noise variance can be estimated as

$$\hat{\sigma}^2 = \frac{1}{m - p} \sum_{j=p+1}^{m} \hat{\lambda}_j \tag{7.69}$$

Since $\mathbf{\Psi} = \sigma^2 \mathbf{I}$, owing to the assumption of second-order stationarity we have the homoscedastic residuals model of Section 6.3.1, that is, the principal components model with $m - p$ isotropic roots.

Once the dimensionality of the AR process is known, its coefficients can be estimated with little extra effort. Letting $\mathbf{Q}_t = \mathbf{E}_t + \mathbf{M}_t$, where $E(\mathbf{Q}_t \mathbf{Q}_t^{\mathrm{T}}) = \mathbf{D}$, we can write $\mathbf{D} = \mathbf{D}^{1/2} \mathbf{D}^{1/2}$ so that the model becomes

$$\mathbf{V}_t = \tilde{\mathbf{B}} \mathbf{F}_t + \mathbf{W}_t \tag{7.70}$$

where $\mathbf{V}_t = \mathbf{D}^{-1/2} \mathbf{Y}_t$, $\mathbf{W}_t = \mathbf{D}^{-1/2} \mathbf{Q}_t$, and $E(\mathbf{W}_t \mathbf{W}_t^{\mathrm{T}}) = \mathbf{I}$. Also using Eq. (7.58) we have $\mathbf{\Gamma} = E(\mathbf{S}_t\mathbf{S}_t^T) = \mathbf{B\Phi B}^T$ where $\mathbf{\Phi} = E(\mathbf{F}_t\mathbf{F}_t^T)$. Let $\mathbf{\Phi} = \mathbf{\Phi}^{1/2}\mathbf{\Phi}^{1/2}$ where

$$\mathbf{\Phi}^{1/2} = \begin{bmatrix} 1 & \theta_{12} & \cdots & \theta_{1p} \\ \theta_{21} & 1 & \cdots & \theta_{2p} \\ \vdots & \vdots & \ddots & \vdots \\ \theta_{p1} & \theta_{p2} & \cdots & 1 \end{bmatrix}$$

and the model can be expressed in general factor form as

$$V_t = \mathbf{HG}_t + \mathbf{W}_t \tag{7.71}$$

where $E(\mathbf{G}_t) = \mathbf{0}$, $E(\mathbf{G}_t\mathbf{G}_t^T) = \mathbf{I}$, and

$$\mathbf{H} = \begin{bmatrix} \theta_{21} & 1 & \theta_{23} & \cdots & \theta_{2p} \\ \theta_{31} & \theta_{32} & 1 & \cdots & \theta_{3p} \\ \vdots & \vdots & \vdots & & \vdots \\ \theta_{p1} & \theta_{p2} & \theta_{p3} & \cdots & 1 \\ \hat{\alpha}_p & \hat{\alpha}_{p-1} & \hat{\alpha}_{p-2} & \cdots & \hat{\alpha}_1 \end{bmatrix} \tag{7.72}$$

such that $\mathbf{\Phi}^{1/2}\boldsymbol{\beta}^* = \boldsymbol{\alpha}$ where $\boldsymbol{\beta}^{*T} = (\beta_p^*, \beta_{p-1}^*, \ldots, \beta_1^*)$ and $\boldsymbol{\alpha}^T = (\alpha_p, \alpha_{p-1}, \ldots, \alpha_1)$. Note that the coefficient matrix has lost its special (Frobenius) form, so that $\mathbf{H}$ is now equivalent to a general factor loadings matrix where the factors $\mathbf{G}$ are unit variance and are uncorrelated. The coefficients of the AR process are then given by

$$\hat{\boldsymbol{\beta}} = \mathbf{\Phi}^{1/2}\hat{\boldsymbol{\alpha}} \tag{7.73}$$

where $\hat{\boldsymbol{\alpha}}$ is the last row of matrix $\mathbf{H}$. Note that the parameters $\boldsymbol{\beta}^*$ are not the same as $\boldsymbol{\beta}$ because of the standardization of the factors $\mathbf{F}_t$. The following example is taken from Rao (1976a).

Example 7.9. A set of 2,000 $N(0, 1)$ random variables are generated, and from this we obtain the time series $\mathbf{X}_t^*$ using the AR process of order $p = 2$,

$$X_{t+1}^* = .80X_t^* - .40X_{t-1}^* + \epsilon_{t+1} \qquad (t = 1, 2, \ldots, 2000) \tag{7.74}$$

Here 2000 additional standard normal deviates $\mathbf{N}_t$ are generated yielding the corrupted series $\mathbf{X}_t = \mathbf{X}_t^* + \mathbf{N}_t$. The estimation algorithm proceeds as follows. First, the observed series $\mathbf{X}_t$ is prewhitened to eliminate serial correlation—perhaps by fitting an AR process of adequate order by least squares. The maximum order of lag for Eq. (7.74) is decided upon, say $m = 10$. This yields the vector

$$\mathbf{Y}_i^{\mathrm{T}} = \{[X_{t+1} - \hat{X}_t(1)], [X_{t+2} - \hat{X}_t(2)], \dots, [X_{t+m} - \hat{X}_t(m)]\}$$

Second, since $\mathbf{Y}_t$ may not represent an independent sequence, the actual analysis is carried out on a subsequence with elements sufficiently apart to guarantee independence. Choosing observations 20 units apart Rao (1976) retains $n = 97$ observations for the estimation process, based on the estimated covariance matrix

$$\hat{\mathbf{\Sigma}} = \frac{1}{97} \sum_{i=1}^{97} (\mathbf{Y}_t - \bar{\mathbf{Y}})(\mathbf{Y}_t - \bar{\mathbf{Y}})^{\mathrm{T}} \tag{7.75}$$

Third, a principal component analysis is carried out on $\hat{\mathbf{\Sigma}}$ to determine the order p of the true AR process, using the chi-squared criterion to test for isotropic roots (Section 4.3). The fourth step consists of letting $m = 4$ and repeating the principal components analysis on the (4×4) matrix $\hat{\mathbf{\Sigma}}$. The actual values are given in Tables 7.12 and 7.13. Now only the last two roots are insignificantly different, and we conclude the AR process is of order $p = 2$.

Table 7.12   Values of $m = 10$ Chi-Squared Statistics and Degrees of Freedom for Testing Latent Roots of $\hat{\mathbf{\Sigma}}^a$

Number of Roots Excluded	Chi-Squared	Degrees of Freedom
0	205.7371	54
1	158.4052	44
2	97.6864	35
3	68.6000	27
4	37.4402	20
5	16.5803	14
6	10.0031	9
7	3.3210	5
8	1.2506	2
9	0.0000	0

Source: Rao, 1976; reproduced with permission.

[a] The last six roots are not significantly different at the 1% level.

Table 7.13 Principal Component Chi-Squared Values for Latent Roots of $\hat{\Sigma}$ with $m = 4$

Number of Roots Excluded	Chi-Squared	Degrees of Freedom
0	39.3093	9
1	25.9924	5
2	4.3215	2
3	0.0000	0

Source: Rao, 1976; reproduced with permission.

The parameters $\boldsymbol{\beta}^*$ are estimated as follows. We have

$$\hat{\mathbf{D}}^{-1/2} = \begin{bmatrix} 1.2045 & 0 \\ 0 & 1.5655 \end{bmatrix} \qquad \mathbf{H} = \begin{bmatrix} \theta_{21} & 1 \\ \alpha_2 & \alpha_1 \end{bmatrix}$$

where

$$\mathbf{Y}\mathbf{Y}^{\mathrm{T}} = \begin{bmatrix} 221.250 & 65.226 \\ 65.226 & 105.843 \end{bmatrix}$$

with latent roots $\hat{\lambda}_1 = 250.638$ and $\hat{\lambda}_2 = 76.459$ and corresponding latent vectors $\mathbf{P}_1^{\mathrm{T}} = (1.466, .660)$ and $\mathbf{P}_2^{\mathrm{T}} = (.365, -.810)$. Equating elements of $\mathbf{H}$ with the elements of the latent vectors we obtain the values $\theta_{12} = 1.466$, $\hat{\beta}_1^* = 1.367$, and $\hat{\beta}_2^* = 1.070$.

7.6.4 Kalman Filtering Using Factor Scores

Consider again the state-space equations of Section 7.6.2,

$$X_{t+1}^* = \beta X_t^* + \theta u_t \tag{7.76}$$

$$X_t = \alpha X_t^* + \epsilon_t \tag{7.77}$$

where X_t and X_t^* are jointly stationary normal processes and u_t and ϵ_t are jointly normal such that

$$E(u_t) = E(\epsilon_t) = 0$$

$$E(u_t u_s^{\mathrm{T}}) = \sigma_u^2 \mathbf{I}$$

$$E(\epsilon_t \epsilon_s^{\mathrm{T}}) = \sigma_v^2 \mathbf{I}$$

$$E(u_t \epsilon_s^{\mathrm{T}}) = 0 \tag{7.78}$$

and

$$E(X_{t+1}^*) = 0$$

$$E(X_t^* u_s^{\mathrm{T}}) = 0 \qquad (s \geq t)$$

$$E(X_t^* \boldsymbol{\epsilon}_s^\mathrm{T}) = 0 \qquad (s \geq t)$$

$$E(X_t u_s^\mathrm{T}) = 0 \qquad (s \geq t) \tag{7.79}$$

Since the main use of state-space models lies in signal extraction and forecasting it is important to develop an efficient algorithm for the estimation of the model. One such algorithm is the so-called Kalman filter, introduced in control engineering by Kalman (1960) and Stratonovich (1960, 1968), which consists of an application of the recursive least squares algorithm to the state-space system (e.g., see Ganin, 1977 and Basilevsky, 1983). An alternative method for deriving the Kalman filter is by the use of factor scores as indicated by Priestley and Rao (1975; see also Priestly, 1981) who have used factor scores to obtain optimal estimates of the unobserved values X_t^* given previous observations X_{t-1}, where α is identified with a matrix of fixed factor scores.

Owing to the presence of serial correlation among the observations, a factor model cannot be used for an autoregressive specification as it stands, and the observed series X_t must first be "prewhitened" (by least squares autoregression), which yields uncorrelated residuals or "white noise" of the form

$$\mathbf{Z}_t = \mathbf{X}_t - E(\mathbf{X}_t \,|\, \mathbf{X}_{(t-1)}) \tag{7.80}$$

where $\mathbf{X}_{(t-1)}$ represents the set of lagged values $\mathbf{X}_{t-1}$, $\mathbf{X}_{t-2}$, and so on. Taking conditional expectation on both sides of Eq. (7.77), we have

$$E(\mathbf{X}_t \,|\, \mathbf{X}_{(t-1)}) = \alpha E(\mathbf{X}_t^* \,|\, \mathbf{X}_{(t-1)}) + E(\boldsymbol{\epsilon}_t \,|\, \mathbf{X}_{(t-1)})$$

$$= \alpha E(\mathbf{X}_t^* \,|\, \mathbf{X}_{(t-1)})$$

$$= \alpha \hat{\mathbf{X}}_{t|t-1}^* \tag{7.81}$$

using Eq. (7.78) and (7.79) (Exercise 7.6) where $\hat{\mathbf{X}}_{t|t-1}^* = E(\mathbf{X}_t^* \,|\, \mathbf{X}_{(t-1)})$. Equation (7.81) represents the linear least squares predictor of the unknown values $\mathbf{X}_t^*$ given a record of past observations $\mathbf{X}_{t-1}$, $\mathbf{X}_{t-2}$, and so on. Subtracting Eq. (7.81) from Eq. (7.77) we then have

$$\mathbf{Z}_t = \alpha(\mathbf{X}_t^* - \hat{\mathbf{X}}_t^* \,|\, \mathbf{X}_{(t-1)}^*) + \boldsymbol{\epsilon}_t$$

$$= \alpha \boldsymbol{\Phi}_t + \boldsymbol{\epsilon}_t \tag{7.82}$$

where both series $\mathbf{Z}_t$ and $\boldsymbol{\Phi}_t$ consist of independent observations such that $E(\boldsymbol{\Phi}_t) = \mathbf{0}$ and $E(\boldsymbol{\Phi}_t \boldsymbol{\Phi}_t^\mathrm{T}) = \boldsymbol{\Omega}$, and $(r \times r)$ covariance matrix of r common factors. Equation (7.82) can be viewed as representing a factor model

(Chapter 6) where the unknown factor scores are given by the estimator

$$\hat{\boldsymbol{\Phi}}_t = E(\boldsymbol{\Phi}_t \mid \mathbf{Z}_{(t)})$$

$$= E[(\mathbf{X}_t^* - \hat{\mathbf{X}}_{t\mid t-1}^*) \mid \mathbf{Z}_{(t)}]$$

$$= E(\hat{\mathbf{X}}_t^* \mid \mathbf{Z}_{(t)}) - E(\hat{\mathbf{X}}_t^* \mid \mathbf{Z}_{(t)})$$

$$= \hat{\mathbf{X}}_{t\mid t}^* - \hat{\mathbf{X}}_{t\mid t-1}^* \tag{7.83}$$

Since $\boldsymbol{\Phi}_t$ is not a linear function of the prewhitened series $\mathbf{Z}_t$, using the regression estimator of Section 6.8.1 yields the $(r \times 1)$ vector

$$\hat{\boldsymbol{\Phi}}_t = \boldsymbol{\Omega}\boldsymbol{\alpha}^{\mathrm{T}}\boldsymbol{\Sigma}^{-1}\mathbf{Z}_t \tag{7.84}$$

Since the quantities $\boldsymbol{\Omega}$, $\boldsymbol{\alpha}$, and $\boldsymbol{\epsilon}$ are not usually known, Eq. (7.84) is replaced by its sample equivalent. Using Eqs. (7.83) and (7.84) we then obtain the Kalman filter

$$\hat{\mathbf{X}}_{t\mid t}^* = \hat{\mathbf{X}}_{t\mid t-1}^* + \hat{\boldsymbol{\Omega}}\hat{\boldsymbol{\alpha}}^{\mathrm{T}}\hat{\boldsymbol{\Sigma}}^{1}\mathbf{Z}_t$$

$$= \hat{\mathbf{X}}_{t\mid t-1}^* + \hat{\boldsymbol{\Omega}}\hat{\boldsymbol{\alpha}}^{\mathrm{T}}\hat{\boldsymbol{\Sigma}}^{-1}(\mathbf{X}_t - \hat{\boldsymbol{\alpha}}\hat{\mathbf{X}}_{t\mid t-1}^*) \tag{7.85}$$

which is a one-step recursive predictor for $\mathbf{X}_{t\mid t}^*$ where the quantity $\hat{\boldsymbol{\Omega}}\hat{\boldsymbol{\alpha}}^{\mathrm{T}}\hat{\boldsymbol{\Sigma}}^{-1}$ is the so-called "Kalman gain" which determines the correction term for the predicted values $\mathbf{X}_{t\mid t}^*$, given the previous values $\mathbf{X}_{t\mid t-1}^*$. The oblique regression factor scores are means of the posterior distribution of the unobserved random factors (Section 6.8) and thus model the adaptive control mechanism expressed by Eqs. (7.76) and (7.77). As is the case for factor analysis, the Kalman filter can also be developed within a Bayesian framework (Harrison and Stevens, 1976; Meinhold and Singpurwalla, 1983).

The equivalence of the regression factor estimator to the recursive Kalman filtering procedure is a theoretical formality and does not necessarily imply any computational simplification. It is however of interest since it makes more evident something which is not readily apparent from the more standard approach—that the Kalman filter is biased. This follows directly from the theory of estimation of factor scores (Section 6.8) where it can be seen that the regression estimator is biased (Rao, 1976b). It follows that an unbiased estimator is obtained by the use of the Bartlett minimum distance estimator which in terms of the state-space model can be expressed as

$$\hat{\mathbf{X}}_{t\mid t}^* = \hat{\mathbf{X}}_{t\mid t-1}^* + \hat{\boldsymbol{\alpha}}^{\mathrm{T}}(\hat{\sigma}_v^2\mathbf{I})^{-1}\hat{\boldsymbol{\alpha}}\hat{\boldsymbol{\alpha}}^{\mathrm{T}}(\hat{\sigma}_v^2\mathbf{I})^{-1}(\mathbf{X}_t - \hat{\boldsymbol{\alpha}}\hat{\mathbf{X}}_{t\mid t-1}^*) \tag{7.86}$$

which differs from Eq. (7.85) in the gain function where $\mathbf{E}(\boldsymbol{\epsilon}_t\boldsymbol{\epsilon}_s^{\mathrm{T}}) = \sigma_v^2\mathbf{I}$. Although Eq. (7.86) is derived by Rao (1976b) in the context of Kalman filtering, it may be inappropriate since (1) the Kalman filter minimizes mean squared error and (2) the Bartlett estimator is only valid, strictly speaking, for fixed rather than random factors, and may cause difficulties when using maximum likelihood estimation (see Section 6.8).

7.7 MULTIPLE TIME SERIES IN THE FREQUENCY DOMAIN

Analyses of time series described in the previous sections possess a common feature in that both the factor loadings and the scores can be identified in terms of time-dependent quantities, that is, they are computed in what is known as the time domain. The objective for time domain analysis is to be able to identify time series interdependence, components of a time series, or lead-lag relationships between points in time. This is often an attractive option because of the additional ease of interpretation when estimators are computed directly in the time domain. Occasionally, however, it is necessary to transform series into what is known as the frequency domain, and the result is then what is generally known as a frequency domain (spectral analysis) of a time series, which enables us to determine periodic components of a series, lead-lag relationship between several series at various periodic frequencies, and so forth

Consider two jointly stationary stochastic processes $Y_1(t)$ and $Y_2(t)$ with autocovariance functions

$$C_{11}(k) = E\{[Y_1(t) - \mu_1][Y_1(t + k) - \mu_1]\}$$

$$C_{22}(k) = E\{[Y_2(t) - \mu_2][Y_2(t + k) - \mu_2]\} \tag{7.87}$$

and cross-covariance functions

$$C_{12}(k) = E\{(Y_1(t) - \mu_1][Y_2(t + k) - \mu_2]\}$$

$$C_{21}(k) = E\{[Y_2(t) - \mu_2][Y_1(t + k) - \mu_1]\} \tag{7.88}$$

where $E[Y_i(t)] = \mu_i$ $(i = 1, 2)$. It can be shown that $C_{ii}(k) = C_{ii}(-k)$ and $C_{ij}(k) = C_{ij}(-k)$ $(i, j = 1, 2)$ where the time displacement k is known as the lead (lag) of the series $(k = 0, \pm 1, \ldots)$. Sample estimates of the moments are then given by

$$\hat{C}_{ij}(k) = \frac{1}{n} \sum_{t=1}^{n-k} (y_{it} - \bar{y}_i)(y_{jt+k} - \bar{y}_j) \tag{7.89}$$

When $\Sigma_{k=-\infty}^{\infty} |C_{ij}(k)| < \infty$ we can define the first- and second-order spectra as

$$S_{ii}(\omega) = (2\pi)^{-1} \sum_{k=-\infty}^{\infty} C_{ii}(k) \exp(-\omega k \sqrt{-1}) \tag{7.90}$$

$$S_{ij}(\omega) = (2\pi)^{-1} \sum_{k=-\infty}^{\infty} C_{ij}(k) \exp(-\omega k \sqrt{-1}) = \overline{S_{ji}(\omega)} \tag{7.91}$$

respectively where $-\infty < \omega < \infty$ is frequency and $\overline{S_{ji}(\omega)}$ is the complex conjugate of $S_{ij}(\omega)$. Equation (7.90) is real, whereas the second order or cross spectrum (Eq. 7.91) is complex. The Fourier series yields a decomposition of variance of the time series at frequency ω, and is known as the Fourier transform of the covariance function (Eq. 7.89). When the series (processes) are continuous, the summations (Eqs. 7.90 and 7.91) are replaced by integrals. The interpretation of a power spectrum of an observed series bears a close resemblance to that of the latent roots of a lagged autocovariance matrix (Section 7.5.2) whereas the cross-power spectrum contains information on the interdependence between the two series at varying frequencies. The following quantities play an important role in a cross-spectral analysis:

$\text{Re } S_{ij}(\omega) = \text{cospectrum}$
$\text{Im } S_{ij}(\omega) = \text{quadrature spectrum}$
$\text{Arg } S_{ij}(\omega) = \text{phase spectrum}$
$|S_{ij}(\omega)| = \text{amplitude spectrum}$

The real part (cospectrum) measures the covariance between the in-phase frequency components of $Y_1(t)$ and $Y_2(t)$ and the imaginary part (quadrature spectrum) measures covariance between the out-of-phase frequency components, that is, the lead-lag relations between the two series at varying frequencies.

7.7.1. Principal Components in the Frequency Domain

Cross-spectral analysis can be generalized to the multiseries case by defining a $(p \times p)$ cross-spectral matrix $\mathbf{S}(\omega)$ with real power spectra (Eq. 7.90) lying on the main diagonal and complex cross spectra (Eq. 7.91) lying on the off-diagonals. Since $\mathbf{S}(\omega)$ is Hermitian, a principal components analysis of $\mathbf{S}(\omega)$ (Section 5.7) reveals the interrelationships between the time series.

Let $\mathbf{Y}(t)$ be a p-vector valued series ($t = 0, \pm 1, \ldots$) with mean $E[\mathbf{Y}(t)] = \boldsymbol{\mu}$ and absolutely summable autocovariance matrix

$$E[\mathbf{Y}(t+k) - \boldsymbol{\mu}][\mathbf{Y}(t) - \boldsymbol{\mu}]^{\mathrm{T}} = \mathbf{C}(k) \tag{7.92}$$

The $(p \times p)$ spectra density matrix can be expressed as

$$\mathbf{S}(\omega) = (2\pi)^{-1} \sum_{k=-\infty}^{\infty} \mathbf{C}(k) \exp(-\omega k \sqrt{-1}) \tag{7.93}$$

where, since $\mathbf{C}(k)$ is real valued, we have

$$\overline{\mathbf{S}(\omega)} = \mathbf{S}(-\omega) = \mathbf{S}^{\mathrm{T}}(\omega) \tag{7.94}$$

that is, $\mathbf{S}(\omega)$ is Hermitian (Section 2.11.1). Under general conditions Eq. (7.93) can be inverted as

$$\mathbf{C}(k) = \int_{-\pi}^{\pi} \exp(\omega k \sqrt{-1}) \, \mathbf{S}(\omega) \, d\omega \tag{7.95}$$

which represents an expansion of the autocovariance function in terms of frequency ω.

Given a zero-mean, p-valued second-order stationary series $\mathbf{X}(t)$, the objective of a principal components analysis in the frequency domain is to find linear combinations

$$\zeta(t) = \sum_{k} \mathbf{b}(t-k)\mathbf{X}(k) \tag{7.96}$$

$(t = 0, \pm 1, \ldots)$ such that

$$\hat{\mathbf{X}}(t) = \sum_{k} \mathbf{c}(t-k)\zeta(k) \tag{7.97}$$

provides a good approximation to $\mathbf{X}(t)$. This is accomplished through a latent vector decomposition of the spectral density matrix $\mathbf{S}(\omega)$.

THEOREM 7.7 (Brillinger, 1969).   Let $\mathbf{X}(t)$ be a p vector-valued second-order stationary series with mean zero, absolutely summable autocovariance function $\mathbf{C}(k)$ and spectral density matrix $\mathbf{S}(\omega)$ $(-\infty < \omega > \infty)$. Then the coefficients $\mathbf{b}(k)$ and $\mathbf{c}(k)$ that minimize

$$\overline{E[\mathbf{X}(t) - \sum_{k} \mathbf{c}(t-k)\zeta(k)]}^{\mathrm{T}} [\mathbf{X}(t) - \sum_{k} \mathbf{c}(t-k)\zeta(k)] \tag{7.98}$$

are given by

$$\mathbf{b}(k) = (2\pi)^{-1} \int_{0}^{2\pi} \mathbf{\Pi}(\theta) \exp(k\omega\sqrt{-1}) \, d\omega \tag{7.99}$$

$$\mathbf{c}(k) = (2\pi)^{-1} \int_{0}^{2\pi} \overline{\mathbf{\Pi}(\omega)}^{\mathrm{T}} \exp(k\omega\sqrt{-1}) \, d\omega \tag{7.100}$$

where

$$\mathbf{\Pi}(\omega) = \begin{bmatrix} \overline{\mathbf{\Pi}_1(\omega)} \\ \overline{\mathbf{\Pi}_2(\omega)} \\ \vdots \\ \overline{\mathbf{\Pi}_p(\omega)} \end{bmatrix}$$

and $\mathbf{\Pi}_i(\omega)$ is the ith latent vector of $\mathbf{S}(\omega)$ at frequency ω.

Theorem 7.7 provides a decomposition of $\mathbf{S}(\omega)$ at each frequency ω, and provides a complex analogue to the real-valued decomposition of a co-variance matrix. The theorem applies to jointly stationary processes, that is processes for which the covariance matrix (Eq. 7.92) is constant (stationary) over time. In practice, when time series are nonstationary—for example, because of the presence of trend and/or seasonal components— they may be "filtered" to remove such terms. As before, the degree of approximation of $\hat{\mathbf{X}}(t)$ to $\mathbf{X}(t)$ is a direct function of how close the latent roots $\lambda(\omega)$ are to zero $(-\infty < \omega < \infty)$. Thus under conditions of Theorem 7.7, the ith principal components series $\zeta_i(t)$ has power spectrum $\lambda_i(\omega)$, with any two components $\zeta_i(t)$ and $\zeta_j(t)$ possessing zero coherency at all frequencies. For greater detail concerning principal components in the complex domain see Brillinger (1981). Bartels (1977) has used frequency domain principal components analysis to analyze Dutch regional cross-sectional/time series economic data, in order to estimate the lead-lag effects of regional economic well-being. A one-factor model of wage rates is also given by Engle and Watson (1981).

7.7.2 Factor Analysis in the Frequency Domain

The use of Fourier transforms can be extended to the case of a frequency domain maximum likelihood factor analysis (Section 6.4). As an example consider the single common-factor model

$$\mathbf{X}(t) = \boldsymbol{\alpha}\phi(t) + \boldsymbol{\epsilon}(t) \tag{7.101}$$

(Geweke, 1977) where $\mathbf{X}(t) = [X_1(t), X_2(t), \ldots, X_p(t)]^{\mathrm{T}}$ is a multivariate series such that $E[\mathbf{X}(t)] = \mathbf{0}$ and $\boldsymbol{\alpha} = [\alpha_1, \alpha_2, \ldots, \alpha_p]^{\mathrm{T}}$ is a $(p \times 1)$ time-invariant vector of loadings, $\phi(t)$ is a single time dependent unobserved factor, and $\boldsymbol{\epsilon}(t)$ is the $(p \times 1)$ vector of independent error terms. We suppose that all assumptions (Section 6.1) hold, except that in addition to being intercorrelated the series $X_1(t), X_2(t), \ldots, X_p(t)$ also possess intercorrelated observations. It is well known from the theory of stochastic processes that the series $\boldsymbol{\alpha}\phi(t)$ and $\boldsymbol{\epsilon}(t)$ can be expressed as infinite linear combinations (moving averages) of time-uncorrelated "white-noise" series $u(t)$ and $v(t)$, that is,

$$\alpha_1\phi(t) = a_1(0)u(t) + a_1(1)u(t-1) + \cdots$$
$$\alpha_2\phi(t) = a_2(0)u(t) + a_2(1)u(t-1) + \cdots$$
$$\alpha_p\phi(t) = a_p(0)u(t) + a_p(1)u(t-1) + \cdots$$

and

$$\epsilon_1(t) = b_{11}(0)v_1(t) + b_{11}(1)v_1(t-1) + \cdots$$
$$\epsilon_2(t) = b_{22}(0)v_2(t) + b_{22}(1)v_2(t-1) + \cdots$$
$$\epsilon_p(t) = b_{pp}(0)v_p(t) + b_{pp}(1)v_p(t-1) + \cdots$$

or

$$\alpha_j\phi(t) = \sum_{s=0}^{\infty} a_j(s)u(t-s) = a_j(s)u(t) \tag{7.102}$$

$$\epsilon_j(t) = \sum_{s=0}^{\infty} b_{jj}(s)v_j(t-s) = b_{jj}(s)v_j(t)(j = 1, 2, \ldots, p) \tag{7.103}$$

so that Eq. (7.101) can be expressed as

$$\mathbf{X}(t) = \mathbf{A}(s)u(t) + \mathbf{B}(s)\mathbf{v}(t) \tag{7.104}$$

where $\mathbf{A}(s) = [a_1(s), a_2(s), \ldots, a_p(s)]^{\mathrm{T}}$, $u(t)$ is a scalar, $\mathbf{B}(s) = \mathrm{diag}\,(b_{jj})$ and $\mathbf{v}(t) = [v_1(t), v_2(t), \ldots, v_p(t)]^{\mathrm{T}}$.

The reduction of signal $\alpha\phi(t)$ and residual error (noise) $\epsilon(t)$ to white noise results in a lack of serial correlation in the two series, thus restoring the initial conditions of the factor model. A potential difficulty still remains, however, since $u(t)$ and $\mathbf{v}(t)$ may be correlated within themselves at different frequencies. Such interdependence can be eliminated by decomposing the series into frequencies. Let $\mathbf{C}(k) = E(\mathbf{X}(t+k)\mathbf{X}(t)]$ be the autocovariance matrix of $\mathbf{X}(t)$. Then from Eq. (7.104)

$$\mathbf{X}(t+k)\mathbf{X}(t)^{\mathrm{T}} = [\mathbf{A}(s+k)u(t+k) + \mathbf{B}(s+k)\mathbf{v}(t+k)][\mathbf{A}(s)u(t) + \mathbf{B}(s)\mathbf{v}(t)]^{\mathrm{T}}$$

$$= \mathbf{A}(s+k)\mathbf{A}^{\mathrm{T}}(s) + \mathbf{B}(s+k)\mathbf{B}^{\mathrm{T}}(s) \tag{7.105}$$

since $u(t)$ and $\mathbf{v}(t)$ are mutually uncorrelated at all leads/lags $k = 0$,

$\pm 1, \ldots,$ by assumption. Let

$$S(\omega) = \sum_{k=-\infty}^{\infty} C(k) \exp(k\omega\sqrt{-1}) \tag{7.106}$$

$$A(\omega) = \sum_{s=-\infty}^{\infty} A(s) \exp(s\omega\sqrt{-1}) \tag{7.107}$$

$$B(\omega) = \sum_{s=-\infty}^{\infty} B(s) \exp(s\omega\sqrt{-1}) \tag{7.108}$$

be the spectral densities (Fourier transforms) of $C(k)$, $A(s)$, and $B(s)$ respectively where diagonal elements of $S(\omega)$ are real and the off-diagonals are complex. Also since $B(s+k)B(t)^T$ is diagonal, it must contain real elements. The analogue of Eq. (7.105) in the frequency domain is then

$$S(\omega) = A(\omega)A^T(\omega) + B(\omega)B^T(\omega) \tag{7.109}$$

(Sargent and Sims, 1977; Geweke, 1977) assuming that $C(k)$ is nonsingular. Since Eqs. (7.105) and (7.109) are invertible transforms of each other they contain the same information, albeit in different guise. Note also that since the processes must be (covariance) stationary, observed series that do not exhibit stationarity must first be transformed by a filtering procedure such as detrending or autoregressive fitting. This is important if we are to preserve the assumptions of the model and achieve unbiased estimation of $S(\omega)$—after the factor model has been estimated, however, the effect(s) of the filter(s) can always be removed if so desired by a procedure known as "recoloring".

It is also possible to derive maximum likelihood estimators for the complex factor model. When $X(t)$ is a multivariate complex process the Fourier transform $S(\omega)$ of $X(t)$ possesses the complex normal distribution

$$f(X) = \frac{1}{\pi^p |S(\omega)|} \exp[-X^H(\omega)S^{-1}(\omega)X(\omega)] \tag{7.110}$$

(Section 2.11). Given n independent observations the likelihood function is then

$$L = \frac{1}{\pi^{np}|S(\omega)|^n} \exp\left[-\sum_{i=1}^{n} x_i(\omega)S^{-1}(\omega)x_i^H(\omega)\right] \tag{7.111}$$

with log likelihood

$$L = -n[p \ln \pi + \ln|S(\omega)| + \operatorname{tr} SS^{-1}(\omega)] \tag{7.112}$$

where

$$S = \frac{1}{n} \sum_{i=1}^{n} \mathbf{x}_i(\omega)\mathbf{x}_i^{H}(\omega) \tag{7.113}$$

Maximization of Eq. (7.112) is equivalent to minimizing

$$\ln|\mathbf{S}(\omega)| + \operatorname{tr} \mathbf{SS}^{-1}(\omega) \tag{7.114}$$

which is the complex analogue of Eq. (6.43). Under the null hypothesis of $r < p$ common factors we therefore minimize

$$\ln|\mathbf{A}(\omega)\mathbf{A}^{T}(\omega) + \mathbf{B}(\omega)\mathbf{B}^{T}(\omega)| + \operatorname{tr} \mathbf{S}|\mathbf{A}(\omega)\mathbf{A}^{T}(\omega) + \mathbf{B}(\omega)\mathbf{B}^{T}(\omega)| \tag{7.115}$$

which is minimized with respect to $\mathbf{A}(\omega)$ and $\mathbf{B}(\omega)\mathbf{B}^{T}(\omega)$. A derivation parallel to the real-domain model (Section 6.4.2) then leads to normal equations which can be solved by adapting Joreskog's program based on the Fletcher–Powel algorithm (see Geweke, 1977). The frequency domain factor model can also be extended to restricted confirmatory factor analysis (Geweke and Singleton, 1981). An extensive numerical example using US economic data may be found in Sargent and Sims (1977).

7.8 STOCHASTIC PROCESSES IN THE SPACE DOMAIN: KARHUNEN–LOÈVE DECOMPOSITION

Serial realizations of a stochastic process occur, physically, as one-dimensional time (space) series (sequences) and can thus be plotted as functions of a single variable (time, directional location, etc.). This however is not the only instance of a stochastic process. For example, when dealing with spatially distributed data we can also encounter correlated observations which are distributed over physical area or volume. Thus any measurement of an object in space will generally be correlated with neighboring measurements. Stochastic processes of this nature are common in geography, geology, engineering, and related areas (Barnett, 1978; Pratt, 1978; Knudson et al., 1977; Kittler and Young, 1973; Fukunaga, 1972). Note that a stochastic process view of spatial data differs from that encountered in Section 7.4, where correlation among observations is ignored since no prior distinction is made between a stochastic (random) variable and a stochastic process. For the sake of concreteness, in this section we confine ourselves only to two-dimensional digital areal (surface) data, for example, as encountered when dealing with picture processing, pattern recognition, radar (sonar) maps, and so forth.

Consider a real function of two variables $f(x, y)$ defined over a region $R = [a_1 < x < b_1,\ a_2 < y < b_2]$ of the x, y plane. Assuming $f(x, y)$ is square-integrable in the interval, that is, assuming

$$\int_R \int [f(x, y)]^2 \, dx \, dy < \infty$$

we wish to find real, orthonormal functions $\phi_{ij}(x, y)$ such that

$$\int_R \int \phi_{ij}(x, y)\phi_{kl}(x, y) \, dx \, dy = \begin{cases} 1 & (i, j) = (k, l) \\ 0 & (i, j) \neq (k, l) \end{cases} \tag{7.116}$$

and

$$f(x, y) = \sum_{i=0}^{\infty} \sum_{j=0}^{\infty} c_{ij}\phi_{ij}(x, y) \tag{7.117}$$

where

$$c_{ij} = \int_R \int f(x, y)\phi_{ij}(x, y) \, dx \, dy \tag{7.118}$$

These equations are two-dimensional analogues of the theory encountered in Section 7.5.1 so that the functions $\phi_{ij}(x, y)$ are assumed to be complete, that is, they form an orthonormal basis in R.

Two-dimensional Karhunen–Loève (KL) analysis is obtained by considering a stochastic process $X(t, s)$ defined over a random field, as encountered for example in digital image processing. Here we consider the problem of transmitting a picture such that storage space is minimized. A digital picture consists of discrete picture elements or pixels, often coded as 6-bit words which can be represented as decimal numbers between 1 and 64 (0 and 63), that is, in 64 levels. The pixels measure light or radiant energy, usually over a wavelength band of about 350–780 mm and this energy is then processed in $(n \times n)$ blocks or matrices. Any pixel in the tth row, sth column can be represented as a sample from a continuous stochastic process $X(t, s)$. The pixel measurements are generally intercorrelated over the random field because of the existence of a nonrandom picture, and a two-dimensional KL model can then be used to compress the image. Since an image will also consist of relatively minor random patterns, a principal components (factor) decomposition will usually result in a substantial reduction in electronic storage requirements as well as in an enhanced image.

Consider a continuous process $X(t, s)$ with autocovariance function $C(t_1, s_1, t_2, s_2)$ for any two points (t_1, s_1) and (t_2, s_2) in R. Following Section 7.5.1 we can write

$$X(t, s) = \sum_{i=1}^{\infty} \sum_{j=1}^{\infty} c_{ij}\phi_{ij}(t, s) \tag{7.119}$$

where $\phi_{ij}(t,s)$ are orthonormal eigen functions, and the coefficients

$$c_{ij} = \int\int_R C(t_1, s_1, t_2, s_2)\phi_{ij}(t, s)\, dt\, ds \qquad (7.120)$$

minimize the mean-square error criterion. The eigenfunctions satisfy the integral equation.

$$\int_R\int c(t_1, s_1, t_2, s_2)\phi_{ij}(t_2, s_2)\, dt_2\, ds_2 = \lambda_{ij}\phi_{ij}(t_1, s_1) \qquad (7.121)$$

where $\lambda_{ij} = E(c_{ij}^2)$ are eigenvalues of the autocovariance function $C(t_1, s_1, t_2, s_2)$. Clearly both eigenfunctions and eigenvalues are real-valued, and $\lambda_{ij} \geq 0$. The coefficients c_{ij} are assumed to be random, but the eigenfunctions $\phi_{ij}(t, s)$ are viewed as fixed. When $X(t, s)$ represents a continuous image, the higher the intercorrelation between the pixels, the less terms are required in Eq. (7.119) to transmit the picture or the consistent part of the image. Since the eigenvalues λ_{ij} are ordered by magnitude and represent the power of the transmission, the first terms of the expansion clearly represent the main features of the image, whereas the last terms contribute to the detail and/or "noise" of the picture. If an excessive number of terms are omitted, this leads to a loss of resolution, whereas an excessive number of terms contribute little, if anything, to the visible picture as far as the observer is concerned. Such a reassemblage of an image may be viewed in terms of transparancy overlays employed in police work when attempting to reconstruct the appearance of a suspect, where each overlay portrays some essential feature of the suspect's appearance. Several overlays are generally required in order to yield a recognizable approximation to the true image. Unlike our analogy of police transparancy overlays, however, each term of the expansion (Eq. 7.119) need not in itself represent an identifiable feature, as long as their total effect results in a recognizable picture. Interestingly, when a picture or image is contaminated by blurs or two-dimensional "noise," the omission of the last terms can result in a clearer image.

In practice, continuous pictures are never transmitted and instead an image is digitized during a sampling process whereby light intensities are stored in pixels which form a $(n \times n)$ lattice or grid. This can be done in a number of ways. Thus with p-channel remote sensing data a common approach is to compute $(p \times p)$ covariance matrices by treating the pixels as replicates of a p-variate observation. A common form of analysis is then to compute principal components of the covariance matrix, although as Switzer and Green (1984) point out this does not necessarily take into account spatial correlation. An alternative method is to use the Kronecker product of two covariance matrices. Consider a $(n \times n)$ image matrix $\mathbf{Y}$ which consists of n discrete picture elements or pixels. First the n^2 elements of $\mathbf{Y}$

are placed into a single column vector by stacking the columns of $\mathbf{Y}$. Call this column vector vec($\mathbf{Y}$). Then the element y_{ij} of $\mathbf{Y}$ is placed into the $[(j-1)n + i]$th position of vec($\mathbf{Y}$). The process is termed column scanning in the engineering literature (e.g., see Hunt, 1973). Now assume that all columns of $\mathbf{Y}$ possess the same covariance matrix $\mathbf{S}_c$ and that the rows possess the same covariance matrix $\mathbf{S}_r$. Then the covariance matrix of vec($\mathbf{Y}$) can at times be written as

$$\mathbf{S} = \mathbf{S}_r \otimes \mathbf{S}_c \tag{7.122}$$

where $\mathbf{S}$ is $(n^2 \times n^2)$. A matrix which can be written in the form of Eq. (7.122) is known as a separable matrix. Separability implies that the column variance/covariance structure of $\mathbf{Y}$ is independent of its row variance/covariance structure and is equivalent to the assumption of stationarity used for time series. The covariance matrices $\mathbf{S}_r$ and $\mathbf{S}_c$ are estimated in the usual way, by using the elements of the rows and columns of $\mathbf{Y}$. The advantage of separability is that $\mathbf{S}$ can be estimated from a single image realization, much in the same way as the moments of a stationary time process can be estimated from a single time series of sufficient length. The covariance matrix $\mathbf{S}$ can be decomposed into a Kronecker product of latent roots and latent vectors as

$$\mathbf{S} = (\mathbf{P}_r \otimes \mathbf{P}_c)(\mathbf{L}_r \otimes \mathbf{L}_c)(\mathbf{P}_r \otimes \mathbf{P}_c) \tag{7.123}$$

(Section 2.9). Equation (7.123) may be viewed as a discrete analogue of the Karhunen–Loève decomposition (Eq. 121). Notice however that the decomposition (Eq. 7.123) does not take into account in an explicit way the spatial autocorrelation between the pixels, which is analogous to the serial correlation of a time series. More recently Switzer and Green (1984) have developed an analysis that resembles factor and canonical corelation analysis and takes explicit account of spatial autocorrelation. Visual examples dealing with engineering problems may be found in Andrews and Patterson (1976a), Huang and Narendra, (1975), Andrews and Patterson (1976b), and Jain and Angel (1974).

7.9 PATTERNED MATRICES

It was seen in Section 3.3 that owing to its special pattern a covariance matrix of the form (3.29) possesses latent roots and latent vectors which exhibit a simplified structure that can be expressed in exact analytic form. This is also true of covariance matrices associated with special types of stochastic processes, whose latent roots and latent vectors can be computed or approximated by algebraic functions without necessarily having to have recourse to solving characteristic equations by numerical means.

7.9.1 Circular Matrices

Consider matrix $\mathbf{C}$ of the form

$$\mathbf{C} = \begin{bmatrix} b_0 & b_1 & b_2 & \cdots & b_{n-1} \\ b_{n-1} & b_0 & b_1 & \cdots & b_{n-2} \\ b_{n-2} & b_{n-1} & b_0 & \cdots & b_{n-3} \\ \vdots & \vdots & \vdots & \ddots & \vdots \\ b_1 & b_2 & b_3 & \cdots & b_0 \end{bmatrix} \tag{7.124}$$

with equal diagonal elements b_0 and periodic rows and columns. A matrix of the form Eq. (7.124) is known as a circular matrix. Since the elements of $\mathbf{C}$ occur periodically with period n, they can be expanded into a finite complex Fourier series with frequency k/n where $k = 0, 1, \ldots, n-1$. It follows that $\mathbf{C}$ is similar to a diagonal matrix $\mathbf{\Lambda}$, with nonzero elements are given by the Fourier transform

$$\lambda_k = \sum_{j=0}^{n-1} b \exp[-\sqrt{-1}(2\pi k/n)j] \tag{7.125}$$

and corresponding column latent vector

$$\mathbf{P}_k = n^{-1/2}\{\exp[-\sqrt{-1}(2\pi k/n)], \exp[-\sqrt{-1}(4\pi k/n)], \ldots ,$$

$$\exp[-\sqrt{-1}(2(n-1)\pi k/n]\}^{\mathrm{T}} \qquad k = 0, 1, \ldots, n-1 \tag{7.126}$$

Let $\mathbf{P}$ be the matrix of latent vectors as in (Eq. 7.126). Then the matrix of Eq. (7.124) can be expressed as

$$\mathbf{C} = \bar{\mathbf{P}}^{\mathrm{T}}\mathbf{\Lambda}\mathbf{P} \tag{7.127}$$

(Section 2.11). Since $\mathbf{C}$ is not generally symmetric the expansion (Eq. 7.127) is complex rather than real. When $b_j = b_{n-j}$ $(i = 0, 1, \ldots, n-1)$ however the latent roots and vectors can be expressed as

$$\lambda_k = \sum_{j=0}^{n-1} b_i \cos\left(\frac{2\pi k_i}{n}\right) \tag{7.128}$$

and

$$\mathbf{P}_k = n^{-1/2}\left\{\left[\cos\left(\frac{2\pi k}{n}\right) + \sin\left(\frac{2\pi k}{n}\right)\right], \left[\cos\left(\frac{4\pi k}{n}\right) + \sin\left(\frac{4\pi k}{n}\right)\right], \ldots ,\right.$$

$$\left.\left[\cos\left(\frac{2(n-1)\pi k}{n}\right) + \sin\left(\frac{2(n-1)\pi k}{n}\right)\right]\right\}^{\mathrm{T}} \tag{7.129}$$

for $k = 0, 1, \ldots, n-1$. Note that the latent vectors of a circular matrix are not direct functions of its elements. Symmetric circular covariance matrices occur, for example, in stochastic processes when equally spaced temporal (spatial) points are equally correlated, that is, when $\mathbf{X}(t) = \mathbf{X}(t + n)$ so that the process has period n and the autocorrelation satisfies the constraint $\rho(k) = \rho(n - k)$ $(k = 1, 2, \ldots, n)$. For example, when $n = 4$ the circular correlation structure is given by the matrix

$$\mathbf{C} = \begin{bmatrix} 1 & \rho_1 & \rho_2 & \rho_1 \\ \rho_1 & 1 & \rho_1 & \rho_2 \\ \rho_2 & \rho_1 & 1 & \rho_1 \\ \rho_1 & \rho_2 & \rho_1 & 1 \end{bmatrix}$$

7.9.2 Tridiagonal Matrices

A matrix of the form

$$\mathbf{G} = \begin{bmatrix} b_1 & c_1 & 0 & 0 & \cdots & 0 \\ a_1 & b_2 & c_2 & 0 & \cdots & 0 \\ 0 & a_2 & b_3 & c_3 & \cdots & 0 \\ \vdots & \vdots & \ddots & & \ddots & c_{n-1} \\ 0 & 0 & \cdots & a_{n-1} & & b_n \end{bmatrix} \tag{7.130}$$

for which $a_i c_i > 0$ $(1, 2, \ldots, n - 1)$ is known as a Jacobi matrix. All elements of a Jacobi matrix are zero except those on the three main diagonals. Tridiagonal matrices of the form of Eq. (7.130), together with several additional special forms, occur in conjunction with stochastic processes (e.g., see, Jain and Angel, 1974) and always possess real spectra, as is shown in the following theorem.

THEOREM 7.8.   A tridiagonal matrix of the form of Eq. (7.130) always possesses real latent roots and latent vectors.

PROOF.   Let $\mathbf{A} = \mathbf{DGD}^{-1}$ where $\mathbf{D} = \mathrm{diag}(d_1, d_2, \ldots, d_n)$. Then matrix $\mathbf{A}$ is of the tridiagonal form

$$\mathbf{A} = \begin{bmatrix} b_1 & c_2(d_1/d_2) & 0 & \cdots & 0 \\ a_2(d_2/d_1) & b_2 & c_3(d_2/d_3) & \cdots & 0 \\ 0 & a_3(d_3/d_2) & b_3 & \cdots & 0 \\ \vdots & \vdots & \ddots & \ddots & \vdots \\ & & & & c_n(d_n/d_{n-1}) \\ 0 & 0 & \cdots & a_{n-1}(d_{n-1}/d_n) & b_n \end{bmatrix}$$

Since $a_i c_i > 0$ we can always find numbers $d_1, d_2, \ldots, d_n$ such that

$$a_2 \frac{d_2}{d_1} = c_2 \frac{d_1}{d_2}, \quad a_3 \frac{d_3}{d_2} = c_3 \frac{d_2}{d_3}, \quad \ldots, \quad a_n \frac{d_{n-1}}{d_n} = c_n \frac{d_n}{d_{n-1}}$$

so that $\mathbf{A}$ is symmetric with nonzero diagonal elements. Since similar matrices possess equal latent roots, it follows that the latent roots of $\mathbf{G}$ are real.

When $c_1 = c_2 = \cdots = c_{n-1} = c$, $b_1 = b_2 = \cdots = b_n = b$, $a_1 = a_2 = \cdots = a_{n-1} = a$, we have what is known as a common tridiagonal matrix. It can be shown that the latent roots of a common tridiagonal Jacobi matrix are given by

$$\lambda_k = b + 2\sqrt{ac}\,\cos\!\left(\frac{k\pi}{n+1}\right) \qquad (k = 1, 2, \ldots, n) \tag{7.131}$$

(see Hammarling, 1970) and the latent vectors can also be expressed in terms of trigonometric functions. For the special case $a = c$ we have a symmetric common tridiagonal matrix with latent roots

$$\lambda_k = b + 2a\,\cos\!\left(\frac{k\pi}{n+1}\right)$$

and latent vectors

$$\mathbf{P}_k = \left(\frac{2}{n+1}\right)^{1/2}\left[\sin\!\left(\frac{k\pi}{n+1}\right), \sin\!\left(\frac{2k\pi}{n+1}\right), \ldots, \sin\!\left(\frac{nk\pi}{n+1}\right)\right]$$
$$(k = 1, 2, \ldots, n) \tag{7.132}$$

which do not depend on the elements of the matrix. A symmetric common tridiagonal matrix is a special case of the symmetric circular matrix $\mathbf{C}$.

7.9.3 Toeplitz Matrices

A matrix that occurs in conjunction with stationary stochastic processes is the Toeplitz matrix. Consider a sequence of elements $b_{-(n-1)}, \ldots, b_{-2}, b_{-1}, b_0, b_1, b_2, \ldots, b_{n-1}$. Then a Toeplitz matrix is obtained by taking b_{i-j} as the (i, j)th element, that is, a Toeplitz matrix contains the elements

$$\mathbf{T} = \begin{bmatrix} b_0 & b_1 & b_2 & \cdots & b_{n-1} \\ b_{-1} & b_0 & b_1 & \cdots & b_{n-2} \\ b_{-2} & b_{-1} & b_0 & \cdots & b_{n-3} \\ \vdots & \vdots & & \ddots & \vdots \\ & & & & b_1 \\ b_{-n+1} & \cdots & b_{-2} & b_{-1} & b_0 \end{bmatrix} \tag{7.133}$$

with equal entries on the main diagonals. When $b_i = b_{-i}$, $\mathbf{T}$ becomes symmetric. Thus for a covariance matrix which possess Toeplitz structure, the covariance elements depend only on the index differences rather than on the indices themselves. Since in practice the indices usually represent time points or space locations, a Toeplitz form frequently characterizes stationary processes. Consider a finite stochastic process with autocovariance $c_k = E(\mathbf{X}_t \mathbf{X}_{t+k})$ ($k = 0, \pm 1, \pm 2, \ldots$) and Toeplitz (stationary) autocovariance matrix

$$\mathbf{C} = \begin{bmatrix} c_0 & c_1 & c_2 & \cdots & c_{T-1} \\ c_1 & c_0 & c_1 & \cdots & c_{T-2} \\ c_2 & c_1 & c_0 & \cdots & c_{T-3} \\ \vdots & \vdots & & \ddots & \vdots \\ & & & & c_1 \\ c_{t-1} & c_{t-2} & & \cdots & c_0 \end{bmatrix} \tag{7.134}$$

where the autocovariance depends only on time differences. A question arises as to whether it is possible to express the latent roots and vectors of a Toeplitz matrix in terms of analytical expressions. This unfortunately turns out not to be the case, for a finite process. However, the latent roots and latent vectors of $\mathbf{C}$ can be approximated (asymptotically) by periodic functions using the following theorem from Toeplitz (1911).

THEOREM 7.9.   Let $\mathbf{C}$ be a matrix as defined by Eq. (7.134). Then as $n \to \infty$ the latent roots and latent vectors of $\mathbf{C}$ approach those of the circular matrix (Eq. 7.124).

The proof of the theorem, together with other related results, can be found in Grenander and Szego (1958), Widom (1965), and Brillinger (1981).

The finite Fourier transform can also be expressed in terms of a $(n \times n)$ periodic orthogonal matrix F with typical element

$$f_{st} = \{\exp[-\sqrt{-1}(2\pi st/n)]\} \tag{7.135}$$

in row $(s + 1)$ and column $(t + 1)$ ($s, t = 0, 1, \ldots, n - 1$). The matrix $\mathbf{F}$ is identical to the matrix of latent vectors $\mathbf{P}$ in Eq. (7.126). The Fourier matrix can then be expressed in terms of sine and cosine functions (see Fuller, 1976). Let $\mathbf{\Lambda} = \text{diag}(\lambda_k)$, where λ_k is given by Eq. (7.125). Then as $t \to \infty$ the contribution from the top and bottom corners of Eq. (7.124) approaches zero, and the complex latent roots and latent vectors of the circular matrix $\mathbf{C}$ provide an approximation to the Toeplitz matrix. When $\mathbf{C}$ is symmetric the latent roots and vectors become real. A principal components analysis of a symmetric Toeplitz matrix can therefore be expressed, asymptotically, in terms of a real-valued Fourier series; that is, given a time series $X(-n), \ldots, X(0), X(1), \ldots, X(n)$ the principal components of the series can be expressed as

$$(2n + 1)^{-1/2} \sum_{t=-n}^{n} \exp\{-\sqrt{-1}[2\pi st/2n + 1]\}X(t) \qquad (7.136)$$

known as the Cramer representation of a time series. For further detail see Brillinger (1981) and Durbin (1984). Also for an application to the two-dimensional space domain in digital image processing, see Andrews and Patterson (1976).

When a finite process is Markov (first order), the Toeplitz autocorrelation matrix assumes the simple form

$$\mathbf{R} = \begin{bmatrix} 1 & \rho & \rho^2 & \cdots & \rho^{n-1} \\ \rho & 1 & \rho & \cdots & \rho^{n-2} \\ \rho^2 & \rho & 1 & & \\ \vdots & \vdots & & \ddots & \rho \\ \rho^{n-1} & \rho^{n-2} & & \cdots & \rho & 1 \end{bmatrix} \qquad (7.137)$$

a symmetric Toeplitz matrix which depends on the single parameter $|\rho| \leq 1$. The inverse of $\mathbf{R}$ can be shown to be the tridiagonal matrix

$$\mathbf{R}^{-1} = \frac{1}{1-\rho^2} \begin{bmatrix} 1 & -\rho & 0 & \cdots & 0 \\ -\rho & 1+\rho^2 & -\rho & \cdots & 0 \\ 0 & -\rho & 1+\rho^2 & \cdots & 0 \\ \vdots & \vdots & & \ddots & -\rho \\ 0 & 0 & \cdots & -\rho & 1 \end{bmatrix} \qquad (7.138)$$

which is almost a common tridiagonal Jacobi matrix, and can therefore be expanded (approximately) into a cosine series (Section 7.9.2).

7.9.4 Block-Patterned Matrices

Consider a block-patterned matrix of the form

$$\mathbf{G} = \begin{bmatrix} \mathbf{A} & \mathbf{B} & \cdots & & \mathbf{B} \\ \mathbf{B} & \mathbf{A} & \mathbf{B} & & \mathbf{B} \\ & & \ddots & & \\ & & & \ddots & \mathbf{B} \\ \mathbf{B} & \mathbf{B} & \cdots & \mathbf{B} & \mathbf{A} \end{bmatrix} \qquad (7.139)$$

consisting of m^2 $(n \times n)$ submatrices

$$
\mathbf{A} = \begin{bmatrix}
a_1 & b_1 & & \cdots & b_1 \\
b_1 & a_1 & b_1 & \cdots & b_1 \\
\cdot & & \ddots & & \cdot \\
\cdot & & & \ddots & b_1 \\
b_1 & b_1 & \cdots & b_1 & a_1
\end{bmatrix}
\qquad
\mathbf{B} = \begin{bmatrix}
a_2 & b_2 & & \cdots & b_2 \\
b_2 & a_2 & b_2 & \cdots & b_2 \\
\cdot & & \ddots & & \cdot \\
\cdot & & & \ddots & b_2 \\
b_2 & b_2 & \cdots & b_2 & a_2
\end{bmatrix}
$$

$\mathbf{G}$ is thus a $(mn \times mn)$ matrix consisting of blocks of equal elements on the main diagonal, and represents a generalization of the equicorrelation matrix (Eq. 3.29). When $\mathbf{G}$ consists of correlation coefficients it is known as the intraclass correlation matrix, and occurs commonly in problems where time-series and cross-sectional data are combined. It can be shown (see Press, 1972) that the latent roots of $\mathbf{G}$ are given by

$$
\lambda_1 = a_1 + (n-1)b_1 + (m-1)[a_2 + (n-1)b_2]
$$

$$
\lambda_2 = \cdots = \lambda_n = (a_1 - b_1) + (m-1)(a_2 - b_2)
$$

$$
\lambda_{n+1} = \lambda_{m+n-1} = (a_1 - a_2) + (n-1)(b_1 - b_2)
$$

$$
\lambda_{n+m} = \lambda_{nm} = (a_1 - a_2) + (b_2 - b_1)
\tag{7.140}
$$

which are an extension of those of Eq. (3.29).

Block-patterned covariance matrices also occur in association with branching or tree-patterned processes, such as those found in biological evolution. Consider seven terminal populations (numbered 1–7) which evolved from populations numbered 8–13, as indicated by the branching process of Figure 7.16. Assuming constant evolution rates the associated covariance between the seven terminal populations is given by the block-patterned array of matrix $\mathbf{E}$ (Cavalli-Sforza and Piazza, 1975; Piazza and Cavalli-Sforza, 1978). The repetitive values for covariances are a consequence of constant evolution rates.

$$
\mathbf{E} = \begin{bmatrix}
a & & & & & & \\
b & a & & & & & \\
c & c & a & & & & \\
d & d & d & a & & & \\
d & d & d & e & a & & \\
d & d & d & f & f & a & \\
d & d & d & f & f & g & a
\end{bmatrix}
$$

For example, population 1 (row 1) is equally correlated with populations 4–7, and so forth, as indicated in matrix $\mathbf{E}$. Let $a = 1.0$, $b = .80$, $c = .50$, $d = .20$, $e = .90$, $f = .40$, and $g = .70$. Then matrix $\mathbf{E}$ becomes

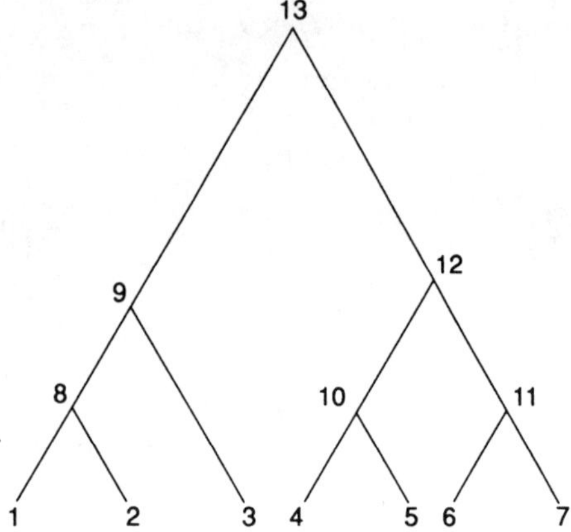

Figure 7.16 A hypothetical biological evolutionary branching process (Cavalli-Sforza and Piazza, 1975).

$$
\mathbf{E} = \begin{bmatrix}
1.00 & & & & & & \\
.80 & 1.00 & & & & & \\
.50 & .50 & 1.00 & & & & \\
.20 & .20 & .20 & 1.00 & & & \\
.20 & .20 & .20 & .90 & 1.00 & & \\
.20 & .20 & .20 & .40 & .40 & 1.00 & \\
.20 & .20 & .20 & .40 & .40 & .70 & 1.00
\end{bmatrix}
$$

The latent vectors of $\mathbf{E}$ exhibit a block-patterned structure (Table 7.14) which indicate the bifurcation points of Figure 7.16. The first split occurs at population (node) 13, and this is indicated by negative values for the first three populations and positive values for the remaining four. Signs for remaining latent vectors are interpreted in a similar fashion. Although latent vectors for matrices of order $2 < p < 4$ can be expressed in analytic form, those of order $p > 4$ are more difficult. Cavalli-Sforza and Piazza (1975) provide an example for actual human populations and indicate how the null hypothesis of constant evaluation rates can be tested employing actual data (see also Anderson, 1973).

Table 14   Latent Vectors (Rows) and Latent Roots of Matrix E

Split at Node	Populations							Latent Roots
	1	2	3	4	5	6	7	
—	.40	.40	.30	.40	.40	.40	.40	3.127
13	−.50	−.50	−.40	.30	.30	.30	.30	1.696
12	0	0	0	.50	.50	−.50	−.50	.992
9	−.30	−.30	.90	0	0	0	0	.585
11	0	0	0	0	0	−.70	.70	.300
8	−.70	.70	0	0	0	0	0	.200
10	0	0	0	.70	−.70	0	0	.100

Source: Cavalli-Sforza and Piazza, 1975; reproduced with permission.

EXERCISES

7.1 Prove that for Eq. (7.5) all elements of the best linear indices $\mathbf{p}$ and $\mathbf{q}$ are strictly positive.

7.2 Show that given $\mathbf{p}$, vector $\mathbf{q}$ can be computed using

$$\mathbf{q} = \left(\frac{1}{\lambda}\right)\mathbf{C}^{\mathrm{T}}\mathbf{p}$$

7.3 Prove Theorem 7.4

7.4 Show that Parseval's equation (Eq. 7.19) holds.

7.5 Using Eq. (7.20) derive the integral equation (Eq. 7.14).

7.6 Prove the relation of Eq. (7.81).

7.7 Show that for the circular matrix (Eq. 7.124)

$$|\mathbf{C}| = \sum_{i=0}^{n-1} b_i^n - n \prod_{i=0}^{n-1} b_i$$

7.8 Show that the determinant of the block-patterned matrix (Eq. 7.139) is given by

$$|\mathbf{G}| = |\mathbf{A} - \mathbf{B}|^{m-1}|\mathbf{A} + \mathbf{B}(m-1)|$$

7.9 Show that the Karhunen–Loève expansion maximizes the entropy measure (Eq. 7.22c).

7.10 Core samples taken from ocean floors are at times analyzed for plankton content in order to infer approximate climatological conditions which existed on earth. Since many species coexist at different times (core depths), their abundance is generally correlated, and a factor analysis is frequently performed on such data. Table 7.15, represents an excerpt of data given by Imbrie and Kipp (1971) for $n = 110$ core depths of a single core, where each depth represents 10 cm and where only $p = 10$ plankton species are considered,

(a) Carry out a principal component analysis of the $p = 10$ species, plotting component scores for the dominant component.

(b) Rotate the loadings and scores using the varimax rotation. Is there a change in the interpretation?

(c) Repeat parts a and b using a maximum likelihood factor model. Horizontal versions of the analysis are also common in geology where

Table 7.15 Percent Abundance of $p = 10$ Species of Plankton at $n = 110$ Depth of an Ocean Core Drilling

Sample Depths	Y_1	Y_2	Y_3	Y_4	Y_5	Y_6	Y_7	Y_8	Y_9	Y_{10}
1	1.792	0.489	43.485	0.814	25.570	0.651	0.0	0.163	0.0	0.163
2	3.203	0.712	37.722	0.356	30.961	0.712	0.0	0.356	0.0	0.0
3	2.564	1.709	47.009	0.855	20.513	1.709	0.0	1.282	0.427	0.0
4	1.124	0.562	47.191	1.124	12.360	2.247	0.0	3.933	0.562	0.562
5	0.671	1.007	43.624	3.020	15.436	1.007	0.0	0.336	0.671	0.336
6	1.149	0.766	52.874	0.766	12.261	0.0	0.0	0.383	2.299	0.0
7	1.990	0.498	53.234	3.980	6.965	0.0	0.0	0.498	0.995	0.0
8	2.222	2.222	45.926	2.222	13.333	2.963	0.0	1.481	1.481	1.481
9	1.786	1.190	49.405	1.786	10.714	1.786	0.0	0.595	0.595	0.0
10	0.621	0.621	36.025	2.484	10.559	0.621	0.0	1.242	1.863	0.0
11	1.418	0.0	46.099	2.837	9.220	4.255	0.0	0.709	2.836	0.0
12	0.0	0.0	38.298	0.709	11.348	2.837	0.0	1.418	5.674	0.0
13	0.498	0.498	48.756	0.0	5.970	1.990	0.498	0.498	2.985	0.0
14	1.379	1.034	42.069	0.690	8.621	2.069	0.0	2.759	1.724	0.690
15	0.662	0.0	46.358	0.0	11.921	0.0	0.0	1.987	3.311	0.0
16	3.429	1.143	45.714	1.143	14.286	1.714	0.0	0.571	3.429	0.571
17	2.899	2.899	42.995	0.0	14.010	1.449	0.0	2.415	2.415	0.483
18	1.198	1.796	50.299	1.198	8.383	2.994	0.0	0.599	0.599	0.599
19	1.887	2.516	38.994	3.145	7.547	2.516	0.0	1.258	1.258	0.0
20	5.143	2.857	38.286	0.0	13.714	1.143	0.0	1.143	1.143	0.0
21	3.067	0.613	37.423	1.227	13.497	2.761	0.0	1.227	0.0	0.307
22	1.961	2.614	41.830	3.268	11.765	1.307	0.654	1.307	0.654	0.0
23	1.515	2.020	37.374	1.010	12.626	2.020	0.0	0.0	0.505	0.0
24	1.422	2.844	38.389	1.422	16.114	0.948	0.0	0.0	0.474	0.0
25	1.630	1.630	36.957	2.174	10.870	2.174	0.0	0.0	0.0	0.0
26	1.571	1.571	37.696	1.571	10.995	4.188	0.0	2.094	2.618	1.047
27	1.826	3.196	36.073	0.913	12.329	2.283	0.0	0.457	0.913	0.457
28	0.926	3.241	28.241	0.463	12.037	0.926	0.0	0.463	1.852	0.463
29	1.379	2.414	35.517	0.345	11.679	0.345	0.0	0.0	4.828	0.0
30	1.036	6.218	34.197	1.036	14.508	0.518	0.0	0.0	1.554	0.518
31	0.649	3.896	39.610	3.896	13.636	1.299	0.0	0.543	0.649	0.0
32	1.485	7.426	29.208	2.475	15.842	1.485	0.0	2.970	1.485	0.0
33	1.087	0.0	42.391	1.630	15.761	1.630	0.0	2.174	1.087	0.0
34	3.404	0.426	32.766	4.255	13.191	2.128	0.0	3.830	0.851	1.70
35	1.429	0.476	42.381	2.857	10.952	1.905	0.0	0.476	0.952	1.90
36	1.449	3.623	36.957	0.0	15.942	3.623	0.0	0.725	1.449	0.72
37	1.685	1.685	48.315	2.809	10.674	1.124	0.0	1.124	1.124	0.0
38	0.772	0.386	40.927	0.772	15.444	2.703	0.0	0.0	0.772	0.38
39	1.266	1.266	37.975	2.532	18.143	3.376	0.0	2.110	0.422	0.0
40	3.627	0.518	41.451	1.554	16.580	0.518	0.0	2.591	1.554	0.0
41	1.869	1.402	37.850	2.804	12.617	2.336	0.0	9.813	0.467	0.93
42	3.509	2.456	42.105	2.105	12.281	1.053	0.351	2.456	0.0	0.0
43	0.904	0.904	44.578	1.205	14.759	0.602	0.301	1.506	0.602	0.0
44	1.449	0.483	43.961	3.865	12.560	1.449	0.0	2.899	0.0	0.0
45	1.299	0.649	38.961	0.325	17.208	1.948	0.0	4.545	1.948	0.0
46	0.0	0.741	33.333	2.222	22.222	2.222	0.0	0.741	0.0	0.0
47	2.513	4.523	35.176	1.005	20.603	0.0	0.0	0.0	0.0	0.0
48	1.026	0.513	42.051	2.051	16.410	2.051	0.0	0.513	2.051	0.0
49	0.565	0.565	44.068	3.955	10.169	1.695	0.0	9.605	3.390	0.0
50	1.523	0.0	34.518	2.030	20.305	2.030	0.0	1.523	1.015	0.0
51	0.508	0.0	40.609	0.508	21.827	0.508	0.0	3.046	0.0	0.0
52	0.0	2.703	28.649	1.622	24.324	3.784	0.0	2.162	3.243	0.0
53	0.629	4.403	39.623	0.629	10.063	3.145	0.0	5.660	5.031	0.0
54	0.800	2.400	50.400	1.600	11.200	2.400	0.0	4.800	0.0	0.0

Sample Depths	Species									
	Y_1	Y_2	Y_3	Y_4	Y_5	Y_6	Y_7	Y_8	Y_9	Y_{10}
55	1.630	0.543	54.348	2.174	7.609	3.804	0.0	1.630	2.717	0.0
56	0.0	0.543	32.609	1.087	11.413	4.891	0.0	3.804	2.717	0.0
57	1.622	1.081	32.973	2.162	11.892	3.784	0.0	9.780	0.541	0.0
58	1.762	0.0	33.921	0.0	16.740	2.643	0.0	9.251	2.643	0.0
59	1.418	0.0	36.879	0.709	11.348	4.255	0.0	4.965	4.965	0.709
60	1.136	2.841	49.432	2.273	11.932	2.273	0.0	0.568	0.0	0.0
61	0.893	3.561	33.036	5.357	13.393	2.679	0.0	4.464	0.893	0.893
62	3.636	1.212	35.758	2.424	6.061	6.061	0.0	3.030	0.0	0.0
63	3.448	1.478	29.064	3.448	14.778	4.433	0.0	2.955	0.0	0.0
64	1.342	2.685	34.228	3.356	12.081	2.685	0.0	2.685	4.027	0.0
65	4.435	2.419	33.468	0.806	17.742	3.226	0.0	0.0	4.032	0.0
66	2.158	2.158	34.532	2.158	15.826	5.036	0.0	0.719	2.158	0.0
67	0.0	4.545	38.636	0.0	15.152	1.515	0.0	2.273	2.273	0.758
68	1.235	0.0	41.975	0.0	12.346	1.852	7.407	0.617	2.469	0.0
69	1.508	1.508	38.191	0.503	3.518	1.508	4.523	1.508	2.010	0.503
70	3.550	2.367	47.337	2.367	5.917	10.059	0.0	0.0	0.592	0.0
71	5.344	0.0	39.695	1.527	13.740	6.870	0.0	0.763	0.0	0.0
72	5.455	0.606	43.636	1.818	10.303	7.273	1.212	0.605	0.0	0.0
73	0.0	0.0	38.095	3.571	4.762	9.524	0.0	3.571	0.0	1.190
74	2.609	1.304	33.043	1.739	9.130	3.913	0.870	3.478	0.435	0.0
75	1.604	1.604	33.690	0.0	19.251	2.139	0.0	3.209	3.209	0.535
76	1.899	0.0	34.177	2.532	12.025	4.430	0.633	2.532	1.266	0.0
77	2.041	0.816	36.327	2.041	20.000	2.449	0.0	2.449	1.224	0.408
78	0.595	2.976	50.000	0.0	7.738	6.548	0.0	2.381	0.595	0.0
79	0.0	6.130	35.249	0.0	10.728	0.0	0.0	0.383	0.383	0.0
80	0.372	5.576	37.918	0.372	15.613	0.743	0.0	0.0	0.372	0.0
81	3.582	5.373	38.209	0.896	17.015	0.896	0.0	0.0	0.896	0.299
82	2.362	2.362	36.220	3.150	14.173	1.969	0.0	0.787	1.575	0.0
83	2.105	4.211	26.842	1.053	13.684	4.737	0.526	5.263	2.105	0.0
84	2.381	3.175	32.143	1.190	17.460	1.587	0.0	0.397	1.190	0.0
85	0.455	0.909	37.273	0.455	24.091	3.182	0.0	0.455	0.455	0.909
86	0.858	3.863	31.760	1.717	21.888	7.296	0.0	4.721	0.858	0.0
87	2.769	1.231	43.385	1.231	2.769	4.000	0.0	6.462	3.077	0.0
88	0.658	1.316	52.632	0.0	3.289	1.974	0.0	3.947	0.658	0.0
89	3.448	0.575	35.632	1.149	14.368	0.0	0.0	4.598	0.575	0.0
90	1.689	0.676	26.689	2.027	8.108	4.392	0.338	13.176	2.027	1.689
91	1.533	0.0	35.249	0.383	9.195	2.682	1.533	13.793	1.533	0.0
92	1.064	0.0	40.957	1.596	6.915	2.660	0.0	3.723	2.660	0.0
93	1.394	0.348	36.585	1.045	8.014	3.833	0.0	6.969	1.394	0.0
94	0.00	0.0	35.533	1.015	13.706	7.614	0.0	3.553	0.0	0.0
95	1.970	2.463	39.901	0.493	15.764	3.941	0.0	0.985	0.493	0.493
96	1.471	2.206	34.559	2.941	15.441	1.471	0.0	0.0	0.735	0.0
97	1.613	0.403	42.742	1.210	16.129	2.823	0.0	2.823	0.403	0.0
98	0.0	0.498	44.776	2.488	19.900	0.995	0.0	1.990	0.995	0.498
99	0.448	0.448	40.359	4.484	12.556	2.242	0.0	6.278	0.897	0.0
100	2.717	0.0	32.065	3.261	15.761	1.087	0.0	6.522	1.087	0.0
101	1.887	1.887	34.906	1.415	12.264	1.415	0.0	3.302	1.415	0.472
102	1.342	2.013	24.161	3.356	11.409	1.342	0.0	9.396	0.0	0.671
103	1.633	0.816	24.898	2.449	6.531	0.408	0.0	12.245	2.041	0.0
104	1.548	0.310	31.269	1.548	9.288	0.0	0.0	9.288	4.644	0.0
105	1.093	0.546	31.694	1.639	14.208	0.0	0.0	19.672	4.372	0.0
106	2.183	1.747	33.188	0.437	13.974	0.437	0.0	4.367	1.747	1.747
107	1.878	0.469	24.883	1.878	14.085	1.408	0.0	9.390	0.939	0.0
108	2.286	2.286	37.143	1.714	8.000	1.714	0.0	8.000	4.571	0.0
109	3.911	2.793	32.961	1.117	14.525	1.117	0.0	2.793	0.559	0.0
110	0.658	0.658	34.868	4.605	15.789	1.316	0.0	3.947	1.974	0.0

Source: Imbrie and Kipp, 1971; reproduced with permission.

they are known as trend surface analyses—see, for example, Harbaugh and Demirmen (1964) and Cameron (1968).

7.11   Carry out a factor analysis of the vehicle fatalities data of Table 7.6 by replacing Provinces by years, that is, by treating the years as variables. Interpret the loadings and scores in terms of vehicle fatalities. Is any additional insight provided by the Q-model?

7.12   Consider the criminological data of Table 7.1 (Example 7.1):

(a) Repeat the analysis using (i) the matrix $\mathbf{Y}^T\mathbf{Y}$ and (ii) the cosine matrix (Section 2.4). Compare your results.

(b) Ahamad (1967) gives the following population values of the 13- to 19-year age group, in thousands:

Year	1950	1951	1952	1953	1954	1955	1956	1957
Values	1914	1912	1920	1927	1899	1935	1980	2057

Year	1958	1959	1960	1961	1962	1963
Values	2105	2173	2330	2455	2542	2601

Express the data of Table 7.1 per thousand and repeat the analysis using the correlation matrix. Compare your results—is there a significant change in the interpretation? Explain.

CHAPTER 8

Ordinal and Nominal Random Variables

8.1 INTRODUCTION

When confronted with multivariate data, it is generally assumed that the observations are drawn from continuous distributions, that is, they represent ratio or interval scales (Section 1.5). This is the case, for example, in the preceding chapters, which deal almost exclusively with factor models in the context of continuous data. Not all observations however are of this type. Thus we frequently encounter situations where variables can only be represented by an absolute count, an ordinal ranking, or else by purely qualitative or nominal categories. Since ordinal and nominal observations do not possess quantitative information (as such), they are usually represented by integers, and this alters certain interpretative aspects of factor analysis. In addition, discrete observations may possess particular configurations in multivariate space— for example, dichotomous dummy variables represent apices of a hypercube (Fig. 9.6), certain rank values lie on the surface of a multidimensional sphere, and so forth. In the following sections we consider classical multivariate analyses of discrete data where the random variables are intercorrelated and the (independent) observations are drawn from a single population, that is the observations are identically distributed.

8.2 ORDINAL DATA

It was seen in Section 1.5 that ordinal or rank-ordered random variables do not possess the property of distance between their particular values. Since ordinal scales are only intended to reflect monotonically increasing (decreasing) sequences of magnitudes they, together with nominal variables, are at times referred to as "qualitative" or "nonmetric." They may occur, for example, in sample surveys of attitudes or opinions (Westley and Jacobson, 1962; Lever and Smooha, 1981), perception or cognition of power (Lusch and Brown, 1982), or judgments of, say, scientists concerned with the biological behavior of chemicals (Mager, 1984) or soil types (Hathout and Hiebert, 1980). Here ordinal data may represent a purely qualitative ordering of categories or else they can reflect a rank ordering of values of a continuous or quantitative variable whose values cannot be observed directly because of physical impediments or perhaps large errors of measurement. Specifically, three situations can be discerned when considering ordinal values. First, the multidimensional continuity is not observed; we observe instead integer ranks from which we wish to estimate the underlying continuity, usually using the factor scores. Here the observed variables are said to be intrinsically continuous. Second, the continuity may be observed but cannot be used in a factor model because of excessive errors of measurement or nonlinearities of unknown form. Here replacing the original values by their rank orders can restore linearity and eliminate much of the error. The cost of such a move is of course the elimination of metric information from the sampled values. Third, a continuity may simply not exist at all, or may be a purely speculative or hypothetical nature. The situation is then described as being purely qualitative or nonmetric. Since ordinal scales do not possess physical units of measure, a factor analysis of such data is normally based either on the correlation or the cosine matrix (Section 2.4).

8.2.1 Ordinal Variables as Intrinsically Continuous: Factor Scaling

A relatively straightforward approach to a factor analysis of ordinal data is to assume an underlying continua for the population, which cannot be observed directly in a sample and which is consequently approximated by an ordinal scale of fixed length. Since ordinal scales are invariant with respect to continuous monotonic transformations, the positive integers $1, 2, \ldots, k$ are normally used for convenience. Also, in order to have the mean (median) of the scale values itself a scale value, the integer k is often chosen to be odd. A typical situation may occur, for example, when a sample of n respondents are asked to express a preference (agreement) on the scale $1, 2, \ldots, k$, concerning a set of p statements. Note that since each respondent may chose any of the k integers per question, there are no constraints on the rows or columns of the data matrix. The objective of factor analysis is then to provide a ratio (or an interval) scale estimate of the multidimensional continua which has generated the observed rankings. Since it is frequently unnecessary to ascribe major measurement error to integer ranks, the model employed is usually that of principal components, although a maximum likelihood factor model can also be used if uneven error is thought to be present. Nothing new then arises in the execution and interpretation of a factor analysis of rank scales. Thus if the analysis is for explanatory purposes, the loading and scores may be rotated using the usual algorithms (Section 5.3), and so forth.

Another major use of a factor analysis of ordinal variables is as a measuring or scaling device of concepts which are inherently multidimensional in nature, for example, socioeconomic or political affiliation. Here, since the concept is initially defined in terms of theory, or some a priori reasoning process, the observed or manifest variables are chosen to reflect the underlying multidimensional scale. If the manifest responses to the questionnaire are further correlated, a reduced number of dimensions can be used to construct a single scale (index) which estimates the relative position of each individual on the scale, and which excludes inconsistencies due purely to individual or residual factors. The main concepts are illustrated by the following examples.

Example 8.1. During the international "oil crisis" in 1982, the federal Canadian government commissioned a survey of 1500 employees of a centrally located company in Winnipeg, Manitoba. The objective of the survey was to construct, mathematically, features of a public transportation mode which would have sufficient appeal (demand) for individuals to forego the use of personal vehicles. The subsequent reduction in the number of such commuters, particularly single-occupant drivers, should then result in a significant decrease of petroleum fuel consumption. The respondents were asked to rank, in order of importance, their "ideal" preference concerning a range of desirable features of public transportation, as represented by the following variables (Basilevsky and Fenton, 1992)

Y_1 = Door to door transportation between home and work
Y_2 = Direct transportation without stops
Y_3 = Short travel time
Y_4 = Freedom to make stops on the way to or from work
Y_5 = Freedom to chose to go at different time and on different days
Y_6 = Freedom from having to drive
Y_7 = Preference for traveling with other people
Y_8 = Preference to have travel arrangements handled by someone else
Y_9 = Importance of low transportation expenses
Y_{10} = Availability of vehicle
Y_{11} = Freedom of responsibility for vehicle maintenance and operation
Y_{12} = The need for space to carry packages
Y_{13} = Freedom from having to obtain parking space
Y_{14} = The need for transportation to and from work during off-hours (nonrush hours)
Y_{15} = Importance of physical disability (if any)
Y_{16} = Preference for traveling alone

The scale used ranges from 5 (very important) to 1 (not at all important), with the digit 3 representing indifference. Since the survey is

Table 8.1   Orthogonal Factors of the "Ideal Preference" Transportation Mode, Winnipeg, Manitoba[a]

Variable	F_1	F_2	F_3	F_4	F_5	F_6	F_7	F_8
X_1			.651	.335				
X_2			.297	.669				
X_3					.847			
X_4			.884					
X_5			.665		.258		.280	
X_6	.399	.387			.470			
X_7		.781						.326
X_8		.834						
X_9	.673							.284
X_{10}				.695				
X_{11}	.736							
X_{12}	.252		.431	.322		.236	.256	
X_{13}	.787							
X_{14}						.968		
X_{15}								.980
X_{16}							.907	

[a] Loadings smaller than .200 are omitted.

Source: Basilevsky and Fenton, 1992.

exploratory in nature, the variables are chosen to reflect a number of different dimensions or components of a public transportation system, the main idea being that once relevant dimensions are known, they can be identified in terms of ingredients of a mixture of a transportation mode which would elicit high consumer use. Owing to the presence of unequal residual error, a maximum likelihood factor analysis was carried out and the rotated factor loadings of the $p = 16$ variables appear in Table 8.1, where the factors may be given the following interpretation. The first factor seems to reflect a consumer convenience—cost effectiveness dimension—the second indicates group-oriented convenience, the third identifies a flexibility dimension which accommodates shopping, while the fourth and fifth factors reflect demand for door-to-door transportation and availability of vehicle. If the extent or strength of preference held by an individual is considered to be continuous, then the factor scores provide estimates of the underlying continuous preference dimensions or scales which have generated the common response to the questionnaire. Further exploratory analysis of ordinal data may be found in Stoetzel (1960), Petrinovich and Hardyck (1964), Mukherjee (1965), Vavra (1972), Heeler et al. (1977), and Bruhn and Schutz (1986).

Example 8.2.   As an illustration of factor scaling consider a survey consisting of 10 questions that pertain to Canada's immigration policy and general economic conditions. The first five variables are intended to measure attitudes towards the degree of economic insecurity. The factor

Table 8.2   Rotated Factor Loadings of 10 Hypothetical Questions

Variable		Factor 1	Factor 2
Y_1:	More white immigration	+.703	+.301
Y_2:	More "boat people"	−.741	−.267
Y_3:	Nonwhites are good citizens	−.705	+.184
Y_4:	Whites are good citizens	−.750	+.371
Y_5:	Whites are more hard working	+.682	−.173
Y_6:	Economic depression imminent	−.271	+.871
Y_7:	Dollar is very unstable	+.103	+.787
Y_8:	Situation is improving	+.080	−.880
Y_9:	Government is doing good job	−.371	+.784
Y_{10}:	Jobs are plentiful	−.291	−.850

Table 8.3   Observed Response for Individual i on the 10 Questions

Question	Observed Response
Y_1	5
Y_2	2
Y_3	4
Y_4	3
Y_5	4
Y_6	4
Y_7	5
Y_8	2
Y_9	4
Y_{10}	1

loadings are given in Table 8.2. The individuals are asked to respond on the scale $5 \equiv$ strongly agree; $4 \equiv$ agree; $3 \equiv$ neutral; $2 \equiv$ disagree; $1 \equiv$ strongly disagree. Consider a hypothetical individual who responds in the manner of Table 8.3. To compute this individual's position on the factor 1 scale, we first reverse the negative signs for the first five loadings that characterize factor 1. Thus for question Y_2 the respondent can also be understood as "not disagreeing," that is, "agreeing," and he/she therefore receive a rank of $(6-2)=4$, which is the "agreed" rank. Similarly Y_3 obtains a rank of $(6-4)=2$, and so forth. The total score for the individual of Table 8.3 on factor 1 is then $.703(5) + .741(2) + .682(3) + .837(4) = 13.375$ or $13.375/5 = 2.675$ (per question), indicating a virtually neutral attitude on the dimension represented by the factor. Likewise the individual's position on the second factor is $[.871(4) + .787(5) + .880(4) + .784(4) + .850(5)]/5 = 3.665$, also indicating an almost neutral position for this scale.   □

When the underlying continua posses a normal distribution, it is possible to define a more appropriate correlation coefficient, the so-called polychoric correlation. Suppose a continuous multivariate population has a normal distribution. The true population values x^* and y^* however are not known and instead we observe the ranked integers $0, 1, 2, \ldots$, which we assume to be related to the true values by the monotone relations (Figs. 8.1 and 8.2).

$$x_i = \begin{cases} 0 & x_i^* < s_1 \\ 1 & s_1 \le x_i^* < s_2 \\ 2 & s_2 \le x_i^* < s_3 \\ \vdots & \vdots \\ k & s_k \le x_i^* \end{cases} \qquad y_i = \begin{cases} 0 & y_i^* < t_1 \\ 1 & t_1 \le y_i < t_2 \\ 2 & t_2 \le y_i^* < t_3 \\ \vdots & \vdots \\ r & t_r \le y_i^* \end{cases} \qquad (8.1)$$

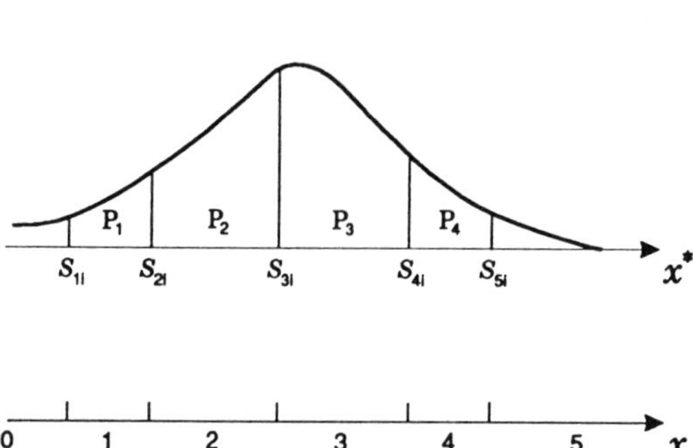

Figure 8.1 Relation between true values x_i^* and observed ranks x_i.

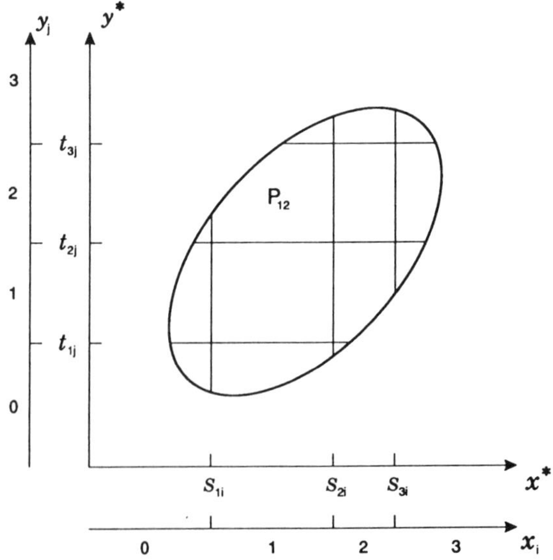

Figure 8.2 Relation between two unobserved continuous variables x^* and y^* in terms of their ranks x and y.

The inequalities implied by Eq. (8.1) are also known as "crude" classifications of the observations in the applied literature and may arise due to lack of data or unacceptably large errors of measurement. Given normality of the population we then wish to estimate the Pearson product-moment correlation between the continuities x^* and y^* using the observed ranks x and y, and then to use the correlations to construct a matrix of polychoric correlation coefficients which could be subjected to a factor analysis.

Consider the bivariate normal distribution $f(x^*, y^*)$. The probability that x^* and y^* lie in the area between s_a, s_{a+1} and t_b, t_{b+1} is given by

$$p_{ab} = \int_{s_a}^{s_{a+1}} \int_{t_b}^{t_{b+1}} f(x^*, y^*) \, dx^* \, dy^* \tag{8.2}$$

(Fig. 8.2), for some correlation value ρ. Given ρ and the thresholds of equation 8.1, the distribution $f(x^*, y^*)$ is completely specified, and the estimated correlation $\hat{\rho}$ between x^* and y^* can be computed by the formula

$$\hat{\rho} = \frac{\sum\limits_{a=1}^{s}\sum\limits_{b=1}^{t} abp_{ab} - \sum\limits_{a=1}^{s}\sum\limits_{b=1}^{t} ap_{ab} \sum\limits_{a=1}^{s}\sum\limits_{b=1}^{t} bp_{ab}}{\left[\sum\limits_{a=1}^{s}\sum\limits_{b=1}^{t} a^2 p_{ab} - \left(\sum\limits_{a=1}^{s}\sum\limits_{b=1}^{t} ap_{ab}\right)^2\right]^{1/2} \left[\sum\limits_{a=1}^{s}\sum\limits_{b=1}^{t} b^2 p_{ab} - \left(\sum\limits_{a=1}^{s}\sum\limits_{b=1}^{t} bp_{ab}\right)^2\right]^{1/2}} \tag{8.3}$$

Using Eqs. (8.2) and (8.3) it is then possible to compare numerically the correlation using ranks with the normal correlation $\hat{\rho}$. A question may also arise as to the robustness of the ranking of the normal values in a maximum likelihood factor model. Olsson (1979b; see also Olsson, 1980) has carried out a simulation of robustness of the factor model with respect to ranking (Eq. 8.1), with the conclusion that the fit of a factor model is not heavily influenced by the number of thresholds k, r, but is influenced to a higher degree by nonnormality of the variables, as expressed by univariate (marginal) skewness.

At times ranked categories are also observed in terms of counts, as for a multicategory two-way contingency table. The situation is illustrated in Figure 8.3 for $k = 3$ and $r = 4$. Employing the Hermite–Chebyshev polynomial expansion of the bivariate normal (Section 8.4.3), Lancaster and Hamdan (1964) have shown that the maximum likelihood estimator of the polychoric correlation coefficient ρ can also be expressed as

$$\varphi^2 = [\chi^2 - (r-1)(k-1)]/n = \sum_{i=1}^{r-1}\sum_{j=1}^{k-1}\left[\sum_{q=1}^{\infty} a_{iq} b_{jq} \hat{\rho}^q\right]^2 \tag{8.4}$$

where φ^2 is the corrected version of Pearson's contingency table correlation. Note also that the thresholds are fixed. Equation (8.4) is considered in

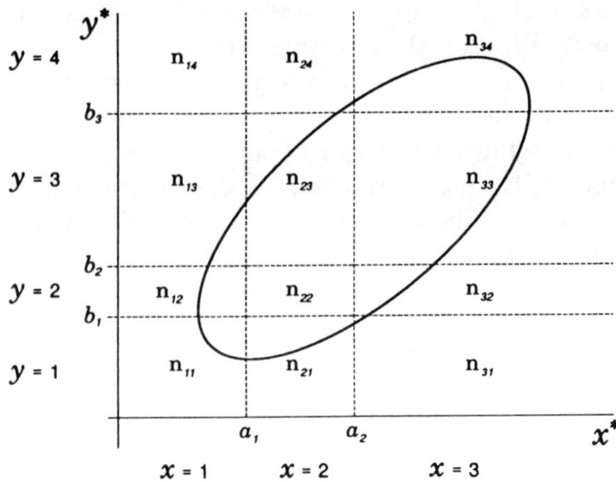

Figure 8.3 The concept of estimating polychoric correlation in terms of a $(r \times k)$ contingency table of counts.

greater detail in Section 8.4.3. The polychoric correlation coefficient is a generalization of Pearson's tetrachoric correlation and is similar to the correlation obtained by a canonical correlation analysis of a two-way contingency table. An algorithm for computing $\hat{\rho}$ has been given by Martinson and Hamdan (1972). It should be kept in mind that a correlation matrix consisting of the estimates $\hat{\rho}$ may not be positive (semi)definite so that a principal components rather than a maximum likelihood factor model should be used in the initial stages of the analysis.

8.2.2 Ranks of Order Statistics

At times even though the data are observed on a continuous scale, the ratio or "quantitative" information inherent in such scales cannot be used, because large errors of measurement or unknown forms of nonlinearities may exist among the variables. Also, a part of the data may be missing and is replaced by, say, the relative positions or the ranks of the missing values should this be possible. Let $y_{(i)}$ be the ith order statistic of a continuous random variable y (Section 7.5.3). Then the rank of $y_{(i)}$ is defined as its position among the order statistics, that is,

$$\text{rank}(y_{(i)}) = i \qquad (i = 1, 2, \ldots, n) \qquad (8.5)$$

Note that there are as many ranks as observations, that is $k = n$. For a multivariate data matrix $\mathbf{Y}$, ranks are usually assigned to the observations on a univariate basis without permuting the rows of $\mathbf{Y}$, that is, for each column $\mathbf{Y}_j$ $(j = 1, 2, \ldots, k)$ we have $\min(y_{ij}) = 1$ and $\max(y_{ij}) = n$. Alternatively, the rows of $\mathbf{Y}$ may be rearranged so that one of the variables (say the first) has its observations in the natural order $1, 2, \ldots, n$, with remaining variables assuming corresponding rank values depending on the magnitudes of the original observations. Thus although the rank-order transformations of the n observations generally do not result in all variables having their observations in the natural order, the sums of the columns are now equal, resulting in closed arrays similar to those encountered in Section 5.9.1. Replacing the original observations by ranks also results in a significant reduction of measurement error (if it exists) and introduces linear relationships between the variable ranks even though the original variables may have been nonlinear. It should be kept in mind that rank-order transformations may result in loss of information if applied to errorless and/or linear data, and the procedure should be used only as a salvage operation of data which is otherwise of limited value.

The rank orders implied by Eq. (8.5) make use of all observations of the data matrix, so that ties cannot occur if the original data are sampled from a continuous distribution. For intrinsically continuous variables the usual correlation (Spearman rho) coefficient is then the appropriate measure of association. A significant number of ties however can be introduced when only a subset of the order statistics is used, for example, deciles or quartiles. The decision to use a smaller number of rank-order statistics often hinges on whether we wish to explore the "main features" of multivariate data, perhaps as a preliminary step to a more complete analysis. Again it is important to stress the exploratory nature of such a strategy, as well as the fact that results of an analysis will depend in part on the ranking scale chosen. Nevertheless, replacing data by quartiles or other broad order statistics may reveal qualitative features which are otherwise submerged by quantitative information, and which can provide a safeguard against becoming "lost" in any subsequent analysis.

Example 8.3. A simulated comparison between principal components analyses of nonlinear continuous variables and rank-transformed data of the same variables has been conducted by Woodward and Overall (1976). The objective of the simulation is to explore the efficacy of rank transformations in being able to simultaneously linearize the variables and remove measurement error from the data. The variables are nonlinear functions of two measurements using a millimeter rule, which represent the north–south and east-west distances of $n = 29$ points on a road map of the state of Texas measured from the bottom, right-hand margins of the map. The original and estimated points appear in Figure 8.4, indicating close agreement between the data and the recovered two-dimensional space spanned by the first two principal components of the (12×12) correlation matrix. Rank transformations thus appear to be effective for the analysis of error-prone continuous, nonlinear data. When measurement error is (approximately) normal, however, a maximum-likelihood factor model may represent a more optimal approach. $\square$

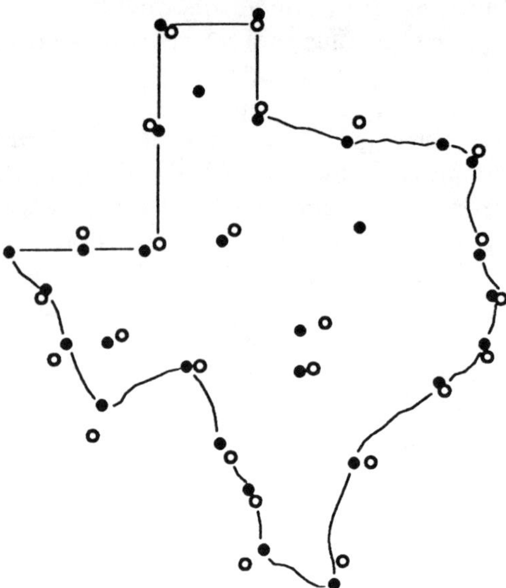

Figure 8.4 Measured configuration (dots) and estimated configuration (circles) using $r = 2$ principal components of locations on a road map of the State of Texas (Woodward and Overall, 1976; reproduced with permission).

Example 8.4. Goldstein (1982) presents data for social and disease variables obtained for $n = 21$ wards of Hull, England (Table 8.4). His objective for using quartiles rather than the original observations is to carry out a preliminary inspection of the data, by means of permuting the rows of the matrix and observing any subsequent patterns that may emerge. Since rank orders tend to linearize relationships among the variables, however, using the principal components model provides a more complete analysis, particularly when the data matrix is large. An examination of the correlation matrices of both the original and quartile data, together with rotated loadings and scores (Tables 8.5–8.7) reveals the close similarity between the two sets of analyses, indicating that the relationships among the variables are approximately linear. Vectors Y_1, Y_3, and Y_4 are seen to form a cluster (Z_1), and consulting the score coefficients it is seen that the cluster Z_1 is underscored, that is, underrepresented in wards 1, 3, 5, 7, and 16. Likewise Z_3 represents another cluster which consists only of the vector Y_2 and which is underrepresented in wards 9, 10, 12, 20, and 21. Components Z_2 and Z_4 can be interpreted in a similar fashion. Both sets of analyses are seen to be similar, so that replacing continuous data by order statistics results in surprisingly little loss of information, and indicates a fair degree of robustness of the principal components model against rank transformations.

$\square$

Table 8.4   Quantitative and Quartile Data for $n = 21$ Wards of Hull, England, 1968–1973

Ward	Crowding (Y_1)	No Toilet (Y_2)	No Car (Y_3)	Unskilled (Y_4)	Jaundice (Y_5)	Measles (Y_6)	Scabies (Y_7)
			Quantitative Data: counts				
1	28	222	627	86	139	96	20
2	53	258	584	137	479	165	31
3	31	39	553	64	88	65	22
4	87	389	759	171	589	196	84
5	29	46	506	76	198	150	86
6	96	385	812	205	400	233	123
7	46	241	560	83	80	104	30
8	83	629	783	255	286	87	18
9	112	24	729	255	108	87	26
10	113	5	699	175	389	79	29
11	65	61	591	124	252	113	45
12	99	1	644	167	128	62	19
13	79	276	699	247	263	156	40
14	88	466	836	283	469	130	53
15	60	443	703	156	339	243	65
16	25	186	511	70	189	103	28
17	89	54	678	147	198	166	80
18	94	749	822	237	401	181	94
19	62	133	549	116	317	119	32
20	78	25	612	177	201	104	42
21	97	36	673	154	419	92	29
			Quartile Data				
1	1	3	2	1	1	2	1
2	2	3	2	2	4	4	2
3	1	2	1	1	1	1	1
4	3	4	4	3	4	4	4
5	1	2	1	1	2	3	4
6	4	4	4	4	4	4	4
7	1	3	1	1	1	2	2
8	3	4	4	4	3	1	1
9	4	1	4	4	1	1	1
10	4	1	3	3	3	1	2
11	2	2	2	2	2	3	3
12	4	1	2	3	1	1	1
13	3	3	3	4	3	3	3
14	3	4	4	4	4	3	3
15	2	4	3	3	3	4	4
16	1	3	1	1	2	2	2
17	3	2	3	2	2	4	4
18	4	4	4	4	4	4	4
19	2	2	1	2	3	3	3
20	2	1	2	3	2	2	3
21	4	1	3	2	4	2	2

Table 8.5a   Correlation Matrix of Social and Disease Variables in the Wards of Hull, England

	X_1	X_2	X_3	X_4	X_5	X_6	X_7
X_1	1.000						
X_2	-1.105	1.000					
X_3	0.779	0.374	1.000				
X_4	0.779	0.263	0.853	1.000			
X_5	0.447	0.484	0.558	0.484	1.000		
X_6	0.005	0.558	0.226	0.116	0.595	1.000	
X_7	0.079	0.374	0.226	0.189	0.521	0.853	1.000

Table 8.5b   Correlation Matrix of Quartile Ranks of Social and Disease Variables in the Wards of Hull, England

	X_1	X_2	X_3	X_4	X_5	X_6	X_7
X_1	1.000						
X_2	0.084	1.000					
X_3	0.733	0.641	1.000				
X_4	0.768	0.522	0.879	1.000			
X_5	0.380	0.480	0.543	0.442	1.000		
X_6	0.055	0.522	0.375	0.209	0.542	1.000	
X_7	0.181	0.378	0.409	0.201	0.425	0.823	1.000

Table 8.6a   Varimax-Rotated Loadings of Original Social and Disease Variables in the Wards of Hull, England

	Z_1	Z_2	Z_3	Z_4
X_1	.955	.045	$-.165$	.164
X_2	.179	.258	.918	.191
X_3	.816	.241	.445	.167
X_4	.874	.035	.386	.130
X_5	.273	.285	.203	.894
X_6	.010	.873	.281	.281
X_7	.152	.956	.081	.090
Latent roots	2.471	1.885	1.344	.994

8.2.3   Ranks as Qualitative Random Variables

Finally, ranks may represent purely qualitative categories which are related by a hierarchical monotonic ordering. Here the only valid relationship among the rank values is "less than" or "greater than." Three broad approaches can be used to factor analyze such data. First, the nonmetric nature of the data can be ignored and Euclidian measures such as Sperman's rho used to summarize relationships among the multivariate rankings. The approach has an interesting interpretation since it can be considered as a

Table 8.6b Varimax-Rotated Loadings of Quartile Ranks of Social and Disease Variables in the Wards of Hull, England

	Z_1	Z_2	Z_3	Z_4
X_1	.920	.004	−.217	.208
X_2	.079	.254	.948	.169
X_3	.904	.104	.257	.165
X_4	.916	.062	.157	.117
X_5	.356	.364	.235	.828
X_6	.009	.869	.308	.240
X_7	.086	.956	.105	.141
Latent root	2.643	1.898	1.199	.875

Table 8.7a Varimax-Rotated Factor Scores of Original Social and Disease Variables in Hull, England

Z_1	Z_2	Z_3	Z_4
−1.173	−0.719	0.767	−0.726
−1.051	−0.346	0.138	2.106
−1.217	−0.763	−0.200	−0.929
0.148	0.986	0.061	1.981
−1.477	1.336	−0.814	−0.580
0.866	2.462	0.007	−0.174
−1.010	−0.325	0.388	−1.195
0.831	−1.440	0.228	−0.290
1.644	−0.585	−0.753	−1.435
1.049	−0.878	−1.375	1.086
−0.376	−0.049	−0.744	0.038
0.860	−0.938	−0.957	−0.912
0.628	−0.059	0.525	−0.368
1.155	−0.402	1.227	0.820
−0.529	1.258	0.954	0.186
−1.690	−0.516	0.266	−0.127
0.483	1.353	−1.124	−0.950
0.830	1.037	1.847	−0.137
−0.760	−0.424	−0.485	0.787
0.308	−0.177	−0.814	−0.508
0.482	−0.810	−1.144	1.419

generalization of Kendall's well-known coefficient of concordance. Secondly, we may define a purely nonparametric correlation coefficient such as Kendall's tau and proceed to decompose a correlation matrix constructed from such coefficients. Third, a nonmetric algorithm can be employed where the factor structure itself is invariant under monotone transformations of the data within each column.

Consider the situation where we have n observers, each of whom ranks k objects on the scale $1, 2, \ldots, k$. The scaling is similar to that used for order

Table 8.7b   Varimax-Rotated Factor Scores of Quartile Ranks of Social and Disease Variables in Hull, England

Z_1	Z_2	Z_3	Z_4
−0.975	−1.035	0.906	−1.036
−0.870	−0.186	0.121	1.837
−1.312	−1.278	0.028	−0.505
0.491	1.007	0.882	0.585
−1.464	1.259	−0.666	−0.303
1.195	1.032	0.906	0.271
−1.283	−0.379	0.840	−1.128
0.928	−1.913	1.542	0.428
1.750	−1.062	−1.040	−1.650
0.713	−0.808	−1.392	0.924
−0.452	0.628	−0.575	−0.524
0.813	−1.062	−0.921	−1.236
0.660	0.311	0.231	−0.059
0.767	−0.032	1.053	0.860
−0.015	1.185	1.018	−0.368
−1.497	−0.579	0.676	0.135
0.294	1.692	−0.761	−1.250
1.195	1.032	0.906	0.271
−0.983	0.428	−0.699	0.876
−0.169	0.386	−1.533	−0.271
0.214	−0.625	−1.621	2.141

statistics (Section 8.2.2) except the ranking is now for variables rather than observations, that is, the row sums rather than the column sums are constant. Since the rows of the $(n \times k)$ data matrix simply consist of permutations (not necessarily distinct) of the integers, $1, 2, \ldots, k$, the rows have the constant sum $[1/2k(k+1)]$ and constant sum of squares, and consequently lie on the surface of a $p-1$ dimensional hypersphere with center at the origin. Overall (nonnull) correlation is suspected for the population, and we wish to estimate the magnitude of such correlation. Kendall has suggested the coefficient

$$W = \frac{12S}{n^2(k^3 - k)} \tag{8.6}$$

as a measure of overall agreement or "concordance" among the n individuals, where S is the sum of squares of deviations of the column sums from the mean column sum (see Kendall, 1970). It can be shown that W is related to the arithmetic mean of the Spearman correlation coefficients amongst the $C\binom{n}{2}$ possible pairs of observers (Exercise 8.1). Also, $0 \le W \le 1$, with unity indicating total agreement among the individuals (Exercise 8.2).

Kendall's statistic of concordance however is a global measure of association and as such possesses two shortcomings. First, it does not shed light on the local structure of agreement/disagreement so that the existence of negative or zero correlation, for example, can be obscured by a value of W which is significantly different from zero. This is especially true for moderate or large data matrices. Second, it does not provide information concerning the correlation among the objects or columns of the matrix, or which clusters of objects (if any) are more (less) preferred by certain groups of individuals. Consequently, there is no opportunity to observe whether all individuals use the same criteria when judging. Note also that even an insignificant value of W does not necessarily imply the absence of local agreement (disagreement).

A more complete understanding of the data can be achieved by a principal components analysis of the rankings, using a $(n \times n)$ matrix of Spearman rho correlations among the observers. In addition the $(k \times k)$ correlation matrix of objects can also be analyzed to uncover preferred (unpreferred) groupings of the objects. The analysis proceeds in the usual fashion, with an examination of the correlation loadings obtained from a Q analysis of the observers (Section 5.4.1). A one-dimensional positive isotropic structure implies uniform agreement amongst the individuals, whereas multidimensional structures imply the existence of groupings or clusters. Linear transformations (rotations) can also be applied to locate clusters of observers with homogeneous behavior, and significance tests can be performed to determine the number of significant components, under the assumption of (approximate) normality (Section 4.3).

Principal components and Kendall's W share a common feature in that both represent metric Euclidian analyses or measures of nonmetric qualitative data. In addition, it is possible to establish a theoretical link between W and the first latent root λ_1 of the correlation matrix in the special case of a one-dimensional isotropic structure. Consider a correlation matrix of the form of Eq. (3.29) where $\sigma^2 = 1$. Then using Eq. (3.32) we have

$$\rho = \frac{\lambda_1 - 1}{n - 1} \tag{8.7}$$

for n observers. Since all correlations are equal we also have $\rho = \rho_a$, where ρ_a is the arithmetic mean, and consequently

$$\rho = \frac{\lambda_1 - 1}{n - 1}$$
$$= \frac{nW - 1}{n - 1} \tag{8.8}$$

which implies

$$W = \frac{\lambda_1}{n} \tag{8.9}$$

using Eq. (8.7) (see also Exercise 8.1). Equation (8.9) is mainly of theoretical interest since a one-dimensional isotropic structure is probably not common for ranked data. It can also be misleading (e.g. see Gorman, 1976) since in the general case a principal components analysis is not related to W, so that the two procedures are generally distinct and the latent root λ_1 should not be confused with W.

Example 8.5. Consider Kendall's example of $n = 4$ observers which rank $k = 6$ objects (Table 8.8) and correlation (Spearman rho) matrix

$$\mathbf{R} = \begin{bmatrix} 1.000 & & & \\ .314 & 1.000 & & \\ -.543 & -.543 & 1.000 & \\ .314 & .029 & .257 & 1.000 \end{bmatrix}$$

for the four rankings, Kendall's coefficient of concordance W equals .229, indicating a somewhat low degree of overall agreement. Principal component loadings and score coefficients (Table 8.9), however, reveal a more complex structure. First, the main source of variance ($\lambda_1/n = .4855$) is due to agreement between the first two observers, who show low preference for object C but higher preference for D. The third individual however disagrees with his first two colleagues' ranking, with observer 4 indicating

Table 8.8 Four Sets of Rankings of Six Objects

	Objects					
Observers	A	B	C	D	E	F
1	5	4	1	6	3	2
2	2	3	1	5	6	4
3	4	1	6	3	2	5
4	4	3	2	5	1	6

Source: Kendall, 1971a–c, p. 94; reproduced with permission.

Table 8.9 Principal Components Analysis of Spearman's Rho Correlation Matrix R for Kendall's Example (Table 8.8)

	Loadings					Scores			
	Z_1	Z_2	Z_3	Z_4		Z_1	Z_2	Z_3	Z_4
1	.772	.417	-.418	.235	A	-.110	.620	-1.163	.605
2	.762	-.096	.628	.121	B	.596	-.461	-.909	-1.619
3	-.873	.332	.188	.304	C	-1.675	-.621	-.273	.432
4	.042	.964	.179	-.194	D	.984	.953	.061	.870
Latent					E	.750	-1.468	.882	.573
roots	1.942	1.222	.637	.200	F	-.545	.977	1.402	-.861

independence from everyone else. In fact, individual 4 appears as "odd man out," showing a relative dislike for E and C and preference for D and F ($\lambda_2/n = .3055$). The third component seems to represent a residual contrast between the first two individuals, with the fourth component representing total individual disagreement which is mainly due to object B. Note that if the objects are of main interest, a principal components analysis can also be performed on the (6×6) object correlation matrix (Exercise 8.4). $\square$

Although the Spearman rho correlation coefficient is frequently used for ranked data, it can lose much of its rationale when the rank orders are purely qualitative, that is, when there is no continuity underlying the ranks. A rank order correlation coefficient however can be defined which only uses rank order information, that is, one that only employs the relations of "greater than" or "less than." Let P and Q represent the number of concordant and discordant pairs respectively. Then Kendall's rank-order correlation coefficient, known as "Kendall's tau" is defined as

$$\tau = \frac{P - Q}{1/2n(n-1)} \tag{8.10}$$

for n objects in the rankings (Kendall, 1970). An alternative form can also be found in Gibbons (1971; see also Exercise 8.3). The coefficient (Eq. 8.10) is distinct from Spearman's rho and often tends to be smaller in magnitude. A factor analysis can be based on a correlation matrix using Kendall's tau. Nothing new of major importance arises with respect to the interpretation of the results—see Marquardt (1974) for an archaeological application.

The use of Kendall's tau in place of Spearman's rho corrects for the nonmetric nature of noncontinuous rank-order variables. When using a factor model in conjunction with Kendall's tau, however, it can be argued that the correction is only partial, since it does not involve the factor algorithm used to compute the loading and the score coefficients of the model. An alternative approach to a factor analysis of ordinal data therefore may be to only recover that minimum dimensional factor representation which is invariant under monotonic transformations of the observations. Such an analysis is frequently termed "nonmetric," in contrast to the usual factor models which assume "metric," that is, intrinsically continuous observations defined on the real line. Kruskal and Shepard (1974) have developed an algorithm to perform nonmetric analysis employing a least squares monotone (isotonic) regression (see also Kruskal, 1964 and Barlow et al., 1972). The algorithm preserves for the jth variable, the same rank orders for the observed and predicted values. Let x_{ij} and x_{kj} be any two observations for the jth variable with predicted values given by

$$x_{ij}^* = \sum_{h=1}^{r} z_{ih} a_{jh} \,, \qquad x_{kj}^* = \sum_{h=1}^{r} z_{kh} a_{jh} \tag{8.11}$$

Then x_{ij}^* and x_{kj}^* are chosen such that

$$x_{ij}^* \leq x_{kj}^* \qquad \text{whenever } x_{ij} \leq x_{kj} \tag{8.12}$$

Least squares monotone regression then consists of finding numbers x_{ij}^* and x_{jk}^* such that $x_{ij}^* \leq x_{kj}^*$ whenever $x_{ij} \leq x_{kj}$ and also which minimize

$$\sum_{i=1}^{n} (x_{ij}^* - x_{ij})^2, \qquad (j = 1, 2, \ldots, p) \tag{8.13}$$

The minimization of Eq. (8.13) cannot be expressed in closed form and is therefore carried out by iterative numerical methods. Although the nonmetric monotone model has theoretical appeal, it seems to possess two practical drawbacks. First, it requires lengthy computation even for moderate data sets, as compared to principal components. Second, the nonmetric model can fare poorly against principal components when variables are nonlinear, especially when in addition they possess errors of measurement (Woodbury and Overall, 1976).   □

8.2.4 Conclusions

Rank-order data lack much of the ratio (interval) information present in continuous random variables. If intrinsic continuity can be assumed, either as a working hypothesis or by invoking a priori theoretical reasoning, most factor models carry through just as if the sample had been taken directly from a continuous population. If the assumption of continuity is tenuous, however, a nonparametric correlation coefficient such as Kendall's tau can be used, and a factor analysis then provides a fictitious but perhaps useful summary of the data. A common condition here is that a $(n \times p)$ data matrix represents a multivariate sample much in the same way as for continuous variables, that is, the observations must be independent. Since rank orders usually consist of integer scales, however, they also share common features with purely qualitative nominal scales, considered in the following sections.

8.3 NOMINAL RANDOM VARIABLES: COUNT DATA

Although ranked data consist of a sequence of discrete, ordered categories, they do preserve to some extent the notion of a quantity (Section 1.5). A nominal scale on the other hand normally expresses purely qualitative categorical attributes which do not contain quantitative information unless the data are expressed as counts or proportions—see e.g., Fleiss (1973). A factor analysis of nominal data possesses certain features that are not shared by ordinal forms of categorical data or other integer-valued variables, and usually require a different interpretational approach.

8.3.1 Symmetric Incidence Matrices

A square matrix $\mathbf{A}$ is said to be a Boolean incidence (or simply a Boolean) matrix when every element a_{ij} of $\mathbf{A}$ consists of either 1 or 0. Such matrices are frequently used as qualitative indicators of the presence or absence of some attribute since we can always let

$$a_{ij} = \begin{cases} 1 & \text{if attribute present} \\ 0 & \text{if attribute absent} \end{cases} \tag{8.14}$$

Furthermore, if $a_{ij} = a_{ji}$, $\mathbf{A}$ is said to be a symmetric Boolean matrix. Although symmetric Boolean matrices are more commonly studied within the context of nonnegative matrices, they can also occur in conjunction with a principal components analysis of undirected graphs, such as transportation networks that possess symmetric incidence matrices. In the geographic literature symmetric matrices are also known as "connectivity" matrices. Undirected graphs occur in structural analyses of the so-called nodal accessibilities of road or airline networks between cities, regions, or countries (see Taaffe and Gauthier, 1973). More generally, undirected graphs occur in any setting where "communication" is two-way.

Consider a graph consisting of n nodes, together with an associated incidence matrix $\mathbf{A}$ where $a_{ij} = 1$ if nodes i and j are connected and zero otherwise. Since connectivity is a symmetric relation, the incidence matrix $\mathbf{A}$ is symmetric. For example, for $n = 5$, the graph of Figure 8.5 has the associated incidence matrix

$$\mathbf{A} = \begin{array}{c} \\ A_1 \\ A_2 \\ A_3 \\ A_4 \\ A_5 \end{array} \begin{array}{c} \begin{array}{ccccc} A_1 & A_2 & A_3 & A_4 & A_5 \end{array} \\ \left[\begin{array}{ccccc} 1 & 1 & 0 & 1 & 0 \\ 1 & 1 & 1 & 0 & 0 \\ 0 & 1 & 1 & 1 & 0 \\ 1 & 0 & 1 & 1 & 1 \\ 0 & 0 & 1 & 1 & 1 \end{array} \right] \end{array}$$

At times the convention of using zeroes on the main diagonal is also employed in which case $\mathbf{A}$ may be thought of as a type of "distance" matrix. Altering the diagonal terms in this way however does not alter significantly a principal component analysis of the matrix since this affects only the latent roots (but not the latent vectors) in a fairly straightforward fashion. Note that since incidence matrices of undirected graphs are symmetric but not necessarily positive definite, the latent roots can be negative, and in order to overcome the difficulty, principal components analysis of $\mathbf{A}$ is at times

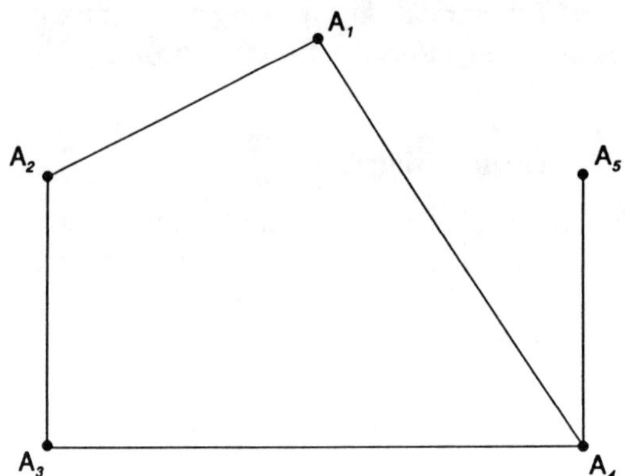

Figure 8.5   An undirected graph consisting of $n = 5$ nodes.

carried out on the matrix $\mathbf{A}^T\mathbf{A}$ (Gould, 1967; Tinkler, 1972). This, however, is unnecessary since for a square symmetric matric we have

$$\mathbf{A}^T\mathbf{A} = \mathbf{A}\mathbf{A}^T = \mathbf{A}^2$$

$$= \mathbf{P}\boldsymbol{\Lambda}^2\mathbf{P}^T \tag{8.15}$$

so that $\mathbf{A} = \mathbf{P}\boldsymbol{\Lambda}\mathbf{P}^T$, that is, $\mathbf{A}^T\mathbf{A}$, $\mathbf{A}\mathbf{A}^T$, and $\mathbf{A}^2$ possess the same latent vectors, the latent roots being squares of those of $\mathbf{A}$.

Example 8.6.   Consider the main road network of Uganda, linking 18 locations (nodes) as indicated in Figure 8.6 (Gould, 1967). The associated incidence matrix $\mathbf{A}$ is symmetric since the roads imply two-way communication between nodes (Table 8.10). The matrix $\mathbf{A}$ however is sparse and virtually diagonal, indicating poor road connections between the locations.

$$\mathbf{A} = \begin{bmatrix}
1 & 1 & 1 & 0 & 0 & 0 & 1 & 1 & 1 & 1 & 0 & 0 & 0 & 1 & 0 & 0 & 0 & 0 \\
1 & 1 & 0 & 0 & 0 & 0 & 0 & 0 & 0 & 0 & 0 & 0 & 0 & 0 & 0 & 0 & 0 & 0 \\
1 & 0 & 1 & 1 & 0 & 0 & 0 & 0 & 0 & 0 & 0 & 0 & 0 & 0 & 0 & 0 & 0 & 0 \\
0 & 0 & 1 & 1 & 1 & 0 & 0 & 0 & 0 & 0 & 0 & 0 & 0 & 0 & 0 & 0 & 0 & 0 \\
0 & 0 & 0 & 1 & 1 & 1 & 0 & 0 & 0 & 0 & 0 & 0 & 0 & 0 & 0 & 0 & 0 & 0 \\
0 & 0 & 0 & 0 & 1 & 1 & 0 & 0 & 0 & 0 & 0 & 0 & 0 & 0 & 0 & 0 & 0 & 0 \\
1 & 0 & 0 & 0 & 0 & 0 & 1 & 0 & 0 & 0 & 0 & 0 & 0 & 0 & 0 & 0 & 0 & 0 \\
1 & 0 & 0 & 0 & 0 & 0 & 0 & 1 & 1 & 0 & 0 & 0 & 0 & 0 & 0 & 0 & 0 & 0 \\
1 & 0 & 0 & 0 & 0 & 0 & 0 & 1 & 1 & 0 & 0 & 0 & 0 & 0 & 0 & 0 & 0 & 0 \\
1 & 0 & 0 & 0 & 0 & 0 & 0 & 0 & 0 & 1 & 1 & 0 & 1 & 0 & 0 & 0 & 0 & 0 \\
0 & 0 & 0 & 0 & 0 & 0 & 0 & 0 & 0 & 1 & 1 & 1 & 0 & 0 & 0 & 0 & 0 & 0 \\
0 & 0 & 0 & 0 & 0 & 0 & 0 & 0 & 0 & 0 & 1 & 1 & 0 & 0 & 0 & 0 & 0 & 0 \\
0 & 0 & 0 & 0 & 0 & 0 & 0 & 0 & 0 & 1 & 0 & 0 & 1 & 0 & 0 & 0 & 0 & 0 \\
1 & 0 & 0 & 0 & 0 & 0 & 0 & 0 & 0 & 0 & 0 & 0 & 0 & 1 & 1 & 0 & 0 & 0 \\
0 & 0 & 0 & 0 & 0 & 0 & 0 & 0 & 0 & 0 & 0 & 0 & 0 & 1 & 1 & 1 & 0 & 0 \\
0 & 0 & 0 & 0 & 0 & 0 & 0 & 0 & 0 & 0 & 0 & 0 & 0 & 0 & 1 & 1 & 0 & 0 \\
0 & 0 & 0 & 0 & 0 & 0 & 0 & 0 & 0 & 0 & 0 & 0 & 0 & 0 & 0 & 0 & 1 & 0 \\
0 & 0 & 0 & 0 & 0 & 0 & 0 & 0 & 0 & 0 & 0 & 0 & 0 & 0 & 0 & 0 & 0 & 1
\end{bmatrix} \tag{8.16}$$

Figure 8.6 A schematic map of the main roads in Uganda, 1921 and 1935 (Gould, 1967; reproduced with permission).

The first four latent vectors, together with their latent roots, are given in Table 8.10. Since an incidence matrix is not generally Grammian, it is more convenient to use latent vectors rather than the correlation loading coefficients. The latent vector elements provide a convenient topological quantification of a road map. The first latent vector is associated with the cluster of towns connected to Kampala (the capital), indicating the city possesses the greatest connectivity or accessability, both direct and indirect. The remaining coefficients of the latent vector indicate satellite towns such as Entebbe, Masaka, Mityana, and Luweru. The remaining three latent vectors indicate the existence of autonomous subclusters of towns which are weakly connected to Kampala (or to each other). Thus the second latent vector groups the Masindi cluster; the third contrasts the two linear branches of the Fort Portal–Mityana string against towns lying on the Kampala–Mbarabara road; and the fourth vector contrasts these two linear branches with the Busenbatia–Jinja–Kampala/Entebbe–Kiboga loop. The remaining latent vectors are ignored since these are either associated with negative latent roots or with features that are unique to particular towns and are thus of little general interest. For an application to transportation networks see Garrison and Marble (1963). □

Table 8.10 The First Four Latent Roots and Latent Vectors of the Incidence Connectivity Matrix Eq. 8.16 Indicating Nodal Structure (Fig. 8.6)[a]

Node	$\lambda_1 = 3.65$	$\lambda_2 = 2.83$	$\lambda_3 = 2.62$	$\lambda_4 = 2.45$
Kampala	.64			−.20
Luweru	.24			−.14
Mityana	.29		.32	
Mubende	.13		.51	.25
Kyenjojo			.51	.33
Fort Portal			.32	.23
Entebbe	.24			−.14
Jinja	.24			−.27
Busenbatia	.11			−.19
Masaka	.35	−.14	−.32	.36
Mbirizi	.15	−.11	−.32	.48
Mbarabara			−.20	.33
Bukata	.13		−.20	.25
Kiboga	.29	.22		−.13
Hoima	.14	.47		
Masindi		.64		.15
Butiaba		.35		.10
Masindi Port		.35		.10

[a] *Coefficient magnitudes less than .10 are omitted.*

Source: Gould, 1967; reproduced with permission.

8.3.2 Asymmetric Incidence Matrices

Not all incidence matrices need be symmetric since at times a graph may represent asymmetric relationships between the nodes in terms of directed, that is, one-way, line segments. An example lies in the context of the general problem of trying to obtain a unique and complete ranking of a set of n objects given pairwise comparisons between two objects. Say, for example, we have a round-robin elimination tournament between n players such that ties are not permitted and each player registers either a win or a loss. The problem is to convert the pairwise elimination results into a complete ranking of the players so that the best player is first, followed by the second best, and so forth. Denoting the players as $A_1, A_2, \ldots, A_n$, the win–loss results of a tournament can be represented by an asymmetric incidence matrix of the form

$$
\mathbf{A} = \begin{bmatrix} 0 & a_{12} & \cdots & a_{1n} \\ a_{21} & 0 & \cdots & a_{2n} \\ \vdots & \vdots & & \vdots \\ a_{n1} & a_{n2} & \cdots & 0 \end{bmatrix}
$$

where

$$a_{ij} = \begin{cases} 1 \text{ if } \mathbf{A}_i \to \mathbf{A}_j \ (\mathbf{A}_j \text{ loses to } \mathbf{A}_i) \\ 0 \text{ if } \mathbf{A}_j \to \mathbf{A}_i \ (\mathbf{A}_i \text{ loses to } \mathbf{A}_j) \end{cases}$$

and $a_{ij} + a_{ji} = 1$ for $i \neq j$. The diagonal elements are again arbitrary, in the sense that they may be coded either by zeroes or ones.

The most straightforward method of assigning scores is to rank players according to the total number of wins, that is, according to the row totals of the incidence matrix $\mathbf{A}$. Such a method of scoring however ignores the transitivities involved, since two players may tie even though one of the two has defeated several stronger players, that is, players who in turn had defeated his tied opponent(s). To take transitivities into account, Kendall (1955) has proposed to rank players according to their "power," that is, according to the elements of the latent vector $\mathbf{P}_1$ that corresponds to the dominant (real) root λ_1 of $\mathbf{A}$. Since for $n > 3$ the nonnegative incidence (the so-called tournament) matrix is primitive, a unique dominant root λ_1 always exists, and a unique and complete ranking can be obtained from the relative magnitudes of the elements of $\mathbf{P}_1$. For a more detailed treatment of nonnegative matrices see Basilevsky (1983).

A weakness of the method is that wins and losses are not treated in a symmetric fashion. Thus rather than rank players based on wins we can also rank them using losses, that is, we can obtain a ranking using the matrix $\mathbf{A}^T$. Since the spectra of the two matrices are not identical, this does not necessarily lead to a reversal of the original ranks, as is demonstrated by the following example (David, 1971).

Example 8.7. Consider the matrix

$$\mathbf{A} = \begin{array}{c} \\ A_1 \\ A_2 \\ A_3 \\ A_4 \\ A_5 \end{array} \begin{array}{ccccc} A_1 & A_2 & A_3 & A_4 & A_5 \\ \begin{bmatrix} 0 & 0 & 1 & 1 & 0 \\ 1 & 0 & 0 & 1 & 0 \\ 0 & 1 & 0 & 1 & 1 \\ 0 & 0 & 0 & 0 & 1 \\ 1 & 1 & 0 & 0 & 0 \end{bmatrix} \end{array}$$

where columns indicate losses and rows indicate wins, that is, A_1 loses to A_2, and so forth (Fig. 8.7). The leading latent vector of $\mathbf{A}$ is $\mathbf{P}_1 = (.4623, .3880, .5990, .2514, .4623)^T$ and that of $\mathbf{A}^T$ is $\mathbf{Q}_1 = (.4623, .3880, .2514, .5990, .4623)^T$. Thus using $\mathbf{A}$ we obtain the ranking A_3, A_1, and A_5 which tie for second place, and A_2 and A_4; using $\mathbf{A}^T$ however we have the ranking A_3, A_2, A_1, and A_5 which tie for third place, followed by A_4. $\qquad\square$

Although the 0-1 dummy code normally denotes a nominal variable, such

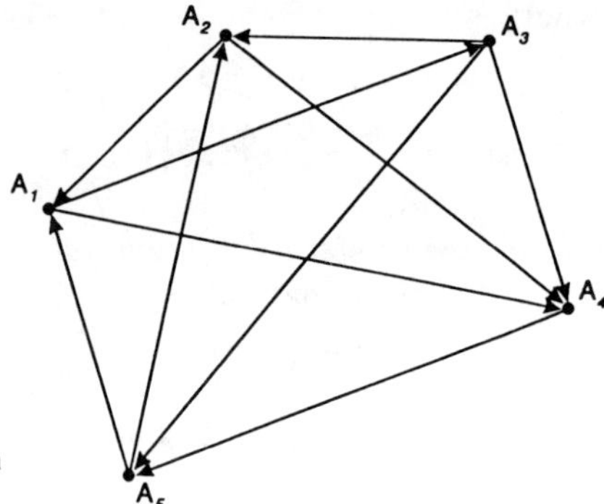

Figure 8.7 The directed graph associated with the incidence matrix of Example 8.7.

a code can also be interpreted as a special case of an ordinal variable, possessing only two levels. Thus a win–loss situation, for example, can be viewed as representing an ordinal variable since a win (code 1) gives a player greater strength than a loss (code 0). The ordinal nature of the variable is further reinforced when we introduce ties, that is, when we have

$$
b_{ij} = \begin{cases} 2 & (B_i \text{ beats } B_j) \\ 1 & (B_i \text{ ties with } B_j) \\ 0 & (B_i \text{ loses to } B_j) \end{cases}
$$

where $b_{ij} + b_{ji} = 2$ for $i \neq j$. Again an overall general ranking is provided by elements of the latent vector that correspond to the largest root. Clearly the method can be generalized to the case of k rankings, and we return to the situation of Section 8.2. Reversing the order does not necessarily result in a reversal of rank orders of the elements of the dominant latent vector.

8.3.3 Multivariate Multinomial Data: Dummy Variables

Square dummy (rank order) matrices of the previous section essentially occur in conjunction with problems of the paired comparisons type. A more general situation occurs for rectangular 0-1 dummy variable matrices which can occur in empirical studies dealing with the presence/absence of k multivariate attributes given n samples. As for the case of ranks (Section 8.2), the 0-1 dummies may represent either purely qualitative categories or else dichotomies defined over difficult-to-observe continuities. First, consider the case of purely qualitative categories where the nominal variables are not given in numeric form and must thus be coded in a particular fashion. As was observed in preceding sections, the 0-1 code is frequently employed. Although from a measurement perspective the 0-1 scale is strictly arbitrary it does lead, in a natural way, to well-known discrete probability distributions and their moments. Consider k dummy variables observed for

Table 8.11 The k Binomial Dummy Variables Observed for n Fixed, Independent Sample Points

Sample Points	$\mathbf{Y}_1$	$\mathbf{Y}_2$	$\cdots$	$\mathbf{Y}_k$
1	1	0	$\cdots$	1
2	0	0	$\cdots$	0
3	1	1	$\cdots$	0
$\vdots$	$\vdots$	$\vdots$		$\vdots$
n	1	0	$\cdots$	1
Totals	n_1	n_2	$\cdots$	n_k

n occasions or sample points (Table 8.11). The variables may represent k sets of "yes–no" answers to a questionnaire, quantal responses in a multivariate bioassay, the occurrence–nonoccurrence of plant species in a certain ecological environment, or whether archaeological artifacts are found in certain closed assemblages. Variables of this type are known as dichotomous or binary. Other names include incidence variables, indicator variables, and pseudo variables. Frequently n is fixed and the probabilities of outcome are constant across the n (independent) sample points. Here the k dummy variables $Y_1, Y_2, \ldots, Y_k$ may then be viewed as defining a set of k binomial random variables, possibly correlated. Although in the statistical literature the term random variable conventionally refers to a count or incidence of occurrence of the code "1," in what follows the term is also used to refer to a 0-1 dummy variable, depending on the context. As is well known, all first and second moments of the dummy variables are functions of the frequencies, either relative or absolute. Thus $\bar{Y}_i = n_i/n$, and the sums-of-squares and products of the dummy variables are as in Table 8.12 where n_{ij} denotes joint bivariate occurrence. The second moments are then given by

$$\frac{\mathbf{Y}_i^T \mathbf{Y}_j}{n} = \frac{n_{ij}}{n} = p_{ij} \qquad (8.17)$$

Table 8.12 The Sum-of-Squares and Products Matrix for k Correlated Binomial Variables

		$\mathbf{Y}_1$	$\mathbf{Y}_2$		$\mathbf{Y}_k$
	$\mathbf{Y}_1$	n_{11}	n_{12}	$\cdots$	n_{1k}
	$\mathbf{Y}_2$	n_{21}	n_{22}	$\cdots$	n_{2k}
$\mathbf{Y}^T\mathbf{Y} =$	$\vdots$	$\vdots$	$\vdots$		$\vdots$
	$\mathbf{Y}_k$	n_{k1}	n_{k2}	$\cdots$	n_{kk}

and

$$\text{var}(\mathbf{Y}_i) = \frac{1}{n}[\mathbf{Y}_i^T\mathbf{Y}_i - n\bar{Y}_j]$$

$$= \frac{n_{ii}}{n} - \left(\frac{n_{ii}}{n}\right)^2$$

$$= p_{ii} - p_{ii}^2$$

$$= p_{ii}(1 - p_{ii})$$

$$= p_{ii}q_{ii} \tag{8.18}$$

$$\text{cov}(\mathbf{Y}_i, \mathbf{Y}_j) = \frac{1}{n}[n_{ij} - n\bar{Y}_i\bar{Y}_j]$$

$$= \frac{n_{ij}}{n} - \left(\frac{n_{ii}}{n}\right)\left(\frac{n_{jj}}{n}\right)$$

$$= p_{ij} - p_{ii}p_{jj} \tag{8.19}$$

where $q_{ii} = (1 - p_{ii})$. Using Eqs. (8.18) and (8.19) the binomial correlation coefficient can then be expressed as

$$r_{ij} = \frac{p_{ij} - (p_{ii})(p_{jj})}{(p_{ii}q_{ii}p_{jj}q_{jj})^{1/2}} \tag{8.20}$$

which corresponds to the usual Pearson product–moment correlation coefficient. Six different analyses of a data matrix of dummy variables are possible, depending on the type of dispersion matrix employed (Section 2.4). The six dispersion matrices are summarized in Table 8.13, where $\mathbf{D} = \text{diag}(n_{ii})$, $\mathbf{C} = \text{diag}(p_{ii}q_{ii})$. Of course when categories represent an unobserved underlying normal distribution, Pearson's tetrachoric correlation coefficient can be used instead (Section 8.2.2). Other measures are also possible and these are discussed further below.

Table 8.13 Six Types of Dispersion Matrices for Binomial Dummy Variables Where the Equality Sign Indicates a Typical Element of the Respective Matrix

	About Origin	About Mean
Absolute frequency	$\mathbf{Y}^T\mathbf{Y} = (n_{ij})$	$\mathbf{X}^T\mathbf{X} = (n_{ij} - n_{ii}n_{jj})$
Relative frequency	$\dfrac{1}{n}\mathbf{Y}^T\mathbf{Y} = \left(\dfrac{n_{ij}}{n} = (p_{ij})\right)$	$\dfrac{1}{n}\mathbf{X}^T\mathbf{X} = (p_{ij} - p_{ii}p_{jj})$
Standardized frequency	$\mathbf{D}^{-1/2}\mathbf{Y}^T\mathbf{Y}\mathbf{D}^{-1/2} = \dfrac{n_{ii}}{(n_{ii}n_{jj})^{1/2}}$	$\mathbf{C}^{-1/2}\mathbf{X}^T\mathbf{X}\mathbf{C}^{-1/2} = \dfrac{p_{ij} - p_{ii}p_{jj}}{(p_{ii}q_{ii}p_{jj}q_{jj})^{1/2}}$

The binomial dichotomy can easily be generalized to a multinomial polytomy (multichotomy) consisting of l categories (levels) $A_1, A_2, \ldots, A_l$. Let $\mathbf{I}_1, \mathbf{I}, \ldots, \mathbf{I}_n$ denote a set of n $(1 \times l)$ observation vectors such that

$$\mathbf{I}_i = \begin{cases} 1 & \text{if } \mathbf{A}_j & (j = 1, 2, \ldots, l) \\ 0 & \text{otherwise} & (i = 1, 2, \ldots, n) \end{cases} \tag{8.21}$$

A single multichotomous nominal variable can then be represented by n dummy variables say $\mathbf{I}_1 = (1, 0, 0, \ldots, 0)^T$, $\mathbf{I}_2 = (0, 1, 0, \ldots, 0)^T$, $\ldots$ $\mathbf{I}_n = (0, \ldots, 1, \ldots, 0)^T$, each consisting of a single "1" and $l - 1$ zeroes (Table 8.14). Since the l categories are mutually exhaustive and exclusive, they cannot be independent. To remove the dependence however we can simply delete a category, for example, the last. Note also that the digit "1" may appear only once for each row of Table 8.14. At times such dummy variables are also termed as "pseudovariables" to distinguish them from nonmultinomial dichotomies which can possesses any number of ones.

Table 8.14 represents a different situation than that depicted in Table 8.11, since in the latter we encounter k binomial dummy variables and in the former we have a single (multinomial) variable consisting of l cataegories or "levels." For fixed $n > l$ let n_i denote the number of times "1" is observed for category $\mathbf{A}_i$. Then $\sum_{i=1}^{n} n_i = n$, and the joint density for n independent sample points can be written as

$$f(\mathbf{N}) = f(n_1, n_2, \ldots, n_l) = \frac{n!}{n_1! n_2! \ldots n_l!} pn_1 pn_2 \cdots pn_l \tag{8.22}$$

It is easy to verify that for $l = 2$ we obtain the binomial probability function and the multinomial is thus the multivariate generalization of the binomial distribution. Also all marginal distributions of Eq. (8.22) are binomial and consequently $E(n_j) = np_j$ and $\text{var}(n_j) = np_j (1 - p_j)$ $(j = 1, 2, \ldots, l)$. Multinomial counts are generally correlated since it can be shown that for any two counts n_i and n_j we have $\text{cov}(n_i, n_j) = -np_i p_j$, and the covariance

Table 8.14　A $(n \times l)$ Dummy Variable Matrix for an l-Category Multinomial Variable

	Multinomial Categories			
	$\mathbf{A}_1$	$\mathbf{A}_2$	$\cdots$	$\mathbf{A}_l$
$\mathbf{I}_1$	1	0	$\cdots$	0
$\mathbf{I}_2$	1	0		0
$\vdots$	$\vdots$	$\vdots$		$\vdots$
$\mathbf{I}_n$	0	0	$\cdots$	1

matrix is therefore

$$\Sigma = \begin{bmatrix} np_1q_1 & -np_1p_2 & \cdots & -np_1p_l \\ -np_1p_2 & np_2q_2 & \cdots & -np_2p_l \\ \vdots & \vdots & & \vdots \\ -np_1p_l & -np_2p_l & \cdots & np_lq_l \end{bmatrix}$$

The correlation coefficient between n_i and n_j can also be expressed as

$$r_{ij} = \frac{-np_ip_j}{(n^2 p_iq_ip_jq_j)^{1/2}}$$

$$= -\frac{(p_ip_j)^{1/2}}{(q_iq_j)^{1/2}} \qquad (i, j = 1, 2, \ldots, l) \tag{8.23}$$

which results in a corresponding correlation matrix for the l subcategories.

A different situation emerges when we consider a set of k interrelated nominal variables, each consisting of $l_1, l_2, \ldots, l_k$ mutually exclusive multichotomies. The special case of $l_1 = l_2 = \cdots = l_k = l$ is shown in Table 8.15. A distribution of this type is known as a multivariate multinomial distribution (see Johnson and Kotz, 1969), and provides a generalization for the k intercorrelated binomial variables of Table 8.11. When $k = 2$ we obtain a further special case, that of the bivariate binomial distribution. Owing to the mutual exclusiveness of the categories there is a single "1" per variable (per sample point) so that the rows of Table 8.15 possess the constant sum k. Since the same trait may be observed for more than one sample point however, the column sums are unrestricted. Let $K = l_1 + l_2 + \cdots + l_k$. Then Table 8.15 can be represented as a $(n \times K)$ matrix $\mathbf{Y}$, and forming the $(K \times K)$ Grammian matrix of the sums-of-squares and products yields the partitioned matrix

Table 8.15   The k Multivariate Multinomial Dummy Variables Observed for n Independent Sample Points

Sample Points	$\mathbf{Y}_1$				$\mathbf{Y}_2$				$\cdots$	$\mathbf{Y}_K$				
	$\mathbf{J}_{11}$	$\mathbf{J}_{12}$	$\cdots$	$\mathbf{J}_{1l}$	$\mathbf{J}_{21}$	$\mathbf{J}_{22}$	$\cdots$	$\mathbf{J}_{2l}$		$\mathbf{J}_{k1}$	$\mathbf{J}_{k2}$		$\mathbf{J}_{kl}$	Total
1	1	0	$\cdots$	0	0	0	$\cdots$	1	$\cdots$	0	1	$\cdots$	0	k
2	0	1	$\cdots$	0	1	0	$\cdots$	0	$\cdots$	1	0	$\cdots$	0	k
.							$\cdots$							
.														$\cdots$
.														
n	0	1	$\cdots$	0	1	0	$\cdots$	0	$\cdots$	0	0	$\cdots$	1	k
Total	N_{11}	N_{12}	$\cdots$	N_{1l}	N_{21}	N_{22}	$\cdots$	N_{2l}	$\cdots$	N_{K1}	N_{K2}	$\cdots$	N_{Kl}	

$$\mathbf{Y}^{T}\mathbf{Y} = \begin{bmatrix} \mathbf{Y}_{1}^{T}\mathbf{Y}_{1} & \mathbf{Y}_{1}^{T}\mathbf{Y}_{2} & \cdots & \mathbf{Y}_{1}^{T}\mathbf{Y}_{K} \\ \mathbf{Y}_{2}^{T}\mathbf{Y}_{1} & \mathbf{Y}_{2}^{T}\mathbf{Y}_{2} & \cdots & \mathbf{Y}_{2}^{T}\mathbf{Y}_{K} \\ \vdots & \vdots & & \vdots \\ \mathbf{Y}_{K}^{T}\mathbf{Y}_{1} & \mathbf{Y}_{K}^{T}\mathbf{Y}_{2} & \cdots & \mathbf{Y}_{K}^{T}\mathbf{Y}_{K} \end{bmatrix}$$

$$= \left[\begin{array}{cccc|c|c} N_{11} & & & 0 & & \\ & N_{12} & & & \mathbf{Y}_{1}^{T}\mathbf{Y}_{2} & \cdots & \mathbf{Y}_{1}^{T}\mathbf{Y}_{K} \\ & & \ddots & & & \\ 0 & & & N_{1l} & & \\ \hline & & & & N_{21} & \\ & & & & N_{22} & \\ & \mathbf{Y}_{2}^{T}\mathbf{Y}_{1} & & & \ddots & \cdots & \mathbf{Y}_{2}^{T}\mathbf{Y}_{K} \\ & & & & & N_{2l} & \\ \hline & & \vdots & & \vdots & \ddots & \vdots \\ & & & & & & N_{K1} \\ & \mathbf{Y}_{K}^{T}\mathbf{Y}_{1} & & & \mathbf{Y}_{K}^{T}\mathbf{Y}_{2} & \cdots & N_{K2} \\ & & & & & & \ddots \\ & & & & & & N_{Kl} \end{array} \right]$$

$$(8.24)$$

where again for simplicity of notation we assume $l_{1} = l_{2} = \cdots = l_{K} = l$. The categories that comprise $\mathbf{Y}_{1}, \mathbf{Y}_{2}, \ldots, \mathbf{Y}_{K}$ are exhaustive and mutually exclusive so that $\mathbf{Y}^{T}\mathbf{Y}$ is symmetric and positive semidefinite of rank $K - k + 1$. The submatrices on the main (block) diagonal are square and diagonal whereas the (generally) rectangular off-diagonals represent nonsymmetric $(l_i \times l_j)$ two-way contingency tables; for example, for $\mathbf{Y}_{1}^{T}\mathbf{Y}_{2}$ we have the $(l_1 \times l_2)$ matrix

$$\mathbf{Y}_{1}^{T}\mathbf{Y}_{2} = \begin{bmatrix} n_{11} & n_{12} & \cdots & n_{1c} \\ n_{21} & n_{22} & \cdots & n_{2c} \\ \vdots & \vdots & \vdots & \vdots \\ n_{r1} & n_{r2} & \cdots & n_{rc} \end{bmatrix}$$

where $l_1 = r$ and $l_2 = c$ the number of rows and columns respectively.

For multivariate multinomial observations however the appropriate measures of association are based on Grammian matrices and this provides the appropriate rationale for a factor analysis of nominal data. Also for large n the multivariate multinomial approaches the K-dimensional normal distribution, which permits the use of normal sampling theory for significance testing.

The categories of Table 8.15 are generally purely qualitative. A special case arises which is of considerable interest in practice, where the multinomial categories represent an observed polytomy of an otherwise continuous (but unobserved) variable. In such a case interest lies in estimating the underlying continuity by using the observed frequencies, much in the same way as in Section 8.2.1. The correlation coefficient between any two underlying continuities may be estimated using the polychoric correlation coefficient (Eq. 8.3), which is also applicable to nominal data. A correlation matrix of such coefficients may then be built up and factor analyzed by an appropriate procedure. When all the variables are nominal, a more commonly used procedure however is to perform a principal components analysis of a matrix of $(0, 1)$ dummy variables. As is seen in the following section, this is equivalent to several other procedures which have been commonly used to analyze discrete data. The appropriate analysis of Table 8.15 is based on the uncentered matrix of direction cosines or relative frequencies $\mathbf{D}^{-1/2}\mathbf{Y}^{T}\mathbf{Y}\mathbf{D}^{-1/2}$, that is, principal components are obtained as solutions of the normal equations

$$(\mathbf{D}^{-1/2}\mathbf{Y}^{T}\mathbf{Y}\mathbf{D}^{-1/2} - \lambda_{i}\mathbf{I})\mathbf{P}_{i} = \mathbf{0} \tag{8.25}$$

or, in alternate form,

$$(\mathbf{Y}^{T}\mathbf{Y} - \lambda_{i}\mathbf{D})\mathbf{P}_{i}^{*} = \mathbf{0} \qquad (i = 1, 2, \ldots, K) \tag{8.26}$$

where $\mathbf{P}_{i}^{*} = \mathbf{D}^{-1/2}\mathbf{P}_{i}$ (see Exercise 8.5) and $\mathbf{D} = \mathrm{diag}(\mathbf{D}_{1}, \mathbf{D}_{2}, \ldots, \mathbf{D}_{k})$ where $\mathbf{D}_{i} = \mathbf{Y}_{i}^{T}\mathbf{Y}_{i}$ are diagonal submatrices of Eq. (8.24). Although $\mathbf{Y}$ consists of discrete random variables, the loadings and the scores are real valued and may be used as estimates of the underlying continuities in the population. The normal equations (Eq. 8.26) contain the maximal latent root $\lambda = \lambda_{1} = k$, equal to the constant row sum, and its corresponding latent vector $\mathbf{1} = (1, 1, \ldots, 1)^{T}$. Since elements of $\mathbf{Y}^{T}\mathbf{Y}$ are not centered about the means, the trivial solution corresponds to the nonzero expected values of the dummy variables. From Eq. (8.26) we have

$$\mathbf{Y}^{T}\mathbf{Y}\mathbf{P}_{i}^{*} = \lambda_{i}\mathbf{D}\mathbf{P}_{i}^{*}$$

and replacing $\mathbf{1}_{K} = \mathbf{P}_{i}^{*}$ and $\lambda_{i} = k$ leads to the expression

$$\mathbf{Y}^{T}\mathbf{Y}\mathbf{1}_{K} = k\mathbf{D}\mathbf{1}_{K} \tag{8.27}$$

where $\mathbf{Y}\mathbf{1}_{K} = k\mathbf{1}_{n} = (k, k, \ldots, k)^{T}$ the $(n \times 1)$ vector or row sums of $\mathbf{Y}$, and $\mathbf{Y}^{T}\mathbf{1}_{n} = \mathbf{D}\mathbf{1}_{k} = (\mathbf{N}_{11}, \mathbf{N}_{12}, \mathbf{N}_{31}, \cdots \mathbf{N}_{Kl})^{T}$ the $(K \times 1)$ vector of column sums of $\mathbf{Y}$. Substituting $\mathbf{Y}\mathbf{1}_{n}$ into the left-side of Eq. (8.27) then leads to

$$k\mathbf{Y}^{T} = k\mathbf{D}\mathbf{1}K$$

the right-hand side of eq. (8.27) and $\lambda_1 = k$, $\mathbf{1}_K = (1, 1, \ldots, 1)^T$ must be a solution of the normal equations (Eq. 8.26), where

$$\mathbf{P}_i^* = \mathbf{D}^{1/2}\mathbf{1}_k$$

$$= (\sqrt{N_{11}}, \sqrt{N_{12}}, \ldots, \sqrt{N_{Kl}})$$

Normalizing we obtain the $(k \times k)$ latent vector

$$\mathbf{E} = \left(\sqrt{\frac{N_{11}}{N}}, \ \sqrt{\frac{N_{12}}{N}}, \ldots, \sqrt{\frac{N_{Kn}}{N}}\right)^T \tag{8.28}$$

where $N = N_{11} + N_{12} + \cdots + N_{Kl}$. Once the effect of the maximal (trivial) solution is removed, the remaining latent roots and vectors are those of the centered solution, that is, the solution of the centered matrix $\mathbf{C}^{-1/2}\mathbf{X}^T\mathbf{X}\mathbf{C}^{-1/2}$ is the same as that of the (first) residual matrix

$$\mathbf{R}_1 = (\mathbf{D}^{-1/2}\mathbf{Y}^T\mathbf{Y}\mathbf{D}^{-1/2} - \lambda_1 \mathbf{P}_1 \mathbf{P}_1^T) \tag{8.29}$$

where $\lambda_1 = k$ and $\mathbf{P}_1 = \mathbf{1}_K$. Finally, since $\mathbf{Y}^T\mathbf{Y}$ is singular, other trivial solutions of Eq. (8.25) are given by $\lambda_i = 0$. In fact, since $\mathbf{Y}^T\mathbf{Y}$ has rank $K - k + 1$ we only obtain that number of nonzero latent roots of the matrix $\mathbf{D}^{-1/2}\mathbf{Y}^T\mathbf{Y}\mathbf{D}^{-1/2}$ (Exercise 8.6).

Example 8.8. Burt (1950) presents physiological human genetic data obtained for $n = 100$ individuals in Liverpool, England: color of hair (Fair, Red, Dark), color of eyes (Light, Mixed, Brown), shape of head (Narrow, Wide), and stature (Tall, Short) in the form of a $k = 4$ dimensional multinomial distribution so that we have $l_1 = 3$, $l_2 = 3$, $l_3 = 2$, and $l_4 = 2$ categories respectively. The (10×10) matrix $\mathbf{Y}^T\mathbf{Y}$ is given in Table 8.16. The adjusted matrix of direction cosine (relative frequencies) $\mathbf{D}^{-1/2}\mathbf{Y}^T\mathbf{Y}\mathbf{D}^{-1/2}$ is presented in Table 8.17 where removing the effect of the trivial solution yields the residual matrix $\mathbf{R}_1$ (Eq. (8.29) of Table 8.18. Once the effect of the sample means is removed, the subsequent loadings are those of the usual (centered) correlation matrix of Table 8.19. The principal components correlation loadings may be interpreted in the usual manner. Also, in addition the first two components can be used to yield a joint mapping of the four traits in a common subspace (Fig. 8.8; see also Section 5.4.4). We thus perceive a tendency for light pigmentation to go together with tall stature and darker pigmentation to be associated with a wider head. A narrow head structure, on the other hand, appears to be equally associated with the first two clusters. The relative groupings of categories is

Table 8.16   Sums of Squares and Products of a (100 × 10) Dummy Variable Matrix Y

Trait	F	R	D	L	M	B	N	W	T	S
Y_1: Hair										
Fair	22	0	0	14	6	2	14	8	13	9
Red	0	15	0	8	5	2	11	4	10	5
Dark	0	0	63	11	25	27	44	19	20	43
Y_2: Eyes										
Light	14	8	11	33	0	0	27	6	29	4
Mixed	6	5	25	0	36	0	20	16	10	26
Brown	2	2	27	0	0	31	22	9	4	27
Y_3: Head										
Narrow	14	11	44	27	20	22	69	0	30	39
Wide	8	4	19	6	16	9	0	31	13	18
Y_4: Stature										
Tall	13	10	20	29	10	4	30	13	43	0
Short	9	5	43	4	26	27	39	18	0	57

Source: Burt, 1950.

Table 8.17   Matrix of Relative Frequencies $D^{-1/2}Y^TYD^{-1/2}$ of Table 8.16

Trait	F	R	D	L	M	B	N	W	T	S
Hair										
Fair	1.000	.000	.000	.519	.213	.077	.359	.306	.423	.254
Red	.000	1.000	.000	.359	.215	.093	.342	.185	.394	.171
Dark	.000	.000	1.000	.241	.525	.611	.667	.430	.384	.718
Eyes										
Light	.519	.359	.241	1.000	.000	.000	.566	.188	.770	.092
Mixed	.213	.215	.525	.000	1.000	.000	.401	.479	.254	.574
Brown	.077	.093	.611	.000	.000	1.000	.476	.290	.110	.642
Head										
Narrow	.359	.342	.667	.566	.401	.476	1.000	.000	.551	.622
Wide	.306	.185	.430	.188	.479	.290	.000	1.000	.356	.428
Stature										
Tall	.423	.394	.384	.770	.254	.110	.551	.356	1.000	.000
Short	.254	.171	.718	.092	.574	.642	.622	.428	.000	1.000

Source: Burt, 1950.

then presumably accounted for in terms of the underlying latent continuum which is related to the genetic and/or environmental factors.

Further examples of principal components analysis of nominal data may be found in David et al. (1977) and Tenenhaus (1977).   □

Table 8.18 Residual Matrix R_1 after the Removal of the "Trivial" Solution

Trait	F	R	D	L	M	B	N	W	T	S
Hair										
Fair	.780	−.182	−.372	.250	−0.68	−.185	−.030	.045	.115	−.100
Red	−.182	.850	−.370	.137	−.017	−.123	.020	−.030	.140	−.121
Dark	−.372	−.370	.370	−.215	.049	.169	.008	−.012	−.136	.118
Eyes										
Light	.250	.137	−.215	.670	−.345	−.320	.089	−.132	.393	−.341
Mixed	−.068	−.017	.049	−.345	.640	−.334	−.097	.145	−.139	.121
Brown	−.185	−.123	.169	−.320	−.334	.690	.130	−.020	−.255	.222
Head										
Narrow	−.030	.020	.008	.089	−.097	.013	.310	−.462	.006	−.005
Wide	.045	−.030	−.012	−.132	.145	−.020	−.462	.690	−.009	.008
Stature										
Tall	.115	.140	−.136	.393	−.139	−.255	.006	−.009	.570	−.495
Short	−.100	−.121	.118	−.341	.121	.222	−.005	.008	−.495	.430

Source: Burt, 1950.

Table 8.19 Centered Principal Components Loadings of Matrix R_1

	Factors						
Traits	I	II	III	IV	V	VI	VII
Hair							
Fair	.469	.420	.231	−.602	−.100	−.404	.119
Red	.387	.333	−.019	.751	.182	−.376	.012
Dark	.794	−.411	−.128	−.011	−.030	.422	−.076
Eyes							
Light	.574	.719	−.164	−.104	.042	.084	−.325
Mixed	.600	−.214	.563	.227	−.442	.048	.125
Brown	.557	−.482	−.437	−.138	.443	−.183	.201
Head							
Narrow	.831	.081	−.428	.039	−.338	−.044	.056
Wide	.557	−.121	.638	−.058	.504	.065	−.084
Stature							
Tall	.656	.619	.010	.047	.120	.335	.241
Short	.755	−.537	−.009	−.041	−.104	−.292	−.209

Source: Burt, 1950.

8.4 FURTHER MODELS FOR DISCRETE DATA

Principal components analysis of multivariate multinomial data of the previous section can be derived in several alternative ways. Although such models are at times considered to be distinct, and historically have been

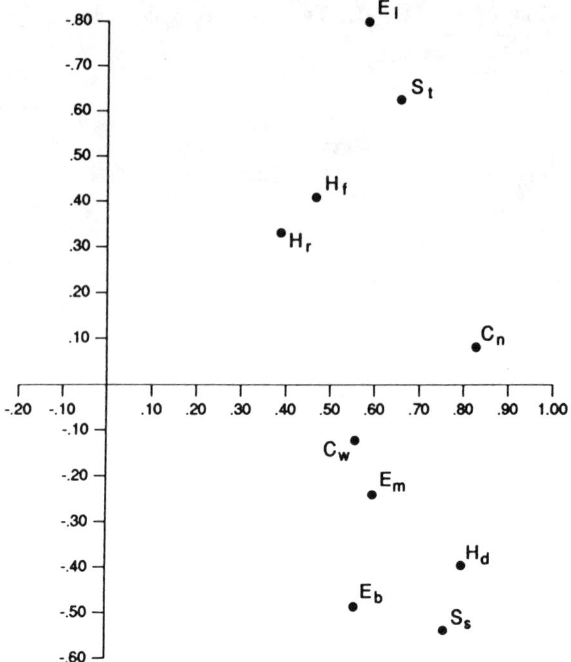

Figure 8.8   Principal component loadings of the first two axes of Table 8.19. H_f = fair hair, H_r = red hair, H_d = dark hair, E_l = light eyes, E_m = mixed eyes, E_b = brown eyes, C_n = narrow head, C_w = wide head, S_t = tall stature, and S_s = short stature.

ascribed to different authors, they are in fact related and can be considered as variants of the model described in Section 8.3.3. In what follows we describe several of the better known alternative derivatives of the multivariate multinomial model. In addition, as a rule the models can also be extended to the rank-order data of Section 8.2.

8.4.1   Guttman Scaling

An alternative rationale to maximizing correlation among nominal variables is to maximize the correlation ratio η^2. This is in fact the approach adopted by Guttman (1941, 1946), who seems to be the first to consider the problem explicitly in terms of a principal components analysis of 0–1 dummy variables. Although the procedure leads to the normal equations (Eq. 8.25), Guttman's derivation provides an additional insight into the model. His derivation does not require directly the assumption of multivariate multinomialness and has as its objective the quantification or scaling of discrete 0–1 multivariate data.

Consider the incidence data matrix of Table 8.15 and assume the dummy variables represent discrete indicators of an underlying continuous subspace. We wish to estimate a set of real-valued loading coefficients and score vectors that maximize a nonlinear coefficient of correlation, known in the context of one-way classification ANOVA as the correlation ratio. Guttman (1941) uses what he terms the "consistency principle" (principle of internal consistency) in deriving the model; in Guttman's (1941, p. 321 words, "in the social sciences we are often confronted with a set of acts of a population of individuals that we would like to consider as a single class of behaviour." The intent of the analysis is thus to maximize consistency between individual sample points, which can be achieved by minimizing variation within the individuals, that is, minimizing the relative variability of the rows of Table 8.15. Let SS_t, SS_w, and SS_b denote the total, within, and between groups sum-of-squares of one-way classification ANOVA, respectively. Then $SS_t = SS_w + SS_b$ and since minimizing SS_w is the same as maximizing SS_b, our task can be defined as maximizing the expression

$$\eta^2 = \frac{SS_b}{SS_t} \tag{8.30}$$

The correlation ratio η^2 provides a general measure of correlation not specifically tied down to linearity of regression. By maximizing Eq. (8.30) we are in fact seeking continuous principal components which are maximally related to the observed dummy variables. Since

$$F = \frac{MS_b}{MS_w} = \frac{\eta^2 \, df_b}{(1 - \eta^2) \, df_w}$$

in usual notation maximizing η^2 is equivalent to maximizing the F statistic.

First consider the correlation ratio for the component loadings. Let $\mathbf{u}$ be a $(K \times 1)$ vector of continuous weights assigned to the columns of $\mathbf{Y}$ and let

$$\bar{\mathbf{u}} = \frac{1}{K} \mathbf{Y} \mathbf{u} \tag{8.31}$$

that is $\bar{\mathbf{u}} = (\bar{u}_1, \bar{u}_2, \ldots, \bar{u}_n)^T$, the $(n \times 1)$ vector of mean weights for individuals $1, 2, \ldots, n$. Also let $\bar{\bar{u}}$ be the overall mean of the elements of $\mathbf{Yu}$. Then the correlation ratio for the loadings can be expressed as

$$\eta_u^2 = \frac{SS_b}{SS_t} = \frac{K\bar{\mathbf{u}}^T\bar{\mathbf{u}} - Kn\bar{\bar{u}}^2}{\mathbf{u}^T\mathbf{D}\mathbf{u} - Kn\bar{\bar{u}}^2}$$

$$= \frac{\frac{1}{K}\mathbf{u}^T\mathbf{Y}^T\mathbf{Y}\mathbf{u} - Kn\bar{\bar{u}}^2}{\mathbf{u}^1\mathbf{D}\mathbf{u} - Kn\bar{\bar{u}}^2}$$

using Eq. (8.31) where $\mathbf{D} = \mathrm{diag}(\mathbf{Y}^T\mathbf{Y})$. The numerator accounts for variation between the rows of $\mathbf{Y}$ (the sample points) and the denominator represents total variation of the elements of $\mathbf{Y}$ when ones are replaced by the weights $u_1, u_2, \ldots, u_K$. Since η^2 is invariant under translation of axes, we can center $\mathbf{u}$ about the overall mean $\bar{\bar{u}}$ by setting $\bar{\bar{u}} = 0$, which yields

$$\eta_u^2 = \frac{\mathbf{u}^T\mathbf{Y}^T\mathbf{Y}\mathbf{u}}{K\mathbf{u}^T\mathbf{D}\mathbf{u}} \tag{8.32}$$

We now have the standard problem of maximizing a ratio of quadratic forms, which can be achieved by maximizing the numerator $\mathbf{u}^T\mathbf{Y}^T\mathbf{Y}\mathbf{u}$ for some fixed value of $\mathbf{u}^T\mathbf{D}\mathbf{u}$. Letting $\eta_u^2 = \lambda$, Eq. (8.32) can then be expressed as

$$\lambda K\mathbf{u}^T\mathbf{D}\mathbf{u} = \mathbf{u}^T\mathbf{Y}^T\mathbf{Y}\mathbf{u}$$

and differentiating Eq. (8.32) with respect to $\mathbf{u}$ yields

$$2\lambda K\mathbf{D}\mathbf{u} + \frac{\partial\lambda}{\partial\mathbf{u}}\mathbf{u}^T\mathbf{D}\mathbf{u} = 2\mathbf{Y}^T\mathbf{Y}\mathbf{u}$$

where the necessary condition for a maximum is

$$\frac{2\mathbf{Y}^T\mathbf{Y}\mathbf{u} - 2\lambda K\mathbf{D}\mathbf{u}}{\mathbf{u}^T\mathbf{D}\mathbf{u}} = 0$$

or

$$\frac{2(\mathbf{Y}^T\mathbf{Y} - \lambda K\mathbf{D})\mathbf{u}}{\mathbf{u}^T\mathbf{D}\mathbf{u}} = 0$$

The maximum value of the correlation coefficient $\lambda = \eta_u^2$ is then a solution of the normal equations

$$\left(\frac{1}{K}\mathbf{Y}^T\mathbf{Y} - \lambda\mathbf{D}\right)\mathbf{u} = 0 \tag{8.33}$$

An alternative derivation is to set $\mathbf{u}^T\mathbf{D}\mathbf{u} = 1$ and maximize the numerator by using Lagrange multipliers (Tjok-Joe, 1976). This procedure however tends to obscure the fact that extrema of ratios of quadratic forms are absolute rather than constrained, a point which is frequently of some importance. Equation (8.33) is the same as Eq. (8.25), with the exception of the constant $1/K$ which can be initially omitted since it only affects the values of the latent roots and not of the latent vectors. We have

$$(\mathbf{D}^{-1/2}\mathbf{Y}^T\mathbf{Y}\mathbf{D}^{-1/2} - \lambda\mathbf{I})\mathbf{P}_i = 0 \tag{8.34}$$

or

$$(\mathbf{Y}^{*T}\mathbf{Y}^* - \lambda\mathbf{I})\mathbf{P}_i = 0 \tag{8.35}$$

where $\mathbf{P}_i = \mathbf{D}^{1/2}\mathbf{u}$ and $\mathbf{Y}^* = \mathbf{Y}\mathbf{D}^{-1/2}$. The vector of weights $\mathbf{u}$ is thus identical to the latent vector $\mathbf{P}_i^*$ of Eq. (8.26), and the latent root is equal (proportional) to η_u^2. Equation (8.35) possesses the same trivial solutions as Eq. (8.26) so that the maximal nontrivial correlation ratio corresponds to the second root of Eq. (8.35).

We can also maximize the correlation ratio for the scores of $\mathbf{Y}^{*T}\mathbf{Y}^*$ which are simply the latent vectors of $\mathbf{Y}^*\mathbf{Y}^{*T}$, by assigning to each sample point a real-valued score, say v_i. Denoting the $(n \times 1)$ vector of scores by $\mathbf{v}$ we have the $(K \times 1)$ vector of mean scores

$$\bar{\mathbf{v}} = \mathbf{D}^{-1}\mathbf{Y}^T\mathbf{v} \tag{8.36}$$

so that the correlation ratio for scores is

$$\eta_v^2 = \frac{\bar{\mathbf{v}}^T\mathbf{D}\bar{\mathbf{v}} - Kn\bar{\bar{v}}}{K\mathbf{v}^T\mathbf{v} - Kn\bar{\bar{v}}}$$

$$= \frac{\mathbf{v}^T\mathbf{Y}\mathbf{D}^{-1}\mathbf{Y}^T\mathbf{v}}{K\mathbf{v}^T\mathbf{v}} \tag{8.37}$$

where we set the overall mean of the scores $\bar{\bar{v}}$ equal to zero. Differentiating and setting to zero then yields the dual of Eq. (8.33), namely,

$$\left(\frac{1}{K}\mathbf{Y}^*\mathbf{Y}^{*T} - \lambda\mathbf{I}\right)\mathbf{v} = \mathbf{0} \tag{8.38}$$

(Exercise 8.7)

where $\lambda = \eta_v^2$. The optimal scores (weights) are then contained in the latent vector $\mathbf{v}$ which is associated with the first nontrivial root of $(1/K)\mathbf{Y}^*\mathbf{Y}^{*T}$, where $\eta_u^2 = \eta_v^2 = \lambda$ (Section 5.4). Guttman's theory of scaling thus leads to a standard principal components analysis of direction cosines, based on the matrix of dummy variables rather than on the contingency table. As for the general method described in Section 8.3, the normal equations (Eqs. 8.35 and 8.38) also possess trivial solutions which can be eliminated by a deflation factor. Thus the matrix

$$\mathbf{X}^T\mathbf{X} = \mathbf{Y}^T\mathbf{Y} - \frac{1}{K}\mathbf{D}\mathbf{1}_K\mathbf{1}_K^T\mathbf{D} \tag{8.39}$$

contains roots $0 \leq \lambda < 1$ and the deflation factor $(1/K)\mathbf{D}\mathbf{1}_K\mathbf{1}_K^T\mathbf{D}$ is equivalent to the matrix of sample means encountered in the continuous case (Section 2.4). Since the matrix $\mathbf{X}^T\mathbf{X}$ has rank $K - k$ it can possess at most this number of nonzero latent roots.

Several extensions of the method are possible. Thus the matrix of dummy variables $\mathbf{Y}$ can be viewed as a contingency table (Shiba, 1965a) or the dummy codes may be replaced by ranks (Guttman, 1950; Shiba, 1965b). It turns out however that the procedure has a dual method of analysis based on contingency tables, which historically has preceded Guttman's principal components decomposition of dummy variables (Section 8.4.3). In fact there has been some debate, particularly in the psychological literature, as to whether a factor analysis (principal components analysis) is suitable for dummy variables and whether the analysis should be considered in terms of so-called scalogram analysis (Guttman, 1953; Burt, 1953). Certainly the surface objectives of the two approaches appear to be different—Burt (1950) seeks to factor analyze a set of dummy variables using principal components, whereas Guttman's (1941) aim is to construct measurement scales of the latent traits. This would seem to correspond to the factor and the multidimensional scaling models, respectively, since the objective of unrestricted factor analysis is exploratory and that of multidimensional scaling is to provide measurements of unobservable phenomena or traits. Although from the application-oriented viewpoint the distinction may be of some importance, from a statistical or mathematical viewpoint the models are identical and are based on a principal components analysis of nominal variables.

8.4.2 Maximizing Canonical Correlation

Consider the multivariate multinomial table of 0-1 dummy variables (Table 8.15). Since each category can be viewed as a distinct set of variables, the matrix $\mathbf{Y}$ can also be viewed as representing K categories. First consider the case for $K = 2$, a situation which is formally equivalent to Hotelling's (1936b) canonical correlation model for continuous variables (Section 5.5.1). Using sample covariances the canonical correlation model can be written as

$$(\mathbf{S}_{11}^{-1}\mathbf{S}_{12}\mathbf{S}_{22}^{-1}\mathbf{S}_{21} - \lambda_i^2)\mathbf{A}_i = \mathbf{0}$$

$$(\mathbf{S}_{22}^{-1}\mathbf{S}_{21}\mathbf{S}_{11}^{-1}\mathbf{S}_{12} - \mu_i^2)\mathbf{B}_i = \mathbf{0} \tag{8.40}$$

with standardization $\mathbf{A}_i^{\mathrm{T}}\mathbf{S}_{11}\mathbf{A}_i = 1$ and $\mathbf{B}_i^{\mathrm{T}}\mathbf{S}_{22}\mathbf{B}_i = 1$ where

$$\mathbf{S} = \begin{bmatrix} \mathbf{S}_{11} & \mathbf{S}_{12} \\ \hline \mathbf{S}_{21} & \mathbf{S}_{22} \end{bmatrix} \tag{8.41}$$

is the sample covariance matrix in partitioned form. To see how $\mathbf{S}$ is related to the sums of squares and products matrix $\mathbf{Y}^{\mathrm{T}}\mathbf{Y}$ (Section 8.3.3) we have, for

$K = 2$,

$$\mathbf{Y}^{\mathrm{T}}\mathbf{Y} = \left[\begin{array}{cccc|cccc} N_{11} & & & & n_{11} & n_{12} & \cdots & n_{1c} \\ & N_{12} & & \mathbf{0} & n_{21} & n_{22} & \cdots & n_{2c} \\ & & \ddots & \vdots & \vdots & \vdots & \ddots & \\ & \mathbf{0} & & N_{1l_1} & n_{r1} & n_{r2} & \cdots & n_{rc} \\ \hline n_{11} & n_{21} & \cdots & n_{r1} & N_{21} & & & \\ n_{12} & n_{22} & \cdots & n_{r2} & & N_{22} & & \mathbf{0} \\ \vdots & \vdots & & \vdots & & \mathbf{0} & \ddots & \\ n_{1c} & n_{2c} & \cdots & n_{rc} & & & & N_{2l_2} \end{array}\right]$$

$$= \left[\begin{array}{c|c} \mathbf{D}_r & \mathbf{N} \\ \hline \mathbf{N}^{\mathrm{T}} & \mathbf{D}_c \end{array}\right] \tag{8.42}$$

where $r = l_1$, $c = l_2$, and $\mathbf{N}$, $\mathbf{N}^{\mathrm{T}}$ consist of joint occurrences, $\mathbf{D}_r$ is the diagonal matrix of totals for $\mathbf{Y}_1$, and $\mathbf{D}_c$ is the diagonal matrix of totals for $\mathbf{Y}_2$. Let $\mathbf{r}$ and $\mathbf{c}$ denote the mean vectors of columns of $\mathbf{Y}_1$ and $\mathbf{Y}_2$, respectively, that is, $\mathbf{r}$ and $\mathbf{c}$ consist of relative frequencies based on the columns of $\mathbf{Y}_1$ and $\mathbf{Y}_2$. Then

$$\mathbf{S}_{11} = \mathbf{Y}_1^{\mathrm{T}}\mathbf{Y}_1 - \mathbf{r}\mathbf{r}^{\mathrm{T}}, \qquad \mathbf{S}_{22} = \mathbf{Y}_2^{\mathrm{T}}\mathbf{Y}_2 - \mathbf{c}\mathbf{c}^{\mathrm{T}}$$

$$= \mathbf{D}_r - \mathbf{r}\mathbf{r}^{\mathrm{T}} \qquad\qquad = \mathbf{D}_c - \mathbf{c}\mathbf{c}^{\mathrm{T}}$$

$$\mathbf{S}_{12} = \mathbf{Y}_1^{\mathrm{T}}\mathbf{Y}_2 - \mathbf{r}\mathbf{c}^{\mathrm{T}}$$

$$= \mathbf{N} - \mathbf{r}\mathbf{c}^{\mathrm{T}} \tag{8.43}$$

where since $\mathbf{S}_{11}\mathbf{1} = \mathbf{S}_{22}\mathbf{1} = \mathbf{0}$, the matrices $\mathbf{S}_{11}$ and $\mathbf{S}_{22}$ are singular of rank $(l_1 - 1)$ and $(l_2 - 1)$, respectively, owing to the mutual exclusiveness of the categories of $\mathbf{Y}_1$ and $\mathbf{Y}_2$. A standard canonical correlation analysis however breaks down in the presence of singularity. Several options exist to remedy the dilemma. First, we can employ unique generalized inverses of $\mathbf{S}_{11}$ and $\mathbf{S}_{22}$ (e.g., see McKeon, 1966). Second, we can omit a column for each nominal variable $\mathbf{Y}_i$, for example, the last. A similar expedient however is to carry out a classical canonical correlation analysis using $\mathbf{Y}_1^{\mathrm{T}}\mathbf{Y}_1 = \mathbf{D}_r$, $\mathbf{Y}_2^{\mathrm{T}}\mathbf{Y}_2 = \mathbf{D}_c$, and $\mathbf{Y}_1^{\mathrm{T}}\mathbf{Y}_2$, that is, without centering the data about the means. The analysis again yields a trivial maximal solution associated with the canonical correlation $\rho^2 = \lambda^2 = 1$, after which the remaining canonical solutions are equivalent to those centered about the means. In practice the solutions are obtained using the matrix of direction cosines. The normal equations can then be written as

$$(\mathbf{D}_r^{-1}\mathbf{N}\mathbf{D}_c^{-1}\mathbf{N}^\mathrm{T} - \lambda_i^2)\mathbf{A}_i = \mathbf{0} \tag{8.44a}$$

$$(\mathbf{D}^{-1}\mathbf{N}^\mathrm{T}\mathbf{D}_r^{-1}\mathbf{N} - \mu_i^2)\mathbf{B}_i = \mathbf{0} \tag{8.44b}$$

with standardization

$$\mathbf{A}_i^\mathrm{T}\mathbf{D}_r\mathbf{A}_i = \mathbf{B}_i^\mathrm{T}\mathbf{D}_c\mathbf{B}_i = 1 \tag{8.44c}$$

As for the general canonical correlation model of Section 5.5.1, it suffices to solve only one of the equations given above.

For $k > 2$ the classical interpretation of canonical correlation as maximizing correlation between two linear combinations (bilinear form) breaks down. This is because correlation is inherently a binary concept and cannot be generalized to more than two variables or vector spaces. Generalizations that seek objectives similar to classical canonical correlation analysis however are possible (McKeon, 1966; Carroll, 1968; Tomassone and Lebart, 1980). Consider k data matrices $\mathbf{Y}_1, \mathbf{Y}_2, \dots, \mathbf{Y}_k$ as in Table 8.15 where the column means are set to zero. We seek an n-component vector $\mathbf{Z}$ and k transformation vectors $\mathbf{A}_j$ ($j = 1, 2, \dots, k$) each with l_j components such that correlation between $\mathbf{Z}$ and $\mathbf{A}_1\mathbf{Y}_1 + \mathbf{A}_2\mathbf{Y}_2 + \cdots + \mathbf{A}_k\mathbf{Y}_k$ is maximized. Although for categorical data this is equivalent to a principal components analysis of multinomial multivariate data (Table 8.15), the comparison is nevertheless interesting since it indicates the inherent limitations of classical canonical correlation analysis. The principal components model however suffers from no such difficulties, and this enables us to generalize to analyses of multivariate multinomial dummy variables and multiple contingency tables. As for principal components, the canonical variates can be rotated to alternative positions (Cliff and Krus, 1976). For further reading concerning canonical correlation of discrete data see also Nishisato (1980) and Greenacre (1984).

8.4.3 Two-Way Contingency Tables: Optimal Scoring

As was seen above an alternative method of developing the topic of a factor analysis of nominal categorical data is through the concept of a two-way contingency table of counts, that is, through the submatrix $\mathbf{N}$ of $\mathbf{Y}^\mathrm{T}\mathbf{Y}$ (Eq. 8.42). The advantage of such an approach is that it obviates the storage and manipulation of large sparse dummy variable matrices. Historically, this is also the original approach, apparently from Hirschfeld (1935; later known as H.O. Hartley), who considered the problem of transforming a discrete bivariate distribution such that regressions between the two variables are linear and yield maximal correlation. By maximizing the correlation ratio he obtained the maximal squared correlation coefficient as the largest latent root of Eq. (8.44), that is, the largest squared canonical correlation between two sets of discrete multinomial categories. In this respect Hirschfeld's work precedes that of Hotelling's paper (1936) on canonical correlation theory, as well as Fisher's (1940) "method of scoring" for discrete variables (see also Maung, 1941) and Guttman's (1941) scalogram analysis. Indeed, to this day the approach seems to be relatively unknown outside of the statistical literature and still continues to elicit publications on the topic (e.g., see Kaiser and Cerny, 1980).

For ease of interpretation we modify our notation to conform to that of contingency table analysis. Consider matrix $\mathbf{N}$ of Eq. (8.42) but in the form of an asymmetric contingency table (Table 8.20). The columns correspond to a nominal (multinomial) variable $\mathbf{A}$ with c categories, and the rows represent a nominal (multinomial) variable $\mathbf{B}$ with r categories. Also $n.. = n$ is the total sample size, and the subtotals $n_{i.}$ and $n_{.j}$ denote marginal frequencies of the rows and columns respectively. Although the rows and columns of $\mathbf{N}$ represent purely nominal categories, they can also consist of ordered categories of the type considered in Section 8.2, particularly when the rank orders are considered to represent unobserved continuities. In any case, although the categories represent observable (but possibly arbitrary) groupings, it is at times not unrealistic to assume the existence of an underlying bivariate continuum $f(x, y)$, which cannot be observed in practice but which may hopefully be estimated from the frequency counts. The purpose for estimating the continuum $f(x, y)$ may be threefold: (1) to remove the artificiality of evenly spaced categories; (2) to estimate the correlation coefficients (regressions) between the variables A and B; (3) to obtain a joint mapping of the two sets of categories in a joint space of lower dimension (Section 5.4.4). The objectives are therefore similar to those of the preceding sections, and the analysis may again be understood either in term of canonical correlation or else in terms of the principal components model. As already pointed out however the main disadvantage of the former

Table 8.20 The Submatrix N in the form of a Two-Way Contingency Table Where y_i and x_j are the Unobserved Scores

			Variable A				
			y_1	y_2	$\cdots$	y_c	
			A_1	A_2	$\cdots$	A_c	Totals
	x_1	B_1	n_{11}	n_{12}	$\cdots$	n_{1c}	$n_{1.}$
	x_2	B_2	n_{21}	n_{22}	$\cdots$	n_{2c}	$n_{2.}$
Variable B	$\vdots$	$\vdots$	$\vdots$	$\vdots$		$\vdots$	$\vdots$
	x_r	B_r	n_{r1}	n_{r2}	$\cdots$	n_{rc}	$n_{r.}$
	Totals		$n_{.1}$	$n_{.2}$	$\cdots$	$n_{.c}$	$n_{..}$

viewpoint is that the canonical correlation model cannot be generalized for $k > 2$, that is, for more than two sets of variables.

Formally, the objective is to estimate the unobserved values x_1, $x_2, \ldots, x_r$ and $y_1, y_2, \ldots, y_c$ of x and y respectively (Table 8.20) such that correlation between them is maximized. This is the prbolem of finding "optimum scores" posed by Hirschfeld (1935), Fisher (1940), and Maung (1941). Assuming large n, the scores are chosen so that they are in standard form—zero means, that is,

$$\frac{1}{n} \sum_{i=1}^{r} n_{i.} x_i = \frac{1}{n} \sum_{j=1}^{c} n_{.j} y_j = 0 \tag{8.45}$$

and unit variance, that is,

$$\mathrm{var}(x) = \frac{1}{n} \sum_{i=1}^{r} n_{i.} x_i^2 = \mathrm{var}(y) = \frac{1}{n} \sum_{j=1}^{c} n_{.j} y_j^2 = 1 \tag{8.46}$$

The correlation between x and y can then be expressed as

$$\rho = \frac{1}{n} \sum_{i=1}^{r} \sum_{j=1}^{c} n_{ij} x_i y_j \tag{8.47}$$

which corresponds to the usual correlation coefficient for grouped data. In matrix form we have

$$\rho = \mathbf{X}^\mathrm{T} \mathbf{N} \mathbf{Y} = [x_1, x_2, \ldots, x_r] \begin{bmatrix} n_{11} & n_{12} & \cdots & n_{1c} \\ n_{21} & n_{22} & \cdots & n_{2c} \\ \vdots & \vdots & & \vdots \\ n_{r1} & n_{r2} & \cdots & n_{rc} \end{bmatrix} \begin{bmatrix} y_1 \\ y_2 \\ \vdots \\ y_c \end{bmatrix} \tag{8.48}$$

a bilinear form in the unknowns $\mathbf{X}$ and $\mathbf{Y}$. Also, using Eq. (8.46) the sums of squares can be expressed as

$$n = \sum_{i=1}^{r} n_{i.} x_i^2, \qquad n = \sum_{j=1}^{c} n_{.j} y_j^2$$

$$= \mathbf{X}^\mathrm{T} \mathbf{N}_i \mathbf{X} \qquad\qquad = \mathbf{Y}^\mathrm{T} \mathbf{N}_j \mathbf{Y} \tag{8.49}$$

where

$$\mathbf{N}_i = \mathrm{diag}(n_{1.}, n_{2.}, \ldots, n_{r.}) \tag{8.50b}$$

$$\mathbf{N}_j = \mathrm{diag}(n_{.1}, n_{.2}, \ldots, n_{.c}) \tag{8.50b}$$

We wish to maximize the correlation (Eq. 8.48) subject to the constraints (Eq. 8.49). Using Lagrange multipliers we have

$$\phi = \mathbf{X}^\mathrm{T} \mathbf{N} \mathbf{Y} - \lambda(\mathbf{X}^\mathrm{T} \mathbf{N}_i \mathbf{X} - n) - \mu(\mathbf{Y}^\mathrm{T} \mathbf{N}_j \mathbf{Y} - n) \tag{8.51}$$

and differentiating with respect to $\mathbf{X}$ and $\mathbf{Y}$ and setting to zero yields the normal equations

$$\frac{\partial \phi}{\partial \mathbf{X}} = \mathbf{NY} - \lambda \mathbf{N}_i \mathbf{X} = \mathbf{0} \tag{8.52a}$$

$$\frac{\partial \phi}{\partial \mathbf{Y}} = \mathbf{N}^{\mathrm{T}} \mathbf{X} - \mu \mathbf{N}_j \mathbf{Y} = \mathbf{0} \tag{8.52b}$$

Premultiplying by $\mathbf{N}_i^{-1/2}$ and $\mathbf{N}_j^{-1/2}$ and solving for $\mathbf{X}$ and $\mathbf{Y}$ yields the two systems

$$(\mathbf{N}_i^{-1/2} \mathbf{NN}_j^{-1/2}) \mathbf{N}_j^{1/2} \mathbf{Y} = \lambda \mathbf{N}_i^{1/2} \mathbf{X} \tag{8.53a}$$

$$(\mathbf{N}_j^{-1/2} \mathbf{N}^{\mathrm{T}} \mathbf{N}_i^{-1/2}) \mathbf{N}_i^{1/2} \mathbf{X} = \mu \mathbf{N}_j^{1/2} \mathbf{Y} \tag{8.53b}$$

where λ and μ are latent roots of $\mathbf{N}$ and $\mathbf{N}^{\mathrm{T}}$ respectively and $\mathbf{N}_j^{1/2} \mathbf{Y}$ and $\mathbf{N}_i^{1/2} \mathbf{X}$ are right and left latent vectors of $\mathbf{N}$. Setting $\lambda^2 = \mu^2 = \rho^2$ and substituting $\mathbf{N}_i^{1/2} \mathbf{X} = (1/\lambda)(\mathbf{N}_i^{1/2} \mathbf{NN}_j^{-1/2}) \mathbf{N}_j^{1/2} \mathbf{Y}$ into Eq. (8.53b) yields

$$\frac{1}{\rho} (\mathbf{N}_j^{-1/2} \mathbf{N}^{\mathrm{T}} \mathbf{N}_i^{-1/2})(\mathbf{N}_i^{-1/2} \mathbf{NN}_j^{-1/2}) \mathbf{N}_j^{1/2} \mathbf{Y} = \rho \mathbf{N}_j^{1/2} \mathbf{Y} \tag{8.54}$$

Let $\mathbf{M} = \mathbf{N}_i^{-1/2} \mathbf{NN}_j^{-1/2}$. Then

$$(\mathbf{M}^{\mathrm{T}} \mathbf{M}) \mathbf{N}_j^{1/2} \mathbf{Y} = \rho^2 \mathbf{N}_j^{1/2} \mathbf{Y} \tag{8.55a}$$

Similarly it can be shown that

$$(\mathbf{MM}^{\mathrm{T}}) \mathbf{N}_i^{1/2} \mathbf{X} = \rho^2 \mathbf{N}_i^{1/2} \mathbf{X} \tag{8.55b}$$

(Exercise 8.8) where $\mathbf{M}^{\mathrm{T}} \mathbf{M}$ and $\mathbf{MM}^{\mathrm{T}}$ are positive semidefinite and symmetric. Equations (8.55a and b) can be viewed as representing a principal components analysis of $\mathbf{M}^{\mathrm{T}} \mathbf{M}$ and $\mathbf{MM}^{\mathrm{T}}$, where $\mathbf{M}$ is the matrix of direction cosines between the two sets of categories and $\mathbf{N}_j^{1/2} \mathbf{Y}$ and $\mathbf{N}_i^{1/2} \mathbf{X}$ are the latent vectors. Both sets are therefore formally equivalent to the "loadings" and "scores" of a continuous data matrix (Section 3.4). Alternatively, Eq. (8.54) can be expressed in canonical correlation format. We have

$$(\mathbf{N}_j^{-1/2} \mathbf{N}^{\mathrm{T}} \mathbf{N}_i^{-1/2} \mathbf{N}_i^{-1/2} \mathbf{NN}_j^{-1/2}) \mathbf{N}_j^{1/2} \mathbf{Y} = \rho^2 \mathbf{N}_j^{1/2} \mathbf{Y}$$

and premultiplying by $\mathbf{N}_j^{-1/2}$ and simplifying yields

$$(\mathbf{N}_j^{-1} \mathbf{N}^{\mathrm{T}} \mathbf{N}_i^{-1} \mathbf{N}) \mathbf{Y} = \rho^2 \mathbf{Y} \tag{8.56a}$$

Also, using Eq. (8.55b) we have

$$(\mathbf{N}_i^{-1} \mathbf{NN}_j^{-1} \mathbf{N}^{\mathrm{T}}) \mathbf{X} = \rho^2 \mathbf{X} \tag{8.56b}$$

which are the same as the canonical correlation equations (Eqs. 8.44a and b). The latent vectors are also usually standardized as $\mathbf{Y}^T\mathbf{N}_j\mathbf{Y} = \mathbf{I}$ and $\mathbf{X}^T\mathbf{N}_i\mathbf{X} = \mathbf{I}$. The analysis of a two-way contingency table is therefore equivalent to a principal components analysis of $\mathbf{M}^T\mathbf{M}$ with possible differences in the scaling of the latent vectors. Again, it must be kept in mind that Eqs. (8.56a and b) cannot be generalized to higher level tables, so that when $k > 2$ the analysis must be conducted along the lines of the principal components model. Also, since the rank of $\mathbf{M}^T\mathbf{M}$ and $\mathbf{M}\mathbf{M}^T$ is equal to $\min(r, c)$, the matrix used in the analysis usually depends on the relative magnitudes of r and c since it is easier to work with matrices of full rank. The contingency table analysis also has the trivial solution $\rho^2 = 1$.

Equations 8.56 maximize correlation between two sets of categories and form a useful tool for an optimal exploratory analysis of nominal and rank order variables. It may be asked, however, to what extent is the analysis distribution free. The problem has been considered by Lancaster (1966; see also Kendall and Stuart, 1979; Naouri, 1970; Hamdan, 1970) in the context of constructing bivariate normality from grouped sample counts n_{ij} by employing Hermite–Chebyshev polynomial expansions. The procedure is analogous to the use of Taylor series expansions of a probability function to obtain the moments of a distribution. Hermite–Chebyshev polynomials however turn out to be particularly useful for the purpose of studying bivariate normality. Let $f(x, y)$ represent the standard bivariate normal distribution with correlation $0 < \rho < 1$ and consider transformations $x' = f_1(x)$ and $y' = f_2(y)$ such that

$$E(x')^2 = (2\pi)^{-1/2} \int_{-\infty}^{\infty} (x') \exp(-1/2(x')^2)\, dx$$

$$E(y')^2 = (2\pi)^{-1/2} \int_{-\infty}^{\infty} (y') \exp(-1/2(y')^2)\, dy \qquad (8.57)$$

Then we can write

$$x' = a_0 + \sum_{i=1}^{\infty} a_i H_i(x), \qquad y' = b_0 + \sum_{i=1}^{\infty} b_i H_i(y) \qquad (8.58)$$

such that $\sum_{i=1}^{\infty} a_i^2 = \sum_{i=1}^{\infty} b_i^2 = 1$, where we can set $a_0 = b_0 = 0$ because of the independence of ρ to change of origin (scale). The functions $H_i(x)$ and $H_i(y)$ are known as Hermite–Chebyshev polynomials and have the property that for any two integers m and n ($m \leq n$)

$$\int_{-\infty}^{\infty} f_1(x)H_n(x)H_m(x)\, dx = \begin{cases} n! & n = m \\ 0 & n \neq m \end{cases} \qquad (8.59)$$

and similarly for $f_2(y)$. The Hermite–Chebyshev polynomials are thus orthogonal over the (normal) weighting functions $f_1(x)$ and $f_2(y)$. Since the polynomials have the generating function $\exp(sx - 1/2s^2)$ and since for the bivariate normal

$$E\left[\exp\left(sx - \frac{s^2}{2} + ty - \frac{t^2}{2}\right)\right] = \exp(st\rho)$$

we have

$$\int_{-\infty}^{\infty} \int_{-\infty}^{\infty} f(x, y) H_n(x) H_m(y)\, dx\, dy = \begin{cases} \rho & n = m \\ 0 & n \neq m \end{cases} \tag{8.60}$$

Using Eq. (8.58) it can then be shown (Exercise 8.9) that the correlation between x' and y' can be expressed as

$$\operatorname{corr}(x', y') = \sum_{i=1}^{\infty} a_i b_i \rho^i \tag{8.61}$$

Hermite–Chebyshev polynomials can also be used to establish the following theorems (Lancaster, 1957; see also Lancaster, 1969).

THEOREM 8.1. Let x and y be bivariate normal $f(x, y)$ with correlation coefficient ρ. If a transformation $x' = f_1(x)$ and $y' = f_2(x)$ is carried out such that $E(x')^2$ and $E(y')^2$ are finite, then the correlation between the transformed variables x' and y' cannot exceed $|\rho|$.

PROOF. From Eq. (8.61) the correlation between x' and y' is $\sum_{i=1}^{\infty} a_i b_i \rho^i$, which is less than $|\rho|$ unless $a_1 = b_1 = 1$, in which case the remaining a_i and b_i must vanish as a result of Eq. (8.58). Thus $\operatorname{corr}(x', y') \leq |\rho|$ and correlation between the transformed variables x' and y' cannot exceed in magnitude the original correlation between the untransformed variables x and y.

THEOREM 8.2. The values to be assigned to the canonical variates are $H_i(x)$ and $H_j(y)$, where the canonical correlations are powers of ρ, in descending order.

PROOF. From Theorem 8.1 the maximum correlation corresponds to the transformations $H_1(x) = x' = x$ and $H_1(y) = y' = y$ using the definition of Hermite–Chebyshev polynomials. Consider a second pair of standardized transformations x'' and y'' which are uncorrelated with the pair x' and y', that is,

$$E(x'') = E(y'') = 0, \qquad E(x', x'') = E(y', y'') = 0,$$

$$E(x'')^2 = E(y'')^2 = 1$$

We can write

$$x'' = \sum_{i=1}^{\infty} c_i H_i(x)$$

$$y'' = \sum_{i=1}^{\infty} d_i H_i(y) \tag{8.62}$$

where $\sum_{i=1}^{\infty} c_i^2 = \sum_{i=1}^{\infty} d_i^2 = 1$. Since $E(x', x'') = 0$, we have

$$E(x', x'') = E[c_1 x' H_1(x) + c_2 x' H_2(x) + c_3 x' H_3(x) + \cdots]$$

$$= E[c_1 H_1^2(x) + c_2 H_1(x)H_2(x) + c_3 H_1(x)H_3(x) + \cdots]$$

$$= c_1 E[H_1^2(x)] + c_2 E[H_1(x)H_2(x)] + c_3 E[H_1(x)H_3(x)] + \cdots$$

$$= c_1$$

so that $c_1 = 0$ since the $H_i(x)$ form an orthonormal system. Similarly it can be shown that $d_1 = 0$. The expansions (Eq. 8.62) can thus be expressed as

$$x'' = \sum_{i=2}^{\infty} c_i H_i(x)$$

$$y'' = \sum_{i=2}^{\infty} d_i H_i(y) \tag{8.63}$$

and following Eq. (8.61) we have

$$\text{corr}(x'', y'') = \sum_{i=2}^{\infty} c_i d_i \rho^i \tag{8.64}$$

which is again maximized for $c_2^2 = d_2^2 = 1$, with the remaining c_i and d_i zero. The maximal correlation is thus for $i = 2$, that is, for ρ^2. Continuing in the same vein we obtain transformed pairs (x''', y''') and (x^{iv}, y^{iv}), and so on with maximal correlations $|\rho|^3$, ρ^4, ...

Theorem 8.1 was previously proved by Maung (1941) using a more lengthy method. Both Theorems 8.1 and 8.2 are of some theoretical interest since they indicate that optimal canonical scores are not strictly distribution-free, a point which is not always realized in practice (see Benzécri, 1970; David et al., 1977). Several other conclusions emerge from the theorems.

1. Although Theorems 8.1 and 8.2 pertain to any data, in general they can also be used for grouped two-way contingency tables where the categories are either ordinal or nominal. If a choice of variables can be made such that a bivariate normal distribution results from a contingency table then ρ^1, ρ^2, $|\rho|^3$, ρ^4, . . . are the latent roots (canonical correlations) and the canonical variables are the orthonormal Hermite–Chebyshev

polynomials

$$(x^{(i)}, y^{(i)}) = [H_i(x), H_i(y)] \tag{8.65}$$

2. The roots of (Eq. 8.56) are powers of the maximal correlation $0 < \rho < 1$, any difference being due to sampling error and the method of grouping.

3. In seeking optimal scoring systems for the rows and columns that maximize correlation we are implicitly attempting to construct a bivariate normal distribution by transforming the margins of the table to univariate normality. It is not sufficient however to transform the marginal categories to univariate normality in order to achieve bivariate normality. The contingency table counts must represent a bivariate normal distribution if correlation is to be maximized.

4. From general latent vector theory, if $\rho_1, \rho_2, \ldots, \rho_l$ are the roots of the system (Eq. 8.56), then the (i, j)th entry of the table can be expressed as

$$n_{ij} = \frac{n_{i.}n_{.j}}{n}\left[1 + \sum_{i=1}^{m-1} \rho_i x_i y_i\right] \tag{8.66}$$

where x_i and y_i are elements of the vectors x and y. The series is analogous to a principal component expansion of continuous data and can therefore be used to obtain predicted counts $\hat{n}_{ij}$. As $m \to \infty$ and the groupings become finer, the expansion (Eq. 8.66) approaches, in the limit, the well-known tetrachoric series

$$f(x, y) = (2\pi)^{-1} \exp[-1/2(x^2 + y^2)]\left[1 + \sum_{i=1}^{\infty} H_i(x)H_i(y)\rho^i\right] \tag{8.67}$$

used by Pearson (1900), but originally from Mehler (1866).

5. The Chebyshev–Hermite polynomials provide a basis for extending the theory to continuous distributions. When a contingency table reflects an underlying bivariate normal distribution with correlation coefficient ρ, it is seen that the latent roots and latent vectors are given by ρ^r and $H_r(x_1)$ and $H_r(x_2)$ respectively. For a p-variate normal distribution with all correlations equal, it can be shown that the first two nontrivial solutions for 0–1 dichotomous data are

$$f_1(x) = a(x_1 + x_2 + \cdots + x_p)$$

$$f_2(x) = b[H_2(x_1) + H_2(x_2) + \cdots + H_2(x_p)]$$

The scores on the second axis are quadratic functions of the first axis, a phenomenon which at times leads to the so-called "horseshoe effect" considered in Section 9.3. Canonical analysis of an $(r \times c)$ table is closely related to the chi-squared statistic and certain other measures of association.

We have

$$\mathrm{tr}(\mathbf{M}^\mathrm{T}\mathbf{M}) = \mathrm{tr}(\mathbf{M}\mathbf{M}^\mathrm{T}) = \sum_{i=1}^{r}\sum_{j=1}^{c} \frac{n_{ij}^2}{n_{i.}n_{.j}} \tag{8.68}$$

since

$$\chi^2 = n \sum_{i=1}^{r}\sum_{j=1}^{c} \frac{[n_{ij} - 1/n(n_{i.}n_{.j})]^2}{n_{i.}n_{.j}}$$

$$= n\left[\sum_{i=1}^{r}\sum_{j=1}^{c} \frac{n_{ij}^2}{n_{i.}n_{.j}} - 1\right] \tag{8.69}$$

and since the latent roots of $\mathbf{M}^\mathrm{T}\mathbf{M}$ and $\mathbf{M}\mathbf{M}^\mathrm{T}$ are $1, \rho_1^2, \rho_2^2, \ldots, \rho_{m-1}^2$ we have from Eq. (8.66)

$$\chi^2 = n(\rho_1^2 + \rho_2^2 + \cdots + \rho_{m-1}^2) \tag{8.70}$$

for $0 < \rho_i < 1$. It is easy to show that the square of Cramer's measure of association is simply the mean squared canonical correlation. We have (see also Kendall and Stuart 1979, vol. 2)

$$C^2 = \frac{\chi^2}{n\min(r-1, c-1)}$$

$$= \frac{n\sum_{i=1}^{m-1}\rho_i^2}{n\min(r-1, c-1)}$$

$$= \frac{\sum_{i=1}^{m-1}\rho_i^2}{m-1} \tag{8.71}$$

which provides a more basic interpretation of correlation within an $(r \times c)$ contingency table. The following properties of C^2 are immediately evident from Eq. (8.71).

(a) C^2 is nonnegative and achieves its lower bound of zero if and only if the polytomies representing the rows and columns are stochastically independent. Thus $C^2 = 0$ if and only if $\rho_i = 0$ $(i = 1, 2, \ldots, m-1)$.

(b) C^2 cannot exceed unity since it equals the arithmetic mean of correlation coefficients, and clearly attains its upper bound $C^2 = 1$ when $\rho_i = 1$ $(i = 1, 2, \ldots, m-1)$.

Other global measures of association which may be used are the geometric mean of the canonical correlations and the so-called squared vector multiple correlation, discussed by Srikantan (1970).

We now turn to tests of significance of the latent roots (canonical correlations) issued from multinomial data. The area seems to be frought with peril, as witnessed by incorrect conjectures which at times are made concerning the distributional properties of the latent roots, particularly in the psychological/educational literature. We first turn to Eq. (8.70). Since the expansion represents a partition of the asymptotic chi-squared statistic it might be assumed that each individual component $n\lambda_i = n\rho_i$ ($i = 1, 2, \ldots, m - 1$) is also asymptotically distributed as chi-squared, with appropriately partitioned corresponding degrees of freedom. As Lancaster (1963) has demonstrated, however, such a conjecture is incorrect, and in practice tends to yield a larger value of $n\rho_1$ (smaller value of $n\rho_{m-1}$) than would be expected from the chi-squared distribution. As pointed out by Greenacre (1984), this generally leads to optimistic significance levels for ρ_1 and pessimistic levels for ρ_{m-1}. The use of the chi-squared statistic discussed above for testing significance of the canonical correlations however does not appear to be in wide use at present. A somewhat different chi-squared test on the other hand still continues to be in use (see Bock, 1960; Butler et al., 1963; Nishisato, 1980) although it also is incorrect for contingency tables. Thus owing to the formal similarity of contingency table analysis to canonical correlation Bock (1960) has suggested that Bartlett's (1951b) approximate large sample chi-squared likelihood ratio test (Section 4.3) be used. The test however requires the assumption of multivariate normality, a condition which generally is not met by the data. Fisher's F test for equality of group means is also at times employed, although in the present context it is incorrect as well. For further details see Butler et al. (1963) and Nishisato (1980). Another test inspired by multivariate normal canonical correlation theory (see Lawley, 1959) is based on the assumption that the variances of $\lambda_i = \rho_i$ can be approximated by the expression $(1/n)(1 - \rho_i^2)^2$ (Kaiser and Cerny, 1980; Kshirsagar, 1972). The approximation however appears to be unsatisfactory and the test should probably not be used for multinomial data (Lebart et al., 1977).

A correct theory proceeds along the lines of Lebart (1976), which employs the multivariate normal asymptotic approximation to the multinomial distribution of the rc entries of the contingency table. The matrix to be diagonalized in the analysis has approximately the same latent roots as a Wishart matrix of some lower order. Thus when the rows and columns are independent, that is, the ρ_i are identically zero, the asymptotic distribution of the $n\rho_i^2$ is the same as the distribution of the latent roots of a central Wishart matrix of order $c - 1$ with $r - 1$ degrees of freedom, assuming, without loss of generality, that $c \leq r$. When the rows and columns are dependent, an asymptotic theory has also been given by O'Neill (1978a,

1978b, 1980, 1981). As noted by Lebart (1976), the percentages of variance of a central Wishart matrix are distributed independently of their sum (trace), and when the usual chi-squared test does not reject the null hypothesis of independence, the major percentages of variance could still be significantly high. Conversely, when the null hypothesis is rejected, it may still be the case that the percentages of variance are not significant.

Example 8.9. An example from the field of dentistry of the use of optimal scoring of contingency tables is provided by Williams (1952), where particular interest lies in detecting the correlational structure between the rows and columns of Table 8.21. The first two nontrivial latent vectors are given by

Column scores y	Row scores x
.8397	-1.3880
.4819	-1.0571
-1.5779	.6016
-1.1378	.9971

where the largest canonical correlation is $\rho = -.5627$ ($\chi^2 = 44.35$), indicating a negative relationship between calcium intake and severity of condition. The scores are scaled to zero mean and weighted sum of squares are equal to $n = 135$, as is customary for this type of analysis. Note that the last score for x reverses an otherwise decreasing trend, indicating a threshold effect for the efficacy of calcium intake after .70 g/day. Merging the last two columns of Table 8.21 results in the correlation of $\rho = -.5581$ ($\chi^2 = 42.05$) with an insignificant second correlation of .0779. The new (recalculated)scores are then

Table 8.21   **Peridontal Condition of $n = 135$ Women as a Function of Calcium Intake: A = best; D = worst.**

Peridontal Condition	$-.40$	$.40-.55$	$.55-.70$	$.70-$	Total
A	5	3	10	11	29
B	4	5	8	6	23
C	26	11	3	6	46
D	23	11	1	2	37
Total	58	30	22	25	135

Source: Williams, 1952; reproduced with permission.

| Column scores | Row scores |
y	*x*
.8448	-1.4072
.4906	-1.0243
-1.3557	.5906
	1.0054

Canonical correlation analysis of contingency tables can therefore be used as an exploratory tool for the underlying correlational structure, as well as for grouping rows and columns. For recent work see Gilula (1986), who also discusses the relationship between canonical and structural models for contingency table analysis. $\square$

Example 8.10.   Another example is Maung's (1941) analysis of Tocher's (1908) eye and hair pigmentation data of Scottish school children. The latent roots and vectors are given in Table 8.23. The latent vectors are again standardized so that their weighted sums of squares equal *n*, the total sample size (Table 8.22). Three questions may be asked of the data: (1) What is (are) the principal underlying dimension(s) of the data? (2) What is (are) the correlation(s) among the dimensions? (3) How closely do the row and column categories cluster together? $\square$

The first underlying dimension (continuum) X_1 and Y_1 has a direct physical interpretation in terms of the amount of pigment present in the outer layer of the retina and in the hair, and thus presumably represents the genetic pigmentation factor. Since X_1 and Y_1 have a correlation of $\rho_1 = .38$, both hair and eye pigment tend to go together, although clearly not to an overwhelming extent. The second dimension X_2 and Y_2 seems to indicate a "lack of extremes" dimension since it relates positively to children with a medium degree of pigmentation, while the third dimension is difficult to interpret and seems to indicate random individual variation among the children. The analysis however is not free of the artificial quadratic

Table 8.22   Two-Way Classification of the Hair and Eye Color of School Children from the City of Aberdeen

Eye Color	Hair Color					
	Fair	Red	Medium	Dark	Black	Total
Blue	1368	170	1041	398	1	2978
Light	2577	474	2703	932	11	6697
Medium	1390	420	3826	1842	33	7511
Dark	454	255	1848	1506	112	5175
Total	5789	1319	9418	5678	157	22361

Source: Maung, 1941; reproduced with permission.

Table 8.23 Latent Roots and Vectors of the Contingency Table 8.22[a]

Categories Variables		Canonical Vectors		
		Y_1	Y_2	Y_3
Hair color	F	-1.3419	$-.9713$	$-.3288$
	R	-0.2933	$.0237$	3.7389
	M	-0.0038	1.1224	$-.1666$
	D	1.3643	$-.7922$	$-.3625$
	B	2.8278	-3.0607	3.8176
		X_1	X_2	X_3
Eye color	B	-1.1855	-1.1443	-1.9478
	L	$-.9042$	$-.2466$	1.2086
	M	$.2111$	1.3321	-1.3976
	D	1.5459	$-.9557$	$.1340$
Cannonical correlation		$\rho_1 = .3806$	$\rho_2 = .1393$	$\rho_3 = .0221$

[a] The latent vectors are standardized such that $\mathbf{X}_i^T \mathbf{D}_c \mathbf{X}_i = \mathbf{Y}_i^T \mathbf{D}_r \mathbf{Y}_i = 22361$.

Source: Maung, 1941; reproduced with permission.

"horseshoe" affect which may influence the interpretation of the canonical variates (Section 9.3).

The scaling of the latent vectors which is customary for canonical correlation of discrete data may be inappropriate since dimensions that account for different correlation (latent roots) are scaled equally. This is analogous to the situation in principal components analysis where the latent vectors are initially scaled to unit length (Sections 3.2–3.4). To account for the differences between the latent roots, however, the weighted sum of squares should be rescaled so that they equal their respective percentages of variance (see Table 8.24 and Fig. 8.9). Alternatively, they may be scaled such that their weighted sum of squares equal the latent roots, but this is not usually done in order to avoid excessively small fractions. The change in the scaling however does not seem to influence the interpretation, since it may be seen again that the first axis measures hair and eyes pigmentation while the second seems to represent a "lack of extremes" dimension. Most of the variation however is accounted for by the first canonical variable. For alternative analyses of the data see Becker and Clogg (1989) and references therein.

8.4.4 Extensions and Other Types of Discrete Data

Other forms of analysis of discrete data are also possible. A method closely related to contingency table analysis is to employ a multinomial-type covariance matrix with prior weights (probabilities) together with constraints (Healy and Goldstein, 1976). Let $\mathbf{X}$ be a vector of scores to be determined

Table 8.24 Rescaled Latent Vector Elements of Table 8.22 such that Their Weighted Sum of Squares $X^T N_i X = X^T D_r X$, $Y^T N_j$, $Y = Y^T D_c Y$ Equal the Percentage of Variance Accounted for[a]

Categories (Variables)		Canonical Vectors	
		Y_1	Y_2
Hair color	F	−.084	−.022
	R	−.018	+.0
	M	−.0	.026
	D	.085	−.018
	B	.178	−.070
		X_1	X_2
Eye color	B	−.074	−.026
	L	−.057	−.006
	M	.013	.031
	D	.097	−.022
Latent root		.14485	.01940
Variance		87.95%	11.78%

[a] X_3 is omitted since it represents random pigment variation among the children.

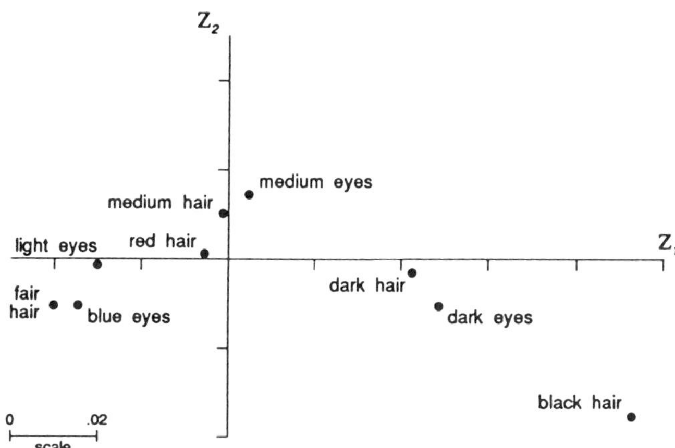

Figure 8.9 The first two canonical axes of Table 8.24 representing the canonical vectors of Table 8.22.

and let **A** be a symmetric matrix with diagonal elements $n_{ij} p_i (1 - p_i)$ and off-diagonal elements $-n_{ijkl} p_i p_k$ where n_{ij} and n_{ijkl} are counts representing the number of elements in level j of attribute i, and both level j of attribute i and level l of attribute k, respectively. The p_i are weights such that $\Sigma_i p_i = 1$ and can thus be viewed in terms of prior probabilities. Also let **Z** be a symmetric matrix with diagonal elements $n_{ij} p_i^2$ and off-diagonal elements $n_{ijkl} p_i p_k$ and let $\mathbf{S} = \mathbf{A} + \mathbf{Z}$. The objective is to minimize disagreement within elements (here subjects) across their respective attribute scores and their weighted average score, that is, to maximize internal consistency. Let λ and μ be Lagrange multipliers. Then we minimize $\mathbf{X}^T\mathbf{AX}$ subject to the constraints $\mathbf{X}^T\mathbf{ZX} = 1$ and $\mathbf{X}^T\mathbf{S1} = 0$, that is, we find $\mathbf{X}$, λ, and μ such that

$$\phi = \mathbf{X}^T\mathbf{AX} - \lambda(\mathbf{X}^T\mathbf{ZX} - 1) - \mu(\mathbf{X}^T\mathbf{S1}) \qquad (8.72)$$

is a minimum. Differentiating with respect to $\mathbf{X}$ and setting to zero yields

$$\frac{\partial \phi}{\partial \mathbf{X}} = 2\mathbf{AX} - 2\lambda\mathbf{ZX} - \mu\mathbf{S1} = 0$$

and multiplying on the left by $\mathbf{1}^T$ we obtain

$$(\mathbf{A} - \lambda\mathbf{Z})\mathbf{X} = 0 \qquad (8.73)$$

since $\mathbf{1}^T\mathbf{A} = \mathbf{1}^T\mathbf{SX} = 0$, which leads to $\mu\mathbf{1}^T\mathbf{S1} = 0$ and thus $\mu = 0$. Since the smallest root of Eq. (8.73) is zero (with latent vector of unities), the smallest latent root of Eq. (8.73) minimizes the criterion, and its latent vector $\mathbf{X}$ provides the optimum scores. Healy and Goldstein (1976) discuss other constraints which, however, do not seem to lead to an optimal solution (Greenacre, 1984). Recently extensions of contingency table canonical analysis have also been given by Gulula et. al. (1986), who consider relationships in terms of "structural" log-linear models (see also Daudin and Trecourt, 1980).

Finally, discrete data may also be observed in terms of integer counts themselves, that is, the $(n \times k)$ data matrix consists of counts n_{ij} which are observed for j random variables across i sample points. Data of this type differ from dummy variables in that they represent a unique absolute scale (Section 1.5.1). It may be analyzed by two methods, depending on the underlying objective. First, we can treat the $(n \times k)$ observed matrix of counts in the same way as any other data matrix and proceed with a principal components analysis. Second, since the entries are counts (or frequencies), we can consider the data matrix in terms of a $(n \times k)$ contingency table and proceed with a canonical analysis such as described in the previous section. The two approaches differ in terms of the weighting scheme employed. A contingency-type analysis is also more conducive to a joint mapping of the random variables and the sample points in a joint vector subspace, and is considered in the following section under the title of correspondence analysis (Section 5.4.4). The following example considers a principal components analysis of counts using human physiological data.

Example 8.11. An analysis of count data is presented by Siervogel et al. (1978) in relation to radial and ulnar digital ridge counts of human finger prints for a sample of $n = 441$ individuals (167 males and 196 females). Since

Table 8.25 Variance of the First 10 PCs of Radial and Ridge Counts for Male and Female Digits

	Males			Females		
Component	Latent root	Percentage of Variance	Cumulative Percentage of Variance	Latent root	Percentage of Variance	Cumulative Percentage of Variance
1	8.589	42.9	42.9	8.021	40.1	40.1
2	1.830	9.2	52.1	1.808	9.0	49.1
3	1.547	7.7	59.8	1.590	8.0	57.1
4	1.236	6.2	66.0	1.431	7.1	64.2
5	1.064	5.4	71.4	1.154	5.8	70.0
6	0.768	3.8	75.2	1.013	5.1	75.1
7	0.634	3.1	78.3	0.714	3.5	78.6
8	0.583	3.0	81.3	0.592	3.0	81.6
9	0.519	2.5	83.8	0.565	2.8	84.4
10	0.472	2.4	86.2	0.495	2.5	86.9

Source: Siervogel et al., 1978; reproduced with permission.

ridge counts are highly correlated between fingers, the authors carry out a principal components analysis of the matrix of counts, and using orthogonal (varimax) rotation obtain biologically meaningful results (Table 8.26) which are consistent for both males and females and for both hands. The results suggest the presence of three distinct digital regions: digit I; digits II and III; digits IV and V (Table 8.26) which could be the focus of further research, for example, for electronic storage of fingerprint information. At times count data are also analyzed in the form of relative frequencies expressed as proportions of the total sample size n (Westley and Lynch, 1962). A further modification is also provided by the use of Haberman's (1973) standardized residuals obtained by considering a matrix of counts as a contingency table. The residuals are defined by

$$e_{ij} = \frac{O_{ij} - E_{ij}}{E_{ij}} \tag{8.74}$$

where $O_{ij} = n_{ij}$ and $E_{ij} = E(n_{ij})$. When the row and column totals $n_{i.}$ and $n_{.j}$ are not very small compared to the overall sample size n, the variance of e_{ij} will not necessarily equal unity. To produce standardized unit variance residuals Haberman (1973) further divides Eq. (8.74) by

$$\sqrt{\text{var}(n_{ij})} = \left[1 - \frac{n_{i.}}{n}\right]^{1/2}\left[1 - \frac{n_{.j}}{n}\right]^{1/2}$$

to obtain the standardized residuals

$$d_{ij} = \frac{e_{ij}}{[\text{var}(n_{ij})]^{1/2}} \tag{8.75}$$

Table 8.26　Rotated Principal Component Loadings for the First Five Components

		Males Components					Females Components				
Digit	Hand	Z_1	Z_2	Z_3	Z_4	Z_5	Z_1	Z_2	Z_3	Z_4	Z_5
				Radical Ridge Counts							
I	R	0.47	0.60	−0.04	0.04	0.32	0.65	0.20	−0.04	0.10	0.45
	L	0.38	0.64	−0.02	0.10	0.34	0.67	0.09	−0.03	0.12	0.46
II	R	0.70	0.22	0.06	0.28	0.10	0.69	0.13	0.12	0.19	−0.02
	L	0.77	0.15	0.31	0.05	0.13	0.60	0.40	0.32	0.09	−0.05
III	R	0.76	0.30	0.18	0.13	0.11	0.63	0.45	0.29	0.05	0.03
	L	0.76	0.30	0.25	0.08	0.03	0.56	0.57	0.28	0.03	0.01
IV	R	0.44	0.65	0.31	0.15	0.01	0.21	0.79	0.21	0.05	0.06
	L	0.50	0.61	0.33	0.02	0.06	0.24	0.82	0.14	0.04	0.03
V	R	0.24	0.77	0.24	0.22	0.02	0.20	0.78	0.12	0.18	0.21
	L	0.08	0.78	0.27	0.14	0.05	0.21	0.76	0.05	0.31	0.20
				Vinar Ridge Count							
I	R	0.12	0.07	0.25	0.08	0.81	0.12	0.15	0.19	0.01	0.83
	L	0.10	0.16	0.17	0.21	0.81	0.06	0.13	0.11	0.06	0.84
II	R	0.14	0.34	0.74	0.00	0.22	0.02	0.61	0.51	−0.01	0.32
	L	0.11	0.29	0.69	0.08	0.18	0.01	0.44	0.56	0.07	0.18
III	R	0.25	0.04	0.78	0.10	0.16	0.18	0.13	0.79	0.05	0.01
	L	0.15	0.09	0.70	0.17	0.03	0.15	0.03	0.80	0.23	0.05
IV	R	0.43	0.20	0.50	0.47	0.10	0.18	0.33	0.62	0.23	0.23
	L	0.36	0.08	0.37	0.66	0.10	0.24	0.25	0.49	0.46	0.20
V	R	0.10	0.16	0.28	0.76	0.25	0.17	0.11	0.31	0.80	−0.05
	L	0.06	0.14	−0.09	0.89	0.06	0.09	0.13	0.07	0.91	0.10
Percent of variance accounted for		17.9	16.8	16.1	11.7	8.9	14.0	20.3	15.0	10.0	10.7

Source: Siervogel et al., 1978; reproduced with permission.

which can then be used in place of n_{ij} in a principal components analysis. Using the standardized residuals d_{ij} also helps to overcome the problem of varying row and column totals by placing relatively rarer categories on the same footing as the more abundant ones. This can be useful, for example, in uncovering whether plant species have certain environmental preferences. □

Example 8.12. Strahler (1978) uses the standardized residuals (Eq. 8.75) in place of the observed frequencies n_{ij} to carry out a Q-mode principal components analysis (Section 5.4.1) of a data matrix where the columns are plant species and the rows are environmental conditions. Since the objective is to obtain a correspondence between the two sets of variables, both the component loadings and the scores are considered (Table 8.27).
Unlike the canonical correlation-type analysis, however, it is not possible to

Table 8.27 Factor Loadings and Scores for Rotated Factors of Rock/Soil Type and Plant Species

Species	Factor				
	Z_1	Z_2	Z_3	Z_4	Z_5
Liquidambar styraciflua L.	.986	.034	−.033	−.118	−.105
Quercus phellos L.	.984	.057	.127	−.104	−.035
Clethra alnifolia L.	.983	−.004	−.135	−.094	−.087
Vaccinium atrococcum (Gray) Heller	.978	−.192	−.081	.027	−.019
Prunus virginiana L.	−.969	−.148	−.019	−.197	−.022
Pinus ridiga Mill.[a]	.966	−.024	−.220	−.089	−.103
Ilex opaca Ait.	.960	−.109	−.110	−.182	−.144
Carya ovata (Mill.) K. Koch	−.960	−.178	.148	−.118	−.107
Juglans nigra L.	−.948	.293	.014	−.121	−.007
Cornus florida L.	−.906	.058	.189	−.293	.231
Vaccinium corymbosum L	.887	.050	−.311	.328	−.076
Ouercus palustris Muenchh.	.878	.179	.333	−.211	−.202
Rhododendron maximum L[a]	.870	−.081	−.458	−.070	−.144
Carya Glabra (Mill.) Sweet	−.849	−.370	−.068	.258	.265
Carya Ovalis (Wang.) Sarg.	−.743	.313	−.306	−.500	.073
Carya tomentosa Nutt.	−.727	−.297	.091	−.179	.586
Quercus falcata Michx.	.727	−.095	−.593	.252	−.218
Rhododendron viscosum (L.) Torr.[a]	.690	−.139	−.687	−.043	−.178
Vitis spp.	−.673	−.103	.609	.103	.393
Liriodendron tulipifera L.	−.616	.153	.503	−.537	.235
Lindera benzoin (L.) Blume	−.569	.455	.462	−.504	.047
Corylus americana Walt.	−.151	.947	.027	−.283	.001
Fagus gradifolia L.	.165	.932	−.283	.133	−.085
Carpinus caroliniana Walt.	−.141	.928	.145	−.308	.047
Quercus Prinus L.	−.241	−.911	−.203	.006	.264
Kalmia latifolia L.	−.133	−.874	−.447	.117	−.066
Castanea pumila (L.) Mill.	−.398	−.873	−.054	−.205	.187
Castanea dentata (Marsh) Borkh.	−.449	−.850	−.095	−.186	.179
Quercus coccinea Muenchh.	.458	−.842	−.266	.049	.095
Lonicera japonica Thunb.	−.354	.791	.337	−.345	−.131
Fraxinus americana L.	−.554	.759	−.215	−.148	.219
Ulmus rubra Muhl.	−.240	.729	.095	−.353	.526
Carya cordiformis (Wang.) K. Koch[a]	.269	.628	.523	−.203	−.467
Sambucus canadensis L.	.095	−.014	.988	−.073	−.095
Rhus radicans L.	−.337	−.020	.920	.182	−.075
Parthenocissus quinquefolia (L.) Planch.	−.548	.274	.783	.016	.104
Vaccinium stamineum L.	.383	−.280	−.726	.498	−.014
Ilex verticellata (1.) Gray	.628	.185	.724	−.206	−.068
Betula nigra L.[a]	.650	.184	.723	−.094	.103
Sassafras albidum (Nutt.) Nees	−.119	−.465	−.668	.309	.478
Gaylussacia dumosa (Andr.) T. & G.	.473	−.533	−.630	−.229	−.205

Table 8.27 *(Continued)*

	Factor				
Species	$\mathbf{Z}_1$	$\mathbf{Z}_2$	$\mathbf{Z}_3$	$\mathbf{Z}_4$	$\mathbf{Z}_5$
Quercus stellata Wang[a]	.025	−.018	−.077	.997	−.002
Quercus marylandica Muench	.163	.213	.040	.957	−.016
Juniperus communis L.	−.269	−.087	.086	.955	−.008
Rhododendron nudiflorum (L.) Torr.	−.367	−.320	−.003	.871	.065
Vaccinium vaccillans Torr.	−.004	−.431	−.510	.737	.105
Pinus virginiana Mill.	.507	−.507	−.329	.600	−.131
Quercus velutina LaM.	−.422	−.523	.014	−.113	.732
	Variance Proportion				
Environment	.441	.290	.121	.115	.034
	Factor scores				
Acid igneous rocks	−.798	.955	.050	−.560	−1.517
Basic igneous rocks	−.348	1.083	−.594	−.390	1.539
Ultramafic rocks	−.173	.048	.057	2.031	−.078
Phyllites and schists	−1.116	−1.584	.306	−.465	.321
Sand and gravel	1.229	−.624	−1.416	−.267	−.438
Interbedded sand and silt	1.207	.121	1.595	−.349	.173

[a]Present in <5% of the sample plots.

plot both the loadings and the scores using common axes. A comparison of high magnitude loadings and scores however reveals which soil types (environmental condition) are conductive to which species. □

Another closely related version of the analysis of two-way discrete tables is Williams' (1978) Q-mode factor-analytic decomposition of families or sets of discrete distribution functions. Let $p_{ij} = p_i(\mathbf{X} = x_j)$ $(i = 1, 2, \ldots, m; \ j = 1, 2, \ldots, n)$ be a family of m probability functions defined on a discrete set $x_1, x_2, \ldots, x_n$. We wish to decompose the finite set of probabilities p_{ij} using a finite orthogonal basis as

$$p_{ij} = \sum_{k=1}^{n} a_{kj} z_{ik} \tag{8.75a}$$

where the a_{kj} are fixed and z_{ik} are random. The variance of the random terms can then be expressed as

$$\text{var}(z_{ik}) = \lambda_k = \sum_{i=1}^{m} f_i z_{ik}^2 - \left(\sum_{i=1}^{m} f_i z_{ik} \right)^2 \tag{8.75b}$$

where $\text{cov}(z_{ik}, z_{il}) = 0$ owing to the orthogonality of Eq. (8.75a) and where f_i are appropriate frequencies. Without loss of generality we can assume that Eq. (8.75a) represents a principal components decomposition of the probabilities p_{ij}. In matrix form we have (Section 3.4)

$$\mathbf{P} = \mathbf{Z}\mathbf{A}^{\mathrm{T}}$$ (8.75c)

where $\mathbf{P}$ and $\mathbf{Z}$ are $(m \times n)$ and $\mathbf{A}$ is $(n \times n)$. Note that because of the present needs of notation $\mathbf{P}$ represents the probability (data) matrix, not the latent vectors which are now given by the columns of $\mathbf{A}$. $\mathbf{Z}$ is therefore the matrix of loadings which correspond to the Q-mode analysis. Also, since Williams' (1978) development of the theory is in terms of Q-mode analysis, the columns of $\mathbf{Z}$ span the m-dimensional "variable" or probability function space while the latent vectors or columns of $\mathbf{A}$ span the n-dimensional space defined by the "observation" or the discrete set $x_1, x_2, \ldots, x_n$. It follows from Eq. (8.75c) that the kth vector $\mathbf{Z}_k$ can be expressed as

$$\mathbf{Z}_k = \mathbf{P}\mathbf{A}_k \qquad (k = 1, 2, \ldots, n)$$ (8.75d)

Also the variances (Eq. 8.75b) can be written in matrix form as

$$\lambda_k = \mathbf{Z}_k^{\mathrm{T}}[\mathrm{diag}(\mathbf{f}) - \mathbf{f}\mathbf{f}^{\mathrm{T}}]\mathbf{Z}_k$$

$$= \mathbf{A}_k^{\mathrm{T}}\mathbf{P}^{\mathrm{T}}[\mathrm{diag}(\mathbf{f}) - \mathbf{f}\mathbf{f}^{\mathrm{T}}]\mathbf{P}\mathbf{A}_k$$ (8.75e)

where

$$[\mathrm{diag}(\mathbf{f}) - \mathbf{f}\mathbf{f}^{\mathrm{T}}] = \begin{bmatrix} f_1(1-f_1) & -f_1 f_2 & \cdots & -f_1 f_m \\ -f_1 f_2 & f_2(1-f_2) & \cdots & -f_2 fm \\ \vdots & \vdots & \ddots & \vdots \\ -f_1 f_m & f_2 f_m & \cdots & f_m(1-f_m) \end{bmatrix}$$

(see also Section 8.3) The f_i $(i = 1, 2, \ldots, m)$ are weights for the m distribution functions such that $\Sigma_{i=1}^{m} f_i = 1$. Williams also introduce weights w_j for the discrete set $x_i, x_2, \ldots, x_n$ such that $\Sigma_{n}^{j=1} w_j = 1$, so that both rows and columns of $\mathbf{P}$ are weighted. Equation (8.75e) can then written as

$$\lambda_k = \mathbf{A}_k^{\mathrm{T}}\mathbf{W}\{\mathbf{W}^{-1}\mathbf{P}^{\mathrm{T}}[\mathrm{diag}(\mathbf{f}) - \mathbf{f}\mathbf{f}^{\mathrm{T}}]\mathbf{P}\mathbf{W}^{-1}\}\mathbf{W}\mathbf{A}_k$$ (8.75f)

where $\mathbf{W} = \mathrm{diag}(w_j)$ and the orthogonality constraint on the weighted latent vectors is now given by

$$\mathbf{A}\mathbf{W}^2\mathbf{A}^{\mathrm{T}} = \mathbf{I}$$ (8.75g)

Thus for the weighted expression (Eq. 8.75f) the $\lambda_1, \lambda_2, \ldots, \lambda_r$ are (nonzero) latent roots of the doubly weighted matrix $\mathbf{W}^{-1}\mathbf{P}^{\mathrm{T}}[\mathrm{diag}(\mathbf{f}) - \mathbf{f}\mathbf{f}^{\mathrm{T}}]\mathbf{P}\mathbf{W}^{-1}$ and $\mathbf{W}\mathbf{A}_1$ and $\mathbf{W}\mathbf{A}_2, \ldots, \mathbf{W}\mathbf{A}_r$ are the corresponding latent vectors. For the last $n - r$ roots we have $\lambda_{r+1} = \lambda_{r+2} = \cdots = \lambda_n = 0$ for which $\mathbf{W}\mathbf{A}_{r+1}, \mathbf{W}\mathbf{A}_{r+2}, \ldots, \mathbf{W}\mathbf{A}_n$ are any corresponding orthonormal latent vectors.

The rank of the matrix $\mathbf{W}^{-1}\mathbf{P}^{\mathrm{T}}[\mathrm{diag}(\mathbf{f}) - \mathbf{ff}^{\mathrm{T}}]\mathbf{PW}^{-1}$ is thus r, which can be shown to be equal to one less than the rank of $\mathbf{P}$ (Exercise 8.13). It can also be shown that the unit vector is always a latent vector which corresponds to a zero latent root, and the analysis of the doubly weighted matrix $\mathbf{W}^{-1}\mathbf{P}^{\mathrm{T}}[\mathrm{diag}(\mathbf{f}) - \mathbf{ff}^{\mathrm{T}}]\mathbf{PW}^{-1}$ is thus similar to the principal components analysis of compositional data (Section 5.9.1).

Example 8.13.   Williams (1978) provides a set of age-distribution histograms of the US population classified by sex, race, and the decade in which the census was taken (Table 8.28). The data are given in percentages and represent the $(n \times m)$ matrix $\mathbf{P}^{\mathrm{T}}$, scaled by 100. Note that the array $\mathbf{P}^{\mathrm{T}}$ can also be taken as an r-way contingency table, which indicates the similarity of the method to procedures described in the preceding sections of the chapter. Since the discrete set $x_1, x_2, \ldots, x_n$ is here represented by time-ordered (age) categories, the elements of the latent vectors $\mathbf{A}_k$ (Eq. 8.75f) of the Q-mode analysis of $\mathbf{P}$ can be plotted on a time axis and the global age–time behavior of the racial groups, by sex, can be observed. Note also that Williams chooses the total 1930 age distribution as the weights w_j and equal frequencies f_i rather than the row and column totals of the Table 8.28 (Exercise 8.14). $\square$

Table 8.28   The Age Distribution of the US Population (millions) in the Decades 1930, 1940, and 1950 Classified by Race and Sex

	1930				1940				1950				
	White		Nonwhite		White		Nonwhite		White		Nonwhite		W_j
Age	M	F	M	F	M	F	M	F	M	F	M	F	
0–4	9.2	9.2	10.4	10.4	7.8	7.7	9.9	9.6	10.8	10.2	12.9	12.3	9.3
5–9	10.1	10.1	11.6	11.6	7.9	7.8	10.2	10.0	8.8	8.4	10.4	10.0	10.25
10–14	9.7	9.7	10.5	10.5	8.8	8.7	10.5	10.2	7.4	7.0	9.3	8.8	9.8
15–19	9.2	9.4	10.1	10.8	9.3	9.3	10.0	10.3	7.0	6.8	8.1	8.2	9.4
20–24	8.5	9.0	9.5	10.6	8.7	8.9	8.8	9.8	7.5	7.6	7.8	8.7	8.9
25–29	7.7	8.1	8.6	9.4	8.4	8.5	8.4	9.3	8.0	8.2	8.1	8.6	8.0
30–34	7.4	7.5	7.2	7.4	7.8	7.9	7.5	7.9	7.6	7.8	7.1	7.7	7.4
35–39	7.6	7.5	7.3	7.6	7.2	7.3	7.4	7.9	7.4	7.5	7.3	7.8	7.5
40–44	6.7	6.4	5.9	5.7	6.7	6.7	6.4	6.3	6.8	6.8	6.4	6.4	6.5
45–49	5.9	5.6	5.5	5.0	6.4	6.3	5.5	5.2	6.1	6.0	5.8	5.6	5.7
50–54	5.1	4.8	4.8	3.7	5.7	5.5	4.6	4.0	5.6	5.6	4.8	4.5	4.9
55–59	4.0	3.8	3.0	2.2	4.6	4.5	3.3	2.9	5.0	4.9	3.6	3.2	3.8
60–64	3.2	3.1	2.3	1.8	3.7	3.7	2.5	2.1	4.2	4.2	2.7	2.5	3.0
65–69	2.4	2.4	1.4	1.2	3.0	3.0	2.4	2.2	3.3	3.5	2.6	2.7	2.3
70–74	1.7	1.7	0.9	0.8	2.0	2.1	1.3	1.2	2.3	2.5	1.5	1.4	1.6
75–	1.5	1.7	1.0	1.1	2.1	2.2	1.2	1.3	2.4	2.9	1.6	1.6	1.55
No record	0.1	0.1	0.1	0.1	0.0	0.0	0.0	0.0	0.0	0.0	0.0	0.0	0.1
Population (M)	55,922,528		6,214,552		59,448,548		6,613,044		67,129,192		7,704,047		
total (F)	54,364,212		6,273,754		58,766,322		6,841,361		67,812,836		8,051,286		
$100\,f_i$	13.80	13.42	1.53	1.55	14.67	14.51	1.63	1.63	16.57	16.74	1.90	1.99	

Source: Williams 1978; reproduced by permission.

8.5 RELATED PROCEDURES: DUAL SCALING AND CORRESPONDENCE ANALYSIS

As was pointed out at the outset (Section 8.4.1), optimal scoring or scaling of contingency tables had an early equivalent in Guttman's (1941) work on principal components analysis of multinomial dummy variables, although the full development of the theory lies elsewhere. Nevertheless, a tradition has developed in the educational and psychological literature of analyzing nominal and rank order data using canonical correlations or principal components e.g., see Skinner and Sheu, 1982). The method is known as "dual scaling" and represents a straightforward adaption of the model(s) discussed above, together with their conventions concerning normalization of canonical variates, data weighting, and so forth. The applications are often restricted to using the canonical variates as optimal "discriminant" scores, to scale the rows and columns of a multinomial or rank-order data matrix representing responses of n subjects to k "stimuli" such as questionnaires, visual experiments, and so forth. For a comprehensive treatment of dual scaling see Nishisato (1980).

A more recent variant, developed in France in the 1960s, is as "analyse factorielle des correspondences" or "correspondence analysis" (Benzecri, 1969; Lebart et al., 1977; Greenacre, 1984; Lebart et al., 1984; Van Rijckevorsel and deLeeuw, 1988). The procedure, which is oriented toward graphical displays, represents an adaptation of the rationale of factor analysis to the objectives of a canonical analysis of a contingency table (see Greenacre and Hastie, 1987). Its principal objective is therefore exploratory, and the aim of the method lies in uncovering joint clusters of the row and column categories. At times attempts are also made to give correspondence analysis greater generality by extending its use to continuous random variables (Section 5.4.4). This however simply reduces correspondence analysis to a multidimensional scaling procedure such as principal axes analysis (Section 5.4.1) and nothing new of essence arises. The method however is at times considered to be novel, no doubt due in part to a certain amount of confusion generated by the original literal translation from the French (Benzecri, 1969; 1973), which seems to have been continued by subsequent English-language publications (Hill, 1974, 1982; Theil, 1975; Greenacre, 1984). Nevertheless, this particular application of principal component/factor analysis of discrete data has revived a perhaps neglected method of multivariate data analysis. The procedure is particularly popular in continental Western Europe, as witnessed by journals such as *Les Cahiers de L'Analyse des Donnees* (Dunod, Paris; 1976–), and it may be useful to give a brief account of correspondence analysis as considered by recent authors.

Let $\mathbf{A}_i$ and $\mathbf{B}_i$ be as defined in Section 8.4.2 where

$$\mathbf{A}_i^\mathsf{T} \mathbf{D}_r \mathbf{A}_i = \mathbf{B}_i^\mathsf{T} \mathbf{D}_c \mathbf{B}_i = 1 \tag{8.76}$$

Correspondence analysis is then simply the system (Eqs. 8.44a and b) where the latent vectors are rescaled, that is, the normalization (Eq. 8.44c) is replaced by

$$\mathbf{A}_i^{*\mathrm{T}}\mathbf{D}_r\mathbf{A}_i^* = \mathbf{B}_i^{*\mathrm{T}}\mathbf{D}_c\mathbf{B}_i^* = \lambda_i^2 \qquad (8.77)$$

In the parlance of correspondence analysis the quantities $\mathbf{A}_i^*$ and $\mathbf{B}_i^*$ are known as the principal coordinates of the row and column "profiles," where profiles are the rows and columns divided by their respective marginal totals. The unit vectors $\mathbf{A}_i$ and $\mathbf{B}_i$ are then referred to as "standard coordinates." A description of correspondence analysis as a factor model may also be found in Valenchon (1982). If the analysis is based on a discrete data matrix, a correspondence analysis often proceeds by using Gower's (1966) principal coordinates where each observation is adjusted by the overall mean and the row and column means, that is, by replacing each y_{ij} by $x_{ij} = y_{ij} + \bar{y}_{..} - \bar{y}_{i.} - \bar{y}_{.j}$ in standard notation (Section 5.9.1). Note that a correspondence (principal axis) analysis essentially ignores the difference between a sample space and a variable subspace and treats a discrete data matrix as a contingency table. Although the practice may be objected to on theoretical grounds it does permit a joint plotting of the sample points and variables in a common (usually two-dimensional) subspace. In particular this permits an alternate approach to the analysis of 0–1 data matrices, termed by Hill (1974) as a zero-order correspondence analysis in contrast to a first-order correspondence analysis of a contingency table.

Definition 8.1   (Hill, 1974). A triple $(\lambda, \mathbf{X}, \mathbf{Y})$ is a solution of a zero-order correspondence analysis if and only if

$$\lambda\mathbf{X} = \mathbf{N}_j^{-1}\mathbf{N}\mathbf{Y}, \qquad \lambda\mathbf{Y} = \mathbf{N}_i^{-1}\mathbf{N}^{\mathrm{T}}\mathbf{X}$$

where $\mathbf{N}$ is a $(r \times c)$ contingency table and $\mathbf{N}_i$, and $\mathbf{N}_j$ are defined by Eq. (8.50). The main idea of a first-order correspondence analysis is that in Eq. (8.52) $\mathbf{N}$ can be taken as a matrix of a 0–1 dummy variable, which essentially leads us back to the analysis considered in the early part of Section 8.3.3. For a two-way contingency table there is thus a direct relationship between the two approaches, which however breaks down in the general k-way analysis.

THEOREM 8.3.   Let $\mathbf{N}$ be a two-way contingency table. Then the solutions of a first-order analysis for $k = 2$ can be put into a 2-1 correspondence with those of a zero-order analysis. The canonical variates are the same, and the latent roots are related by the function $2\lambda^2 - 1$.

For a proof the reader is referred to Hill (1974) and Greenacre (1984). The point is that when $(\lambda, \mathbf{X}, \mathbf{Y})$ is the solution of the normal equations (Eq. 8.52) for a 0–1 matrix $\mathbf{N}$, then $(2\lambda^2 - 1, \mathbf{X}, \mathbf{Y})$ is the solution when $\mathbf{N}$ is a contingency table, where $\mathbf{U}$ is partitioned as

$$\mathbf{U} = \begin{bmatrix} \mathbf{X} \\ \text{-} \\ \mathbf{Y} \end{bmatrix} \tag{8.78}$$

corresponding to the columns and rows of the contingency table-see Greenacre (1984). Hill (1974) also considers other relations between the model and principal components analysis, where variates are rescaled elastically to be as concordant as possible. The procedure is also at times viewed as a multidimensional scaling method. The method can also be used to detect reducibility (near reducibility) of a 0–1 data matrix (see also Benzecri, 1980). Applications using ranked and mixed data are given in Marubini et al. (1979), Francis and Lauro (1982), and Nakache et al. (1978). Fortran programs may be found in David et al. (1977), Nishisato (1980), Greenacre (1984), and Lebart et al. (1977). A review of PCIBM programs is given by Hoffman (1991).

Example 8.14. Consider an example from geology (Theil and Cheminée, 1975; see also Theil, 1975; Valenchon, 1982). The data are compositional rather than discrete (Section 5.4.1), and this illustrates the peculiarity of correspondence analysis—it can also be viewed as an attempt to find a common mapping for both the variables as well as the sample points of continuous data consisting of proportions.

The samples consist of rocks from a certain geographic areas which are then analyzed for trace elements. The objective is to uncover relationships between the elements and the rock samples. The first three factors account for 89.5% of the total variance and can be identified as follows (Table 8.29. The first distinguishes the trace elements found in the extreme members of

Table 8.29 The First Three Axes of the Trace Elements Found in Rocks in Ethiopia

Factor 1	Factor 2	Factor 3
Bn = −28.3	Cr = −31.2	Cu = −7.8
Zn = −6.7	Ba = −20.7	Cr = −6.1
Cr = 51.7	V = 26.3	Ni = 82.3
Ni = 5.0	Cu = 6.9	
	Sr = 7.3	
57.2%	23.2%	9.1%

Source: Theil and Cheminée, 1975; reproduced with permission.

the series—chromium (in picrite basalt) and barium (in the trachytic rhyolites and rhyolites). The second factor separates the extreme trace elements from the trace elements found in greater abundance in the intermediate rocks (such as vanadium), and the third contrasts high Ni against small traces of Cu and Cr. A joint distribution of the trace elements and rock type is displayed in Figure 8.10. A genetic application has been given by Greenacre and Degos (1977). $\qquad\square$

8.6 CONCLUSIONS

Most of the models considered in Chapters 3 and 5 are seen to be readily applicable to rank-order and nominal dummy variables, although adjustments are usually required in the measures of association and in the interpretation of the final results. The flexibility of factor models is further underlined when we come to consider applications of a more specialized nature (Chapter 9). Note that units of measure do not pose a problem as is sometimes the case with continuous variables, since an analysis of qualitative data is based on the correlation or cosine matrix, depending on the centering requirement of the analysis. A potential pitfall nevertheless exists, which may catch the unwary user off guard—the reversal of a particular code used will generally alter the analysis. Thus reversing zeros and ones of a dichotomous code, for example, can yield a factor structure with little resemblance to the original solution. The decision of which digit is to be attached to what category must therefore constitute an integral part of the analysis of qualitative data, and should form a basic reference point for any subsequent interpretation of the output.

EXERCISES

8.1 Prove that Kendall's coefficient of concondance (Eq. 8.6) is given by

$$W = \frac{\rho_a(n-1)+1}{n-1}$$

where ρ_a is the mean value of the Spearman rho correlation between $C(\genfrac{}{}{0pt}{}{n}{2})$ possible pairs of observers.

8.2 Show that for the coefficient W we have $0 \le W \le 1$.

8.3 Let c be the number of crossings linking the same objects in the two

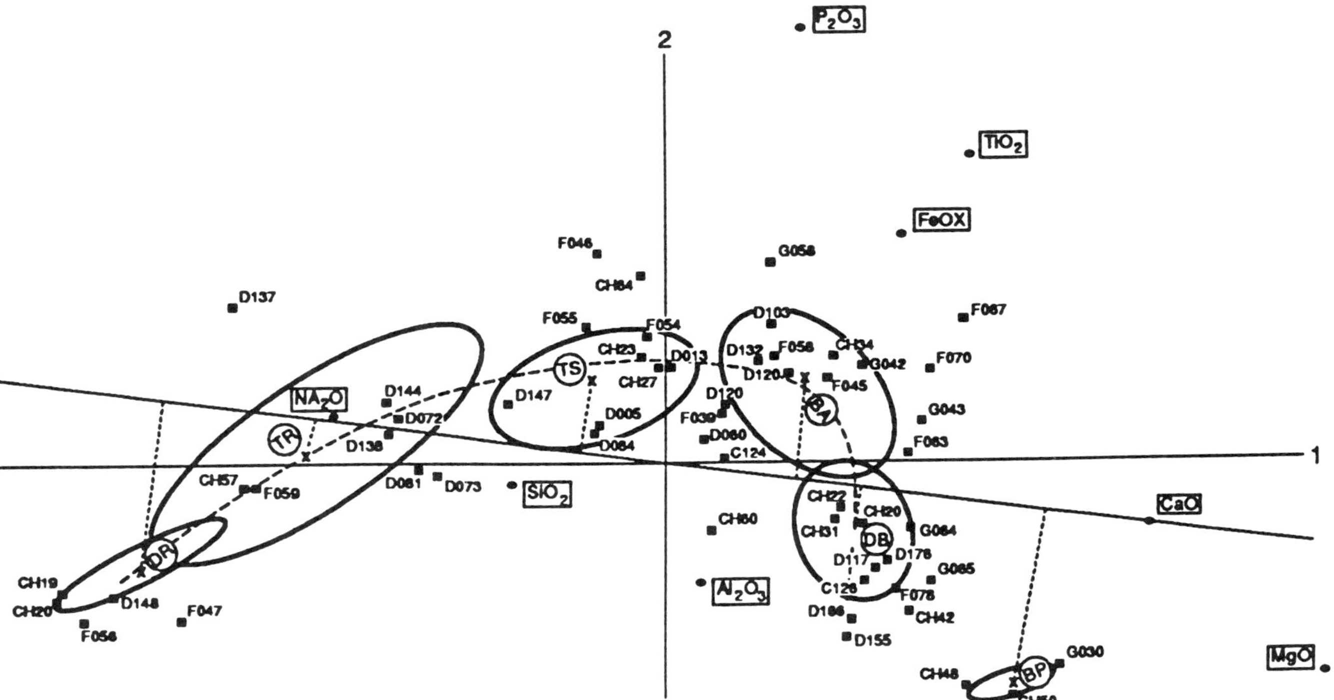

Figure 8.10 The joint distribution of rock type-trace element in the Erta ale chain (Theil and Cheminée, 1975; reproduced with permission).

565

rankings. Show that Kendall's tau can be expressed as

$$\tau = \frac{C\binom{n}{2} - 2c}{C\binom{n}{2}}$$

8.4 Using the data of Table 8.8 carry out a principal components analysis of the object correlation matrix using (a) Spearman's rho and (b) Kendall's tau correlation coefficients. What do you conclude?

8.5 Prove that the normal equations (Eqs. 8.25 and 8.26) are equivalent.

8.6 Show that the rank of matrix $\mathbf{Y}^T\mathbf{Y}$ (Eq. 8.24) is $K - k + 1$.

8.7 Show that maximizing expression (Eq. 8.37) yields the normal equations (Eq. 8.38).

8.8 Prove Eq. (8.55b).

8.9 Show that for a bivariate normal distribution the correlation coefficient between x' and y' can be expressed in terms of the infinite expansion (Eq. 8.61).

8.10 The following data are provided by Pearson (1904) concerning joint father's and son's occupations:

Father's Occupation	Son's Occupation														
	(1)	(2)	(3)	(4)	(5)	(6)	(7)	(8)	(9)	(10)	(11)	(12)	(13)	(14)	Total
(1)	28	0	4	0	0	0	1	3	3	0	3	1	5	2	50
(2)	2	51	1	1	2	0	0	1	2	0	0	0	1	1	62
(3)	6	5	7	0	9	1	3	6	4	2	1	1	2	7	54
(4)	0	12	0	6	5	0	0	1	7	1	2	0	0	10	44
(5)	5	5	2	1	54	0	0	6	9	4	12	3	1	13	115
(6)	0	2	3	0	3	0	0	1	4	1	4	2	1	5	26
(7)	17	1	4	0	14	0	6	11	4	1	3	3	17	7	88
(8)	3	5	6	0	6	0	2	18	13	1	1	1	8	5	69
(9)	0	1	1	0	4	0	0	1	4	0	2	1	1	4	19
(10)	12	16	4	1	15	0	0	5	13	11	6	1	7	15	106
(11)	0	4	2	0	1	0	0	0	3	0	20	0	5	6	41
(12)	1	3	1	0	0	0	1	0	1	1	1	6	2	1	18
(13)	5	0	2	0	3	0	1	8	1	2	2	3	23	1	51
(14)	5	3	0	2	6	0	1	3	1	0	0	1	1	1	32
Total	84	108	37	11	122	1	15	64	69	24	57	23	74	86	775

The categories are: (1) army; (2) art; (3) teacher, clerk, or civil servant; (4) crafts; (5) divinity; (6) agriculture; (7) land ownership; (8) law; (9) literature; (10) commerce; (11) medicine; (12) navy; (13) politics and court; (14) scholarship and science.

(a) Carry out a correspondence analysis of the contingency table (matrix).

(b) Plot the scores in a common two-dimensional subspace. Which occupations have changed more (less) during the generations?

8.11 Le Bras (1974) has given the following (artificial) data from neuro-psychology concerning certain functions which are hypothetically performed in specific areas of the brain's right hemisphere:

Individual	Y_1	Y_2	Y_3	Y_4	Y_5	Y_6	Y_7	Y_8	Y_9	Y_{10}	Y_{11}	Y_{12}	Y_{13}	Y_{14}	Y_{15}	Y_{16}	Y_{17}	Y_{18}	Y_{19}	Y_{20}	Y_{21}	Y_{22}	Y_{23}	Y_{24}	Y_{25}
No. 1	1	1	1	1	1	1	1	0	1	0	1	1	0	0	0	1	1	1	1	0	1	0	0	0	0
No. 2	0	0	0	0	1	0	0	0	1	1	0	1	1	1	1	0	0	1	1	1	1	1	1	1	1
No. 3	0	0	0	0	0	0	1	1	1	0	1	1	1	1	1	0	0	1	1	0	1	1	1	1	1
No. 4	0	1	1	1	0	0	1	1	1	1	0	1	1	1	1	0	0	0	1	1	0	0	0	1	1
No. 5	0	1	1	1	1	0	0	1	1	1	0	0	1	1	1	0	1	1	1	1	0	1	0	0	0
No. 6	0	0	1	0	0	0	1	1	0	0	1	1	1	0	0	0	1	1	1	1	1	1	1	1	1
No. 7	0	1	1	1	0	0	0	0	1	1	0	1	0	1	1	0	1	1	1	1	0	0	1	1	1
No. 8	1	1	1	1	1	0	1	1	1	1	0	1	0	1	1	0	0	1	1	1	0	0	0	0	0
No. 9	0	0	0	1	1	0	0	1	1	1	1	1	1	1	1	1	1	1	1	1	0	0	0	0	0
No. 10	1	1	0	0	0	1	1	1	1	0	1	1	0	0	0	1	1	1	1	1	1	1	0	0	0
No. 11	0	0	0	0	0	1	1	0	1	1	1	1	1	1	1	1	1	0	1	1	1	0	0	0	1
No. 12	1	1	1	1	0	0	1	1	1	0	1	1	0	0	0	0	1	1	0	0	1	1	1	1	0
No. 13	0	0	1	0	0	0	0	1	0	1	1	1	1	1	1	0	1	1	1	1	0	0	1	1	1
No. 14	1	1	1	1	0	1	1	1	1	0	1	1	1	1	1	0	0	1	0	0	0	0	1	0	0
No. 15	1	1	1	1	1	0	1	1	1	1	1	1	1	1	0	0	1	1	0	0	0	0	0	0	0
No. 16	0	1	0	1	1	0	1	1	1	1	0	0	1	1	1	0	0	1	1	0	0	1	1	1	0
No. 17	0	1	1	1	1	1	1	1	1	1	0	0	1	1	1	0	0	0	1	1	0	0	0	0	1
No. 18	1	1	1	1	0	1	1	1	1	0	1	1	0	0	0	0	1	0	0	0	1	1	1	1	0
No. 19	1	0	0	1	1	1	1	0	1	1	0	1	1	1	1	0	0	1	0	1	0	0	0	1	1
No. 20	0	0	0	1	1	0	0	0	1	1	0	0	1	1	1	0	1	1	1	1	0	1	1	1	1
No. 21	0	0	1	1	1	0	0	0	1	0	0	0	1	1	1	0	0	1	1	1	1	1	1	1	1
No. 22	0	0	0	0	0	1	1	0	1	1	1	1	0	1	1	0	1	1	1	1	1	1	0	1	0
No. 23	0	1	1	1	1	0	1	1	1	1	1	1	1	0	0	0	1	1	1	0	0	1	0	0	0
No. 24	0	0	1	0	0	0	0	1	1	0	0	1	1	1	1	1	1	1	1	1	0	1	0	1	1
No. 25	0	0	0	0	0	1	0	0	0	0	1	1	1	1	1	1	1	1	1	1	0	1	1	1	1
No. 26	1	0	0	1	0	1	1	1	1	0	1	1	1	1	0	1	1	1	1	0	0	0	1	0	0
No. 27	0	0	0	0	0	0	0	1	0	0	1	1	1	1	1	0	1	1	1	1	1	1	1	1	1
No. 28	1	1	1	1	1	0	0	1	1	1	0	1	1	1	1	0	0	0	1	0	0	0	0	1	1
No. 29	0	0	0	1	0	0	1	1	1	0	0	0	1	1	1	0	1	1	1	1	1	1	1	1	0
No. 30	0	1	1	1	0	0	0	1	0	0	1	1	1	1	0	0	1	1	1	1	0	0	1	1	1
No. 31	1	1	1	1	1	0	0	1	1	1	0	1	1	1	1	0	1	1	0	1	0	0	0	0	0
No. 32	0	1	1	0	0	1	1	1	1	1	0	1	1	1	1	0	0	1	1	1	0	0	1	0	0
No. 33	1	1	0	0	0	1	1	1	1	0	1	1	1	1	0	1	1	1	0	0	1	1	0	0	0
No. 34	1	1	1	1	1	1	1	1	1	0	1	1	0	1	1	1	0	0	0	0	1	0	0	0	0
No. 35	0	0	0	1	1	0	0	0	1	1	1	1	1	1	1	0	0	1	0	1	1	1	1	0	1
No. 36	0	1	1	1	1	0	1	1	1	1	0	0	1	1	1	0	0	0	1	1	0	0	0	1	1
No. 37	0	1	0	0	0	0	1	1	1	0	1	1	1	1	0	0	1	0	1	0	1	1	1	1	1
No. 38	0	0	1	1	0	0	0	1	1	1	0	1	1	1	0	0	0	1	1	1	0	1	1	1	1
No. 39	0	1	1	0	0	1	1	1	1	1	1	1	1	1	1	1	1	0	1	0	0	0	0	0	0
No. 40	1	1	1	0	0	1	1	1	0	0	1	1	1	1	1	0	1	1	0	1	0	0	0	0	1
No. 41	0	0	0	0	0	0	1	0	0	1	0	1	1	1	1	1	1	1	1	1	1	1	1	1	0
No. 42	1	0	1	1	1	1	1	1	1	1	1	1	1	1	1	0	0	1	0	0	0	0	0	0	0
No. 43	0	0	1	1	0	1	1	1	1	1	0	1	1	1	1	1	1	1	0	0	0	1	0	0	0
No. 44	0	0	1	1	1	0	0	0	1	1	0	0	1	1	1	0	1	1	1	1	0	1	1	1	0
No. 45	0	0	1	1	1	0	1	1	1	1	0	0	0	1	1	0	0	1	1	1	0	0	1	1	1
No. 46	1	0	1	0	0	1	1	1	1	1	1	0	1	1	1	1	0	0	1	0	0	0	0	1	1
No. 47	1	1	0	0	0	1	1	1	0	0	1	1	1	1	0	1	1	1	1	0	0	1	1	0	0
No. 48	1	0	0	0	0	1	0	0	1	0	1	1	1	1	0	1	1	1	1	1	0	1	1	1	0
No. 49	1	1	1	1	1	1	1	1	1	0	1	1	1	1	0	1	1	0	0	0	0	0	0	0	0
No. 50	0	0	0	0	1	0	0	0	1	1	0	1	1	1	1	0	1	1	1	1	0	1	1	1	1

	Variables																								
No. 51	0	0	1	1	1	0	0	1	1	1	0	1	1	1	1	0	0	0	1	1	0	0	1	1	1
No. 52	0	1	0	0	0	0	1	0	1	0	1	0	1	0	1	1	1	1	1	1	1	0	1	1	1
No. 53	1	1	1	0	0	1	1	1	1	1	0	1	1	1	1	0	0	1	1	0	0	0	1	0	0
No. 54	1	1	1	0	0	1	1	1	1	1	1	1	1	1	0	1	0	0	0	0	1	1	0	0	0
No. 55	0	0	0	0	0	0	0	1	0	1	1	0	1	1	1	1	1	1	1	1	0	1	1	1	1
No. 56	0	0	0	1	1	0	0	1	1	1	0	1	1	1	1	0	0	1	1	1	0	0	1	1	1
No. 57	1	1	1	1	0	1	1	1	1	1	1	1	1	0	0	1	1	0	0	0	0	0	0	0	0
No. 58	1	1	1	1	1	1	1	1	1	1	0	1	1	1	0	1	1	0	0	0	0	0	0	0	0
No. 59	0	1	1	1	0	0	1	1	1	1	0	0	1	1	1	0	0	1	1	0	0	0	1	1	1
No. 60	1	1	1	1	0	0	1	1	1	1	0	1	1	1	1	0	1	0	1	1	0	0	0	0	0
No. 61	1	1	1	1	0	1	1	1	1	1	1	1	0	1	1	0	0	0	1	1	0	0	0	0	0
No. 62	1	1	1	1	0	0	1	1	1	0	1	1	1	0	0	1	1	1	0	0	1	1	0	0	0
No. 63	1	1	1	1	1	1	1	1	1	1	0	1	1	0	1	0	0	0	1	1	0	0	0	0	0
No. 64	1	1	0	1	0	1	1	1	1	0	1	0	1	1	0	0	0	0	1	1	0	0	1	1	1
No. 65	1	1	1	1	1	1	0	1	1	0	1	1	0	1	0	0	1	1	1	0	0	1	0	0	0
No. 66	0	0	1	0	0	0	1	1	0	1	0	1	1	0	1	1	1	1	1	1	0	1	1	1	0
No. 67	0	1	1	1	1	0	1	1	1	1	0	1	0	1	1	0	0	0	0	1	0	0	1	1	1
No. 68	1	1	1	1	1	1	0	1	1	0	1	0	1	1	1	0	0	1	1	0	0	0	0	1	0
No. 69	1	1	1	0	0	1	1	1	1	0	0	0	1	1	0	0	0	0	1	1	0	1	1	1	1
No. 70	1	1	1	0	0	1	1	1	0	0	1	1	1	1	0	1	1	1	0	0	1	1	0	0	0

To discover if a function is not performed, presumably due to a tumor, hemmorhage, and so on, individuals are given 25 tests where "pass" $= 0$ and "fail" $= 1$. The areas where the functions are performed however are not known—all we have are the outcomes of the tests. Assuming the right hemishpere consists of 25 specific connected areas located on a grid, we wish to assign each of the 25 tests to a location on the grid.

(a) Using principal components estimate the underlying grid structure of the $p = 25$ variables. How many dimensions are required to estimate the structure? Confirm by plotting the variables in the common factor space.

(b) Using principal axes (Section 8.5) repeat the exercise of part a. Plot the variables and individuals in the common factor space.

8.12 An important area of biochemical medical research is in the development of antitumor drugs, where interest lies in being able to predict a drug's activity using properties of its chemical structure. Since the resultant biological measures possess large variability they are often analyzed using order statistics. The following data are given by Mager (1980a) where

$Y_1 =$ Harmmett's electronic constant

$Y_2 =$ Electronic constant in a different position

$Y_3 =$ Logarithm of molar volume

Y_1	Y_2	Y_3	Y_1	Y_2	Y_3
0.78	0.00	1.76	−0.04	0.00	1.76
0.60	0.00	1.76	−0.04	0.00	1.76
0.25	0.00	1.76	0.15	0.00	1.76
0.78	0.00	1.76	0.10	0.00	1.76
0.50	0.00	1.76	0.14	0.00	1.76
0.66	0.00	1.76	0.18	0.00	1.76
0.65	0.00	1.76	0.16	0.00	1.76
0.00	0.00	1.76	0.18	0.00	1.76
−0.24	0.00	1.76	0.11	0.00	1.76
−0.04	0.00	1.76	−0.04	0.00	1.76
−0.04	0.00	1.76	−0.06	0.78	3.67
−0.04	0.00	1.76	−0.09	0.00	1.76
0.21	0.00	1.76	−0.04	−0.17	3.03
−0.08	0.00	1.76	−0.56	0.00	1.76
−0.04	0.07	2.49	−0.04	−0.24	3.67
−0.04	0.00	1.76	−0.04	−0.18	3.23
0.12	0.00	1.76	−0.66	0.00	1.76
−0.04	0.00	1.76	−0.04	−0.17	3.03
−0.13	0.00	1.76	−0.24	0.00	1.76
−0.13	0.25	2.93	−0.04	−0.66	3.36
−0.10	0.00	1.76	−0.04	−0.66	3.36
−0.04	0.00	1.76	−0.13	−0.66	3.36
−0.04	0.25	2.93	−0.04	−0.66	3.36
−0.13	0.00	1.76	−0.04	−0.66	3.36
−0.44	0.00	1.76	−0.13	−0.66	3.36

(a) Using the original observations carry out a principal components analysis of the variables.

(b) Perform a principal components analysis using quartiles in place of the original observations. What do you conclude?

8.13 Consider the matrix $\mathbf{W}^{-1}\mathbf{P}^{\mathrm{T}}[\mathrm{diag}(\mathbf{f}) - \mathbf{ff}^{\mathrm{T}}]\mathbf{PW}^{-1}$ of Eq. (8.75f) of Section (8.44). Show that the rank of the matrix is one less than the rank of $\mathbf{P}$.

8.14 Refer to Example 8.13 and Table 8.28

(a) Repeat the Q-mode analysis of Table 8.28 using the weights w_j and f_i and compare to Williams (1978). What do you conclude?

(b) Reanalyze the data using R-mode analysis, that is, using the age distribution as the rows of the data matrix (Table 8.28). Can you determine the historical (time) behavior of the different age groups across race and sex?

CHAPTER 9

Other Models for Discrete Data

9.1 INTRODUCTION

The methods introduced in Chapter 8 represent what may be called classical procedures of analyzing discrete data. Several additional questions arise however when considering nominal and rank order variables, for example, what can be done in the presence of nonindependence in the sample observations, or how is it possible to analyze a $(n \times p)$ data matrix which consists of both discrete and continuous random variables. Also the latent space itself may be discrete rather than continuous, that is, a given set of discrete variables may be generated by a set of unobserved classes or categories which themselves are discrete. These problems, together with extensions of techniques described in Chapter 8, are considered in the following sections.

9.2 SERIALLY CORRELATED DISCRETE DATA

Generally speaking a sample is drawn randomly and independently from some population, and this is usually sufficient to ensure that points in the sample space are not correlated. As was seen in Chapter 7 however, observations can, in practice, be correlated. Also, given a $(n \times p)$ matrix of observations for n objects and p attributes we may wish to discover which objects are more similar to each other in terms of the attributes, in other words which objects tend to be characterized more by which attributes, and vice versa. This implies the uncovering of object/attribute clusters, not unlike those considered in Section 8.5, which could provide useful information concerning the nature of the objects and/or the attributes. This is especially common when the observations are either ordered over time or are distributed in physical space, since this induces serial correlations among the discrete observations. In this section we examine two applications of optimal scoring, described in Sections 8.3 and 8.4, when the discrete data reflect a sequential ordering over time (space).

9.2.1 Seriation

The general question posed by seriation may be stated as follows. Given a set of "objects," how can we place these objects (items) in an approximately correct serial order, so that the position of each object best reflects the degree of similarity between it and the remaining items. The problem may be further complicated by the presence of uncertainty due to the partial unreliability of the data. Seriation originated in the late 19th century in archaeology with sequence dating, where interest lay in ordering n large sets of closed assemblages (such as graves) into an ordered chronological series. Kendall (1975; see also Doran and Hodson, 1975) has also referred to such activity as one-dimensional map-making, since seriation can also be applied to the mapping of diverse objects such as genes and viruses (Benzer, 1961), as well as to ecological species that are distributed over a certain area. The latter activity is known as ordination and is considered in Section 9.2.2. In some respects the situation is similar to that encountered in Chapter 7 when estimating time functions, except the data matrix is integer-valued rather than continuous. For the sake of concreteness in what follows we consider seriation in the archaeological/historical setting.

Consider a set of n closed assemblages examined for the presence or absence of p preselected artifacts that are thought to be of importance. If the assemblages span a period of time and if an order can be discerned in the distribution of the 0–1 binary code values, then the ordering may be ascribed to a directed unilinear time dimension. This is because if the known time-sensitive variables are chosen judiciously, they will be highly inter-correlated, and can then be represented by a single factor which is attributed to "time." Other variables that correlate highly with the time factor may then be inferred to be good candidates for additional time-dependent variables. In any case, as a rule of thumb the first two factors are usually expected to account for at least 50% of the variance, as a check on the time-discriminating ability of the variables (Marquardt, 1978). The time direction of the seriated objects cannot be determined in the same analysis *in situ* without, of course, the presence of external information provided by the archaeologist. Any nonnegative data such as rank orders, absolute counts, and proportions can be used, although in archaeological practice seriation is usually based on 0–1 dummy variables to denote the absence or presence of some attribute. The main reason for preferring 0–1 absence–presence data as opposed to counts or proportions seems to be that the time similarity between assemblages of similar function does not necessarily increase with the number or abundance of the attributes they have in common (Goldman, 1971), but rather whether the attributes are simply present or not. For example, a greater number of ornamental broaches in certain graves may be due to a relative wealth differential rather than to temporal change. The end product of such coding is then a matrix of 0–1 dummy variables where rows represent assemblages and columns the artifacts. If the ones (presence of attribute) can be grouped together by permuting the rows and columns of the matrix, then this is sufficient to establish the presence of serial order among the artifacts and assemblages. A matrix that can be so permuted has been termed by Kendall (1975) to be in "Petrie form," and upon further permutation of the rows and columns can be put in quasi-diagonal form. The process may be illustrated by means of an artificial numerical example.

Example 9.1.   Consider a cemetary consisting of $n = 13$ graves, where an archaeologist selects $k = 15$ artifacts as being of importance (Graham et al., 1976). The occurrence/nonoccurrence of an artifact in a given grave is denoted by a 1 or 0 (Tables 9.1–9.3) where the zeroes are omitted for convenience. Consulting Table 9.1 it may be concluded that the occurrence of artifacts is distributed randomly across the graves. The impression however is mainly due to the arbitrariness of the initial numbering or listing of the graves and artifacts in the matrix, and disappears once the rows and columns of the matrix are permuted in a judicious fashion. Thus Table 9.2 reveals the presence of distinct clusters, and further permutation of columns yields the quasi-diagonal structure of Table 9.3, which is characteristic of a seriated matrix. Both the graves and artifacts are now in temporal order, and with further knowledge of direction it is possible to determine which graves are oldest and which are most recent. The rearrangement of rows thus determines the ordinal chronology of the graves, and this in turn

Table 9.1   An Unordered Incidence Matrix

Artifact types

Graves	1	2	3	4	5	6	7	8	9	10	11	12	13	14	15
1	1	1						1					1		
2	1					1	1	1			1			1	
3				1				1					1		
4				1			1	1			1				1
5			1	1								1			
6				1									1		
7	1	1					1	1						1	
8		1		1				1					1		
9			1	1								1			
10			1	1					1	1		1			1
11			1	1						1	1				1
12			1	1					1	1					1
13	1	1				1	1	1						1	

Table 9.2 The Incidence Matrix of Table 9.1 with the Rows Ordered such that it is in Petrie Form

Artifact types

		1	2	3	4	5	6	7	8	9	10	11	12	13	14	15
G	6					1								1		
	3					1			1					1		
	8		1			1			1					1		
	1	1	1						1					1		
r	7	1	1					1	1						1	
a	13	1	1				1	1	1						1	
v	2	1					1	1	1						1	
e	4				1			1	1			1				1
s	11			1	1						1	1				1
	12			1	1					1	1	1				1
	10			1	1					1	1		1			1
	5			1	1								1			
	9			1	1								1			

Table 9.3 The Petrie Matrix of Table 9.2 with the Columns and Rows Ordered Along the Main diagonal

Artifact types

		5	13	2	8	1	14	6	7	11	15	10	4	9	3	12
G	6	1	1													
	3	1	1		1											
	8	1	1	1	1											
	1		1	1	1	1										
r	7			1	1	1	1									
a	13		1	1	1	1	1	1								
v	2			1	1	1	1	1	1							
e	4				1				1	1	1		1			
s	11								1	1	1	1	1		1	
	12										1	1	1	1	1	
	10										1	1	1	1	1	1
	5												1		1	1
	9												1	1		

assigns a range of "sequential dates" to each artifact. Note that the identification of the direction of the graves with time must further be inferred since direction could presumably also be due to spatial variables such as physical or geographic position of the graves and/or artifacts. □

In Example 9.1 we are able to determine the existence of the quasi-diagonal structure simply by a trial and error permutation of the rows and columns of the matrix. In a realistic situation, however, where the matrix may possess hundreds of rows and/or columns such an approach is clearly not feasible for a systematic seriation of an incidence matrix. Moreover, a matrix may not have a diagonal structure since there is no a priori guarantee that we can locate the temporal order, perhaps owing to a lack of suitable artifacts. It is of some interest therefore to investigate mathematical conditions for the existence of a diagonal structure, as well as computational procedures which could be used to locate such a structure. A number of algorithms that rearrange the rows/columns of nonnegative integer matrices can be used in order to uncover such informative clusters (see e.g., McCormick et al., 1972; Exercise 9.5). It turns out that necessary and sufficient conditions for a correct ordering can be obtained from general properties of Hamiltonian graph-theoretic circuits, which must pass through the rows (graves) of the matrix (see Shuchat, 1984). The problem of determining whether a given incidence matrix can be rearranged by rows (columns) so as to bring together all the unities in each column appears to have been first considered by Fulkerson and Gross (1965) in a genetic problem of testing whether experimental data are consistent with a linear model of the gene. Both approaches however have a drawback in that they are only applicable to deterministic or near deterministic situations, which is not the case for seriation in archaeology and other related disciplines. Kendall (1969) has shown that for a $(0, 1)$ incidence matrix $\mathbf{Y}$ a principal components analysis of either $\mathbf{Y}^{\mathrm{T}}\mathbf{Y}$ or $\mathbf{Y}\mathbf{Y}^{\mathrm{T}}$ yields sufficient information to decide on the possibility of a diagonal arrangement, as well as the actual arrangement itself. Although Kendall uses multidimensional scaling, a principal components (factor) analysis is probably preferable in the presence of noisy data (see Graham et al., 1976). A factor analytic model also provides a more unified approach to the seriation problem, and can handle a fairly large matrix if principal components are used. Figure 9.1 illustrates a seriation of actual graves from the Munsinger-Rains region. The relationship between diagonalization using row/column permutations and latent vector decomposition is stated in the following theorem.

THEOREM 9.1. Let $\mathbf{Y}$ be a $(0, 1)$ incidence matrix. If the rows and columns of $\mathbf{Y}$ can be permuted so that all the ones in every row and column come together, there exists (degeneracies aside) a unique ordering of the rows and columns, generated by the first (nontrivial) axis of the optimal scaling of $\mathbf{Y}$.

An outline of the proof is given by Hill (1974). The idea is to consider an incidence matrix $\mathbf{Y}$ as a contingency table and to compute scores that maximize the rows (assemblages) and columns (artifacts) as described in Sections 8.4 and 8.5. If there exists an underlying continuum for the rows and columns, it is estimated by the axis corresponding to the largest (nontrivial) latent root of the decomposition. Another approach is to use the general theory of nonnegative matrices (e.g., see Basilevsky, 1983). For

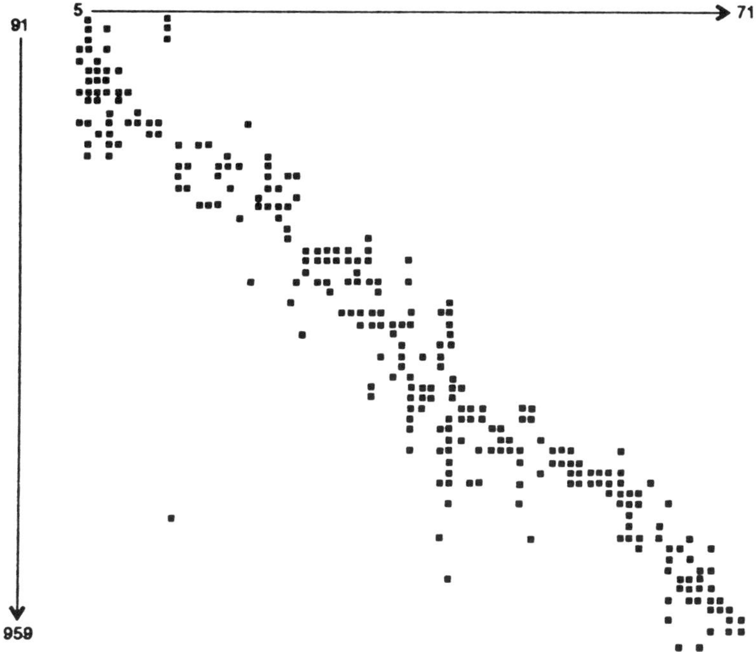

Figure 9.1 A seriated archaelogical incidence matrix representing the graves from the Munsinger–Rains region.

archaeological or historical time-sequence data the continuum can clearly be interpreted in terms of a time trend due to the serially correlated nature of the artifacts and assemblages. The method can also be extended in a straightforward manner to matrices containing integer counts of artifacts, the so-called "abundance" matrices (Kendall, 1971a). As seen in Section 9.3, however, non-Euclidean measures of association may have to be used, owing to the existence of the quadratic or the so-called "horseshoe" effect.

Seriation has also been applied to classical writings such as the works of Plato (Cox and Brandwood, 1959) using absolute counts. The following example is taken from Boneva (1971).

Example 9.2. The problem may be posed as follows. Given a count of Plato's writing style (sentence-endings or the so-called clausula), is it possible to place his works in chronological order? The input matrix consists of the books (rows) and types of clausulae (columns), there being in all 45 books and 32 different types of clausulae. The entries of the matrix are then counts of clausulae in the various works. The matrix can again be considered as a contingency table, in conformity with Sections 8.4 and 8.5, and a seriation of the rows using the first nontrivial axes yields the coordinates of Table 9.4 and the corresponding graph of Figure 9.2. Since the ordering is based on grammatical structure (the clausulae), it clearly does not necessari-

Table 9.4   Two-Dimensional Coordinates of Figure 9.2 of the Works of Plato

	Final Configuration		Name
1	0.179	1.020	Charmides
2	−0.292	0.662	Laches
3	1.513	0.108	Lysis
4	0.706	0.483	Euthyphro
5	0.248	0.463	Gorgias
6	1.177	0.395	Hippas Minor
7	0.443	1.144	Euthydemus
8	0.346	0.167	Cratylus
9	0.310	0.990	Meno
10	1.119	1.344	Menexenus
11	0.811	1.046	Phaedrus
12	0.195	0.774	Symposium
13	0.323	0.524	Phaedo
14	0.237	0.428	Theaetetus
15	0.874	0.613	Parmenides
16	0.339	0.812	Protagorus
17	−0.335	1.125	Crito
18	0.641	1.137	Apology
19	0.194	0.645	R_1
20	0.561	0.597	R_2
21	−0.295	0.956	R_3
22	0.647	0.826	R_4
23	0.403	0.635	R_5 Republic
24	0.301	1.238	R_6
25	−0.019	1.494	R_7
26	0.178	1.681	R_8
27	0.316	1.639	R_9
28	−0.009	0.892	R_{10}
29	−0.566	−1.158	L_1
30	−0.799	−1.893	L_2
31	−1.053	−1.449	L_3
32	−0.348	−2.040	L_4
33	−0.489	−2.301	L_5
34	−1.404	−1.643	L_6 Laws
35	−0.851	−1.569	L_7
36	−1.323	−1.707	L_8
37	−1.438	−1.351	L_9
38	−1.176	−1.899	L_{10}
39	−1.824	−1.040	L_{11}
40	−1.096	−1.257	L_{12}
41	0.945	−1.009	Critias
42	−0.584	−1.802	F (Philebus)
43	−0.335	−1.162	Politicus
44	0.232	−0.327	Sophist
45	0.997	−0.232	Trimaeus

Source: Boneva, 1971; reproduced with permission.

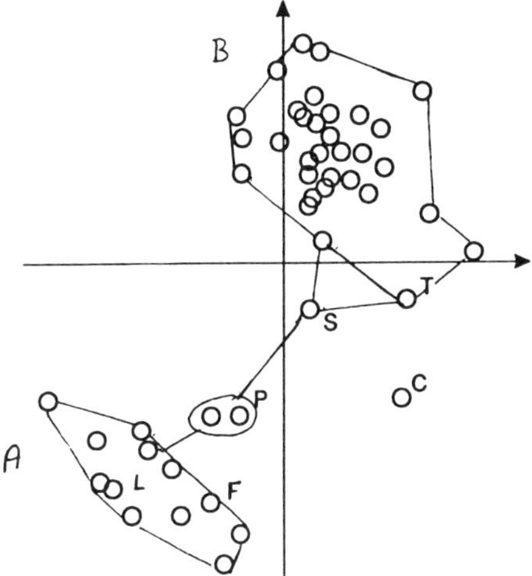

Figure 9.2 A chronological seriation of the works of Plato.

ly have to reflect a temporal evolution of Plato's philosophical concepts unless grammatical style itself is time-dependent. Also, the ordering is estimated using only the first two principal axes, which may not provide enough local information since subsequent axes can well reveal the existence of subgroupings which contrast finer aspects of Plato's style, and which distort the two-dimensional sequence. Nevertheless, the ordering may be of some interest since it reveals that the books fall into two well-marked groups: (1) those for which both coordinates are negative (cluster A; Philebus, Politicus, the Laws), and (2) those for which the second coordinate is positive and which contains the Republic and the first 18 works (cluster B). Critias however is a member of neither group and, indeed, forms a cluster all by itself (Table 9.4; Figure 9.2). □

9.2.2 Ordination

The spatial analogue of temporal seriation is the ordination of spatially correlated incidence or count data in one; two; or three-dimensional space – for background see Gittins (1969). The procedure originated in ecological studies of plant species and is known as reciprocal averaging, weighted averaging, or gradient analysis. The objective is to identify the causes or process(es) that underlie and determine the behavior of the individual species which collectively form the vegetation of a given area. More generally ordination considers spatial distributions of objects across various conditions or subareas, in an attempt to uncover which types of objects have an affinity for what kind of an environment. As in the case of seriation the initial data matrix (contingency table) may consist either of (0, 1) incidences of occurrence/nonoccurrence or of abundance measurements such as counts, percentage of area covered, and so forth. Since the fundamental objective of such research is often qualitative in nature (existence–nonexistence of species), most work is conducted using $(0, 1)$ incidence matrices, although exceptions exist. Since the objective or ordination is a joint plot of the rows and columns of a data matrix it can be achieved using either a principal components analysis or a canonical correlation analysis when the data matrix is considered as a contingency table.

In ecology, gradient analysis is often combined with a scheme of successive approximations, as follows. For example, from an approximate floristic gradient such as water relations the species are divided into "wet" and "dry" species. An initial gradient analysis of sites can be obtained by scoring 0 for stands containing only "wet" species, 1.00 for stands containing only "dry" species, and .50 for those containing equal amounts of both. Provided that the stand scores are a reasonable initial indication of water retention, they can be used to derive an improved second-order new calibration of the species. The new species scores are the averages of the scores of the stands in which they occur. Thus species that occur mainly in wet stands, for example, receive a low score; species that occur mainly in dry stands receive a high score; and intermediate species are given an intermediate score. Thus if the new species scores are rescaled from 0 to 1.00 the process can be repeated and the stands recalibrated.

The process of repeated cross-calibration represents a traditional method in plant ecology and in the limit yields a unique one-dimensional ordination of both species and stands, since after sufficient iterations the scores stabilize. The process is known as "reciprocal averaging" by practitioners and can be applied to a matrix of counts, proportions (probabilities), or $(0, 1)$ absence–presence data. It can be shown however that the iteration procedure is in fact identical to the computation of the dominant (nontrivial) latent root and latent vector of the discrete data matrix used.

THEOREM 9.2.   (Hill, 1973). Let $\mathbf{A} = (a_{ij})$ denote a $(0, 1)$ matrix where rows represent species and columns represent the stands. Then the cross calibration of the rows/columns by reciprocal averaging is equivalent to the extraction of the first nontrivial latent root and its corresponding latent vector.

PROOF.   Let

$$r_i = \sum_{j=1}^{n} a_{ij} \qquad c_j = \sum_{i=1}^{m} a_{ij}$$

be the row and column totals of $\mathbf{A}$. Then the reciprocal averaging process can be represented by the equations

$$x_i = \frac{1}{r_i} \sum_{j=1}^{n} a_{ij} y_j \qquad y_j = \frac{1}{c_j} \sum_{i=1}^{m} a_{ij} x_i$$

or, in matrix form, as

$$\mathbf{X} = \mathbf{R}^{-1}\mathbf{A}\mathbf{Y} \qquad \mathbf{Y} = \mathbf{C}^{-1}\mathbf{A}^{\mathrm{T}}\mathbf{X} \qquad (9.1)$$

where $\mathbf{X}$ and $\mathbf{Y}$ are the row and column scores and $\mathbf{R}$ and $\mathbf{C}$ are diagonal matrices containing the row and column totals. Substituting, say, for $\mathbf{Y}$ in Eq. (9.1) the first step of the iteration is

$$\mathbf{X}' = \mathbf{R}^{-1}\mathbf{A}\mathbf{C}^{-1}\mathbf{A}^{\mathrm{T}}\mathbf{X} \qquad (9.2)$$

where $\mathbf{X}'$ are the new values for the row scores. When there is no rescaling, the largest latent root of Eq. (9.2) is 1, which corresponds to the unit latent vector. The matrix expression preceding $\mathbf{X}$ in Eq. (9.2) however is not symmetric. Let $\mathbf{Z} = \mathbf{R}^{1/2}\mathbf{X}$. Then Eq. (9.2) may be rewritten as

$$\begin{aligned}
\mathbf{Z}' &= \mathbf{R}^{-1/2}\mathbf{A}\mathbf{C}^{-1}\mathbf{A}^{\mathrm{T}}\mathbf{R}^{-1/2}\mathbf{Z} \\
&= (\mathbf{R}^{-1/2}\mathbf{A}\mathbf{C}^{-1/2})(\mathbf{C}^{-1/2}\mathbf{A}^{\mathrm{T}}\mathbf{R}^{-1/2})\mathbf{Z} \\
&= \mathbf{B}\mathbf{B}^{\mathrm{T}}\mathbf{Z} \qquad\qquad\qquad\qquad\qquad\qquad (9.3)
\end{aligned}$$

where $\mathbf{Z}' = \mathbf{R}^{1/2}\mathbf{X}'$ and $\mathbf{B} = \mathbf{R}^{-1/2}\mathbf{A}\mathbf{C}^{-1/2}$. The matrix preceding $\mathbf{Z}$ in Eq. (9.3) is the same as in Eq. (8.54), which is also of the same form as in Eq. (8.44) but with different scaling.

An artificial numerical example illustrating the similarity between reciprocal averaging and principal components is given by Hill (1973). Hatheway (1971) has used contingency table optimal scoring to study rain forest vegetation. For a comparison of reciprocal averaging with other ordination techniques such as standardized and unstandardized principal components and other latent vector methods see Gauch et al. (1977) and Orloci (1966). Again, a choice is available as to the location of the origin—centering by columns, rows, or both—for a discussion see Noy-Meir (1971). The following example is drawn from Orloci (1966).

Example 9.3. The data consist of the presence or absence of plant species from dune and slack vegetation in Newborough Warren, Anglesey. The random sample consists of 131 plots (2×2 m each) where plant shoot presence–absence is determined for all species encountered in 50 random 20×20 cm quadrants within each sample plot. A total of 100 plots is used in the ordination, using 101 species. □

The (100×101) incidence data matrix is subject to a principal components ordination to determine specie richness and locate specie communities. A plot for the first two components is given in Figure 9.3. Because species-rich communities differ in more species than the species-poor communities, the analysis results in a dense clustering of the dune plant

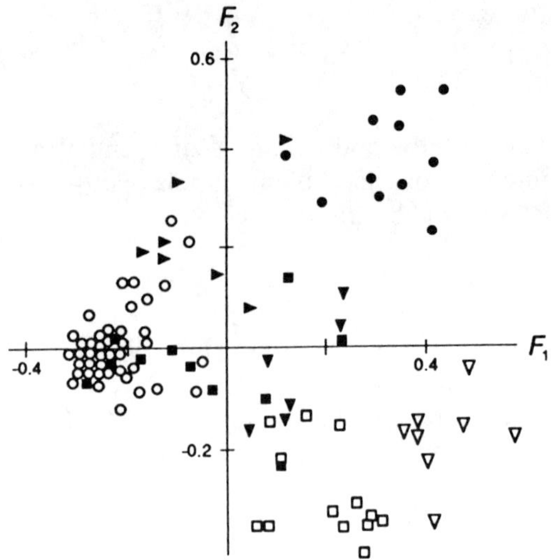

Figure 9.3 Presence–absence ordination by principal component analysis (○ mobile dunes; ▶ fixed dunes; ● fixed slacks (leached soils); ■ hummocks; ▼ raised ground in slacks; □ alkaline wet slacks with semi-open vegetation; ▽ alkaline wet slacks with closed vegetation) (Orloci 1966, reproduced with permission).

communities and a gradual quadratic-like fanning out of the points (Sections 8.4.3 and 9.3). The first principal axis can be identified as corresponding to a vegetational gradient starting from the Ammophila arenaria-type of mobile dunes and leading to plant communities of the wet alkaline slacks (Orloci, 1966).   □

Finally, it must be noted that both seriation and ordination are special cases of a more general problem. Given a nonnegative object–attribute matrix whose (i, j)th element measure (on an ordinal scale) the degree to which object j possesses attribute i, can we cluster the rows/columns such that similar groups of attributes characterize similar groups of objects (Exercise 9.5).

9.2.3 Higher-Dimensional Maps

Both seriation and ordination seek to uncover a one-dimensional ordering of objects, either in temporal or in physical space. The idea can also be extended to a higher-dimensional ordering of objects, for example, in physical or in geographic space. Thus by using integer counts (or functions of integer counts) as measures of similarity, it is possible to construct or estimate a higher-dimensional map from what may appear to be meager data. Kendall (1971b) has aptly described the procedure as "construction of maps from odd bits of information," and using a similarity measure based on intermarriage rates computed from parish registers has estimated the relative location of eight lost Otmoor parishes in Oxfordshire, England.

Also, in a problem of determining fine genetic structure Benzer (1959) utilized similar mapping concepts using a $(0, 1)$ recombination matrix, while Levine (1972) produced a three-dimensional map indicating the structure of bank-industrial corporate interlocked directorates. The algorithms which can be employed for mapping purposes are those of multidimensional scaling including principal components and principal axes analysis. The conversion of non-metric into metric information is also considered by Abelson and Tukey (1959; 1963). The following example can serve as an illustration in the area of archaeological exploration.

Example 9.4. Tobler and Wineburg (1971), using cunieform tablets uncovered in Kültepe (Turkey) containing cross references to Assyrian towns (1940–1740 B.C.) have attempted to estimate the location of these lost Bronze Age centers. The basic reasoning employed is that towns that are closer to each other engage (on the average) in more frequent trade than those that are far away, and thus can be expected to be mentioned more frequently on their neighbor's records or tablets. In addition, it is expected that interaction between the Assyrian towns also depended on the relative sizes of the towns, so that in total it is posited that interaction increases with size and decreases with distance. The authors thus use the well-known gravity model index

$$I_{ij} = \frac{kP_iP_j}{d_{ij}^2} \tag{9.4}$$

where I_{ij} is the interaction between place i and j; P_i and P_j are populations of towns i and j, d_{ij} is distance; and k is a constant of proportionality. The estimated map of the towns' locations is given in Figure 9.4. The fit of the configuration is reported to be in excess of 80%. Since any solution only gives the relative coordinates of the towns, at least two points (towns) must be known in terms of absolute coordinates in order to determine the scale, and a third point to determine the absolute orientation. Of course the more towns that are known in advance, the more precise and stable will be the statistical solution. It must be kept in mind however that a statistical resolution of a mapping problem is at best an initial approximation which must usually be adjusted by an experienced archaeologist. Nevertheless a first approximation based on a mathematical algorithm can result in a saving of human labor and may provide greater objectivity in the comparison or search process.

Example 9.5. The objective of the method of Example 9.4 can be described as follows. Given a square matrix of interpoint distances (similarities), is it possible to estimate, in a space of lower dimension, the coordinates of the physical locations of the points which have given rise to these distances (similarities). An alternative method of obtaining a two-

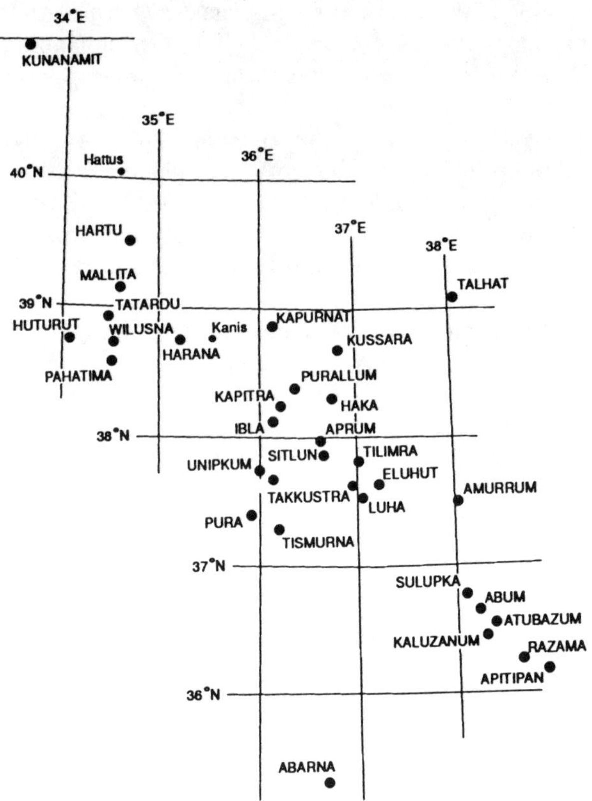

Figure 9.4　Predicted location of 33 pre-Hittite Assyrian towns (Tobler and Wineburg, 1971; reproduced with permission).

dimensional map is one that makes use of the spatial distribution which often exists among the nominal observations. The procedure is to use the loadings from a Q-mode (scores from an R-mode) factor analysis to obtain an estimated spatial distribution or map of the factors, which hopefully reveal the essential structure(s) of the original variables. The purpose of the plot is exploratory and is mainly intended to detect groupings or clusters among the original variables or sample points. Rowlett and Pollnac, (1971), for example, have used factor clusters to observe whether archaeological variables can be clustered arealy. The data consist of $p = 104$ grave assemblages (the variables) which are distributed, in varying degrees, among $n = 77$ archaeological sites in an area which is approximately 100×75 km, located in the Marne Valley of northern Champagne. The objective is to detect whether the so-called Marne Culture (480–400 B.C.), long recognized as a distinctive variant of the widespread La Tène Culture, can itself be subdivided meaningfully into subcultures. The map of Figure 9.5, constructed using scores of the correlation matrix, indicates three distinct

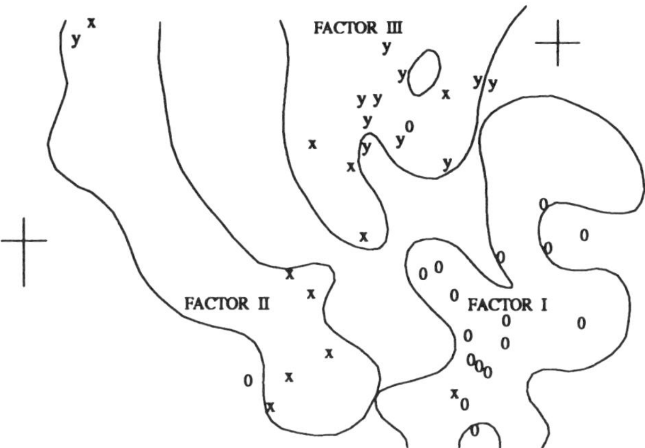

Figure 9.5 Spatial distribution of factors. Boundaries are factor loadings of $+.30$. $0 \equiv$ graves among houses; $y \equiv$ houses adjacent to cemetery; $x \equiv$ houses distance from cemetery.

clusters of the $n = 77$ sites, suggesting the presence of three distinct subcultures. Since the nominal data are not too highly correlated, however, the $r = 3$ factors only account for 20.5% of the variance (9.3, 6.0, and 5.2% respectively). The results appear to be meaningful, and the authors conclude that factor mapping should prove to be useful in anthropology and archaeology since it appears to yield superior results to more traditional forms of clustering such as hierarchical dendograms (see also Christenson and Read 1977). The existence of the three distinct clusters or subcultures however may very well be spurious. First, it is not clear that the three common factors are statistically significant. Second, the factor method used is the so-called "common factor analysis" or the image factor model whereby unities on the main diagonal are replaced by squared multiple correlations. As was seen in Section (6.3.3) however, the model is inappropriate since it introduces bias. Actually, with dichotomous variables it is probably best to use the principal components decomposition rather than an errors-in-variables factor model of the type described in Sections 6.4 and 6.5. Nevertheless, the present numerical example illustrates the type of analysis which can be performed with discrete data using the Q-mode approach (Section 5.4) (Table 9.5).

9.3 THE NONLINEAR "HORSESHOE" EFFECT

The mathematical validity of factor analysis does not depend on the type of particular configuration of points in multivariate space. Owing to the very nature of $(0, 1)$ data however, the observations cannot lie in any position in the vector space but must lie on the apices of a hypercube (Fig. 9.6), and this generally modifies the implementation and interpretation of factor

Table 9.5   Factor Scores of a (75 × 75) Correlation Matrix of Nominal Dummy Variables Describing Archaeological Artifacts Representing the La Tène Culture

Assemblage Items	Factors		
	I	II	III
1. Vases chiefly to right, at foot	−1.22	0.28	0.39
2. Lances chiefly to the left	−0.50	−0.34	0.26
3. a°-la bowls	−0.37	−0.07	−1.76
4. A°-la jar	−1.12	3.48	−1.30
5. A° jar with neck cordon	0.04	−0.53	0.65
6. B vase with flat, everted rim	−0.82	−0.06	−0.40
7. B vase with vertical rim	−0.11	−0.18	0.20
8. Rimless B and b vases	−0.41	−0.13	0.44
9. Biconic plates with foot	−0.29	−0.70	0.66
10. Footless carinated cup	−0.19	−0.71	0.63
11. Footless ovoid cup c°-2b	0.06	−0.72	0.68
12. Rectilinear conical cist predominant form	−0.89	−0.87	0.74
13. Orange-yellow pottery at least 10% of ceramic colors	−4.32	−0.25	0.07
14. Thin red paint	−4.21	0.22	−0.10
15. Wide-band painting technique	−4.56	0.21	−0.00
16. Triple chevron ceramic motif	−0.88	0.52	0.12
17. Inverted chevron ceramic motif	−1.29	0.12	0.14
18. Zig-zag line ceramic motif	−0.58	−0.45	0.09
19. Reticular ceramic motif	−1.17	0.16	0.05
20. Circular ceramic motif	0.33	0.20	0.54
21. Vertical wavy combmarks on ceramics	−0.35	0.75	0.57
22. Less than 50% twisted torcs	−1.50	−0.31	0.21
23. Majority of twisted torc hooks in the plane of the torc	−1.54	−0.50	−0.48
24. Bird torcs, bird vases, and other bird images (except on fibulae)	−0.19	−0.57	0.36
25. Torcs with exterior annelets	−0.38	−0.21	0.59
26. Bracelet with continuous series of incised lines	−0.61	−0.09	0.62
27. Bracelet with serpentine decoration around the exterior	0.33	−0.67	0.17
28. Pin-and-socket bracelet with flattened wire section	−0.56	0.05	0.48
29. Multiple-node bracelet	0.03	−0.99	0.43
30. Fibula terminal semihemispherical (as at La Gorge Meillet)	0.48	−0.70	0.48
31. Fibula with false spring on foot	0.24	−0.98	0.81
32. Disc fibula	−0.23	−0.69	0.69
33. Bronze sword scabbards	0.62	−0.86	0.56
34. Predominantly high arc on scabbard mouth	0.00	−1.16	0.09
35. Knife with complete handle and rectangular pommel	−0.35	−0.48	−0.08

Table 9.5 *(Continued)*

Assemblage Items	Factors		
	I	II	III
36. Knives with convex dorsal lines and short riveted handles (D − 1 and D3e)	−1.15	−0.20	0.48
37. Narrow felloe clamp	−0.00	−1.16	0.71
38. Trapezoidal chariot burial pit	0.81	−1.06	0.82
39. Vases chiefly to the left side, not at foot	0.29	−0.75	−1.92
40. Black piriform wheelmade urns	0.20	−0.62	−1.57
41. A-3c urn with short, flat upper shoulder	0.48	−1.06	−0.66
42. Spheroid jars A°-3	−0.17	0.52	−3.82
43. b°-3 spherical pots	0.72	−0.18	−1.09
44. Ta° chalice with flaring rim	0.80	−0.25	−1.02
45. Tc°-3 conical chalice	0.25	−0.17	−1.43
46. Black pottery predominant (50% or more)	0.21	−0.41	−5.09
47. Relief decorative technique	0.45	−1.01	−1.59
48. Nested serial lozenge ceramic motif	0.78	−0.27	−0.60
49. A vases with rounded bellies	0.64	−0.35	−1.05
50. Concentric semicircle decorative motif	0.60	−0.81	0.06
51. Vertical comb marks with top section curved	0.46	0.09	−0.39
52. Punctate decoration with relief margins	0.30	−0.57	−0.96
53. Pyramidal ceramic motif	0.35	−0.74	−0.89
54. La Tène curvilinear designs on wheel-made pottery	0.41	−0.76	−0.73
55. Bracelet of flattened wire with square-chapped ends (B2a)	0.03	0.22	0.09
56. Knife with complete handle with splayed butt	0.65	−0.80	−0.67
57. Knife with concave dorsal line, short handle	0.71	−0.28	−0.20
58. Wide felloe clamp	0.83	−0.48	−0.03
59. Square or rectangular chariot pit	0.43	−0.34	−0.32
60. Cremations as well as inhumations	0.63	0.82	0.34
61. Pots chiefly at the head	0.99	0.71	0.68
62. A-3b urn with short cylindrical neck, wide flat rim	0.67	0.56	−0.10
63. A-2b vase with dropping upper shoulder	0.57	1.22	0.67
64. b°-2 vase	0.65	0.56	0.60
65. Cylindrical cist predominant cist form	0.66	0.99	0.57
66. Many grey pots (over 33%)	0.51	1.81	0.44
67. Triplet parallel lines on ceramic cists	0.72	0.30	0.66
68. Single and stacked lozenges ceramic motif	0.01	1.70	0.95
69. Left-oriented step ceramic motif	−0.41	1.59	0.46
70. Solid dot ceramic motif	0.15	0.65	0.46
71. Over 50% twisted torcs	0.67	3.00	−0.41
72. Plaque catch-plate on torcs	0.59	2.02	0.68
73. Majority of torc hooks perpendicular to the plane of the torc	0.48	3.29	−0.28
74. Bracelet with alternate band decoration with lines at right angles to the bands	0.36	0.55	−0.27
75. Rectangle and triangles bracelet motif	0.04	2.09	1.00

Table 9.5 (*Continued*)

Assemblage Items	Factors		
	I	II	III
76. Pointed-ended bracelet with flattened section, overlapping ends	0.79	0.94	0.36
77. Predominantly low arc scabbard mouth	0.70	1.94	−1.18
78. Knife with arched back, stepped pommel	0.84	−0.13	0.32
79. Knife with short, stepped handle	0.86	0.55	0.57
80. Red piriform wheel-made urns	0.56	−1.15	0.56
81. Piriform flasks	0.06	−0.75	0.50
82. B°-3 jar (high rounded shoulder)	0.20	−0.08	0.48
83. b°-1 bowl (high rounded shoulder)	−0.35	−0.72	0.31
84. Rimless chalice Tc°-2a	0.23	−0.21	0.95
85. Triangle ceramic motif	−0.38	−0.70	0.26
86. Asymmetrical rectangular meander ceramic motif	−0.04	−0.93	0.50
87. La Tène curvilinear designs on handmade pottery	0.46	−0.80	0.40
88. Bracelet of ribbon twist with pointed ends	0.32	−0.77	0.62
89. Pointed oval design on fibula bow complemented by tick marks	0.56	−0.57	0.83
90. Knife with pointed handle	0.09	−1.01	0.71
91. Spear with long socket	0.18	−0.83	0.34
92. Lances chiefly at the feet	0.82	0.34	0.24
93. B° vase with incurvate upper shoulder (B°-3)	0.84	−0.20	−1.20
94. B° rimless ceramic situla	0.23	−0.79	−0.79
95. Cross-hatched decoration on fibula bows	0.11	−0.33	−0.57
96. Lances chiefly at the right	0.61	0.47	0.48
97. Symmetrical rectangular meander ceramic motif	0.30	0.15	−0.20
98. Pin-and-socket bracelets with round socket	0.43	−0.31	−0.14
99. Flat rectangular fibula terminal with X-design incised	0.82	0.53	0.68
100. Vases chiefly at foot, no side preference	0.68	−0.27	0.55
101. Orange-brown pottery over 10%	0.42	0.45	0.77
102. Thick paint predominant technique of ceramic decoration (50% plus)	0.37	0.03	0.57
103. Double chevron design	−0.66	0.87	0.34
104. Series of small circles decorating bronzes	0.18	1.06	0.46

Source: Rowlett and Pollnac, 1971; reproduced with permission.

analysis. In this section we consider a particular problem that occurs when using dummy variables in a canonical or a factor model. It was seen in Section 8.4.3 that latent vectors associated with nominal data consist of Chebyshev–Hermite polynomials when the (0. 1) observations are generated by a normal distribution. In particular, for a bivariate normal distribution the first two principal axes may be related by a quadratic function even though the eigenvectors are orthogonal. The phenomenon manifests itself in terms of a quadratic configuration of points in the plane, which can

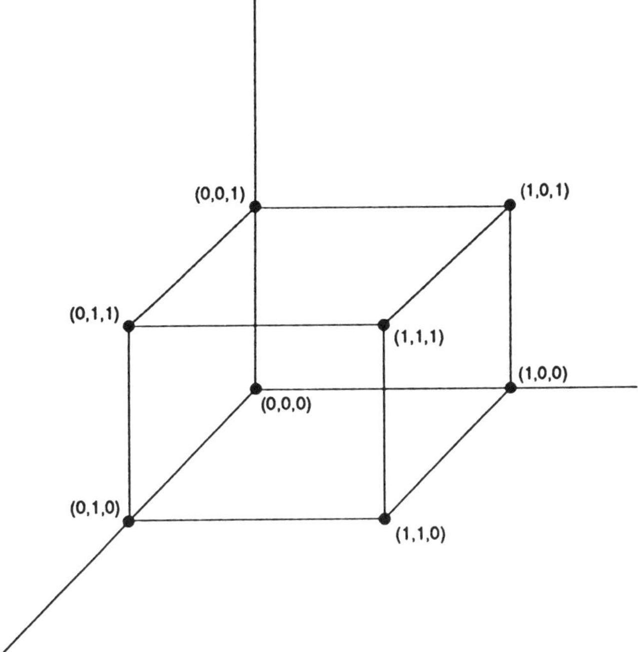

Figure 9.6 Hypercube in three-dimensional space.

frequently be observed in the applied literature and which has been termed by Kendall as the "horseshoe effect" (Kendall, 1971a, 1975; Hill and Gauch, 1980). Research workers are frequently surprised by the phenomenon, although it can be expected in practice under widespread and general conditions because of the very nature of the data and measures of association used. The main difficulty with a quadratic configuration of points is that it renders interpretation more difficult, particularly when conducting a seriation of discrete time series or an ordination of spatially distributed points (Section 9.2). This is because the "horseshoe" is a purely mathematical artifact and does not necessarily correspond to any real structure of the data, and cannot be interpreted in a meaningful, substantive sense. A way out of the difficulty is to consider alternative non-Euclidean measures of association which do not rely on linearity to maximize correlation. Before considering non-Euclidean measures of association it is useful to take a closer look at the quadratic horseshoe effect and the reason for its presence.

Consider the matrix

$$\mathbf{A} = \begin{bmatrix} 1 & 1 & 1 & 1 & 1 & 0 & 0 & 0 & 0 & 0 & 0 & 0 & 0 & 0 & 0 \\ 0 & 0 & 1 & 1 & 1 & 1 & 1 & 0 & 0 & 0 & 0 & 0 & 0 & 0 & 0 \\ 0 & 0 & 0 & 0 & 1 & 1 & 1 & 1 & 1 & 0 & 0 & 0 & 0 & 0 & 0 \\ 0 & 0 & 0 & 0 & 0 & 0 & 1 & 1 & 1 & 1 & 1 & 0 & 0 & 0 & 0 \\ 0 & 0 & 0 & 0 & 0 & 0 & 0 & 0 & 1 & 1 & 1 & 1 & 1 & 0 & 0 \\ 0 & 0 & 0 & 0 & 0 & 0 & 0 & 0 & 0 & 0 & 1 & 1 & 1 & 1 & 1 \end{bmatrix} \tag{9.5}$$

which represents serially ordered objects corresponding to the rows and columns of **A**. The horseshoe effect is a direct product of the fact that the data points are restricted to lie on the corners of a hypercube (Fig. 9.6). Thus in terms of the similarity measures between the rows of the matrix we observe a quadratic effect in spite of the fact that the ordering is linear (diagonal). Consider the matrix of inner products between the rows of **A**

Table 9.6 Sample Counts of Euphausia Species taken at Discovery II and Deutsche Südpolar Expedition Stations[a]

	Y_1	Y_2	Y_3	Y_4	Y_5	Y_6	Y_7	Y_8	Y_9	Y_{10}	Y_{11}	Y_{12}	Y_{13}	Y_{14}	Y_{15}	Y_{16}	Y_{17}
00°54′	12	2	7	6	1	–	–	–	–	–	–	–	–	–	–	–	–
02°00′	44	11	103	176	–	–	–	–	–	–	–	–	–	–	–	–	–
04°06′	9	4	36	–	–	–	–	–	–	–	–	–	–	–	–	–	–
05°02′	–	4	18	11	–	–	–	–	–	–	–	–	–	–	–	–	–
07°03′	4	16	1	4	–	–	–	–	–	–	–	–	–	–	–	–	–
07°52′	4	6	41	9	–	–	–	–	–	–	–	–	–	–	–	–	–
10°51′	12	4	26	21	1	–	–	–	–	–	–	–	–	–	–	–	–
12°23′	2	–	8	2	–	–	–	–	–	–	–	–	–	–	–	–	–
13°37′	2	14	94	21	–	–	–	–	–	–	–	–	–	–	–	–	–
16°37′	10	2	170	21	–	–	–	–	–	–	–	–	–	–	–	–	–
19°35′	–	–	29	–	–	–	–	–	–	–	–	–	–	–	–	–	–
20°19′	1	–	27	19	8	21	–	–	–	–	–	–	–	–	–	–	–
22°38′	–	2	39	2	1	–	–	–	–	–	–	–	–	–	–	–	–
23°51′	–	–	12	–	–	–	8	–	–	–	–	–	–	–	–	–	–
25°51′	–	–	46	–	–	–	12	–	–	–	–	–	–	–	–	–	–
26°24′	–	–	8	–	–	5	–	–	–	–	–	–	–	–	–	–	–
29°42′	–	–	8	–	–	2	–	–	–	–	–	–	–	–	–	–	–
30°28′	–	–	–	–	1	21	8	3	–	5	–	–	–	–	–	–	–
32°07′	–	–	7	–	–	20	165	–	–	–	8	–	–	–	–	–	–
32°49′	–	–	–	4	–	50	116	176	10	27	94	–	–	–	–	–	–
34°19′	–	–	–	–	3	56	459	2	8	11	–	–	–	–	–	–	–
36°08′	–	–	–	–	–	33	223	49	83	57	16	–	–	–	–	–	–
37°09′	–	–	–	–	–	47	252	84	23	–	2	–	–	–	–	–	–
39°21′	–	–	–	–	–	–	1	9	–	15	12	–	–	–	–	–	–
39°53′	–	–	–	–	–	–	–	3	4	52	9	–	–	–	–	–	–
42°24′	–	–	–	–	–	–	–	–	–	1	55	24	24	3	–	–	–
45°11′	–	–	–	–	–	–	–	–	–	1	50	83	218	1	–	–	–
46°02′	–	–	–	–	–	–	–	–	–	–	–	21	20	–	–	–	–
48°04′	–	–	–	–	–	–	–	–	–	–	39	16	62	18	–	–	–
49°53′	–	–	–	–	–	–	–	–	–	–	–	8	5	–	–	–	–
50°43′	–	–	–	–	–	–	–	–	–	–	–	–	388	19	–	–	–
53°23′	–	–	–	–	–	–	–	–	–	–	–	–	46	90	–	–	–
54°13′	–	–	–	–	–	–	–	–	–	–	–	–	6	1	134	10	–
57°55′	–	–	–	–	–	–	–	–	–	–	–	–	10	2	2154	4	–
60°33′	–	–	–	–	–	–	–	–	–	–	–	–	–	1	89	–	–
62°49′	–	–	–	–	–	–	–	–	–	–	–	–	–	–	68	8	–
64°53′	–	–	–	–	–	–	–	–	–	–	–	–	–	–	–	4	–
65°30′	–	–	–	–	–	–	–	–	–	–	–	–	–	–	–	–	1
66°02′	–	–	–	–	–	–	–	–	–	–	–	–	–	–	–	4	17

[a]The counts from Discovery II Stations are corrected for a 20 minN100B haul. Stations' positions are shown in Figure 9.9.

Source: Baker, 1965; reproduced with permission.

given by the symmetric Toeplitz matrix

$$\mathbf{S} = \begin{bmatrix} 5 & 3 & 1 & 0 & 0 & 0 \\ 3 & 5 & 3 & 1 & 0 & 0 \\ 1 & 3 & 5 & 3 & 1 & 0 \\ 0 & 1 & 3 & 5 & 3 & 1 \\ 0 & 0 & 1 & 3 & 5 & 3 \\ 0 & 0 & 0 & 1 & 3 & 5 \end{bmatrix} \tag{9.6}$$

Since the similarity between the first row and the last three rows is zero (even though the rows are highly disimilar) and since the ordering between the first and last rows is reversed this results in the quadratic horseshoe effects such as that of Fig. 9.7 when a conventional principal components analysis is performed on the matrix. Also related procedures discussed in the preceding section such as reciprocal averaging, correspondence analysis, and canonical correlation of a contingency table are also subject to the quadratic effect. In addition the horseshoe effect may introduce difficulties concerning the rotation of axes since the principal components are no longer independent.

A closer examination of the problem reveals that the zeroes of the matrix (Eq. 9.6) are mainly responsible for the quadratic effect of an ordination. A solution which is therefore sometimes used is to transform the inner product matrix in such a way that the resulting similarity matrix is free of the effect of the zeroes. This implies the use of alternative measures of similarity between the rows (columns) of the original data matrix. Two such measures are in common use; the "step-across' method of Williamson (1978) developed for ecologic ordination of plant species, and the "circle" measure of Kendall (1971c, 1975), used in archaeological seriation. First consider the "step-across" method used in ordination. The procedure consists of replacing inner products by the well-known Minkowski L_1 distances d_{kl} where

$$d_{kl} = \sum_{i=1}^{n} |x_{ki} - x_{li}| \tag{9.7}$$

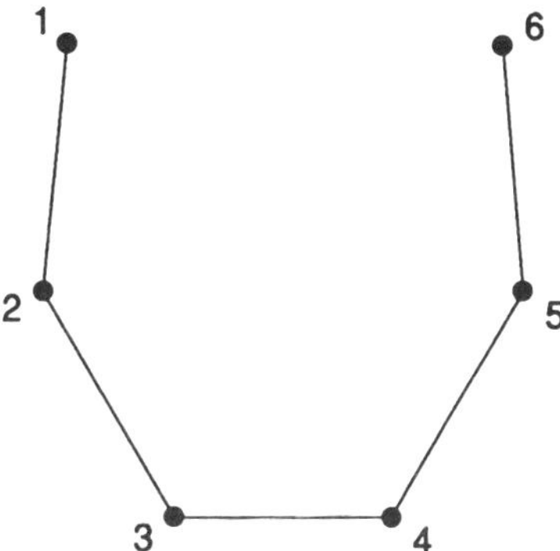

Figure 9.7 Ordination resulting from a principal components analysis of the incidence matrix (Eq. 9.6).

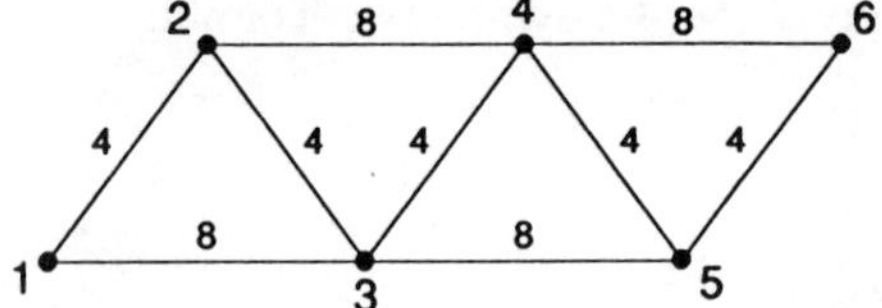

Figure 9.8 The network of distances (Eq. 9.9) between the rows of the incidence matrix (Eq. 9.6).

for some kth and lth rows (see below). The L_1 or the "city block" distance is a more natural measure of association, owing to the hypercube structure of $(0, 1)$ data. For the data matrix (Eq. 9.5) we then have the L_1 distance matrix

$$
\mathbf{D} = \begin{bmatrix}
0 & 4 & 8 & 10 & 10 & 10 \\
4 & 0 & 4 & 8 & 10 & 10 \\
8 & 4 & 0 & 4 & 8 & 10 \\
10 & 8 & 4 & 0 & 4 & 8 \\
10 & 10 & 8 & 4 & 0 & 4 \\
10 & 10 & 10 & 8 & 4 & 0
\end{bmatrix}
\tag{9.8}
$$

which however is still not free of the quadratic horseshoe effects as may be verified, by example, by comparing the first and the last three rows. Williamson (1978) next proposes to replace the equal entries in $\mathbf{D}$ by a graph-theoretic shortest distance between the points, as illustrated in Figure 9.8. The resultant matrix of non-Euclidean distances is then given by

$$
\mathbf{G} = \begin{bmatrix}
0 & 4 & 8 & 12 & 16 & 20 \\
4 & 0 & 4 & 8 & 12 & 16 \\
8 & 4 & 0 & 4 & 8 & 12 \\
12 & 8 & 4 & 0 & 4 & 8 \\
16 & 12 & 8 & 4 & 0 & 4 \\
20 & 16 & 12 & 8 & 4 & 0
\end{bmatrix}
\tag{9.9}
$$

The linear ordination of points can now be carried out using $\mathbf{G}$ in place of the usual Euclidean association matrices. A minor drawback in using Eq. (9.9) however is that although all the latent roots and latent vectors are real, owing to the symmetry of the matrix, some roots can be negative because of the nonpositive (semi)definiteness of $\mathbf{G}$. Components that correspond to negative latent roots however can usually be ignored owing to their lack of interpretability.

Example 9.6. Using Baker's (1965) data, which conform closely to a diagonal structure, Williamson (1978) carries out an ordination of stations and species found in the Indian Ocean (Table 9.6, Fig. 9.9). The data consist of presences (counts) and absences of $p = 17$ species of euphausiids

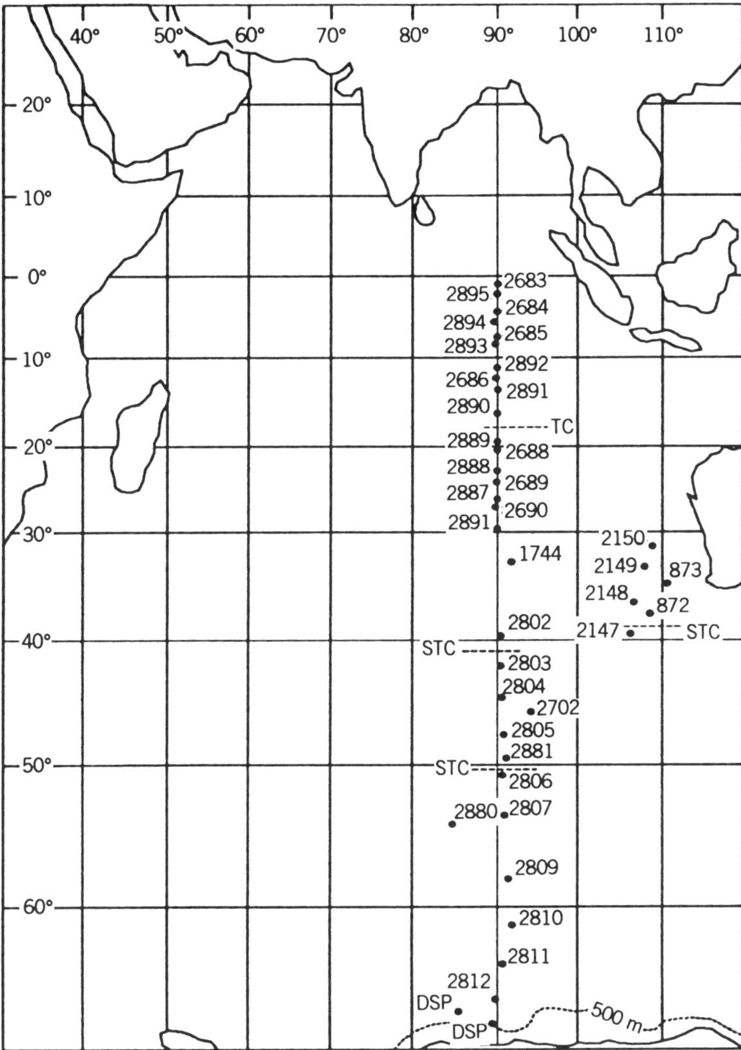

Figure 9.9 Chart showing the positions of R.R.S. "Discovery II" stations (used). The approximated positions of the Tropical (TC), Subtropical (STC) and Antarctic (AC) convergences are also given (DSP are Deutsche Sudpolar Expedition stations; Baker, 1965; reproduced with permission).

(planktonic crustacea) taken at various stations in the Indian ocean along a north–south axis. The species are defined as follows:

$$Y_1 = E.\ diomedeae\ (DI)$$
$$Y_2 = E.\ paragibba\ (PA)$$
$$Y_3 = E.\ brevis\ (BR)$$

$Y_4\ = E.\ tenera$ (TE)

$Y_5\ = E.\ mutica$ (MU)

$Y_6\ = E.\ hemigibba$ (HE)

$Y_7\ = E.\ recurva$ (RE)

$Y_8\ = E.\ spinifera$ (SP)

$Y_9\ = E.\ lucens$ (LU)

$Y_{10} = E.\ similis$ var.armata (SA)

$Y_{11} = E.\ similis$ (SI)

$Y_{12} = E.\ longirostris$ (L)

$Y_{13} = E.\ vallentini$ (VA)

$Y_{14} = E.\ triacantha$ (TR)

$Y_{15} = E.\ frigida$ (FR)

$Y_{16} = E.\ superba$ (SU)

$Y_{17} = E.\ crystallorophias$ (CR)

Since the abundance of species depends in part on water temperatures, it is suspected that there exists a north–south abundance gradient which can be estimated by a statistical ordination. As can be seen from Figure 9.10 however, a conventional principal components analysis of the data of Table 9.6 produces a quadratic rather than a linear configuration of points. Using the step-across similarity matrix we obtain a more linear distribution of the Euphausia species, which indicates a north–south gradient (Fig. 9.11).

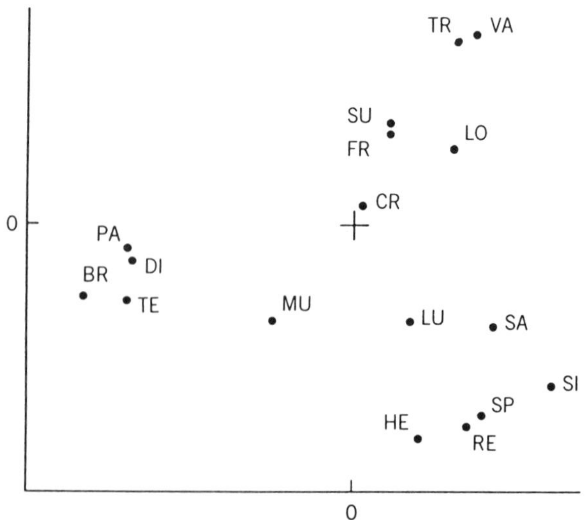

Figure 9.10   Ordinary principal components analysis of the *E. crystallorophias* species counts of Table 9.6.

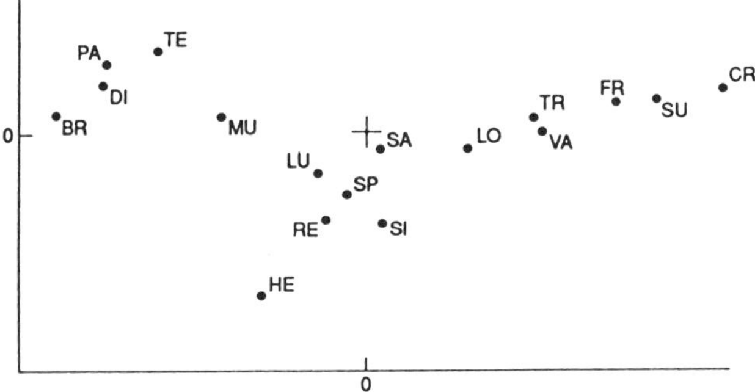

Figure 9.11 A step-across analysis of the *E. crystallorophias* species counts of Table 9.6.

□

Several authors also find the "horseshoe" or the "arch effect" to cause difficulties in optimal scaling, where it also has no meaningful interpretation. For example, Hill and Gauch (1980) have proposed an alternative method for eliminating the horseshoe effect. Since the effect is caused by the first two principal axes being highly related to each other (while at the same time being restricted to be orthogonal), a "detrending" procedure can be used, whereby axes are obtained in such a way that at any point along the first axis the mean value of all subsequent axes is zero. This has the effect of "flattening out" the horseshoe, resulting in a linear ordering of the points. The procedure, however, does not seem to attack the root cause of the problem. In archaeological seriation Kendall (1971c, 1975) has introduced his "circle product" which can be used as a measure of similarity either between elements of an incidence matrix or a matrix of counts (see also Wilkinson, 1974) and which appears to work well in practice. Let n_{ij} be the ith and jth element of a count matrix (0, 1 for an incidence matrix). Then the ith and jth element of Kendall's similarity matrix is defined as

$$(S \circ S)_{ij} = K(i, j) = \sum_{h=0}^{k} \min(n_{ih}, n_{jh}) \tag{9.10}$$

The procedure of adding the smallest pairwise elements of any two rows, implied by Eq. (9.10), may have to be repeated several times until an association matrix results which does not yield a horseshoe effect. A comparison of various approaches can be found in Graham et al. (1976), who have conducted a Monte Carlo study to test the effectiveness of various seriation procedures and measures of association.

9.4 MEASURES OF PAIRWISE CORRELATION OF DICHOTOMOUS VARIABLES

Section 9.3 illustrates the point that at times it may be desirable to employ nonstandard measures of association when considering nominal dummy

variables. Actually a wide choice is available since several such measures
have been developed over the years which embody Euclidian as well as
non-Euclidian properties. In the following section we consider, in a general
context, some of the better known coefficients of association which can be
used for nominal variables when carrying out a factor analysis.

9.4.1 Euclidean Measures of Association

When the categories represent binomial dichotomies, Euclidean measures of
association provide appropriate inputs to a factor analysis, and as is seen in
Section 8.3.3 it becomes particularly convenient to employ $(0, 1)$ dummy
variables as indicators of the presence or absence of some given trait.
Consider the (2×2) contingency table (Table 9.7) which expresses the
essential binary relationships between the dichotomous categories. A well-
known and commonly used measure is the Pearson product correlation
coefficient, which can be expressed as

$$
\begin{aligned}
r &= \frac{n_{11} - (n_{11} + n_{12})(n_{11} + n_{21})}{(n_{11} + n_{12})(n_{21} + n_{22})(n_{11} + n_{21})(n_{12} + n_{22})} \\[2mm]
&= \frac{n_{11} - n_{1.}n_{.1}}{n_{1.}n_{2.}n_{.1}n_{.2}} \\[2mm]
&= \frac{n_{11}n_{22} - n_{12}n_{21}}{n_{1.}n_{2.}n_{.1}n_{.2}}
\end{aligned}
\tag{9.11}
$$

The coefficient is also known as the "phi" coefficient or simply as Pearson's
ϕ. It can be shown that r^2 is related to the chi-squared statistic by the
relationship

$$
r^2 = \frac{\chi^2}{n}
\tag{9.12}
$$

where χ^2 is the chi-squared statistic from the (2×2) contingency table

**Table 9.7   The (2×2) Contingency Table for $(0, 1)$
Binary Data**

Variable B \\ Variable A		1	0	Totals
	1	n_{11}	n_{12}	$n_{1.}$
	0	n_{21}	n_{22}	$n_{2.}$
Totals		$n_{.1}$	$n_{.2}$	n

(Table 9.7), with 1 degree of freedom (Exercise 9.1). The coefficient r^2 is commonly used, for example, in plant ecology when performing a cluster analysis of species (sites) (Williams and Lambert, 1959) and can also be employed in a factor analysis in situations where the distinction between positive and negative association is not of importance. Where such a distinction is of importance, for example, in marketing research or opinion poll questionnaires (Heeler et al., 1977; Jones and Siller, 1978) then r^2 is clearly the appropriate measure of association. Both maximum likelihood factor analysis and principal component models can be employed in conjunction with r^2, although it must be kept in mind that multivariate normality can only hold approximately at best. Another well-known measured based on r^2 is the Pearson coefficient of contingency, defined as

$$P = \left[\frac{r^2}{1 + r^2} \right]^{1/2}$$

$$= \left[\frac{\chi^2}{n + \chi^2} \right]^{1/2} \tag{9.13}$$

Although using the common $(0, 1)$ coding scheme (Table 9.7) is at times convenient, other coding systems can also be used. Thus the $(0, 1)$'s can be replaced by $(-1, +1)$ codes where -1 signifies absence and $+1$ indicates the presence of a trait. Euclidian measures of association can then be computed using such alternative codes. Thus replacing $1, 0$ in Table 9.7 by $+1, -1$, it can be shown that the cosine of the angle between the two random variables is given by

$$r_c = \frac{(n_{11} + n_{22}) - (n_{21} + n_{12})}{n} \tag{9.14}$$

In the psychological and educational literature (Eq. 9.14) is also known as the G coefficient. Note, however, that since it does not correct for the means (marginal differences), r_c is not a true Pearson-type correlation coefficient. Actually Eq. (9.14) has the interesting interpretation as the difference in probabilities between the two sets of equal scores, that is, the probability of scoring the same on both variables minus the probability of scoring differently on both variables. Because of this property it is frequently employed for marking true–false multiple choice questions in an examination in an attempt to eliminate the effect of random guessing. Since Eq. (9.14) can also be defined for sample points it can be used in a Q-mode factor analysis of individuals, and in this context is at times understood as the "correlation" between students' true or real knowledge and the test criteria used to evaluate the students' knowledge (see Ives and Gibbons, 1967).

9.4.2 Non-Euclidean Measures of Association

Because of its symmetry the Euclidean correlation coefficient (Eq. 9.11) requires that both off-diagonal entries of Table 9.7 vanish in order that all those who are A be also B, that is, both nondiagonal terms must vanish to obtain perfect (positive) association. This may at times be an undesirable property and an alternative to Eq. (9.11) which can be used is the coefficient

$$Q = \frac{n_{11}n_{22} - n_{12}n_{21}}{n_{11}n_{22} + n_{12}n_{21}} \tag{9.15}$$

proposed by Yule, where only one off-diagonal frequency need vanish in order to achieve total positive association. Thus $Q = 1$ when all A are B (all B are A), that is, when either $n_{12} = 0$ or $n_{21} = 0$. Kendall and Stuart (1979) have termed this type of association as "complete association" as opposed to "absolute association" represented by $r_r = 1$ in Eq. (9.11). Another measure of complete association is provided by the proportion

$$X = \frac{n_{12}n_{21}}{n_{11}n_{22}} \tag{9.16}$$

where $X = 1$ only when $(n_{11}n_{22} - n_{12}n_{21}) = 0$.

More specialized non-Euclidean measures of association can also be used. Owing to the structure of $(0, 1)$ data, the L_1 distance, for example, can at times be more appropriate than either r or P (Williamson, 1978). Both Euclidean and L_1 measures however utilize all cells of the (2×2) contingency table. In certain applications, such as biological ecology, it is at times inappropriate to use all the cells. Thus when the codes $(0, 1)$ correspond to absence/presence data it may be argued that nothing can be said concerning unobserved events (absences of a trait) and thus not all of the cells should be used when matching occurrence/nonoccurrence of objects or events. If this is deemed to be the case several measures, known collectively as matching coefficients, are available. A well-known coefficient is the Russell–Rao coefficient defines as

$$R = \frac{n_{11}}{n_{11} + n_{21} + n_{12} + n_{22}}$$

$$= \frac{n_{11}}{n} \tag{9.17}$$

which omits those counts from the numerator that correspond to unobserved events, that is, those that correspond to the code 0. When the $(0, 0)$ cell is also removed from the denominator we obtain

$$J = \frac{n_{11}}{n_{11} + n_{12} + n_{21}} \tag{9.18}$$

known as the Jaccard coefficient. Removing also the $(1, 1)$ matches from the denominator yields the further modification

$$K = \frac{n_{11}}{n_{12} + n_{21}} \tag{9.19}$$

or the Kulczynski coefficient. Other combinations designed to fit specific conditions are also possible and may be found, for example, in Anderberg (1973). When a Q-analysis of persons or sample points is of interest, Kendall's measure of association can also be used (see Vegelius, 1982). On the other hand when a bivariate normal continuity is suspected to underlie the dichotomies, a special case of the polychoric correlation coefficient (Eq. 8.3), known as tetrachoric correlation, becomes appropriate. When the two continuous variables are dichotomized at some thresholds h and k, the estimate of ρ is then the tetrachoric estimate obtained by solving

$$f(h, k, \rho) = [2\pi(1 - \rho^2)^{1/2}]^{-1} \int_{-\infty}^{h} \int_{-\infty}^{k} \exp\left\{ -\left[\frac{x^2 - \partial \rho xy + y^2}{2(1 - \rho)^2} \right] \right\} dx \, dy$$

$$\tag{9.20}$$

where the bivariate normal can be expanded using the Hermite–Chebyshev polynomials (Section 8.4.3). Again, a correlation matrix can be built up consisting of tetrachoric correlations and factor analyzed by means of an appropriate factor model. Note again that the resultant correlation matrix need not be positive (semi)definite and its elements may exceed the bounds $(-1, 1)$ for highly nonnormal data. This implies the possibility of negative latent roots, in which case the maximum likelihood factor model cannot be used. Finally, discrete data can also be used to compute proportions or probability estimates, much in the same way as is done for continuous data (Section 5.9.1). The input data for an analysis can then consist of a $(n \times k)$ multivariate data matrix, or else a symmetric matrix of observed proportions which can be viewed as a similarity (association) matrix. Although similarity matrices of this type are normally considered within the framework of multidimensional scaling analysis, factor models can also be used to uncover spaces of minimal dimension which are capable of accounting for the observed proportions (Exercise 9.4).

9.5 MIXED DATA

Frequently data will be mixed. Thus in many applications it is not uncommon to observe data sets consisting of nominal, ranked, and continuous random variables. When ranked data can be viewed as being intrinsically continuous, no major difficulties arise when comparing ordinal, interval, or ratio scales. We can simply treat the ranks as if they represent values on a continuous scale and compute the usual Pearson product-moment correlation coefficient. Alternatively, continuous variables can be converted into ranks if we suspect the presence of nonlinearities or large errors in variables (Section 8.2.1).

A more fundamental issue arises when we wish to compare nominal and ratio or interval scale information. Thus given a set of mixed $(0, 1)$ and continuous random variables a question arises as to whether it is legitimate to include both types of variables in a factor analysis. Two types of attitudes seem to prevail. On the one hand the "purist school," guided perhaps by an excessive degree of theoretical rigidity, is usually wont to caution against the practice of "mixing apples and oranges." On the other end of the spectrum both types of variables are often thrown in together without taking heed of the differences in their matrices, and "churning them through" the usual computer software packages. A compromise between the two attitudes is however both possible and necessary, and lies in the recognition that the use of standard algorithms and procedures with mixed data must be accompanied by specific objectives, an appropriate measure of association and an interpretation of the final results which is in accordance with the type and composition of the data.

First, we observe that in most applications it is probably not feasible to convert continuous variables into dichotomies since this results in a loss of information. The way out of the impasse thus usually lies in the selection of an appropriate correlation coefficient which accords with the structure of the data and which agrees with the origins or genesis of the dichotomous variables. Two well-known correlation coefficients are available, depending on whether the dummy variables represent purely qualitative categories or are simply manifestations of unobserved continuities.

9.5.1 Point Biserial Correlation

Let x be a continuous random variable and y a $(0, 1)$ dummy, say a binomial random variable that represents two mutually exclusive qualitative categories. Then the Pearson product moment correlation between x and y is given by

$$\rho = \frac{E(xy) - E(x)E(y)}{\sigma_x \sigma_y}$$

$$= \frac{E(xy) - E(x)p}{\sigma_x (pq)^{1/2}} \tag{9.21}$$

where $E(y) = p = (1 - q)$ is the true proportion of ones in the population and $\sigma_y^2 = pq$ is the binomial variance (Section 8.3). Let x_i and y_i be sample values and let $\bar{x}_1$ and $\bar{x}_2$ be the sample means of x in the two groups which correspond to the codes $1, 0$ of the dichotomous variable y. Then the sample equivalent of Eq. (9.21) can be expressed as

$$r = \frac{\left(\frac{1}{n_1 + n_2}\sum_{i=1}^{n_1} x_i\right) - \left(\frac{n_1}{n_1 + n_2}\bar{x}\right)}{(s_x^2 \hat{p}\hat{q})^{1/2}}$$

$$= \frac{(\bar{x}_1 - \bar{x}_2)(\hat{p}\hat{q})^{1/2}}{s_x} \tag{9.22}$$

where $\hat{p} = n_1/(n_1 + n_2)$, $\bar{x} = \hat{p}\bar{x}_1 + \hat{q}\bar{x}_2$, and n_1 and n_2 represent the number of observations in the two groups (corresponding to the dummies 0, 1). Equation (9.22) is known as the (sample) point biserial correlation coefficient. The point biserial correlation between a dichotomous dummy and a continuous random variable is therefore equal to the mean difference of the continuous variable in the two groups, adjusted for the standard deviations. A factor analysis of such mixed data is therefore not interpretable in the same way as in the case of a homogeneous set of variables, since we are now comparing correlation coefficients and mean values. Often such a comparison cannot be meaningfully justified and an alternative approach at times proves fruitful. Consider the partitioned sample covariance matrix

$$\mathbf{S} = \begin{bmatrix} \mathbf{S}_{11} & \mathbf{S}_{12} \\ \mathbf{S}_{21} & \mathbf{S}_{22} \end{bmatrix} \tag{9.23}$$

where block diagonals represent covariance matrices of the continuous and dummy variables, respectively, and off diagonals contain covariances between the two sets, that is, terms containing mean differences. We may then wish to cast the analysis in the form of the canonical correlation model (Eq. 8.40). The normal equations are $(\mathbf{S}_{11}^{-1}\mathbf{S}_{12}\mathbf{S}_{22}^{-1}\mathbf{S}_{21} - \lambda_i^2)\mathbf{A}_i = \mathbf{0}$, which now represent a decomposition of the mean differences of the continuous set, adjusted for the covariance matrices $\mathbf{S}_{11}$ and $\mathbf{S}_{22}$ of the two sets, much in the same way as for the bivariate correlation coefficient (Eq. (9.22). Note that the canonical correlation model can also be used in the event the continuities are dichotomized (say, using the median as cut-off point), in which case we end up with a contingency table-type analysis discussed in Section 8.4. The use of the point biserial correlation coefficient in a principal component analysis and the advantage of the method over other procedures such as correspondence analysis in biological ordination is given by Hill and Smith (1976).

9.5.2 Biserial Correlation

The point biserial r does not assume an underlying continuity for the dummy variable. When the dichotomy represents an underlying (normal) continuity, however, Eq. (9.22) becomes inappropriate. In this case the estimator of ρ

is

$$r = \frac{(\bar{x}_1 - \bar{x}_2)(\hat{p}\hat{q})}{s_x z_k} \qquad (9.24)$$

where z_k is the standard normal deviate corresponding to the point of dichotomy k of the normal distribution of y. Equation (9.24) is known as the (sample) biserial correlation coefficient. As for the case of the tetrachoric correlation coefficient the biserial r may exceed the bounds $(-1, 1)$ for highly nonnormal data, and can yield non-Grammian matrices. For approximately normal data, however, the point biserial r can be used to construct correlation matrices, and a factor analysis may then proceed in the usual way since Eq. (9.24) is comparable to the usual Pearson product-moment correlation coefficient. Note however that for correlations between the dummy variables we should now use the tetrachoric correlation rather than the ϕ coefficient (Section 9.4). When the continuous variable is represented by dichotomous counts within a contingency table an alternative form for the biserial r must also be used (see Kendall and Stuart, 1979).

When the continuous normal variate is observed as a multistage polytomy, that is, in the form of rank orders, the notion of biserial correlation can be extended to that of a polyserial correlation coefficient, much in the same way as is done for the tetrachoric correlation of Section 8.2.1 (Olsson et al., 1982). The correlation of a polytomy with a continuous variable can also be understood in terms of the correlation which would be registered if the continuous variable has been approximated as closely as possible by an additive effects model based on the observed character states of the discrete variable (Hill and Smith, 1976). Mixed rank-order and nominal data however are best analyzed in terms of optimal scoring and related techniques discussed in Sections 8.3 and 8.4. For mixed nominal and continuous observations Escofier (1979) has also proposed a recoding of the continuous variables where each is replaced by a set of two codes $(1 - x_i)/2$ and $(1 + x_i)/2$.

Example 9.7. Mixed data frequently occur in exploratory medical research. Thus Nakache et al. (1978), for example, consider data obtained for a set of patients suffering from acute myocardial infarction, which is complicated by pump failure. Since a technique known as intraaortic balloon pumping (IABP) can be beneficial, the authors attempt to uncover high-risk patients who have a high probability of nonsurvival and for whom IABP is particularly necessary. The following 24 variables, which are thought to be good predictors of survival/nonsurvival, are selected for the analysis, with the continuous variables being reduced to discrete categories. The final data used

Original Variables	Unit of Measurement
Sex (M/F)	1 for male, 2 for female
Age (AG)	Numerical value (years)
Previous myocardial infarction (PM)	1 for absence, 2 for presence
Time of study from probable onset of MI (TS)	Numerical value (day)
Location(A/I)	1 for anterior infarction 2 for inferior infarction
Heart rate (HR)	Numerical value
Cardiac index (CI)	Numerical value
Stroke index (SI)	Numerical value
Diastolic aortic pressure (DAP)	Numerical value
Mean aortic pressure (MAP)	Numerical value
Peripheral vascular resistance (PVR)	Numerical value
Ventricular filling pressure (LVFP)	Numerical value
Mean pulmonary artery pressure (MPAP)	Numerical value
Right ventricular filling pressure (RVFP)	Numerical value
Total pulmonary resistance (TPR)	Numerical value
Left ventricular stroke work index (LVSWI)	Numerical value
Right ventricular stroke work index (RVSWI)	Numerical value
Left ventricular minute work index (LVSWI)	Numerical value
Right ventricular minute work index (RVSWI)	Numerical value
LVSWI : LVFP ratio (LI)	Numerical value
RVSWI : RVFP ratio (RIO)	Numerical value
Survival/Death	1 for survival, 2 for death
Conduction disturbances (CD)	1 for absence, 2 for presence
Right ventricular pressure generation index (RVPGI)	Numerical value

consist of binary codes, where the interval of each physical variable is split into approximately equal-size classes by taking into account the physio-pathological significance of the bounds (Table 9.8). The procedure generally results in a loss of information, but the objective of the authors is to use "correspondence analysis" (Section 8.5) to determine which variables characterize survival. The situation is portrayed in Figure 9.12 using an initial "learning" sample of $n = 101$ individuals, and the classification is then tested on a further sample of 55 patients who subsequently entered the Coronary Care Unit. Figure 9.12 indicates the clustering or joint mapping of patients/variables in a joint two-dimensional factor space which is seen to be

Table 9.8   Redefined Categories of Original Variables Obtained for Patients Suffering from Acute Myocardial Infarction Complicated by Pump Failure

Variables	Number of Classes	Code	Classes				
			1	2	3	4	5
Y_1	2	M/F	Male	Female			
Y_2	3	AG	57	58–69	70		
Y_3	2	PM	Absence	Presence			
Y_4	3	TS	1	2	3		
Y_5	2	A/I	Anterior	Inferior			
Y_6	3	HR	82	84–96	99		
Y_7	5	CI	1.23	1.24–1.56	1.58–1.92	1.93–2.40	2.41
Y_8	5	SI	13.8	13.9–17.4	17.5–21.7	21.8–26.9	27.2
Y_9	3	DAP	58	59–67	68		
Y_{10}	3	MAP	69	70–81	82		
Y_{11}	5	PVR	2156	2172–2809	2846–3164	3173–3897	4031
Y_{12}	5	LVFP	14	15–17	18–20	21–24	25
Y_{13}	5	MPAP	19	20–23	24–27	28–31	32
Y_{14}	3	RVFP	7	8–11	12		
Y_{15}	4	TPR	797	807–1108	1131–1608	1610	
Y_{16}	4	LVSWI	10.49	10.72–14.83	15.29–24.00	34.70	
Y_{17}	4	RVSWI	2.74	2.82–4.15	4.16–6.15	6.22	
Y_{18}	4	LVSWI	1022	1031–1325	1328–2098	2131	
Y_{19}	4	RMRVSWI	242	246–387	388–564	573	
Y_{20}	5	LI	0.44	0.45–0.64	0.67–0.93	0.95–1.65	
Y_{21}	5	RIO	0.20	0.21–0.35	0.36–0.53	0.54–1.14	1.16
Y_{22}	2	S/D	Survivor	Non survivor			
Y_{23}	2	CD	Absence	Presence			
Y_{24}	4	RV	0.94	1.05–1.82	1.83–2.88	3.00	

Source: Nakache et al. (1978); reproduced with permission.

composed of a survival zone, a death zone, and the indeterminate zone of overlap.

9.6   THRESHOLD MODELS

It was observed in Section 9.4 that when a set of p dichotomous dummy variables are assumed to represent p underlying normal variates, it is more appropriate to consider tetrachoric correlation coefficients. A principal components analysis can then be performed on the matrix of tetrachoric correlations, with little change in the final interpretation of the results. Such a heuristic approach however suffers from the drawback that a tetrachoric correlation matrix need not be positive—definite, with the result that it is not possible to use weighted factor models such as those described in Chapter 6.

An alternative approach is to introduce the concept of a threshold point, as traditionally employed in probit or logit bioassay experiments when estimating quantal-response effective dosages. Consider a set of p continuous random variables $x_1, x_2, \ldots, x_p$ which are of main interest. For various

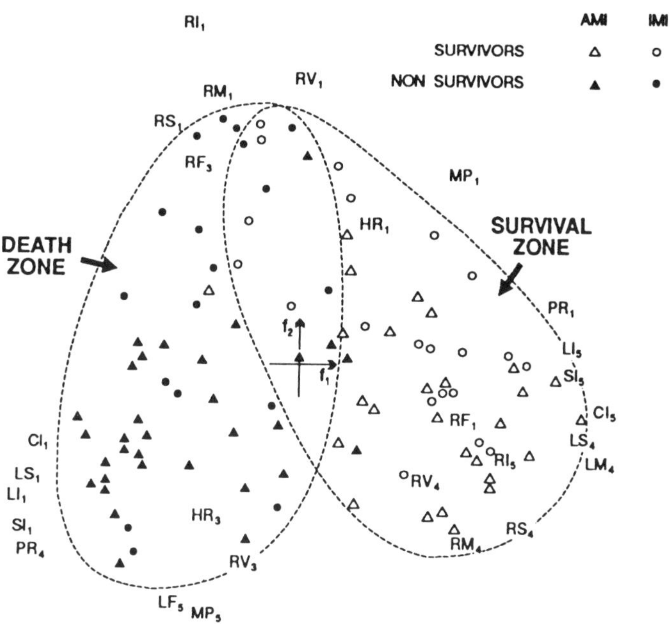

Figure 9.12 Simultaneous scatter configuration of learning sample patients and binary variables on the factorial plane (F_1 and F_2) issued from correspondence analysis.

reasons we are not able to observe the continuities, but can only observe corresponding quantative changes in terms of the dummy variables, that is, we can only observe

$$y = \begin{cases} 1 & \text{if } x_i \geq t_i \\ 0 & \text{if } x_i < t_i \end{cases} \tag{9.25}$$

Equation (9.25) is equivalent to assuming that an observed $2 \times 2 \times \cdots \times 2 = 2^p$ contingency table arises from grouping each dimension of a p-variate multivariate normal distribution into categories or dichotomous ranks. The t_i are referred to as critical points or threshold values and are in general assumed to vary randomly from individual to individual. Fixed threshold points occur commonly in the physical sciences, for example, such as melting (freezing) points of liquids. Here the underlying variable (heat) is readily observed and can be measured directly. In many other areas such as encountered in the social, biological, or life sciences, however, the critical points are random, and in addition cannot be observed directly. Thus the decision for a housewife to reenter the labor force, for example, may be a function of her spouse's income. As long as the income is above a certain threshold point for that individual, the housewife will not join the labor force. Once it descends below this critical value, however, she may well feel the necessity to rejoin the labor force in order to maintain family income above this value. An economic threshold point will in addition vary randomly between individuals depending on their life styles, position in the life cycle, and so on. Also, when applying a dosage of insecticide to an insect species, for example, we cannot observe directly the insect's intrinsic tolerance level to the poison. Rather we observe, after a sufficient period of time, two states. Live insects for whom the dosage was below their threshold point t, and dead insects whose resistance was insufficient to overcome the dosage.

When the dummy variables are functions of unobservable threshold points, it becomes inappropriate to compute factors solely in terms of the observed dummy variables, since the factors of interest are those that account for the variation in the unobservable continuities $X_1, X_2, \ldots, X_p$. The general factor model is then

$$\mathbf{X} = \boldsymbol{\mu} + \boldsymbol{\alpha}\boldsymbol{\Phi} + \boldsymbol{\epsilon} \tag{9.26}$$

where $\mathbf{X}$ is the $(p \times 1)$ vector of unobserved variables and $\boldsymbol{\Phi}$ represents the $(r \times 1)$ vector of r common factors (Chapter 6). Assuming $E(\mathbf{X}) = \boldsymbol{\mu} = \mathbf{0}$, the unobserved covariance matrix $\boldsymbol{\Sigma}$ can be decomposed, generally, as

$$\boldsymbol{\Sigma} = \boldsymbol{\alpha}\boldsymbol{\Omega}\boldsymbol{\alpha}^{\mathrm{T}} + \boldsymbol{\Psi} \tag{9.27}$$

where $\boldsymbol{\Omega}$ is the $(r \times r)$ common factor covariance (correlation) matrix. Note that in Eq. (9.26) both the dependent as well as independent variables are latent, that is, not observed. The model may thus be viewed as incorporating a "two-tier" latent structure, where the observed dichotomies have underlying latent continuities which in turn are functions of other latent factor continuities. The multivariate probit maximum likelihood factor analysis model assumes further that the underlying continuity is multivariate normal, which permits the estimation of the so-called "probits." The population probability of observing a p-dimensional dummy variable, say $\mathbf{Y} = (1, 0, 1, \ldots, 0)^{\mathrm{T}}$, is

$$g(y) = \int_{t_1}^{\infty} \int_{-\infty}^{t_2} \int_{t_2}^{\infty} \cdots \int_{-\infty}^{t_p} |\boldsymbol{\Sigma}|^{-1/2} (2\pi)^{-p/2} \exp[-1/2(\mathbf{u}^{\mathrm{T}}\boldsymbol{\Sigma}\mathbf{u})] \, d\mathbf{u} \tag{9.28}$$

where $\mathbf{u}$ is a normal vector of integration variables and $\boldsymbol{\Sigma}$ is as in Eq. (9.27). The integral (Eq. 9.28) on which maximum likelihood estimation is based does not exist in closed form, and must consequently be evaluated by numerical integration. This places a heavy burden on the limits of computation since the full maximum likelihood method implied by Eq. (9.28) uses all 2^p cell frequencies of the implied contingency table, that is, it uses 2-way, 3-way, $\ldots$, p-way associations between the dichotomous variables. As pointed out by Bock and Lieberman (1970; see also Lawley, 1944), however, for a one-factor model the practical limit for such an exercise appears to be (at present) at most a dozen or so variables. The situation therefore differs from the usual maximum likelihood factor analysis where only pairwise associations or correlations are considered. If for the probit model third and higher order associations can be ignored, this results in a significant simplification in the computations since we at most consider two-dimensional marginals, that is, we consider the integrals

$$P_i^* = P(x_i \equiv 1) = \int_{t_i}^{\infty} (2\pi)^{-1/2} \exp\left(-\frac{u^2}{2}\right) du \tag{9.29}$$

$$P_{ij}^* = P(x_i \equiv 1, x_j \equiv 1) = \int_{t_i}^{\infty} \int_{t_j}^{\infty} |\mathbf{\Sigma}_{ij}|^{-1/2}(2\pi)^{-1} \exp[-1/2(u^{\mathrm{T}}\mathbf{\Sigma}_{ij}u) \, du$$

$$\tag{9.30}$$

where

$$\mathbf{\Sigma}_{ij} = \begin{bmatrix} 1 & \sigma_{ij} \\ \sigma_{ij} & 1 \end{bmatrix}, \qquad \sigma_{ij} = \sum_{k=1}^{r} \sum_{l=1}^{r} \alpha_{ik}\alpha_{jl}\Omega_{kl} \tag{9.31}$$

and Ω_{kl} is the (k, l)th element of the factor covariance matrix $\mathbf{\Omega}$. This is the method adopted by Christofferson (1975, 1977), and has the advantage of only using the first- and second-order marginals of the 2^p contingency table. The simplification permits the use of a greater number of variables and enables the extraction of more common factors. The numerical integration implied by Eq. (9.30) is further simplified by using the Hermite–Chebyshev polynomial expansion of Section 8.4.2, that is, by expressing P_{ij}^* as the infinite series

$$P_{ij}^* = \sum_{k=0}^{\infty} \sigma_{ij}^2 \tau_k(t_i)\tau_k(t_j) \tag{9.32}$$

where τ_k is the kth tetrachoric function

$$\tau_k(x) = \frac{H_{k-1}(x)f(x)}{(k!)^{1/2}} \tag{9.33}$$

We can then define, for the p random variables,

$$P_i = P_i^* + \epsilon_i \qquad (i = 1, 2, \ldots, p)$$

$$P_{ij} = P_{ij}^* + \epsilon_{ij} \qquad (i = 1, 2, \ldots, p-1; j = i+1, \ldots, p)$$

which expresses the observed sample proportions P_i and P_{ij} in terms of the threshold levels. To estimate the parameters of the factor model Christofferson (1975, 1977) minimizes the generalized least squares criterion

$$F(t, \alpha, \Phi) = \Sigma^{\mathrm{T}} \mathbf{S}^{-1} \Sigma$$

$$= (\mathbf{P} - \mathbf{P}^*) \mathbf{S}^{-1} (\mathbf{P} - \mathbf{P}^*) \tag{9.34}$$

where $\mathbf{S}$ is a consistent estimator of the residual covariance matrix. When the loadings are not rotated, min $F(t, \alpha, \Phi)$ is asymptotically distributed as chi-squared with $(p/2)(p + 1)$ degrees of freedom.

This method requires repeated computation of integrals of the form of Eqs. (9.29) and (9.30), together with their derivatives. More recently Muthén (1978) has suggested an iterative algorithm based on the tetrachoric correlation coefficients which avoids the iterated computation of the integrals. Since the probit maximum likelihood factor model depends in part of nonlinear expressions Muthén's strategy is to first linearize the relationship between the observed proportions and the vector $\Theta = (\theta_1^{\mathrm{T}}, \theta_2^{\mathrm{T}})$, where θ_1 is the vector of population thresholds t_i and θ_2 is the vector of population tetrachoric correlations. The algorithm also permits the extraction of multiple common factors, but because of its greater simplicity it can also be used with a larger number of variables. In effect Muthén's (1978) procedure uses tetrachoric correlations from two-way contingency tables to estimate correlations between the unobserved normal continuities, and then estimates the threshold levels themselves using weighted least squares to increase efficiency. However, for a larger set of 20–25 variables, the author reports that little difference can be observed between the weighted and unweighted least squares criteria. Both the weighted and unweighted criteria can be tested using an asymptotic chi-squared statistic.

More generally, and computational aspects aside, the Bock and Lieberman (1970), Christofferson (1975), and Muthén (1978) models all share common features in that they consider the same population model, for which the continuities underlying the 0–1 dummy variables (as well as the latent common factors) are assumed to be normally distributed. This allows the use of the well-known probit function defined as $\mathrm{probit}(u) = \phi^{-1}(u)$ where ϕ is the normal cumulative probability distribution function. More recently Bartholomew (1980, 1983) has proposed another well-known cumulative function, the so-called logit function of a probability p defined as $\mathrm{logit}(p) = \ln[p/(1 - p)]$ which can be used in place of probits for nonnormal data. The logistic is similar to the probit function, but accepts more outlying observations. Indeed, Batholomew (1980) shows that both the probit and logistic functions satisfy a list of six conditions of a well-behaved response function, which probably explains in part why both tend to yield fairly similar results. The logistic function however has the advantage in that it can be expressed in algebraic or closed form, and this makes it easier to manipulate terms. Let $p_i(\phi)$ be a response function defined as the probability that a positive response ($y_i = 1$) occurs for an individual whose latent position is given by ϕ. Bartholomew (1980) then defines the logistic factor model as

$$G[p_i(\phi)] = \alpha_{io} + \sum_{j=1}^{r} \alpha_{ij} H(\phi_j) \qquad (i = 1, 2, \ldots, p) \qquad (9.35)$$

where α_{ij} are the factor loadings and the expressions G and H represent logistic functions, which then yields

$$\text{logit}[p_i(\phi)] = \ln[p_i/(1 - p_i)] + \sum_{j=1}^{q} \alpha_{ij} \ln[\phi_j/(1 - \phi_j)]$$

$$(i = 1, 2, \ldots, p) \qquad (9.36)$$

The theory is developed within a Baysian framework and can be generalized to more than a single threshold level when ranked or polytomous data are used, but for the case $r > 1$ the logit factor model appears to suffer from computational difficulties. Probit and logit threshold models are also described by Andersen (1980), who classifies them generally under the heading "latent structure models," which should not be confused with the latent class: (structure) models of the following section.

9.7   LATENT CLASS ANALYSIS

The classical factor models described in the previous chapters, together with their extensions and embellishments, all possess one main common characteristic. In all cases the underlying factor scores are defined in a continuous vector space, and the factor loadings for both continuous and discrete manifest variables can then be considered in terms of Pearson product-moment correlation coefficients. An exception to this general model was, however, already noted in Section 6.13 where Lazarsfeld's (1950) latent profile model is defined in terms of a discrete factor structure which is considered to be responsible for generating the set of continuous observed or manifest variables. A further departure from the usual factor models is the so-called latent class analysis model where both manifest and latent "factors" represent discrete categories (Lazarsfeld, 1950). The general situation is depicted in Table 9.9 in terms of type of manifest and latent random variables, where it may be seen that the latent class model requires the input variables to be in nominal form. Since ratio or rank order information is not required the latent class model is well suited for the analysis or reduction of purely qualitative categories into a smaller set of qualitative latent classes. The need for such a reduction may occur, for example, when the observed multivariate classifications represent a mixture

Table 9.9   A Cross Classification of Factor Model Versus Type of Random Variable

Factors (Latent Variables)	Observed Variables	
	Metrical (Continuous; Ranks)	Categorical (Nominal)
Metrical continuous; ranks	Standard PCA; FA	PCA, FA of multivariate multinomial data
Categorical (nominal)	Latent profile analysis (LPA)	Latent class analysis (LCA)

of a smaller number of unobserved categories, and the objective is then to "unmix" the sample into its basic discrete constituents.

Since latent class analysis uses categorical, that is, nominal factors, consider a hypothetical example provided by Lazarsfeld and Henry (1968) where we have three dichotomies: "education" (high, low); "read newspaper A" (yes, no); "read newspaper B" (yes, no). Forming the (2×2) contingency table of newspaper readership we may conclude that some individuals who read A also tend to read B, that is, newspaper readership is not independent across the sample of individuals. The interdependency however may be due to a third underlying variable such as education, in the sense that once the (2×2) table is stratified on this variable, "newspaper readership" may well become independent (Table 9.10), since the dependence may be due to the nonhomogeneous nature of the observed sample; that is, the sample may represent a mixture of binomial (multinomial) distributions. In this example the third category is observable and may thus be readily taken into account. More generally the intervening categories may not be observable or may be difficult to observe, thus constituting a latent category or class. The principal objective of latent class analysis is then to estimate membership and the number of such classes on condition

Table 9.10   Stratification of a (2×2) Contingency Table on a Third Category (Education) such that Rows and Columns are Independent Within the High Education and Low Education Branches

Education (High, Low)

High Education

	Read A	Do not read A
Read B Do not read B	p_{11}	p_{12}
	p_{21}	p_{22}

Low Education

	Read A	Do not read A
Read B Do not read B	p_{11}	p_{12}
	p_{21}	p_{22}

that once latent classes are estimated the observed interdependencies of the (2×2) contingency tables vanish. Lazarsfeld has termed this "the axiom of local independence" (see Lazarsfeld and Henry, 1968) which corresponds to the conditional independence assumption of continuous factor models (Chapter 6) where the observed correlations vanish once the common factors are taken into account.

The latent class model can be understood and developed from standard probability theory making use of Bayes' theorem. Consider any two dichonomous random variables X_1 and X_2 defined as

$$x_i = \begin{cases} 1 & \text{if event } E_i \text{ is observed} \\ 0 & \text{if event } E_i \text{ not observed} \end{cases}$$

for $i = 1$ and 2. The outcome corresponding to the code "1," which is defined arbitrarily, is known as the "positive" outcome or the "success" of the trial and the code "0" is termed as the "negative" or the "failure" outcome of the trial. For example, x_i may represent independent binomial trials or outcomes for any other discrete distribution. The following theorem is well known from classical probability theory.

THEOREM 9.2. Let A_1, A_2, ..., A_m represent a mutually exclusive partition of a sample space S, such that $P(A_s) \neq 0$, $s = 1, 2, \ldots, m$, and $\Sigma_{s=1}^{m} P(A_s) = 1$. Let E_i be some arbitrary event defined in S. Then

$$P(E_i) = \sum_{s=1}^{m} P(A_s)P(E_i \mid A_s) \tag{9.37}$$

PROOF. We have

$$P(E_i) = P(E_i \cap S)$$

$$= P[E_i \cap (A_1 \cup A_2 \cup \cdots \cup A_m)]$$

$$= P[(E_i \cap A) \cup (E_i \cap A_2) \cup \cdots \cup (E_i \cap A_m)]$$

$$= P(E_i \cap A_1) + P(E_i \cap A_2) + \cdots + P(E_i \cap A_m)$$

$$= \sum_{s=1}^{m} P(E_i \cap A_s)$$

and using the definition of a conditional probability we have

$$P(E_i \mid A_s) = \frac{P(E_i \cap A_s)}{P(A_s)}$$

so that

$$P(E) = \sum_{s=1}^{m} P(A_s)P(E_i \mid A_s) \tag{9.38}$$

The result of Theorem 9.2 can be generalized to any finite number of independent events. Thus for two events we have

$$P(E_i \cap E_j) = \sum_{s=1}^{m} P(E_i \cap E_j \cap A_s)$$

where

$$P(E_i \cap E_j \mid A_s) = \frac{P(E_i \cap E_j \cap A_s)}{P(A_s)}$$

and cross multiplying yields

$$P(E_i \cap E_j) = \sum_{s=1}^{m} P(A_s)P(E_i \cap E_j \mid A_s)$$

$$= \sum_{s=1}^{m} P(A_s)P(E_i \mid A_s)P(E_j \mid A_s) \tag{9.39}$$

It is easy to show by induction that for any finite number of k arbitrary events $E_1, E_2, \ldots, E_k$, we have the relation

$$P(E_1 \cap E_2 \cap \cdots \cap E_k) = \sum_{s=1}^{m} P(A_s)P(E_1 \mid A_s)P(E_2 \mid A_s) \cdots P(E_k \mid A_s)$$

$$\tag{9.40}$$

In a typical application of Eq. (9.40) the probabilities $P(A_s)$ of the partitioning or "background" events are given, and $k < m$. In the latent class model however we assume the events $A_1, A_2, \ldots, A_m$ to be unobservable and the probabilities $P(A_s)$ to be unknown. The problem then poses itself as follows. Given the observed probabilities $P(E_1), P(E_2), \ldots, P(E_k)$ is it possible to compute the conditional probabilities $P(E_i \mid A_s)$ $(i = 1, 2, \ldots, k;$ $s = 1, 2, \ldots, m)$ together with the partition probabilities $P(A_s)$ such that $m < k$? Consider k dichotomous random variables (the observed classes) and $m < k$ unobserved classes, where

$$\hat{P}(E_i) = \hat{p}_i =$$ Observed proportion of sample points that respond positively to the ith dichotomy $(i = 1, 2, \ldots, k)$.

$$\hat{P}(E_i \cap E_j) = \hat{p}_{ij} =$$ Observed proportion of sample points which respond positively to both the ith and jth dichotomies $(i \neq j,\ p_{ij} = p_{ji})$.

$$\hat{P}(E_i \cap E_j \cap \cdots \cap E_k) = \hat{p}_{ij\ldots k} =$$ Observed proportion of sample points that respond positively to the ith, jth, $\ldots, k$th dichotomies $(i \neq j \neq \cdots \neq k)$, where permutations of indices are excluded.

$$P(A_s) = \pi_s =$$ The unobserved probability of being in the sth latent class, $s = 1, 2, \ldots, m$.

$$P(E_i \mid A_s) = \nu_{is} =$$ The unobserved conditional probability that a sample point in the sth latent class is also in the ith observed category (probability of sample point in latent class A_s responding favorably to dichotomy E_i).

Since the sets $A_1, A_2, \ldots, A_m$ represent a partition of the sample space, $\Sigma_{s=1}^{m} \pi_s = 1$. When conditions for multinomial sampling are met, the π_s can also be considered as multinomial probabilities. Using the notation shown above we have the normal equations

$$1 = \sum_{s=1}^{m} \pi_s \tag{9.41a}$$

$$\hat{p}_i = \sum_{s=1}^{m} \pi_s \nu_{is} \tag{9.41b}$$

$$\hat{p}_{ij} = \sum_{s=1}^{m} \pi_s \nu_{is} \nu_{js} \qquad (i, j = 1, 2, \ldots, k) \tag{9.41c}$$

$$\hat{p}_{ij\cdots l} = \sum_{s=1}^{m} \pi_s \nu_{is} \nu_{js}, \ldots, \nu_{ls} \qquad (i, j, \ldots, l = 1, 2, \ldots, k) \tag{9.41d}$$

where $i \neq j \neq \cdots \neq l$ and permuted subscripts do not appear. Equations (9.41) express observed probabilities in terms of unknown probabilities, and represent the general system of normal equations for a latent class model. They are also known in the literature as the "accounting" equations. In practice, given a sample the observed joint frequencies or the manifest probabilities are substituted on the left-hand side of Eq. (9.41), and assuming the existence of a unique solution (identifiability) the system can be solved to yield estimates of the unobserved parameters π_s, ν_{is}, $\nu_{js}, \ldots, \nu_{ls}$, known as the latent probabilities.

The system of equations (Eq. 9.41) resembles the normal equations of factor analysis, except for two major differences. First, the notion of a continuous latent variable or factor is replaced by the corresponding notion of a discrete latent set of classes or categories. Second, the accounting or normal equations of latent class analysis contain third, fourth, $\ldots, l$th higher moments or "interactions" in terms of the joint probabilities, which do not exist in classical factor models which only make use of second order covariance terms. For $k > m$ observed dichotomies, the largest number of joint probabilities possible is when $l = k$, although in practice very few higher order moments are normally used. Taking $p_0 \equiv 1$ as the "null subscript" probability, the maximum number of equations which is possible is then $\Sigma_{l=0}^{k} C(_l^p) = 2^k$. Since for k observed categories and $m < k$ latent classes the total number of unknown latent parameters is $m + mk = m(k + 1)$, a necessary condition for identifiability is that $2^k \geq m(k + 1)$, that is,

$$\frac{2^k}{k + 1} \geq m \qquad (9.42)$$

Equation (9.42) can always be satisfied however since given m we can always take k sufficiently large assuming a large sample. Conversely, if the system is identifiable and if there are more equations (i.e., joint probabilities) than parameters, a solution can always be found by using a subset of the normal equations. Also, not all of the k dichotomous variables need be used. Thus, given the somewhat large degree of latitude which emerges from the model specification, a number of different models have been developed, depending on the type and amount of information employed (i.e., depending on the number of variables) and the number of joint probabilities which are used.

Since Eq. (9.41) is generally nonlinear, an added difficulty arises when we attempt to express the normal equations in matrix form. The most straightforward solution of the normal equations is that of Anderson (1954, 1959; see also Lazarsfeld and Henry 1968; Fielding, 1978), which extends Lazarsfeld's (1950) original work (see also Koopmans, 1951). The simplified model uses only $2m - 1$ variables, and only second- and third-order joint probabilities are employed to define the set of (identified) equations from which the latent probabilities are estimated. In addition, as for the maximum likelihood factor model, the latent class model requires a prior specification of m. The latent class model can then be expressed in modified form as

$$1 = \sum_{s=1}^{m} \pi_s \qquad (9.42a)$$

$$p_i = \sum_{s=1}^{m} \pi_s \nu_{is} \qquad (i = 1, 2, \ldots, k) \qquad (9.42b)$$

$$p_{ij} = \sum_{s=1}^{m} \pi_s \nu_{is} \nu_{js} \qquad (i = 1, 2, \ldots, k) \qquad (9.42c)$$

$$p_{ijl} = \sum_{s=1}^{m} \pi_s \nu_{is} \nu_{js} \nu_{ls} \qquad (i, j, l = 1, 2, \ldots, k) \qquad (9.42d)$$

for all possible combinations of the indices i, j, k, where $p_0 \equiv 1$. The Anderson (1954) model requires a selection of $2m - 2$ observed dichotomies in order to form two distinct sets of $m - 1$ variables, each such that $2m - 1 < k$ plus an additional (observed) variable called the stratifier variable. This permits us to express Eq. (9.42) in relatively straightforward matrix form, where the stratifier variable (which receives the subscript l) is used to compute the third-order joint probabilities of Eq. (9.42d). Which of the $2m - 2$ observed variates and which stratifier are chosen for the analysis is arbitrary, so that if an initial choice proves unsatisfactory, a different portion of the variables may be used to improve the solution.

Following Anderson, define the following matrices:

$$\mathbf{N} = \begin{bmatrix} \pi_1 & & & \mathbf{0} \\ & \pi_2 & & \\ \mathbf{0} & & \ddots & \\ & & & \pi_. \end{bmatrix} \qquad \mathbf{\Delta} = \begin{bmatrix} \nu_{k1} & & & \mathbf{0} \\ & \nu_{k2} & & \\ \mathbf{0} & & \ddots & \\ & & & \nu_{km} \end{bmatrix}$$

$$\mathbf{\Lambda} = \begin{bmatrix} 1 & \nu_{11} & \nu_{21} & \cdots & \nu_{2m-2,1} \\ 1 & \nu_{12} & \nu_{22} & \cdots & \nu_{2m-2,2} \\ \hline 1 & \nu_{1m} & \nu_{2m} & \cdots & \nu_{2m-2,m} \end{bmatrix}$$

$$\mathbf{P}_1 = \begin{bmatrix} 1 & \bar{p}_m & \cdots & p_{2m-2} \\ p_1 & p_{1,m} & \cdots & p_{1,2m-2} \\ p_2 & p_{2,m} & \cdots & p_{2,2m-2} \\ \vdots & \vdots & & \vdots \\ p_{m-1} & p_{m-1,m} & \cdots & p_{m-1,m-2} \end{bmatrix}$$

$$\mathbf{P}_2 = \begin{bmatrix} p_k & p_{m,k} & \cdots & p_{2m-2,k} \\ p_{1k} & p_{1,m,k} & \cdots & p_{2,2m-2,k} \\ \vdots & \vdots & & \vdots \\ p_{m-1,k} & p_{m-1,m,k} & \cdots & p_{m-1,2m-2,k} \end{bmatrix}$$

$$\mathbf{\Lambda}_1 = \begin{bmatrix} 1 & \nu_{11} & \nu_{21} & \cdots & \nu_{m-1,1} \\ 1 & \nu_{12} & \nu_{22} & \cdots & \nu_{m-1,2} \\ \hline 1 & \nu_{1m} & \nu_{2m} & \cdots & \nu_{m-1,m} \end{bmatrix} \qquad \mathbf{\Lambda}_2 = \begin{bmatrix} 1 & \nu_{m,1} & \nu_{m+1,1} & \cdots & \nu_{2m-2,1} \\ 1 & \nu_{m,2} & \nu_{m+1,2} & \cdots & \nu_{2m-2,2} \\ \hline 1 & \nu_{m,m} & \nu_{m+1,m} & \cdots & \nu_{2m-2,m} \end{bmatrix}$$

using the selected $2m - 2$ dichotomies (without the stratifier variable) where the first column corresponds to the dummy item $p_0 \equiv 1$, $\mathbf{P}_1$ is the $(m \times m)$ matrix of observed first- and second-order joint probabilities, and $\mathbf{P}_2$ is the $(m \times m)$ matrix of first-, second- and third-order probabilities involving the stratifier dichotomy. The matrices $\mathbf{\Lambda}_1$ and $\mathbf{\Lambda}_2$ correspond to the latent conditional probabilities of the two sets of observed variables containing $m - 1$ dichotomies each, together with a column of unities (but excluding the stratifier), that is, they consist of the latent marginal probabilities bordered by a column of ones. The normal equations (Eq. 9.42) can then be expressed in matrix form as

$$\mathbf{P}_1 = \mathbf{\Lambda}_1^{\mathrm{T}}\mathbf{N}\mathbf{\Lambda}_2 \tag{9.43a}$$

$$\mathbf{P}_2 = \mathbf{\Lambda}_1^{\mathrm{T}}\mathbf{N}\mathbf{\Delta}\mathbf{\Lambda}_2 \tag{9.43b}$$

where Eq. (9.43a) contains the first- and second-order joint probabilities and Eq. (9.43b) contains the first-, second-, and third-order probabilities involving the stratifier dichotomy.

Example 9.8. As an example consider the case where $k = 3$ and $m = 2$. We have

$$\mathbf{P}_1 = \mathbf{\Lambda}_1^{\mathrm{T}}\mathbf{N}\mathbf{\Lambda}_2$$

that is,

$$\begin{bmatrix} 1 & p_2 \\ p_1 & p_{12} \end{bmatrix} = \begin{bmatrix} 1 & 1 \\ \nu_{11} & \nu_{12} \end{bmatrix} \begin{bmatrix} \pi_1 & 0 \\ 0 & \pi_2 \end{bmatrix} \begin{bmatrix} 1 & \nu_{21} \\ 1 & \nu_{22} \end{bmatrix}$$

$$= \begin{bmatrix} (\pi_1 + \pi_2) & (\pi_1 \nu_{21} + \pi_2 \nu_{23}) \\ (\pi_1 \nu_{11} + \pi_2 \nu_{12}) & (\pi_1 \nu_{11} \nu_{21} + \pi_2 \nu_{12} \nu_{22}) \end{bmatrix}$$

or, in equation form,

$$1 = \pi_1 + \pi_2$$

$$p_1 = \pi_1 \nu_{11} + \pi_2 \nu_{12}$$

$$p_2 = \pi_1 \nu_{21} + \pi_2 \nu_{22}$$

$$p_{12} = \pi_1 \nu_{11} \nu_{21} + \pi_2 \nu_{12} \nu_{23}$$

Likewise, for $\mathbf{P}_2 = \mathbf{\Lambda}_1^{\mathrm{T}}\mathbf{N}\mathbf{\Delta}\mathbf{\Lambda}_2$ we have

$$\begin{bmatrix} p_3 & p_{23} \\ p_{13} & p_{123} \end{bmatrix} = \begin{bmatrix} 1 & 1 \\ \nu_{11} & \nu_{12} \end{bmatrix} \begin{bmatrix} \pi_1 & 0 \\ 0 & \pi_2 \end{bmatrix} \begin{bmatrix} \nu_{31} & 0 \\ 0 & \nu_{32} \end{bmatrix} \begin{bmatrix} 1 & \nu_{21} \\ 1 & \nu_{22} \end{bmatrix}$$

$$= \begin{bmatrix} (\pi_1 \nu_{31} + \pi_2 \nu_{32}) & (\pi_1 \nu_{31} \nu_{21} + \pi_2 \nu_{32} \nu_{22}) \\ (\pi_1 \nu_{11} \nu_{31} + \pi_2 \nu_{12} \nu_{32}) & (\pi_1 \nu_{11} \nu_{31} \nu_{21} + \pi_1 \nu_{11} \nu_{32} \nu_{22}) \end{bmatrix}$$

or, in equation form,

$$p_3 = \pi_1 \nu_{31} + \pi_2 \nu_{32}$$

$$p_{13} = \pi_1 \nu_{11} \nu_{31} + \pi_2 \nu_{12} \nu_{32}$$

$$p_{23} = \pi_1 \nu_{31} \nu_{21} + \pi_2 \nu_{32} \nu_{23}$$

$$p_{123} = \pi_1 \nu_{11} \nu_{31} \nu_{21} + \pi_1 \nu_{11} \nu_{32} \nu_{22}$$

where the third subscript (3) corresponds to the stratifier dichotomy. $\square$

The normal equations (Eq. 9.43) can be solved by finding the generalized left and right latent vectors of the matrices $\mathbf{P}_1$ and $\mathbf{P}_2$ (Anderson, 1954, 1959; see also Basilevsky, 1983). The objective here is to find a suitably small number $m < k$ of latent classes for which the observed dichotomies (contingency tables) become independent. We first consider the following lemma:

Lemma 9.1. Let

$$(\mathbf{P}_2 - \phi_i\mathbf{P}_1)\mathbf{X}_i = \mathbf{0} \tag{9.44a}$$

$$\mathbf{Y}_i^\mathrm{T}(\mathbf{P}_2 - \mu_i\mathbf{P}_1) = \mathbf{0} \tag{9.44b}$$

where $\mathbf{X}_i$ and $\mathbf{Y}_i$ are the right and left latent vectors of $\mathbf{P}_2$ with respect to $\mathbf{P}_1$, respectively. Then $\phi_i = \mu_i$. $\square$

Proof. The proof follows by observing that for the homogeneous equations (Eq. 9.44) we have

$$|\mathbf{P}_2 - \phi_i\mathbf{P}_1| = |\mathbf{P}_2 - \mu_i\mathbf{P}_1| = 0$$

so that ϕ_i and μ_i satisfy the same characteristic equation. $\square$

The solution of the system (Eq. 9.43) proceeds by first finding $\mathbf{\Delta}$, which contains the latent roots of Eq. (9.44). We have

$$
\begin{aligned}
|\mathbf{P}_2 - \phi_i\mathbf{P}_1| &= |\mathbf{\Lambda}_1^\mathrm{T}\mathbf{N}\mathbf{\Delta}\mathbf{\Lambda}_2 - \phi_i\mathbf{\Lambda}_1^\mathrm{T}\mathbf{N}\mathbf{\Lambda}_2| \\
&= |\mathbf{\Lambda}_1^\mathrm{T}||\mathbf{N}||\mathbf{\Delta} - \phi\mathbf{I}||\mathbf{\Lambda}_2| \\
&= 0 \tag{9.45}
\end{aligned}
$$

so that the elements $v_{k1}, v_{k2}, \ldots, v_{km}$ of $\mathbf{\Delta}$ are the latent roots $\phi_1 > \phi_2 > \cdots > \phi_m$. In addition since $\mathbf{N}$, $\mathbf{\Lambda}_1$, and $\mathbf{\Lambda}_2$ are nonsingular, this implies the elements of $\mathbf{N}$ are nonzero.

Next, the elements of $\mathbf{\Lambda}_1$ and $\mathbf{\Lambda}_2$ can be obtained as the latent vectors of the product

$$
\begin{aligned}
\mathbf{P}_2\mathbf{P}_1^{-1} &= (\mathbf{\Lambda}_1^\mathrm{T}\mathbf{N}\mathbf{\Delta}\mathbf{\Lambda}_2)(\mathbf{\Lambda}_1^\mathrm{T}\mathbf{N}\mathbf{\Lambda}_2)^{-1} \\
&= \mathbf{\Lambda}_1^\mathrm{T}\mathbf{N}\mathbf{\Delta}\mathbf{\Lambda}_2\mathbf{\Lambda}_2^{-1}\mathbf{N}^{-1}(\mathbf{\Lambda}_1^\mathrm{T})^{-1} \\
&= \mathbf{\Lambda}_1^\mathrm{T}\mathbf{\Delta}(\mathbf{\Lambda}_1^\mathrm{T})^{-1}
\end{aligned}
$$

since both $\mathbf{N}$ and $\boldsymbol{\Delta}$ are diagonal. Also

$$\mathbf{P}_1^{-1}\mathbf{P}_2 = (\boldsymbol{\Lambda}_1^{\mathrm{T}}\mathbf{N}\boldsymbol{\Lambda}_2)^{-1}(\boldsymbol{\Lambda}_1^{\mathrm{T}}\mathbf{N}\boldsymbol{\Delta}\boldsymbol{\Lambda}_2)$$

$$= \boldsymbol{\Lambda}_2^{-1}\mathbf{N}^{-1}(\boldsymbol{\Lambda}_1^{\mathrm{T}})^{-1}\boldsymbol{\Lambda}_1^{\mathrm{T}}\mathbf{N}\boldsymbol{\Delta}\boldsymbol{\Lambda}_2$$

$$= \boldsymbol{\Lambda}_2^{-1}\boldsymbol{\Delta}\boldsymbol{\Lambda}_2$$

which further shows that $\boldsymbol{\Delta}$ is the diagonal matrix of latent roots and $\boldsymbol{\Lambda}_1^{\mathrm{T}}$ and $\boldsymbol{\Lambda}_2$ are the latent vectors of Eq. (9.44), that is, of the products $\mathbf{P}_2\mathbf{P}_1^{-1}$ and $\mathbf{P}_1^{-1}\mathbf{P}_2$, respectively. Specifically, $\boldsymbol{\Lambda}_2$ corresponds to the right latent vectors of the system

$$(\mathbf{P}_2 - \phi_i\mathbf{P}_1)\mathbf{X}_i = \mathbf{0} \qquad\qquad (9.46a)$$

and $\boldsymbol{\Lambda}_1^{\mathrm{T}}$ to the left latent vectors $\mathbf{Y}_i^{\mathrm{T}}$ where

$$\mathbf{Y}_i^{\mathrm{T}}(\mathbf{P}_2 - \phi_i\mathbf{P}_1) = \mathbf{0} \qquad\qquad (9.46b)$$

Finally, the matrix $\mathbf{N}$ can be found from Eq. (9.43a). Note that although the latent vectors are determined up to constants of proportionality, the first column of $\boldsymbol{\Lambda}_1$ and $\boldsymbol{\Lambda}_2$ consists of unities and are thus unique.

Several difficulties may be encountered when applying Anderson's model. First, for a sample of observed probabilities of $2m - 1$ dichotomies using different $(m \times m)$ observed matrices will not generally yield the same solutions. Thus which partition is used can have a marked effect on the results. This is partly due to the fact that Anderson's method does not use all of the sample data, and also because in a sample the normal equations need not be consistent, unlike for the population values. The estimates of the latent probabilities moreover need not be admissible. Thus it is possible, for example, to obtain complex solutions since the matrices of the normal equations are not generally symmetric. Even in the case of real-valued solutions the estimated latent probabilities may lie outside of the 0–1 interval. The procedure is nevertheless computationally straightforward and yields convenient closed-form solutions, and provides an interesting example of how latent root and latent vector models can be used for an exploratory multivariate analysis of dichotomous random variables.

Other formulations of latent class analysis are also possible, but these may be computationally cumbersome and may also depend on imaginary entities (e.g., see Green 1951). Gibson (1955) has extended Anderson's solution by (1) involving all of the observed dichotomies for estimating third order moments, (2) using more than a single stratifier, and (3) augmenting the probability matrices $\mathbf{P}_1$ and $\mathbf{P}_2$. The analysis then follows that of Anderson, except that it is based on a sum of matrices of the form $\mathbf{P}_2$ where each matrix term in the sum involves a different stratifier. The augmentation process also uses a latent vector analysis, in conjunction with least squares orthogonal projections (least squares regression coefficients). As in factor analysis, however, it is also important to consider conditions under which a latent class model is identifiable, that is, necessary and sufficient conditions under which a set of manifest parameters corresponds to a unique set of latent parameters such that both sets can be related by the accounting equations of the form of Eq. (9.41). Conditions for identification have been described by Madansky (1960) who extends Anderson's model using higher moments but, unlike Gibson (1955), retains a single stratifier dichotomy. Madansky's (1960) method is to use sufficient sample information in the form of higher order product moments in order to resolve identification difficulties. Moreover the model can also be made to include the extensions proposed by Gibson (1955), which makes it an attractive option for empirical application.

A more direct link between latent class and factor analysis is provided by Green's (1951) model, which also represents an alternative method of matricizing the normal accounting equations (Eq. 9.41). An interesting feature of the model is that it avoids some of the difficulties associated with Anderson's estimators since the observed or manifest probabilities are represented by symmetric matrices and thus yield real-valued solutions. in this respect the model also bears a greater similarity to a principal components decomposition of a covariance matrix since it only makes use of second-order product moments. Let

$$\mathbf{P} = \begin{bmatrix} 1 & p_1 & p_2 & \cdots & p_k \\ p_1 & p_{11} & p_{12} & \cdots & p_{1k} \\ p_2 & p_{12} & p_{22} & \cdots & p_{2k} \\ \vdots & \vdots & \vdots & & \vdots \\ p_k & p_{1k} & p_{2k} & \cdots & p_{kk} \end{bmatrix} \quad \Lambda = \begin{bmatrix} 1 & \nu_{11} & \nu_{21} & \cdots & \nu_{m1} \\ 1 & \nu_{12} & \nu_{22} & \cdots & \nu_{m2} \\ \hline 1 & \nu_{1k} & \nu_{2k} & \cdots & \nu_{mk} \end{bmatrix}$$

$$\mathbf{N} = \begin{bmatrix} \pi_1 & & & \mathbf{0} \\ & \pi_2 & & \\ & & \ddots & \\ \mathbf{0} & & & \pi_m \end{bmatrix}$$

where the diagonal elements p_{ii} remain undefined in the initial stage of the analysis—for example, they may be considered as missing values, although it is clear that they do not correspond to any substantively concrete aspect of the data. The accounting equations can now be expressed in matrix form as

$$\mathbf{P} = \Lambda^{\mathrm{T}} \mathbf{N} \Lambda \tag{9.47}$$

where Λ and $\mathbf{N}$ are the latent vectors and latent roots of $\mathbf{P}$, respectively.

Since the matrix **P** contains probabilities, additional constraints are also used in practice to simplify the rotation of the "factor axes." The interpretation of the output proceeds in the usual way, keeping in mind the discrete nature of the "factor space." For an application of the model to hospital mortality data see Miller et al. (1962).

When the observed dichotomies are binomial the method of maximum likelihood can be used to estimate the latent probabilities. A computer program to maximize the likelihood function is given by Henry (1975). Formann (1982) uses the logistic function to obtain maximum likelihood estimators of the latent parameters (see also Dayton and Macready, 1988). Also Goodman (1974a,b) extends the latent class model to deal with contingency tables involving multinomial categories, which are assumed to be explained by a discrete latent variable composed of k classes (see also Haberman, 1979; Clogg, 1981; Bergan, 1983). The analysis has also been generalized to several different types of multidimensional contingency tables (Clogg and Goodman, 1984). For an application of Goodman's model see Madden and Dillon (1982). Finally, a recent development is to employ the EM algorithm (Section 4.7.3) to yield maximum likelihood estimators of the latent parameters. The algorithm seems to be particularly appropriate when some observations are missing randomly (see Aitkin et al., 1981).

Before attempting to interpret the maximum likelihood analysis it would be desirable to test the null hypothesis that there exist m latent classes. This is because like most cluster models such as factor analysis and hierarchical cluster analysis the latent class model can produce relatively homogeneous clusters even from random data. A natural candidate for the test would seem to be the asymptotic chi-squared distribution for the likelihood ratio test statistic, but it is known that it does not apply to mixture models. Aitkin et al. (1981) discuss the difficulties involved, and propose a test together with graphical analysis, based on normal mixture theory. The ability of the procedure to discriminate between the latent classes however does not seem to be known, and simulation studies are perhaps in order to assess the approach. Mixture theory nevertheless seems to provide a useful rationale or framework for the latent class model, not unlike that employed for a Q-mode factor model for unmixing samples (Section 5.9.2). This is because latent class models can be used to classify sample points into discrete classes, each possessing a probability distribution. This is done by computing the so-called recruitment probabilities, which are analogues of the factor scores. Let

$P(R) = $ Probability of an observed response to a particular $(0, 1)$ configuration (response pattern)

$P(S/R) = $ the observed probability of belonging to the sth latent class, given a response R

Then using Bayes' theorem we have

$$P(S/R) = \frac{P(S)P(R/S)}{P(R)}$$

$$= \frac{\hat{\pi}_s P(R/S)}{P(R)} \tag{9.48}$$

where the $\hat{\pi}_s$ are estimates given by the latent class model.

Example 9.8. Latent class analysis has been used by Henry (1974) to classify individuals during a socioeconomic experiment held in Indiana (see Basilevsky and Hum, 1984) which grew out of the "war on poverty" program in the United States. The analysis is performed on the following dichotomized variables obtained through a sample survey:

$Y_1 =$ Parents should teach their children that there isn't much you can do about the way things are going to turn out in life.

$Y_2 =$ Parents should teach their children that planning only makes a person unhappy, since your plans hardly every work out anyway.

$Y_3 =$ Nowadays the wise parents will teach the child to live for today and let tomorrow take care of itself.

$Y_4 =$ Parents should teach their children not to expect too much out of life so they won't be disappointed.

$Y_5 =$ Parents should teach their children that when a man is born the success he is going to have is already in the cards, so he might as well accept it and not fight it out.

The sample consists of $n = 1762$ low income members of the labor force, where each question is coded as

$$Y_j = \begin{cases} 1 & \text{if individual disagrees} \\ 0 & \text{if agrees} \end{cases}$$

The positive score of unity indicates an "optimistic" response and zero denotes "pessimism." A latent model consisting of three classes is found to provide the best fit (Table 9.12), where the recruitment probabilities denoting class size are given in Table 9.13.

The class size probabilities $\hat{\pi}_s$ can be considered as analogues to the latent roots of principal components analysis, whereas conditional prob-

Table 9.12   Partitioning Latent Probabilities $P(A_s) = \hat{\pi}_s$ and Latent Conditional Probabilities $P(E_i \mid A_s) = \hat{\nu}_{is}$ for $m = 3$ (Theorem 9.2)

	$\hat{\nu}_{i1}$	$\hat{\nu}_{i2}$	$\hat{\nu}_{i3}$
Y_1: Not much can do	.920	.510	.232
Y_2: Plans do not work	.989	.685	.266
Y_3: Live for today	.952	.746	.269
Y_4: Do not expect too much	.915	.515	.070
Y_5: Accept life	.923	.877	.268
Class size $(\hat{\pi}_s)$	.666	.197	.137

Source: Henry, 1974.

abilities play a role equivalent to the factor loadings. It can be seen that the largest latent class A_1 can be understood as representing an "optimism" element in parents' attitudes since the high probabilities indicate a high frequency of disagreement (codes 1). Many of the working poor therefore seem to indicate a counter-defeatist attitude with respect to their children.

Table 9.13   Recruitment Probabilities for Type of Response to Questionnaire

Response Pattern					Class			Observed Frequency
Y_1	Y_2	Y_3	Y_4	Y_5	I	II	II	
0	0	0	0	0	.000	.012	.988	71
1	0	0	0	0	.000	.040	.960	19
0	1	0	0	0	.001	.067	.932	25
0	0	1	0	0	.000	.088	.912	24
0	0	0	1	0	.000	.145	.855	6
0	0	0	0	1	.000	.189	.811	33
1	1	0	0	0	.035	.192	.773	11
1	0	1	0	0	.008	.247	.746	15
1	0	0	1	0	.017	.362	.621	1
0	1	1	0	0	.039	.352	.609	13
1	0	0	0	1	.003	.444	.553	9
0	1	0	1	0	.078	.465	.457	6
0	0	1	1	0	.016	.566	.418	5
0	1	0	0	1	.016	.574	.410	21
0	0	1	0	1	.003	.649	.348	27
0	0	0	1	1	.005	.762	.232	9
0	0	1	1	1	.044	.921	.035	22
1	0	1	0	1	.043	.828	.129	21
1	0	0	1	1	.068	.856	.076	5
0	1	1	0	1	.145	.785	.070	44
1	1	0	0	1	.203	.660	.137	26

Table 9.13 (*Continued*)

Response Pattern					Class			Observed Frequency
Y_1	Y_2	Y_3	Y_4	Y_5	I	II	II	
1	0	1	1	0	.206	.654	.140	4
0	1	0	1	1	.212	.750	.038	17
1	0	1	1	1	.345	.648	.007	29
1	1	1	0	0	.449	.367	.185	17
0	1	1	1	0	.499	.446	.055	9
1	1	0	1	0	.592	.318	.091	5
0	1	1	1	1	.652	.345	.002	117
1	1	1	0	1	.668	.324	.008	116
1	1	0	1	1	.756	.241	.004	57
1	1	1	1	0	.923	.074	.003	79
1	1	1	1	1	.955	.045	.000	899
								1762

Source: Henry, 1974.

The second latent class A_2 also exhibits optimism but to a lesser extent, and seems to indicate a "realistic" modification to the first latent class. The third class A_3 is more difficult to identify, but seems to represent "pessimistic" individuals. Note that the lack of a clear-cut distinction between the three classes may indicate that the underlying "optimism–pessimism" scale is essentially continuous so that a principal component model may be more appropriate. In any case, as with any exploratory model, the results of a latent class analysis should be treated as tentative and subject to further corroboration. Latent class analysis has also been used to estimate ethnicity using census data (Johnson, 1990); effect of response errors is considered by Bye and Schechter (1986).

EXERCISES

9.1 Prove Eq. 9.12.

9.2 Prove that the G coefficient can be derived as given by Eq. (9.14).

9.3 Using an appropriate model of Section 9.2 seriate (ordinate) the following Boolean matrix:

```
1   1   0   1   1   0   0   0   0   0
0   0   0   0   0   1   0   1   1   1
1   1   0   0   0   0   1   0   0   0
0   0   0   0   1   0   1   1   1   1
0   0   0   0   1   0   0   1   1   1
0   0   1   0   0   1   0   0   1   0
0   0   0   0   1   0   1   1   0   0
1   1   0   0   0   0   0   0   0   0
0   1   0   0   1   0   1   1   0   1
0   1   0   0   1   0   1   1   0   0
0   0   1   0   0   1   0   0   0   0
1   1   0   0   1   0   1   1   0   0
0   0   1   0   0   1   0   1   1   0
1   1   0   1   0   0   1   0   0   0
```

9.4 The following data was obtained by Rothkopf (1957) using 598 individuals who were asked to judge whether a pair of consecutively presented Morse code singals were identical or not. Using the pairwise comparisons similarity data matrix of proportions, we wish to discern a possible ordering, in a space of smaller dimension, of correctly identified (incorrectly confused) Morse code signals:

(a) Carry out a principal components analysis of the similarity matrix.

	A	B	C	D	E	F	G	H	I	J	K	L	M	N	O	P	Q	R	S	T	U	V	W	X	Y	Z	1	2	3	4	5	6	7	8	9	Ø
A	92	04	06	13	03	14	10	13	46	05	22	03	25	34	06	06	09	35	23	06	37	13	17	12	07	03	02	07	05	05	08	06	05	06	02	03
B	05	84	37	31	05	28	17	21	05	19	34	40	06	10	12	22	25	16	18	02	18	34	08	84	30	42	12	17	14	40	32	74	43	17	04	04
C	04	38	87	17	04	29	13	07	11	19	24	35	14	03	09	51	34	24	14	06	06	11	14	32	82	38	13	15	31	14	10	30	28	24	18	12
D	08	62	17	88	07	23	40	36	09	13	31	56	08	07	09	27	09	45	29	06	17	20	27	40	15	33	03	09	06	11	09	19	08	10	05	06
E	06	13	14	06	97	02	04	04	17	01	05	06	04	04	05	01	05	10	07	67	03	03	02	05	06	05	04	03	05	03	05	02	04	02	03	03
F	04	51	33	19	02	90	10	29	05	33	16	50	07	06	10	42	12	35	14	02	21	27	25	19	27	13	08	16	47	25	26	24	21	05	05	05
G	09	18	27	38	01	14	90	06	05	22	33	16	14	13	62	52	23	21	05	03	15	14	32	21	23	39	15	14	05	10	04	10	17	23	20	11
H	03	45	23	25	09	32	08	87	10	10	09	29	05	08	08	14	08	17	37	04	36	59	09	33	14	11	03	09	15	43	70	35	17	04	03	03
I	64	07	07	13	10	08	06	12	93	03	05	16	13	30	07	03	05	19	35	16	10	05	08	02	05	07	02	05	08	09	08	08	05	02	04	09
J	07	09	38	09	02	24	18	05	04	85	22	31	08	03	21	63	47	11	02	07	09	09	09	22	32	28	67	66	33	15	07	11	28	29	26	23
K	05	24	38	73	01	17	25	11	05	27	91	33	10	12	31	14	31	22	02	02	23	17	33	63	16	18	05	09	17	08	08	18	14	13	05	06
L	02	69	43	45	10	24	12	26	09	30	27	86	06	02	09	37	36	28	12	05	16	19	20	31	25	59	12	13	17	15	26	29	36	16	07	03
M	24	12	05	14	07	17	29	08	08	11	23	08	96	62	11	10	15	20	07	09	13	04	21	09	18	08	05	07	06	06	05	07	11	07	10	04
N	31	04	13	30	08	12	10	16	13	03	16	08	59	93	05	09	05	28	12	10	16	04	12	04	06	11	05	02	03	04	04	06	02	02	10	02
O	07	07	20	06	05	09	76	07	02	39	26	10	04	08	86	37	35	10	03	04	11	14	25	35	27	27	19	17	07	07	06	18	14	11	20	12
P	05	22	33	12	05	36	22	12	03	78	14	46	05	06	21	83	43	23	09	04	12	19	19	19	41	30	34	44	24	11	15	17	24	23	25	13
Q	08	20	38	11	04	15	10	05	02	27	23	26	07	06	22	51	91	11	02	03	06	14	12	37	50	63	34	32	17	12	09	27	40	58	37	24
R	13	14	16	23	05	34	26	15	07	12	21	33	14	12	12	29	08	87	16	02	23	23	62	14	12	13	07	10	13	04	07	12	07	09	01	02
S	17	24	05	30	11	26	05	59	16	03	13	10	05	17	06	06	03	18	96	09	56	24	12	10	06	07	08	02	02	15	28	09	05	05	05	02
T	13	10	01	05	46	03	06	06	14	06	14	07	06	05	06	11	04	04	07	96	08	05	04	02	02	06	05	05	03	03	03	08	07	06	14	06
U	14	29	12	32	04	32	11	34	21	07	44	32	11	13	06	20	12	40	51	06	93	57	34	17	09	11	06	06	16	34	10	09	09	07	04	03
V	05	17	24	16	09	29	06	39	05	11	26	43	04	01	09	17	10	17	11	06	32	92	17	57	35	10	10	14	28	73	44	36	25	10	01	05
W	09	21	30	22	09	36	25	15	04	25	29	18	15	06	26	20	25	61	12	04	19	20	86	22	25	22	10	22	19	16	05	09	11	06	03	07
X	07	64	45	19	03	28	11	06	01	35	50	42	10	08	24	32	61	10	12	03	12	17	21	91	48	26	12	20	24	27	16	57	29	16	17	06
Y	09	23	62	15	04	26	22	09	01	30	12	14	05	06	14	30	52	05	07	04	06	13	21	44	86	23	26	44	40	15	11	26	22	33	23	16
Z	03	46	45	18	02	22	17	10	07	23	21	51	11	02	15	59	72	14	04	03	09	11	12	36	42	87	16	21	27	09	10	25	66	47	15	15
1	02	05	10	03	03	05	13	04	02	02	05	14	09	07	14	30	28	09	04	02	03	12	14	17	19	22	84	63	13	08	10	08	19	32	57	55
2	07	14	22	05	04	20	13	03	25	26	09	14	02	03	17	37	28	06	05	03	06	10	11	17	30	13	62	89	54	20	05	14	20	21	16	11
3	03	08	21	05	04	32	06	12	02	23	06	13	05	02	05	37	19	09	07	06	04	16	06	22	25	12	18	64	86	31	23	41	16	17	08	10
4	06	19	19	12	08	25	14	16	07	21	13	19	03	03	02	17	29	11	09	03	17	55	08	37	24	03	05	26	44	89	42	44	32	10	03	03
5	08	45	15	14	02	45	04	67	07	14	04	41	02	00	04	13	07	09	27	02	14	45	07	45	10	10	14	10	30	69	90	42	24	10	06	05
6	07	80	30	17	04	23	04	14	02	11	11	27	06	02	07	16	30	11	14	03	12	30	09	58	38	39	15	14	26	24	17	88	69	14	05	14
7	06	33	22	14	05	25	06	04	06	24	13	32	07	06	07	36	39	12	06	02	03	13	09	30	30	50	22	29	18	15	12	61	85	70	20	13
8	03	23	40	06	03	15	15	06	02	33	10	14	03	06	14	12	45	02	06	04	05	07	05	24	35	50	42	29	15	16	09	30	60	89	61	25
9	03	14	23	03	01	06	14	05	02	30	06	07	16	11	10	31	32	05	06	07	06	03	08	11	21	24	57	39	09	12	04	11	42	56	91	78
Ø	09	03	11	02	05	07	14	04	05	30	08	03	02	03	25	21	29	02	03	04	05	03	02	12	15	20	50	26	09	11	05	22	17	52	81	94

(b) Carry out a maximum likelihood factor analysis using information
obtained from part a.

(c) Compare the two analyses with each other, as well as to Shepard's
(1963) multidimensional scaling analysis of the data.

9.5 McCormick et al. (1972) give a ranking, on the scale of 0 to 2, of the
degree (subjectively determined) to which each of a total of 53 military
aircraft can perform one or more of a total of 27 functions. This yields
the following object/attribute matrix:

Type of Aircraft

Application	1	2	3	4	5	6	7	8	9	10	11	12	13	14	15	16	17	18	19	20	21	22	23	24	25	26	27	28	29	30	31	32	33	34	35	36	37	38	39	40	41	42	43	44	45	46	47	48	49	50	51	52	53
1	0	0	0	0	0	0	0	0	0	0	0	0	0	0	0	0	0	0	0	0	0	0	1	1	2	1	1	2	1	1	1	1	2	1	1	2	2	2	1	0	0	0	0	0	0	0	0	0	0	0	0	0	0
2	0	0	0	2	0	0	0	0	0	0	2	0	0	0	0	0	2	2	0	2	2	0	2	0	0	0	0	0	0	0	0	0	0	0	0	0	0	0	0	1	0	0	0	0	0	0	1	0	0	0	0	0	1
3	0	1	0	1	0	2	0	0	1	0	0	0	0	2	0	2	2	2	0	1	0	0	2	0	1	2	2	2	0	0	2	2	2	0	2	2	2	2	0	0	0	0	0	1	1	0	1	1	1	1	1	1	1
4	1	1	1	0	1	1	0	1	1	0	0	0	0	1	0	1	0	1	0	0	0	0	0	1	1	2	2	2	0	0	0	0	0	0	0	0	0	0	0	0	0	1	1	0	1	1	1	1	1	1	1	1	1
5	0	0	0	2	0	0	1	0	0	0	1	0	1	0	1	1	2	1	2	2	1	0	2	0	0	0	0	0	0	0	0	0	0	0	0	0	0	0	0	0	0	0	0	0	0	0	0	0	0	0	0	0	0
6	1	2	1	2	0	2	0	0	0	0	0	0	0	2	0	2	2	2	0	1	0	0	2	0	1	0	0	1	0	0	0	1	1	0	1	1	1	1	0	0	0	0	0	0	0	0	0	0	0	0	0	0	0
7	0	1	0	0	0	1	0	0	0	0	0	0	0	2	0	1	1	1	0	0	0	0	0	0	1	1	1	2	0	1	2	2	2	0	1	2	2	2	0	0	0	0	0	0	0	0	0	0	0	0	0	0	0
8	1	2	2	1	1	2	2	1	2	2	2	2	2	2	2	2	2	2	1	1	2	1	1	1	1	1	1	1	1	1	1	1	1	1	1	1	1	1	0	1	2	1	1	2	2	1	1	1	1	1	1	1	0
9	1	2	0	0	0	0	0	0	0	0	0	0	0	0	0	0	0	0	0	0	0	0	1	2	1	1	1	1	1	1	1	1	1	1	1	1	1	1	0	0	0	0	0	0	0	0	0	0	0	0	1	0	0
10	2	1	1	0	2	1	0	2	1	0	0	1	0	0	0	0	0	0	0	0	0	0	2	1	2	2	1	2	1	2	2	2	2	1	1	2	1	1	1	1	2	2	0	1	0	1	2	2	2	2	1	2	2
11	1	1	1	2	1	1	2	1	1	2	2	1	2	2	2	2	2	2	2	2	2	2	1	1	1	1	1	1	1	1	1	1	1	1	1	1	1	1	1	2	1	1	1	1	1	1	1	1	1	1	1	1	1
12	0	0	0	0	0	0	0	0	0	0	0	0	0	0	0	0	0	0	0	0	0	0	0	0	0	0	0	0	0	0	0	0	0	0	0	0	0	0	0	0	0	0	0	0	0	0	0	0	0	0	0	0	0
13	0	1	0	2	0	1	0	0	0	1	2	0	1	1	1	1	1	2	1	0	2	0	1	0	0	1	0	1	0	1	0	1	0	0	0	0	0	1	2	0	1	0	0	0	0	0	0	0	0	0	0	0	0
14	1	1	2	1	1	1	1	1	1	1	1	1	2	2	1	2	2	1	2	1	2	1	1	1	1	1	1	1	1	1	1	1	1	1	1	1	2	1	1	1	1	1	1	1	1	1	1	1	1	1	1	1	1
15	1	1	1	2	0	1	1	1	1	2	1	1	1	2	1	1	2	1	1	2	1	2	2	2	0	0	0	0	0	0	0	0	0	0	0	0	0	0	0	0	2	2	1	1	0	0	0	0	0	0	0	0	0
16	1	1	0	0	0	1	0	0	0	0	1	0	0	1	1	1	1	1	0	0	0	0	0	1	1	0	1	0	1	0	0	1	1	1	1	1	0	0	0	0	1	0	0	0	0	0	0	0	0	0	0	0	0
17	1	2	1	1	1	2	1	1	2	2	2	2	1	2	2	2	2	1	1	1	1	1	2	1	1	1	1	1	0	1	1	1	1	1	1	1	1	1	1	1	2	1	1	1	1	1	1	1	1	1	1	1	1
18	1	2	2	2	1	2	1	2	2	2	2	2	2	2	2	2	2	1	2	1	2	1	1	1	0	1	1	1	0	0	2	2	1	1	0	1	1	0	0	0	2	2	1	2	1	1	1	1	0	0	1	0	1
19	1	2	2	2	1	2	1	1	1	2	2	2	2	1	2	2	2	2	2	1	2	1	2	1	1	1	1	1	1	0	0	0	0	0	0	0	0	0	0	0	2	2	1	2	1	1	0	1	1	0	0	0	1
20	1	2	2	2	1	2	1	2	2	2	2	2	2	2	2	2	2	2	1	2	0	0	0	1	0	0	0	0	0	0	0	0	0	0	0	0	0	0	0	0	2	2	1	2	1	0	0	0	0	0	0	0	0
21	1	2	2	1	1	2	1	2	2	2	1	2	1	2	1	2	1	2	2	1	1	1	1	1	1	2	2	2	2	1	2	2	2	1	2	2	2	1	0	0	0	0	0	0	0	0	0	0	0	0	1	1	2
22	0	0	0	0	0	0	0	0	0	0	0	0	0	0	0	0	0	0	0	0	0	0	0	0	0	2	2	2	2	0	0	0	0	0	0	0	0	0	0	1	0	0	0	0	0	0	1	0	1	0	2	2	0
23	1	1	1	1	1	1	2	1	1	1	1	1	1	1	1	1	1	1	1	2	2	1	1	1	0	0	0	0	0	0	0	0	0	0	0	0	0	0	0	1	1	1	1	2	1	1	1	1	1	1	1	1	1
24	0	0	0	0	0	0	0	0	0	0	0	0	0	0	0	0	0	0	0	0	0	0	0	0	0	2	1	0	1	2	1	1	2	1	1	1	2	1	1	1	2	2	0	0	0	0	2	0	0	1	2	0	2
25	0	1	0	0	0	1	0	0	0	0	0	0	1	0	1	0	1	0	0	0	0	0	0	2	1	0	2	2	0	1	0	1	0	1	0	0	0	0	0	0	0	0	0	0	0	0	0	0	0	0	0	0	0
26	2	1	1	1	1	1	1	1	1	1	1	1	1	1	1	1	1	1	1	1	1	1	1	1	1	1	1	1	1	1	1	1	1	1	1	1	1	1	1	1	2	2	2	2	1	1	1	1	1	1	1	1	1
27	1	1	0	0	0	0	0	0	0	0	0	0	0	0	0	0	0	0	0	0	0	0	0	1	1	0	1	1	2	2	2	2	2	2	2	2	2	2	1	1	1	0	0	1	0	0	1	1	1	0	1	1	1
28	0	1	0	0	0	1	0	0	0	0	0	0	0	1	0	1	0	1	0	0	0	0	0	0	0	2	2	2	2	2	2	2	2	2	2	2	2	2	1	0	0	1	0	0	1	0	0	1	1	1	0	1	1
29	0	1	0	0	0	1	0	0	0	0	0	0	0	1	0	1	0	1	0	0	0	0	0	0	0	0	0	0	0	1	0	0	0	1	0	1	0	1	0	0	0	0	0	0	0	0	0	0	0	0	0	0	0
30	1	1	1	0	1	1	0	1	1	0	0	0	0	0	0	0	0	0	0	0	0	0	0	0	0	1	1	1	1	1	1	1	1	0	1	1	1	0	1	1	0	1	0	0	1	0	1	1	1	1	1	1	1
31	2	2	2	2	1	2	2	2	2	2	2	2	2	2	2	2	2	2	2	2	2	2	2	2	2	2	2	2	2	2	2	2	2	2	2	2	2	2	1	0	0	2	2	2	2	2	1	1	1	1	1	2	1
32	1	0	0	0	0	0	0	0	0	0	0	0	0	0	0	0	0	0	0	0	0	0	0	0	1	1	0	1	0	0	2	2	2	2	2	2	2	2	2	2	2	2	2	1	0	1	0	0	1	0	0	1	1
33	0	1	1	1	0	1	2	0	0	1	1	1	1	1	1	1	1	2	1	2	1	1	1	0	1	0	0	0	0	0	0	0	0	0	0	0	0	0	0	0	1	1	0	1	1	0	0	0	0	0	0	0	0
34	0	0	0	2	0	0	0	0	0	0	0	0	0	0	0	1	1	0	2	0	0	2	0	0	0	0	0	0	0	0	0	0	0	0	0	1	1	0	2	0	0	2	0	0	0	0	0	0	0	0	0	0	0
35	0	1	1	2	0	0	2	0	0	1	1	0	1	1	1	1	1	1	1	2	1	2	0	0	0	0	0	0	0	0	0	0	0	0	0	0	0	0	0	1	0	1	1	0	0	0	0	0	0	0	0	0	0
36	1	1	0	0	0	0	0	0	0	0	0	0	0	0	0	0	0	0	0	0	0	0	1	1	0	0	0	0	0	0	0	0	0	0	0	0	0	0	0	0	2	2	0	0	0	0	0	0	0	0	0	0	0
37	0	0	0	1	0	0	0	0	0	0	1	0	0	0	2	0	1	1	0	1	1	0	1	0	0	0	0	0	0	0	0	0	0	0	0	2	0	1	1	0	1	1	0	1	0	0	0	0	0	0	0	0	0

Using the ordination of Section 9.2.2 (see also Section 8.5), determine
which aircraft are similar in terms of their application (use).

CHAPTER 10

Factor Analysis and Least Squares Regression

10.1 INTRODUCTION

The previous chapters dealt with factor analysis in terms of a set of statistical data-analytic procedures which may be utilized to estimate unobserved latent variables, influential dimensions, or clusters within the variable (sample) space. This is the traditional conceptual framework within which factor analytic models have been viewed, and which constitutes much of their application today. Another major area of relevance for factor models also exists, which recently has captured greater attention—the use of factors in the estimation of least squares planes and other functional specifications. Generally speaking this takes two forms. First, factor models can be utilized directly to estimate multivariate functions, for example, when dealing with the so-called functional and structural relationships between the dependent and independent variables, biological growth curves, chemical free-energy relations, and so forth. This occurs when all variables are subject to measurement or observational error or when the explanatory variables are not observed. Second, the predictor variables may be subject to multicollinearity and factor models can then be employed in conjunction with Gauss–Markov least squares in order to augment precision. In the following sections we describe several factor analysis-based least squares regression estimators which can be used to estimate linear functional forms or to reduce bias and inefficiency in least squares regression.

10.2 LEAST SQUARES CURVE FITTING WITH ERRORS IN VARIABLES

In this section, we return to the mathematical and statistical origins of principal components—the use of orthogonal ellipsoidal principal axes to

624

estimate a least squares regression plane. Actually the relevant model is the weighted principal components model (Section 5.6) since generally speaking errors may be correlated and may possess uneven variance. At first glance this seems a surprising origin for factor analysis given the well-developed nature of the Gauss–Markov least squares model. The puzzle disappears however once we note the restricted nature of least squares regression. A crucial assumption of Gauss–Markov analysis is that the independent or predictor variables are free of measurement error. Thus although the predictor variables can be allowed to vary randomly their true values must be known in advance. In this sense least squares regression can be viewed as utilizing a priori knowledge in the form of zero restrictions placed upon the error terms of the independent variables. In the presence of measurement error however, sufficient prior knowledge (or replications) must also be available to at least estimate the error variance, since in the absence of such information the Gauss–Markov model yields inconsistent and inefficient estimators.

When measurement error is present in the predictor variables their true values become, in effect, unobserved latent variables and consequently can be estimated using the weighted principal components model (Section 5.6). A further advantage of viewing regression in terms of weighted principal components is that the requirement for prior information is recognized in an explicit manner when no replications are available. Such an approach seems to have been first considered by Adcock (1878), Kummel (1879), and Pearson (1901). The latter author used cross sections of an ellipsoid in what represents the first de facto principal components analysis of a correlation matrix, but in terms of latent vectors associated with the smallest rather than the largest latent root(s). The problem can also be posed within the context of multivariate normal theory by considering rotations of normal ellipsoids to independent form, as described originally in Bravais' (1846) work on the multivariate normal distribution. The theory has since been considered by Gini (1921), Rhodes (1937), Frisch (1929), Van Uven (1930), Dent (1935), Lindley (1947), Anderson (1951a), Sprent (1966), and others. Summary reviews of the material can be found in Moran (1971), Mak (1978), Mandel (1984) and Anderson (1984b).

Specifically, the situation is as follows. Let $y = \xi + \epsilon$, $x_1 = \chi_1 + \Delta_1$, $x_2 = \chi_2 + \Delta_2, \ldots, x_{k-1} = \chi_{k-1} + \Delta_{k-1}$ where ξ, χ_1, $\chi_2, \ldots, \chi_{k-1}$ are true values related by the exact linear equation

$$\xi = \beta_0 + \beta_1\chi_1 + \beta_2\chi_2 + \cdots + \beta_{k-1}\chi_{k-1} \tag{10.1}$$

The true values, however, are not observed owing to the presence of residual errors, and in their place we have the variables $y, x_1, x_2, \ldots, x_{k-1}$ such that $E(\epsilon) = E(\Delta_1) = \cdots = E(\Delta_{k-1}) = 0$. The predictor or explanatory variables can contain error for a number of reasons. When the predictor variables are allowed to vary randomly, they are often observed in very much the same way as the dependent variable, and can then be subject to measurement error and other residual variation. For example, in large sample surveys of human populations all variables are generally affected by recall error, intentional misreporting, data recording and entry errors, loss of data, and so forth. Also when variables contain missing values they are frequently estimated by a number of different techniques (e.g., see Basilevsky et al., 1985) and this introduces error terms into the predictors even in the case when the predictors are fixed. The difficulty is also not obviated when the original values are replaced by ranks (Section 8.2) since these will also be subject to error, and what is required is an estimation technique which takes into explicit account error in all the variables of the equation.

Before considering curve-fitting methods when all variables contain measurement error it is perhaps worthwhile to first point out the effects of such errors on regression analysis and why the usual least squares linear regression is not optimal. Consider the true values $\xi, \chi_1, \chi_2, \ldots, \chi_{k-1}$ which are related as in Eq. (10.1). To estimate parameters $\beta_0, \beta_1, \ldots, \beta_{k-1}$, n independent observations $y_i, x_{i1}, x_{i2}, \ldots, x_{ik-1}$ $(i = 1, 2, \ldots, n)$ are obtained such that $y_i = \xi_i + \epsilon_i$, $x_{i1} = \chi_{i1} + \Delta_{i1}$, $x_{i2} = \chi_{i2} + \Delta_{i2}, \ldots, x_{ik-1} = \chi_{ik-1} + \Delta_{ik-1}$. Equation (10.1) then becomes

$$(y_i - \epsilon_i) = \beta_0 + \beta_1(x_{i1} - \Delta_{i1}) + \beta_2(x_{i2} - \Delta_{i2}) + \cdots + \beta_{k-1}(x_{ik-1} - \Delta_{ik-1})$$

or

$$y_i = \beta_0 + \beta_1 x_{i1} + \beta_2 x_{i2} + \cdots + \beta_{k-1} x_{ik-1} + (\epsilon_i - \beta_1 \Delta_{i1} - \cdots - \beta_{k-1} \Delta_{ik-1})$$

$$= \beta_0 + \beta_1 x_{i1} + \beta_2 x_{i2} + \cdots + \beta_{k-1} x_{ik-1} + \delta_i \qquad (i = 1, 2, \ldots, n) \qquad (10.2)$$

Equation (10.2) does not represent a simple regression equation since the variables $x_1, x_2, \ldots, x_{k-1}$ are correlated with the overall error term δ. If Eq. (10.2) is estimated by Gauss–Markov least squares, the estimator of $\boldsymbol{\beta}$ in $\mathbf{Y} = \mathbf{X}\boldsymbol{\beta} + \boldsymbol{\delta}$ is inconsistent, and the variance of the error term is inflated owing to the errors of measurement in the predictor variables (Section 6.3.3). Thus the Gauss–Markov least squares fit becomes progressively poorer as the error variance increases, which is also accompanied by a corresponding increase in bias (Example 10.2). Also linearity is generally not preserved, that is, even when the true equation (Eq. 10.1) is linear, it does not follow that the observed regression (Eq. 10.2) is necessarily linear (Kendall and Stuart, 1979, pp. 438–440).

Two broad strategies are available when all variables in a regression equation contain error. In the simple case when the errors can be estimated with some accuracy (e.g., when replications are available), the variables can be corrected for error and least squares estimates may be computed in the usual manner. When insufficient information exists for the error terms, however, prior information must be available, at least for their relative variances. For example, we may know that all variables possess equal error variance (or nearly so), or else a ranking of the error variance magnitudes may be indicated using prior information (should this be available). In this situation we would wish to fit the plane so that the error variance is no longer minimized only in the direction of the dependent variable. A generalized (weighted) least squares estimator of the coefficients of Eq. (10.1) can be derived which minimizes error variance in any arbitrary direction, and which makes use of weighted principal components associated with the smallest latent root(s) and their latent vector(s).

10.2.1 Minimizing Sums of Squares of Errors in Arbitrary Direction

The following procedure can be used to estimate Eq. (10.1) in a direction determined by the variance/covariance structure of the errors. Let $\mathbf{X} = (x_1, x_2, \ldots, x_k)^{\mathrm{T}}$ be a vector of observed random variables where $y = x_k$, $\mathbf{X} = \boldsymbol{\chi} + \boldsymbol{\Delta}$, and $\boldsymbol{\chi}$ and $\boldsymbol{\Delta}$ represent the true and error parts respectively such that $E(\boldsymbol{\Delta}) = \mathbf{0}$. Although $\boldsymbol{\Delta} = (\Delta_1, \Delta_2, \ldots, \Delta_k)^{\mathrm{T}}$, and thus $\mathbf{X}$, are assumed to be random the true parts $\boldsymbol{\chi} = (\chi_1, \chi_2, \ldots, \chi_k)^{\mathrm{T}}$ may be either random or fixed. In the former case the linear form (Eq. 10.1) is referred to as a structural relationship; in the latter case it is known as a functional relationship. In the structural relationship case it is assumed that $\boldsymbol{\chi}$ has finite mean and finite covariance matrix. For the present we suppose that the true values $\boldsymbol{\chi}$ are random. The variance/covariance matrix of $\mathbf{X}$ can then be expressed as

$$E(\mathbf{X}\mathbf{X}^{\mathrm{T}}) = E[(\boldsymbol{\chi} + \boldsymbol{\Delta})(\boldsymbol{\chi} + \boldsymbol{\Delta})^{\mathrm{T}}]$$

or
$$\boldsymbol{\Sigma} = E(\boldsymbol{\chi}\boldsymbol{\chi}^{\mathrm{T}}) + E(\boldsymbol{\Delta}\boldsymbol{\Delta}^{\mathrm{T}})$$

$$= \boldsymbol{\Sigma}^* + \boldsymbol{\Psi} \tag{10.3}$$

assuming $E(\boldsymbol{\Delta}\boldsymbol{\chi}^{\mathrm{T}}) = \mathbf{0}$, that is, the errors and the true values are uncorrelated. The condition is usually imposed when replications are not available since in this case the variance/covariance structure of the error terms cannot be estimated. The error covariance matrix $\boldsymbol{\Psi}$ however may be either diagonal or otherwise depending on whether the error terms are correlated, although in practice errors can be expected to be independent of each other. Again, in the event the errors are correlated replications are required to estimate the off-diagonal elements of $\boldsymbol{\Psi}$.

Since all variables are subject to error they are treated in a symmetric fashion, and the equation is estimated in the implicit rather than in the explicit form, that is, the equation is expressed as $\alpha_1 x_1 + \alpha_2 x_2 + \cdots + \alpha_k x_k = \delta$ or

$$\boldsymbol{\alpha}^{\mathrm{T}}\mathbf{X} = \boldsymbol{\delta} \tag{10.4}$$

where $E(\mathbf{\Delta}) = E(\mathbf{\delta}) = \mathbf{0}$ and $\mathbf{\delta} = \mathbf{\alpha}^T\mathbf{\Delta}$ is the joint error term. Also, the coefficients $\mathbf{\alpha}$ are assumed to be fixed. The objective is to estimate Eq. (10.1) using only Eq. (10.4), that is, by estimating $\mathbf{\alpha}$ such that

$$\mathrm{var}(\mathbf{\delta}) = E[(\mathbf{\alpha}^T\mathbf{X})(\mathbf{\alpha}^T\mathbf{X})^T]$$

$$= \mathbf{\alpha}^T E(\mathbf{X}\mathbf{X}^T)\mathbf{\alpha}$$

$$= \mathbf{\alpha}^T\mathbf{\Sigma}\mathbf{\alpha} \tag{10.5}$$

is minimized. Since the error variance (Eq. 10.5) can also be expressed as $\mathbf{\alpha}^T\mathbf{\Psi}\mathbf{\alpha}$ (Exercise 10.1), the minimization of Eq. (10.5) cannot be carried out unless $\mathbf{\alpha}^T\mathbf{\Psi}\mathbf{\alpha}$ is fixed. This implies the well-known Lagrangian expression

$$\phi = \mathbf{\alpha}^T\mathbf{\Sigma}\mathbf{\alpha} - \lambda(\mathbf{\alpha}^T\mathbf{\Psi}\mathbf{\alpha} - c) \tag{10.6}$$

where c and λ are positive constants. Differentiating (Eq. 10.6) with respect to $\mathbf{\alpha}$ and setting to zero yields

$$\frac{\partial \phi}{\partial \mathbf{\alpha}} = \mathbf{\Sigma}\hat{\mathbf{\alpha}} - \lambda\mathbf{\Psi}\hat{\mathbf{\alpha}} = \mathbf{0}$$

or

$$(\mathbf{\Sigma} - \lambda\mathbf{\Psi})\hat{\mathbf{\alpha}} = \mathbf{0} \tag{10.7}$$

We thus arrive at the same expression as for the weighted principal components model (Eq. 5.104) except that the optimum solution $\hat{\mathbf{\alpha}}$ now corresponds to the *smallest* latent root of $\mathbf{\Sigma}$ in the metric $\mathbf{\Psi}$ (Exercise 10.2). Thus the lower-order principal components represent residual error measured at an arbitrary angle to the plane whereas components corresponding to dominant roots estimate the plane itself. The weighted least squares model therefore takes the weighted principal component $\mathbf{\zeta}_k$ as the optimal estimator of the error term $\mathbf{\delta}$.

Since the population matrixes $\mathbf{\Sigma}$ and $\mathbf{\Psi}$ are usually not known, they must be estimated using sample values. Let $\mathbf{X}$ denote a $(n \times k)$ matrix of sample observations for the dependent and independent variables such that $\bar{\mathbf{X}} = \mathbf{0}$. The sample counterpart of Eq. (10.7) is then the set of normal equations

$$(\mathbf{X}^T\mathbf{X} - l\hat{\mathbf{\Psi}})\hat{\mathbf{\alpha}} = \mathbf{0} \tag{10.8}$$

where $\hat{\mathbf{\Psi}}$ denotes the sample estimator of $\mathbf{\Psi}$. Premultiplying Eq. (10.8) by $\hat{\mathbf{\alpha}}^T$ and setting $\hat{\mathbf{\alpha}}^T\hat{\mathbf{\Psi}}\hat{\mathbf{\alpha}} = 1$ we obtain

$$\hat{\mathbf{\alpha}}^T(\mathbf{X}^T\mathbf{X} - l\hat{\mathbf{\Psi}})\hat{\mathbf{\alpha}} = \mathbf{0}$$

or

$$\hat{\boldsymbol{\alpha}}^{\mathrm{T}}(\mathbf{X}^{\mathrm{T}}\mathbf{X})\hat{\boldsymbol{\alpha}} = l\hat{\boldsymbol{\alpha}}^{\mathrm{T}}\hat{\boldsymbol{\Psi}}\hat{\boldsymbol{\alpha}}$$

$$= l \tag{10.9}$$

so that the estimated sample regression plane is given by

$$\mathbf{Z}_k = \hat{\boldsymbol{\alpha}}^{\mathrm{T}}\mathbf{X}$$

$$= \hat{\alpha}_1\mathbf{X}_1 + \hat{\alpha}_2\mathbf{X}_2 + \cdots + \hat{\alpha}_k\mathbf{X}_k$$

$$= \hat{\boldsymbol{\delta}} \tag{10.10}$$

where $\mathbf{Z}_k$ is the weighted principal component corresponding to the smallest root l_k such that $\mathbf{Z}_k^{\mathrm{T}}\mathbf{Z}_k = l_k$. An alternative standardization is to let $\hat{\boldsymbol{\alpha}}^{\mathrm{T}}\hat{\boldsymbol{\Psi}}\hat{\boldsymbol{\alpha}} = l_k^{-1}$ in which case $\mathbf{Z}_k^{\mathrm{T}}\mathbf{Z}_k = 1$. To obtain the explicit form solution of Eq. (10.10) let $\mathbf{X}_k = \mathbf{Y}$, be the dependent variable of the regression equation. Then

$$\hat{\alpha}_k\mathbf{X}_k = \hat{\alpha}_k\mathbf{Y} = -\hat{\alpha}_1\mathbf{X}_1 - \hat{\alpha}_2\mathbf{X}_2 - \cdots - \hat{\alpha}_{k-1}\mathbf{X}_{k-1} + \hat{\boldsymbol{\delta}}$$

or

$$\mathbf{Y} = -\frac{\hat{\alpha}_1}{\hat{\alpha}_k}\mathbf{X}_1 - \frac{\hat{\alpha}_2}{\hat{\alpha}_k}\mathbf{X}_2 - \cdots - \frac{\hat{\alpha}_{k-1}}{\hat{\alpha}_k}\mathbf{X}_{k-1} + \frac{\hat{\boldsymbol{\delta}}}{\hat{\alpha}_k}$$

$$= \tilde{\beta}_1\mathbf{X}_1 + \tilde{\beta}_2\mathbf{X}_2 + \cdots + \tilde{\beta}_{k-1}\mathbf{X}_{k-1} + \hat{\boldsymbol{\delta}}^* \tag{10.11}$$

The intercept term is obtained from the usual condition that the regression plane must pass through the mean point.

Since all variables are subject to error, the term $\hat{\boldsymbol{\delta}}$ consists of a linear combination of the individual error terms. We have $\hat{\boldsymbol{\delta}} = \boldsymbol{\Delta}\hat{\boldsymbol{\alpha}}$ or

$$\hat{\boldsymbol{\delta}} = \begin{bmatrix} \Delta_{11} & \Delta_{12} & \cdots & \Delta_{1k} \\ \Delta_{21} & \Delta_{22} & \cdots & \Delta_{2k} \\ . & . & . & . \\ . & . & . & . \\ \Delta_{n1} & \Delta_{n2} & \cdots & \Delta_{nk} \end{bmatrix} \begin{bmatrix} \hat{\alpha}_1 \\ \hat{\alpha}_2 \\ . \\ . \\ \hat{\alpha}_k \end{bmatrix} \tag{10.12}$$

so that

$$\hat{\delta}_i^* = \frac{\hat{\delta}_i}{\hat{\alpha}_k} = \frac{1}{\hat{\alpha}_k}(\Delta_{i1}\hat{\alpha}_1 + \Delta_{i2}\hat{\alpha}_2 + \cdots + \Delta_{ik}\hat{\alpha}_k) \tag{10.13}$$

and the fitted plane is given by

$$y_i = \tilde{\beta}_1 x_{i1} + \tilde{\beta}_2 x_{i2} + \cdots + \tilde{\beta}_{k-1} x_{ik-1}$$

$$+ \left(\Delta_{i1} \frac{\hat{\alpha}_1}{\hat{\alpha}_k} + \Delta_{i2} \frac{\hat{\alpha}_2}{\hat{\alpha}_k} + \cdots + \Delta_{ik-1} \frac{\hat{\alpha}_{k-1}}{\hat{\alpha}_k} + \Delta_{ik} \frac{\hat{\alpha}_k}{\hat{\alpha}_k} \right)$$

$$= \tilde{\beta}_1 x_{i1} + \tilde{\beta}_2 x_{i2} + \cdots + \tilde{\beta}_{k-1} x_{ik-1}$$

$$+ (-\tilde{\beta}_1 \Delta_{i1} - \tilde{\beta}_2 \Delta_{i2} - \cdots - \tilde{\beta}_{k-1} \Delta_{ik-1} + \Delta_{ik})$$

$$= \tilde{\beta}_1 (x_{i1} - \Delta_{i1}) + \tilde{\beta}_2 (x_{i2} - \Delta_{i2}) + \cdots + \tilde{\beta}_{k-1} (x_{ik-1} - \Delta_{ik-1}) + \Delta_{ik}$$

or

$$\hat{y}_i = \tilde{\beta}_1 \hat{x}_{i1} + \tilde{\beta}_2 \hat{x}_{i2} + \cdots + \tilde{\beta}_{k-1} \hat{x}_{ik-1} \qquad (i = 1, 2, \ldots, n) \quad (10.14)$$

where $\hat{\mathbf{X}} = \hat{\chi}_1, \hat{\mathbf{X}}_2 = \hat{\chi}_2, \ldots, \hat{\mathbf{X}}_{k-1} = \hat{\chi}_{k-1}$, and $\hat{\mathbf{Y}} = \hat{\chi}_k$ are the estimated true values of $\mathbf{X}_1, \mathbf{X}_2, \ldots, \mathbf{X}_{k-1}$ $\mathbf{Y}$ and $\tilde{\beta}_j = -\hat{\alpha}_j/\hat{\alpha}_k$ $(j = 1, 2, \ldots, k - 1)$ as in (Eq. 10.11). The predicted values can also be estimated in terms of the weighted principal components of $\mathbf{X}$ as

$$\begin{aligned} \hat{\mathbf{X}}_1 &= p_{11}\mathbf{Z}_1 + p_{12}\mathbf{Z}_2 + \cdots + p_{1k-1}\mathbf{Z}_{k-1} \\ \hat{\mathbf{X}}_2 &= p_{21}\mathbf{Z}_1 + p_{22}\mathbf{Z}_2 + \cdots + p_{2k-1}\mathbf{Z}_{k-1} \\ \hline \hat{\mathbf{X}}_k &= p_{k1}\mathbf{Z}_1 + p_{k2}\mathbf{Z}_2 + \cdots + p_{kk-1}\mathbf{Z}_{k-1} \end{aligned} \qquad (10.15)$$

where $\tilde{\boldsymbol{\delta}} = \mathbf{Z}_k$, that is, the predicted values are the weighted nonorthogonal projections of the columns of $\mathbf{X}$ into the subspace spanned by the first $k - 1$ weighted principal components.

The weighted least squares regression plane selects the smallest latent root as the optimal error variance of the equation on the assumption that there is only a single regression plane passing through the variables. This is so, however, only when the smallest latent root is significantly different from one or more of the remaining roots. When this is not the case, there exist more than one regression plane and the choice of which equation to use becomes less clear. When $k - r$ smallest roots are insignificantly different (Section 4.3.2) the estimates of the elements of $\hat{\alpha}$ can be defined as sample averages of the $k - r$ latent vector elements. The main difficulty here however is that the latent roots by themselves are unable to pick out the appropriate explicit-form regression equation since small roots will also correspond to nonpredictive multicollinearities which are present among the predictors (Section 10.3.2). This must be so, for $r < k - 1$.

Whichever plane is selected as the appropriate estimate of the linear equation, another major assumption is that the variance/covariance matrix of the errors is known, or at least can be estimated from the data. This is not always the case and in this situation a priori conditions have to be imposed on the structure of $\boldsymbol{\Psi}$ in order to solve the normal equations (Eq. 10.9). Such information may be available in the sample in terms of replicated observations, or else can be given in the form of prior knowledge about the residual errors. The first simplifying assumption commonly used is that the error terms are uncorrelated, in which case $\boldsymbol{\Psi}$ becomes a diagonal matrix. Under regularity conditions the diagonal error variance matrix may then be estimated by an iterative procedure. Alternatively, if this is not possible the unknown error variance terms can at times be replaced by the rank orders $1, 2, \ldots, k$, if the relative error magnitudes of the variance terms are known. Also, as a special case, when the observed variables are further deemed to possess equal error variance the weighted regression model reduces to finding that (unweighted) principal component which corresponds to the smallest latent root of $\mathbf{X}^T\mathbf{X}$. Here the errors are fitted perpendicularly to the estimated plane, and the model is then known as the orthogonal-norm least squares model (Fig. 10.1). Since the orthogonal-norm model is a principal components model it retains all optimality properties of the latter (Chapter 3). Chen (1974) has given a Bayesian optimality property of the model.

The weighted regression model (Eq. 10.8) assumes that the true values $\chi_1, \chi_2, \ldots, \chi_k$ are random. The model also holds for the case when the true values are fixed so that the distinction between deterministic and random variables need not generally be made when using weighted least squares. The situation is identical to the case when principal components can be considered either fixed or random (Sections 3.6 and 6.7). Note also that the development assumes that errors are independent and homoscedastic in the

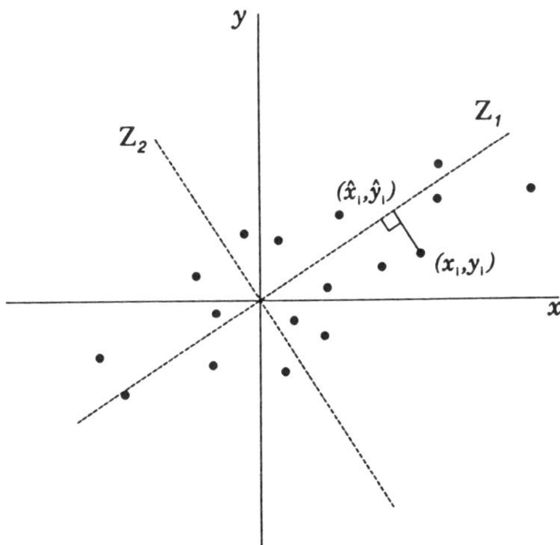

Figure 10.1 Orthogonal-norm least squares regression by rotation of axes.

sample space, that is, the error specification is that of the ordinary least squares model. The model can be generalized however to the case where the errors are heteroscedastic and correlated in the sample space (Sprent, 1966).

Example 10.1.   First consider a bivariate example (Table 10.1) using data from Sprent (1966), where $x = $ logarithm of wood growth and $y = $ logarithm of girth increment of apple trees. Sprent's model for the data is somewhat more general since, as pointed out above, the observations for time series need not be independent or homoscedastic. Ignoring the serial correlation we note that six replications are available per year so that the residual errors for both variables can be estimated, and corrected values $X^* = X - \Delta$ and $Y^* = Y - \epsilon$ may be used in a Gauss–Markov regression. To illustrate the weighted least squares model however we have the normal

Table 10.1   Logarithms of Wood Growth and Girth Increment of Apple Trees with Six Replications per Year

	X	Y	Δ	ϵ
1954	3.44	.49	$-.002$	.013
	3.62	.51	.178	.033
	3.42	.43	$-.022$	$-.047$
	3.47	.43	.028	$-.047$
	3.28	.48	$-.162$	.003
	3.42	.52	$-.022$	.043
Means	3.442	.477	.00	.00
1956	3.97	.56	.027	.055
	3.97	.52	.027	.015
	3.79	.43	$-.153$	$-.075$
	4.04	.52	.097	.015
	3.94	.49	$-.003$	$-.015$
	3.95	.51	.007	.005
Means	3.943	.505	.00	.00
1958	4.06	.63	$-.008$	.020
	4.04	.57	$-.028$	$-.040$
	4.07	.59	$-.002$	$-.020$
	4.11	.64	.042	.030
	3.93	.59	$-.138$	$-.020$
	4.20	.64	.132	.030
Means	4.068	.610	.00	.00
Overall means	3.818	.531	.00	.00

Source: Sprent, 1966; reproduced with permission.

equations

$$(\mathbf{X}^T\mathbf{X} - l\hat{\boldsymbol{\Psi}})\hat{\boldsymbol{\alpha}} = (\hat{\boldsymbol{\Psi}}^{-1}\mathbf{X}^T\mathbf{X} - l\mathbf{I})\hat{\boldsymbol{\alpha}} = \mathbf{0} \tag{10.16}$$

Since $\bar{\boldsymbol{\Delta}} = \bar{\boldsymbol{\epsilon}} = \mathbf{0}$, the first and second moments of the system are given by

$$\sum_{i=1}^{18} x_i = 68.7200\,, \qquad \sum_{i=1}^{18} y_i = 9.5500$$

$$\sum_{i=1}^{18} x_i^2 = 263.8108\,, \qquad \sum_{i=1}^{18} y_i^2 = 5.1475\,, \qquad \sum_{i=1}^{18} x_i y_i = 36.7095$$

$$\sum_{i=1}^{18} \Delta_i^2 = .133102\,, \qquad \sum_{i=1}^{18} \epsilon_i^2 = .021484\,, \qquad \sum_{i=1}^{18} \Delta_i \epsilon_i = .028014$$

and

$$\sum_{i=1}^{18} (x_i - \bar{x})^2 = 1.453111\,, \qquad \sum_{i=1}^{18} (y_i - \bar{y})^2 = .080694\,,$$

$$\sum_{i=1}^{18} (x_i - \bar{x})(y_i - \bar{y}) = .249722$$

Since the errors are correlated ($r_{\Delta\epsilon} = .524$) we have

$$\hat{\boldsymbol{\Psi}}^{-1} = \begin{bmatrix} .133102 & .028014 \\ .028014 & .021484 \end{bmatrix}^{-1} = \frac{1}{.002075}\begin{bmatrix} .021484 & -.028014 \\ -.028014 & .133102 \end{bmatrix}$$

$$= \begin{bmatrix} 10.353735 & -13.500723 \\ -13.500723 & 64.145542 \end{bmatrix}$$

so that

$$\hat{\boldsymbol{\Psi}}^{-1}\hat{\boldsymbol{\Sigma}} = \begin{bmatrix} 10.353735 & -13.500723 \\ -13.500723 & 64.145542 \end{bmatrix}\begin{bmatrix} 1.453111 & .249722 \\ .249722 & .080694 \end{bmatrix}$$

$$= \begin{bmatrix} 11.673699 & 1.496128 \\ -3.599496 & 1.804732 \end{bmatrix}$$

whose latent roots are solutions of the determinantal equation

$$\begin{vmatrix} (11.673699 - l) & 1.496128 \\ -3.599496 & (1.804732 - l) \end{vmatrix} = 0$$

or

$$l^2 - 13.478431l + 26.453205 = 0$$

that is, $l_1 = 11.093962$ and $l_2 = 2.384469$. The latent vectors corresponding to the two roots are then given by Eq. (10.16), that is, corresponding to l_1 we have

$$(\hat{\boldsymbol{\Psi}}^{-1}\hat{\boldsymbol{\Sigma}} - l_1\mathbf{I})\hat{\boldsymbol{\alpha}}_1 = \mathbf{0}$$

or

$$\begin{bmatrix} (11.673699 - 11.093962) & 1.496128 \\ -3.599496 & (1.804732 - 11.093962 \end{bmatrix} \begin{bmatrix} \hat{\alpha}_{11} \\ \hat{\alpha}_{21} \end{bmatrix} = \mathbf{0}$$

subject to the constraint $\hat{\boldsymbol{\alpha}}_1^{\mathrm{T}}\hat{\boldsymbol{\Psi}}\hat{\boldsymbol{\alpha}}_1 = 1$. This yields the system of equations

$$\left. \begin{array}{l} .579737\,\hat{\alpha}_{11} + 1.496128\,\hat{\alpha}_{21} = 0 \\ -3.599496\,\hat{\alpha}_{11} - 9.28923\,\hat{\alpha}_{21} = 0 \end{array} \right\} \quad \hat{\alpha}_{11} = -2.580701\,\hat{\alpha}_2$$

where

$$[\hat{\alpha}_{11}, \hat{\alpha}_{21}] \begin{bmatrix} .133102 & .0280140 \\ .280140 & .021484 \end{bmatrix} \begin{bmatrix} \hat{\alpha}_{11} \\ \hat{\alpha}_{21} \end{bmatrix} = 1$$

so that

$$.133102\,\hat{\alpha}_{11}^2 + .056028\,\hat{\alpha}_{11}\,\hat{\alpha}_{21} + .021488\,\hat{\alpha}_{21}^2 = 1$$

the unique solution to which is (aside from rotational indeterminacy and sign change) $\hat{\alpha}_{11} = -2.9538$ and $\hat{\alpha}_{21} = 1.1446$. Similarly, corresponding to $l_2 = 2.384469$ we have the latent vector $\hat{\alpha}_{12} = -1.276775$ and $\hat{\alpha}_{22} = 7.927326$ so that the implicit form equation is given by

$$\hat{\alpha}_{12}(\hat{\mathbf{X}} - \bar{\mathbf{X}}) + \hat{\alpha}_{22}(\hat{\mathbf{Y}} - \bar{\mathbf{Y}}) = \mathbf{0}$$

or

$$-1.276775(\hat{\mathbf{X}} - \bar{\mathbf{X}}) + 7.927326(\hat{\mathbf{Y}} - \bar{\mathbf{Y}}) = \mathbf{0}$$

where $\bar{Y} = .5306$ and $\bar{X} = 3.8178$. Solving for $\tilde{\beta}_0$ and $\tilde{\beta}_1$ we have

$$(\hat{\mathbf{Y}} - \bar{\mathbf{Y}}) = \frac{1.276775}{7.927326}(\hat{\mathbf{X}} - \bar{\mathbf{X}})$$

$$= .16106(\hat{\mathbf{X}} - \bar{\mathbf{X}})$$

so that

$$\hat{\mathbf{Y}} = \tilde{\beta}_0 + \tilde{\beta}_1\hat{\mathbf{X}}$$

$$= -.0843 + .16106\hat{\mathbf{X}}$$

The Gauss–Markov ordinary least squares estimate for the slope is $\hat{\beta}_1 = .17185$ so that the effect of the error in $\mathbf{X}$ is small. Actually the main influence here is exerted by the correlation between the two error terms since if the correlation is assumed to be zero we obtain $\tilde{\beta}_1 = .5433$, a much larger value.

Example 10.2. Under certain environmental conditions a protein known as vicilin, when isolated from legume (faba bean) seeds, undergoes self-association to form a micelle-type structure, a relatively unique arrangement for protein molecules. A micelle is a stable spherical aggregation of molecules in an aqueous system, in which nonpolar or hydrophobic ends of the molecules are oriented inwards and the polar portions are exposed to the water medium. The capacity of vicilin to form micelles seems to be related to several molecular structural parameters which, in turn, are influenced by the specific type of environment. Experimental data (Table 10.2; Ismond, 1984; personal communication) is obtained for the following variables:

Y_1 = Surface hydrophobicity, in terms of how many hydrophobic amino acid residues were on the surface of the protein
Y_2 = Temperature of denaturation
Y_3 = Micelle rating; the (subjective) ordinal scale 0–9
Y_4 = Logarithmic (base 10) transformation of Y_1
Y_5 = Logarithmic (base 10) transformation of Y_2

A part of the total data set is given in Table 10.2 for $n = 56$ experimental observations. Using the entire data set of 130 observations we obtain the following correlation matrix together with its latent roots and latent vectors for the first three variables (Table 10.3).

Taking $\mathbf{X}_3$ as the dependent variable we have the estimated equation (Eq. 10.14)

$$\hat{X}_3 = \frac{.2837}{.5797}\,\hat{X}_1 - \frac{.7638}{.5797}\,\hat{X}_2$$

or

$$\hat{Y}_3 = 229.152 - .4894\,\hat{Y}_1 - 1.3176\,\hat{Y}_2 \tag{10.17}$$

The data are used only for illustrative purposes since the assumption of equal error variance is doubtful.

10.2.2 The Maximum Likelihood Model

The weighted least squares regression model can be viewed as a generalization of the usual Gauss–Markov ordinary least squares where the residual errors are fitted at an arbitrary angle to the plane. At times it is also

Table 10.2 Part of the Original $n = 130$ Observations for $k = 5$ Experimental Variables in a Study of Micelle Formation

Y_1	Y_2	Y_3	Y_4	Y_5
315	89.0	9	2.49831	1.94939
299	89.0	9	2.47567	1.94939
266	86.0	9	2.42488	1.93450
235	86.0	9	2.37107	1.93450
271	88.5	9	2.43297	1.94694
248	89.0	9	2.39445	1.94939
352	88.5	9	2.54654	1.94694
264	88.0	9	2.42160	1.94448
255	89.0	9	2.40654	1.94939
263	91.0	9	2.41996	1.95904
269	91.0	9	2.42975	1.95904
225	90.5	9	2.35218	1.95665
225	91.0	9	2.35218	1.95904
288	95.0	6	2.45939	1.97772
202	95.0	6	2.30535	1.97772
198	95.0	6	2.29667	1.97772
167	90.0	9	2.22272	1.95424
167	91.0	9	2.22272	1.95904
153	92.5	8	2.18469	1.96614
156	92.5	9	2.19312	1.96614
168	92.5	0	2.22531	1.96614
162	92.5	0	2.20952	1.96614
127	94.5	0	2.10380	1.97543
112	94.5	0	2.04922	1.97543
174	98.0	5	2.24055	1.99123
162	98.0	5	2.20952	1.99123
160	99.0	5	2.20412	1.99564
153	100.0	5	2.18469	2.00000
172	103.0	0	2.23553	2.01284
151	104.0	0	2.17898	2.01703
144	104.5	0	2.15836	2.01912
136	105.0	0	2.13354	2.02119
233	87.0	9	2.36736	1.93952
228	87.5	9	2.35793	1.94201
194	87.5	9	2.28780	1.94201
176	88.0	9	2.24551	1.94448
226	91.0	9	2.35411	1.95904
197	91.0	9	2.29447	1.95904
190	91.0	9	2.27875	1.95904
192	92.0	9	2.28330	1.96379
153	100.5	9	2.18469	2.00217
186	97.0	8	2.26951	1.98677
211	96.0	8	2.32428	1.98227
219	105.0	0	2.34044	2.02119
172	105.0	0	2.23553	2.02119

Table 10.2   (Continued)

174	105.0	0	2.24055	2.02119
175	106.5	0	2.24304	2.02735
228	86.0	2	2.35793	1.93450
188	86.0	2	2.27416	1.93450
183	87.0	2	2.26245	1.93952
165	88.0	2	2.21748	1.94448
214	89.0	2	2.33041	1.94939
178	90.5	2	2.25042	1.95665
177	92.5	2	2.24797	1.96614
166	93.0	2	2.22011	1.96848
143	100.0	0	2.15534	2.00000

Source: Ismond, 1984, personal communication.

Table 10.3   Correlation Matrix, Mean Values, Latent Roots, and Latent Vectors of the First Three Variables Using the Original $n = 130$ Observations of a Micelle-Rating Experiment

	X_1	X_2	X_3		Z_1	Z_2	Z_3
X_1	1.0000				.5176	.8072	.2837
X_2	$-.5047$	1.0000			$-.6309$	.1361	.7638
X_3	.3385	$-.6322$	1.0000		.5780	$-.5743$	.5797
	204.49	94.63	4.39	l_i	1.9932	.6732	.3328

Source: Ismond, 1984; personal communication.

convenient to assume normality of the errors, for example, for purposes of significance testing. For the moment we continue not to make a distinction between functional and structural forms, although, as will be seen shortly, such a distinction is of some importance from the theoretical perspective when considering normal errors. Consider the errors-in-variables model of the previous section where the errors are distributed as the multivariate normal. Then for any ith observation Δ_i we have

$$f(\Delta_i) + \frac{1}{(2\pi)^{k/2}|\Psi|^{1/2}} e^{-1/2(\Delta_i^T \Psi^{-1} \Delta_i)} \qquad (i = 1, 2, \ldots, n) \qquad (10.18)$$

where Ψ is nonsingular and Δ_i is the ith row of Δ. The likelihood of the sample is then given by

$$L(\Delta) = \frac{1}{(2\pi)^{nk/2}|\Psi|^{n/2}} e^{-1/2 \sum_{i=1}^{n} \Delta_i^T \Psi^{-1} \Delta_i} \qquad (10.19)$$

It is easy to show that when error is present only in the dependent variable $Y = X_1$, Eq. (10.19) reduces to the usual Gauss–Markov ordinary least squares likelihood function

$$L(\boldsymbol{\epsilon}) = \frac{1}{(\sigma_\epsilon^2 2\pi)^{n/2}}\, e^{-\frac{1}{\sigma^2}\boldsymbol{\epsilon}^T\boldsymbol{\epsilon}} \tag{10.20}$$

where $\boldsymbol{\epsilon}$ is the error term in $\mathbf{Y}$. Since $\boldsymbol{\Delta} \sim N(\mathbf{0}, \boldsymbol{\Psi})$ we have $\boldsymbol{\delta} \sim N(\mathbf{0}, \boldsymbol{\alpha}^T\boldsymbol{\Psi}\boldsymbol{\alpha})$. Let $\mathbf{x}_i^T$ denote the ith row of $\mathbf{X}$. Then Eq. (10.4) can be written as

$$\mathbf{x}_i^T\boldsymbol{\alpha} = \boldsymbol{\Delta}_i^T\boldsymbol{\alpha} = \boldsymbol{\delta}_i \qquad (i = 1, 2, \ldots, n) \tag{10.21}$$

and the likelihood function of $\delta_1, \delta_2, \ldots, \delta_n$ is given by

$$
\begin{aligned}
\mathbf{L}(\boldsymbol{\delta}) &= ke^{-\frac{1}{2}\sum_{i=1}^{n} \mathbf{x}_i^T\boldsymbol{\alpha}(\boldsymbol{\alpha}^T\boldsymbol{\Psi}\boldsymbol{\alpha})^{-1}\boldsymbol{\alpha}^T\mathbf{x}_i} \\
&= ke^{-\frac{1}{2}\sum_{i=1}^{n} \frac{\mathbf{x}_i^T\boldsymbol{\alpha}\boldsymbol{\alpha}^T\mathbf{x}_i}{\boldsymbol{\alpha}^T\boldsymbol{\Psi}\boldsymbol{\alpha}}} \\
&= ke^{-\frac{1}{2}\frac{\boldsymbol{\alpha}^T(\mathbf{x}^T\mathbf{x})\boldsymbol{\alpha}}{\boldsymbol{\alpha}^T\boldsymbol{\Psi}\boldsymbol{\alpha}}}
\end{aligned}
\tag{10.22}
$$

where k is the constant of proportionality. Also, it is easy to verify by direct expansion that $\mathbf{x}_i^T\boldsymbol{\alpha}\boldsymbol{\alpha}^T\mathbf{x}_i = \boldsymbol{\alpha}^T\mathbf{x}_i\mathbf{x}_i^T\boldsymbol{\alpha}$ and consequently

$$
\sum_{i=1}^{n} \mathbf{x}_i^T\boldsymbol{\alpha}\boldsymbol{\alpha}^T\mathbf{x}_i = \sum_{i=1}^{n} \boldsymbol{\alpha}^T\mathbf{x}_i\mathbf{x}_i^T\boldsymbol{\alpha}
$$

$$
= \boldsymbol{\alpha}^T(\mathbf{X}^T\mathbf{X})\boldsymbol{\alpha} \tag{10.23}
$$

The likelihood function (Eq. 10.23) can be maximized by minimizing the exponent, subject to the constraint $\boldsymbol{\alpha}^T\boldsymbol{\Psi}\boldsymbol{\alpha} = 1$. The Lagrangian expression is

$$\phi = \boldsymbol{\alpha}^T(\mathbf{X}^T\mathbf{X})\boldsymbol{\alpha} - \lambda(\boldsymbol{\alpha}^T\boldsymbol{\Psi}\boldsymbol{\alpha} - 1) \tag{10.24}$$

where λ is a Lagrangian scalar multiplier. Differentiating Eq. (10.24) with respect to the parameters $\boldsymbol{\alpha}$ and equating to zero yields

$$\frac{\partial\phi}{\partial\boldsymbol{\alpha}} = 2(\mathbf{X}^T\mathbf{X})\hat{\boldsymbol{\alpha}} - 2l\hat{\boldsymbol{\Psi}}\hat{\boldsymbol{\alpha}} = \mathbf{0}$$

or

$$[(\mathbf{X}^T\mathbf{X}) - l_k\hat{\boldsymbol{\Psi}}]\hat{\boldsymbol{\alpha}} = \mathbf{0} \tag{10.25}$$

where l is the sample estimator of λ.

The normal equations (Eq. 10.25) are the same as Eq. (10.8) and $\hat{\boldsymbol{\alpha}}$ again denotes the latent vector of $\mathbf{X}^T\mathbf{X}$ which corresponds to the smallest latent root l_k. For purposes of computation it may also be more convenient to write Eq. (10.25) as $[\hat{\boldsymbol{\Psi}}^{-1}(\mathbf{X}^T\mathbf{X}) - l_k\mathbf{I}]\hat{\boldsymbol{\alpha}} = \mathbf{0}$ and to solve the resulting characteristic equation (Section 2.10). Although both $\hat{\boldsymbol{\Psi}}^{-1}$ and $\mathbf{X}^T\mathbf{X}$ are symmetric and positive definite, their product need not be symmetric, and the $\hat{\boldsymbol{\alpha}}$ therefore need not represent an orthogonal coordinate system. Both the eigenvectors $\hat{\boldsymbol{\alpha}}_i$ and the eigenvalues l_i however are real, and $l_i \geq 0$.

The derivation of the maximum likelihood model does not distinguish between the case when the true values are random or fixed. It was pointed out in Section 6.7 however that maximum likelihood estimators do not exist for fixed principal components. In fact Solari (1969) has shown that for functional forms these values correspond to a saddle point rather than to a true maximum of likelihood surface. Although such a result is indeed surprising, since we would more readily expect difficulties with the more general structural form, this does not necessarily imply that good approximations cannot be obtained from the functional form as well. this is because, as pointed out by Copas (1972), continuous population probability densities are only continuous hypothetical approximations to likelihood functions when using actual data, which are never recorded as a pure continuum. Rather, we observe data that are rounded off to some decimal point, that is, data grouped within intervals, and this is sufficient to ensure (with unit probability) that error variances will not be zero, which is what produces saddlepoint solutions. The least squares estimator is thus often used with normally distributed errors for both structural and functional forms, and although "maximum likelihood" estimators will still be inconsistent with grouped data, such inconsistency will usually be small in most cases encountered in practice (Sprent, 1970). Maximum likelihood estimation of the general nonlinear functional relationship has also been considered by Dolby and Lipton (1972).

10.2.3 Goodness of Fit Criteria of Orthogonal-Norm Least Squares

Several criteria are available with which to estimate the sum of squares resulting from the least squares plane. First we note that Eq. (10.11) yields biased predicted values of the dependent variable $\mathbf{Y}$ since in

$$\hat{\mathbf{Y}} = \mathbf{X}\tilde{\boldsymbol{\beta}} \tag{10.26}$$

the variables $\mathbf{X}$ contain error and $E(\hat{\mathbf{Y}}) \neq \boldsymbol{\chi}\boldsymbol{\beta}$. A consistent estimator of the conditional regression equation is provided by Eq. (10.14), that is, $\hat{\mathbf{Y}} = \hat{\mathbf{X}}\hat{\boldsymbol{\beta}}$ where values of $\hat{\mathbf{X}}$ are given by Eq. (10.15). We have

$$E(\hat{\mathbf{X}}^T\hat{\boldsymbol{\delta}}) = E[(\mathbf{Z}_{k-1}\mathbf{P})^T\hat{\boldsymbol{\delta}}]$$

$$= E[\mathbf{P}^T\mathbf{Z}_{k-1}^T\hat{\boldsymbol{\delta}}]$$

$$= \mathbf{0} \tag{10.27}$$

so that the predicted values and the error term are independent. Since residual variance of the implicit-form regression equation is provided by the smallest root l_k, the proportion of variance explained by the first $(k-1)$ major axes is then

$$R_A^2 = \frac{\displaystyle\sum_{i=1}^{k-1} l_i}{\displaystyle\sum_{i=1}^{k} l_i}$$

$$= 1 - \frac{l_k}{\displaystyle\sum_{i=1}^{k} l_i} \qquad (10.28)$$

which may be used as a goodness-of-fit criterion. Note, however, that R_A^2 is not bounded by zero from below—for example, in a bivariate regression we obtain

$$R_A^2 = \frac{l_1}{l_1 + l_2} = \frac{1}{2}$$

in the case of independence between the variables, since in this case $l_1 = l_2$. Also note that Eq. (10.28) is always larger (in a given sample) than the ordinary least squares R^2 criterion since for the former case the residuals are fitted orthogonally to the hyperplane and thus represent minimum distance between observations and the hyperplane. A bivariate measure which is bounded from below by zero is

$$R_B^2 = \frac{l_1 - l_2}{l_1 + l_2} \qquad (10.29)$$

which can be shown to equal the bivariate correlation coefficient between the dependent and independent variables. Other measures that generalize to the multivariate case are also available and may be found in Kloek and Banink (1962).

Finally, a useful measure of multicollinearity is given by

$$S_k^2 = \frac{l_k}{a_{ik}^2} \qquad (10.30)$$

since Hawkins (1973) has shown that when a_{lk} is small (large) relative to l_k, multicollinearity will also be high (low) (Section 10.3.3).

10.2.4 Testing Significance of Orthogonal-Norm Least Squares

Once a structural (functional) form has been fitted to the data, it is of some interest to test for significance and estimate confidence intervals of the slope coefficients. In what follows we describe two possible approaches for deriving standard errors of the orthogonal norm coefficients, depending on whether errors are normal or otherwise. Owing to the difference between the two models (Section 10.2.2), this also largely coincides with whether a structural or functional form is considered.

When the true values and the error terms follow the normal distribution, the coefficients of the orthogonal-norm equation can be tested using Girshick's (1939) principal components distribution theory, together with the theory of ratio estimators (Basilevsky, 1980). Let $\hat{\alpha}_{ik}$ be any ith element of the latent vector $\hat{\boldsymbol{\alpha}}_k$ which is associated with the smallest latent root l_k. Then for multivariate normal data we know (Section 4.3.4) that $\hat{\alpha}_{ik}$ has asymptotic variance

$$\hat{\sigma}_i^2 = \frac{l_k}{n} \sum_{s=1}^{k} \frac{l_s}{(l_s - l_k)^2} \, \hat{\alpha}_{is}^2 \qquad (s \neq k; i = 1, 2, \ldots, k) \qquad (10.31)$$

If the k regression variables are assumed to be distributed as a multivariate normal it follows that each marginal distribution is normal, and $(\hat{\alpha}_{ik} - \alpha_{ik})$ are thus asymptotically $N(o, \sigma_i^2)$ normal variates $(i = 1, 2, \ldots, k)$. In large samples the null hypothesis $\alpha_{ik} = 0$ can therefore be tested by means of the normal distribution.

A more frequent requirement is to be able to test the significance of the conditional regression parameters $\tilde{\beta}_i = -(\hat{\alpha}_i/\hat{\alpha}_k)$, that is, ratios of normal variates (Section 10.2.1). This may be done by using the general theory of ratio estimators. It is known that when $x \sim N(\mu_x, \sigma_x^2)$ and $y \sim N(\mu_y, \sigma_y^2)$ the probability density function of $z = x/y$ is the Fieller distribution (Fieller, 1932), which depends on μ_x, μ_y, σ_x/σ_y, and ρ, the correlation coefficient between x and y. Although the theoretical distribution of z is known, no tables of its probabilities are available, no doubt because of the complexity of the function (see Hinkley, 1969). Another difficulty arises when $\mu_x = \mu_y = 0$. In this case Curtiss (1940) has shown that (see also Marsaglia, 1965) z follows the Cauchy distribution (Fig. 10.2)

$$C(\mu, \lambda) = \frac{\lambda}{\pi[(z - \mu)^2 + \lambda^2]} \qquad (10.32)$$

where

$$\lambda = \frac{\sigma_x}{\sigma_y} \sqrt{1 - \rho^2} \quad \text{and} \quad \mu = \rho \frac{\sigma_x}{\sigma_y} = \frac{\sigma_{xy}}{\sigma_y^2}$$

In its general form the Cauchy distribution (Eq. 10.32) depends on the parameters μ and λ which determine location and scale, respectively. Although the Cauchy distribution possesses no moments, confidence inter-

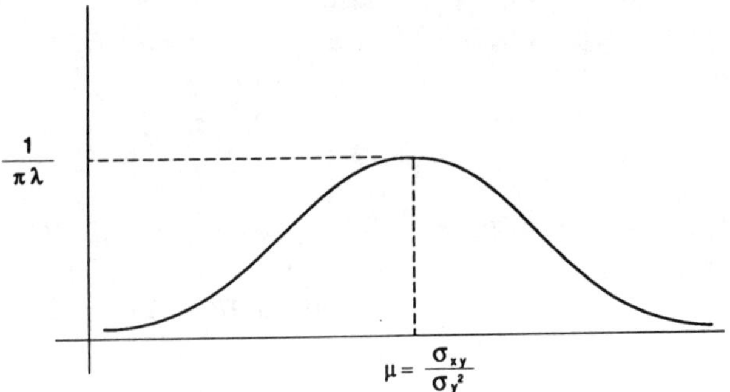

Figure 10.2   The non-central Cauchy distribution.

vals and critical values for μ are available, and ML estimators exist (Haas et al., 1970) so that hypotheses concerning μ and λ can be tested.

In the orthogonal regression context, however, the Cauchy distribution is only relevant in the degenerate case when both numerator and denominator of the ratio coefficients are distributed normally with zero mean. More generally, Lianos and Rausser (1972) have estimated an empirical cumulative distribution of a ratio of normal variates, as well as the critical values for various levels of significance, and conclude that the standard normal distribution cannot be used for testing the significance of a ratio in samples of size $n = 125$ or smaller. It is nevertheless advantageous to be able to use a well-tabulated distribution such as the standard normal to test for significance of $\tilde{\beta}_i$, if only in an approximate fashion. As noted above, a way out of the difficulty is to consider the $\tilde{\beta}_i$ as ratio estimators, well known in sampling theory. Let

$$r = \frac{a}{b}, \qquad R = \frac{\alpha}{\beta} \qquad (10.33)$$

where r is a sample value of the ratio and R its population value such that $a \sim N(\alpha, \sigma_a^2)$ and $b \sim N(\beta, \sigma_b^2)$. It is known that r is a consistent but biased estimator of R unless $r = [\text{cov}(a, b)/\text{var}(b)]$. However, consider the quantity

$$q = a - Rb \qquad (10.34)$$

which is normally distributed since a and b are normal, that is, $q \sim N(0, \sigma_q^2)$ since $E(q) = \alpha - R\beta = 0$. The variance of q can be obtained (approximately) by the first-order Taylor series expansion about the mean, as

$$\sigma_q^2 \simeq (\sigma_a^2 + R^2\sigma_b^2 - 2R\sigma_{ab})/\beta^2 \qquad (10.35)$$

so that approximately,

$$\frac{a - Rb}{1/\beta\sqrt{\sigma_a^2 + R^2\sigma_b^2 - 2R\sigma_{ab}}} \sim N(0, 1) \qquad (10.36)$$

Also, by expanding r into a bivariate Taylor series it can be shown that the square of the denominator of Eq. (10.36) is the first-order approximation of σ_r^2 so that we have (approximately),

$$\frac{r - R}{\sigma_r} \sim N(0, 1) \qquad (10.37)$$

On the null hypothesis $R = 0$ we then have

$$\frac{r}{\sqrt{\dfrac{S_a^2}{b^2}}} \sim N(0, 1) \qquad (10.38)$$

where σ_a^2 has been replaced by its sample value s_a^2. Equation (10.38) may be used as a large sample test of significance of the ratio coefficients $\hat{\beta}_i = r$ where we replace b by $\hat{\alpha}_k$ and s_a^2 by $\hat{\sigma}_i^2$ using Eq. (10.31). Equation (10.38) is similar to the usual statistic employed in Gauss–Markov ordinary least squares and is not difficult to compute. Its use however is conditional on the roots of the quadratic (Eq. 10.35) being real, which is the case most of the time when $3\sigma_b \leq b$ and $3\sigma_a \leq a$ (Geary, 1930). Also the bias of the ratio estimator r can be shown to satisfy the inequality

$$\frac{|\text{Bias } r|}{\sigma_r} \leq \frac{\sigma_b}{b} \qquad (10.39)$$

The bias can usually be ignored however when $\sigma_b/b \leq .1$, that is, when the sample size is large.

Example 10.3. As a numerical comparison between ordinary least squares and orthogonal-norm regression consider the true equation

$$\mathbf{Y}^* = .50\mathbf{\chi}_1 - .10\mathbf{\chi}_2 \qquad (10.40)$$

where the observed values are given by $\mathbf{X}_1 = \mathbf{\chi}_1 + \mathbf{\Delta}_1$, $\mathbf{X}_2 = \mathbf{\chi}_2 + \mathbf{\Delta}_2$, and $\mathbf{Y} = \mathbf{Y}^* + \mathbf{\epsilon}$. The equation is defined for a population size of $n = 100$ where the values are randomly drawn from the integers in the range 1–100 to avoid multicollinearity. The correlational structure and the least squares estimates are given in Tables 10.4 and 10.5.

The ratio estimator test is based on the assumption of approximate multivariate normality of the variables. Nonparametric methods that do not

Table 10.4 The Correlational Structure Between True Values and Error Measurements for a Simulated Population

	X_1	X_2	Y^*	Δ_1	Δ_2	ϵ
X_1	1.00					
X_2	$-.1664$	1.00				
Y^*	.9260	$-.5263$	1.00			
Δ_1	$-.0382$	$-.0174$	$-.0262$	1.00		
Δ_2	$-.0055$	.0408	$-.0204$	$-.0573$	1.00	
ϵ	$-.0456$	.0361	$-.0531$	$-.0234$	.0013	1.00

Table 10.5 The Estimated Equation (Eq. 10.40) Using All $n = 100$ Values for Ordinary (OLS) and Orthogonal-Norm (ON) Least Squares[a]

	OLS		ONLS		Error SS		
					Δ_1	Δ_2	ϵ
Coefficients	.4811	$-.0893$	.4965	$-.0899$			
Standard deviation	.0203	.0224	.0168	.0230	17.4%	19.9%	162%
t-Statistic	23.70	3.99	29.55	3.91			
R^2		.8661		.9133			

[a] The R^2 coefficient for orthogonal-norm least squares is obtained using Eq. (10.26).

require prior distributional assumptions, however, are also possible. Anderson (1975), for example, uses the jackknife estimator to test the angle of rotation associated with the bivariate functional form, using small samples. Assuming (1) bivariate normality, (2) bivariate uniformness, and (3) uniform and exponential errors, Anderson (1975) uses Monte Carlo sampling to investigate the influence of nonnormality. A more conclusive study is carried out by Kelly (1984) who uses Monte Carlo methodology to compare sampling behavior of the jackknife, bootstrap, normal-theory, and influence function estimators of variability. Kelly (1984), concludes that while the bootstrap seems to perform well in a variety of sampling situations, the influence function estimator of the standard error is consistently biased downward, as is the normal theory estimator in nonnormal situations. The conclusions are that although no estimator is globally best, the influence function method should be avoided. Also, the jackknife estimator is conservative in that on the average it overestimates the standard error. The bootstrap seems to do better than the jackknife in terms of mean squared error, but it can apparently give biased results.

10.2.5 Nonlinear Orthogonal Curve Fitting

Unlike the usual least squares regression model it is not possible to generalize the orthogonal-norm plane to nonlinear functions such as polynomials because in order to maintain orthogonality to the tangent the residuals must alter their orientation with respect to the axes. An iterative algorithm however has been developed by Hastie and Stuetzle (1989) which fits "principal curves" to data points in an orthogonal manner. This is achieved by an index which at each fitting stage minimizes the distance from the point to the curve. The nonparametric algorithm begins with the first (largest) principal component and checks if this curve is "self-consistent" by projecting and averaging. The iteration continues until convergence, which appears to take place in practice most of the time, although this is not guaranteed.

10.3 LEAST SQUARES REGRESSION WITH MULTICOLLINEARITY

The previous sections deal with the weighted least squares errors-in-variables model, which is the inverse of weighted principal components (Section 5.6), where least squares optimality is associated with the smallest latent root(s) of both the dependent and independent variables. In recent years, with the advent of electronic computers, another specification error of major importance has been recognized for least squares—that of multicollinearity. The term multicollinearity refers to the situation (and the difficulties created by it) when in a multiple least squares regression equation the correlation between any two (or more) predictors arbitrarily tends to unity in the limit, reaching unity (exact correlation) when the affected predictors lie in a common subspace. In this section we concern ourselves with linear multicollinearity, so that high correlation between terms of a quadratic equation, for example, are not considered. In the situation of unit correlation the predictors are related by one (or more) exact linear relationship(s), which are usually due to a badly specified model, that is, a predictor is inadvertently defined as a linear combination of some (or all) other predictors. Here, the difficulty can be resolved by removing explanatory variables which are responsible for the multicollinearity.

When correlation is high but not perfect, the Gauss–Markov estimator is still unique but $(\mathbf{X}^T\mathbf{X})$ tends to singularity, thus inflating the diagonal elements of its inverse. The result is high inefficiency, and a paradoxical situation arises whereby we tend to accept the null hypothesis for all least squares coefficients of the equation even though the F-statistic indicates a high overall fit. Thus a consequence for using regression as an exploratory tool is that it becomes difficult to distinguish between significant and insignificant predictor variables. Furthermore, multicollinearity can be enhanced (masked) by the presence of outlier observations, which contribute further to serious difficulties of estimation, interpretation, and prediction. By the very nature of the problem it is generally not possible to predict beforehand exactly how low correlation must be in order to identify statistically significant predictors, although diagnostic statistics may be used to alert the analyst to the problem (see Hocking and Pendleton, 1983).

When the explanatory variables can be predetermined (held fixed), for example, in a scientific experiment, multicollinearity will usually not be a problem even when the regressors are not orthogonal. When all variables vary randomly and not much a priori information is available for the structure of the population, (for the use of a priori information see Fomby and Hill, 1979) multicollinearity emerges as a distinct possibility. Although multicollinearity is common for data such as time series, it can also appear with random independent samples. There are two distinct ways of considering multicollinearity. First, the excessively high correlation can be viewed as a sampling deficiency whereby the sample contains insufficient information to estimate a linear model. This view of multicollinearity has been advocated by Farrar and Glauber (1967), whereby multicollinearity is defined in its strict sense as departure from orthogonality. There are two difficulties associated with such a viewpoint, particularly when dealing with random rather than fixed regressors. (1) a departure from the orthogonality definition of multicollinearity is too stringent in practice, and (2) it is inconsistent with the concept of multivariate regression, which is intended to generalize univariable regression precisely through the notion of correlated predictor variables.

The second, and probably more productive view of multicollinearity, is in terms of population correlation, where although the observed variables are linearly independent, owing to errors in variables and/or sampling error, their true population values are perfectly correlated because of the existence of common underlying dimensions. Here multicollinearity is viewed as a model specification error rather than a data difficulty. Also, whereas the sampling approach treats multicollinearity as a problem due to lack of information, the population approach emphasizes the presence of supplementary information which is available in terms of the high correlation between the regressors. It is only when such information is ignored that multicollinearity emerges as a misspecification of the regression equation. Population multicollinearities are usually due to inherent characteristics of the population since the difficulty cannot be removed by resampling or by an appropriate choice of experimental or observational units (see also Gunst, 1983; Basilevsky, 1973).

The causal structures that may exist among a set of random regressors can be uncovered by a factor analysis of the variables, since it is well known that although some coefficients cannot be estimated precisely, their linear combinations often can. The situation is similar to the concept of reparametrization commonly encountered in discrete ANOVA-type least squares regression. The first suggestion of this approach to multicollinearity was from Tintner (1945), who suggested that latent roots of a sample dispersion matrix be used to estimate the number of estimable linear combinations between the regressors, and proposed that multicollinearity be tested on the basis of these latent roots. Somewhat later Kendall (1957) reiterated the idea of carrying out a principal components analysis of the independent variables of a regression equation while Kloek and Mennes

(1960) used principal components to estimate equation systems. In what follows we consider several approaches developed for using principal components to solve the multicollinearity problem

10.3.1 Principal Components Regression

When nonexperimental data are used in a least squares regression context the predictor variables are frequently stochastic rather than fixed, and the outcome may be high intercorrelation among the variables. Even with fixed experimental variables, high correlation may become a problem, for example, in calibration experiments using a near infrared reflectance instrument (Fearn, 1983) or in experiments that require random covariates. The matrix $\mathbf{X}^T\mathbf{X}$ of the predictor variables becomes ill-conditioned and this results in inefficiency of the least squares estimators. A common strategy to improve efficiency is to replace the unbiased ordinary least squares estimator by one which is biased but more efficient, and which minimizes the mean squared error criterion. Several such estimators exist (see Hocking, 1976). First, the predictor variables can be tested for individual contribution to the sum of squares using the F statistic. This includes stepwise regression procedures such as backward and forward elimination. Since step-wise methods do not make use of all possible combinations of the predictors, they may fail to locate the optimal subset, and at times a better strategy is to perform all $2^k - 1$ regressions if the number of predictors is not too high. Second, we can introduce prior "information" in the form of "phoney data" and use the so-called ridge estimators, which by themselves do not result in a reduced set of predictor variables but which are nevertheless less subject to multicollinearity. A possibly unsatisfactory feature of ridge regression however is that it represents a purely technical "quick fix" and does not attempt to uncover the structure and/or cause(s) of multicollinearity. A third option is to use linear combinations of the predictor variables as the new regressors. The linear combinations play the role of linear restrictors and may either be suggested by prior theory (Theil, 1963) or else may be estimated directly from the sample by using a factor model such as principal components. An advantage of the approach is that it will reveal the correlational structure of the predictors, which can yield important information (see, e.g., Chatterjee and Price, 1977). The principal components regression model seems to have been first proposed by Tintner (1945), Butchatzsch (1947), and Stone (1947) who used principal components to eliminate multicollinearity in sociodemographic and economic time series data. The use of the method grew somewhat and similar applications appeared in White et al. (1958), Versace (1960), Spurrell (1963), Fiering (1964), Hashiguchi and Morishima (1969), Daling and Tamura (1971), Janssen and Jobson (1980), Kung and Sharif (1980), and Mager (1988), to name but a few. More theoretical aspects of the model can also be found in Massy (1965), Basilevsky (1973), Silvey (1969) Oman (1978), Fomby et al.

(1978), Trenkler (1980), Park (1981), and Mandel (1982). An interesting use of the complex decomposition (Section 5.7) can also be found in Doran (1976), in conjunction with the generalized least squares estimator of a time series regression model. Cohen and Jones (1969) and Basilevsky (1973) also consider time series regression in terms of the continuous Karhunen–Loève decomposition (Section 7.5).

There are two broad objectives when employing principal components of the predictor variables. First, we may have a large number of highly correlated explanatory variables, not all of which represent unique or well-defined influences. Indeed, the number of explanatory variables may exceed the sample size, thus precluding the estimation of a regression plane. A principal components analysis of the explanatory variables can then be carried out to explore the correlational structure of the regressors and to locate correlated subsets or clusters, should such exist, and the component scores may then be used in place of the original variables. We call this procedure Model I. If all of the explanatory variables are highly intercorrelated, a single dimension (component) suffices to capture all of the relevant information in the variable subspace (see Section 3.3). The regression of the dependent variable on the dominant principal component then provides a global test of whether the independent variables play a significant role in explaining the behavior of the dependent variable. In general, more than a single principal component is usually required. Note that since principal components replace the original explanatory variables, the use of Model I implies a substantive identification of the components, including rotation of the components should this be required (Section 5.3). The difference between such an approach and ridge regression is now clear. Whereas the latter requires an increase in (artificial) information, the former reduces dimensionality by omitting low-variance dimensions. There are three advantages to using Model I principal components regression: (1) it simplifies the structure of the model and thus brings out the essential information contained in the independent variables; (2) it increases the efficiency of the regression estimators by eliminating or greatly reducing multicollinearity, *which is assumed to exist in the population*; (3) it produces unbiased (consistent) estimators (in the event the lower case principal components represent measurement error). Thus the essential feature of a factor approach to multicollinearity is that it attempts to model the correlational structure of the predictor variables. The cost or disadvantage of such an approach is that the original predictors are not used directly in the regression equation and thus their coefficients cannot be estimated in the model.

The second major approach to using principal components within a regression model is to employ the components as a reduced set of "instrumental" variables and to use them to expand the original least squares estimators into a finite linear series. We refer to this approach as Model II. This implies a third step in the analysis after the dependent variable has been regressed on all of the k principal components—inversion of the original ordinary least squares coefficients in terms of the principal components regression coefficients. Here the original regression variables are of interest rather than the principal components themselves, which are only used as convenient instruments with which to carry out an expansion of the ordinary least squares coefficients. A substantive identification of the principal components is therefore not required, although such information is always of interest. Also, when both error and multicollinearity coexist with the k predictor variables, it is possible to assume that some (or all) of the true predictor values would be perfectly correlated were it not for the presence of random errors. When only multicollinearity is present, the removal of small roots results in biased but more efficient regression estimators, which are expected to lie closer to the true population values.

The principal components regression model is best thought of as a special case of the canonical correlation model (Section 5.5). Let $\mathbf{Z} = \mathbf{XP}$ and $\mathbf{V} = q\mathbf{Y}$ be two sets of principal components where $\mathbf{X}$ is the $(n \times k)$ matrix of k standardized explanatory variables and $\mathbf{Y}$ is a $(n \times 1)$ standardized vector of n dependent observations. Then if $\mathbf{P} = (p_{ij})$ is a $(k \times k)$ matrix of latent vectors associated with $\mathbf{X}$, and q is an arbitrary scalar (set equal to 2), the principal components regression model can be derived as follows: maximize the correlation

$$\mathbf{Z}^T\mathbf{V} = 2\mathbf{Z}^T\mathbf{Y} = 2\mathbf{P}^T(\mathbf{X}^T\mathbf{Y}) \tag{10.41}$$

subject to the multivariate constraint $\mathbf{P}^T(\mathbf{X}^T\mathbf{X})\mathbf{P} = l\mathbf{I}$, where $\mathbf{X}^T\mathbf{X}$ is the correlation matrix of predictor variables and l is a scalar quantity. The Lagrangian expression can be written as

$$\phi = 2\mathbf{P}^T(\mathbf{X}^T\mathbf{Y}) - [\mathbf{P}^T(\mathbf{X}^T\mathbf{X})\mathbf{P} - l\mathbf{I}]\boldsymbol{\gamma}^* \tag{10.42}$$

where $\boldsymbol{\gamma}^*$ is a $(k \times 1)$ vector of Lagrange multipliers. Differentiating Eq. (10.42) with respect to $\mathbf{P}$ and setting the vector of partial derivatives to zero yields the following normal equations:

$$\frac{\partial \phi}{\partial \mathbf{P}} = 2(\mathbf{X}^T\mathbf{Y}) - 2(\mathbf{X}^T\mathbf{X})\mathbf{P}\hat{\boldsymbol{\gamma}}^* = \mathbf{0}$$

or

$$\hat{\boldsymbol{\gamma}}^* = \mathbf{P}^T(\mathbf{X}^T\mathbf{X})^{-1}\mathbf{X}^T\mathbf{Y}$$

$$= \mathbf{P}^T\hat{\boldsymbol{\beta}} \tag{10.43}$$

since $\mathbf{P}$ is an orthogonal matrix and $\hat{\boldsymbol{\beta}}$ is the usual vector of ordinary least squares regression coefficient. Thus $\hat{\boldsymbol{\gamma}}^*$ is the vector of principal components regression coefficients obtained by regressing $\mathbf{Y}$ on the matrix of unstandardized principal components $\mathbf{Z}^* = \mathbf{XP}$ since from Eq. (10.43) we have

$$\hat{\boldsymbol{\gamma}}^* = \mathbf{P}^{-1}(\mathbf{X}^T\mathbf{X})^{-1}\mathbf{PP}^T\mathbf{X}^T\mathbf{Y}$$

$$= (\mathbf{P}^T\mathbf{X}^T\mathbf{XP})^{-1}\mathbf{P}^T\mathbf{X}^T\mathbf{Y}$$

$$= (\mathbf{Z}^{*T}\mathbf{Z}^*)^{-1}\mathbf{Z}^{*T}\mathbf{Y}$$

$$= \mathbf{L}^{-1}\mathbf{Z}^{*T}\mathbf{Y} \tag{10.44}$$

that is,

$$
\begin{bmatrix} \hat{\gamma}_1^* \\ \hat{\gamma}_2^* \\ \vdots \\ \hat{\gamma}_r^* \\ \hdashline \hat{\gamma}_{r+1}^* \\ \vdots \\ \hat{\gamma}_k^* \end{bmatrix}
=
\left[\begin{array}{ccccc:ccc}
\frac{1}{l_1} & & & & & & \mathbf{0} & \\
 & \frac{1}{l_2} & & & & & & \\
 & & \ddots & & & & & \\
\mathbf{0} & & & \frac{1}{l_r} & & & & \\
\hdashline
 & & & & \frac{1}{l_{r+1}} & & \mathbf{0} & \\
 & & & & & \ddots & & \\
 & \mathbf{0} & & & & \mathbf{0} & & \frac{1}{l_k}
\end{array}\right]
\begin{bmatrix} \mathbf{Z}_1^{*T}\mathbf{Y} \\ \mathbf{Z}_2^{*T}\mathbf{Y} \\ \vdots \\ \mathbf{Z}_r^{*T}\mathbf{Y} \\ \hdashline \mathbf{Z}_{r+1}^{*T}\mathbf{Y} \\ \vdots \\ \mathbf{Z}_k^{*T}\mathbf{Y} \end{bmatrix}
$$

or

$$
\begin{bmatrix} \hat{\boldsymbol{\gamma}}_{(r)}^* \\ \hdashline \hat{\boldsymbol{\gamma}}_{(k-r)}^* \end{bmatrix}
=
\begin{bmatrix} \mathbf{L}_{(r)}^{-1} & \mathbf{0} \\ \hdashline \mathbf{0} & \mathbf{L}_{(k-r)}^{-1} \end{bmatrix}
\begin{bmatrix} \mathbf{Z}_{(r)}^{*T}\mathbf{Y} \\ \hdashline \mathbf{Z}_{(k-r)}^{*T}\mathbf{Y} \end{bmatrix}
\tag{10.45}
$$

where

$$\hat{\boldsymbol{\gamma}}_1^* = \frac{1}{l_1}\mathbf{Z}_1^{*T}\mathbf{Y},\ \hat{\gamma}_2^* = \frac{1}{l_2}\mathbf{Z}_2^{*T}\mathbf{Y}, \ldots, \hat{\gamma}_k^* = \frac{1}{l_k}\mathbf{Z}_k^{*T}\mathbf{Y} \qquad (l_i \neq 0; i = 1, 2, \ldots, k)$$

For very small roots the regression coefficients cannot be estimated with precision, and normally only the first r components will be used. It does not necessarily follow however that only the first r principal components are statistically significant. The significance of principal components regression coefficients is not only dependent on the latent roots of $(\mathbf{X}^T\mathbf{X})$ but also is a function of the sum of products (correlation) of $\mathbf{Y}$ and the unstandardized components $\mathbf{Z}_i^*$, since when both the principal components and the variables are standardized we have

$$\hat{\gamma}^* = \mathbf{P}^T(\mathbf{X}^T\mathbf{X})^{-1}\mathbf{X}^T\mathbf{Y}$$
$$= \mathbf{P}^T[\mathbf{PL}^{1/2}\mathbf{Z}^T\mathbf{ZL}^{1/2}\mathbf{P}^T]^{-1}\mathbf{PL}^{1/2}\mathbf{Z}^T\mathbf{Y}$$
$$= \mathbf{P}^T[\mathbf{PLP}^T]^{-1}\mathbf{PL}^{1/2}\mathbf{Z}^T\mathbf{Y}$$
$$= \mathbf{L}^{-1/2}\mathbf{Z}^T\mathbf{Y}$$
$$= \mathbf{L}^{-1/2}\hat{\gamma} \tag{10.46}$$

or $\hat{\gamma} = \mathbf{L}^{1/2}\hat{\gamma}^* = \mathbf{Z}^T\mathbf{Y}$ where $\mathbf{Y} = \mathbf{Z}\gamma + \epsilon$ for standardized $\mathbf{Z}$ (and $\mathbf{Y}$). The regression coefficients therefore do not depend only on the latent roots, and principal components that correspond to large latent roots are therefore not necessarily optimal in the regression when it comes to accounting for the sum of squares of the dependent variable. Thus in general all regression coefficients $\hat{\gamma}_1$, $\hat{\gamma}_2, \ldots, \hat{\gamma}_k$ should be tested for significance if we wish to maximize the predictive power of the $\mathbf{Z}_i$. If the components are orthogonal this can be done simply by using the t-test for each $\hat{\gamma}_i$. When the predictor variables are also subject to error however, the low-order components should not prove significant since they usually correspond to measurement error in the predictors.

The main sampling properties of the principal components regression Model I can be summarized by the following theorem.

THEOREM 10.1. Let $\mathbf{Y} = \mathbf{X}\boldsymbol{\beta} + \epsilon$ where $\mathbf{X}$ is a $(n \times k)$ matrix of random predictor variables, and let $\mathbf{Z}^* = \mathbf{XP}$ be the $(n \times k)$ matrix of unstandardized principal components of the correlation matrix $\mathbf{R} = \mathbf{X}^T\mathbf{X}$. Then the Model I principal component regression estimator has the following properties:

(i) Replacing $\mathbf{X}$ by all of the principal components $\mathbf{Z}$ does not alter the predicted values or the error sum of squares.

(ii) $\hat{\gamma}^*$ is a consistent estimator of $\gamma^* = \mathbf{P}^T\boldsymbol{\beta}$ where $\mathbf{P}$ is a fixed matrix of coefficients.

(iii) The asymptotic covariance matrix of $\hat{\gamma}^*$ is $\sigma^2\mathbf{L}^{-1}$ where $\sigma^2 = \text{var}(\epsilon)$.

(iv) When $\mathbf{Y} \sim N(\mathbf{X}\boldsymbol{\beta}, \sigma^2)$ the distribution of

$$t_i = l_i^{1/2}\frac{\hat{\gamma}_i^*}{\hat{\sigma}}$$

approaches the standard normal distribution.

PROOF

(i) We have $\hat{\mathbf{Y}} = \mathbf{P}_x\mathbf{Y}$ where $\mathbf{P}_x$ is the projection matrix that projects $\mathbf{Y}$ onto the basis of the column space of $\mathbf{X}$. The columns of $\mathbf{Z}$ span the same space as those of $\mathbf{X}$ since

$$\hat{\mathbf{Y}} = \mathbf{P}_x \mathbf{Y} = \mathbf{Z}^* \mathbf{P}^{\mathrm{T}} (\mathbf{P} \mathbf{Z}^{*\mathrm{T}} \mathbf{Z}^* \mathbf{P}^{\mathrm{T}})^{-1} \mathbf{P} \mathbf{Z}^{*\mathrm{T}} \mathbf{Y} = \mathbf{P}_{\mathbf{Z}^*}$$

so that replacing $\mathbf{X}$ by $\mathbf{Z}^*$ does not alter the predicted values $\hat{\mathbf{Y}}$. It follows that the error sum of squares also remains unchanged.

(ii) From Eq. (10.44) the least squares estimator is given by $\hat{\boldsymbol{\gamma}}^* = \mathbf{L}^{-1} \mathbf{Z}^{*\mathrm{T}} \mathbf{Y}$, and taking probability limits we have

$$
\begin{aligned}
p \lim \hat{\boldsymbol{\gamma}}^* &= p \lim \mathbf{L}^{-1} \mathbf{Z}^{*\mathrm{T}} \mathbf{Y} \\
&= p \lim \mathbf{L}^{-1} \mathbf{Z}^{*\mathrm{T}} (\mathbf{X}\boldsymbol{\beta} + \boldsymbol{\epsilon}) \\
&= p \lim[\mathbf{L}^{-1} \mathbf{Z}^{*\mathrm{T}} \mathbf{X}\boldsymbol{\beta}] + p \lim[\mathbf{L}^{-1} \mathbf{Z}^{*\mathrm{T}} \boldsymbol{\epsilon}] \\
&= p \lim[\mathbf{L}^{-1} \mathbf{L} \mathbf{P}^{\mathrm{T}} \boldsymbol{\beta}] + p \lim[\mathbf{L}^{-1} \mathbf{P}^{\mathrm{T}} \mathbf{X}^{\mathrm{T}} \boldsymbol{\epsilon}] \\
&= \mathbf{P}^{\mathrm{T}} \boldsymbol{\beta} + \mathbf{L}^{-1} \mathbf{P}^{\mathrm{T}} p \lim(\mathbf{X}^{\mathrm{T}} \boldsymbol{\epsilon}) \\
&= \mathbf{P}^{\mathrm{T}} \boldsymbol{\beta}
\end{aligned}
$$

since by assumption $p \lim \mathbf{X}^{\mathrm{T}} \boldsymbol{\epsilon} = 0$, $\hat{\boldsymbol{\gamma}}^*$ therefore represent consistent estimators of the linear combinations $\mathbf{P}^{\mathrm{T}} \boldsymbol{\beta}$.

(iii) We have

$$
\begin{aligned}
p \lim[(\hat{\boldsymbol{\gamma}}^* - \boldsymbol{\gamma}^*)(\hat{\boldsymbol{\gamma}}^* - \boldsymbol{\gamma}^*)^{\mathrm{T}}] &= p \lim[(\mathbf{P}^{\mathrm{T}}\hat{\boldsymbol{\beta}} - \mathbf{P}^{\mathrm{T}}\boldsymbol{\beta})(\mathbf{P}^{\mathrm{T}}\hat{\boldsymbol{\beta}} - \mathbf{P}^{\mathrm{T}}\boldsymbol{\beta})^{\mathrm{T}}] \\
&= p \lim[\mathbf{P}^{\mathrm{T}}(\hat{\boldsymbol{\beta}} - \boldsymbol{\beta})(\hat{\boldsymbol{\beta}} - \boldsymbol{\beta})^{\mathrm{T}}\mathbf{P}] \\
&= \mathbf{P}^{\mathrm{T}}[p \lim(\hat{\boldsymbol{\beta}} - \boldsymbol{\beta})(\hat{\boldsymbol{\beta}} - \boldsymbol{\beta})^{\mathrm{T}}]\mathbf{P} \\
&= \mathbf{P}^{\mathrm{T}} \sigma^2 (\mathbf{X}^{\mathrm{T}}\mathbf{X})^{-1} \mathbf{P} \\
&= \sigma^2 \mathbf{L}^{-1}
\end{aligned}
$$

(iv) Since $\hat{\boldsymbol{\beta}} \sim N[\boldsymbol{\beta}, \sigma^2(\mathbf{X}^{\mathrm{T}}\mathbf{X})^{-1}]$ it follows, using parts ii and iii of the theorem, that the distribution of $\hat{\boldsymbol{\gamma}}^*$ approaches the normal $N[\hat{\boldsymbol{\gamma}}^*, \sigma^2 \mathbf{L}^{-1}]$. Since $\hat{\gamma}_1^*, \hat{\gamma}_2^*, \ldots, \hat{\gamma}_k^*$ are uncorrelated random variables the expression

$$t_i = \frac{\hat{\gamma}_i^*}{\sqrt{\hat{\sigma}^2/l_i}} = l_i^{1/2} \frac{\hat{\gamma}_i^*}{\hat{\sigma}} \qquad (i = 1, 2, \ldots, k)$$

tends to be distributed as $N(0, 1)$ for increasing n.

Part iv of Theorem 10.1 indicates that given fixed residual error variance, the magnitudes of the t_i statistic depend on both the latent roots as well as on $\hat{\gamma}_i^*$. When standardized components $\mathbf{Z}$ are used, the regression coefficients (Eq. 10.46) do not depend on l_i, and the t statistic can then be expressed as $t_i = \hat{\gamma}_i / \hat{\sigma}$ $(i = 1, 2, \ldots, k)$. When the dependent variable $\mathbf{Y}$ is also standardized to unit length, the orthogonal regression coefficients $\hat{\gamma}_1, \hat{\gamma}_2, \ldots, \hat{\gamma}_k$ estimate the correlation between $\mathbf{Y}$ and the principal components.

Multicollinearity is not just a problem of correlated predictor variables, since multiple regression is designed precisely to handle nonorthogonality of the regressors. Rather multicollinearity enters the stage when two predictors (or more) are so highly correlated as to render the regression coefficients both highly unstable and statistically insignificant owing to the inefficiency of the least squares estimator. In this case a principal components analysis will generally yield a reduced number of principal components, which account for most of the observed variance in the regressors. Nevertheless it is interesting to note that principal components can also be used in the presence of perfect linear dependence among the predictor variables, as was indicated by Rao (1962). Let $\mathbf{X}^T\mathbf{X}$ have rank $1 \le r < k$. Then the last $k - r$ latent roots are precisely zero, and only r linear combinations of the original least squares coefficients are estimable. The estimator (Eq. 10.44) then becomes

$$\left[\begin{array}{c} \hat{\boldsymbol{\gamma}}^*_{(r)} \\ \hline \mathbf{0} \end{array}\right] = \left[\begin{array}{c:c} \mathbf{L}^{-1}_{(r)} & \mathbf{0} \\ \hdashline \mathbf{0} & \mathbf{0} \end{array}\right] \left[\begin{array}{c} \mathbf{Z}^{*T}_{(r)}\mathbf{Y} \\ \hline \mathbf{0} \end{array}\right] \tag{10.47a}$$

or

$$\hat{\boldsymbol{\gamma}}^*_{(r)} = \mathbf{L}^{-1}_{(r)}\mathbf{Z}^{*T}_{(r)}\mathbf{Y} \tag{10.47b}$$

omitting the zero parts of the partitioned matrix/vectors.

When $\rho(\mathbf{X}^T\mathbf{X}) = k$ all $\hat{\gamma}^*_i$ are estimable, although some may turn out to be statistically insignificant. Further results may be found in Eubank and Webster (1985).

The second major objective of principal components regression, as stated at the outset, is to use the components as a convenient orthogonal basis by which to expand the original regression coefficients into a finite linear series (Model II). The expansion may easily be obtained by noting that Eq. (10.43) can be inverted to express $\hat{\beta}_i$ in terms of $\hat{\gamma}^*_i$. We have

$$\hat{\boldsymbol{\beta}} = \mathbf{P}\hat{\boldsymbol{\gamma}}^*$$

$$= \mathbf{P}\mathbf{L}^{-1/2}\hat{\boldsymbol{\gamma}} \tag{10.48}$$

so that each coefficient can be expanded as the linear combination

$$\hat{\beta}_i = \mathbf{P}_i\mathbf{L}^{-1/2}\hat{\boldsymbol{\gamma}}$$

$$= l_1^{-1/2}p_{i1}\hat{\gamma}_1 + l_2^{-1/2}p_{i2}\hat{\gamma}_2 + \cdots + l_k^{-1/2}p_{ik}\hat{\gamma}_k \tag{10.49}$$

where $\mathbf{P}_i$ is the ith row vector of $\mathbf{P}$. The rth order principal components regression estimator is obtained by replacing the last $k - r$ terms (which correspond to "small" latent roots $l_{r+1}, l_{r+2}, \ldots, l_k$) by zeros, that is, we have for $1 \le r < k$

$$
\begin{bmatrix} \tilde{\beta}_1 \\ \tilde{\beta}_2 \\ \vdots \\ \tilde{\beta}_k \end{bmatrix} =
\begin{bmatrix}
p_{11} & p_{12} & \cdots & p_{1r} & 0 & \cdots & 0 \\
p_{21} & p_{22} & \cdots & p_{2r} & 0 & \cdots & 0 \\
\vdots & \vdots & & \vdots & \vdots & & \vdots \\
p_{k1} & p_{k2} & \cdots & p_{kr} & 0 & \cdots & 0
\end{bmatrix}
$$

$$
\times
\begin{bmatrix}
l_1^{-1/2} & & & & & \\
 & l_2^{-1/2} & & & \mathbf{0} & \\
 & & \ddots & & & \\
 & & & l_r^{-1/2} & & \\
\hline
 & & & & 0 & \\
 & \mathbf{0} & & & & \ddots \\
 & & & & & 0
\end{bmatrix}
\begin{bmatrix} \hat{\gamma}_1 \\ \hat{\gamma}_2 \\ \vdots \\ \hat{\gamma}_r \\ 0 \\ \vdots \\ 0 \end{bmatrix}
$$

or

$$
\tilde{\beta} = [\mathbf{P}_{(r)} \vdots \mathbf{0}]
\begin{bmatrix} \mathbf{L}_{(r)}^{-1/2} & \mathbf{0} \\ \mathbf{0} & \mathbf{0} \end{bmatrix}
\begin{bmatrix} \hat{\gamma}_{(r)} \\ \mathbf{0} \end{bmatrix}
\tag{10.49a}
$$

Alternatively, since the principal components regression estimator is computed in a fixed r-dimensional subspace, Eq. (10.49a) can also be written as

$$
\tilde{\beta} = \mathbf{P}_{(r)} \mathbf{L}_{(r)}^{-1/2} \hat{\gamma}_{(r)}
\tag{10.49b}
$$

where the zeros are omitted. The ith principal components regression estimator can then be expressed as the truncated expression

$$
\tilde{\beta}_i = l_1^{-1/2} p_{i1} \hat{\gamma}_1 + l_2^{-1/2} p_{i2} \hat{\gamma}_2 + \cdots + l_r^{-1/2} p_{ir} \hat{\gamma}_r
\tag{10.50}
$$

Although $\tilde{\beta}_i$ depends on the latent vector elements p_{ij}, the greatest influence on $\tilde{\beta}_i$ comes from the last (omitted) $k - r$ latent roots, whose magnitudes approach zero as the degree of multicollinearity increases. In the limit when some predictors are perfectly correlated, at least one of the roots becomes identically zero and $\hat{\beta}$ does not exist (but $\tilde{\beta}$ does). The main sampling properties of $\tilde{\beta}$ can be summarized as follows.

THEOREM 10.2. Let $\mathbf{Y} = \mathbf{X}\beta + \epsilon$ where $\mathbf{P}^{T}\mathbf{X}^{T}\mathbf{X}\mathbf{P} = \mathbf{L}$ and $\tilde{\beta} = \mathbf{P}_{(r)} \mathbf{L}_{(r)}^{-1/2} \hat{\gamma}_{(r)}$ is the r-component regression estimator of β. Then

    (i) $\tilde{\beta}$ is a consistent estimator of β only when for the last $k - r$ terms of Eq. (10.49) we have

$$p \lim \hat{\gamma}_{r+1} = p \lim \hat{\gamma}_{r+2} = \cdots = p \lim \hat{\gamma}_k = 0$$

(ii) The variance–covariance matrix of $\hat{\boldsymbol{\beta}}$ can be expressed as

$$E[(\hat{\boldsymbol{\beta}} - \boldsymbol{\beta})(\hat{\boldsymbol{\beta}} - \boldsymbol{\beta})^{\mathrm{T}}] = \sigma^2 \mathbf{P} \mathbf{L}^{-1} \mathbf{P}^{\mathrm{T}}$$

where

$$E[(\tilde{\boldsymbol{\beta}} - \boldsymbol{\beta})(\tilde{\boldsymbol{\beta}} - \boldsymbol{\beta})^{\mathrm{T}}] = \sigma^2 \mathbf{P}_{(r)} \mathbf{L}_{(r)}^{-1} \mathbf{P}_{(r)}^{\mathrm{T}} \quad \text{when } p \lim \tilde{\boldsymbol{\beta}} = \boldsymbol{\beta}$$

PROOF

(i) From Eq. (10.49) we have

$$p \lim \hat{\beta}_i = l_1^{-1/2} p_{i1} p \lim \hat{\gamma}_1 + l_2^{-1/2} p_{i2} p \lim \hat{\gamma}_2 + \cdots$$

$$+ l_r^{-1/2} p_{ir} p \lim \hat{\gamma}_r + l_{r+1}^{-1/2} p_{i,r+1} p \lim \hat{\gamma}_{r+1} + \cdots$$

$$+ l_k^{-1/2} p_{i,k} p \lim \hat{\gamma}_k$$

$$= l_1^{-1/2} p_{i1} p \lim \hat{\gamma}_1 + l_2^{-1/2} p_{i2} p \lim \hat{\gamma}_2 + \cdots$$

$$+ l_r^{-1/2} p_{ir} p \lim \hat{\gamma}_r$$

for

$$p \lim \hat{\gamma}_i = 0 \qquad (i = r+1, r+2, \ldots, k)$$

It follows that $p \lim \hat{\beta}_i = p \lim \tilde{\beta}_i = \beta_i$ $(i = 1, 2, \ldots, k)$ that is, $\tilde{\beta}$ is a consistent estimator of β only when the Model I regression coefficients for the last $k - r$ principal components are insignificantly different from zero i.e. their non-zero values are due to error. Since this does not hold in general, we conclude that $\tilde{\boldsymbol{\beta}}$ is an inconsistent estimator of $\boldsymbol{\beta}$.

(ii) We have, for large samples,

$$E[(\hat{\boldsymbol{\beta}} - \boldsymbol{\beta})(\hat{\boldsymbol{\beta}} - \boldsymbol{\beta})^{\mathrm{T}}] = \sigma^2 (\mathbf{X}^{\mathrm{T}} \mathbf{X})^{-1}$$

$$= \sigma^2 \mathbf{P} \mathbf{L}^{-1} \mathbf{P}^{\mathrm{T}}$$

so that the variance-covariance matrix of $\tilde{\boldsymbol{\beta}}$ is given by

$$E[(\tilde{\boldsymbol{\beta}} - \boldsymbol{\beta})(\tilde{\boldsymbol{\beta}} - \boldsymbol{\beta})^{\mathrm{T}}] = \sigma^2 \mathbf{P}_{(r)} \mathbf{L}_{(r)}^{-1} \mathbf{P}_{(r)}^{\mathrm{T}} \tag{10.51}$$

The (i, j)th element of the matrix $E[(\hat{\boldsymbol{\beta}} - \boldsymbol{\beta})(\hat{\boldsymbol{\beta}} - \boldsymbol{\beta})^{\mathrm{T}}]$ can thus be expanded as

$$E[(\hat{\beta}_i - \beta_i)(\hat{\beta}_j - \beta_j)^{\mathrm{T}}] = \sigma^2 [p_{j1}, p_{j2}, \ldots, p_{jk}] \begin{bmatrix} p_{i1} \, l_1^{-1} \\ p_{i2} \, l_2^{-1} \\ \vdots \\ p_{ik} \, l_k^{-1} \end{bmatrix}$$

$$= \sigma^2 (p_{i1} p_{j1} l_1^{-1} + p_{i2} p_{j2} l_2^{-1} + \cdots + p_{ik} p_{jk} l_k^{-1})$$

and for $i = j$ we have

$$\mathrm{var}(\hat{\beta}_i) = E[(\hat{\beta}_i - \beta_i)^2]$$

$$= \sigma^2 \left(\frac{p_{i1}^2}{l_1} + \frac{p_{i2}^2}{l_2} + \cdots + \frac{p_{ik}^2}{l_k} \right) \tag{10.52}$$

Omitting the last $k - r$ terms for some fixed r we obtain a more efficient estimator since $\mathrm{var}(\tilde{\beta}_i) < \mathrm{var}(\hat{\beta}_i)$, and a modified t statistic can be computed as

$$t_m = \frac{\tilde{\beta}_i}{\mathrm{SD}(\tilde{\beta}_i)} \tag{10.53}$$

Since σ^2 is generally not known, it is replaced by the sample estimator s^2 where degrees of freedom are increased to $n - r$ from $n - k$. Also the notion of variance is undefined when $\tilde{\beta}$ is a biased (inconsistent) estimator of β.

Equation (10.51) expresses the "variance/covariance" matrix of $\tilde{\beta}$ in terms of the first $1 \le r < k$ latent roots and latent vectors of the matrix $(\mathbf{X}^{\mathrm{T}}\mathbf{X})$, for some constant σ^2. Since $\tilde{\beta}$ is an inconsistent (biased) estimator of β, however, the term "variance/covariance" is incorrent, strictly speaking and the expression $E[(\tilde{\beta} - \beta)(\tilde{\beta} - \beta)^{\mathrm{T}}]$ is referred to as the mean squared error (MSE) criterion. This is because the notion of variance and covariance is undefined for a biased estimator. Since the latent roots of $(\mathbf{X}^{\mathrm{T}}\mathbf{X})$ are ranked in decreasing order the estimator $\tilde{\beta}$ must clearly minimize the MSE criterion when a fixed number of $k - r$ principal components are deleted from the expansion (10.52). Another way of seeing this is to note that the $\tilde{\beta}_i$ represent linear combinations of the OLS estimators $\hat{\beta}_i$ (equation 10.49b) and then to ask for what set of constants $\mathbf{C}^{\mathrm{T}} = (c_1, c_2, \ldots, c_k)$ is the variance of the linear combination $\mathbf{C}^{\mathrm{T}}\hat{\beta}$ minimized (see also Theorem 10.1). This can be easily achieved by the method of Lagrange multipliers, as was the case for principal components (Section 3.2) and nothing new arises in principle (Exercise 10.12).

Theorem 10.2 establishes that when the last $k - r$ principal components do not explain significant variance of $\mathbf{Y}$, that is, when $p \lim \hat{\gamma}_i = 0$ ($i = r + i$, $r + 2, \ldots, k$), the estimator $\tilde{\beta}$ is optimal in the sense of being consistent and possessing minimal variance. This is the case when the last $(k - r)$ principal components of $\mathbf{X}^T\mathbf{X}$ represent random error variation which is irrelevant for explaining the variance of $\mathbf{Y}$. The proper procedure for deleting principal components is then based on testing for isotropic latent roots (Section 4.3.2) since these can be expected to correspond to measurement error in the predictors, on the assumption of multivariate normality. On the other hand, when the explanatory variables do not possess random error, $\tilde{\boldsymbol{\beta}}$ is no longer a consistent estimator of $\boldsymbol{\beta}$ although it minimizes the MSE criterion (Eq. 10.51). This is because omitting a component which is associated with a small latent root (but which at the same time is strongly related to $\mathbf{Y}$) can lead to a significant decrease in variance. Note that a more general principal components regression estimator can also be used, one based on deleting only those principal components that are insignificantly different from zero, as determined by the t statistic (Theorem 10.1). Since this does not necessarily imply deleting components with least variance, the preservation of consistency necessarily results in the risk of higher imprecision of the estimator. Nevertheless, when a large number of multicollinear predictors exist, the use of a smaller set of components results in an increase in the degrees of freedom, thus mitigating to a certain extent the effect of small latent roots. For greater detail concerning the selection of components in a principal components regression see Mansfield et al. (1977), Fomby and Hill (1978), Saxena (1980)), Park (1981), and Rao and Oberhelman (1982). Jackson (1991) provides an overview of principal components regression. Finally, note that no matter which terms are deleted, a reduced number of components necessarily leads to linear dependence between the original k regression coefficients.

The Model II principal components regression estimator possesses two further important optimality criteria, as demonstrated by the following theorem.

THEOREM 10.3. Let $\tilde{\boldsymbol{\beta}}$ be the vector of principal components regression coefficients as defined in Theorem 10.2. Then

    (i) $\tilde{\boldsymbol{\beta}}$ minimizes the determinant of the MSE criterion.

    (ii) $\tilde{\boldsymbol{\beta}}$ minimizes the trace of the MSE criterion.

PROOF

    (i) From well-known properties of the determinant (e.g., see Basilevsky, 1983) we have

$$|(\tilde{\boldsymbol{\beta}} - \boldsymbol{\beta})(\tilde{\boldsymbol{\beta}} - \boldsymbol{\beta}^T| = |\sigma^2 \mathbf{PL}^{-1}\mathbf{P}^T|$$

$$= \sigma^2 |\mathbf{P}||\mathbf{L}^{-1}||\mathbf{P}^T|$$

$$= \sigma^2 |\mathbf{L}^{-1}|$$

which is minimized for $l_1 > l_2 > \cdots > l_r$, assuming constant σ^2.

(ii) We have

$$\operatorname{tr}[(\tilde{\boldsymbol{\beta}} - \boldsymbol{\beta})(\tilde{\boldsymbol{\beta}} - \boldsymbol{\beta})^{\mathrm{T}}] = \operatorname{tr} \sigma^2 \mathbf{P} \mathbf{L}^{-1} \mathbf{P}^{\mathrm{T}}$$

$$= \sigma^2 \operatorname{tr}[\mathbf{P}^{\mathrm{T}} \mathbf{L}^{-1} \mathbf{P}^{\mathrm{T}}]$$

$$= \sigma^2 \operatorname{tr}(\mathbf{L}^{-1})$$

which again is minimized for $l_1 > l_2 > \cdots > l_r$.

Note again that for biased $\tilde{\boldsymbol{\beta}}$ the minimization is of the mean squared error criterion rather than variance.

The estimator $\tilde{\boldsymbol{\beta}}$ has two interesting interpretations which may be used to relate it more closely to standard least squares regression theory.

THEOREM 10.4.   Let $\tilde{\boldsymbol{\beta}}$ be the vector of principal components regression coefficients as defined by Theorem 10.2. Then
  (i) $\tilde{\boldsymbol{\beta}}$ is a restricted least squares regression estimator, obtained by setting $\mathbf{P}_{(k-r)}^{\mathrm{T}} \boldsymbol{\beta} = \mathbf{0}$.
  (ii) $\tilde{\boldsymbol{\beta}}$ is an adjusted instrumental variables estimator of $\boldsymbol{\beta}$.

PROOF
  (i) Let $\hat{\boldsymbol{\beta}}^*$ denote the restricted least squares estimator given by (e.g., see Goldberger, 1964),

$$\hat{\boldsymbol{\beta}}^* = \hat{\boldsymbol{\beta}} - (\mathbf{X}^{\mathrm{T}}\mathbf{X})^{-1}\mathbf{R}^{\mathrm{T}}[\mathbf{R}(\mathbf{X}^{\mathrm{T}}\mathbf{X})^{-1}\mathbf{R}^{\mathrm{T}}](\mathbf{R}\hat{\boldsymbol{\beta}} - \mathbf{r}) \qquad (10.54)$$

where $\mathbf{r} = \mathbf{R}\hat{\boldsymbol{\beta}}$ are the linear restrictions placed upon the least squares estimators $\hat{\boldsymbol{\beta}}$ for some fixed matrix $\mathbf{R}$. We also have $\hat{\boldsymbol{\beta}} = \mathbf{P}_{(r)}\hat{\boldsymbol{\gamma}}_{(r)}^* + \mathbf{P}_{(k-r)}\hat{\boldsymbol{\gamma}}_{(k-r)}^*$, where for the sake of convenience we use unstandardized latent vectors. Setting $\mathbf{r} = \mathbf{0}$ and $\mathbf{R} = \mathbf{P}_{(k-r)}^{\mathrm{T}}$ in Eq. (10.54) we have

$$\hat{\boldsymbol{\beta}}^* = \hat{\boldsymbol{\beta}} - (\mathbf{X}^{\mathrm{T}}\mathbf{X})^{-1}\mathbf{P}_{(k-r)}[\mathbf{P}_{(k-r)}^{\mathrm{T}}(\mathbf{X}^{\mathrm{T}}\mathbf{X})^{-1}\mathbf{P}_{(k-r)}]^{-1}\mathbf{P}_{(k-r)}^{\mathrm{T}}\hat{\boldsymbol{\beta}}$$

$$= \hat{\boldsymbol{\beta}} - \mathbf{P}\mathbf{L}^{-1}\mathbf{P}^{\mathrm{T}}\mathbf{P}_{(k-r)}\mathbf{L}_{(k-r)}\mathbf{P}_{(k-r)}^{\mathrm{T}}\mathbf{P}\hat{\boldsymbol{\gamma}}^*$$

$$= \hat{\boldsymbol{\beta}} - \mathbf{P}\mathbf{L}^{-1}\begin{bmatrix} \mathbf{0} \\ \cdots \\ \mathbf{I}_{(k-r)} \end{bmatrix}\mathbf{L}_{(k-r)}[\mathbf{0} \; \vdots \; \mathbf{I}_{(k-r)}]\hat{\boldsymbol{\gamma}}^*$$

$$= \hat{\boldsymbol{\beta}} - \mathbf{P}\begin{bmatrix} \mathbf{0} \\ \cdots \\ \hat{\boldsymbol{\gamma}}_{(k-r)}^* \end{bmatrix}$$

$$= \hat{\boldsymbol{\beta}} - \mathbf{P}_{(k-r)}\hat{\boldsymbol{\gamma}}_{(k-r)}^*$$

where

$$\mathbf{P}_{(k-r)}\hat{\boldsymbol{\gamma}}^*_{(k-r)} = \mathbf{P}_{(r+1)}\hat{\boldsymbol{\gamma}}^*_{(r+1)} + \mathbf{P}_{(r+2)}\hat{\boldsymbol{\gamma}}^*_{(r+2)} + \cdots + \mathbf{P}_{(k)}\hat{\boldsymbol{\gamma}}^*_{(k)}$$

Thus using Eq. (10.54) we have

$$\hat{\boldsymbol{\beta}}^* = \mathbf{P}_{(r)}\hat{\boldsymbol{\gamma}}^*_{(r)}$$

$$= \tilde{\boldsymbol{\beta}}$$

so that the principal components regression estimator $\tilde{\boldsymbol{\beta}}$ is a restricted least squares estimator.

(ii) For $r = k$ the instrumental variables estimator of $\boldsymbol{\beta}$ is given by

$$\tilde{\boldsymbol{\beta}}_I = (\mathbf{Z}^{*\mathrm{T}}\mathbf{X})^{-1}\mathbf{Z}^{*\mathrm{T}}\mathbf{Y}$$

$$= [(\mathbf{XP})^{\mathrm{T}}\mathbf{X}]^{-1}(\mathbf{XP})^{\mathrm{T}}\mathbf{Y}$$

$$= (\mathbf{X}^{\mathrm{T}}\mathbf{X})^{-1}\mathbf{X}^{\mathrm{T}}\mathbf{Y}$$

$$= \hat{\boldsymbol{\beta}}$$

since the number of components is the same as the number of predictor variables. We can therefore write

$$\hat{\boldsymbol{\beta}}_I = \mathbf{P}_{(r)}\hat{\boldsymbol{\gamma}}^*_{(r)} + \mathbf{P}_{(k-r)}\hat{\boldsymbol{\gamma}}^*_{(k-r)}$$

$$= \tilde{\boldsymbol{\beta}} + \mathbf{P}_{(k-r)}\hat{\boldsymbol{\gamma}}^*_{(k-r)}$$

or

$$\tilde{\boldsymbol{\beta}} = \hat{\boldsymbol{\beta}}_I - \mathbf{P}_{(k-r)}\hat{\boldsymbol{\gamma}}^*_{(k-r)}$$

so that $\tilde{\boldsymbol{\beta}}$ can be taken as an "adjusted" instrumental variables estimator.

Since instrumental variables estimators are normally used when the predictors are subject to measurement error the use of principal components in this context is not optimal and a more natural development for $r < k$ instrumental variables lies in the context of factor analysis regression, considered in Section (10.4.1). Note also that the restricted least squares interpretation of $\tilde{\boldsymbol{\beta}}$ in the first part of the theorem allows us to interpret (10.54) in terms of an optimal restricted least squares estimator, which minimizes the mean squared error criterion of $\tilde{\boldsymbol{\beta}}$ (see also Fomby et al., 1978). In this view principal components regression can be considered as an exploratory search for optimal zero restrictions when $\mathbf{R}$ is not provided ahead of time.

Example 10.4   As an example of Model I PC regression consider the car data of Example 6.2 where $\mathbf{X}_3(\mathbf{Y}_3)$ is omitted and $\mathbf{Y}$ represents mileage per gallon. A PC analysis of the correlation matrix of the predictors yields the following loadings and latent roots

	$\mathbf{Z}_1$	$\mathbf{Z}_2$	$\mathbf{Z}_3$	$\mathbf{Z}_4$
$\mathbf{X}_1$	.9471	.2383	$-.0744$	.2015
$\mathbf{X}_2$	.9367	$-.1902$	.2922	$-.0329$
$\mathbf{X}_3$	.8332	.5173	$-.1097$	$-.1618$
$\mathbf{X}_4$	$-.6416$	.7459	.1744	.0394
l_i	2.88026	.91698	.13336	.06941

where $\mathbf{Z}_1$ represents heavy, powerful cars and $\mathbf{Z}_2$ denotes high acceleration automobiles with smaller engines and lower horsepower. The regression of mileage ($\mathbf{Y}$) on all four components (Table 10.6) indicates that the main determinant of gasoline consumption is component $\mathbf{Z}_1$, that is, engine size ($\mathbf{X}_1$), horsepower ($\mathbf{X}_2$), and weight of car ($\mathbf{X}_3$) which is capable of high acceleration ($\mathbf{X}_4$). Since the principal components are orthogonal, the t-test possesses maximal efficiency, and $\hat{\gamma}_i$ are unbiased from the usual least squares regression theory.

To see the effect of the individual predictors on $\mathbf{Y}$ we turn to Model II which yields efficient but biased estimators. The correlation matrix $\mathbf{R}$ is given by

Table 10.6   Regression Coefficients of Mileage on the Standardized Principal Components Where $\hat{\sigma} = .07965$

	$\hat{\gamma}$	$|t|$
$\mathbf{Z}_1$	$-.86612$	10.87
$\mathbf{Z}_2$	$-.21560$	2.71
$\mathbf{Z}_3$	.09847	1.24
$\mathbf{Z}_4$	.14944	1.87

$$
\mathbf{R} = \begin{array}{c}
\mathbf{Y} \\
\mathbf{X}_1 \\
\mathbf{X}_2 \\
\mathbf{X}_4 \\
\mathbf{X}_5
\end{array}
\begin{array}{ccccc}
\mathbf{Y} & \mathbf{X}_1 & \mathbf{X}_2 & \mathbf{X}_3 & \mathbf{X}_4 \\
\left[\begin{array}{ccccc}
1.0000 & & & & \\
-.8489 & 1.0000 & & & \\
-.7464 & .8134 & 1.000 & & \\
-.8682 & .8880 & .6553 & 1.000 & \\
.4180 & -.4350 & -.6932 & -.1742 & 1.000
\end{array}\right]
\end{array}
$$

and the latent vectors of the predictors by

$$
\mathbf{P} = \begin{array}{c} \\ \mathbf{X}_1 \\ \mathbf{X}_2 \\ \mathbf{X}_3 \\ \mathbf{X}_4 \end{array}
\begin{array}{cccc}
\mathbf{Z}_1 & \mathbf{Z}_2 & \mathbf{Z}_3 & \mathbf{Z}_4 \\
\left[\begin{array}{cccc}
.55808 & .24887 & -.20384 & .76489 \\
.55191 & -.19863 & .80023 & -.12480 \\
.49093 & .54025 & -.30025 & -.61398 \\
-.37806 & .77894 & .47742 & .14963
\end{array}\right]
\end{array}
$$

When all variables are standardized to zero mean and unit sum of squares, the least squares regression of mileage ($\mathbf{Y}$), engine size ($\mathbf{X}_1$), horsepower ($\mathbf{X}_2$), weight ($\mathbf{X}_3$), and inverse of acceleration ($\mathbf{X}_4$) is

	$\hat{\boldsymbol{\beta}}$	SD	$|t|$	
$\mathbf{X}_1$	.038070	.237843	.160	$F = 32.66$
$\mathbf{X}_2$	$-.091955$	.181194	.507	$R^2 = .8287$
$\mathbf{X}_3$	$-.801427$	.203215	3.944	
$\mathbf{X}_4$	.231188	.131918	1.753	

Here we see the effects of multicollinearity since in spite of a high F value only one variable appears significant as judged by the t statistic ($\alpha = .05$). Also the coefficient for $\mathbf{X}_1$ is of the wrong sign and there exists a strong likelihood that the magnitudes of both the least squares coefficients and their standard deviations are distorted. Using correlations (least squares coefficients) between $\mathbf{Y}$ and $\mathbf{Z}_1$, $\mathbf{Z}_2$, $\mathbf{Z}_3$, and $\mathbf{Z}_4$ the $\hat{\beta}_i$ can be reproduced using Eq. (10.44) as $\hat{\boldsymbol{\beta}} = \mathbf{P}\mathbf{L}^{-1/2}\,\tilde{\boldsymbol{\gamma}}$ or

$$
\begin{bmatrix} \hat{\beta}_1 \\ \hat{\beta}_2 \\ \hat{\beta}_3 \\ \hat{\beta}_4 \end{bmatrix}
=
\begin{bmatrix}
.550808 & .24887 & -.20384 & .76489 \\
.55191 & -.19863 & .80023 & -.12480 \\
.49093 & .54025 & -.30025 & -.61398 \\
-.37806 & .77894 & .47742 & .14963
\end{bmatrix}
$$

$$
\times
\begin{bmatrix}
1.69713 & & & \mathbf{0} \\
 & .95758 & & \\
\mathbf{0} & & .36518 & \\
 & & & .26345
\end{bmatrix}^{-1}
\begin{bmatrix}
-.86612 \\
-.21560 \\
.09847 \\
.14944
\end{bmatrix}
$$

The individual least squares coefficients can thus be expanded as

$$
\hat{\beta}_1 = \frac{-(.55808)(.86612)}{1.69713} - \frac{(.24887)(.21560)}{.95758} - \frac{(.20384)(.09847)}{.36518}
$$

$$
+ \frac{(.76489)(.14944)}{.26345}
$$

$$
= -.28481 - .056033 - .054965 + .433870
$$

$$
= .0380
$$

$$\hat{\beta}_2 = -.28166 + .04478 + .21578 - .07079$$

$$= -.0919$$

$$\hat{\beta}_3 = -.25054 - .12163 - .08096 - .34827$$

$$= -.8014$$

$$\hat{\beta}_4 = .19294 - .17536 + .12873 + .084876$$

$$= .2312$$

and their standard deviations as

$$\mathrm{var}(\hat{\beta}_1) = \hat{\sigma}^2\left(\frac{p_{11}^2}{l_1} + \frac{p_{12}^2}{l_2} + \frac{p_{13}^2}{l_3} + \frac{p_{14}^2}{l_4}\right)$$

$$= .07965^2\left(\frac{.31145}{2.88026} + \frac{.06194}{.91698} + \frac{.04155}{.13336} + \frac{.58505}{.06941}\right)$$

$$= .07965^2(.10813 + .06755 + .31156 + 8.42890)$$

$$= .07965^2(8.91614)$$

$$= .05665$$

and so forth so that

$$\mathrm{SD}(\hat{\beta}_1) = .2380$$

$$\mathrm{SD}(\hat{\beta}_2) = .07965(.10575 + .043025 + 4.8018 + .22439)^{1/2}$$

$$= .18119$$

$$\mathrm{SD}(\hat{\beta}_3) = .07965(.08368 + .31829 + .67599 + 5.43108)^{1/2}$$

$$= .2032$$

$$\mathrm{SD}(\hat{\beta}_4) = .07965(.04962 + .66168 + 1.70913 + .32256)^{1/2}$$

$$= .1319$$

using Eq. (10.52). The effect and source of the multicollinearity is now clear since examining the terms of the expansion indicates that $\mathbf{Z}_3$ and $\mathbf{Z}_4$ contribute disproportionately to the overall magnitudes of the coefficients and their standard deviations, even though they account for a small percentage of the overall variance of the predictors. Using Eqs. (10.50), (10.51), and (10.53) the biased but efficient PCR coefficients, standard deviations, and modified t values can readily be computed by omitting the last two terms ($r = 2$), and these appear in Table 10.7. Note that the

Table 10.7 Principal Components Regression Coefficients, Standard Deviations, and Values of the Modified t Statistic

	$\tilde{\beta}$	$|t|$	$|t_m|$
X_1	$-.3408$	.0334	10.21
X_2	$-.2369$	.0307	7.71
X_3	$-.3722$	.0505	7.37
X_4	.0176	.0067	2.62

coefficient for X_1 is now negative, as is expected on theoretical grounds, and all four predictors are significant.

10.3.2 Comparing Orthogonal-Norm and Y-Norm Least Squares Regression

In Section 10.2 the orthogonal-norm least squares regression model is considered in its own right without any reference to Gauss–Markov regression. Several relationships however can be established between the two models which are of interest when dealing with multicollinearity. Consider Figure 10.3 where the straight line through the origin is given by the equation $a_1 y_1 + a_2 x = 0$ or, alternatively $y_1 - b_2 x = 0$ where $b_2 = (-a_2/a_1)$. Any point off the line can then be represented as $y_1 - b_2 x = e$ for some value e. From Euclidean geometry we know that the perpendicular distance

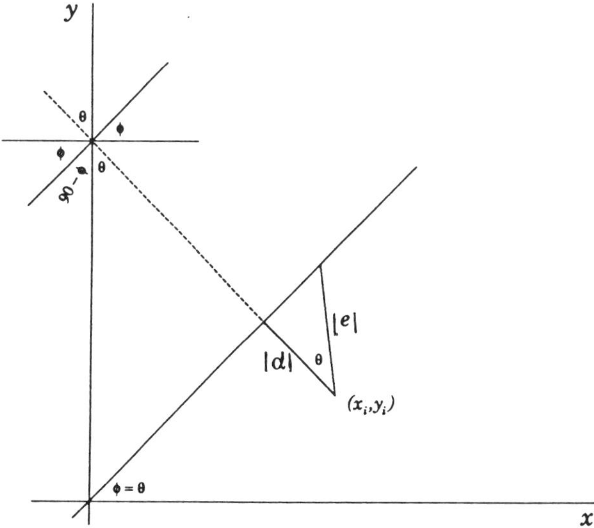

Figure 10.3 Relationship between y-norm and orthogonal-norm least squares regression in terms of the residual errors of the two models.

from some arbitrary point to the straight line is given by

$$|d| = \frac{|y_1 - b_2 x|}{|(1 + b_2^2)^{1/2}|}$$

$$= \frac{|e|}{|(1 + b_2^2)^{1/2}|}$$

$$= \frac{|e|}{|(1 + a_2^2/a_1^2)^{1/2}|}$$

$$= \frac{|a_1||e|}{|(a_1^2 + a_2^2)^{1/2}|}$$

and using the normalization $(a_1^2 + a_2^2) = 1$ we obtain the simplified expression

$$|d\| = |a_1||e| \tag{10.55}$$

where a_1 and a_2 are direction cosines, that is, $a_1 = \cos\theta$, $a_2 = \sin\theta$, and θ is the angle between the straight line and the horizontal axis, which in turn must be the same as the angle between the y-norm and the perpendicular distance to the straight line (Exercise 10.7). We thus have $|d| = \cos\theta|e| = a_1|e|$ and when $a_1 \to 1$ we have $\theta = 0$ in which case $|d| \to |e|$, that is, both the orthogonal-norm and y-norm regressions approach each other in the degenerate case. Since the degenerate case also coincides with multicollinearity between x and the constant term, this provides a method for diagnosing the problem (see Hawkins, 1973). The relationship between y-norm and orthogonal-norm regression of Figure 10.4 provides the basis of

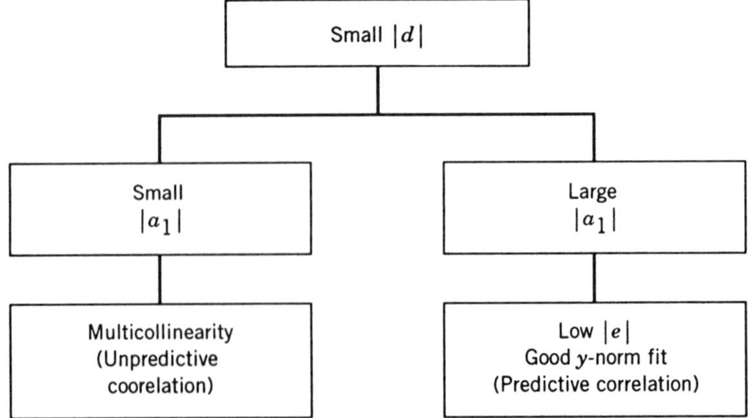

Figure 10.4   The identification of y-norm multicollinearity using orthogonal-norm regression.

yet another model for principal components regression, described in the following section.

10.3.3 Latent Root Regression

Equation (10.55) implies that orthogonal-norm least squares can be used as a diagnostic tool for the y-norm regression plane, and this completes the two-way table of possible outcomes as shown in Table 10.8. The method is due to Hawkins (1973), Webster and Mason (1974) and Gunst et al. (1976). The data matrix for latent root regression is as in Section 10.2 where the first column represents the dependent variable and the next $k-1$ columns represent predictor variables. Let the data matrix be denoted by $\mathbf{V}$. Then for n observations we have

$$
\mathbf{V} = \begin{bmatrix}
y_1 & x_{11} & x_{12} & \cdots & x_{1k} \\
y_2 & x_{21} & x_{22} & \cdots & x_{2k} \\
\vdots & \vdots & \vdots & & \vdots \\
y_n & x_{n1} & x_{n2} & \cdots & x_{nk}
\end{bmatrix}
$$

with the associated matrix of column latent vectors

$$
\begin{array}{ccccc}
\mathbf{P}_0 & \mathbf{P}_1 & \mathbf{P}_2 & \cdots & \mathbf{P}_k
\end{array}
$$

$$
\mathbf{P} = \begin{bmatrix}
p_{00} & p_{01} & p_{02} & \cdots & p_{0k} \\
p_{10} & p_{11} & p_{12} & \cdots & p_{1k} \\
p_{20} & p_{21} & p_{22} & \cdots & p_{2k} \\
\vdots & \vdots & \vdots & & \vdots \\
p_{k0} & p_{k1} & p_{k2} & \cdots & p_{kk}
\end{bmatrix}
$$

partitioned along the first row. We assume that all variables are standardized to unit length so that the latent vectors are those of the correlation matrix. Note that the dependent variable receives the subscript zero rather than unity, in the notation of Gunst et al. (1976). The $k+1$ unstandardized

Table 10.8 Four Possible Least Squares Models Depending on Direction of Fit and Whether the Dependent Variable is Included or Excluded from the Analysis

Direction of Residuals	Dependent Variable	
	Included	Excluded
Orthogonal norm	Pearson's orthogonal regression	Ordinary principal components
y-norm	Latent root regression	Principal components regression

principal components are then given by

$$
\mathbf{Z}_0 =
\begin{bmatrix}
p_{00}y_1 + \sum_{h=1}^{k} p_{h0}x_{1h} \\
p_{00}y_2 + \sum_{h=1}^{k} p_{h0}x_{2h} \\
\hline
p_{00}y_n + \sum_{h=1}^{k} p_{h0}x_{nh}
\end{bmatrix}
\quad
\mathbf{Z}_1 =
\begin{bmatrix}
p_{01}y_1 + \sum_{h=1}^{k} p_{h1}x_{1h} \\
p_{01}y_2 + \sum_{h=1}^{k} p_{h1}x_{2h} \\
\hline
p_{01}y_n + \sum_{h=1}^{k} p_{h1}x_{nh}
\end{bmatrix}
\quad \cdots
$$

$$
\mathbf{Z}_k =
\begin{bmatrix}
p_{0k}y_1 + \sum_{h=1}^{k} p_{hk}x_{1h} \\
p_{0k}y_2 + \sum_{h=1}^{k} p_{hk}x_{2h} \\
\hline
p_{0k}y_n + \sum_{h=1}^{k} p_{hk}x_{nh}
\end{bmatrix}
$$

or $\mathbf{Z} = \mathbf{VP}$ where $\mathbf{Z}_0 = \mathbf{VP}_0$, $\mathbf{Z}_1 = \mathbf{VP}_1, \ldots, \mathbf{Z}_k = \mathbf{VP}_k$ for latent roots $l_0 \geq l_1 \geq l_2 \geq \cdots \geq l_k$. Again recall that although the inequality is not strict, the probability of obtaining equal roots is arbitrarily close to zero for continuous data. For $l_k = 0$ we have the linear combination

$$
p_{00}\mathbf{Y} + p_{10}\mathbf{X}_1 + p_{20}\mathbf{X}_2 + \cdots + p_{k0}\mathbf{X}_k = \mathbf{0}
$$

or

$$
\mathbf{Y} = -p_{00}^{-1}(p_{10}\mathbf{X}_1 + p_{20}\mathbf{X}_2 + \cdots + p_{k0}\mathbf{X}_k) = \mathbf{0} \tag{10.56}
$$

as in Section 10.2. Thus for any jth zero-variance principal component and for any ith observation we have

$$
y_i = -p_{0j}^{-1} \sum_{h=1}^{k} p_{hj}x_{ih} \qquad (i = 1, 2, \ldots, n) \tag{10.57}
$$

where the matrix $\mathbf{X}^{\mathsf{T}}\mathbf{X}$ approaches singularity as $p_{0j} \to 0$, that is, the columns of the data matrix for the predictors tend to approach linear dependence and the relationship being, in the limit, $\sum_{h=1}^{k} p_{hi}x_{ih} = 0$.

Now let $\mathbf{P}_j^0$ denote the jth (column) latent vector with the first element p_{0j} omitted. Then Eq. (10.57) can be expressed in vector form as

$$
\hat{\mathbf{Y}}_{(i)} = -p_{0j}^{-1}\mathbf{X}\mathbf{P}_j^0 \qquad (i = 1, 2, \ldots, k)
$$

where $\hat{\mathbf{Y}}_{(i)}$ denotes that part of $\mathbf{Y}$ which is predicted by the jth principal component, assuming $p_{0j} \neq 0$ and $l_j \neq 0$. The residual sum of squares can then be expressed as

$$(\mathbf{Y} - \hat{\mathbf{Y}}_{(j)})^{\mathrm{T}}(\mathbf{Y} - \hat{\mathbf{Y}}_{(j)}) = \frac{l_j}{p_{0j}^2} = l_j^* \tag{10.58}$$

Usually none of the individual components will by itself be a good predictor and an optimal linear combination is sought using the method of Lagrange multipliers (Gunst et al., 1976). Let such a combination be

$$\hat{\mathbf{Y}} = \sum_{j=0}^{k} a_j p_{0j} \hat{\mathbf{Y}}_{(j)}$$

$$= -\sum_{j=0}^{k} a_j \mathbf{X} \mathbf{P}_j^0$$

$$= -\mathbf{X} \sum_{j=0}^{k} a_j \mathbf{P}_j^0 \tag{10.59}$$

for some coefficients $a = (a_1, a_2, \ldots, a_k)^{\mathrm{T}}$. Equation (10.59) is of the usual least squares form where $\hat{\boldsymbol{\beta}} = \Sigma_{i=0}^{k} a_i \mathbf{P}_i^0$. Imposing the arbitrary constraint $\Sigma_{i=0}^{k} a_i p_{0i} = 1$, we wish to minimize the Lagrange expression $\phi(\mathbf{a}) = \mathbf{a}^{\mathrm{T}} \boldsymbol{\Lambda} \mathbf{a} - 2\mu(\mathbf{a}^{\mathrm{T}} \mathbf{P}_{(j)}^{(0)} - 1)$, where the latent roots $\boldsymbol{\Lambda}$ are functions of the residual sums of squares of the least squares fit (see Webster et al., 1974). Differentiating and setting to zero yields the expression $\partial\phi/\partial\mathbf{a} = \boldsymbol{\Lambda}\mathbf{a} - 2\mu\mathbf{P}_i^0 = \mathbf{0}$ or $\mathbf{a} = \mu\mathbf{P}_j^0\boldsymbol{\Lambda}^{-1}$ so that $a_i = l_i^{-1} p_{0i}\mu$. We have $a_i p_{0i} = l_i^{-1} p_{0i}^2 \mu$ or

$$\sum_{j=0}^{k} a_j p_{0j} = \sum_{j=0}^{k} l_j^{-1} p_{0j} \mu$$

$$= \mu \sum_{j=0}^{k} l_j^{-1} p_{0j}^2$$

$$= 1$$

so that the Lagrange multiplier is given by $\mu = (\Sigma_{i=0}^{k} l_i^*)^{-1}$. The ordinary least squares estimator can therefore be expressed as

$$\hat{\boldsymbol{\beta}} = \sum_{i=0}^{k} a_i \mathbf{P}_i^0$$

$$= -\sum_{i=0}^{k} l_i^{-1} p_{0i} \left(\sum_{h=0}^{k} l_h^{*-1} \right)^{-1} \mathbf{P}_i^0$$

$$= \left(\sum_{h=0}^{k} l_j^{*-1} \right) \sum_{i=0}^{k} l_i^{-1} p_{0i} \mathbf{P}_i^0 \tag{10.60}$$

Since $l_k^{*-1} = (p_{0h}^2 / l_h)$, Eq. (10.60) can be written as

$$\hat{\boldsymbol{\beta}} = \frac{\displaystyle\sum_{i=0}^{k} l_i^{-1} p_{0i} \mathbf{P}_i^0}{\displaystyle\sum_{h=0}^{k} \left(\frac{p_{0h}^2}{l_h} \right)} \tag{10.61}$$

which is a function of the latent roots and the latent vectors only. Note that solutions of the normal equations can be obtained regardless of whether $\mathbf{X}^T\mathbf{X}$ is singular or not by simply omitting zero latent roots. Equation (10.61) thus provides an alternative expansion of the y-norm least squares regression coefficients using latent roots and latent vectors of both the dependent and independent variables. The general idea here is to omit terms that correspond to small latent roots, but only those terms for which the latent vector element is small as well, since these terms correspond to multicollinear (nonpredictive) correlation amongst the predictors (Fig. 10.4). As is the case for principal components regression (Eq. 10.50), the latent root least squares estimator is efficient but biased.

Example 10.5   Referring to the car data of the previous example we have the latent roots and latent vectors of the dependent and predictor variables as in Table 10.9 where the matrix of latent vectors is transposed for convenience. The least squares coefficients of Example 10.4 can be reproduced using Eq. (10.61)—the coefficient $\hat{\beta}_1$, for example, can be expanded as

$$-c\hat{\beta}_1 = \sum_{i=0}^{4} l_i^{-1} p_{0i} p_{1j}$$

where

$$c = \sum_{i=0}^{4} \left(\frac{p_{0i}^2}{l_j} \right)$$

Table 10.9   Latent Roots and Latent Vectors of the Dependent and Predictor Variables of Example 10.4 Where p_{hi} ($h = 0, 1, \ldots, 4$) is the Element of the hth Column

l_i	Y	X_1	X_2	X_3	X_4
$l_0 = 3.68875$	$-.48160$	.49458	.47290	.45393	$-.30621$
$l_1 = .95657$	$-.18642$	.15774	$-.27717$	.46272	.80587
$l_2 = .18179$	.69848	.34260	.55554	$-.01321$	.29317
$l_3 = .11719$	.37872	.54712	$-.62099$	.21294	$-.35533$
$l_4 = .05571$	.31938	$-.56017$	.07273	.73097	$-.21117$

We have

$$-c\hat{\beta}_1 = \frac{-(.48160)(.49458)}{3.68875} - \frac{(.18642)(.15774)}{.95657} + \frac{(.69848)(.34260)}{.18179}$$

$$+ \frac{(.37872)(.54712)}{.11719} - \frac{(.31938)(.56017)}{.05571}$$

$$= -.06457 - .03074 + 1.131635 + 1.76811 - 3.21140$$

$$= 3.08446 - 3.30671$$

$$= -.22225,$$

$$c = \frac{(-.48160)^2}{3.68875} + \frac{(-.18642)^2}{.95657} + \frac{(.69848)^2}{.18179} + \frac{(.37872)^2}{.11719} + \frac{(.31938)^2}{.05571}$$

$$= .068875 + .03633 + 2.68372 + 1.22390 + 1.83097$$

$$= 5.843795$$

so that

$$\hat{\beta}_1 = \frac{.22225}{5.843795}$$

$$= .03803$$

The remaining coefficients can be expanded in a similar manner. The main idea here is the same as for principal components regression, that is, we wish to omit terms in the expansion which are responsible for multicollinearity. In view of Section 10.3.2 this corresponds to small l_i and small p_{0i} values. Webster and Mason (1974) recommend deletion of terms for which $l \leq .05$ and $p_{0i} \leq .10$, which is not realized for our example (Table 10.9), although multicollinearity is present. To be consistent with principal components regression (Example 10.4) we omit the last two terms, which yields the latent root regression coefficients $\tilde{\beta}_1^* = -.4378$, $\tilde{\beta}_2^* = .7626$, $\tilde{\beta}_3^* = -.0718$, and $\tilde{\beta}_4^* = 1.0094$. Comparing these with the $\tilde{\beta}_i$ of Example 10.4 it can be seen that latent root regression tends to yield values that are not necessarily close to the principal components regression coefficients—indeed one of the coefficients ($\tilde{\beta}_2^* = .7626$) is of the wrong sign. A disadvantage of latent root regression is that it is difficult to test individual significance of the coefficients by an adjusted t statistic, although a global F value can be computed. Given the more complex nature of latent root regression, it does not appear to be as advantageous as principal components regression in the present context, although in practice it is always a good idea, when dealing with multicollinearity, to examine several different estimators.

10.3.4 Quadratic Principal Components Regression

Principal components regression can also be employed with a quadratic regression equation to more easily identify turning points and multicollinear relationships. Consider a quadratic regression model in k predictor variables

$$y = \beta_0 + \sum_{i=1}^{k} \beta_i x_i + \sum \sum_{i \geq j} \beta_{ij} x_i x_j + \epsilon \tag{10.62}$$

For a sample of n observations Eq. (10.62) can be expressed in matrix form as

$$\mathbf{Y} = \boldsymbol{\beta}_0 + \mathbf{X}\boldsymbol{\beta} + \mathbf{XBX}^{\mathrm{T}} + \boldsymbol{\epsilon} \tag{10.63}$$

where $\boldsymbol{\beta} = (\beta_1, \beta_2, \ldots, \beta_k)^{\mathrm{T}}$ and

$$\mathbf{B} = \begin{bmatrix} \beta_{11} & \frac{1}{2}\beta_{21} & \cdots & \frac{1}{2}\beta_{k1} \\ \frac{1}{2}\beta_{21} & \beta_{22} & \cdots & \frac{1}{2}\beta_{k2} \\ \vdots & \vdots & \ddots & \vdots \\ \frac{1}{2}\beta_{k1} & \frac{1}{2}\beta_{k2} & \cdots & \beta_{kk} \end{bmatrix} \tag{10.64}$$

is a symmetric matrix of coefficients for the purely quadratic terms. Let $\mathbf{P}$ denote the $(k \times k)$ matrix of latent vectors of $\mathbf{X}^{\mathrm{T}}\mathbf{X}$. Then Eq. (10.63) can be rewritten (rotated) as

$$\mathbf{Y} = \boldsymbol{\beta}_0 + (\mathbf{XP}^{\mathrm{T}})(\mathbf{P}\boldsymbol{\beta}) + (\mathbf{XP}^{\mathrm{T}})(\mathbf{PBP}^{\mathrm{T}})(\mathbf{PX}^{\mathrm{T}}) + \boldsymbol{\epsilon}$$

$$= \boldsymbol{\beta}_0 + \mathbf{Z}\boldsymbol{\gamma} + \mathbf{Z}\boldsymbol{\Theta}\mathbf{Z}^{\mathrm{T}} + \boldsymbol{\epsilon} \tag{10.65}$$

where $\mathbf{Z} = \mathbf{XP}^{\mathrm{T}}$, $\boldsymbol{\gamma} = \mathbf{P}\boldsymbol{\beta}$, and $\boldsymbol{\Theta} = \mathbf{PBP}^{\mathrm{T}}$ are the principal components for $\mathbf{X}$, the vector of principal component regression coefficients, and the diagonal matrix of latent roots of $\mathbf{B}$, respectively. Although $\mathbf{B}$ is symmetric, it is not necessarily positive definite, so that $\boldsymbol{\Theta}$ may contain both positive and negative terms. To illustrate the procedure let $k = 2$. Then Eq. (10.65) can be written in vector form as

$$\mathbf{Y} = \boldsymbol{\beta}_0 + \gamma_1 \mathbf{Z}_1 + \gamma_2 \mathbf{Z}_2 + \theta_1 \mathbf{Z}_1^2 + \theta_2 \mathbf{Z}_2^2 + \boldsymbol{\epsilon} \tag{10.66}$$

which is a quadratic form in the principal components $\mathbf{Z}_1$ and $\mathbf{Z}_2$. The canonical form (Eq. 10.66) is often employed in the so-called response surface methodology to analyze optimal experimental conditions (see Box and Draper, 1987). This can be achieved more efficiently using Eq. (10.66) since we do not have to worry about the cross-product interaction terms. Negative θ_j indicate the existence of a maximum. Positive θ_j indicate a minimum, and a positive and negative coefficient reveals the existence of a saddle point. Note also that by rotating the quadratic form we increase the degrees of freedom since we do not have to estimate the cross-product terms. The original linear coefficients can also be expanded as in Section 10.3.1, whereas the quadratic coefficients are given by $\mathbf{B} = \mathbf{P}^T \mathbf{\Theta} \mathbf{P}$ or

$$
\mathbf{B} = \begin{bmatrix} \beta_{11} & \frac{1}{2}\beta_{21} \\ \frac{1}{2}\beta_{21} & \beta_{22} \end{bmatrix} = \begin{bmatrix} p_{11} & p_{12} \\ p_{21} & p_{22} \end{bmatrix} \begin{bmatrix} \theta_1 & 0 \\ 0 & \theta_2 \end{bmatrix} \begin{bmatrix} p_{11} & p_{21} \\ p_{12} & p_{22} \end{bmatrix}
$$

$$
= \begin{bmatrix} (\theta_1 p_{11}^2 + \theta_2 p_{12}^2) & (\theta_1 p_{11} p_{21} + \theta_2 p_{12} p_{22}) \\ (\theta_1 p_{11} p_{21} + \theta_2 p_{12} p_{22}) & (\theta_1 p_{21}^2 + \theta_2 p_{22}^2) \end{bmatrix}
$$

$$(10.67)$$

The extension to $k > 2$ predictors is straightforward.

10.4 LEAST SQUARES REGRESSION WITH ERRORS IN VARIABLES AND MULTICOLLINEARITY

When the predictor variables are highly correlated but free of residual error, all alternatives to ordinary least squares lead to biased estimators, and optimality is considered in terms of the minimum mean square (MMS) criterion. When both multicollinearity and errors in the variables are present, however, the regression hypothesis is altered in a fundamental way. Here, although the observed predictors are not perfectly correlated because of residual and/or measurement error, the true values may be assumed to be linearly dependent owing to the presence of a smaller number of unobserved predictor variables or dimensions. The objective then becomes to estimate the true values together with their least squares regression coefficients with the dependent variable. This can be done in one of two ways, using maximum likelihood factor analysis, depending on whether (1) the dependent variable is excluded (Basilevsky, 1981) or (2) the dependent variable is included (Scott, 1966).

10.4.1 Factor Analysis Regression: Dependent Variable Excluded

The most straightforward approach to factor analysis regression (FAR) is to extend the principal components regression model of Section 10.3.1. Let $\mathbf{X}$ have a k-dimensional normal distribution $N(\mathbf{0}, \mathbf{\Sigma})$ and let $\mathbf{X} = \chi + \mathbf{\Delta}$ where χ denotes the true part of $\mathbf{X}$ and $E(\mathbf{\Delta}^T \mathbf{\Delta})$ is diagonal. Factor analysis regression then consists of the joint hypotheses

$$\mathbf{Y} = \mathbf{X}\boldsymbol{\beta} + \boldsymbol{\eta} \tag{10.68a}$$

$$\mathbf{X} = \boldsymbol{\phi}\boldsymbol{\alpha} + \mathbf{\Delta} \tag{10.68b}$$

such that $E(\mathbf{X}) = \boldsymbol{\chi} = \boldsymbol{\phi}\boldsymbol{\alpha}$ is of rank $r < k$, $E(\boldsymbol{\phi}^T\boldsymbol{\Delta}) = \mathbf{0}$, $\boldsymbol{\phi} \sim N(\mathbf{0}, \mathbf{I})$ is a $(n \times r)$ matrix of standardized and uncorrelated factors (scores), and $\boldsymbol{\alpha}$ represents a $(r \times k)$ matrix of factor coefficients (loadings). To initially simplify the presentation we assume that both $\boldsymbol{\phi}$ and $\boldsymbol{\alpha}$ are known. The regression specification is therefore conditional on the acceptance of the factor model (Eq. 10.68b) so that although the observed explanatory variables $X_1, X_2, \ldots, X_k$ possess a nonsingular distribution, the true values $\chi_1, \chi_2, \ldots, \chi_k$ lie in a subspace of dimension $r < k$. Since $\boldsymbol{\chi}$ is not of full rank, the k population regression parameters also lie in an r-dimensional subspace and we denote these values by $\boldsymbol{\beta}^*$. Our purpose then is to estimate $\boldsymbol{\beta}^*$ on the assumption that the true regression equation is given by

$$\mathbf{Y} = \boldsymbol{\chi}\boldsymbol{\beta}^* + \boldsymbol{\epsilon} \tag{10.69}$$

where $\boldsymbol{\epsilon} \sim N(\mathbf{0}, \sigma^2\mathbf{I})$, $\mathbf{Y}$ is a $(n \times 1)$ vector of sample observations on a dependent random variable Y, $\boldsymbol{\chi}$ is the unobserved matrix of true values, and $\boldsymbol{\epsilon}$ is the residual vector for $\mathbf{Y}$.

When $\boldsymbol{\chi}$ is of full rank, $\boldsymbol{\beta}^* = \boldsymbol{\beta}$ and the normal equations can be solved in the usual way, allowance being made for errors in the predictor variables. Also, the best linear minimum norm (BLMN) estimator can be obtained when $\mathbf{X}$ contains no error but $(\mathbf{X}^T\mathbf{X})$ is singular, by considering the generalized inverse $(\mathbf{X}^T\mathbf{X})^-$ (e.g., Basilevsky, 1983; Graybill, 1983; Searle, 1982). In the factor analysis case where both singularity and measurement error are assumed to be present it is still possible to obtain an unbiased BLMN estimator of $\boldsymbol{\beta}^*$. The unique generalized inverse of $\boldsymbol{\chi}$ is given by

$$\boldsymbol{\chi}^- = \boldsymbol{\alpha}^T(\boldsymbol{\alpha}\boldsymbol{\alpha}^T)^{-1}\boldsymbol{\phi}^T \tag{10.70}$$

(see Albert, 1972), and postmultiplying by $\mathbf{Y}$ yields the FAR estimator

$$\tilde{\boldsymbol{\beta}}^* = \boldsymbol{\chi}^{-1}\mathbf{Y} = \boldsymbol{\alpha}^T(\boldsymbol{\alpha}\boldsymbol{\alpha}^T)^{-1}\boldsymbol{\phi}^T\mathbf{Y}$$

$$= \boldsymbol{\alpha}^T(\boldsymbol{\alpha}\boldsymbol{\alpha}^T)^{-1}\hat{\boldsymbol{\delta}} \tag{10.71}$$

where $\hat{\boldsymbol{\delta}} = \boldsymbol{\phi}^T\mathbf{Y}$ is the $(r \times 1)$ vector of correlations between $\boldsymbol{\phi}_1, \boldsymbol{\phi}_2, \ldots, \boldsymbol{\phi}_r$ and $\mathbf{Y}$. Premultiplying by $\boldsymbol{\alpha}$ then yields the inverse relationship $\boldsymbol{\alpha}\boldsymbol{\beta}^* = \hat{\boldsymbol{\delta}}$. The inverse $\boldsymbol{\chi}^-$, which provides a unique solution to the singular least squares problem, replaces the generalized inverse $(\mathbf{X}^T\mathbf{X})^{-1}\mathbf{X}^T$ normally used in regression, and it is easy to show that in the nonsingular case $\boldsymbol{\chi}^- = (\mathbf{X}^T\mathbf{X})^{-1}\mathbf{X}^T$. Also when $\boldsymbol{\alpha} = \mathbf{A} = \mathbf{L}^{1/2}\mathbf{P}^T$ we obtain the PCR estimator.

Equation (10.71) represents an estimator of the orthogonal projection $\boldsymbol{\beta}^* = \mathbf{P}_\alpha\boldsymbol{\beta}$ where $\mathbf{P}_\alpha = \boldsymbol{\alpha}^T(\boldsymbol{\alpha}\boldsymbol{\alpha}^T)^{-1}\boldsymbol{\alpha}$ is the projection operator on the range space $R(\boldsymbol{\alpha}^T)$ of $\boldsymbol{\alpha}^T$ and $(\mathbf{I} - \mathbf{P}_\alpha)$ is the projection operator on the null space $N(\boldsymbol{\alpha}^T)$ so that $\boldsymbol{\chi}(\mathbf{I} - \mathbf{P}_\alpha) = \mathbf{0}$. Here $\tilde{\boldsymbol{\beta}}^*$ is also the unbiased estimator of

$\boldsymbol{\beta}^* = \mathbf{P}_\alpha \boldsymbol{\beta}$ since

$$
\begin{aligned}
E(\tilde{\boldsymbol{\beta}}^*) &= E[\boldsymbol{\alpha}^{\mathrm{T}}(\boldsymbol{\alpha}\boldsymbol{\alpha}^{\mathrm{T}})^{-1}\boldsymbol{\phi}^{\mathrm{T}}\mathbf{Y}] \\
&= E[\boldsymbol{\alpha}^{\mathrm{T}}(\boldsymbol{\alpha}\boldsymbol{\alpha}^{\mathrm{T}})^{-1}\boldsymbol{\phi}^{\mathrm{T}}(\mathbf{X}\boldsymbol{\beta} + \boldsymbol{\eta})] \\
&= E[\boldsymbol{\alpha}^{\mathrm{T}}(\boldsymbol{\alpha}\boldsymbol{\alpha}^{\mathrm{T}})^{-1}\boldsymbol{\phi}^{\mathrm{T}}\mathbf{X}\boldsymbol{\beta}] + E[\boldsymbol{\alpha}^{\mathrm{T}}(\boldsymbol{\alpha}\boldsymbol{\alpha}^{\mathrm{T}})^{-1}\boldsymbol{\phi}^{\mathrm{T}}\boldsymbol{\eta}] \\
&= E[\boldsymbol{\alpha}^{\mathrm{T}}(\boldsymbol{\alpha}\boldsymbol{\alpha}^{\mathrm{T}})^{-1}\boldsymbol{\phi}^{\mathrm{T}}(\boldsymbol{\phi}\boldsymbol{\alpha} + \boldsymbol{\Delta})\boldsymbol{\beta}] + E[\boldsymbol{\alpha}^{\mathrm{T}}(\boldsymbol{\alpha}\boldsymbol{\alpha}^{\mathrm{T}})^{-1}\boldsymbol{\phi}^{\mathrm{T}}\boldsymbol{\eta})] \\
&= \boldsymbol{\alpha}^{\mathrm{T}}(\boldsymbol{\alpha}\boldsymbol{\alpha}^{\mathrm{T}})^{-1}\boldsymbol{\alpha}\boldsymbol{\beta} + \boldsymbol{\alpha}^{\mathrm{T}}(\boldsymbol{\alpha}\boldsymbol{\alpha}^{\mathrm{T}})^{-1}E(\boldsymbol{\phi}^{\mathrm{T}}\boldsymbol{\Delta})\boldsymbol{\beta} + \boldsymbol{\alpha}^{\mathrm{T}}(\boldsymbol{\alpha}\boldsymbol{\alpha}^{\mathrm{T}})^{-1}E(\boldsymbol{\phi}^{\mathrm{T}}\boldsymbol{\eta}) \\
&= \mathbf{P}_\alpha \boldsymbol{\beta} \\
&= \boldsymbol{\beta}^* \qquad\qquad\qquad (10.72)
\end{aligned}
$$

since $E(\boldsymbol{\phi}^{\mathrm{T}}\boldsymbol{\eta}) = E(\boldsymbol{\phi}^{\mathrm{T}}\boldsymbol{\Delta}) = \mathbf{0}$. Also, $\tilde{\mathbf{Y}} = \boldsymbol{\chi}\tilde{\boldsymbol{\beta}}^*$ is an unbiased estimator of $E(\mathbf{Y}) = \boldsymbol{\chi}\boldsymbol{\beta}^*$ since $E(\tilde{\mathbf{Y}}) = E(\boldsymbol{\chi}\tilde{\boldsymbol{\beta}}^*) = \boldsymbol{\chi}E(\tilde{\boldsymbol{\beta}}^*) = \boldsymbol{\chi}\boldsymbol{\beta}^*$. Note, however, that the projection of the OLS estimator $\hat{\boldsymbol{\beta}}$ onto $R(\boldsymbol{\alpha}^{\mathrm{T}})$ is a biased estimator of $\boldsymbol{\beta}^*$ since

$$
\begin{aligned}
E(\mathbf{P}_\alpha \hat{\boldsymbol{\beta}}) &= E[\mathbf{P}_\alpha (\mathbf{X}^{\mathrm{T}}\mathbf{X})^{-1}\mathbf{X}^{\mathrm{T}}\mathbf{Y}] \\
&= E[\mathbf{P}_\alpha (\mathbf{X}^{\mathrm{T}}\mathbf{X})^{-1}\mathbf{X}^{\mathrm{T}}(\mathbf{X}\boldsymbol{\beta} + \boldsymbol{\eta})] \\
&= \mathbf{P}_\alpha \boldsymbol{\beta} + \mathbf{P}_\alpha E[(\mathbf{X}^{\mathrm{T}}\mathbf{X})^{-1}(\boldsymbol{\phi}\boldsymbol{\alpha} + \boldsymbol{\Delta})^{\mathrm{T}}\boldsymbol{\eta}] \\
&= \mathbf{P}_\alpha \boldsymbol{\beta} + \mathbf{P}_\alpha \boldsymbol{\Sigma}^{-1}\boldsymbol{\alpha}^{\mathrm{T}}E(\boldsymbol{\phi}^{\mathrm{T}}\boldsymbol{\eta}) + \boldsymbol{\Sigma}^{-1}E(\boldsymbol{\Delta}^{\mathrm{T}}\boldsymbol{\eta}) \\
&= \mathbf{P}_\alpha \boldsymbol{\beta} + \boldsymbol{\Sigma}^{-1}E(\boldsymbol{\Delta}^{\mathrm{T}}\boldsymbol{\eta}) \qquad\qquad (10.73)
\end{aligned}
$$

where $E(\boldsymbol{\Delta}^{\mathrm{T}}\boldsymbol{\eta}) \neq \mathbf{0}$. Finally, the covariance matrix of $\tilde{\boldsymbol{\beta}}^*$ can be obtained as

$$
\begin{aligned}
E[(\tilde{\boldsymbol{\beta}}^* - \boldsymbol{\beta}^*)(\tilde{\boldsymbol{\beta}}^* - \boldsymbol{\beta}^*)^{\mathrm{T}}] &= \boldsymbol{\alpha}^{\mathrm{T}}(\boldsymbol{\alpha}\boldsymbol{\alpha}^{\mathrm{T}})^{-1}E(\boldsymbol{\epsilon}\boldsymbol{\epsilon})^{\mathrm{T}}(\boldsymbol{\alpha}\boldsymbol{\alpha}^{\mathrm{T}})^{-1}\boldsymbol{\alpha} \\
&= \boldsymbol{\alpha}^{\mathrm{T}}(\boldsymbol{\alpha}\boldsymbol{\alpha}^{\mathrm{T}})^{-2}\sigma^2\mathbf{I} \qquad\qquad (10.74)
\end{aligned}
$$

When the true model is of full rank it can be shown that Eq. (10.74) reduces to the usual OLS covariance matrix $\sigma^2(\mathbf{X}^{\mathrm{T}}\mathbf{X})^{-1}$, and when $\boldsymbol{\alpha} = \mathbf{A} = \mathbf{L}^{1/2}\mathbf{P}^{\mathrm{T}}$ we obtain the PCR covariance matrix $\sigma^2(\mathbf{PL}^{-1}\mathbf{P}^{\mathrm{T}})$. On the usual assumption of normality of $\mathbf{Y}$ Eq. (10.74) can be used to compute interval estimates of $\tilde{\boldsymbol{\beta}}^*$ and to test hypotheses.

Our development has so far assumed that the true population values of $\boldsymbol{\phi}$ and $\boldsymbol{\alpha}$ are available. When this is not the case, both $\boldsymbol{\phi}$ and $\boldsymbol{\alpha}$ are estimated from sample values and are replaced by their estimates $\hat{\boldsymbol{\phi}}$ and $\hat{\boldsymbol{\alpha}}$. In this case the results given above still hold, on the assumption that the loadings $\hat{\boldsymbol{\alpha}}$ and errors $\hat{\boldsymbol{\Delta}}$ are uncorrelated, which is the case in large samples. The factor scores $\hat{\boldsymbol{\phi}}$ however are usually not required since the FAR estimator can be obtained more directly from Eq. (10.71) as

$$\tilde{\boldsymbol{\beta}}^* = \hat{\boldsymbol{\alpha}}^T(\hat{\boldsymbol{\alpha}}\hat{\boldsymbol{\alpha}}^T)^{-2}\hat{\boldsymbol{\alpha}}\mathbf{X}^T\mathbf{Y} \qquad (10.75)$$

since it is straightforward to verify that

$$\hat{\boldsymbol{\delta}} = \hat{\boldsymbol{\phi}}^T\mathbf{Y} = (\hat{\boldsymbol{\alpha}}\hat{\boldsymbol{\alpha}}^T)^{-1}\hat{\boldsymbol{\alpha}}\mathbf{X}^T\mathbf{Y} \qquad (10.75a)$$

Factor analysis regression can also be considered within the context of optimality of errors-in-variables models, as is demonstrated by the following theorem.

THEOREM 10.5   Let $\hat{\boldsymbol{\beta}}_I$ be an instrumental variable estimator of $\boldsymbol{\beta}$ when (1) there are fewer instruments than variables, and (2) the instruments are maximally correlated with the predictor variables. Then

$$\tilde{\boldsymbol{\beta}}^* = \mathbf{P}_\alpha\hat{\boldsymbol{\beta}}_I$$

PROOF.   Premultiplying Eq. (10.68a) by $\boldsymbol{\phi}^T$ yields the normal equations

$$\boldsymbol{\phi}^T\mathbf{Y} = \boldsymbol{\phi}^T\mathbf{X}\boldsymbol{\beta} + \boldsymbol{\phi}^T\boldsymbol{\eta}$$

$$= \boldsymbol{\phi}^T\mathbf{X}\boldsymbol{\beta}_I$$

$$= \boldsymbol{\alpha}\hat{\boldsymbol{\beta}}_I$$

where by assumption $\boldsymbol{\phi}^T\mathbf{X} \neq \mathbf{0}$. Thus $\hat{\boldsymbol{\beta}}_I$ is the optimal instrumental variables estimator since the common factors $\boldsymbol{\phi}$ are maximally correlated with $\mathbf{X}$. Premultiplying by $\boldsymbol{\alpha}^T(\boldsymbol{\alpha}\boldsymbol{\alpha}^T)^{-1}$ then yields $\boldsymbol{\alpha}^T(\boldsymbol{\alpha}\boldsymbol{\alpha}^T)^{-1}\boldsymbol{\phi}^T\mathbf{Y} = \boldsymbol{\alpha}^T(\boldsymbol{\alpha}\boldsymbol{\alpha}^T)^{-1}\boldsymbol{\alpha}\hat{\boldsymbol{\beta}}_I$ or $\tilde{\boldsymbol{\beta}}^* = \mathbf{P}_\alpha\hat{\boldsymbol{\beta}}_I$.

When the number of instruments is the same as the total number of predictors it is well known that $\hat{\boldsymbol{\beta}}_I$ is an unbiased estimator of $\boldsymbol{\beta}$. This is clearly not the case here since, by the very nature of the ML factor model, the number of common factors is always less than the total number of variables.

10.4.2   Factor Analysis Regression: Dependent Variable Included

An alternative method for factor analysis regression, as was the case for the principal components regression model of Section 10.3.3, is to include the dependent variable Y in the factoring process. The model can be written in partitioned form as

$$
\begin{bmatrix} y \\ \cdots \\ x_1 \\ x_2 \\ \vdots \\ x_k \end{bmatrix} = \begin{bmatrix} \alpha_{11} & \alpha_{12} & \cdots, & \alpha_{1,r} \\ \hline \alpha_{21} & \alpha_{22} & \cdots, & \alpha_{2,r} \\ \vdots & \vdots & & \vdots \\ \alpha_{k+1,1} & \alpha_{k+1,2} & \cdots, & \alpha_{k+1,r} \end{bmatrix} \begin{bmatrix} \phi_1 \\ \phi_2 \\ \vdots \\ \phi_r \end{bmatrix} + \begin{bmatrix} \epsilon_1 \\ \cdots \\ \epsilon_2 \\ \vdots \\ \epsilon_{k+1} \end{bmatrix}
$$

or

$$
Y = \alpha_1 \Phi + \epsilon_1
$$

$$
X = \alpha_2 \Phi + \epsilon_2 \tag{10.76}
$$

where α_1 is the $(1 \times r)$ vector of loadings for the predictor variables and Φ is the $(1 \times r)$ vector of common factors for both the independent variable Y and the k predictor variables X. Since the predictors are subject to measurement error and multicollinearity, the objective is to use a factor model to estimate the true part of X and to use their values in an ordinary least squares regression model. The model was originally considered by Scott (1966) and Lawley and Maxwell (1973)— see also and Isogawa and Okamoto (1980). Using Eq. (10.76) and assuming ϵ_2 represents the errors in the predictors, the true values can be estimated, for a sample of n observations as $\hat{X} = \Phi\hat{\alpha}_2$ where $\hat{\alpha}_2$ are the maximum likelihood estimates of the factor loadings and Φ is an orthogonal matrix of factor scores. The Scott–Lawley factor analysis regression estimator can then be expressed as

$$
\begin{aligned}
\beta^* &= (\hat{X}^T\hat{X})^{-1}\hat{X}^T Y \\
&= (\hat{\alpha}_2^T\Phi^T\Phi\hat{\alpha}_2)^{-1}\hat{\alpha}_2^T\Phi^T Y \\
&= (\hat{\alpha}_2^T\hat{\alpha}_2)^{-1}\hat{\alpha}_2^T\Phi^T(\Phi\hat{\alpha}_1 + \epsilon_1) \\
&= (\hat{\alpha}_2^T\hat{\alpha}_2)\hat{\alpha}_2^T\hat{\alpha}_1
\end{aligned} \tag{10.77}
$$

since the common factors and the errors are uncorrelated. Equation (10.77) represents the vector of regression coefficients of the loadings $\hat{\alpha}_2$ on the loadings vector $\hat{\alpha}_1$ and does not involve the factor scores Φ. Alternative expressions for Eq. (10.77), which require the estimation of Φ, can also be found in Lawley and Maxwell (1973). For the bivariate case with one common factor ($k = 2$; $r = 1$) it is straight-forward to show that $\beta^* = \hat{\alpha}_1/\hat{\alpha}_2$ (Exercise 10.13).

THEOREM 10.6   Let $X = \chi + \Delta$ as in Section (10.4.1). Then β^* is a consistent estimator of the true population coefficient when $\hat{X}$ is a consistent estimator of χ, that is, when the common factors correctly estimate χ.

PROOF.   We have from Eq. (10.77)

$$\boldsymbol{\beta}^* = (\hat{\mathbf{X}}\hat{\mathbf{X}})^{-1}\hat{\mathbf{X}}^{\mathrm{T}}\mathbf{Y}$$

$$= (\hat{\mathbf{X}}^{\mathrm{T}}\hat{\mathbf{X}})^{-1}\hat{\mathbf{X}}^{\mathrm{T}}(\boldsymbol{\Phi}\hat{\boldsymbol{\alpha}}_1 + \hat{\boldsymbol{\epsilon}}_1)$$

so that

$$p \lim \boldsymbol{\beta}^* = p \lim(\hat{\mathbf{X}}^{\mathrm{T}}\hat{\mathbf{X}})^{-1}\hat{\mathbf{X}}^{\mathrm{T}}\boldsymbol{\Phi}\hat{\boldsymbol{\alpha}}_1 + p \lim(\hat{\mathbf{X}}^{\mathrm{T}}\hat{\mathbf{X}})^{-1}\hat{\mathbf{X}}^{\mathrm{T}}\boldsymbol{\epsilon}_1$$

$$= p \lim(\hat{\mathbf{X}}^{\mathrm{T}}\hat{\mathbf{X}})^{-1} p \lim \hat{\mathbf{X}}^{\mathrm{T}}\boldsymbol{\Phi}\hat{\boldsymbol{\alpha}}_1 + p \lim(\hat{\mathbf{X}}^{\mathrm{T}}\hat{\mathbf{X}})^{-1}\hat{\mathbf{X}}^{\mathrm{T}}\hat{\boldsymbol{\epsilon}}_1$$

$$= p \lim(\hat{\mathbf{X}}^{\mathrm{T}}\hat{\mathbf{X}})^{-1} p \lim \hat{\boldsymbol{\alpha}}_2^{\mathrm{T}}\hat{\boldsymbol{\alpha}}_1$$

$$= (\boldsymbol{\alpha}_2^{\mathrm{T}}\boldsymbol{\alpha}_2)^{-1}\boldsymbol{\alpha}_2\boldsymbol{\alpha}_1$$

$$= \boldsymbol{\beta}$$

where $p \lim \hat{\mathbf{X}}^{\mathrm{T}}\hat{\boldsymbol{\epsilon}}_1 = \mathbf{0}$.

Once $\boldsymbol{\beta}^*$ are known, the predicted values of the dependent variable can be obtained as $\hat{\mathbf{Y}} = \hat{\mathbf{X}}\boldsymbol{\beta}^* = \hat{\boldsymbol{\Phi}}\hat{\boldsymbol{\alpha}}_2\boldsymbol{\beta}^*$ where $\boldsymbol{\Phi}$ is an $(n \times r)$ matrix of factor scores (Section 6.8). When the scores $\boldsymbol{\Phi}$ are estimated using Eq. (6.95), however, $\hat{\mathbf{Y}}$ is an inconsistent estimator of the true values of the dependent variable (Lawley and Maxwell, 1973). Also Chan (1977) has shown that Eq. (6.101) leads to inconsistent estimates of the dependent variable, and provides one that is consistent. Since using factor scores (Eq. 6.101) leads to the minimization of $E(\hat{\mathbf{Y}} - \mathbf{Y})^2$ however, inconsistency may not be an overriding factor because of the increased efficiency of the estimator $\boldsymbol{\beta}^*$

10.5   FACTOR ANALYSIS OF DEPENDENT VARIABLES IN MANOVA

A regression equation will usually consist of k predictor variables and a single dependent variable. At times however $p > 1$ dependent variables are available which are correlated with each other, and carrying out a set of p separate regressions will provide incorrect estimators and tests of significance. The situation occurs in experimental designs when more than one response is observed for the same set of experimental conditions and a multivariate analysis of variance (MANOVA) is appropriate (Dempster, 1963), or in the context of observational data when carrying out a path analysis between two sets of variables (Chung et al., 1977). The multivariate multiple regression model (e.g., see Finn, 1974) can be written as

$$\mathbf{Y} = \mathbf{XB} + \mathbf{E} \tag{10.78}$$

where $\mathbf{Y}$ is $(n \times p)$, $\mathbf{X}$ is $(n \times k)$, $\mathbf{B}$ is the $(k \times p)$ matrix of coefficients, and $\mathbf{E}$ is the $(n \times p)$ matrix of residual errors. When the columns of $\mathbf{Y}$ are highly correlated, they cannot be analyzed individually because of the resultant correlation of the residual errors, since this results in wider confidence intervals for the elements of $\mathbf{B}$. Also, the columns of $\mathbf{Y}$ frequently represent alternative measurements on a smaller number of common traits or dimensions, which renders individual interpretation difficult. This is the case, for example, when a set of growth curves are observed under experimental treatment, the so-called repeated measures designs (Chapter 7) described in Rao (1958), Rao and Boudieau (1985), Church (1966), and Snee (1972).

A way out of the difficulty is to carry out a factor analysis of $\mathbf{Y}$ and to use the factor scores as new dependent variables. Since the factors are orthogonal, Eq. (10.78) can be replaced by $r < p$ separate regression equations and standard testing of significance can be carried out, under the usual assumptions. The method was initially proposed by Rao (1958) in the context of growth curves, and a sequential test of significance based on principal components was developed by Demster (1963). Consider Eq. 10.78 where some (or all) columns of $\mathbf{Y}$ are highly correlated, but the predictors $\mathbf{X}$ are not. Let $\mathbf{L}$ and $\mathbf{P}$ be the latent roots and latent vectors, respectively, of $\mathbf{Y}^T\mathbf{Y}$, that is, $\mathbf{P}^T\mathbf{Y}^T\mathbf{Y}\mathbf{P} = \mathbf{L}$. Post multiplying Eq. (10.78) by $\mathbf{P}$ yields

$$\mathbf{YP} = \mathbf{XBP} + \mathbf{EP} \quad \text{or} \quad \mathbf{Z} = \mathbf{X\delta} + \mathbf{\eta} \tag{10.79}$$

where $\mathbf{Z} = \mathbf{YP}$ is the $(n \times p)$ matrix of principal components of $\mathbf{Y}$, $\mathbf{\eta} = \mathbf{EP}$ is the transformed matrix of errors and $\mathbf{\delta}$ is the new matrix of regression coefficients. We have

$$\hat{\mathbf{\delta}} = (\mathbf{X}^T\mathbf{X})^{-1}\mathbf{X}^T\mathbf{Z}$$

$$= (\mathbf{X}^T\mathbf{X})^{-1}\mathbf{X}^T(\mathbf{YP})$$

$$= \hat{\mathbf{B}}\mathbf{P} \tag{10.80}$$

which may be compared to Eq. (10.46). Since multicollinearity among the dependent variables is assumed not to be a problem, it is not usually necessary to expand $\hat{\mathbf{B}}$ in terms of the $\hat{\mathbf{\delta}}$. Also note that the deletion of dependent variables does not affect the remaining values of $\hat{\mathbf{B}}$. Equation (10.79) implies a partition of $\mathbf{Y}^T\mathbf{Y}$ into the predicted and residual matrices of sums-of-squares since

$$\mathbf{L} = \mathbf{P}^T\mathbf{Y}^T\mathbf{YP} = \mathbf{P}^T\hat{\mathbf{B}}^T\mathbf{X}^T\mathbf{X}\hat{\mathbf{B}}\mathbf{P} + \mathbf{P}^T\mathbf{E}^T\mathbf{EP}$$

$$= \hat{\mathbf{L}} + \mathbf{L}_E \tag{10.81}$$

say where $\mathbf{L}$, $\hat{\mathbf{L}}$, and $\mathbf{L}_E$ are diagonal.

The main reason for using principal components in place of the observed variables is to transform the covariance matrix of $\mathbf{E}$ into diagonal form and to replace Eq. (10.78) by $r < p$ separate regressions. Since observed variables are replaced by component (factor) scores, it becomes important to identify the factors in terms of real, substantive behavior. This may imply a full-blown factor analysis of $\mathbf{Y}$, including rotation of the loadings and scores to identifiable form. If an oblique rotation is chosen, however, this may not completely alleviate the original problem of dependence among the $\mathbf{Y}$ variables, but since oblique factors are usually less correlated than the original variables, a factor analysis of $\mathbf{Y}$ may still prove of some value.

10.6 ESTIMATING EMPIRICAL FUNCTIONAL RELATIONSHIPS

It was seen in Section 10.2 that principal components can be used directly to estimate a linear regression function, by using latent vectors that are associated with small latent roots. Another use of the PC expansion is to estimate an intrinsically linear empirical function of the form

$$\hat{x}_{ij} = \sum_{h=1}^{r} a_{ih} z_{hj}^{*} \tag{10.82}$$

where the a_{ih} and z_{hj}^{*} are the loadings and scores of $\mathbf{X}^{\mathrm{T}}\mathbf{X}$ respectively (see Tucker, 1958). Alternatively, a ML factor model can be used on the usual assumptions (Chapter 6), or the analysis may be based on the inner product matrix $\mathbf{Y}^{\mathrm{T}}\mathbf{Y}$ (Sheth, 1969) or any other matrix of Euclidian measures of association. Equation (10.82) thus represents a standard factor analysis of a $(n \times k)$ data matrix, and nothing new of a theoretical statistical nature arises in the context. Several applications however are of particular interest when analyzing growth curves or certain experimental data.

The use of principal components to estimate functional growth curves is due to Rao (1958) and represents a special case of a factor analysis of multivariate data when the rows (columns) of the data matrix $\mathbf{Y}$ are ordered over time and/or space (Section 7.2). The original data can represent sample survey observations or may emanate from controlled experiments, the so-called repeated measures designs, where a sequence of measurements is taken for each experimental unit and the resultant observations are analyzed using a MANOVA model (Section 10.5). Since the data are of the type discussed in Chapter 7, the reader should consult the first four sections of the chapter. Also, since growth curves represent time functions it usually does not make sense to employ the usual rotations of the factor loadings or scores (Section (5.3). A different type of transformation however may at times be appropriate in order to distinguish between true or explanatory functions and random noise. The relevant criterion in the context of correlated observations is that of "smoothness," which may be defined as

$$\mathrm{sm}(y) = \sum_{t=2}^{n} (y_{t-1} - y_{t})^{2} \tag{10.83}$$

as proposed by Arbuckle and Friendly (1977), where $\mathbf{Y} = (y_1, y_2, \ldots, y_n)^{\mathrm{T}}$ is an observation vector ordered over time (or space). In what follows we assume, without loss of generality, that observations are ordered over time. Since a first difference of a series represents a linear transformation we can write

$$\mathbf{\Delta} = \begin{bmatrix} y_1 - y_2 \\ y_2 - y_3 \\ \vdots \\ y_{n-1} - y_n \end{bmatrix} = \begin{bmatrix} 1 & -1 & 0 & 0 & \cdots & 0 & 0 \\ 0 & 1 & -1 & 0 & \cdots & 0 & 0 \\ 0 & 0 & 1 & -1 & \cdots & 0 & 0 \\ & & & & & & \\ 0 & 0 & \cdots & \cdots & & 1 & -1 \end{bmatrix} \begin{bmatrix} y_1 \\ y_2 \\ \vdots \\ y_n \end{bmatrix}$$

$$= \mathbf{DY}$$

say, so that Eq. (10.83) can be expressed in matrix form as $\mathbf{\Delta}^{\mathrm{T}}\mathbf{\Delta} = \mathbf{Y}^{\mathrm{T}}\mathbf{D}^{\mathrm{T}}\mathbf{DY}$ where $\mathbf{Y}^{\mathrm{T}}\mathbf{Y} = 1$. Now, consider a $(n \times k)$ data matrix $\mathbf{X}$ such that $\mathbf{XP} = \mathbf{Z}$ where $\mathbf{P}$ and $\mathbf{Z}$ are the latent vectors and component scores, respectively. To transform the component scores to smoother form we take first differences $\mathbf{\Delta} = \mathbf{DZ}$ and minimize

$$\mathbf{\Delta}^{\mathrm{T}}\mathbf{\Delta} = \mathbf{Z}^{\mathrm{T}}\mathbf{D}^{\mathrm{T}}\mathbf{DZ}$$

$$= \mathbf{P}^{\mathrm{T}}(\mathbf{X}^{\mathrm{T}}\mathbf{D}^{\mathrm{T}}\mathbf{DX})\mathbf{P} \tag{10.84}$$

so that the smoothest function is given by the principal component

$$\mathbf{Z}_k^* = \mathbf{DXP}_k \tag{10.85}$$

where $\mathbf{P}_k$ is the latent vector that corresponds to the smallest latent root l_k of $\mathbf{X}^{\mathrm{T}}\mathbf{D}^{\mathrm{T}}\mathbf{DX}$. The smoothness of the transformed component scores is thus measured by the smallest latent root l_k. Higher-order differences can also be minimized in a similar fashion, for which the reader is referred to Arbuckle and Friendly (1977).

Example 10.6   Sheth (1969) and Arbuckle and Friendly (1977) consider a sample of $n = 154$ families who bought a certain commodity on 60 (or more) occasions, the commodity consisting of several competing brand names. For each family the first 60 purchases are grouped into $k = 10$ blocks or "trials" of six purchases each. Within each block the frequency of purchase of a given brand (say Brand A) having the highest market share is then calculated, resulting in a (154×10) data matrix $\mathbf{Y}$ with typical element $y_{ij} =$ number of times family i purchased Brand A during the jth time block or "trial," so that $0 \le y_{ij} \le 6$.

The PC analysis is carried out on the average inner product $(1/n)(\mathbf{Y}^{\mathrm{T}}\mathbf{Y})$ (Section 2.4.1) in order to preserve information concerning level and variability of purchase (Table 10.10). Since we wish to observe the time

Table 10.10   The Averaged Inner Product Matrix $(1/n)$ (Y^TY) of Brand A Purchases

					Trials					
Trials	1	2	3	4	5	6	7	8	9	10
1	1430	1176	1157	1155	1126	1144	1131	1138	1176	1030
2	1176	1306	1148	1149	1079	1116	1163	1082	1139	1031
3	1157	1148	1318	1197	1122	1144	1150	1136	1182	1034
4	1155	1149	1197	1387	1149	1216	1233	1208	1242	1095
5	1126	1079	1122	1149	1327	1221	1192	1146	1220	1091
6	1144	1116	1144	1216	1221	1516	1318	1317	1350	1200
7	1131	1163	1150	1233	1192	1318	1495	1312	1328	1210
8	1138	1082	1136	1208	1146	1317	1312	1505	1373	1216
9	1176	1139	1182	1242	1220	1350	1328	1373	1595	1315
10	1030	1031	1034	1095	1091	1200	1210	1216	1315	1377

Source: Sheth, 1969; reproduced with permission.

behavior of families, the component loadings rather than the scores are of interest and these are given in Table 10.11, together with their associated latent roots. Also, since the association matrix consists of inner products, the loadings for the first PC reflect the general mean level of purchases and the remaining loadings are the direction cosines between the trials and the principal components. The loadings for the first $r = 4$ components of Tables 10.11 and 10.12 can be plotted and compared for relative smoothness (Exercise 10.10).

Table 10.11   Principal Component Loadings of the Inner Product Matrix $(1/n)$ (Y^TY) and Their Associated Latent Roots

				Principal Component Loadings						
Trials	Z_1	Z_2	Z_3	Z_4	Z_5	Z_6	Z_7	Z_8	Z_9	Z_{10}
Y_1	2.70	.982	−.609	.573	−.383	.212	.144	−.234	−.052	.222
Y_2	2.64	.824	−.014*	−.402	.174	.503	−.330	.217	−.055	−.464
Y_3	2.69	.824	.013*	−.310	.036	−.518	−.125	.442	.329	.390
Y_4	2.79	.370	.410	−.452	−.106	−.435	.053	−.672	−.171	−.112
Y_5	2.71	.161	.298	.594	.676	−.213	.469	.132	.027	−.263
Y_6	2.92	−.455	.530	.568	−.012	.160	−.676	−.134	.114	.090
Y_7	2.91	−.339	.460	−.321	.008	.672	.386	.094	−.214	.362
Y_8	2.89	−.628	.026*	.003	−.828	−.077	.249	.191	.324	−.335
Y_9	3.01	−.658	−.461	.002	.001	−.384	−.159	.250	−.662	.017
Y_{10}	2.70	−.766	−.871*	−.274	.501	.088	.002	−.295	.427	.077
l_i	12058.6	620.4	284.4	253.3	243.9	221.7	163.4	150.8	142.9	116.4

Source: Arbuckle and Friendly, 1977; reproduced with permission.

Table 10.12 The First Six Components of Table 10.11 Using First Differences

Trials	Principal Component Loadings					
	Z_1^*	Z_2^*	Z_3^*	Z_4^*	Z_5^*	Z_6^*
Y_1	.3196	.4518	−.4914	.4282	.1525	.3198
Y_2	.3126	.3498	−.2497	.0557	−.0518	−.5221
Y_3	.3067	.3099	.0232	−.5024	−.1175	.1039
Y_4	.3120	.1943	.2079	−.4484	−.3267	.0371
Y_5	.3081	.1103	.3756	−.1017	.6458	.0871
Y_6	.3198	−.0993	.4899	.3619	.1508	.1418
Y_7	.3252	−.1021	.2639	.3094	−.2494	−.5333
Y_8	.3133	−.2553	−.0010	.2342	−.5258	.3845
Y_9	.3354	−.4052	−.2211	−.2046	.0656	.2621
Y_{10}	.3084	−.5272	−.3909	−.1621	.2660	.2904

Source: Arbuckle and Friendly, 1977; reproduced with permission.

Example 10.7 An example of estimating the dimensionality of energy-free chemical relations is given by Hutton et al. (1986) who consider shifts in resonance frequency of substituted phenols and 2-nitrophenols. The data matrix for 2-substituted nitrophenols is given in Table 10.13. The initial objective of the principal components analysis is to determine the number of factors required to reproduce the substituent parameters (Eq. 10.86), free of

Table 10.13 ^{13}C Chemical Shifts of 2-Substituted Shifts of 2-Substituted Nitrophenols (ppm) from Internal TMS

Substituent	2-Nitrophenol					
	C-1	C-2	C-3	C-4	C-5	C-6
H	152.6	136.6	125.2	119.5	135.6	119.4
NH_2	144.5	135.6	107.4	141.9	124.3	120.2
OCH_3	146.9	135.8	107.7	151.7	123.7	120.4
CH_3	150.4	136.0	24.6	128.8	136.4	119.2
F	148.9	136.2	111.5	153.9	122.8	120.6
Cl	151.2	137.2	124.5	122.5	134.9	120.9
Br	151.4	137.7	127.2	109.3	137.5	121.3
$COCH_3$	155.8	136.8	126.0	128.2	134.4	119.2
CHO	157.1	137.2	128.3	128.0	134.6	120.0
COOH	156.0	136.6	127.2	122.1	135.9	119.6
$COOCH_3$	156.2	136.5	126.9	120.6	135.4	119.6
CN	155.6	137.5	130.2	101.5	137.9	120.3
NO_2	157.6	136.4	122.0	138.5	129.5	119.9
$C(CH_3)_3$	150.3	135.9	120.9	142.1	133.0	119.1
I	151.8	138.2	132.8	79.9	143.2	121.5
COC_6H_5	155.9	136.9	127.8	127.8	136.1	119.4
$NHCOCH_3$	148.5	135.4	114.8	131.5	127.3	119.7
C_6H_5	151.6	137.3	122.7	131.5	133.5	119.8

Source: Hutton et al., 1986; reproduced with permission.

Table 10.14 Varimax Principal Component Loadings of the Correlation Matrix of Data of Table 10.2

	$\mathbf{Z}_1^*$	$\mathbf{Z}_2^*$	$\mathbf{Z}_3^*$	$\mathbf{Z}_4^*$	$\mathbf{Z}_5^*$	$\mathbf{Z}_6^*$
C_1	0.302	0.938	-0.148	0.078	-0.009	-0.006
C_2	0.614	0.355	0.505	0.491	0.005	0.002
C_3	0.812	0.547	0.005	0.164	0.096	0.076
C_4	-0.904	-0.142	-0.337	-0.016	0.222	0.008
C_5	0.935	0.292	0.022	0.155	0.121	-0.025
C_6	0.105	-0.158	0.980	0.051	-0.014	-0.000
l_I	2.830	1.436	1.353	0.301	0.073	0.007

Source: Hutton et al., 1986; personal communication.

experimental error. Using the varimax rotation (Section 5.3.1) the correlation loadings of Table 10.14 indicate the presence of three systematic dimensions and the data of Table 10.13 can therefore be fitted by the equation.

$$\hat{y}_{ij} = \bar{y}_i + \sum_{h=1}^{3} a_{ih} z_{hj} \tag{10.86}$$

The variances are usually set to unity in order to avoid the dominating variance of the ipso and ortho-like portions of the molecule, that is, the analysis is based on the correlation matrix. A number of variants of the method are also available—for a general overview of related applications in chemistry see Malinowski (1977), Malinowski and Howery (1979), Kowalski et al. (1982), Kowalski and Wald (1982), and Johnels et al. (1983; 1985; Strouf, 1986). □

10.7 OTHER APPLICATIONS

The previous sections describe what may be termed the "standard" uses of factor models in estimating least squares planes and functional forms. Other applications and variants are also possible and these may be found in the literature of the various disciplines that employ multivariate data. Several specialized applications however are of particular interest and these are described briefly in the following sections.

10.7.1 Capital Stock Market Data: Arbitrage Pricing

Factor models can be used to analyze rates of returns on portfolio investments if we assume the existence of systematic risk factors that influence some or all of the stock returns over time. (Roll and Ross, 1980; Sharpe, 1982; Kryzanowski and To, 1983). Let $\mathbf{Y}$ represent a $(T \times k)$ matrix of returns for k investment stocks over T time periods. Then if there exist

$r < k$ unobserved risk factors ϕ_1, ϕ_2, ..., ϕ_r, a ML factor model can be used to expand the jth column (asset) of $\mathbf{Y}$ as

$$\mathbf{Y}_j = E(\mathbf{Y}_j) + \alpha_{j1}\phi_1 + \alpha_{j2}\phi_2 + \cdots + \alpha_{jr}\phi_r + \epsilon_j \tag{10.87}$$

where $E(\mathbf{Y}_j)$ is the expected return on security j, α_{ji}, α_{j2}, ..., α_{jr} are ML factor loadings, and ϵ_j ($j = 1, 2, \ldots, k$) are uncorrelated error terms (Chapter 6).

In matrix form we have the equations $\mathbf{X}_j = \mathbf{\Phi}\boldsymbol{\alpha}_j + \epsilon_j$ where $\mathbf{X}_j = \mathbf{Y}_j - \bar{\mathbf{Y}}_j$. The loadings are known as "reaction coefficients" and measure the sensitivity of the jth stock to the particular risk factor. In practice, in the finance literature an unnecessary duplication occurs, since the common factor scores $\mathbf{\Phi}$ are first estimated and then used as regressor variables for $\mathbf{X}_1, \mathbf{X}_2, \ldots, \mathbf{X}_k$ to test the significance of the common factors.

Once the reaction coefficients are estimated, the second state is to compute the risk premium λ_i for the ith common risk factor. This is done by computing T cross-sectional regressions of the rows of $\mathbf{X}$ (or $\mathbf{Y}$) on the $(k \times r)$ matrix of loadings $\hat{\boldsymbol{\alpha}}$ (or the regression coefficients of factor scores on the stocks) as

$$\begin{aligned} \mathbf{X}_1 &= \hat{\boldsymbol{\alpha}}\boldsymbol{\lambda}_1 + \hat{\boldsymbol{\delta}}_1 \\ \mathbf{X}_2 &= \hat{\boldsymbol{\alpha}}\boldsymbol{\lambda}_2 + \boldsymbol{\delta}_2 \\ \hline \mathbf{X}_T &= \hat{\boldsymbol{\alpha}}\boldsymbol{\lambda}_T + \hat{\boldsymbol{\Delta}}_T \end{aligned} \tag{10.88}$$

where $\mathbf{X}_j$ ($j = 1, 2, \ldots, T$) are row vectors of $\mathbf{X}$, that is, we estimate $\mathbf{X}^{\mathrm{T}} = \hat{\boldsymbol{\alpha}}\boldsymbol{\lambda} + \boldsymbol{\delta}$ where $\boldsymbol{\lambda}$ is the $(r \times T)$ matrix of coefficients to be estimated. The columns of $\boldsymbol{\lambda}$ are then averaged to obtain the final estimates of risk premia $\bar{\lambda}_1, \bar{\lambda}_2, \ldots, \bar{\lambda}_r$, one for each common factor. Using Eq. (10.88) the risk premia can be used to calculate the cost of capital of each asset.

Example 10.8 Arthur et al. (1988) (Table 10.15) use the ML factor model to price $k = 24$ agricultural assets over $T = 35$ quarterly time periods. The $k = 24$ assets are then regressed on the common factors to yield the reaction (regression) coefficients shown in Table 10.16. To facilitate interpretation the rotated varimax correlation loadings are given in Table 10.17. Using the reaction coefficients of Table 10.16 the (average) premia are $\bar{\lambda}_1 = .089$, $\bar{\lambda}_2 = .013$, $\bar{\lambda}_3 = .073$, and $\bar{\lambda}_4 = .123$. Quarterly premia for each asset can then be estimated by using the $\bar{\lambda}_j$ in Eq. (10.88)—for example, the quarterly premium for gold is

$$.12148(.089) - .04072(.013) + .02489(.073) + .02563(.123) = 0.151$$

or 1.51%, and so forth.

Table 10.15   Standardized $r = 4$ Common Factor Scores of a (35×24) Matrix of Rates of Returns for 24 Assets

Quarter	$\hat{\phi}_1$	$\hat{\phi}_2$	$\hat{\phi}_3$	$\hat{\phi}_4$
1	.	.	.	.
2	0.2206	0.6448	0.3988	0.6947
3	0.0831	−0.5684	0.0940	−1.5416
4	−0.5848	−1.5549	0.2851	−1.4378
5	−0.0691	0.4814	−0.4741	0.7246
6	−0.5386	−0.3885	0.0588	0.5358
7	−0.4713	−2.8349	0.8227	0.4169
8	−1.0202	0.6940	1.2561	−0.4904
9	−0.9484	1.1693	1.6122	1.6140
10	1.1061	1.0887	−0.0246	1.8018
11	0.9823	−1.0558	1.9362	0.2047
12	−1.4182	0.2694	0.3441	0.4935
13	0.2973	0.4914	−0.3952	1.9335
14	0.1527	1.2488	−2.0068	0.0381
15	0.6642	0.2491	2.0883	−1.3126
16	−0.6465	0.2170	0.0888	−0.3827
17	0.4279	−0.4399	−0.2547	0.0143
18	−0.4817	−0.2739	−0.2310	−1.3495
19	2.2910	1.5568	0.2729	1.3130
20	0.9592	1.4059	−0.8678	−0.6440
21	−0.5042	0.4364	−1.4520	−1.0298
22	0.1073	−0.0645	−0.9355	0.2244
23	−1.3470	−2.3355	−0.2553	0.2789
24	−1.1140	−0.2886	1.8936	−1.5368
25	−1.2706	0.0991	−0.2689	−0.4798
26	−0.3125	0.2347	−1.0074	0.9138
27	0.0289	−1.1595	−0.7505	0.0182
28	2.5445	−0.6526	0.4453	−1.2220
29	1.5062	0.1996	0.1875	0.7910
30	1.2818	0.9733	−0.4978	−0.3396
31	−0.0533	0.1469	−1.6843	−1.3172
32	−0.1132	0.5274	1.1267	−0.6952
33	−0.9482	0.2101	−0.3624	1.3805
34	−1.0461	0.4899	−0.3618	−0.2986
35	0.2351	−1.2169	−1.0808	−0.2737

Table 10.16 Estimated Reaction (Regression) Coefficients Using Standardized Common Factors and Unstandardized Dependent Variables

Asset	Factor 1	Factor 2	Factor 3	Factor 4	R^{2} [a]
1 (Gold)	0.121480[b]	−0.04072	0.02489	0.02563	.47
2 (S & P)	0.051119[b]	0.00060	−0.00538	0.00472	.83
3. (Dow-Jones)	0.058190[b]	0.00312	−0.00235	−0.00813[b]	.95
4 (NY Exchange)	0.053539[b]	−0.00036	−0.00086	−0.00548[b]	.97
5 (American Exchange)	0.078357	−0.00625	0.03969[b]	0.04535[b]	.44
6 (Municipal bonds)	0.017848	−0.01592	0.00200	−0.00370	.13
7 (Vegetables)	−0.010425	0.03610	0.00739	0.00225	.05
8 (Cotton)	0.044173[b]	0.03173[b]	−0.01563	0.01108	.34
9 (Feed & hay)	0.008098	0.07916[b]	−0.05848[b]	0.01949	.66
10 (Food grains)	0.025862	0.03764[b]	0.01562	0.01219	.27
11 (Fruit)	−0.041710	−0.01240	−0.00529	0.00292	.09
12 (Tobacco)	0.214586	−0.24347[b]	−0.15751	−0.15448	.22
13 (Dairy, eggs)	0.040689[b]	−0.00681	0.01412	0.07048[b]	.32
14 (Meat)	−0.004446[b]	0.00859[b]	−0.00138[b]	0.07578[b]	1.00
15 (Corn)	0.017764[b]	0.07848[b]	−0.01524	0.01304	.74
16 (Barley)	−0.005655[b]	0.09886[b]	0.00854[b]	−0.02264[b]	1.00
17 (Steer)	−0.002875	0.00986	−0.00672	0.05749[b]	.66
18 (Hogs)	0.003376	−0.01544	−0.00576	0.09544[b]	.53
19 (Wheat)	0.007072	0.04595[b]	0.03702[b]	−0.00032	.54
20 (Mark)	−0.003271	0.00300	0.4702[b]	0.00363	.72
21 (Swiss franc)	0.000430[b]	0.00498[b]	0.07198[b]	0.00010[b]	1.00
22 (Yen)	0.015895	0.00138	0.04448[b]	−0.01699	.32
23 (Farmland)	−0.000858	0.01011	0.01094	0.00345	.13
24 (Farmland and dividends)[c]	−0.001071	0.01033	0.01118	0.00356	.13

[a] $R^{2} = 1.00$ for cases wherein one factor nearly perfectly explains all variations in that asset (factor 2 for asset 16, factor 3 for asset 21, factor 4 for asset 14).
[b] Significant at a 90% confidence level.
[c] Annual returns to farmland included.

Source: Arthur et al., 1988; reproduced with permission.

10.7.2 Estimating Nonlinear Dimensionality: Sliced Inverse Regression

Consider a general equation of the form

$$Y = f(\boldsymbol{\beta}_1\mathbf{X}, \boldsymbol{\beta}_2\mathbf{X}, \ldots, \boldsymbol{\beta}_k\mathbf{X}, \boldsymbol{\epsilon}) \tag{10.89}$$

where Y is the dependent variable, $\mathbf{X}$ is a column vector of p predictor variables, the $\boldsymbol{\beta}_j$ are row vectors, and $\boldsymbol{\epsilon}$ is the error term. Equation (10.88) is assumed to be linear in the unknowns $\boldsymbol{\beta}_j$ but not necessarily in the predictors $\mathbf{X}$. The functional form f is unknown and we wish to estimate the dimensionality of the space generated by the values of $\boldsymbol{\beta}_j$, that is, we wish to estimate whether f is linear (unit dimensionality), quadratic (dimensionality two), or generally k dimensional. Li (1991) has termed this space as the effective dimension reduction (edr) space and has proposed a procedure to estimate it based on inverse regression. Li shows that if the p-dim column vector is standardized to zero mean and unit variance, the inverse regression

Table 10.17 Varimax Rotated Correlation Loading for $k = 24$ Assets[a]

Asset	Factor 1	Factor 2	Factor 3	Factor 4	$R^{2\,a}$
4 (New York)	97[a]	5	0	−6	
3 (Dow-Jones)	95[a]	10	−3	−9	
2 (S & P)	88[a]	9	−9	11	
1 (Gold)	62[a]	−15	15	13	
5 (American)	51[a]	3	27	31	
8 (Cotton)	43	32	−15	14	
11 (Fruit)	−27	−9	−2	0	
16 (Barley)	11	99	2	−12	
15 (Corn)	28	76[a]	−19	21	
9 (Feed and hay)	15	61[a]	−46	21	
19 (Wheat)	18	57[a]	44	5	
10 (Food grains)	29	38	12	15	
7 (Vegetables)	−2	22	3	4	
6 (Municipal bonds)	24	−23	5	−7	
12 (Tobacco)	22	−30	−16	−21	
21 (Swiss franc)	1	2	100[a]	0	
20 (Mark)	−5	2	84[a]	5	
22 (Yen)	17	−1	51[a]	−18	
23 (Farmland)	2	24	24	10	
14 (Meat)	−2	24	−4	97[a]	
17 (Steers)	0	24	−11	77[a]	
18 (Hogs)	1	0	−4	72[a]	
13 (Dairy, eggs)	27	4	9	48	

[a] Values are multiplied by 100. The square of a pattern coefficient represents the direct contribution of the factor to the variance of the variable. The sum of these squares across any variable represents the commonality or total variance explained.
[b] Values greater than 50.

Source: Arthur et al., 1988; reproduced with permission.

curve $E(\mathbf{X}/y)$ will fall into this space. The procedure is based on the idea of inverse regression whereby instead of regressing Y on $\mathbf{X}$ we regress $\mathbf{X}$ on Y, that is, we have an equation with p dependent variables and a single independent variable Y. As $\mathbf{Y}$ varies $E(\mathbf{X}/Y)$ draws out a curve called the inverse regression curve which lies in the p-dimensional space, but which also lies close to the k-dimensional subspace. A principal component analysis of $E(\mathbf{X}/Y)$ (or of the standardized version $E(\mathbf{Z}/Y)$) will then recover the subspace. Li (1991) has developed an algorithm which he calls the "sliced inverse regression" (SIR) algorithm, which provides sample estimates of the dimensionality k of the subspace. Given data $(Y_i, \mathbf{X}_i)$ $(i = 1, 2, \ldots, n)$ the algorithm operates as follows:

1. Standardize $\mathbf{X}$ to zero mean and unit variance.
2. Divide the range of Y into H partitions or "slices" $\mathbf{I}_1, \mathbf{I}_2, \ldots, \mathbf{I}_H$
3. Within each partition compute the sample mean $\bar{X}_h$ $(h = 1, 2, \ldots, H)$.

4. Carry out a principal component analysis of the weighted sample covariance matrix

$$\hat{\mathbf{V}} = \sum_{h=1}^{H} \frac{1}{n_h} \bar{\mathbf{X}}_h^{T} \bar{\mathbf{X}}_h \tag{10.90}$$

where n_h is the number of observations in partition h. Let $\mathbf{P}_k$ ($k = 1, 2, \ldots, K$) be the (row) latent vectors associated with the largest k latent roots. Then estimates of the $\boldsymbol{\beta}_j$ are given by

$$\hat{\boldsymbol{\beta}}_k = \mathbf{P}_k \hat{\boldsymbol{\Sigma}}^{-1/2} \tag{10.91}$$

where $\hat{\boldsymbol{\Sigma}}$ is the sample covariance matrix of $\mathbf{X}$. For further detail and simulated examples the reader is referred to Li (1991).

10.7.3 Factor Analysis and Simultaneous Equations Models

In this chapter we described applications of factor models when fitting a single multivariate function to a set of n observations. Other systems are also possible, whereby we have n observations for k independent variables which influence a set of p dependent variables, for example, the MANOVA model considered in Section 10.5. Here the exogenous treatment variables determine the behavior of the dependent experimental variables which are observed endogenously (internally) to the experimental system. More general models also exist whereby exogenous variables have an indirect causal effect on a set of endogenous variable. The initial model, known as path analysis, originated with Wright (1934, 1960) in the context of genetics, and seems to have been discovered independently by research workers of the Cowles Commission (see Koopmans, 1950; Malinvaud, 1966) working with econometric models, known as simultaneous equations models which deal with two sets of variates. Relations between two sets of variables can also be posited in terms of latent traits as estimated by common factors such as encountered in psychology, sociology, and other areas of the social sciences (e.g., see Jöreskog, 1979). Although such uses of factor models are of growing importance, (e.g., see Woodward et al., 1984) they lie outside the scope of the book and are not considered in the present volume.

EXERCISES

10.1 Prove that for Eq. (10.5) we have $\operatorname{var}(\boldsymbol{\delta}) = \boldsymbol{\alpha}^{T}\boldsymbol{\Psi}\boldsymbol{\alpha}$ where $\boldsymbol{\delta}$ is given by Eq. (10.4)

10.2 Using Theorem 3.3 show that the optimal latent vector $\hat{\boldsymbol{\alpha}} = \hat{\boldsymbol{\alpha}}^{*}$ corresponds to the smallest latent root of Eq. (10.7).

10.3 Show that for the normal equations (Eq. 10.7), when $\mathbf{\Psi} = c\mathbf{I}$ for some constant c the model reduces to finding orthogonal principal components of $\mathbf{\Sigma}$.

10.4 In Example 10.1, use Sprent's data to compute corrected values of $\mathbf{X}^*$ and $\mathbf{Y}^*$ and estimate the ordinary least squares line $\mathbf{Y}^* = \beta_0 + \beta_1 \mathbf{X}^*$. How does it differ from the errors-in-variables equation? The ordinary least squares line?

10.5 Let $\mathbf{b}$ be an estimator of $\mathbf{\beta}$ such that for the bias vector $B(\mathbf{b})$ we have $B(\mathbf{b}) = E(\mathbf{b}) - \mathbf{\beta} \neq \mathbf{0}$. Show that for the mean squared error criterion MSE $(\mathbf{b})$ we have

$$\text{MSE}(\mathbf{b}) = \text{var}(\mathbf{b}) + B(\mathbf{b})B(\mathbf{b})^{\text{T}}$$

where $\text{var}(\mathbf{b}) = E[\mathbf{b} - E(\mathbf{b})][\mathbf{b} - E(\mathbf{b})]^{\text{T}}$.

10.6 Show that when the last $k - r$ latent roots are set to zero the principal components Model II regression coefficients $\hat{\mathbf{\beta}}_{(r)} = \mathbf{P}_{(r)}^{\text{T}} \hat{\mathbf{\gamma}}^*$ are solutions of the following constrained minimization problem (Hocking, 1976):

$$\text{minimize } (\hat{\mathbf{\beta}} - \mathbf{\beta})^{\text{T}} \mathbf{X}^{\text{T}} \mathbf{X}(\hat{\mathbf{\beta}} - \mathbf{\beta})$$

$$\text{subject to } \mathbf{P}_{(k-r)} \mathbf{\beta} = \mathbf{0}$$

10.7 Show that θ is the angle between the y norm and the perpendicular distance to the straight line (Fig. 10.3).

10.8 Consider Equation (10.48) where $\hat{\mathbf{\gamma}}^*$ and $\hat{\mathbf{\gamma}}$ are $(k \times 1)$ vectors of regression coefficients obtained by regressing a dependent variable $\mathbf{Y}$ on unstandardized and standardized principal components, respectively. Derive the following expressions for $\hat{\mathbf{\beta}} = (\mathbf{X}^{\text{T}} \mathbf{X})^{-1} \mathbf{X}^{\text{T}} \mathbf{Y}$ where $(\mathbf{X}^{\text{T}} \mathbf{X}) = \mathbf{R}$ the sample correlation matrix of explanatory variables.
a) $\hat{\mathbf{\beta}} = \mathbf{P} \mathbf{L}^{-1} \mathbf{Z}^{*\text{T}} \mathbf{Y}$
b) $\hat{\mathbf{\beta}} = \mathbf{P} \mathbf{L}^{-1/2} \mathbf{Z}^{\text{T}} \mathbf{Y}$
c) $\hat{\mathbf{\beta}} = (\mathbf{X}^{\text{T}} \mathbf{X})^{-1} \mathbf{P} \mathbf{L} \hat{\mathbf{\gamma}}^*$

10.9 In a principal components regression model the matrix $(\mathbf{X}^{\text{T}} \mathbf{X})$ is usually taken to be the correlation matrix of the explanatory variables. Is it still valid to use the covariance matrix when carrying out a principal components regression? When would it not make sense to do so?

10.10 Compare the smoothness of the first $r = 4$ components of Tables

10.11 and 10.12 of Example 10.6 by plotting the loading coefficients on a graph.

10.11 Consider the principal components regression estimator $\tilde{\boldsymbol{\beta}}$ as defined by Equation (10.49b). Prove that $\tilde{\boldsymbol{\beta}}$ minimizes the sum of squared residuals $\phi(\mathbf{b}) = (\mathbf{Y} - \mathbf{Xb})^{\mathrm{T}}(\mathbf{Y} - \mathbf{Yb})$ for all $\mathbf{b}$ contained in the r-dimensional subspace spanned by the first r latent vector of $(\mathbf{X}^{\mathrm{T}}\mathbf{X})$ (Marquardt, 1970).

10.12 The variance/covariance matrix of the linear combinations $\mathbf{C}^{\mathrm{T}}\hat{\boldsymbol{\beta}}$ is given by

$$E[(\mathbf{C}^{\mathrm{T}}\hat{\boldsymbol{\beta}} - \mathbf{C}^{\mathrm{T}}\boldsymbol{\beta})(\mathbf{C}^{\mathrm{T}}\hat{\boldsymbol{\beta}} - \mathbf{C}^{\mathrm{T}}\boldsymbol{\beta})^{\mathrm{T}} = \mathbf{C}^{\mathrm{T}}E[(\hat{\boldsymbol{\beta}} - \boldsymbol{\beta})(\hat{\boldsymbol{\beta}} - \boldsymbol{\beta})^{\mathrm{T}}]\mathbf{C}$$

$$= \sigma^2 \mathbf{C}^{\mathrm{T}}(\mathbf{X}^{\mathrm{T}}\mathbf{X})^{-1}\mathbf{C}$$

Show, by minimizing the above expression (subject to the constraint $\mathbf{C}^{\mathrm{T}}\mathbf{C} = \mathbf{I}$), that $\mathbf{C} = \mathbf{P}_{(r)}$, the latent vectors of $(\mathbf{X}^{\mathrm{T}}\mathbf{X})^{-1}$ (i.e. latent vectors of $(\mathbf{X}^{\mathrm{T}}\mathbf{X})$) associated with the first r latent roots (see also Greenberg, 1975).

10.13 Show that for the bivariate case estimator (10.77) reduces to $\hat{\alpha}_1/\hat{\alpha}_2$.

References

Abelson, R. P. and J. W. Tukey (1959). "Efficient Conversion of Non-Metric Information into Metric Information," *Proc. Stat. Assoc., So. Stat. Sec.* 226–230.

Abelson, R. P. and J. W. Tukey (1963). "Efficient Utilization of Non-Numerical Information in Quantitative Analysis; General Theory and the Case of Simple Order," *Assoc. Math. Stat.*, **34**; 1347–1369.

Abizadeh, F., S. Abizadeh, and A. Basilevsky, (1990). "Potential for Economic Development: A Quantiative Approach," *Soc. Indic. Res.*, **22**; 97–113.

Abizadeh, S. and A. Basilevsky (1986). "Socioeconomic Classification of Countries: A Maximum Likelihood Factor Analysis Technique," *Soc. Sci. Res.*, **15**; 97–112.

Abizadeh, S. and A. Basilevsky (1990). "Measuring the Size of Government," *Public Finance*, **45**; 353–377.

Acito, F. and R. D. Anderson (1980). "A Monte Carlo Comparison of Factor Analytic Methods," *J. Mark. Res.*, **17**; 228–236.

Acito, F., R. D. Anderson, and J. L. Engledow, (1980). "A Simulation Study of Methods for Hypothesis Testing in Factor Analysis," *J. Consum. Res.*, **7**; 141–150.

Adams, E. W., R. F. Fagat and R. E. Robinson (1965). "A Theory of Appropriate Statistics," *Psychometrika*, **30**; 99–127.

Adcock, R. J. (1878). "A Problem in Least Squares" *The Analyst*, **5**; 53–54.

Adelman, I. and C. T. Morris (1965). "A Factor Analysis of the Interrelationship Between Social and Political Variables and Per Capita Gross National Product," *Q. J. Econ.*, **79**; 555–662.

Adelman, I. and C. T. Morris, (1967). Society, Politics, and Economic Development: A Quantitative Approach, John Hopkins University Press, Baltimore.

Adelman, I. and C. T. Morris (1970). "Factor Analysis and Gross National Product: A Reply," *Q. J. Econ.*, **84**; 651–662.

Ahamad, B. (1967). "An Analysis of Crimes by the Method of Principal Components," *Appl. Stat.*, **16**; 17–35.

Ahmed, N. and K. R. Rao, (1975). Orthogonal Transforms for Digital Signal Processing, Springer Verlag, New York.

Aitchison, J. (1983). "Principal Component Analysis of Compositional Data," *Biometrika*, **70**; 57–65.

Aitchison, J. (1984). "Reducing the Dimensionality of Compositional Data Sets," *Math. Geol.*, **16**; 617–635.

Aitchison, J. and J. Brown (1957). *The Log-Normal Distribution*, Cambridge University Press, Cambridge.

Aitkin, M., D. Anderson, and J. Hinde (1981). Statistical Modelling of Data on Teaching Styles," *J. R. Stat. Soc. (A)*, **144**; 419–461.

Akaike, H. (1971a). "Autoregressive Model Fitting for Control," *Ann. Inst. Statis. Math.*, **23**; 163–180.

Akaike H. (1971b). "Determination of the Number of Factors by an Extended Maximum Likelihood Principle," Research Memorandum #44, Institute Statistical Mathematics, Tokyo (unpublished manuscript).

Akaike, H. (1974a). "A New Look at the Statistical Model Identification," *IEEE Trans. Autom. Control*, **AC-19**; 716–723.

Akaike, H. (1974b). "Markovian Representation of Stochastic Processes and its Application to the Analysis of Autoregressive Moving Average Processes, *Ann. Inst. Stat. Math.*, **26**; 363–387.

Akaike, H. (1976). "Canonical Correlation Analysis of Time Series and the Use of an Information Criterion," in R. K. Mehra and D. G. Laissiotis (Eds.), '*System Identification*: *Advances and Case Studies*, Academic Press, New York.

Akaike, H. (1987). "Factor Analysis and AIC," *Psychometrika*, **52**; 317–332.

Albert, A. (1972). '*Regression and the Moore Penrose Pseudoinverse*," Academic Press, New York.

Albert, G. E. and R. L. Tittle (1967). "Nonnormality of Linear Combinations of Normally Distributed Random Variables," *American Math. Mon.*, **74**; 583–585.

Allen, S. J. and Hubbard R. (1986). "Regression Equations for the Latent Roots of Random Data Correlation Matrices with Unities in the Diagonal," *Multivar. Behav. Res.*, **21**; 393–398.

Allison, T. and D. V. Cicchetti (1976). "Sleep in Mammals: Ecological and Constitutional Correlates," *Science*, **194**; 732–734.

Amato, V. (1980). "Logarithm of a Rectangular Matrix with Application to Data Analysis," in E. Diday, L. Lebart, J. P. Pagés, and R. Tomassone (Eds.), *Data Analysis and Informatics*, North-Holland, New York.

Anderberg, M. R. (1973). *Cluster Analysis for Application*, Academic Press, New York.

Anderson, A. B., A. Basilevsky, and D. J. Hum (1983a). "Measurement: Theory and Techniques," in P. H. Rossi, J. D. Wright, and A. B. Anderson (Eds.), *Handbook of Survey Research*, Academic Press, New York, Chapter 7.

Anderson, A. B., A. Basilevsky, and D. J. Hum (1983b). "Missing Data: A Review of the Literature," in P. H. Rossi, J. D. Wright, and A. B. Anderson (Eds.), *Handbook of Survey Research*, Academic Press, New York.

Anderson, E. (1935). "The Irises of the Gaspé Peninsula," *Bull. Am. Iris Soc.*, **59**; 2–5.

Anderson, E. B. (1980). *Discrete Statistical Models with Social Science Applications*, North Holland, New York.

Anderson, J. A. (1976). "Multivariate Methods in Clinical Assessment," *J. R. Statist. Soc. (A)*, **139**; 161–182.

Anderson, N. H. (1961). "Scales and Statistics: Parametric and Non-Parametric," *Psych. Bull.*, **58**; 305–316.

Anderson, T. W. (1951a). "The Asymptotic Distribution of Certain Characteristic Roots and Vectors," Proceedings of the Second Berkeley Symposium on Mathematical Statistics and Probability, University of California Press, pp. 103–130.

Anderson, T. W. (1951b). "Estimating Linear Restrictions on Regression Co-efficients for Multivariate Normal Distributions," *Ann. Math. Stat.*, **22**; 327–351.

Anderson, T. W. (1954). "On Estimation of Parameters in Latent Structure Analysis," *Psychometrika*, **19**; 1–10.

Anderson, T. W. (1958). *An Introduction to Multivariate Statistical Analysis*, Wiley, New York.

Anderson, T. W. (1959). "Some Scaling Models and Estimation Procedures in the Latent Class Model," in U. Grenander (Ed.); Probability and Statistics: *The Harald Cramer Volume*, Wiley, New York. pp. 9–38.

Anderson, T. W. (1963a). "Asymptotic Theory for Principal Components," *Ann. Math. Stat.*, **34**; 122–148.

Anderson, T. W. (1963b). "The Use of Factor Analysis in the Statistical Analysis of Time Series, *Psychometrika*, **28**; 1–25.

Anderson, T. W. (1973). "Asymptotically Efficient Estimates of Covariance Matrices with Linear Structure," *Ann. Stat.*, **1**; 135–141.

Anderson, T. W. (1984a). *An Introduction to Multivariate Statistical Analysis*, Wiley, New York.

Anderson, T. W. (1984b). "The 1982 World Memorial Lectures: Estimating Linear Statistical Relationship," *Ann. Stat.*, **12**; 1–45.

Anderson, T. W. and D. A. Darling (1952). "Asymptotic Theory of Certain 'Goodness of Fit' Criteria Based on Stochastic Processes," *Ann. Math. Stat.*, **23**; 193–212.

Anderson, T. W. and H. Rubin (1956). "Statistical Inference in Factor Analysis," Proceedings of the Third Berkeley Symposium on Mathematical Statistics and Probability, Vol. V, J. Neyman, (Ed.), University of California, Berkeley, pp. 111–150.

Anderson-Sprecher, R. and J. Lederter (1991). "State-Space Analysis of Wildlife Telemetry Data," *J. Am. Stat. Assoc.*, **86**; 596–602.

Andrews, D. F., R. Gnanadesikan, and J. L. Warner (1971). "Transformations of Multivariate Data," *Biometrics* **27**; 825–840.

Andrews, D. F., R. Gnanadesikan, and J. L. Warner (1973). "Methods for

Assessing Multivariate Normality," in P. R. Krishnaiah (Ed.), *Multivariate Analysis-III*, Academic, New York, pp. 95–116.

Andrews, H. C. and C. L. Patterson (1976a). "Outer Product Expansions and Their Uses in Digital Image Processing," *IEEE Trans. Comput.*, **C-25**; 1–47.

Andrews, H. C. and C. L. Patterson (1976b). "Singular Value Decomposition and Digital Image Processing, *IEEE Trans. Acoust., Speech Sign. Proc.*, **ASSP-24**; 26–53.

Andrews, T. G. (1948). "Statistical Studies in Allergy," *J. Allergy*, **19**; 43–46.

Arbuckle, J. A. and M. L. Friendly, (1977). "On Rotating to Smooth Functions," *Psychometrika*, **42**; 127–140.

Archer, C. O. and R. I. Jennrich (1973). "Standard Errors for Rotated Factor Loadings," *Psychometrika*, **38**; 581–592.

Armstrong, J. S. (1967). "Derivation of Theory by Means of Factor Analysis or Tom Swift and His Electric Factor Analysis Machine," *Am. Stat.*, **21**; 17–21.

Arthur, L. M., C. A. Carter, and F. Abizadeh (1988). "Arbitrage Pricing, Capital Asset Pricing, and Agricultural Assets," *Am. J. Agric. Econ.*, **70**; 359–365.

Baker, A. de C. (1965). "The Latitudinal Distribution of Euphausia Species in the Surface Waters of the Indian Ocean," *Discovery Rep.*, **33**; 309–334.

Barlow, R. E., D. J. Bartholomew, J. M. Bremner, and H. D. Brunk (1972). *Statistical Inference Under Order Restrictions*, Wiley, New York.

Barnett, T. P. (1978). "Estimating Variability of Surface Air Temperature in the Northern Hemisphere," *Am. Meteor. Soc.*, **106**; 1353–1366.

Barnett, T. P. (1978). "Estimating Variability of Surface Air Temperature in the Northern Hemisphere," *Mon. Weather Rev.*, **106**; 1353–1367.

Barnett, V. (1975). "Probability Plotting Methods and Order Statistics," *Appl. Stat.*, **24**; 95–108.

Bartels, C. P. A. (1977). *"Economic Aspects of Regional Welfare, Income Distribution and Unemployment,"* Martin Nijhoff, Leiden.

Bartholomew, D. J. (1980). "Factor Analysis for Categorical Data,"*J. R. Stat. Soc. (B)*, **42**; 293–321.

Bartholomew, D. J. (1981). "Posterior Analysis of the Factor Model,' *Br. J. Math. Stat. Psych.*, **34**; 93–99.

Bartholomew, D. J. (1983). "Latent Variable Models For Ordered Categorical Data," *J. Econ.*, **22**; 229–243.

Bartholomew, D. J. (1984). "The Foundations of Factor Analysis," *Biometrika*, **71**; 221–232.

Bartholomew, D. J. (1987). *Latent Variable Models and Factor Analysis*, Oxford University Press, New York.

Bartlett, M. S. (1937). "The Statistical Conception of Mental Factors," *Br. J. Psych.*, **28**; 97–104.

Bartlett, M. S. (1950). "Tests of Significance in Factor Analysis," *Br. J. Psych. (Stat. Sect.)*, **3**; 77–85.

Bartlett, M. S. (1951a). The Goodness of Fit of a Single Hypothetical Discriminant Function in the Case of Several Groups," *Ann. Eugen.*, **16**; 199–214.

Bartlett, M. S. (1951b). "The Effect of Standardization on a χ^2 Approximation in Factor Analysis," *Biometrika*, **38**; 337–344.

Bartlett, M. S. (1954). "A Note on the Multiplying Factors for Various χ^2 Approximations," *J. R. Stat. Soc. (B)*, **16**; 296–298.

Basilevsky, A. (1973). "A Multivariate Analysis of Consumer Demand in the U.K., 1955–1968," Unpublished Ph.D. Thesis, University of Southampton, England.

Basilevsky, A. (1980). "The Ratio Estimator and Maximum Likelihood Weighted Least Squares Regression," *Qual. and Quant.*, **14**; 377–395.

Basilevsky, A. (1981). "Factor Analysis Regression," *Can. J. Stat.*, **9**; 109–117.

Basilevsky, A. (1983). *Applied Matrix Algebra in the Statistical Sciences*, North-Holland/Elsevier, New York.

Basilevsky, A. and R. Fenton (1992). "Estimating Demand for Nonexistent Public Commodities: To-and From Work Travel Experience of Winnipeg Commuters," *J. Urban Res.*, **1**; 1–15.

Basilevsky, A. and D. P. Hum (1979). "Karhunen–Loève Analysis of Historical Time Series with an Application to Plantation Births in Jamaica," *J. Am. Stat. Assoc.*, **74**; 284–290.

Basilevsky, A. and D. P. Hum (1984). *Experimental Social Programs and Analytic Methods*, Academic Press, Orlando.

Basilevsky, A., D. Sabourin, D. P. Hum, and A. Anderson (1985). "Missing Data Estimators in the General Linear Model: An Evaluation of Simulated Data as an Experimental Design," *Commun. Stat.-Simul. Comput.*, **14**; 371–394.

Becker, M. P. and C. C. Clogg (1989). "Analysis of Sets of Two-Way Contingency Tables Using Association Models,' *J. Am. Stat. Assoc.*, **84**; 142–151.

Behboodian, J. (1972). "A Simple Example of Some Properties of Normal Random Variables," *Am. Math. Mon.*, **79**; 632–634.

Behboodian, J. (1990). "Examples of Uncorrelated Dependent Random Variables Using a Bivariate Mixture," *Am. Stat.*, **44**; 218.

Bentler, P. M. (1982). Confirmatory Factor Analysis via Noniterative Estimation: A Fast, Inexpensive Method," *J. Mark. Res.*, **19**; 417–424.

Bentler, P. M. and S. Y. Lee (1979). "A Statistical Development of Three-Mode Factor Analysis," *Br. J. Math. Stat. Psych.*, **32**; 87–104.

Bentler, P. M. and J. S. Tanaka (1983). "Problems with EM Algorithms for ML Factor Analysis," *Psychometrika*, **48**; 247–251.

Benzecri, J. P. (1969). "Statistical Analysis as a Tool to Make Patterns Emerge from Data," in S. Watanabe (Ed.), *Methodologies of Pattern Recognition*, Academic, New York.

Benzecri, J. P. (1970). *Distance Distributionnelle et Métrique du Chi deux en Analyse Factorielle des Correspondence*, 3rd ed., Lab. Stat. Math. Fac. Sci., Paris.

Benzecri, J. P. (1973). *L'analyse des Données, Tome II, L'Analyse des Correspondences*, Dunod, Paris.

Benzecri, F. (1980). "Introduction a la Classification Automatique d'après un Exemple de Données Medicales," *Cah. Anal. Données*, **5**; 311–340.

Benzer, S. (1959). "On the Topology of the Genetic Fine Structure," *Proc. Math Acad. Sci. USA*, **45**; 1607–1624.

Benzer, S. (1961). "On the Topography of the Genetic Fine Structure," *Proc. Natl. Acad. Sci. USA*, **47**; 403–415.

Bergan, J. R. (1983). "Latent Class Models in Educational Research," E. W. Gordon (Ed.), *Review of Research in Education*, Vol. 10; American Education Research Association, 1983.

Bock, R. D. (1960). "Methods and Applications of Optimal Scaling," University of North Carolina Psychometric Laboratory Research Memorandum No. 25.

Bock, R. D. and M. Lieberman (1970). "Fitting a Response Model For n Dichotomously Scored Items," *Psychometrika*, **35**; 179–197.

Boekee, D. E. and H. H. Buss (1981). "Order Estimation of Autoregressive Models," *4th Proceedings of the Aachen Colloquium: Theory and Application of Signal Processing*, pp. 126–130.

Boneva, L. I. (1971). "A New Approach to a Problem of Chronological Seriation Associated with the Works of Plato," in F. R. Hodson, D. G. Kendall, and P. Tautu (Eds.), *Mathematics in the Archaeological and Historical Sciences*, Edinburgh University Press, Edinburgh.

Bookstein, F. L. (1989). "Size and Shape: A Comment on Semantics," *Syst. Zool.*, **38**; 173–180.

Bouroche, J. M. and A. M. Dussaix (1975). "Several Alternatives for Three-Way Data Analysis," *Metra*, **14**; 299–319.

Box, G. E. P. (1949). "A General Distribution Theory for a Class of Likelihood Criteria," *Biometrika* **36**; 317–346.

Box, G. E. P. and N. R. Draper (1987). *Empirical Model-Building and Response Surfaces*, Wiley, New York.

Box, G. E. P. and G. M. Jenkins (1976). *Time Series Analysis; Forecasting and Control*, Holden Day, San Francisco, CA.

Bozdogan, H. (1987). "Model Selection and Akaike's Information Criterion (AIC): The General Theory and its Analytical Extensions," *Psychometrika*, **52**; 345–370.

Bradu, D. and F. E. Grine (1979). "Multivariate Analysis of Diademodontine Crania from South Africa and Zambia," *South Af. J. Sci.*, **75**; 441–448.

Bravais, A. (1846). "Analyse Mathématique sur les Probabilités des Erreurs de Situation d'un Point," Mémoires Prèsentés a L'Académie Royale des Sciences de L'Institut de France, *Sci. Math. Phys.*, **9**; 255–332.

Brier, G. W. and G. T. Meltsen (1976). "Eigenvector Analysis for Prediction of Time Series," *J. Appl. Meteor*, **15**; 1307–1312.

Brillinger, D. R. (1969). "The Canonical Analysis of Stationary Time Series," P. R. Krishnaiah (Ed.), *Multivariate Analysis*, Academic, New York.

Brillinger, D. R. (1981). *Time Series: Data Analysis and Theory*, (Expanded Edition) Holden-Day 331–350.

Brillinger, D. R. and H. K. Preisler (1983). "Maximum Likelihood Estimation in a Latent Variable Problem" in S. Karlin, T. Amemiya, and L. A. Goodman (Eds.), *Studies in Econometrics, Time Series, and Multivariate Statistics*, Academic, New York.

Broffitt, J. D. (1986). "Zero Correlation, Independence and Normality," *Am. Stat.*, **40**; 276–277.

Brookins, O. T. (1970). "Factor Analysis and Gross National Product: A Comment," *Q. J. Econ.*, **84**; 648–650.

Broschat, T. K. (1979). "Principal Component Analysis in Horticultural Research," *Hort. Sci.*, **14**; 114–117.

Brown, R. L., J. Durbin, and J. M. Evans (1975). "Techniques for Testing the Constancy of Regression Relationships over Time," *J. R. Stat. Soc. (B)*, **37**; 149–192.

Browne, M. W. (1974). "Generalized Least-Squares Estimators in the Analysis of Covariance Structures," *South Afr. Stat. J.*, **8**; 1–24.

Browne, M. W. (1979). "The Maximum-Likelihood Solution in Inter-Battery Factor Analysis," *Br. Math. Stat. Psych.*, **32**; 75–86.

Browne, M. W. (1980). "Factor Analysis of Multiple Batteries by Maximum Likelihood," *Br. J. Math. Stat. Psych.*, **33**; 184–199.

Bruhn, C. M. and H. G. Schutz (1986). "Consumer Perceptions of Dairy and Related-Use Foods," *Food Technol.*, **40**; 79–85.

Butchatzch, E. J. (1947). "The Influence of Social Conditions on Mortality Rates," *Popul. Stud.*, **1**; 229–248.

Bulmer, J. T. and H. F. Shurvell (1975). "Factor Analysis as a Complement to Band Resolution Techniques. III. Self Association in Trichloroacetic Acid," *Can. J. Chem.*, **53**; 1251–1257.

Bunch, J. R. and C. P. Nielsen (1978). "Updating the Singular Value Decomposition," *Numer. Math.*, **31**; 111–129.

Bunch, J. R., C. P. Nielsen, and D. C. Sorensen (1978). "Rank-one Modification of the Symmetric Eigenproblem," *Numer. Math.*, **31**; 31–48.

Burgard, D. R., S. P. Perone, and J. L. Wiebers (1977). "Factor Analysis of the Mass Spectra of Oligodeoxyribonucleotides," *Anal. Chem.*, **49**; 1444–1446.

Burt, C. (1950). "The Factorial Analysis of Categorical Data," *Br. J. Psych. (Stat. Sec.)*, **3**; 166–185.

Burt, C. (1953). "Scale Analysis and Factor Analysis; Comments on Dr. Guttman's Paper," *B. J. Stat. Psych.*, **6**; 5–23.

Butler, J. C. (1976). "Principal Components Analysis Using the Hypothetical Closed Array," *Math. Geol.*, **8**; 25–36.

Butler, J. M., L. N. Rice, A. K. Wagstaff, and S. C. Knapp (1963). *Quantiative Naturalistic Research*, Prentice Hall, New Jersey.

Bye, B. V. and E. S. Schechter (1986). "A Latent Markov Model Approach to the Estimation of Response Errors in Multiwave Panel Data," *J. Am. Stat. Assoc.*, **81**; 375–380.

Cadzow, J. A. (1982). "Spectral Estimation: An Overdetermined Rational Model Equation Approach," *Proc. IEEE*, **70**; 907–939.

Cadzow, J. A. (1984). "ARMA Time Series Modelling: A Singular Value Decomposition Approach," in E. J. Wegman and J. G. Smith (Eds.), *Statistical Signal Processing*, Marcel Dekker, New York.

Cameron, E. M. (1968). "A Geochemical Profile of the Swan Hills Reef," *Can. J. Earth Sci.*, **5**; 287–309.

Cammarata, A. and G. K. Menon (1976). "Pattern Recognition. Classification of Therapeutic Agents According to Pharmacophones," *J. Med. Chem.*, **19**; 739–747.

Campbell, N. A. (1980). "Robust Procedures in Multivariate Analysis I: Robust Covariance Estimation," *Appl. Stat.*, **29**; 231–237.

Carroll, J. D. (1968). "Generalization of Canonical Correlation Analysis to Three or More Sets of Variables," *Proceedings, 76th American Psychology Association Convention*, **3**; 227–228.

Cartwright, H. (1986). "Color Perception and Factor Analysis," *J. Chem. Educ.*, **63**; 984–987.

Cattell, R. B. (1949). "The Dimensions of Culture Patterns by Factorization of National Characters," *J. Abnormal Normal Psych.*, **44**; 443–469.

Cattell, R. B. (1965a). "Factor Analysis: An Introduction to Essentials I. The Purpose and Underlying Models," *Biometrics*, **21**; 190–215.

Cattell, R. B. (1965b). "Factor Analysis: An Introduction to Essentials II. The Role of Factor Analysis in Research," *Biometrics*, **21**; 405–435.

Cattell, R. B. (1966). "The Scree Test for the Number of Factors," *Mult. Behav. Res.*, **1**; 245–276.

Cattell, R. B. (1978). *The Scientific Use of Factor Analysis in Behavioral and Life Sciences*, Plenum Press.

Cattell, R. B. and M. Adelson (1951). "The Dimensions of Social Change in the U.S.A. as Determined by P-technique" *Soc. Forces*, **30**; 190–210.

Cavalli-Sforza, L. L. and A. Piazza (1975). "Analysis of Evolution: Evolutionary Rates, Independence and Treeness," *Theor. Popul Biol.*, **8**; 127–165.

Chambers, J. M., W. S. Cleveland, B. Kleiner, and P. A. Tukey (1983). *Graphical Methods for Data Analysis*, Wadsworth International Group, Belmont, CA.

Chan, N. N. (1977). "On an Unbiased Predictor in Factor Analysis," *Biometrika*, **64**; 642–644.

Chang, W. C. (1983). "On Using Principal Components Before Separating a Mixture of Two Multivariate Normal Distributions," *Appl. Stat.*, **32**; 267–275.

Chatfield, C. and A. J. Collins (1980). *Introduction to Multivariate Analysis*, Chapman and Hall, New York.

Chatterjee, S. (1984). "Variance Estimation in Factor Analysis: An Application of the Bootstrap," *Br. J. Math. Stat. Psych.*, **37**; 252–262.

Chatterjee, S. and B. Price (1977). *Regression Analysis by Example*, Wiley, 1977, New York.

Chen, C. W. (1974). "An Optimal Property of Principal Components," *Communications in Statistics*, **3**; 979–983.

Christenson, A. L. and D. W. Read (1977). "Numerical Taxonomy, R-Mode Factor Analysis and Archaeological Classification," *Am. Antiq.*, **42**; 163–179.

Christofferson, A. (1975). "Factor Analysis of Dichotomized Variables," *Psychometrika*, **40**; 5–32.

Christofferson, A. (1977). "Two-Step Weighted Least Squares Factor Analysis of Dichotomized Variables," *Psychometrika*, **42**; 433–438.

Chung, C. S., M. C. W. Kau, S. C. Chung, and D. C. Rao (1977). "A Genetic and Epidemiologic Study of Periodontal Disease in Hawaii. II Genetic and Environmental Influence," *Am. J. Human Genet.*, **29**; 76–82.

Church, A. (1966). "Analysis of Data When the Response is a Curve," *Technometrics*, **8**; 229–246.

Claringbold, P. J. (1958). "Multivariate Quantal Analysis," *R. Stat. Society (B)*, **20**; 398–405.

Clarke, D. L. (Ed.) (1972). *Models in Archaeology*, Methuen, London.

Clarke, T. L. (1978). "An Oblique Factor Analysis Solution for the Analysis of Mixtures," *Math. Geol.*, **10**; 225–241.

Clarkson, D. B. (1979). "Estimating the Standard Errors of Rotated Factor Loadings by Jackknifing," *Psychometrika*, **44**; 297–314.

Cliff, N. (1962). "Analytic Rotation to a Functional Relationship," *Psychometrika*, **27**; 283–295.

Cliff, N. (1966). Orthogonal Rotation to Congruence," *Psychometrika*, **31**; 33–42.

Cliff, N. and D. J. Krus (1976). "Interpretation of Canonical Analysis: Rotated vs. Unrotated Solutions," *Psychometrika*, **41**; 35–42.

Clogg, C. C. (1981). "New Developments in Latent Structure Analysis," D. M. Jackson and E. F. Borgatta (Eds.), *Factor Analysis and Measurement in Sociological Research*, Sage, Beverely Hills, CA.

Clogg, C. C. and L. A. Goodman (1984). "Latent Structure Analysis of a Set of Multidimensional Contingency Tables," *J. Am. Stat. Assoc.*, **79**; 762–771.

Cochran, N., and H. Horne (1977). "Statistically Weighted Principal Component Analysis of Rapid Scanning Wavelength Kinetics Experiments," *Anal. Chem.*, **49**; 846–853.

Cohen, A., and R. H. Jones (1969). "Regression on a Random Field," *J. Am. Stat. Assoc.*, **64**; 1172–1182.

Comrey, A. L. (1973). *A First Course in Factor Analysis*, Academic Press, New York.

Consul, P. C. (1969). "The Exact Distributions of Likelihood Criteria for Different Hypotheses," in P. R. Krishnaiah (Ed.) *Multivariate Analysis II*, Academic, New York pp. 171–181.

Coombs, C. H. and R. C. Kao (1960). "On a Connection Between Factor Analysis and Multidimensional Unfolding," *Psychometrika*, **25**; 219–231.

Coombs, C. H. H. Raiffa, and R. M. Thrall (1954). "Some Views on Mathematical Models and Measurement Theory," *Psych. Rev.*, **61**; 132–144.

Cooper, J. C. B. (1983). "Factor Analysis: An Overview," *Am. Stat.*, **37**; 141–147.

Copas, J. B. (1972). "The Likelihood Surface in the Linear Functional Relationship Problem," *J. Roy. Stat. Soc. (B)*, **34**; 274–278.

Cox, D. R. and L. Brandwood (1959). "On a Discriminatory Problem Connected with the Works of Plato," *J. Roy. Stat. Soc. (B)*, 21; 195–200.

Cox, D. R. and N. J. H. Small (1978). "Testing Multivariate Normality," *Biometrika*, **65**; 263–272.

Craddock, J. M. (1965). "The Meteorological Application of Principal Component Analysis," *Statistician*, **15**; 143–156.

Craddock, J. M. and S. Flintoff (1970). "Eigenvector Representations of Northern Hemispheric Fields," *Q. J. R. Meteor. Soc.*, **96**; 124–129.

Craddock, J. M. and C. R. Flood (1969). "Eigenvectors for Representing the 500 mb. Geopotential Surface Over the Northern Hemisphere," *Q. J. R. Meterol. Soc.*, **95**; 576–593.

Cramér, H. (1937). "Random Variables and Probability Distributions," Cambridge Tracts in Mathematics No. 36, Cambridge University Press, England.

Creer, K. M. (1957). "The Remanent Magnetization of Unstable Keuper Marls," *Philos. Trans., R. Soc. (Great Br.)*, **250**; 130–143.

Critchley, F. (1985). "Influence in Principal Components Anaysis," *Biometrika* **72**; 627–636.

Cureton, T. K. and L. F. Sterling, (1964). "Factor Analysis of Cardiovascular Test Variables," *J. Sports Med. Phys. Fitness*, **4**; 1–24.

Curtiss, J. M. (1940). "On the Distribution of the Quotient of Two Chance Variables, *Ann. Math. Stat.*, **11**; 409–421.

Daling, J. R. and H. Tamura (1971). "Use of Orthogonal Factors for Selection of Variables in a Regression Equation—An Illustration," *Appl. Stat.*, **20**; 260–268.

Datta, A. K. and A. Ghosh (1978). "The Use of Multivariate Analysis in Classification of Regions," *Econ. Affairs*, **23**; 159–168.

Daudin, J. J. and P. Trecourt (1980). "Analyse Factorielle des Correspondences et Modèle LogLineare: Comparaison des Deux Methodes sur un Exemple," *Rev. Stat. Appl.*, **28**; 5–24.

Dauxois, J., A. Pousse, and Y. Romain (1982). "Asymptotic Theory for the Principal Component Analysis of a Vector Random Function: Some Applications to Statistical Inference," *J. Multiv. Anal.*, **12**; 136–154.

David, H. A. (1971). "Ranking the Players in a Round Robin Tournament," *Rev. Int. Stat. Inst.*, **39**; 137–146.

David, M., M. Dagbert, and Y. Beauchemin (1977). "Statistical Analysis in Geology: Correspondence Analysis Method," *Q. Colorado, School Mines*, **72**; 1–60.

Davis, A. W. (1977). "Asymptotic Theory for Principal Component Analysis: Non-normal Case," *Aust. J. Stat.*, **19**; 206–212.

Davis, R. C. (1952) "On the Theory of Prediction of Nonstationary Stochastic Processes," *J. Appl. Phys.*, **23**; 1047–1053.

Dayton, C. M. and G. B. Macready (1988). "Concomitant-Variable Latent-Class Models," *J. Am. Stat. Assoc.*, **83**; 173–178.

De Jong, F. J. (1967). *Dimensional Analysis for Economists*, North-Holland, Amsterdam.

deLigny, C. L., G. H. E. Nieuwdorp, Brederode, W. K. and W. E. Hammers (1981). "An Application of Factor Analysis with Missing Data," *Technometrics*, **23**; 91–95.

Demster, A. P. (1963). "Stepwise Multivariate Analysis of Variance Based on Principle Variables," *Biometrics*, **19**; 478–490.

Demster, A. P., N. M. Laird, and D. B. Rubin (1977). "Maximum Likelihood from Incomplete Data via the EM Algorithm," *J. R. Stat. Soc. (B)*, **39**; 1–38.

Dent, B. (1935). "On Observations of Points Connected by Linear Relation," *Proc. Phys. Soc.*, **47**; 92–106.

Derflinger, G. (1979). "A General Computing Algorithm for Factor Analysis," *Biometrica J.*, **24**; 25–38.

de Sarbo, S., R. E. Hausman, S. Lin, and W. Thomspon (1982). "Constrained Cannonical Correlation," *Psychometrika*, **47**; 489–516.

Deville, J. C. and E. Malinvaud (1983). "Data Analysis in Official Socio-Economic Statistics," *J. R. Stat. Soc. (A)*, **146**; 335–361.

Devlin, S. J., R. Gnanadesikan, and J. R. Kettenring (1981). "Robust Estimation of Dispersion Matrices and Principal Components," *J. Am. Stat. Assoc.*, **76**; 354–362.

Diaconis, P. and B. Efron (1983). "Computer-Intensive Methods in Statistics," *Sci. Am.*, **248** (May); 116–130.

Dobson, R., T. F. Golob, and R. L. Gustafson (1974). "Multidimensional Scaling of Consumer Preferences for a Public Transportation System: An Application of Two Approaches, *Socio-Econ. Plann. Sci.*, **8**; 23–36.

Dolby, G. R. and S. Lipton (1972). "Maximum Likelihood Estimation of the General Nonlinear Functional Relationship with Replicated Observations and Correlated Errors," *Biometrika*, **59**; 121–129.

Doran, H. E. (1976). "A Spectral Principal Components Estimator of the Distributed Lag Model," *Int. Econ. Rev.*, **17**; 8–25.

Doran, J. E. and F. R. Hodson (1975). *Mathematics and Computers in Archaeology*, Harvard University Press, Cambridge MA.

Drury, G. G. and E. B. Daniels (1980). "Predicting Bicycle Riding Performance Under Controlled Conditions," *J. Saf. Res.*, **12**; 86–95.

Dudzinski, M. L., J. T. Chmura, and C. B. H. Edwards (1975). "Repeatability of Principal Components in Samples: Normal and Non-Normal Data Compared," *Multivar. Behav. Res.*, **10**; 109–118.

Duewer, D. L., B. R. Kowalski, K. J. Clayson, and R. J. Roby (1978). "Elucidating the Structure of Some Chemical Data," *Comput. Biomed. Res.*, **11**; 567–580.

Dumitriu, M., C. Dumitriu, and M. David (1980). "Typological Factor Analysis: A New Classification Method Applied to Geology," *Math. Geol.*, **12**; 69–77.

Dunn, W. J. and S. Wold (1978). "Structure-Activity of β-Adrenergic Agents Using the SIMCA Method of Pattern Recognition," *J. Med. Chem.*, **21**; 922–930.

Dunn, W. J. and S. Wold (1980). "Relationships Between Chemical Structure and Biological Activity Modeled by SIMCA Pattern Recognition," *Biorg. Chem.*, **9**; 505–523.

Durbin, J. (1984). "Present Position and Potential Developments: Some Personal Views," *Time Series Anal.*, *J.R. Stat. Soc.* (A): 161–173.

Durbin, J. and M. Knott (1972). "Components of Cramér-Von Mises Statistics I," *J. R. Stat. Soc.*, **34**; 290–307.

Dwyer, P. S. (1967). "Some Applications of Matrix Derivatives in Multivariate Analysis," *J. Am. Stat. Assoc.*, **62**; 607–625.

Dwyer, P. S. and M. S. MacPhail (1948). "Symbolic Matrix Derivatives," *Ann. Math. Stat.*, **19**; 517–534.

Dyer, T. G. J. (1981). "On a Technique to Investigate the Covariation Between Streamflow and Rainfall in Afforested Catchments," *Ecol. Model.*, **13**; 149–157.

Eastment, H. T. and W. J. Krzanowski (1982). "Cross-Validatory Choice of the Number of Components From a Principal Comonent Analysis," *Technometrics*, **24**; 73–77.

Easton, G. S. and R. E. McCulloch (1990). "A Multivariate Generalization of Quartile-Quartile Plots," *J. Am. Stat. Assoc.*, **85**; 376–386.

Eckart, C. and G. Young (1936). "The Approximation of One Matrix by Another of Lower Rank," *Psychometrika*, **1**; 211–218.

Efron, B. (1979). "Bootstrap Methods: Another Look at the Jackknife," *Ann. Stat.*, **7**; 1–26.

Efron, B. (1981). "Nonparametric Estimates of Standard Error: The Jackknife, the Bootstrap, and Other Methods," *Biometrika* **68**; 589–599.

Ehrenberg, A. S. C. (1962). "Some Questions About Factor Analysis," *The Statistician*, **12**; 191–208.

Elffers, H., J. Bethlehem, and R. Gill (1978). Indeterminacy Problems and the Interpretation of Factor Analysis Results," *Stat. Neerandica*, **32**; 181–199.

Engle, R. and M. Watson (1981). "A One-Factor Multivariate Time Series Model of Metropolitan Wage Rates," *J. Am. Stat. Assoc.*, **76**; 774–781.

Erez, J. and D. Gill (1977). "Multivariate Analysis of Biogenic Constituents in Recent Sediments Off Ras Burka, Gulf of Elat, Red Sea," *Math. Geol.*, **9**; 77–98.

Escofier, B. (1979). "Traitement Simultané de Variables Qualitatives et Quantitatives en Analyse Factorielle,' *Cah. Anal. Donnés*, **4**; 137–146.

Escoufier, Y. (1980a). "Exploratory Data Analysis When Data are Matrices," *Recent Developments in Statistical Inference and Data Analysis*, K. Matusita (Ed.), North-Holland, New York.

Escoufier, Y. (1980b). "L'analyse Conjointe de Plusieurs Matrices de Donnees," *Biometrie Temps*, 59–76.

Eubank, R. L. and J. T. Webster (1985). "The Singular Value Decomposition as a Tool for Solving Estimability Problems," *Am. Stat.*, **39**; 64–66.

Eysenck, H. J. (1951). "Primary Social Attitudes as Related to Social Class and Political Party," *Br. J. Soc.*, **2**; 198–209.

Eysenck, H. J. (1952). "Uses and Abuses of Factor Analysis," *Appl. Stat.*, **1**; 45–49.

Falkenhagen, E. R. and S. W. Nash (1978). "Multivariate Classification in Provenance Research," *Silvae Genet.*, **27**; 14–23.

Fomby, T. B. and R. C. Hill (1978). "Deletion Criteria for Principal Components Regression Analysis," *Am. J. Agric. Econ.*, **60**; 524–527.

Fomby, T. B. and R. C. Hill (1979). "Multicollinearity and the Value of A-Priori Information," *Commun. Stat.—Theor. Meth.*, **A8(5)**; 477–486.

Farmer, S. A. (1971). "An Investigation into the Results of Principal Components Analysis of Data Derived from Random Numbers," *Statistician*, **20**; 63–72.

Farrar, D. E. and R. R. Glauber (1967). "Multicollinearity in Regression Analysis: The Problem Revisited," *Rev. Econ. Stat.*, **49**; 92–107.

Fearn, T. (1983). "A Misuse of Ridge Regression in the Calibration of a Near Infrared Reflectance Instrument," *Appl. Stat.*, **32**; 73–79.

Fielding, A. (1978). "Latent Structure Models," in C. A. O'Muircheartaigh and C. Payne (Ed.), *The Analysis of Survey Data*, Vol. 1, Wiley, New York.

Fieller, E. C. (1932). "The Distribution of the Index in a Normal Bivariate Population," *Biometrika* **24**; 428–442.

Fiering, M. F. (1964). "Multivariate Technique for Synthetic Hydrology," *J. Hydrol. Div.*, American Society of Civil Engineers, HY5.

Finn, J. D. (1974). *A General Model for Multivariate Anaysis*, Holt, Rinehart and Winston, New York.

Fisher, R. (1953). "Dispersion on a Sphere," *Proc. R. Soc. (GB)*, **217**; 295–305.

Fisher, R. A. (1936). "The Use of Multiple Measurements in Taxonomic Problems," *Ann. Eugen.*, **7**; 179–188.

Fisher, R. A. (1940). "The Precision of Discriminant Functions," *Ann. Eugen.*, **10**; 422–429.

Fisher, R. A. and W. A. Mackenzie (1923). "Studies in Crop Variation. II. The Manurial Response of Different Potato Varieties," *J. Agric. Sci.*, **13**; 311–320.

Fleiss, J. L. (1973). *Statistical Methods for Rates and Proportions*, Wiley, New York.

Fletcher, R., and M. J. D. Powell (1963). "A Rapidly Convergent Descent Method for Minimization," *Comput. J.*, **2**; 163–168.

Flores, R. M. and G. L. Shideler (1978)., "Factors Controlling Heavy-Mineral Variations on the South Texas Outer Continental Shelf, Gulf of Mexico," *J. Sediment. Pet.*, **48**; 269–280.

Flury, B., (1984). "Common Principal Components in k Groups," *J. Am. Stat. Assoc.* **79**; 892–898.

Flury, B. (1986a). "Asymptotic Theory for Common Principal Component Analysis," *Ann. Stat.*, **14**; 418–430.

Flury, B. (1986b). "Proportionality of k Covariance Matrices," *Stat. Probabl. Lett.*, **4**; 29–33.

Flury, B. (1988). *"Common Principal Components and Related Multivariate Models,"* Wiley, New York.

Fomby, T. B., R. C. Hill, and S. R. Johnson (1978). "An Optimal Property of Principal Components in the Context of Restricted Least Squares," *J. Am. Stat. Assoc.*, **73**; 191–193.

Fordham, B. G. and G. D. Bell (1978). "An Empirical Assessment and Illustration of Some Multivariate Morphometric Techniques," *Math. Geol.*, **10**; 111–139.

Fordham, B. G. and G. D. Bell (1978). "An Empirical Assessment and Illustration of Some Multivariate Morphometric Techniques," *Math. Geol.*, **10**; 111–139.

Formann, A. K. (1982). "Linear Logistic Latent Class Analysis," *Biom. J.*, **24**; 171–190.

Francis, I. (1974). "Factor Analysis: Fact or Fabrication," *Math. Chron.*, **3**; 9–44.

Francis, I. and N. Lauro (1982). "An Analysis of Developer's and User's Ratings of

Statistical Software Using Multiple Correspondence Analysis, *Compstat*; *Proc. Comp. Stat.*, **5**; 212–217.

Frane, J. W. and M. HIll (1976). "Factor Analysis as a Tool for Data Analysis," *Commun. Stat. Theory Methods*, **A5** (6); 487–506.

Fréchet, M. (1951). Généralization de la lai de Probabilité de Laplace," *Ann. Inst. Henri Poincaré*, **13**; 1–29.

Friedman, J. H. and J. W. Tukey (1974). "A Projection Pursuit Algorithm for Exploratory Data Analysis," *IEEE Trans. Comput.*, **C-23**; 881–889.

Frisch, R. (1929). "Correlation and Scatter in Statistical Variables," *Nordic Stat. J.*, **8**; 36–102.

Fujikoshi, Y. (1977). "Asymptotic Expansions for the Distributions of Some Multivariate Tests," in P. R. Krishnaiah, (Ed.), *Multivariate Analysis* Vol. 4, North Holland, New York, pp. 55–71.

Fukunaga, K. (1990). *Introduction to Statistical Pattern Recognition*, Academic, New York.

Fulkerson, D. R. and O. A. Gross (1965). "Incidence Matrices and Interval Graphs," *Pacific J. Math.*, **15**; 835–855.

Full, W. E., R. Ehrlich, and J. E. Klovan (1981), "Extended Q Model-Objective Definition of External End Members in the Analysis of Mixtures," *Math. Geol.*, **13**; 331–344.

Fuller, W. A. (1976). *Introduction to Statistical Time Series*, Wiley, New York.

Gabriel, K. R. (1971). "The Biplot Graphic Display of Matrices with Application to Principal Component Analysis," *Biometrika*, **58**; 453–467.

Gabriel, K. R. (1978). "Least Squares Approximation of Matrices by Additive and Multiplicative Models," *J. R. Stat. Soc. (B)*, **40**; 186–196.

Gabriel, K. R. and M. Haber (1973). "The Moore–Penrose Inverse of a Data Matrix—A Statistical Tool With Some Meteorological Applications," Third Conference of Probability and Statistics in Atmospheric Science; Boulder, CO (American Meteorological Society), pp. 110–117.

Gabriel, K. R. and C. L. Odoroff (1983). "Resistant Lower Rank Approximation of Matrices," in J.E. Gentle (Ed), *Computer Science and Statistics: The Interface*, North Holland, New York.

Gabriel, K. R. and S. Zamir (1979). "Lower Rank Approximation of Matrices by Least Squares With Any Choice of Weights," *Technometrics* **21**; 489–498.

Gaito, J. (1980). "Measurement Scales and Statistics: Resurgence of an Old Misconception," *Psych. Bull.*, **87**; 564–567.

Garrison, W. L. and D. F. Marble (1963). "Factor Analytic Study of the Connectivity of a Transportation Network," *Regional Science Association*, Papers XII: Lund Congress.

Gauch, H. G., R. H. Whittaker, and T. R. Wentworth (1977). "A Comparative Study of Reciprocal Averaging and Other Ordination Techniques," *J. Ecol.*, **65**; 157–174.

Geary, R. C. (1930). "The Frequency Distribution of the Quotient of Two Normal Variables," *J. R. Stat. Soc.*, **93**; 442–446.

Ganin, Y. (1977), "Further Comments in the Derivation of Kalman Filters, Section

II: Gaussian Estimates and Kalman Filtering," in T. Kailath (Ed.), *Linear Least-Squares Estimation*, Dowden, Hutchinson and Ross, Straudsburg, PA, p. 281–289.

Gerbrands, J. J. (1981). "On the Relationships Between SVD, KLT and PCA," *Pattern Recogn.*, **14**; 375–381.

Geweke, J. F. (1977). "The Dynamic Factor Analysis of Economic Time-Series Models" in D. J. Aigner and A. S. Goldberger (Eds.), Latent Variables in Socio-Economic Models, North Holland, New York, pp. 365–383.

Geweke, J.F. and K. J. Singleton (1980). "Interpreting the Likelihood Ratio Statistic in Factor Models When Sample Size is Small," *J. Am. Stat. Assoc.*, **75**; 133–137.

Geweke, J. F. and K. J. Singleton (1981). "Maximum Likelihood 'Confirmatory' Factor Analysis of Economic Time Series," *Int. Econ. Rev.*, **22**; 37–54.

Gibbons, J. D. (1971). *Nonparametric Statistical Inference*, McGraw Hill, New York.

Gibson, A. R., A. J. Baker, and A. Moeed (1984). "Morphometric Variation in Introduced Populations of the Common Myna (acridotheres tristis): An Application of the Jackknife to Principal Components Analysis," *Syst. Zool.*, **33**; 408–421.

Gibson, W. A. (1955). "An Extension of Anderson's Solution for the Latent Structure Equations," *Psychometrika*, **20**; 69–73.

Gilula, A. and S. J. Haberman (1986). "Canonical Analysis of Contingency Tables by Maximum Likelihood," *J. Am. Stat. Assoc.*, **81**; 780–788.

Gilula, Z. (1986). "Grouping and Association in Contingency Tables: An Exploratory Canonical Correlation Approach, *J. Am. Stat. Assoc.*, **81**; 773–779.

Gini, C. (1921). "Sull' interpolazione di Una Retta Quando i Valori della Variable Indipendente Sono affetti da Errori Accidentali," *Metron*, **1**; 63–82.

Girshick, M. A. (1936). "Principal Components," *J. Am. Stat. Assoc.*, **31**; 519–528.

Girschick, M. A. (1939). "On the Sampling Theory of Roots of Determinantal Equations," *Ann. Math. Stat.*, **10**; 203–224.

Gittins, R. (1969). "The Appliction of Ordination Techniques," in I. H. Rorison (Ed.), *Ecological Aspects of the Mineral Nutrition of Plants*, British Ecological Society Symposium 9, Blackwell Scientific Publishers, Oxford, pp. 37–66.

Glahn, H. R. (1968). "Canonical Correlation and its Relationship to Discriminant Analysis and Multiple Regression, *J. Atmos. Sci.*, **25**; 23–31.

Glynn, W. J. and R. J. Muirhead, (1978). "Inference in Canonical Correlation Analysis," *J. Multiv. Anal.*, **8**; 468–478.

Gnanadesikan, R. (1977). *Methods of Statistical Data Analysis of Multivariate Observations*, Wiley, New York.

Gnanadesikan, R. and J. R. Kettenring (1972). "Robust Estimates, Residuals, and Outlier Detection with Multiresponse Data," *Biometrics*, **28**; 81–124.

Gold, R. J. M., H. S. Tenenhouse, and L. S. Adler (1976). "A Method for Calculating the Relative Protein Contents of the Major Keratin Components from their Amino Acid Composition," *Biochem.*, **159**; 157–160.

Goldberger, A. S. (1964). *Econometric Theory*, Wiley, New York.

Goldman, K. (1971). "Some Archaelogical Criteria for Chronological Seriation," in

F. R. Hodson, D. G. Kendall and P. Tautu, (Eds.), *Mathematics in the Archaelogical and Historical Sciences*, Edinburgh University Press, Edinburgh.

Goldstein, M. (1982). "Preliminary Inspection of Multivariate Data," *Am. Stat.*, **30**; 358–362.

Good, I. J. (1969). "Some Applications of the Singular Decomposition of a Matrix," *Technometrics*, **11**; 823–831.

Goodall, D. W. (1954). "Objective Methods for Classification of Vegetation," *Aust. J. Botany*, **2**; 304–324.

Goodman, L. A. (1974a). "The Analysis of Systems of Qualitative Variables when some of the Variables are Unobservable Part I—A Modified Latent Structure Approach," *Am. J. Sociol*, **79**; 1179–1259.

Goodman, L. A. (1974b). "Exploratory Latent Structure Analysis Using both Identifiable and Unidentifiable Models," *Biometrika*, **61**; 215–231.

Goodman, N. R. (1963). "Statistical Analysis Based on a Certain Multivariate Complex Gaussian Distribution (an Introduction)," *Ann. Math. Stat.*, **34**; 152–177.

Gordon, I. R. and R. M. Whittaker (1972). "Indicators of Local Prosperity in the South West Region," *Reg. Stud.*, **6**; 299–313.

Gorman, B. S. (1976). "Principal Components Analysis as an Alternative to Kendall's Coefficient of Concordance W," *Educ. Psych. Measure*, **36**; 627–629.

Gorsuch, R. L. (1974). *Factor Analysis*, Saunders, Toronto.

Gould, J. (1981). *The Mismeasure of Man*, Norton, New York.

Gould, P. R. (1967). "On the Geographical Interpretation of Eigenvalues," *Trans. Inst. Br. Geog.*, **42**; 53–86.

Gower, J. C. (1966). "Some Distance Properties of Latent Root and Vector Methods Used in Multivariate Analysis," *Biometrika*, **53**; 325–338.

Gower, J. C. (1967). "Multivariate Analysis and Multidimensional Geometry," *Statistician*, **17**; 13–28.

Graham, A. (1981). *"Kronecker Products and Matrix Calculus with Applications,"* Ellis Horwood, Chichester.

Graham, I., P. Galloway, and I. Scollar (1976). "Model Studies in Computer Seriation," *J. Arch. Sci.*, **3**; 1–31.

Graybill, F. A. (1983). *"Introduction to Matrices with Applications in Statistics,"* Wadsworth, Belmont, CA.

Green, B. F. (1951). "A General Solution for the Latent Class Model of Latent Structure Analysis," *Psychometrika*, **16**; 151–166.

Greenacre, M. J. and L. G. Underhill (1982). "Scaling a Data Matrix in Low-Dimensional Euclidean Space," in D. M. Hawkins, (Ed.), *Topics in Applied Multivariate Analysis*, Cambridge University Press, Cambridge.

Greenacre, M. and Hastie, T. (1987). "The Geometric Interpretation of Correspondence Analysis," *J. Am. Stat. Assoc.*, **82**; 437–447.

Greenacre, M. J. (1984). *Theory and Applications of Correspondence Analysis*, Academic, London.

Greenacre, M. J. and L. Degos (1977). "Correspondence Analysis of HLA Gene Frequency Data from 124 Population Samples," *Am. J. Human Genet.*, **29**; 60–75.

Greenberg, E. (1975). "Minimum Variance Properties of Principal Component Regression," *J. Am. Stat. Association*, **70**; 194–197.

Gregg, L. W. and R. G. Pearson (1961). "Factorial Structure of Impact and Damage Variables in Lightplane Accidents," *Hum. Factors*, **3**; 237–244.

Grenander, V. and G. Szego (1958). *Toeplitz Forms and their Applications*, University of California Press, Berkeley.

Gunst, R. F. (1983). "Regression Analysis With Multicollinear Predictor Variables: Definition, Detection, and Effects," *Commun. Stat. Theor. Meth.*, **12**; 2217–2260.

Gunst, R. F., J. T. Webster, R. L. Mason (1976). "A Comparison of Least Squares and Latent Root Regression Estimators," *Technometrics*, **18**; 74–83.

Gupta, R. P. (1965). "Asymptotic Theory for Principal Components Analysis in the Complex Case," *J. Indian Stat. Assoc.*, **3**; 97–106.

Guttman, I., D. Y. Kim, and I. Olkin (1985). "Statistical Inference for Constants of Proportionality," in P. R. Krishnaiah (Ed.), *Multivariate Analysis VI: Proceedings of the Sixth International Symposium on Multivariate Analysis*, North Holland, New York.

Guttman, L. (1941). "The Quantification of a Class of Attributes: A Theory and Method of Scale Construction," in P. Host (Eds.), *The Prediction of Personal Adjustment*, Social Science Research Council, New York.

Guttman, L. (1946). "An Approach for Quantifying Paired Comparisons and Rank Order," *Ann. Math. Stat.*, **17**; 144–163.

Guttman, L. (1950). "The Principal Components of Scale Analysis," in Stoufer et al. (Eds.), *Measurement and Prediction*, Princeton University Press, Princeton, NJ.

Guttman, L. (1953). "A Note on Sir Cyril Burt's Factorial Analysis of Qualitative Data," *Br. J. Stat. Psych.*, **6**; 1–4.

Haas, G., L. Bain, and C Antle (1970). "Inference for the Cauchy Distribution Based on Maximum-Likelihood, Estimators," *Biometrika*, **57**; 403–408.

Haberman, S. J. (1973). "The Analysis of Residuals in Cross-Classified Tables," *Biometrics*, **29**; 205–220.

Haberman, S. J. (1979). *Analysis of Qualitative Data*," Volume 2—New Developments, Academic, New York.

Haitovsky, Y. (1966). "A Note on Regression on Principal Components," *Am. Stat.*, **20**; 28–29.

Hamdan, M. A. (1970). "The Equivalence of Tetrachoric and Maximum Likelihood Estimates of ρ in 2×2 Tables," *Biometrika*, **57**; 212–215.

Hammarling, S. J. (1970). "Latent Roots and Latent Vectors," University of Toronto Press, Toronto.

Hannan, E. J. and B. G. Quinn (1979). "The Determination of the Order of an Autoregression," *J. Am. Stat. Assoc. (B)* **41**; 190–195.

Hanson, R. J. and M. J. Norris (1981). "Analysis of Measurements Based on the Singular Value Decomposition," *SIAM J. Sci. Stat. Comput.*, **2**; 363–373.

Harbaugh, J. W. and F. Demirmen (1964). "Application of Factor Analysis to Petrologic Variations of Americas Limestone (Lower Permian), Kansas and

Oklahoma," State Geological Survey, University of Kansas, Publication No. #15.

Hardy, D. M. and J. J. Walton (1978). "Principal Components Analysis of Vector Wind Measurements," *Appl. J. Meteorol.*, **17**; 1153–1162.

Harman, H. H. (1967). *Modern Factor Analysis*, University of Chicago Press, Chicago.

Harrison, P. J. and C. F. Stevens (1976). "Bayesian Forecasting," *J. R. Stat. Soc. (B)*, **38**; 205–247.

Hashiguchi, S. and J. Morishima (1969). "Estimation of Genetic Contribution of Principal Components to Individual Variates Concerned," *Biometrics* **25**; 9–15.

Hastie, T. and W. Stuetzle (1989). "Principal Curves," *J. Am. Stat. Assoc.*, **84**; 502–516.

Hatcher, R. E. M., A. M. Hedges, A. M. Pollard, and P. M. Kenrick (1980). "Analysis of Hellenistic and Roman Fine Pottery from Benghazi," *Archaeometry*, **22**; 133–151.

Hatheway, W. H. (1971). "Contingency-Table Analysis of Rain Forest Vegetation," in G. P. Patil, E. C. Pielou, and W. E. Watson (Eds.), Statistical Ecology, Vol. 3, Penn. State University, University Park, PA.

Hathout, S. and D. Hiebert (1980). "Quantitative Methods in Land Evaluation for Wheat Production in West Kilimanjaro, Tanzania," *Singapore J. Trop. Geog.*, **1**; 47–54.

Hawkins, D. M. (1973). "On the Investigation of Alternative Regression by Principal Components Analysis," *Appl. Stat.*, **22**; 275–286.

Hawkins, D. M. (1974). "The Detection of Errors in Multivariate Data Using Principal Components, *J. Am. Stat. Assoc.*, **69**; 340–344.

Hay, A (1975). "On the Choice of Methods in the Factor Analysis of Connectivity Matrices: A Comment," *Trans. Inst. Br. Geogr.*, **66**; 163–167.

Healy, M. J. R. (1968). "Multivariate Normal Plotting," *Appl. Stat.*, **17**; 157–161.

Healy, M. J. R. and H. Goldstein (1976). "An Approach to the Scaling of Categorized Attributes," *Biometrika*, **63**; 219–229.

Hearnshaw, L. S. (1979). *Cyril Burt Psychologist*, London, Hodder and Staughton.

Heeler, R. M., T. W. Whipple, and T. P. Hustad (1977). "Maximum Likelihood Factor Analysis of Attitude Data," *J. Marketing Res.*, **14**; 42–51.

Henry, N. W. (1974). *"Latent Structure Analysis of a Fatalism Scale,"* Gary Income Maintenance Experiment, Purdue University.

Henry, N. W. (1975). *"Maximum Likelihood Estimation of Parameters of Latent Class Models,"* Gary Income Maintenance Experiment, Purdue University,

Henry, R. C. (1977). "The Application of Factor Analysis to Urban Aerosol Source Identification," *5th Conference on Probability and Statistics in Atmospheric Science*, pp. 134–138.

Henrysson, S. (1962). "The Relation Between Factor Loadings and Biserial Correlations in Item Analysis," *Psychometrika*, **27**; 419–424.

Herson, J. (1980). "Evalution of Toxicity: Statistical Considerations," *Cancer Treat. Rep.*, **54**; 463–468.

Higman, B. (1964). *Applied Group-Theoretic and Matrix Methods*, Dover Publishing, New York (originally published by Oxford University Press, 1955).

Hill, I.D. (1974). "Association Football and Statistical Inference," *Appl. Stat.*, **23**; 203–208.

Hill, M. D. (1982). *Encyclopedia of Statistical Sciences*, Vol. 2, in S. Katz and L. Johnson (Eds.), Wiley, New York, pp. 204–210.

Hill, M. O. (1973). "Reciprocal Averaging: An Eigenvector Method of Ordination," *J. Ecol.*, **61**; 237–249.

Hill, M. O. and H. G. Gauch (1980). "Detrended Correspondence Analysis: An Improved Ordination Technique," *Vegetatio*, **42**; 47–58.

Hill, M. O. and A. J. E. Smith (1976) "Principal Component Analysis of Taxonomic Data with Multi-Stage Discrete Characters," *Taxon*, **25**; 249–255.

Hills, M. (1977). "Book Review," *Appl. Stat.*, **26**; 339–340.

Hinkley, D. V. (1969). "On the Ratio of Two Correlated Normal Random Variables," *Biometrika*, **56**; 635–639.

Hirschfeld, H. O. (1935). "A Connection Between Correlation and Contingency," *Cambridge Philos. Soc. Proc.*, **31**; 520–524.

Hitchon, B., G. K. Billings, and J. E. Klovan (1971). "Geochemistry and Origin of Formation Waters in the Western Canada Sedimentary Basin-III. Factors Controlling Chemical Composition," *Geochim. Cosmochim. Acta.*, **35**; 567–598.

Hocking, R. R. (1976). "The Analysis and Selection of Variables in Linear Regression," *Biometrics*, **32**; 1–49.

Hocking, R. R. and O. J. Pendleton (1983). "The Regression Dilemma," *Comm. Stat.-Theory Meth.*, **12**; 497–527.

Hoffman, D. L. (1991). "Review of Four Correspondence Analysis Programs for the IBM PC," *Am. Stat.*, **45**; 305–310.

Hohn, M. E. (1979). "Principal Components Analysis of Three-Way Tables," *Math. Geol.*, **11**; 611–626.

Hooper, J. W. (1959). "Simultaneous Equations and Canonical Correlation Theory," *Econometrika*, **27**; 245–256.

Hopkins, J. W. (1966). "Some Considerations in Multivariate Allometry," *Biometrics*, **22**; 747–760.

Horn, J. L. (1965). "A Rationale and Test for the Number of Factors in Factor Analysis," *Psychometrika*, **30**; 179–185.

Horst, P. (1965). *Factor Analysis of Data Matrices*, Holt, Rinehart and Winston, New York.

Hotelling, H. (1933). "Analysis of a Complex of Statistical Variables into Principal Components," *J. Educ. Psych.*, **24**; 417–441, 498–520.

Hotelling, H. (1936a). "Simplified Computation of Principal Components," *Psychometrika*, **1**; 27–35.

Hotelling, H. (1936b). "Relations Between Two Sets of Variates," *Biometrika*, **28**; 321–377.

Howery, D. G. (1976). "Factor Anlayzing the Multifactor Data of Chemistry," *Int. Lab.*, March/April; 11–21.

Huang, T. S. and P. M. Narendra (1975). "Image Restoration by Singular Value Decomposition," *Appl. Opt.*, **9**; 2213–2216.

Huber, P. J. (1985). "Projection Pursuit," *Ann. Stat.*, **13**; 435–475.

Hudson, C. B. and R. Ehrlich (1980). "Determination of Relative Provenance Contributions in Samples of Quartz Sand Using a Q-Mode Factor Analysis of Fourier Grain Shape Data," *J. Sediment. Pet.*, **50**; 1101–1110.

Hunt, B. R. (1973). "The Application of Constrained Least Squares Estimation of Image Restoration by Digital Computer, *IEEE Trans. Comput.*, **C-22**; 805–812.

Hunter, A. A. and A. H. Latif (1973). "Stability and Change in the Ecological Structure of Winnipeg: A Multi-Method Approach," *Can. Rev. Soc. Anth.*, **10**; 308–333.

Hurewicz, W. and H. Wallman (1974). *Dimension Theory*, Princeton University Press, Princeton, NJ.

Hutton, H. M., K. R. Kunz, J. D. Bozek, and B. J. Blackburn (1986). "Determination of Substituent Effects by Factor Analysis and Multiple Linear Regression for the Carbon-13 Nuclear Magnetic Resonance Chemical Shifts in 4-Substituted Phenols and 2-Nitrophenols," *Can. J. Chem.*, **65**; 1316–1321.

Imbrie, J. (1963). "Factor and Vector Analysis Programs for Analyzing Geologic Data," *Technical Report No. 6*, Office of Naval Research, Geography Branch, Northwestern University, Evanson, Ill.

Imbrie, J. and N.G. Kipp (1971). "A New Micropaleontological Method of Quantitative Paleo-Climatology: Application to a Late Pleistocene Carribean Core," in K.K. Turekian (Ed.), *The Late Cenozoic Glacial Ages*, Yale University Press, New Haven, CT.

Imbrie, J. and and E. G. Purdy (1962). "Classification of Modern Bahamian Carbonate Sediments," *Memoirs, Am. Assoc. Pet. Geol.*, **1**; 253–272.

Ipsen, D. C. (1960). *Units, Dimensions, and Dimensionless Numbers*, McGraw-Hill, New York.

Isogawa, Y. and M. Okamato (1980). "Linear Prediction in the Factor Analysis Model," *Biometrika*, **67**; 482–484.

Jackson, J. E. (1959). "Quality Control Methods for Several Related Variables," *Technometrics* **1**; 359–377.

Jackson, J. E. (1981). "Principal Components and Factor Analysis: Part II—Additional Topics Related to Principal Components," *J. Qual. Technol.*, **13**; 46–58.

Jackson, J. E. (1991). *A User's Guide to Principal Components*, Wiley-Interscience, New York.

Jackson, J.E. and F. T. Hearne (1973). "Relationships Among Coefficients of Vectors Used in Principal Components," *Technometrics*, **15**; 601–610.

Jackson, J.E. and G.S. Mudholkar (1979). "Central Procedures for Residuals Associated with Principal Component Analysis," *Technometrics*, **21**; 341–349.

Jain, A. K. and E. Angel (1974). "Image Restoration, Modelling, and Reduction of Dimensionality, *IEEE Trans., Comput.*, **C-23**; 470–474.

James, A. T. (1964). "Distribution of Matrix Variates and Latent Roots Derived from Normal Samples," *Ann. Math. Stat.*, **35**; 475–501.

James, A. T. (1969). "Tests of Equality of Latent Roots of the Covariance Matrix," in P. R. Krishnaiah, (Ed.), *Multivariate Analysis II*, Academic, New York, pp. 205–218.

Janssen, C. T. L. and J. D. Jobson (1980). "Applications and Implementation on the Choice of Realtor," *Decis. Sci.*, **11**; 299–311.

Jaswinski, A. H. (1970). *Stochastic Processes and Filtering Theory*, Academic, New York.

Jaumotte, I., J. H. P. Paelinck, J. M. Leheureux, and M. Pietquin (1971). "The Differential Economic Structures of the Belgian Provinces: A Time Varying Factor Analysis," *Region Urban Econ.*, **1**; 41–75.

Jeffers, J. N. R. (1967). Two Case Studies in the Application of Principal Components Analysis," *Appl. Stat.*, **16**; 225–236.

Jennrich, R. I. (1973). "Standard Errors for Obliquely Rotated Factor Loadings," *Psychometrika*, **38**; 593–604.

Jennrich, R. I. and D. T. Thayer (1973). "A Note on Lawley's Formulas for Standard Errors in Maximum Likelihood Factor Analysis," *Psychometrika*, **38**; 571–580.

Johnels, D., V. Edlund, H. Grahn, S. Hellberg, M. Sjostrom, and S. Wold (1983). "Clustering of Aryl Carbon-13 Nuclear Magnetic Resonance Substituent Chemical Shifts. A Multivariate Data Analysis Using Principal Components," *J. Chem. Soc. Perkin Trans. II*, 863–871.

Johnels, D., U. Edlund, and S. Wold (1985). "Multivariate Data Analysis of Carbon-13 Nuclear Magnetic Resonance Substituent Chemical Shifts of 2-Substituted Napthalenes," *J. Chem. Soc. Perkin Trans. II*, 1339–1343.

Johnson, N. L. and S. Kotz (1969). *Discrete Distributions*, Wiley, New York.

Johnson, N. L. and S. Kotz (1970). *Continuous Univariate Distributions*, Vols. 1 and 2, Wiley, New York.

Johnson, R. A. (1990). Measurement of Hispanic Ethnicity in the U.S. Census: An Evaluation Based on Latent-Class Analysis," *J. Am. Stat. Assoc.*, **85**; 58–65.

Johnson, R. A. and D. W. Wichern (1982). *Applied Multivariate Statistical Analysis*, Prentice-Hall, Englewood Cliffs, NJ.

Jolicoeur, P. (1963). "The Multivariate Generalization of the Allometry Equation," *Biometrics*, **19**; 497–499.

Jolicoeur, P. (1984). "Principal Components, Factor Analysis, and Multivariate Allometry: A Small-Sample Direction Test," *Biometrics* **40**; 685–690.

Jolicoeur, P. and J. Mosimann (1960). "Size and Shape Variation in the Painted Turtle: A Principal Component Analysis," *Growth*, **24**; 339–354.

Joliffe, I. T. (1972). "Discarding Variables in a Principal Component Analysis, I. Artificial Data," *Appl. Stat.*, **21**; 160–173.

Joliffe, I. T. (1973). "Discarding Variables in a Principal Component Analysis. II: Real Data," *Appl. Stat.*, **22**; 21–31.

Joliffe, I. T. (1986). *Principal Components Analysis*, Springer-Verlag, New York.

Jones, V. J. and F. H. Siller (1978). "Factor Analysis of Media Exposure Data Using Prior Knowledge of the Medium," *J. Marketing Res.*, **15**; 137–144.

Jöreskog, K. G. (1962). "On the Statistical Treatment of Residuals in Factor Analysis," *Psychometrika*, **27**; 335–354.

Jöreskog, K. G. (1963). *Statistical Estimation in Factor Analysis*, Almqvist & Wiksell, Stockholm.

Jöreskog, K. G. (1970). "A General Method for Analysis of Covariance Structures," *Biometrika*, **57**; 239–251.

Jöreskog, K. G. (1971). "Simultaneous Factor Analysis in Several Populations," *Psychometrika*, **36**; 409–426.

Jöreskog, K. G. (1977). "Factor Analysis by Least Squares and Maximum-Likelihood Methods," in K. Enslein, A. Ralston, and H. S. Wilf (Eds.), *Statistical Methods for Digital Computers*, Vol. III of Mathematical Methods for Digital Computers, Wiley, New York.

Jöreskog, K. G. (1979a). "Basic Ideas of Factor and Component Analysis," J. Magidson (Ed.), *Advances in Factor Analysis and Structural Equation Models*, Abt Books, Cambridge MA.

Jöreskog, K. G. and A. S. Goldberger (1972). "Factor Analysis by Generalized Least Squares," *Psychometrika*, **37**; 243–260.

Jöreskog, K. G., J. E. Klovan and R. A. Reyment (1976). *Geological Factor Analysis*, Elsevier, Amsterdam.

Jöreskog, K. G. and D. N. Lawley (1968). "New Methods in Maximum Likelihood, Factor Analysis," *B. J. Math. Stat. Psych.*, **21**; 85–96.

Kac, M. (1951). "On Some Connections Between Probability Theory and Differential and Integral Equations, Proceedings at the 2nd Berkeley Symposium on Mathematical Statistics and Probability.

Kaiser, H. F. (1958). "The Varimax Criterion for Analytic Rotation in Factor Analysis," *Psychometrika*, **23**; 187–200.

Kaiser, H. G. and B. A. Cerny (1980). "On the Canonical Analysis of Contingency Tables," *Educ. Psych. Meas.*, **40**; 95–99.

Kale, B. K. (1970). "Normality of Linear Combinations of Non-Normal Random Variables," *Am. Math. Monthly*, **77**; 992–995.

Kalman, R. E. (1960). "A New Approach to Linear Filtering and Prediction Problems," *Trans. ASME J. Basic Eng. (D)*, **83**; 95–108.

Kalman, R. E. and R. S. Bucy (1961). "New Results in Linear Filtering and Prediction Theory," *J. Basic Eng.*, **83**; 95–108.

Kalman, R. E., P. L. Falb and M. A. Arbib (1969). *Topics in Mathematical System Theory*, McGraw Hill, New York.

Kano, Y. (1986). "Consistency Conditions on the Least Squares Estimator in Single Common Factor Analysis Model," *Ann. Inst. Stat. Math.*, **38**; 57–68.

Kaplan, W. (1952). *Advanced Calculus*, Addison Wesley, Reading, MA.

Kapteyn, A., H. Neudecker, and T. Wansbeck (1986). "An Approach to n-Mode Components Analysis," *Psychometrika*, **51**; 269–275.

Karhunen, K. (1947). "Uber Linear Methoden in der Wahrscheinlichkeitsrechnung," *Ann. Acad. Sci. Fenn.*, (*AI*), **37**; 1–79.

Karl, T. R., A. J. Koscielny, and H. F. Diaz (1982). "Potential Errors in the

Application of Principal Component (Eigenvector) Analysis to Geophysical Data," *J. Appl. Meteorol.*, **21**; 1183–1186.

Katzner, D. W. (1983). *Analysis without Measurement*, Cambridge University Press, Cambridge, 304 pp.

Kelly, G. (1984). "The Influence Function in the Errors in Variables Problem," *Ann. Stat.*, **12**; 87–100.

Kendall, D. G. (1969). "Incidence Matrices, Interval Graphs and Seriation in Archaeology," *Pacific J. Math.*, **28**; 565–570.

Kendall, D. G. (1971a). "Seriation from Abundance Matrices," in F. R. Hodson, D. G. Kendall, and P. Tautu (Eds.), *Mathematics in the Archaeological and Historical Sciences*, Edinburgh University Press, Edinburgh.

Kendall, D. G. (1971b). "Construction of Maps from 'Odd Bits of Information,'" *Nature*, **231**; 158–159.

Kendall, D. G. (1971c). "Maps from Marriages: An Application of Non-Metric Multi-Dimensional Scaling to Parish Register Data," in F. R. Hodson, D. G. Kendal, and P. Tautu, (Eds.), *Mathematics in the Archaeological and Historical Sciences*, Edinburgh University Press.

Kendall, D. G. (1975). "Recovery of Structure from Fragmentary Information," *Phil Trans. R. Soc. (Series A)*, **279**; 547–582.

Kendall, M. G. (1950), "Factor Analysis," *J. R. Stat. Soc.*, **12**; 60–73.

Kendall, M. G. (1957). *A Course in Multivariate Analysis*, Griffin, London.

Kendall, M. G. (1970). *Rank Correlation Methods*, 4th ed., Griffin, London.

Kendall, M. G. and D. N. Lawley (1956). "The Principles of Factor Analysis," *J. R. Stat. Soc.*, (*B*), **18**; 83–84.

Kendall, M. G. and A. Stuart (1979). The Advanced Theory of Statistics, Vol. 2, 4th ed., Griffin, London.

Kettenring, J. R. (1971). "Canonical Analysis of Several Sets of Variables," *Biometrika* **58**; 433–451.

Kittler, J. and P. Young (1973). "A New Approach to Feature Selection Based on the Karhunen–Loève Expansion," *Pattern Recognition*, **5**; 335–352.

Kloek, T. and R. Bannink (1962). "Principal Components Analysis Applied to Business Test Data," *Stat. Neerlandica* **16**; 57–69.

Kloek, T. and deWit, G. M. (1961). "Best Linear and Best Linear Unbiased Index Numbers," *Econometrica*, **29**; 602–616.

Kloek, T. and L. B. M. Mennes (1960). "Simultaneous Equations Estimation Based on Principal Components of Predetermined Variables," *Econometrica*, **28**; 45–61.

Klovan, J. E. (1966). "The Use of Factor Analysis in Determining Depositional Environments from Grain-Size Distributions," *J. Sediment. Pet.*, **36**; 115–125.

Klovan, J. E. (1975). "R and Q-mode Factor Analysis," in *Concepts in Geostatistics*, R. B. McCammon (Ed.), Springer-Verlag, New York, Chapter 2.

Klovan, J. E. (1981). "A Generalization of Extended Q-Mode Factor Analysis to Data Matrices with Variable Row Sums," *Math. Geol.*, **13**; 217–224.

Klovan, J. E. and A. T. Miesch (1976). "Extended Cabfac and Q-Mode Computer

Programs for Q-Mode Factor Analysis of Compositional Data," *Comput. Geosci.*, **1**; 161–178.

Knudson, E. J., D. L. Duewer and G. D. Christian (1977). "Application of Factor Analysis to the Study of Rain Chemistry in the Puget Sound Region," B. R. Kowalski (Ed.), *American Chemical Society*, ACS Symposium Series 52, Washington, D.C.

Koopman, R. F. (1978). "On Baysian Estimation in Unrestricted Factor Analysis," *Psychometrika*, **43**; 109–110.

Koopmans, T. C. (1950), *Statistical Inference in Dynamic Economic Models*, Cowles Commission Monograph No. 10, Wiley, New York.

Koopmans, T. C. (1951). *"Identification Problems in Latent Structure Analysis,"* Cowles Commission Discussion Paper: Statistics No. 360.

Kosambi, D. D. (1943). "Statistics in Function Space," *J. Indian Math. Soc.*, **7**; 76–88.

Kowalski, B. R., R. W. Gerlach and H. Wold (1982). "Chemical Systems Under Indirect Observation," in K. G. Jöreskog and H. Wold (Ed.), *Systems Under Indirect Observation Part II*, North Holland, New York, pp. 191–207.

Kowalski, B. R. and S. Wold (1982). "Pattern Recognition in Chemistry," in P. R. Krishnaiah and L. N. Kanal (Eds.), *Handbook of Statistics*, **2**; 673–697.

Krantz, D. H., R. D. Luce, P. Suppes, and A. Tversky (1971). *Foundations of Measurement*, Vols. 1–3, Academic, New York.

Kroonenberg, P. M. (1983a). *Three-Mode Principal Component Analysis: Theory and Applications*, DSWO Press, Leiden.

Kroonberg, P. M. (1983b). "Annotated Bibliography of Three-Mode Factor Analysis," *Br. J. Math. Stat. Psych.*, **36**; 81–113.

Kruskal, J. B. (1964). "Nonmetric Multidimensional Scaling: A Numerical Method," *Psychometrika*, **29**; 28–42.

Kruskal, J. B. and R. N. Shepard (1974). "A Nonmetric Variety of Linear Factor Analysis," *Psychometrika*, **39**; 123–157.

Kryzanowski, L. and Minh Chan To (1983). "General Factor Models and the Structure of Security Returns," *J. Finance Quant. Anal.*, **18**; 31–52.

Krzanowski, W. J. (1976). "Canonical Representation of the Location Model for Discrimination or Classification," *J. Am. Stat. Assoc.*, **71**; 845–876.

Krzanowski, W. J. (1979). "Between Groups Comparison of Principal Components." *J. Am. Stat. Assoc.*, **74**; 703–707.

Krzanowski, W. J. (1983). "Cross-Validatory Choice in Principal Component Analysis: Some Sampling Results," *J. Stat. Comput. Simulation*, **18**; 299–314.

Krzanowski, W. J. (1984). "Sensitivity of Principal Components," *J.R. Stat. Soc. (B)*, **46**; 558–563.

Kshirsagar, A. M. (1961). "The Goodness-of-Fit of a Single (Non-Isotropic) Hypothetical Principal Component," *Biometrika*, **48**; 397–407.

Kshirsagar, A. M. (1966). "The Non-Null Distribution of a Statistic in Principal Components Analysis," *Biometrika*, **53**; 490–494.

Kshirsagar, A. M. (1972). *"Multivariate Analysis,"* Marcell Dekker, New York.

Kullback, S. and R. A. Leibler (1951). "On Information and Sufficiency," *Ann. Math. Stat.*, **22**; 79–86.

Kummel, C. H. (1879). "Reduction of Observed Equations Which Contain More than One Observed Quantity," *Analyst*, **6**; 97–105.

Kung, E. C. and T. A. Sharif (1980). "Multi-Regression Forecasting of the Indian Summer Monsoon With Antecedent Patterns of the Large-Scale Circulation," WMO Symposium on Probabilistic and Statistical Methods in Weather Forecasting, pp. 295–302.

Kuz, T. J., E. Baril, D. Hiebert, A. Morrison, and C. Skonberg (1979). "Winnipeg, A Multivariate Analysis 1951, 1961, and 1971," University of Winnipeg and Department of Environmental Planning, City of Winnipeg, Manitoba.

La Marche, V. C. and H. C. Fritts (1971). "Anomaly Patterns of Climate Over the Western United States 1700–1930 Derived From Principal Component Analysis of Tree-Ring Data," *Mon. Weather Rev.*, **99**; 138–142.

Lancaster, H. O. (1957). "Some Properties of the Bivariate Normal Distribution Considered in the Form of a Contingency Table," *Biometrika*, **44**; 289–292.

Lancaster, H. O. (1963). "Canonical Correlations and Partitions," *Quart. J. Math.*, **14**; 220–224.

Lancaster, H. O. (1966). 'Kolmogorov's Remark on the Hotelling Canonical Correlations," *Biometrika*, **53**; 585–588.

Lancaster, H. O. (1969). *The Chi-Squred Distribution*, Wiley, New York.

Lancaster, H. O. and M. A. Hamdan (1964). "Estimation of the Correlation Coefficient in Contingency Tables with Possibly Nonmetric Characteristics," *Psychometrika*, **29**; 381–391.

Lautenschlager, G. J. (1989). "A Comparision of Alternatives to Conducting Monte Carlo Analyses for Determining Parallel Analysis Criteria," *Multivariate Behav. Res.*, **24**; 365–395.

Lawley, D. N. (1940). "The Estimation of Factor Loadings by the Method of Maximum Likelihood," *Proc. R. Soc. Edinburgh* (*A*), **60**; 64–82.

Lawley, D. N. (1941). "Further Investigation in Factor Estimation," *Proc. R. Soc. Edinburgh* (*A*), **61**; 176–185.

Lawley, D. N. (1942). "Further Investigations in Factor Estimation," *Proc. R. Soc. Edinburgh*, **62**; 176–185.

Lawley, D. N. (1944). "The Factorial Analysis of Multiple Item Tests," *Proc. R. Soc. Edinburgh*, **62-A**; 74–82.

Lawley, D. N. (1953). "A Modified Method of Estimation in Factor Analysis and Some Large Sample Results," Uppsala Symposium on Psychological Factor Analysis: Nordisk Psykologi Monograph Series, #3, pp. 35–42.

Lawley, D. N. (1956). "Tests of Significance of the Latent Roots of Covariance and Correlation Matrices," *Biometrika*, **43**; 128–136.

Lawley, D. N. (1959). "Tests of Significance in Canonical Analysis," *Biometrika*, **46**; 59–66.

Lawley, D. N. (1963). "On Testing a Set of Correlation Coefficients for Equality," *Ann. Math. Stat.*, **34**; 149–151.

Lawley, D. N. and A. E. Maxwell (1971). *Factor Analysis as a Statistical Method*, Butterworths, London.

Lawley, D. N. and A. E. Maxwell (1973). "Regression and Factor Analysis," *Biometrika*, **60**; 331–338.

Lawton, W. H. and A. Sylvestre (1971). "Self Modelling Curve Resolution," *Technometrics* **13**; 617–633.

Layard, M. W. J. (1974). "A Monte-Carlo Comparison of Tests for Equality of Covariance Matrices, *Biometrika*, **16**; 461–465.

Lazarsfeld, P. F. (1950). "The Logic and Mathematical Foundtaion of Latent Structure Analysis," S. Stouffer et al. (Eds.), *Studies in Social Psychology in World War II*, Vol. IV. Measurement and Prediction, Princeton University Press, Princeton, NJ.

Lazarsfeld, P. F. and N. W. Henry (1968). *Latent Structure Analysis*, Houghton Mifflin, Boston.

Lebart L. (1976). "The Significance of Eigenvalues Issued from Correspondence Analysis of Contingency Tables," *Proc. Comput. Stat. (Compstat)*. 38–45; Physica-Verlag, Vienna.

Lebart, L., A. Morineau, and N. Tabard (1977). *Techniques de la Description Statistique*, Dunod, Paris.

Lebart, L., A. Morineau, and K. M. Warwick (1984). *Multivariate Descriptive Statistical Analysis: Correspondence Analysis and Related Techniques for Large Matrices*, Wiley, New York.

Le Bras, H. (1974). "Vingt Analyses Multivarieés d'Une Structure Commue," *Math. Sci. Hum.*, **12** (No. 47); 37–55.

Ledermann, S. and J. Breas (1959). "Les Dimensions de la Mortalité," *Population*, **14**; 637–682.

LeMaitre, R. W. (1968). "Chemical Variation Within and Between Volcanic Rock Series – A Statistical Approach," *J. Pet.*, **9**; 220–252.

Lever, H. and S. Smooha (1981). "A Part-Whole Strategy for the Study of Opinions," *Pub. Opin. Q.*, **45**; 560–570.

Levine, J. H. (1972). "The Spheres of Influence," *Am. Soc. Rev.*, **37**; 14–27.

Lewi, P. J. (1982). *Multivariate Data Analysis in Indistural Practice*, Wiley, New York.

Li, G. and Z. Chen (1985). "Projection Pursuit Approach to Robust Dispersion Matrices and Principal Components: Primary Theory and Monte Carlo," *J. Am. Stat. Assoc.*, **80**; 759–766.

Li, Ker-Chau (1991). "Sliced Inverse Regression for Dimension Reduction," *J. Am. Stat. Assoc.* **86**; 316–327.

Lianos, T. P. and G. C. Rausser (1972). "Approximate Distribution of Parameters in a Distributed Lag Model," *J. Am. Stat. Assoc.*, **67**; 64–67.

Linden, M. (1977). "Factor Analytical Study of Olympic Decathlon Data," *Res. Q.*, **48**; 562–568.

Lindley, D. V. (1947). "Regression Lines and the Linear Functional Relationship," *Suppl., J. R. Stat. Soc.*, **9**; 218–244.

Lindsay, S. L. (1986). "Geographic Size Varition in *Tamiasciurus Douglasii*: Significance in Relation to Conifer Core Morphology, *J. Mamm.*, **67**; 317–325.

Loève, M. (1945). "Fonctions Aleatoires de Second Ordre," *CR Acad. Sci. Paris*, **220**; 469–.

Lusch, R. F. and J. R. Brown (1982). "A Modified Model of Power in the *Marketing Channel*," *Market. Res.*, **19**; 312–323.

Lyttkeus, E. (1972). "Regression Aspects of Canonical Correlation," *J. Mult. Anal.*, **2**; 418–439.

Machin, D. (1974). "A Multivariate Study of the External Measurements of the Sperm Whale (*Physeter catodon*)," *J. Zool., London*, **172**; 267–288.

Machin, D. and B. L. Kitchenham (1971). "A Multivariate Study of the External Measurements of the Humpback Whale (*Megaptera Novaeangliae*)," *J. Zool., London*; **165**; 415–421.

Madansky, A. (1960). "Determinantal Methods in Latent Class Analysis," *Psychometrika*, **25**; 183–198.

Madden, T. J. and W. R. Dillon (1982). "Causal Analysis and Latent Class Models. An Application to a Communication Hierarchy of Effects Model," *J. Mark. Res.*, **19**; 472–490.

Mager, P. P. (1980a). "Principal Component Regression Analysis Applied in Structure–Activity Relationships. 1, Selective Dihydrofolic Acid Reductase Inhibitors," *Biom. J.*, **5**; 441–446.

Mager, P. P. (1980b), "Correlation Betwen Qualitatively Distributed Predicting Variables and Chemical Terms in Aeridine Derivatives Using Principal Components," *Biometrical J.*, **22**; 813–825.

Mager, P. P. (1984). *Multidimensional Pharmacochemistry: Design for Safer Drugs*, Academic, Orlando, FL.

Mager, P. P. (1988). *Multivariate Chemometrics in QSAR: A Dialogue*, Wiley, New York.

Mak, T. K. (1978). *Estimation of Linear Structural and Functional Relationships*, Department of Mathematics, University of Western Ontario.

Malinowski, E. R. (1977). "Abstract Factor Analysis–A Theory of Error and its Application to Analytical Chemistry," in B. R. Kowalski (Ed.), Symposium Series 52, American Chemical Society, Washington, DC, 1977.

Malinowski, E. R. and D. G. Howery (1979). *Factor Analysis in Chemistry*, Wiley, New York.

Malinvaud, E. (1966). *Statistical Methods of Econometrics*, North-Holland, Amsterdam.

Malkovich, J. F. and A. A. Afifi (1973). "On Tests for Multivariate Normality," *J. Am. Stat. Assoc.*, **68**; 176–179.

Mallows, C. L. (1960). "Latent Vectors of Random Symmetric Matrices, *Biometrika*, **48**; 133–149.

Mallows, C. L. (1973). "Some Comments on Cp," *Technometrics*, **15**; 661–675.

Mandel, J. (1982). "Use of the Singular Value Decomposition in Regression Analysis," *Am. Stat.*, **36**; 15–24.

Mandel, J. (1984). "Fitting Straight Lines When Both Variables are Subject to Error," *J. Qual. Technol.*, **16**; 1–14.

Mansfield, E. R., J. T. Webster, and R. F. Gunst (1977). "An Analytic Variable Selection Technique for Principal Components Regression," *Appl. Stat.*, **34**; 34–40.

Marcus-Roberts, H. M. and F. S. Roberts (1987). *Meaningless Statistics*, Rutgers Research Report #18–87, Rutgers University, New Jersey.

Mardia, K. V. (1970). "Measures of Multivariate Skewness and Kurtosis with Applications," *Biometrika*, **57**; 519–530.

Mardia, K. V. (1975). "Statistics of Directional Data (with discussion), *J. R. Stat. Soc.* (*B*); 349–393.

Marida, K. V. (1977). "Mahalanabis Distances and Angles," in P. R. Krishnaiah (Ed.), *Multivariate Analysis IV*, North Holland, New York, pp. 495–511.

Mardia, K. V. (1978). "Some Properties of Classical, Multi-Dimensional Scaling," *Commun. Stat., Theory* Methods, **A7(13)**; 1233–1241.

Mardia, K. V. (1980). "Tests of Univariate and Multivariate Normality," in P. R. Krishnaiah (Ed.), *Handbook of Statistics*, Vol 1. pp. 279–320.

Mardia, K. V., J. T. Kent and J. M. Bibby (1979). *Multivariate Analysis*, Academic, London.

Mardia, K. V. and P. J. Zemroch (1975). "Algorithm AS84. Measures of Multivariate Skewness and Kurtosis," *Appl. Stat.*, **24**; 262–265.

Marquardt, W. H. (1974). "A Statistical Analysis of Constituents in Human Paleofecal Specimens from Mammoth Cave," in P. J. Watson (Ed.) *Archaeology of the Mammoth Cave Area*, Academic, New York.

Marquardt, W. H. (1978). "Advances in Archaeological Seriation," in B. Schiffer (Ed.), *Advances in Archaeological Method and Thoery*, Vol. 1, pp. 257–314.

Martin, J. K. and R. McDonald (1975). "Baysian Estimation in Unrestricted Factor Analysis: A Treatment for Heywood Cases," *Psychometrika*, **40**; 505–517.

Martinson, E. O. and M. A. Hamdan (1972), "Algorithm AS 87: Calculation of the Polychoric Estimate of Correlation in Contingency Tables," *Appl. Stat.*, **24**; 272–278.

Marubini, E., M. Rainisio, and V. Mandelli (1979). "A Multivariate Approach to Analyze Ordinal Data Collected in Clinical Trials of Analgesic Agents," *Meth. Inf. Med.*, **18**; 175–179.

Marsaglia, G. (1965). "Ratio of Normal Variables and Ratios of Sums of Uniform Variables," *J. Am. Stat. Assoc.*, **60**; 193–204.

Massy, W. F. (1965). "Principal Component Regression in Exploratory Statistical Research," *J. Am. Stat. Assoc.*, **60**; 234–256.

Matalas, N. C. and B. J. Reiher (1967). "Some Comments on the Use of Factor Analysis," *Water Res. Res.*, **3**; 213–223.

Mathai, A. M. and R. S. Katiyar (1979). "Exact Percentage Points for Testing Independence," *Biometrika*, **66**; 353–356.

Matthews, J. N. S. (1984). "Robust Methods in the Assessment of Multivariate Normality," *Appl. Stat.*, **33**; 272–277.

Matthews, L. H. (1938). "The Humpback Whale, 'Megaptera Nodosa'," *Discovery Rep.*, **17**; 9–93.

Mauchly, J. W. (1940). "Significance Test for Sphericity of a Normal n-variate Distribution," *Ann. Math. Stat.*, **11**; 204–209.

Marquardt, D. W. (1970). "Generalized Inverses, Ridge Regression, Biased Linear Estimation, and Nonlinear Estimations," *Technometrics* **12**; 591–612.

Maung, K. (1941). "Measurement of Association in a Contingency Table with Special Reference to the Pigmentation of Hair and Eye Colours of Scottish School Children," *Ann. Eugen.*, **11**; 189–223.

Maxwell, A. E. (1976). "The Learning of Motor Movements: A Neurostatistical Approach," *Psych. Med.*, **6**; 643–648.

McCabe, G. P. (1984). "Principal Variables," *Technometrics*, **26**; 137–144.

McCammon, R. B. (1966). "Principal Component Analysis and its Application in Large-Scale Correlation Studies," *J. Geol.*, **74**; 721–733.

McCormick, W. T., P. J. Schweitzer, and T. W. White (1972). "Problem Decomposition and Data Reorganization by a Clustering Technique," *Oper. Res.*, **20**; 993–1009.

McDonald, R. P. (1962). "A General Approach to Nonlinear Factor Analysis," *Psychometrika*, **27**; 397–415.

McDonald, R. P. (1970). "Three Common Factor Models for Groups of Variables," *Psychometrika*, **35**; 111–128.

McDonald, R. P. (1975). "A Note on Ripp's Test of Significance in Common Factor Analysis," *Psychometrika*, **40**; 117–119.

McDonald, R. P. (1979). "Simultaneous Estimation of Factor Loadings and Scores," *Br. J. Math. Stat. Psych.*, **32**; 212–228.

McDonald, R. P. (1985). *Factor Analysis and Related Methods*, Lawrence Erlbaum Associates, New Jersey.

McDonald, R. P. and E. J. Burr (1967). "A Comparison of Four Methods of Constructing Factor Scores," *Psychometrika*, **32**; 381–401.

McElroy, M. N. and R. L. Kaesler (1965). "Application of Factor Analysis to the Upper Cambrian Reagen Sandstone of Central and Northwest Kansas," *Compass*, **42**; 188–201.

McGillivray (1985). "Size, Sexual Size Dimorphism and their Measurement in Great Horned Owls in Alberta, *Can. J. Zoo.*, **63**; 2364–2372.

McKeon, J. J. (1966). "Canonical Analysis: Some Relations Between Canonical Correlation, Factor Analysis, Discriminant Analysis, and Scaling Theory," Psychometric Monography, Vol. 13.

McReynolds, W. O. (1970). "Characterization of Some Liquid Phases," *J. Chromatogr. Sci.*, **8**; 685–691.

Mehler, F. G. (1966). "Über die Entwicklung einer Funktion von beliebig vielen Variablen nach Laplaceschen Funktionen Hoeherer Ordnung," *J. Reine Angew. Math.*, **66**; 161–176.

Mehra, R. K. (1979). "Kalman Filters and their applications to Forecasting," in Makridakis, S., Wheelwright, S. C. (eds.) "Studies in the Management Sciences,

12; 75–94 TIMS Studies in the Management Sciences; North Holland, Amsterdam.

Meinhold, R., and Singpurwalla, N. D. (1983). "Understanding the Kalman Filter," *The Am. Statist.*, **37**; 123–127.

Melnick, E. L. and A. Tenenbein (1982). "Misspecifications of the Normal Distribution," *Am. Statistician* **36**; 372–373.

Middleton, G. V. (1964). "Statistical Studies in Scapolites, *Can. J. of Earth Sciences* **1**; 23–34.

Miesch, A. T. (1980). "Scaling Variables and Interpretation of Eigenvaleus in Principal Component Analysis of Geologic Data," *Mathematical Geology* **12**; 523–538.

Miesch, A. T. (1976b). "Q-Mode Factor Analysis of Compositional Data," *Comput. Geos.*, **1**; 147–159.

Miesch, A. T. (1976a). "Q-Mode Factor Analysis of Geochemical and Petrologic Data Matrices with Constant Row-Sums, *Geological Survey Professional Paper* 574-G; U.S. Dept. of the Interior.

Miller, C. R., G. Sabagh and H. F. Dingman (1962). "Latent Class Analysis and Differential Mortality," *J. Amer. Statist. Assoc.*, **57**; 430–438.

Mills, J. T., R. N. Sinha and H. A. H. Wallace (1978). "Multivariate Evaluation of Isolation Techniques for Fungi Associated with Stored Rapeseed," *Phytopathology* **68**; 1520–1525.

Moran, P. A. P. (1971). "Estimating Structural and Functional Relationships," *J. Multi. Anal.*, **1**; 232–255.

Morrison, D. F. (1967). "*Multivariate Statistical Methods*," McGraw-Hill, New York.

Mosteller, F. and D. L. Wallace (1963). 'Inference in an Authorship Problem," *J. Am. Statist. Assoc.*, **58**; 275–309.

Mudholkar, G. S. and P. Subbaiah (1981). "Complete Independence in the Multivariate Normal Distribution," in C. Taillie *et al.* (eds.) *Statistical Distributions in Scientific Work*, **5**; 157–168.

Muirhead, R. J. (1982). "*Aspects of Multivariate Statistical Theory*," Wiley, New York.

Muirhead, R. J. and C. M. Waternaux (1980). Asymptotic Distributions in Canonical Correlation Analysis and other Multivariate Procedures for Nonnormal Populations," *Biometrika* **67**; 31–43.

Mukherjee, B. N. (1965). "A Factor Analysis of Some Qualitative Attributes of Coffee," *Advertising Research* **5**; 35–38.

Mukherjee, B. N. (1974). "A Factor-Analytic Study of Respondent Variability in Demographic Data," *Demography India* **3**; 375–396.

Mulaik, S. A. (1972). "*The Foundations of Factor Analysis*," McGraw-Hill, New York.

Muthén, B. (1978). "Contributions to Factor Analysis of Dichotomous Variables," *Psychometrika*, **43** 551–560.

Nagle, R. K. and E. B. Saff (1986). *Fundamentals of Differential Equations*, Benjamin/Cummings, Menlo Park. CA.

Nakache, J. P., P. Lorente, and C. Chastaing (1978). "Evaluation of Prognosis and

Determination of Therapy Using Multivariate Methods," *Biomedicine* (special issue), *European Journal Clinical and Biological Research*, International Meeting on Comparative Therapeutic Trials, Paris 1978.

Naouri, J. C. (1970). "Analyse Factorielle des Correspondences Continue," *Publ. Inst. Stat. Univ. Paris*, **19**; 1–100.

Nishisato, S. (1980). *Analysis of Categorical Data: Dual Scaling and Its Applications*, University of Toronto Press, Toronto.

Noy-Meir, I. (1971). "Multivariate Analysis of the Semi-Arid Vegetation in South-Easter Australia: Nodal Ordination by Component Analysis," *Proc. Ecol. Soc. Aust.*, **6**; 159–193.

O'Neill, M. E. (1978a). "Asymptotic Distributions of the Canonical Correlations from Contingency Tables," *Aust. J. Stat.*, **20**; 65–82.

O'Neill, M. E. (1978b). "Distributional Expansions for Canonical Correlations from Contingency Tables," *J. R. Stat. Soc. (B)*, **40**; 303–312.

O'Neill, M. E. (1980). "The Distribution of Higher-Order Interactions in Contingency Tables," *J. R. Stat. Soc. (B)*, **42**; 357–365.

O'Neill, M. E. (1981). "A Note on the Canonical Correlations from Contingency Tables," *Aust. J. Stat.*, **23**; 58–66.

Obenchain, R. L. (1972). "Regression Optimality of Principal Components," *Ann. Math. Stat.*, **43**; 1317–1319.

Olsson, U. (1980). "Measuring Correlation in Ordered Two Way Contingency Tables," *J. Mark. Res.*, **17**; 391–394.

Okamoto, M. (1969). "Optimality of Principal Components," in P. R. Krishnaiah (Ed.), *Multivariate Analysis II*, Academic, New York, pp. 673–685.

Okamoto, M. (1973). "Distinctness of the Eigenvalues of a Quadratic Form in a Multivariate Sample," *Ann. Stat.*, **1**; 763–765.

Olsson, U. (1979a). "Maximum Likelihood Estimation of the Polychoric Correlation Coefficient," *Psychometrika*, **44**; 443–460.

Olsson, U. (1979b). "On the Robustness of Factor Analysis Against Crude Classification of the Observations," *Mult. Beh. Res.*, **14**; 485–500.

Olsson, U., F. Drasgow, and N. J. Dorans (1982). "The Polyserial Correlation Coefficient," *Psychometrika*, **47**; 337–347.

Oman, S. D. (1978). "A Baysian Comparison of Some Estimators Used in Linear Regression With Multicollinear Data," *Commun. Stat. Theor. Meth.*, **A7**(6); 517–534.

Orloci K. (1966)."The Theory and Appliction of Some Ordination Methods," *J. Ecol.*, **54**; 193–215.

Orloci L. (1967). "Data Centering: A Review and Evaluation with Reference to Component Analysis," *Syst. Zool.*, **16**; 208–212.

Otter, P. and J. Schur (1982). "Principal Component Analysis in Multivariate Forecasting of Economic Time Series," in O. D. Anderson (Ed.) *Time Series Analysis: Theory and Practice I*, North Holland, New York.

Oudin, J. L. (1970). "Analyse Geochimique de la Matiére Organique Extraite des Roches Sédimentaire I. Composés Extraitible au Chloroforme," *Inst. Francais Pet. Rev.*, **25**; 3–15.

Overall, J. E. and C. M. Williams (1961). "Models for Medical Diagnosis: Factor Analysis. Part One, Theoretical," *Med. Doc.*, **5**; 51–56.

Ozeki, K. (1979). "A Coordinate-Free Theory of Eigenvalue Analysis Related to the Method of Principal Components and the Karhunen–Loève Expansion," *Inf. Control*, **42**; 38–59.

Pagan, A. (1975). "A Note on the Extraction of Components From Time Series," *Econometrica*, **43**; 163–168.

Park, S. H. (1981). "Collinearity and Optimal Restrictions on Regression Parameters of Estimating Responses," *Technometrics*, **23**; 289–295.

Parzen, E. (1983). "Time Series ARMA Model Identification by Estimating Information," in J. E. Gentle (Ed.), *Computer Science and Statistics: The Interface*, North Holland, New York.

Pearson, K. (1900). "Mathematical Contributions to the Theory of Evolution in the Inheritance of Characters not Capable of Exact Quantitative Measurement, VIII, *Philos. Trans. R. Soc. (A)*, **195**; 79–150.

Pearson, K. (1901). "On Lines and Planes of Closest Fit to a System of Points in Space," *Philosophical Magazine* **2**, 6th Series; 557–572.

Pearson, K. (1904). "On the Theory of Contingency and its Relation to Association and Normal Correlation," *Karl Pearson's Early Statistical Papers*, Cambridge, University Press, pp. 443–475.

Pearson, K. and L. N. G. Filon (1898). "On the Probable Errors of Frequency Constants and on the Influence of Random Selection on Variation and Correlation," *Phil. Trans. R. Soc. (A)*, **191**; 229–311.

Pearson, K. and M. Maul (1927). "The Mathematics of Intelligence I. The Sampling Errors in the Theory of a Generalized Factor," Biometrika, **19**; 246–292.

Pennel, R. (1972). "Routinely Computable Confidence Intervals for Factor Loadings Using the Jackknife," *Br. J. Math. Stat. Psyc.*, **25**; 107–114.

Peters, W. S. and J. Q. Butler (1970). "The Construction of Regional Economic Indicators by Principal Components," *Ann. Reg. Sci.*, **4**; 1–14.

Petrinovich, L. and C. Hardyck (1964). "Behavioral Changes in Parkinson Patients Following Surgery," *J. Chronic Dis.*, **17**; 225–233.

Pfanzagl, J. (1968). *Theory of Measurement*, Wiley, New York.

Piazza, A. and L. L. Cavalli-Sforza (1978). "Multivariate Techniques for Evolutionary Analyses," *Riv. Stat. Appl.*, **11**; 239–254.

Pimentel, R. A. (1979). *Morphometrics: The Multivariate Analysis of Biological Data*, Kendall/Hunt, Dubuque, Iowa.

Pirkle, F. L., R. J. Beckman, and H. L. Fleischhauer (1982). A Multivariate Uranium Favourability Index Using Aerial Radiometric Data," *Geol.*, **90**; 109–124.

Pratt, W. K. (1978). Digital Image Processing, Wiley, New York.

Press, S. J. (1972). *Applied Multivariate Analysis*, Holt, Rinehart and Winston, New York.

Priestly, M. B. (1981). *Spectral Analysis and Time Series*, Vol. 2: Multivariate Series, Prediction and Control, Academic, London.

Priestly, M. B. and T. S. Rao (1975). "The Estimation of Factor Scores and Kalman

Filtering for Discrete Parameter Stationary Process," *Int. J. Control*, **21**; 971–974.

Prigonine, I. and I. Stengers (1984). *Order out of Chaos*, Bantam, Toronto.

Quenouille, M. H. (1968). *The Analysis of Multiple Time-Series*, Hafner, New York.

Ram, R. (1982). Composite Indices of Physical Quality of Life, Basic Needs Fulfilment, and Income; *J. Dev. Econ.*, **11**; 227–247.

Ramsey, F. L. (1986). "A Fable of PCA," *Am. Stat.*, **40**; 323–324.

Rao, C. R. (1955). "Estimation and Tests of Significance in Factor Analysis," *Psychometrika*, **20**; 93–111.

Rao, C. R. (1958). Some Statistical Methods for Comparison of Growth Curves," *Biometrics* **14**; 1–17.

Rao, C. R. (1962). "A Note on a Generalized Inverse of a Matrix with Applications to Problems in Mathematical Statistics," *J. R. Stat. Soc. (B)*, **24**; 152–158.

Rao, C. R. (1964). "The Use and Interpretation of Principal Component Anaysis in Applied Research," *Sankhya (A)*, **26**; 329–358.

Rao, C. R. (1965). *Linear Statistical Inference and its Applications*, Wiley, New York, 522 pp.

Rao, C. R. (1983). "Likelihood Ratio Tests for Relationships between Covariance Matrices," in S. Karlin, T. Amemiya, and L. A. Goodman (Ed.), *Studies in Econometrics, Time Series and Multivariate Statistics*, Academic, New York, pp. 529–543.

Rao, C. R. and R. Boudreau (1985). "Prediction of Future Observations in a Factor Analytic Type Growth Model," P. R. Krishnaiah (Ed.), *Multivariate Analysis – VI*, 449-466, Elsevier, Amsterdam.

Rao, K. and H. D. Oberhelman (1982). "A Study on the Proper Selection of Components in a Principal Components Regression," ASA Proc. Bus. Econ. Stat. Sec., 207–209.

Rao, T. S. (1976a). "Canonical Factor Analysis and Stationary Time Series Models," *Sankhya* **38**(B); 256–271.

Rao, T. S. (1976b). "A Note on the Bias in the Kalman-Bucy Filter," *Int. J. Control*, **23**; 641–645.

Rayner, A. C. (1970). "The Use of Multivariate Analysis in Development Theory: A Critique of the Approach Adopted by Adelman and Morris," *Q. J. Econ.*, **84**; 638–647.

Rayner, J. C. W. (1985). "Maximum-Likelihood Estimation of μ and Σ from a Multivariate Normal Distribution," *Am. Stat.*, **39**; 123–124.

Riersol, O. (1950). "On the Identifiability of Parameters in Thurstone's Multiple Factor Analysis," *Psychometrika*, **15**; 121–159.

Relethford, J. H., C. L. Francis, and P.J. Byard (1978). "The Uses of Principal Components in the Analysis of Cross-Sectional Growth Data," *Hum. Biol.*, **50**; 461–475.

Rencher, A. (1992). "Interpretation of Canonical Discriminant Functions, Canonical Variates, and Principal Components," *Am. Stat.*, **46**; 217–225.

Rencher, A. C. (1988). "On the Use of Correlations to Interpret Canonical Functions," *Biometrika*, **75**; 363–365.

Reyment, R. A. (1963). "Multivariate Analytic Treatment of Quantitative Species Associations: An Example from Palaeoecology," *J. Anim. Ecol.*, **32**; 535–547.

Reyment, R. A. (1969). "A Multivariate Palaeontological Growth Problem," *Biometrics*, **25**; 1–8.

Reyment, R. A. (1971). "Multivariate Normality in Morphometric Analysis," *Math. Geol.*, **3**; 357–368.

Reyment, R. A. (1982). "Phenotypic Evolution in a Cretaceous Foraminifer," *Evolution*, **36**; 1182–1199.

Reyment, R. A., R. E. Blackith, and N. A. Campbell (1984). *Multivariate Morphometrics*, Academic, London.

Rhodes, E. C. (1937). "The Construction of An Index of Business Activity," *J. R. Stat. Soc.*, **100**; 18–66.

Richman, M. B. (1981). "Obliquely Rotated Principal Components: An Improved Meteorological Map Typing Technique? *J. Appl. Meteor.*, **20**; 1145–1159.

Rissanen, J. (1978). "Modeling by Shortest Data Description," *Automatica* **14**; 465–471.

Ritter, G. L., S. R. Lowry, and T. L. Isenhour (1976). "Factor Analysis of the Mass Spectra of Mixtures," *Anal. Chem.*, **48**; 591–595.

Roberts, F. S. (1979). "Measurement Theory, With Applications to Decision Making, Utility, and the Social Sciences," in *Encyclopedia of Mathematics*, Vol. 7, Addison-Wesley, Reading, MA.

Roll, R. and Ross, S. A. (1980). "An Empirical Investigation of the Arbitrage Pricing Theory," *J. Financ.*, **35**; 1073–1103.

Rosenberg, L. (1965). "Nonnormality of Linear Combinations of Normally Distributed Random Variables," *Am. Math. Mon.*, **72**; 888–890.

Rosenblatt, M. (1952). "Limit Theorems Associated with Variants of the Von Mises Statistic," *Ann. Math. Stat.*, **23**; 617–623.

Rothkopf, E. Z. (1957). "A Measure of Stimulus Similarity and Errors in Some Paired-Associate Learning Tasks," *J. Exp. Psych.*, **53**; 94–101.

Rousseeuw, P. J. and B. C. Van Zomeren (1990). "Unmasking Multivariate Outliers and Leverge Points, *J. Am. Stat. Assoc.*, **85**; 633–639.

Rowlett, R. M. and R. B. Pollnac (1971). "Multivariate Analysis of Marnian La Téne Cultural Groups," in F. R. Hodson, D. G. Kendall, and P. Tautu (Eds.), Mathematics in the Archaelogical and Historical Sciences, Edinburgh University Press, Edinburgh.

Roy, S. N. (1957). *Some Aspects of Multivariate Anaysis*, Wiley, New York.

Royston, J. P. (1983). "Some Techniques for Assessing Multivariate Normality Based on the Shapiro-Wilk W." *Appl. Stat.*, **32**; 121–133.

Rozett, R. W. and E. M. Petersen (1976). "Classification of Compounds by the Factor Analysis of their Mass Spectra," *Anal. Chem.*, **48**; 817–825.

Rubin, D. B. and D. T. Thayer (1982). "EM Algorithms for ML Factor Analysis," *Psychometrika*, **47**; 69–76.

Rubin, D. B. and D. T. Thayer (1983). "More on EM for ML Factor Analysis," *Psychometrika*, **48**; 253–257.

Russell, B. (1937). *The Principles of Mathematics*, 2nd ed., George Allen & Unwin Ltd., London.

Russell, M. A., J. Peto, and U. A. Patel (1974). "The Classification of Smoking By Factorial Structure of Motives," *J. R. Stat. Soc. (A)*, **137**; 313–346.

Ruymgaart, F. H. (1981). A Robust Principal Component Analysis, *J. Mult. Anal.*, **11**; 485–497.

Ryder, N. B. and C. F. Westoff (1971). *Reproduction in the United States, 1965*, Princeton University Press, 1971.

Sargent, T. J. and C. A. Sims (1977). "Business Cycle Modelling Without Pretending to Have Too Much A-Priori Economic Theory," Conference Proceedings and New Methods in Business Cycle Research, Federal Reserve Bank of Minneapolis.

Saxena, S. K. (1969). "Silicate Solid Solutions and Geothermy 4. Statistical Study of Chemical Data on Garnets and Clinopyroxene," *Contr. Mineral. Petrol.*, **23**; 140–156.

Saxena, S. K. and L. S. Walker (1974). "A Statistical Chemical and Thermodynamic Approach to the Study of Lunar Mineralogy," Geochim. Cosmochem. Acta, **38**; 79–95.

Saxena, A. K. (1980). Principal Components and its Use in Regression Analysis: The Problem Revisited," *Statistica*, **XL**; 363–368.

Schickendanz, P. T. (1977). "Applications of Factor Analysis in Weather Modification Reserach," 5th Conference on Probability and Statistics in: Atmospheric Sciences, pp. 190–195.

Schott, J. R. (1987). "An Improved Chi-Squared Test for a Principal Component," *Stat. Prob. Lett.*, **5**; 361–365.

Schott, J. R. (1991). "A Test for a Specific Principal Component of a Correlation Matrix," *J. Am. Stat. Assoc.*, **86**; 747–751.

Schwarz, G. (1978). "Estimating the Dimension of a Model," *Ann. Stat.*, **6**; 461–464.

Sclove, S. (1987). "Application of Model-Selection Criteria to Some Problems in Multivariate Analysis," *Psychometrika*, **52**; 333–343.

Scott, J. T. (1966). "Factor Analysis and Regression," *Econometrica* **34**; 552–562.

Searle, S. R. (1982). *Matrix Algebra Useful for Statistics*, Wiley, New York.

Sharpe, W. F. (1982). "Factors in New York Stock Exchange Security Returns, 1931–1979," *J. Portfolio Manage.*, **8**; 5–19.

Shea, G. (1978). "Statistical Diagnosis and Test of Factor Hypotheses," *J. Am. Stat. Assoc.*, **73**; 346–350.

Shepard, R. N. (1963). "Analysis of Proximities as a Study of Information Processing in Man," *Hum. Factors.* **5**; 33–48.

Sher, A. M., A. C. Young, and W. M. Meridith (1960). "Factor Analysis of the Electrocardiogram. Test of Electrocardiographic Theory: Normal Hearts," *Circ. Res.*, **8**; 519–526.

Sheth, J. N. (1969). "Using Factor Analysis to Estimate Parameters," *J. Am. Stat. Assoc.*, **64**; 808–822.

Shiba, S. (1965a). "A Method for Scoring Multicategory Items," *Japanese Psych. Res.*, **7**; 75–79.

Shiba, S. (1965b). "The Generalized Method for Principal Components of Scale Analysis," *Japenese Psych. Res.*, **7**; 163–165.

Shier, D. R. (1988). "The Monotonicity of Power Means Using Entropy," *Am. Stat.*, **42**; 203–204.

Shuchat, A. (1984). "Matrix and Network Models in Archaeology," *Math. J.*, **57**; 3–14.

Siervogel, R. M., A. F. Roche, and E. M. Roche (1978). "Developmental Fields for Digital Dermatoglyphic Traits as Revealed by Multivariate Analysis," *Hum. Biol.*, **50**; 541–556.

Silvey, S. D. (1969). "Multicollinearity and Imprecise Estimation, *J. R. Stat. Soc. (B)* **31**; 539–552.

Sinha, R. N. and P. J. Lee (1970). "Maximum Likelihood, Factor Analysis of Natural Arthropod Infestations in Stored Grain Bulks," *Res. Pop. Ecol.*, **12**; 51–60.

Sinha, R. N., H. A. H. Wallace, and F. S. Chebib (1968). "Canonical Correlations of Seed Viability, Seed-Borne Fungi, and Environment in Bulk Grain Ecosystems," *Can. J. Bot.*, **47**; 27–34.

Sinha, R. N., H. A. H. Wallace, and F. S. Chebib (1969). "Canonical Correlation Between Groups of Acarine, Fungal and Environmental Variables in Bulk Grain Ecosystems," *Res. Pop. Ecol.*, **11**; 92–104.

Sinha, R. N., G. Yaciuik, and W. E. Muir (1973). "Climate in Relation to Deterioration of Stored Grain: A Multivariate Study," *Oecologia*, **12**; 69–88.

Skinner, C. J., D. J. Holmes, and M. F. Smith (1986). "The Effect of Sample Design on Principal Components Analysis," *J. Am. Stat. Assoc.*, **81**; 789–798.

Skinner, H. A. and W. J. Sheu (1982). "Dimensional Analysis of Rank-Order and Categorical Data," *Appl. Psych. Meas.*, **6**; 41–45.

Small, N. J. H. (1978). "Plotting Squared Radii," *Biometrika*, **65**; 657–658.

Small, N. J. H. (1980). "Marginal Skewness and Kurtosis in Testing Multivariate Normality," *Appl. Stat.*, **29**; 85–87.

Smirnov, N. V. (1936). "Sur la Distribution de w^2 (Criterium de M. R. V. Mises)." *CR (Paris)*, **202**; 449–452.

Snee, R. D. (1972). "On the Analysis of Response Curve Data," *Technometrics*, **14**; 47–62.

Solari, M. E. (1969). "The 'Maximum Likelihood Solution' of the Problem of Estimating a Linear Functional Relationship," *J. R. Stat. Soc. (B)*, **31**; 372–375.

Somers, K. M. (1986). "Multivariate Allometry and Removal of Size with Principal Components Analysis," *Syst. Zool.*, **35**; 359–368.

Somers, K. M. (1989). "Allometry, Isometry and Shape in Principal Components Analysis," *Syst. Zool.*, **38**; 169–173.

Sparling, D. W. and J. D. Williams (1978). 'Multivariate Analysis of Avion Vocalizations," *J. Theor. Biol.*, **74**; 83–107.

Spearman, C. (1904). "General Intelligence, Objectively Determined and Measured," *Am. J. Psych.*, **15**; 201–293.

Spearman, C. (1913). Correlations of Sums and Differences. *Br. J. Psych.*, **5**; 417–426.

Spencer, D. (1966). "Factors Affecting Element Distributions in a Silurian Graphtalite Band," *Chem. Geol.*, **1**; 221–249.

Spjotvoll, E., H. Martens, and R. Valden (1982). "Restricted Least Squares Estimation of the Spectra and Concentration of Two Unknown Constituents Available in Mixtures," Technometrics, **24**; 173–180.

Sprent, P. (1966). "A Generalized Least Squares Approach to Linear Functional Relationship," *J. R. Stat. Soc. (B)*, **28**; 278–297.

Sprent, P. (1970). "The Saddle Point of the Likelihood Surface for a Linear Functional Relationship," *J. R. Stat. Soc. (B)*, **32**; 432–434.

Sprent, P. (1972). "The Mathematics of Size and Shape," *Biometrics* **28**; 23–37.

Spurrell, D. J. (1963). "Some Metallurgical Applications of Principal Components," *Appl. Stat.*, **12**; 180–188.

Srikantan, K. S. (1970). "Canonical Association Between Nominal Measurements," *J. Am. Stat. Assoc.*, **65**; 284–292.

Srivastava, M. S. and E. M. Carter (1983). *An Introduction to Applied Multivariate Statistics*, North Holland, New York.

Srivastava, M. S. and C. G. Khatri (1979). *An Introduction to Multivariate Statistics*, North-Holland, New York.

Stauffer, D. F., E. O. Garton, and R. K. Steinhort (1985). "A Comparison of Principal Components From Real and Random Data," *Ecology* **66**; 1693–1698.

Stevens, S. S. (1946). "On the Theory of Scales of Measurement," *Science*, **103**; 677–680.

Steward, D. W. (1981). "The Application and Misapplication of Factor Analysis in Marketing Research," *J. Mark. Res.*, **18**; 51–62.

Stitt, F. W., M. Frane, and J. W. Frane (1977). "Mood Change as a Tool in Clinical Research," *J. Chronic Dis.*, **30**; 135–145.

Stoetzel, J. (1960). "A Factor Analysis of the Liquor Presences of French Consumers, *Advert. Res.*, **1**; 7–11.

Stone, M. (1974). "Cross-Validatory Choice and Assessment of Statistical Prediction," *J. R. Stat. Soc. B*, **36**; 111–133.

Stone, R. (1947). "On the Interdependence of Blocks of Transactions," *J. R. Stat. Soc. (Suppl)* **9**; 1–45.

Strahler, A. H. (1978). "Binary Discriminant Analysis: A New Method for Investigating Species-Environment Relationships," E*cology*, **59**; 108–116.

Stratonovich, R. L. (1968), *Conditional Markov Procecesses and their Applications to the Theory of Optimal Control*, Elsevier, New York.

Stratonovich, R. L. (1960). "Application of the Theory of Markoff Processes in Optimal Signal Dissemination," *Radiotekh. Elek.*, **5**; 1751–1763. (in Russian).

Stratonovich, R. L. (1960), "Application of the Theory of Markoff Processes in Optimal Signal Discrimination," *Radio Eng. Electron. Phys.*, **1**; 1–19 (in Russian). Also in R. Kailath (Ed.) (1977). *Linear Least Squares Estimation*, Dowden Hutchinson and Ross, Straudsburg, PA.

Strouf, O. (1986). *Chemical Pattern Recognition*, Wiley, New York.

Sundberg, P. (1980). "Shape and Size-Constrained Principal Components Analysis," *Syst. Zool.*, **38**; 166–168.

Switzer, P. and A. A. Green (1984). "Mini/Max Autocorrelation Factors for Multivarite Spatial Imagery," *Technical Report #6*, Dept. of Statistics, Stanford.

Taaffe, E. J. and H. L. Gauthier (1973). *Geography of Transportation*, Prentice Hall, Englewood Cliffs, NJ.

Takemura, A. (1985). "A Principal Decomposition of Hotellings T^2 Statistic," in R. P. Krinshnaiah (Ed.), *Multivariate Anaysis-VI*; Elsevier, Amsterdam, pp. 583–597.

Takeuchi, K., H. Yanai, and B. N. Mukherjee (1982). *The Foundations of Multivariate Analysis*, Wiley Eastern Ltd., New Delhi.

Temple, J. T. (1978). "The Use of Factor Analysis in Geology," *Math. Geol.*, **10**; 379–387.

Tenenhaus, M. (1977). "Analyse en Composantes Principal d'Un Ensemble de Variables Nominales ou Numeriques, *Rev. Stat. Appl.*, **24**; 39–56.

Theil, H. (1960). "Best Linear Index Numbers of Prices and Quantities," *Econometrica*, **28**; 464–480.

Theil, H. (1963). "On the Use of Incomplete Prior Information in Regression Analysis, *J. Am. Stat. Assoc.*, **58**; 401–414.

Theil, H. (1975). "Correspondence Factor Analysis: An Outline of its Method," *Math. Geol.*, **7**; 3–2.

Theil, H., and J. L. Cheminée (1975). "Application of Correspondence Factor Analysis to the Study of Major and Trace Elements in the Erta Ale Chain (Afar, Ethiopia)" *Math. Geol.*, **7**; 14–30.

Thomas, H. (1985). "Measurement Structures and Statistics," in S. Katz., and N. L. Johnson (Eds.), *Encyclopedia of Statistical Sciences*, Vol. 5; Wiley, New York, pp. 381–386.

Thompson, G. H. (1951). *The Factorial Analysis of Human Ability*, London University Press.

Thorpe, R. S. (1983). "A Review of the Numerical Methods for Recognizing and Analyzing Racial Differentiation," in J. Felsenstein (Ed.), *Numerical Taxonomy*, Springer-Verlag, New York, pp. 404–423.

Thurstone, L. L. (1935). *The Vectors of Mind-*, University of Chicago Press.

Tinkler, K. J. (1972). "The Physical Interpretation of Eigenfunctions of Dichotomous Matrices," *Trans. Inst. Br. Geog.*, **55**; 17–44.

Tintner, G. (1945). "A Note on Rank, Multicollinearity and Multiple Regression," *Ann. Math. Stat.*, **16**; 304–307.

Tobler, W. and S. Wineburg (1971). "A Cappadocian Speculation," *Nature*, 1971.

Tocher, J. F. (1908). "Pigmentation Survey of School Children in Scotland," *Biometrika*, **6**; 130–235.

Toeplitz, O. (1911). "Zur Theorie der Quadaischen und Bilinearen Formen Von Unendlichvillen Veränderlichen," *Math. Ann.*, **70**; 351–376.

Tomassone, R. and L. Lebart (1980). "Survey Analysis: Statistical Management of Large Data Sets," *Proc. Comp. Stat. (Compstat)*, **4**; 32–42.

Torgerson, W. S. (1952). "Multidimensional Scaling: I. Theory and Method," *Psychometrika*, **17**; 401–419.

Torgerson, W. S. (1958). *Theory and Methods of Scaling*, Wiley, New York.

Tracy, D. S. and P. S. Dwyer (1969). "Multivariate Maxima and Minima with Matrix Derivatives," *J. Am. Stat. Assoc.*, **64**; 1576–1594.

Trenkler, G. (1980). "Generalized Mean Squared Error Comparisons of Biased Regression Estimators," *Commun. Stat. – Theor. Meth.*, **A9**(12); 1247–1259.

Tucker, L. R. (1958). "Determination of Parameters of a Functional Relations by Factor Analysis," *Psychometrik*, **23**; 19–23.

Tucker, L. R. (1966). "Some Mathematical Notes On Three-Mode Factor Analysis," *Psychometrika*, **31**; 279–311.

Tucker, L. R. (1967). "Three-Mode Factor Analysis of Parker-Fleishman Complex Tracking Behavior Data," *Multivariate Behav. Res.*, **2**; 139–151.

Tucker, L. R. (1972). "Relations Between Multidimensional Scaling and Three-Mode Factor Analysis," *Psychometrika* **37**; 3–27.

Tukey, J. W. (1979). "Comment," *J. Am. Stat. Assoc.*, **74**; 121–122.

Tyler, D. E. (1981). Asymptotic Inference for Eigenvectors," *Ann. Stat.*, **9**; 725–736.

Tyror, J. G. (1957). "The Distribution of the Directions of Perihelia of Long-Period Comets," *Mon. Notes, R. Astron. Soc.*, **117**; 3–13.

United Nations Organization (1962). "Factor Analysis of Sex-Age Specific Death Rates: A Contribution to the Study of the Dimensions of Mortality," Pop. Bull. UN, No. 6, New York.

Upton, G. J. G., and B. Fingleton (1989). *Spatial Data Analysis by Example*: *Categorical and Directional Data*, Vol. 2, Wiley, New York.

Valenchon, F. (1982). "The Use of Correspondence Analysis in Geochemistry," *Math. Geol.*, **14**; 331–342.

Van Bemmel, J. H. (1982). "Recognition of Electrocardiographic Patterns," in P. R. Krishnaiah and L. N. Kanal (Eds.), Handbook of Statistics, Vol. 2, North-Holland, New York, pp. 501–526.

Van den Wallenberg, A. L. (1977). "Redundacy Analysis: An Alternative for Canonical Correlation Analysis," *Psychometrika*, **42**; 207–219.

VanDriel, O. P. (1978). "On Various Causes of Improper Solutions in Maximum Likelihood Factor Analysis," *Psychometrika*, **43**; 225–243.

Van Rijckevorsel, J. L. A. and J. de Leeuw (1988). *Component and Correspondence Analysis*, Wiley and Son, Chichester.

Van Uven, M. J. (1930), "Adjustment of n Points (in n-Dimensional Space)," II; *K. Acad. Wet. Amsterdam, Proc. Sect. Sci.*, **33**; 307–326.

Vavra, T. G. (1972). "Factor Analysis of Percepetual Change," *J. Mark. Res.*, **9**; 193–199.

Vegelius, J. (1980). "Is the G-Index a Correlation Coefficient," *Educ. Psych. Meas.*, **40**; 349–352.

Vegelius, J. (1982). "A Q-Analysis for Nominal Data," *Educ. Psych. Meas.*, **42**; 105–111.

Velicer, W. F. (1976). "Determining the Number of Components From the Matrix of Partial Correlations," *Psychometrika*, **41**; 321–327.

Velicer, W. F. (1977). "An Empirical Comparison of the Similarity of Principal Component, Image and Factor Patterns," *Multivariate Behav. Res.*, 3–22.

Versace, J. (1960). "Factor Analysis of Roadway and Accident Data," *High. Res. Board Bull.*, **240**; 24–32.

Vierra, R. K. and D. L. Carlson (1981). "Factor Analysis, Random Data, and Patterned Results," *Am. Antiq.*, **46**; 272–283.

Walker, M. A. (1967). "Some Critical Comments on 'An Analysis of Crimes by the Method of Principal Components' by B. Ahamad," *Appl. Stat.*, **16**; 36–39.

Wallis, C. P. and R. Maliphant (1967). "Delinquent Areas in the County of London: Ecological Factors," *Br. J. Criminol.*, **7**; 250–284.

Waternaux, C. M. (1976). "Asymptotic Distribution of the Sample Roots for a Non-normal Population," *Biometrika*, **63**; 639–645.

Watson, G. S. (1966). "The Statistics of Orientation Data," *J. Geol.*, **74**; 786–797.

Webb, W. M. and L. I. Briggs (1966). "The Use of Principal Components Analysis to Screen Mineralogical Data," *J. Geol.*, **74**; 716–720.

Weber, E. and J. Berger (1978). "Two-Way Table Analysis by Biplot Methods," *Proceedings in Computational Statistics*, 3rd Symposium, Physica-Verlag, Leiden.

Webster, J. T. and R. L. Mason (1974). "Latent Root Regression Analysis," *Technometrics*, **16**; 513–522.

Wei, W. W. S. (1990). *"Time Series Analysis: Univariate and Multivariate Methods,"* Addison-Wesley.

Weiner, P. H. (1977). "Solve Problems via Factor Analysis," *Chem. Tech.*, **7**; 321–328.

Weiner, P. H. and D. G. Howery (1972). "Factor Analysis of Some Chemical and Physical Influences in Gas Liquid Chromatography," *Analyt. Chem.*, **44**; 1189–1194.

Wells, M. T. (1990). "The Relative Efficiency of Goodness-of-Fit Statistics in the Simple and Composite Hypothesis-Testing Problem," *J. Am. Stat. Assoc.*, **85**; 459–463.

Westley, B. H. and H. K. Jacobson (1962). "Dimensions of Teachers' Attitudes Towards Instructional Television," *AV Commun. Rev.*, **10**; 179–185.

Westley, B. H. and M. D. Lynch (1962). "Multiple Factor Analysis of Dichotomous Audience Data," *Journalism Q.*, **39**; 369–372.

White, R. M., D. S. Cooley, R. C. Derly, and F. A. Seaver (1958). "The Development of Efficient Liner Statistical Operators for the Prediction of Sea-Level Pressure," *J. Meteorol.*, **15**; 426–440.

Whittle, P. (1953). "On Principal Components and Least Squares Methods of Factor Analysis," *Skand. Aktuarietidskr.*, **35**; 224–239.

Wiberg, T. (1976). "Computation of Principal Components When Data are Missing," *Proceedings in Computational Statistics (Comptstat.)*, Physica Verlag, Vienna, pp. 229–236.

Widom, H. (1965). "Toeplitz Matrices," in I. J. Hirschman (Ed.), *Studies in Real and Complex Analysis*, Vol. 3, M.A.A. Studies in Mathematics, Prentice Hall, Englewood Cliffs, NJ.

Wilkinson, E. M. (1974). "Techniques of Data Analysis: Seriation Theory," *Archaeo-Physica*, **5**; 1–142.

Wilks, S. S. (1946). "Sample Criteria for Testing Equality of Means, Equality of Variances, and Equality of Covariances in a Normal Multivariate Distribution," *Ann. Math. Stat.*, **17**; 257–281.

Williams, E. J. (1952). "Use of Scores for the Analysis of Association in Contingency Tables, *Biometrika*, **39**; 274–289.

Williams, J. S. (1978). "Canonical Analysis: A Factor-Analytic Method of Comparing Finite Discrete Distribution Functions," *J. Am. Stat. Assoc.*, **73**; 781–786.

Williams, W. T., and J. M. Lambert (1959). "Multivariate Analysis in Plant Communities, I: Association Analysis in Plant Communities, *J. Ecol.*, **47**;

Williamson, M. H. (1978). "The Ordination of Incidence Data," *J. Ecol.*, **66**; 911–920.

Wold, H. (1953). *"Some Artificial Experiments in Factor Analysis*," Uppsala Symposium on Psychological Factor Analysis; Nordisk Psykologi Monogrpah Series, No. 3, pp. 43–64.

Wold, H. (1980). "Model Construction and Evaluation When Theoretical Knowledge is Scarce," J. Kmenta, and J. B. Ramsey (Eds.), Evaluation of Econometric Models, Academic, New York, pp. 47–74.

Wold, S. (1978). "Cross-Validatory Estimation of the Number of Components in Factor and Principal Components Models," *Technometrics*, **20**; 397–405.

Wold, S. (1976). "Pattern Recognition by Means of Disjoint Principal Components Models," *Pattern Recognition*, **8**; 127–139.

Wold, S., C. Albano, W. Dunn, U. Edlund, B. Eliasson, E. Johansson, B. Nordén, and M. Sjöstrom (1982). "The Indirect Observation of Molecular Chemical Systems," in K. G. Jöreskog and H. Wold (Eds.), *Systems Under Indirect Observation, Part II*, North Holland, New York, pp. 177–190.

Wolf, J. H. (1970). "Pattern Clustering in Multivariate Mixture Analysis," *Multivar. Behav. Res.*, **5**; 329–350.

Woodbury, M. A. and Hickey, R. J. (1963). "Computers in Behavioural Research," *Behav. Sci.*, **8**; 347–354.

Woodbury, M. A. and Siler, W. (1966). "Factor Analysis with Missing Data," *Ann. NY Acad. Sci.*, **28**; 746–754.

Wooding, R. A. (1956). "The Multivariate Distribution of Complex Normal Variates," *Biometrika*, **43**; 212–215.

Woodward, J. A. and J. E. Overall (1976). "Factor Analysis of Rank-Ordered Data: An Old Approach Revisited," *Psych. Bull.*, **83**; 864–867.

Woodward, J. A., R. L. Retka, and Lin Ng (1984). "Construct Validity of Heroin Abuse Estimation," *Int. J. Addict.*, **19**; 93–117.

Wright, S. (1934). "The Method of Path Coefficients," *Ann. Math. Stat.*, **5**; 161–125.

Wright, S. (1960). "Path Coefficients and Regression: Alternative or Complementary Concepts?" *Biometrics*, **16**; 189–202.

Wu, C. F. J. (1983). "On the Convergence Properties of the EM Algorithm," *Ann. Stat.*, **11**; 95–103.

Yanai, H., Y. Inaba, H. Takagi, H. Toyokawa, and S. Yamamoto (1978). "An Epidemiological Study of Mortality Rates of Various Cancer Sites During 1958 to 1971 by Means of Factor Analysis," *Behaviormetrika*, **5**; 55–74.

Yawo, H., Y. Shinagawa, and Y. Shinagawa (1981). "Principal Component Analysis of Systemic Lupus Erythematosus (SLE): A Proposal for Handling Data with Many Missing Values," *Comput. Biomed. Res.*, **14**; 248–261.

Yin, Y. Q. and P. R. Krishnaiah (1985). "Limit Theorems for the Eigenvalues of a Product of Large Dimensional Random Matrices When the Underlying Distribution is Isotropic," *SIAM: Thoery Probab.*, **31**; 342–346.

Young, G. and A. S. Householder (1938). "A Discussion of a Set of Points in Terms of Their Mutual Distances," *Psychometrika*, **3**; 19–22.

Young, P. (1985). "Recursive Identification, Estimation and Control," in E. J. Hannan, P. R. Krishnaiah, and M. M. Rao, (Eds.), *Handbook of Statistics*, Vol. 5, North Holland, New York, pp. 213–255.

Zegura, S. L. (1978). "Components, Factors and Confusion," *Yearb. Phys. Anthropol.*, **21**; 151–159.

Zhou, D., T. Chang, and J. C. Davis (1983). "Dual Extraction of R-Mode and Q-Mode Factor Solutions," *Math. Geol.*, **15**; 581–606.

Index